24B
$80

2008 VIII 15

CHRONOLOGY AND EVOLUTION OF MARS

The International Space Science Institute is organized as a foundation under Swiss law. It is funded through recurrent contributions from the European Space Agency, the Swiss Confederation, the Swiss National Science Foundation, and the University of Bern. For more information, see the homepage at http://www.issi.unibe.ch/.

The titles published in this series are listed at the end of this volume.

CHRONOLOGY AND EVOLUTION OF MARS

*Proceedings of an ISSI Workshop,
10–14 April 2000, Bern, Switzerland*

Edited by

REINALD KALLENBACH
International Space Science Institute, Bern, Switzerland

JOHANNES GEISS
International Space Science Institute, Bern, Switzerland

WILLIAM K. HARTMANN
Planetary Science Institute, Tucson, USA

Editorial Assistant: URSULA PFANDER
International Space Science Institute, Bern, Switzerland

Reprinted from *Space Science Reviews*, Volume 96, Nos. 1–4, 2001

KLUWER ACADEMIC PUBLISHERS
DORDRECHT / BOSTON / LONDON

Space Sciences Series of ISSI

Library of Congress Cataloging-in-Publication Data.

Chronology and evolution of Mars / edited by R. Kallenbach, J. Geiss, and W.K. Hartmann
p.cm. – (Space science Series of ISSI: v. 12)
Includes index.
ISBN 0-7923-7051-1
1. Mars–Geology-Congresses. I. Kallenbach, R. II. Geiss, Johannes. III. Hartmann, William
K. IV. Space science series of ISSI: v. 12.
QB641.C53 2001
559.9′23–dc21   2001035763

Published by Kluwer Academic Publishers,
P.O. Box 17, 3300 AA Dordrecht, The Netherlands

Sold and distributed in North, Central and South America
by Kluwer Academic Publishers,
101 Philip Drive, Norwell, MA 02061, U.S.A.

In all other countries, sold and distributed
by Kluwer Academic Publishers,
P.O. Box 322, 3300 AH Dordrecht, The Netherlands

Printed on acid-free paper

Printed in the Netherlands

TABLE OF CONTENTS

EPILOGUE

Chronology and Evolution of Mars
ISSI Workshop, 10–14 April 2000, Bern, Switzerland

Group Photographs

1. Jean-Pierre Bibring
2. Gabriela Indermühle
3. Jim Head
4. Ronald Greeley
5. Philippe Masson
6. Kurt Marti
7. Grenville Turner
8. Otto Eugster
9. Dieter Stöffler
10. Jean-Louis Birck
11. Tim Swindle
12. Ansgar Greshake
13. Michael Drake
14. Robert Clayton
15. Paul Warren
16. Heinrich Wänke
17. Larry Nyquist
18. Francis Albarède
19. Tilman Spohn
20. Ernst Hauber
21. John Saxton
22. Don Bogard
23. Gerhard Neukum
24. John Bridges
25. Bill Hartmann
26. Reinald Kallenbach
27. Mario Acuña
28. Ralf Jaumann
29. Uli Ott
30. Graham Ryder

Not in this picture: Gabriele Arnold, Doris Breuer, Therese Encrenaz, Johannes Geiss, Matthew Golombek, Alex Halliday, Elmar Jessberger, Tobias Owen

Picture taken by Urs Lauterburg, Physikalisches Institut, University of Bern

FRIENDS OF THIS BOOK

The following people supported the preparation of this book in various ways, including submitting reviews of various chapters, coordinating meeting and publishing preparations, providing publication advice, maintaining computer infrastructure, and other forms of assistance.

Leslie Baker, Rocky Mt. College, Billings
Nadine G. Barlow, University of Central Florida, Orlando
Daniel C. Berman, Planetary Science Institute, Tucson
Harry Blom, Kluwer Academic Publishers, Dordrecht
Donald R. Davis, Planetary Science Institute, Tucson
Therese Encrenaz , Observatoire de Paris, Meudon
Otto Eugster, University of Bern, Bern
Jamie Gilmour, Manchester University, Manchester
Ron Greeley, Arizona State University, Tempe
Gabriela Indermühle, International Space Science Institute, Bern
Boris Ivanov, Institute for Dynamics of Geospheres, Moscow
Bruce Jakosky, University of Colorado, Boulder
Elmar Jessberger, Institut für Planetologie, Universität Münster
Melissa Lane, Planetary Science Institute, Tempe
Barbara Lucchita, U.S. Geological Survey, Flagstaff
Kurt Marti, University of California at San Diego, San Diego
Philippe Masson, Université Paris-Sud, Paris
Harry Y. McSween, University of Tennessee, Knoxville
Gerhard Neukum, Deutsches Zentrum für Luft- und Raumfahrt, Berlin
Tobias Owen, University of Hawaii, Honolulu
Elaine Owens, Planetary Science Institute, Tucson
Carl Pilcher, NASA Headquarters, Washington, DC
Graham Ryder, Lunar and Planetary Institute, Houston
Xavier Schneider, International Space Science Institute, Bern
Rudolf von Steiger, International Space Science Institute, Bern
Dieter Stöffler, Museum für Naturkunde, Humboldt-Universität, Berlin
Timothy D. Swindle, University of Arizona, Tucson
Ken L. Tanaka, U.S. Geological Survey, Flagstaff
Grenville Turner, Manchester University, Manchester
Silvia Wenger, International Space Science Institute, Bern

U.S. participants of the workshop *Chronology and Evolution of Mars* were in part supported by the NASA Mars Global Surveyor program through the Jet Propulsion Laboratory and the NASA Planetary Geology and Geophysics program and through the office of Dr. Carl Pilcher. The support by the Planetary Science Institute, Tucson, AZ, USA, and its staff is greatly appreciated. The editors wish to express their sincere thanks to Ursula Pfander, editorial assistant at the International Space Science Institute, Bern, Switzerland.

INTRODUCTION: A NEW CHAPTER IN MARS RESEARCH

REINALD KALLENBACH[1], JOHANNES GEISS[1] and WILLIAM K. HARTMANN[2]

[1]*International Space Science Institute, CH-3012 Bern, Switzerland*
[2]*Planetary Science Institute, Tucson, AZ 85705-8331, USA*

February 2001

This book reviews the most recent progress in constraining the timescales and geological processes in the evolution of Mars. It developed from a series of workshops on planetary science at the International Space Science Institute (ISSI), in Bern. The first of these meetings in February, 1999, concentrated on the interdisciplinary assessment of the astronomical, chronological, and geochemical constraints of the formation of the inner solar system $\sim$4.56 Gyr ago (Benz *et al.*, 2000). It appeared natural to continue with a core group meeting, reviewing the knowledge on the subsequent chronology of the inner solar system until the present day on the exemplary cases of Mars and Moon. Among the terrestrial planets, Mars is unique to have undergone all planetary evolutionary steps, without global resets, till the present (Encrenaz *et al.*, 1995; Bibring and Erard, 2001). The discussion on the "red planet" climaxed with a workshop on Chronology and Evolution of Mars in April, 2000, emphasizing communication and collaboration between the geochemical dating community, the crater chronologists, and the photogeologists.

The results reported here open a new chapter of Martian exploration. An early chapter began in 1895 with Percival Lowell's hypothesis, still seriously debated in the mid-20th century, that advanced life might exist on Mars. A new chapter opened in 1965 when the first closeup photos of Mars from Mariner 4 revealed impact craters, suggesting to Leighton *et al.* (1965) that Mars was Moon-like, geologically and biologically dead, but with a thin atmosphere redistributing the dust.

Still another chapter opened in 1971, when Mariner 9 discovered dry riverbeds and towering volcanoes, implying a very non-lunar Mars with a complex geologic history. The first crater count studies yielded ages of a few 100 Myr for massive volcanic constructs, but no substantial magnetic field was observed that would indicate modern geological activity. Viking landers in 1976 revealed basaltic soils laden with sulfates and salts suggesting ancient water exposure, but lacking organics, which almost ended discussion of life on Mars. In 1986, Tanaka developed a relative stratigraphy, dividing Mars into the Noachian (early), Hesperian (middle), and Amazonian (late) periods, with most of the activity in the Noachian and Hesperian, but the absolute timescale of this chronology remained controversial.

On the Earth and Moon, the biggest geologic advances came when the relative timescale was connected to absolute radiometric ages of terrestrial rocks and samples returned by the Apollo and Luna missions. This gave better causal insights into

Space Science Reviews **96**: 3-6, 2001.
© 2001 *Kluwer Academic Publishers. Printed in the Netherlands.*

the fossil record, the development of the crust/mantle structure, cratering histories, and the simultaneity of various events in different regions. From crater counts on Mars it was estimated that most volcanism, water flow, and other forms of geologic activity occurred very early at $\sim$2.5 $-$ 4.5 Gyr ago, but the absolute timescale still remained uncertain. Thus, the latest chapter in Mars research is partly defined by the attempt to convert from relative to absolute timescales, based on the dating of Martian meteorites and conversion of lunar cratering chronology to Mars.

The absolute dating and chemical analysis of the currently known 16 rocks from Mars has led to much discussion. During early meteorite studies, three classes of achondrites, Shergottites, Nakhlites, and Chassignites, were identified on the basis of special petrographic properties. In 1958, Geiss and Hess were surprised to obtain a low K-Ar formation age of only $\sim$580 Myr on the Shergotty meteorite (now dated at $\sim$170 Myr, Nyquist *et al.*, 2001), because virtually all other meteorites had ages of >4 Gyr, the time of planet formation. To dismiss this result by random argon loss would have required 98% losses. By 1974, Papanastassiou and Wasserburg postulated metamorphism on some relatively large but unknown parent body for these young meteorites. The Moon was ruled out by Apollo rock samples. Nyquist *et al.* (1979), Wasson and Wetherill (1979), and Wood and Ashwal (1981) were among the first to suggest that the parent body was Mars – though most impact experts of that time denied that rocks could be blasted intact off Mars.

Acceptance that these rocks were from Mars came with elemental and isotopic identification of Martian atmospheric gases in them (Bogard and Johnson, 1983; Becker and Pepin, 1984; Dreibus and Wänke, 1987). By the 1990s researchers accepted that Mars had produced not only a 4.5 Gyr-old crustal rock, but also basaltic rocks as young as 0.2 $-$ 1.3 Gyr, and that these had come from only four to eight impact sites. The new chapter of Mars research marks a period where we have a handful of samples from unknown locations on Mars. The diverse pieces of information on surface features and chemistry, presently available from the Viking and Pathfinder landing sites and from Mars Global Surveyor (MGS) and earlier NASA and Soviet spacecraft observations (Pellinen and Raudsepp, 2000), have to be integrated to rough out an absolute chronology of Martian geologic evolution.

In Part I of this book, cratered lunar surfaces, which have been dated precisely by the radiometric ages of returned samples, are taken as a reference (Stöffler and Ryder, 2001) to derive ages of cratered geological units on Mars. The lunar production function, i.e. the size-frequency distribution of craters as they form on the Moon (Neukum *et al.*, 2001), is converted to describe impact events on Mars (Ivanov, 2001). The resulting Martian crater retention ages date the youngest detected flows at $\leq$10 Myr (Hartmann and Neukum, 2001). This is consistent with the conclusion from radiometric dating that at least some geologically young igneous rocks with ages of a few 100 Myr exist on Mars (Nyquist *et al.*, 2001).

Part II examines the origin of the planet, the interior structure, and the surface rocks. Tungsten isotope data, Ba/W and time-integrated Re/Os ratios in Martian meteorites affirm that Mars differentiated during the first 20 Myr of solar system history into core, mantle, crust and atmosphere (Halliday *et al.*, 2001). Thus, the oldest Martian surfaces involve primordial crustal material, as can also be concluded from the crystallization age of 4.5 Gyr for meteorite ALH84001 and the high crater densities of some areas. Based on the new MGS topography and gravity data, and the data from Pathfinder on the rotation of Mars, Spohn *et al.* (2001) constrain their models of the present interior structure. Their article also reviews the MGS discovery that the remnant magnetization of the oldest parts of the Martian crust (Acuña *et al.*, 1999) are an order of magnitude stronger than crustal magnetism on Earth. Head *et al.* (2001) give an update on the stratigraphic system and geologic processes of the Martian surface. On the latter, remote sensing identifies two broad groups of igneous rock units, basaltic and andesitic (Bibring and Erard, 2001), as is confirmed by in-situ chemical analyses at the Pathfinder landing site (Wänke *et al.*, 2001).

Part III emphasizes volatile history. Certain Martian meteorites indicate fluvial activity younger than the rocks themselves, in one case ∼670 Myr (Shih *et al.*, 1998; Swindle *et al.*, 2000; Bridges *et al.*, 2001). Youthful water seeps suggest even more recent liquid water mobility (Malin and Edgett, 2000; Hartmann, 2001). The geomorphologic evidence for liquid water on Mars is thoroughly reviewed by Masson *et al.* (2001). The Martian atmosphere's history is outlined by Encrenaz (2001), reporting constraints by remote sensing, and by Bogard *et al.* (2001), discussing meteorite studies and Viking measurements. Mars is not as depleted in moderately volatile elements as Earth, where volatiles may have been lost by a giant early impact (Halliday *et al.*, 2001). Finally, atmospheric processes may in turn limit our access to the Martian chronology, as winds erode impact craters and/or cover them with deposits (Greeley *et al.*, 2001).

In the epilogue, Hartmann *et al.* (2001) summarize the new chapter of Mars research and give directions to open a future chapter. A glossary, a subject index, and a list of acronyms are added for the benefit of the reader.

References

Acuña, M.H., Connerney, J.E.P., Wasilewski, P., Kletetschka, G., Ness, N.F., Rème, H., Lin, R., and Mitchell, D.: 1999, 'Mars Crustal Magnetism – Global Distribution, Morphology and Source Models', *Am. Astron. Soc., DPS meeting*, abstract #31, #59.07.

Becker, R.H., and Pepin, R.O.: 1984, 'The Case for a Martian Origin of the Shergottites: Nitrogen and Noble Gases in EET 79001', *Earth Planet. Sci. Lett.* **69**, 225–242.

Benz, W., Kallenbach, R., and Lugmair, G.W. (eds.): 2000, *From Dust to Terrestrial Planets*, Kluwer Academic Publishers, Dordrecht, The Netherlands.

Bibring, J.-P., and Erard, S.: 2001, this volume.

Bogard, D.D., and Johnson, P.: 1983, 'Martian Gases in an Antarctic Meteorite?', *Science* **221**, 651–654.

Bogard, D.D., Clayton, R.N., Marti, K., Owen, T., and Turner, G.: 2001, this volume.

Bridges, J.C., Catling, D.C., Saxton, J.M., Swindle, T.D., Lyon, I.C., and Grady, M.M.: 2001, this volume.

Dreibus, G., and Wänke, H.: 1987, 'Volatiles on Earth and Mars: A Comparison', *Icarus* **71**, 225–240.

Encrenaz, T.: 2001, this volume.

Encrenaz, T., Bibring, J.-P., Blanc, M., and Dunlop, S.: 1995, *The Solar System*, Springer-Verlag.

Geiss, J., and Hess., D.C.: 1958, 'Argon-potassium Ages and the Isotopic Composition of Argon from Meteorites', *Astrophys. J.* **127**, 224–236.

Greeley, R., Kuzmin, R.O., and Haberle, R.M: 2001, this volume.

Halliday, A.N., Wänke, H., Birck, J.-L., and Clayton, R.N.: 2001, this volume.

Hartmann, W.K.: 2001, this volume.

Hartmann, W.K., and Neukum, G.: 2001, this volume.

Hartmann, W.K., Kallenbach, R., Geiss, J., and Turner, G.: 2001, this volume.

Head, J.W., *et al.*: 2001, 'Geological Processes and Evolution', this volume.

Ivanov, B.A.: 2001, 'Mars/moon Cratering Rate Ratio Estimates', this volume.

Leighton, R.B., Murray, B., Sharp, R., Allen, J., and Sloan, R.: 1965, 'Mariner IV Photography of Mars: Initial Results', *Science* **149**, 627–630.

Malin, M.C., and Edgett, K.S.: 2000, 'Evidence for Recent Groundwater Seepage and Surface Runoff on Mars', *Science* **288**, 2330–2335.

Masson, P., Carr, M.H., Costard, F., Greeley, R., Hauber, E., and Jaumann, R.: 2001, this volume.

McSween, H.Y., Jr.: 1994, 'What Have we Learned about Mars from SNC Meteorites', *Meteoritics* **29**, 757–779.

Neukum, G., Ivanov, B.A., and Hartmann, W.K.: 2001, this volume.

Nyquist, L.E., Bogard, D., Wooden, J., Wiesmann, J., Shih, C.-Y., Bansal, B., and McKay, G.: 1979, 'Early Differentiation, Late Magmatism, and Recent Bombardment on the Shergottite Parent Planet', *Meteoritics* **14**, 502.

Nyquist, L.E., Bogard, D.D., Shih, C.-Y., Greshake, A., Stöffler, D., and Eugster, O.: 2001, this volume.

Papanastassiou, D.A., and Wasserburg, G.J.: 1974, 'Evidence for Late Formation and Young Metamorphism in the Achondrite Nakhla', *Geophys. Res. Lett.* **1**, 23–26.

Pellinen, R., and Raudsepp, P. (eds.): 2000, *Towards Mars*, Raud Publishing, Helsinki, Finland.

Sawyer, D.J., McGehee, M.D., Canepa, J., and Moore, C.B.: 2000, 'Water Soluble Ions in the Nakhla Martian Meteorite', *Met. Planet. Sci.* **35**, 743–748.

Shih, C.-Y., Nyquist, L.E., Reese, Y., and Wiesmann, H.: 1998, 'The Chronology of the Nakhlite, Lafayette: Rb-Sr and Sm-Nd Isotopic Ages', *Proc. 29th Lunar Planet. Sci. Conf.*, LPI, Houston, abstract #11435 (CD-ROM).

Spohn, T., Acuña, M., Breuer, D., Golombek, M., Greeley, R., Halliday, A., Hauber, E., Jaumann, R., and Sohl, F.: 2001, this volume.

Stöffler, D., and Ryder, G.: 2001, this volume.

Swindle, T.D., Treiman, A., Lindstrom, D., Burkland, M., Cohen, B., Grier, J., and Olson, E.: 2000, 'Noble Gases in Iddingsite from the Lafayette Meteorite; Evidence for Liquid Water on Mars in the Last Few Hundred Million Years', *Met. Planet. Sci.* **35**, 107–116.

Tanaka, K.L.: 1986, 'The Stratigraphy of Mars', *J. Geophys. Res.* **91 suppl.**, 139–158.

Wänke, H., Brückner, J., Dreibus, G., Rieder, R., and Ryabchikov, I.: 2001, this volume.

Wasson, J., and Wetherill, G.: 1979, 'Dynamical, Chemical, and Isotopic Evidence Regarding the Formation Locations of Asteroids and Meteorites', *in* T. Gehrels (ed.), *Asteroids*, Univ. Arizona Press, Tucson, pp. 926–974.

Wood, C.A., and Ashwal, C.A.: 1981, 'SNC Meteorites: Igneous Rocks from Mars', *Proc. 12th Lunar Planet. Sci. Conf.*, 1359–1375.

Address for offprints: International Space Science Institute, Hallerstrasse 6, CH-3012 Bern, Switzerland; (reinald.kallenbach@issi.unibe.ch)

I: CHRONOLOGY OF MARS AND OF THE INNER SOLAR SYSTEM

STRATIGRAPHY AND ISOTOPE AGES OF LUNAR GEOLOGIC UNITS: CHRONOLOGICAL STANDARD FOR THE INNER SOLAR SYSTEM

D. STÖFFLER[1] and G. RYDER[2]

[1]*Institut für Mineralogie, Museum für Naturkunde, Humboldt-Universität zu Berlin, Invalidenstrasse 43, D-10099 Berlin, Germany*
[2]*Lunar and Planetary Institute, 3600 Bay Area Boulevard, Houston, TX 77058, USA*

Received: 11 August 2000; accepted: 25 February 2001

Abstract. The absolute ages of cratered surfaces in the inner solar system, including Mars, are derived by extrapolation from the impact flux curve for the Moon which has been calibrated on the basis of absolute ages of lunar samples. We reevaluate the lunar flux curve using isotope ages of lunar samples and the latest views on the lunar stratigraphy and the principles of relative and absolute age dating of geologic surface units of the Moon. The geological setting of the Apollo and Luna landing areas are described as far as they are relevant for this reevaluation. We derive the following best estimates for the ages of the multi-ring basins and their related ejecta blankets and present alternative ages for the basin events (in parentheses): 3.92 ± 0.03 Gyr (or 3.85 ± 0.05 Gyr) for Nectaris, 3.89 ± 0.02 Gyr (or 3.84 ± 0.04 Gyr) for Crisium, 3.89 ± 0.01 Gyr (or 3.87 ± 0.03 Gyr) for Serenitatis, and 3.85 ± 0.02 Gyr (or 3.77 ± 0.02 Gyr) for Imbrium. Our best estimates for the ages of the mare landing areas are: 3.80 ± 0.02 Gyr for Apollo 11 (old surface), 3.75 ± 0.01 Gyr for Apollo 17, 3.58 ± 0.01 Gyr for Apollo 11 (young surface), 3.41 ± 0.04 Gyr for Luna 16, 3.30 ± 0.02 Gyr for Apollo 15, 3.22 ± 0.02 Gyr for Luna 24, and 3.15 ± 0.04 Gyr for Apollo 12. The ages of Eratosthenian and Copernican craters remain: ~ 2.1 (?) Gyr (Autolycus), 800 ± 15 Myr (Copernicus), 109 ± 4 Myr (Tycho), 50.3 ± 0.8 (North Ray crater, Apollo 16), and 25.1 ± 1.2 (Cone crater, Apollo 14). When plotted against the crater densities of the relevant lunar surface units, these data result in a revised lunar impact flux curve which differs from the previously used flux curve in the following respects: (1) The ages of the stratigraphically most critical impact basins are notably younger, (2) the uncertainty of the calibration curve is decreased, especially in the age range from about 4.0 to 3.0 Gyr, (3) any curve for ages older than 3.95 Gyr (upper age limit of the Nectaris ejecta blanket) is abandoned because crater frequencies measured on such surface formations cannot be correlated with absolute ages obtained on lunar samples. Therefore, the impact flux curve for this pre-Nectarian time remains unknown. The new calibration curve for lunar crater retention ages less than about 3.9 Gyr provides an updated standard reference for the inner solar system bodies including Mars.

1. Introduction

Apart from the Earth, the Moon (Figure 1) is the only planetary body for which we have both a detailed stratigraphic history and rock samples that we can relate with varied degrees of confidence to specific geologic - or at least morphologic - units. Furthermore, the Moon has preserved much of its surficial magmatic and impact record of at least the last 4 billion years. While its endogenic history is unique, the Moon has become a calibration plate for the cratering record of the Earth-Moon

Space Science Reviews **96**: 9-54, 2001.
© 2001 *Kluwer Academic Publishers. Printed in the Netherlands.*

Figure 1. Telescopic view of the nearside of the Earth's moon with landing sites of the Apollo and Luna missions.

system, and by extrapolation, of the entire inner solar system by the assumption of a heliocentric origin for impactor populations. These populations range from long and short period comets through asteroids to interplanetary dust and cover a size range from hundreds of kilometers to micrometers.

We summarize the absolute and relative chronology of the endogenic and exogenic history of the Moon as expressed by specific surface units for which cratering rates can be calibrated against time by counting craters on sampled and dated surfaces. Our summary is derived from and revises previous work, in particular the data presented by Neukum and colleagues (e.g., Neukum *et al.*, 1975; Neukum, 1983; Neukum and Ivanov, 1994; Hartmann *et al.*, 1981). Our goals differ from those of the Basaltic Volcanism Study Project (Hartmann *et al.*, 1981) in that we are not attempting to understand the chronology of lunar volcanism, but to constrain impact flux chronologies in the inner solar system as the Moon has recorded them. In particular, we examine how the presently available radiometric and exposure ages for lunar rocks are correlated with surface units, and those in turn with crater density counts. In contrast with the previous reviews, our revised flux curve differs slightly during the period of mare volcanism, and is absent for the period prior to about 3.9 Gyr because there are no dated surfaces to calibrate the flux.

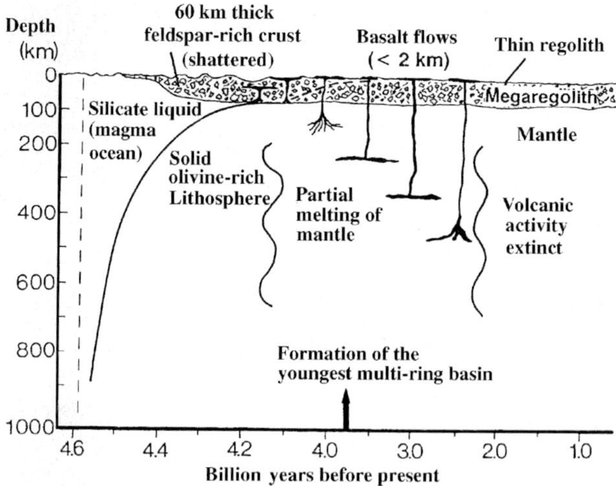

Figure 2. Geologic history of the moon schematically represented by the variation of the constitution of the lunar crust and mantle as a function of time.

2. Geologic History and Basic Stratigraphy of the Moon

2.1. GEOLOGIC HISTORY

The formation of the Moon is part of the formation process of the terrestrial planets, and it is commonly accepted that the Moon accreted from material similar to the material of the Earth as is clearly shown by the oxygen isotopes (Clayton and Mayeda, 1975; Hartmann *et al.*, 1984; Halliday, 2000). Although the origin of the Moon is still subject to debate, there has been growing evidence and consensus that it formed rapidly from material ejected during a tangential collision of a Mars-sized planetary body with the Proto-Earth after the Earth's core had formed (Hartmann and Davis, 1975; Cameron and Ward, 1976; Cameron, 1984), most probably ~4.50 Gyr ago (Halliday, 2000). This rapid accretion theory (Canup and Agnor, 2000) requires that the Moon was initially in a completely or partially molten state, a conclusion independently derived from the observation that the primordial crust of the Moon is highly feldspathic (anorthositic). The latter most probably formed by fractional crystallization and differentiation of a global "magma ocean" (Taylor, 1982) (Figure 2). A complementary ultramafic, dunitic mantle formed at the lower boundary of the magma ocean.

The continued crystallization of the global magma ocean led at about 4.4 Gyr ago to a residual melt layer enriched in incompatible elements (KREEP; K = potassium, REE = rare earth elements, P = phosphorus) including the radioactive elements K, U, and Th, between the mantle and crust (Figure 2). This heat-producing layer was involved in the earliest phase of lunar volcanism producing basaltic intrusives and extrusives of KREEP-rich and high alumina basaltic composition after ~4.3 Gyr ago (Shearer and Papike, 1999), as well as in the formation of ancient plutonic intrusions into the lunar crust. These produced diverse igneous rocks such as rocks of the Mg-suite (troctolites, norites, gabbronorites, dunites) and rocks of the alkali-suite such as alkali anorthosites and granitic lithologies (James, 1980).

The early lunar magmatism led to a crust of varied thickness ranging from about 60 to 110 km which was subjected, simultaneously with the magmatic events, to the intense early collisional bombardment (Figure 2). The early impact processing reworked the crust to some unknown depth (probably several km), destroying some of the primary structure induced by igneous processes, and producing an upper megabreccia layer of mixed feldspathic composition (megaregolith, Hartmann, 1973).

Isotope systematics of mare basalts indicate that the ultramafic cumulate mantle, from which they were extracted by partial melting, existed at about 4.4 Gyr. The extrusion of volcanics began as early as 4.3 Gyr (Shearer and Papike, 1999) but none of the volcanic surfaces formed before about 3.8 Gyr are still preserved because of the heavy bombardment (Figure 2). The mare basalt flows, which cover only about 17% of the lunar surface, formed recognizable maria from 3.9 Gyr to about 2 Gyr or even 1 Gyr ago. Crater densities and crater degradation provide tools for establishing relative ages of the volcanic and crater ejecta surfaces.

2.2. PRINCIPLES OF RELATIVE AGE DATING

Baldwin (1949) made a strong case for the impact origin of most lunar craters and for the volcanic nature of the mare plains. He also obtained general time relationships based on crater densities and superposition. In the decade prior to Apollo, geological mapping of the Moon was pioneered by Shoemaker and Hackman (1962), using principles established by Gilbert (1893). Spacecraft images (mainly Lunar Orbiter and Apollo) established rock-stratigraphic units and enhanced understanding of their formative processes.

The basic methods and the results of lunar mapping and stratigraphic analysis, as derived from telescopic and spacecraft imagery and from the analysis of returned lunar rocks, are described comprehensively in Wilhelms (1987) (Figure 3). Work on defining the relative ages of units, their geological and chemical definition, and their formative processes has continued since then, and progressed with global data obtained from the Clementine and Lunar Prospector missions (e.g., Binder, 1998; Nozette et al., 1994; Staid et al., 1996; Jolliff et al., 2000b). These missions have also helped to understand more of the third dimension of the lunar crust.

The fundamental method used is the application of the law of superposition: Younger units overlie, cut, or overlap older ones. It was apparent that older units recognized this way had more craters than younger units, consistent with the craters being of impact origin. The technique of using this relative crater density as a method of deriving relative ages when superposition relationships were lacking (e.g., non-contacting units) became standard for the Moon and for other planets, including Mars. The principles of using size-frequency distributions as measures

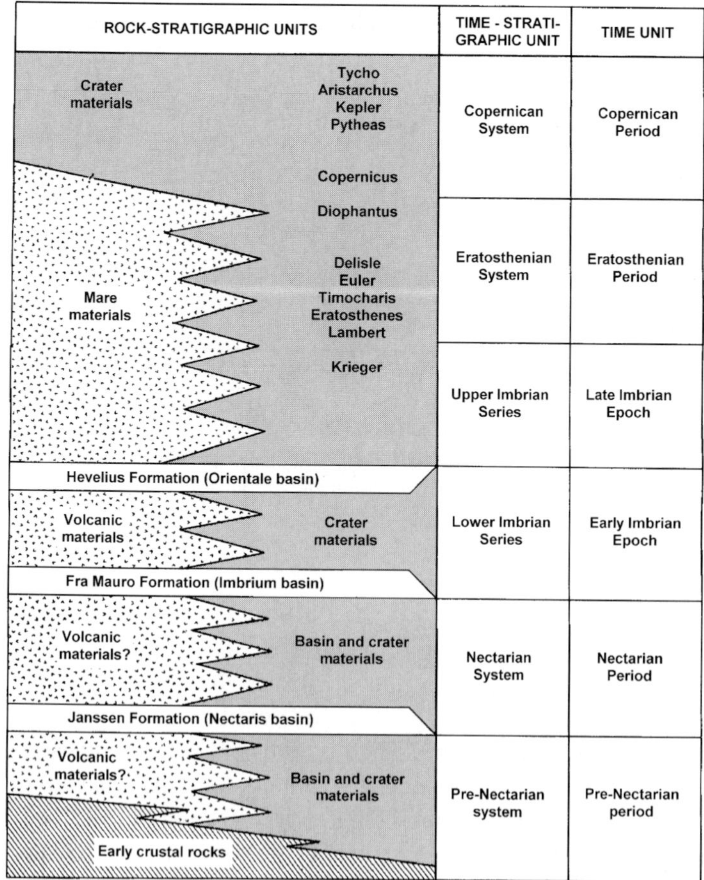

Figure 3. The lunar stratigraphic column with rock-stratigraphic, time-stratigraphic and time units (from Wilhelms, 1987).

of relative ages have been long established (reviewed in Hartmann *et al.*, 1981; Wilhelms, 1984, 1987; Neukum *et al.*, 2001; Figure 4). The degree of crater degradation is also an indicator of relative age: for a given size, a fresh crater is younger than a degraded one (*ceteris paribus*).

An erosion-based method is especially useful where the defined unit is too small for significant crater size-frequency determination. However, there are several different ways of addressing the erosion-caused morphology of a crater, and many have been used. Numerical values such as D_L relate to the size of the largest crater that is nearly destroyed (Soderblom and Lebofsky, 1972; Boyce and Dial, 1975; Wilhelms, 1980). On small surfaces, such a crater will not necessarily actually appear, and D_L is defined as the diameter of craters with the shallow wall slope of 1° (Wilhelms, 1987). Despite some pitfalls, this method has been successful and used extensively (Wilhelms, 1987).

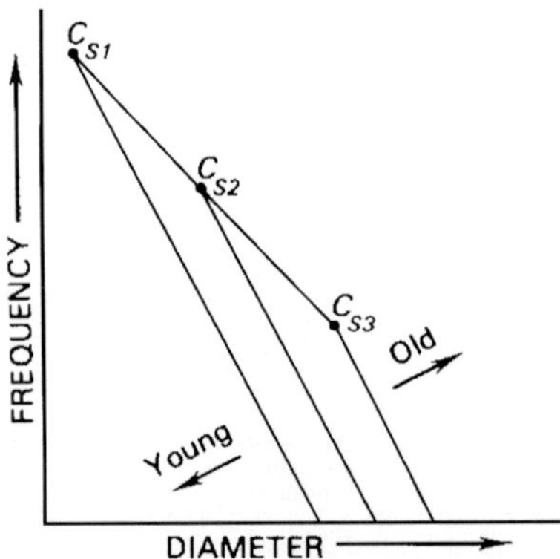

Figure 4. Principle graph for age dating of planetary surfaces by cumulative crater frequencies as a function of crater diameter; the kinks in the curves define the parameter C_S which is the transition diameter between the crater production curve (steep) and the crater saturation curve (flat); C_{S1}, C_{S2}, and C_{S3} represent increasing relative crater retention ages; from Wilhelms (1987).

2.3. PRINCIPLES OF ABSOLUTE AGE DATING

The Moon is the only extraterrestrial body for which we have rock samples that can be related to specific geological units. Most of the lunar rocks were collected by astronauts at six Apollo landing sites, and robotic sample returns from three documented sites were also accomplished by the Soviet Luna missions (Figure 1; Table I). There are also 20 lunar meteorites (not counting paired finds) (Grossman, 2000; Koblitz, 1999); however, these meteorites cannot be ascribed to specific surface units but only to general types such as highland or mare terrain.

According to the different rock types returned from the moon – igneous rocks, crystalline impact melt rocks and impact glasses, thermometamorphic rocks (granulitic lithologies) and polymict clastic matrix breccias (Table I; Stöffler *et al.*, 1980; Heiken *et al.*, 1991) – different types of ages result from radiogenic isotope dating: (1) crystallization ages defining either magmatic, impact melting or recrystallization events, (2) impact breccia formation ages defining the time of the assembly and deposition of a polymict breccia, (3) ages of thermally induced disturbances defining the time of any (mostly impact-induced) event partially or completely resetting a radiogenic isotope system, and (4) exposure ages defining the time since which an impact-displaced rock fragment was exposed to cosmic rays.

The methods of radiogenic isotope dating are well-documented (e.g., Faure, 1986; Dalrymple, 1991). For the purposes of this paper, the methods that yield the last crystallization age of a rock are the most important. The main internal isochron methods used for lunar samples are Rb-Sr and Sm-Nd, especially, though not only, for coarser-grained rocks. ^{40}Ar-^{39}Ar-methods, including laser techniques, have produced considerable data for both igneous and impact-produced rocks; they have the advantage of being applicable to fine-grained crystalline rocks as well as glasses. U,Th-Pb methods, particularly the dating of individual zircon grains using

TABLE I

Classification of lunar rocks (modified after Stöffler *et al.*, 1980) and statistics of samples at the lunar landing sites (data from Heiken *et al.*, 1991); weight percentages of the total weight of all samples of each mission given in italics; * Wt.% of total of rocks > 10 mm

Generation	Rock type	A11	A12	A14	A15	A16	A17	L16	L20	L24
	Number of samples	58	69	227	370	731	741	1 core	1 core	1 core
	Total weight (kg)	21.6	34.3	42.3	77.3	95.7	110.5	0.101	~0.05	0.170
	Wt.% igneous and metamorphic rocks > 10 mm	*44.9*	*80.6*	*67.3*	*74.7*	*72.3*	*65.9*	–	–	–
	Wt.% Fines < 10 mm	*54.6*	*16.8*	*30.6*	*17.0*	*19.3*	*26.7*	*100*	*100*	*100*
	Wt.% Drill cores	*0.4*	*1.2*	*0.9*	*6.0*	*7.4*	*6.6*	*100*	*100*	*100*
First (primary)	IGNEOUS ROCKS									
	Plutonic rocks									
	Anorthosites*				0.4	11	0.5			
	Ferroan anorthosites									
	Magnesian anorthosites									
	Alkali anorthosites									
	Norite									
	Gabbronorites (various types)									
	Troctolites (various types)									
	Dunite									
	"Granite"									
	Volcanic rocks									
	Basalts*	20.0	80.6	9.3	41.3		31.4			
	Aluminous b. (High-Al and Al-)									
	KREEP-basalt									
	Basaltic glasses (various types)									
	Non-classified igneous rocks	2.1	0.2				4.3			
Second (secondary)	METAMORPHIC ROCKS (thermometamorphic)									
	Granulites									
	Granulitic breccias									
	IMPACTITES*	23.0	1.95	59.4	39.7	67.5	34.9			
	Monomict (impact) breccias									
	Cataclastic plutonites									
	Cataclastic metamorphic rocks									
	Polymict (impact) breccias									
	Impact melt breccias, feldspathic									
	Impact melt breccias, mafic									
	Impact melt rocks (clast-free)									
	Impact glass									
	Fragmental breccias									
Third (tertiary)	*Polymict impact breccias*									
	Fragmental breccias									
	Regolith breccias									
	Impact glass									
	IMPACTOCLASTIC SEDIMENT* < 10 mm	54.8	17.2	31.2	18.5	21.1	28.8	100	100	100
	Regolith									

ion probes, have been applied in particular to those more incompatible-element enriched, coarse-grained rocks such as quartz-monzodiorites. Cosmic ray exposure ages are used to date distinct bedrock excavation events such as the young impact craters at the Apollo landing sites whose rim deposits have been sampled (e.g., North Ray, South Ray, Cone, Shorty, Camelot craters). General reviews of the results of lunar chronology have been provided by Turner (1977), Dalrymple (1991), Nyquist and Shih (1992), and Snyder *et al.* (2000), among others.

The geological and stratigraphic interpretation of isotope ages of lunar rocks is not always straightforward. There is the immediate problem when correlating a measured age with a rock-stratigraphic unit that none of the rock samples was collected directly from a bedrock unit (although some came close at the Apollo 15 landing site) as the entire lunar surface is covered with impact-produced regolith several meters thick. Specific problems arise for the different rock types:

Volcanic rocks of mare provenance. Even for the comparatively simple case of a volcanic rock, it is not necessarily easy to relate that rock to a mapped geological unit. At any given mare collection site there is a range of basalt types - brought to the surface by multiple reworking of the regolith - that in some cases covers a distinct range of ages. While the youngest of these is most probably the age of the surface unit, if that unit is thin or discontinuous it might not be the surface that is mapped and which retains the crater density/crater degradation characteristics used to define the age of the unit in question.

Impact melt and clastic breccia lithologies of highland provenance. Radiometric age dating of impact melt rocks is generally possible by direct dating of the glassy or crystalline matrix. However, since datable impact melt rocks are either displaced indiviual clasts within the lunar regolith or displaced clasts residing in polymict breccias, it is not obvious what geologic unit they were excavated from and what impact crater they represent. For polymict clastic impact breccia deposits, the age can only be constrained to be younger than that of the youngest clast, because assembly of the breccia components and their deposition occurs at too low a temperature to reset radiogenic isotopes of the clasts. For ancient clastic breccia deposits produced at times when the impact rate was high, the youngest clast is likely to be very close to the assembly age. Complete or partial resetting of clasts is possible for clasts residing in impact melt breccias. In this case, the oldest clast gives a lower limit for the age of the precursor rocks of the impact melt unit (e.g., Jessberger *et al.*, 1977). Datable impact melts (including glasses) that intrude or coat a breccia also define a minimum age for a breccia. The thermal events that have reset or disturbed the radiogenic isotope systems in lunar rocks can be used to either date units directly (e.g., granulitic rocks) or to constrain their ages, according to geologic constraints. For all types of highland rocks, a meaningful interpretation of their geologic provenance and the correlation with a time-stratigraphic unit can only be made on the basis of photogeologic models of their parent geologic formations (see Section 6).

3. The Lunar Stratigraphic Column

Lunar stratigraphy establishes geologic units and assembles them into a relative time-sequenced column of global significance. A pre-requisite is the identification of rock units or morphological units formed in a single stage process. These are dominantly related to impact basins, impact craters, and lava flows. Morphological units, rather than the exposed bedrock, are necessarily used in photo-based stratigraphy. The units are assembled into higher-order packages. The rock System boundaries are intended to be the same absolute age everywhere. This chronostratigraphic division (Systems, Series) can be converted into a chronometric division (Periods, Epochs), and with application of radiogenic ages into absolute time. Figure 3 shows these stratigraphic columns, following Wilhelms (1987).

The stratigraphic boundaries are reasoned, convenient markers that subdivide lunar history, but they do not generally imply any fundamental changes in geological processes. The older boundaries are defined by the deposition of ejecta from specific basin-forming impact events, ending with the deposition of the Hevelius Formation (Orientale, the last basin-forming event). The boundaries of the later Systems are less precisely defined; although depending on quantitative expressions of the relative degrees of crater degradation, the criteria are not entirely unambiguous in practical application. Lava flows do not define stratigraphic boundaries as they are themselves only regional (Figure 5; Table II).

3.1. PRE-NECTARIAN SYSTEM

The *pre-Nectarian System* comprises all landforms *older* than the Nectaris basin, and includes about 30 recognized impact basins. Some directly underlie deposits of Nectaris, and others are recognized as pre-Nectarian by the size-frequency curves of superimposed craters (Figure 5; Table II). They are chronologically sequence-able at least roughly. The oldest recognized basin is Procellarum, but this may well not be of impact origin. The oldest – also the deepest and largest – basin of almost-certain impact origin is South Pole-Aitken. The pre-Nectarian landforms are dominantly of impact origin; no volcanic landforms, or even faults or folds have been recognized. Pre-Nectarian terrain is predominant on the farside (Figure 6). Any rocks from such ancient terrains in the Apollo or Luna sample collections have been reworked as fragmental material into later impact breccia deposits.

3.2. NECTARIAN SYSTEM

The *Nectarian System* comprises all landforms produced between the formation of the Nectaris impact basin and the formation of the Imbrium impact basin (Figure 5; Table II). Nectaris itself has deposits over a fairly wide area. Eleven other Nectarian basins have been recognized, including Serenitatis and Crisium. Direct superpositional relationships allow some definition of their stratigraphic sequence, but some crater frequency distributions have been affected by later basins, e.g.,

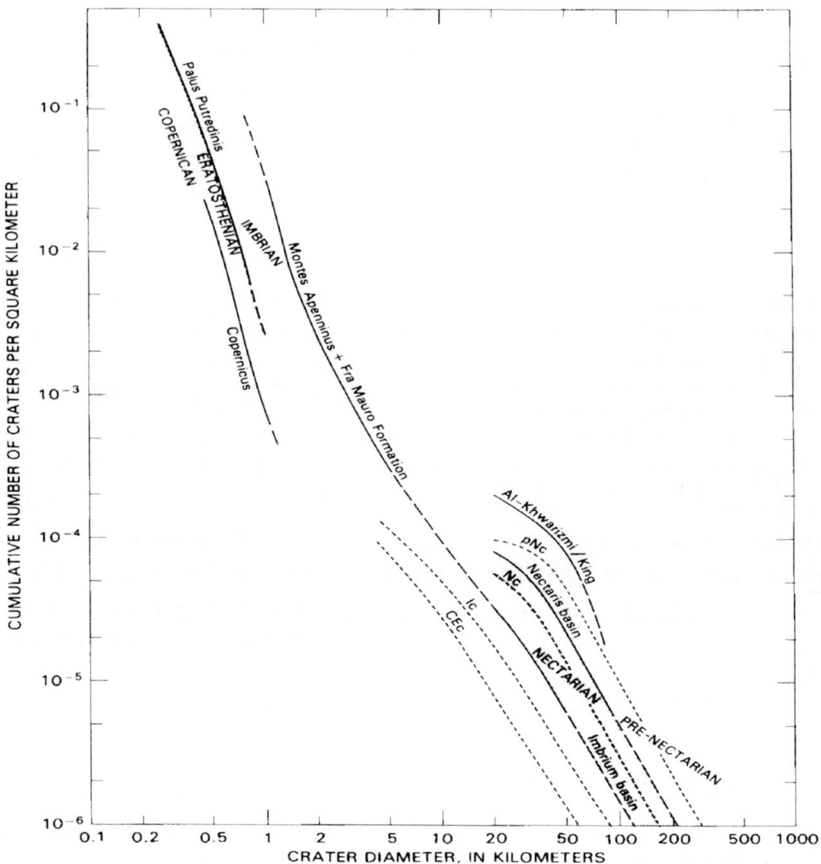

Figure 5. Cumulative crater frequency-crater diameter-curves for the major rock-stratigraphic and time stratigraphic units of the moon's history as listed in Figure 3; dashed curves are average frequencies of impact craters of the pre-Nectarian (pNc), Nectarian (Nc), Imbrian (Ic), and the Copernican/Eratosthenian Periods (CEc); from Wilhelms (1987).

Serenitatis ejecta is badly degraded by Imbrium ejecta. Nectarian "light plains" are more evident than are pre-Nectarian ones, and some of these have been suggested to be volcanic in origin (Figure 7; Wilhelms, 1987). The Nectarian System has been masked by the Imbrian basins and later volcanic activity; thus, it is more common on the lunar farside (Figure 6).

3.3. LOWER IMBRIAN SERIES

The *Lower Imbrian Series* comprises all landforms produced between the formation of the Imbrium impact basin and the formation of the Orientale impact basin (Table II). The deposits of these two basins constitute extensive laterally continuous horizons, although much of the Imbrium basin itself was later flooded with mare lavas. Crater counts as well as its topographic freshness suggest that the

TABLE II

Stratigaphic criteria for lunar time-stratigraphic units (Wilhelms, 1987); D_L: diameter of largest crater eroded to $1°$ interior slopes; C_S: limiting crater diameter for the steady state crater frequency distribution (Figure 4) and from the approximate formula $D_L = 1.7C_S$; n/a = not applicable.

System or Series	Crater frequency (number per km^2)		C_S (m)	D_L (m)
	≥ 1 km	≥ 20 km		
Copernican System	$< 7.5 \times 10^{-4}$ (mare) $< 1.0 \times 10^{-3}$ (crater)	n/a	?	< 165 (mare) $< ca.$ 200 (crater)
Eratosthenian System	7.5×10^{-4} to $\sim 2.5 \times 10^{-3}$ (mare)	n/a	< 100 (mare)	145–250 (mare)
Upper Imbrian Series	$\sim 2.5 \times 10^{-3}$ (mare) to $\sim 2.2 \times 10^{-2}$	2.8×10^{-5}	$80 - 300$ (mare)	$230 - 550$ (mare)
Lower Imbrian Series	$\sim 2.2 - 4.8 \times 10^{-2}$ (basin)	$1.8 - 3.3 \times 10^{-5}$	$320 - 860$ (basin)	n/a
Nectarian System	n/a	$2.3 - 8.8 \times 10^{-5}$	$800 - 4{,}000?$ (basin)	n/a
Pre-Nectarian System	n/a	$> 7.0 \times 10^{-5}$	$> 4{,}000?$ (basin)	n/a

Schrödinger basin is Lower Imbrian, but no other basins are in the Lower Imbrian. The size-frequency distribution curves for Orientale deposits lie slightly below those for Imbrium, and both are distinctly below those for Nectaris (Figure 5). Many "light plains", including the Cayley plains on which Apollo 16 landed, have a Lower Imbrian age (Figure 7), and many may be related to the Imbrium basin. At least some light plains may also be of volcanic origin, despite the absence of volcanic rocks in the Apollo 16 samples. The Apennine Bench formation, a plains unit within the Imbrium basin, is almost certainly a volcanic unit, as it correlates with chemically volcanic KREEP basalts collected at the Apollo 15 landing site.

3.4. UPPER IMBRIAN SERIES

The *Upper Imbrian Series* comprises the landforms produced between the formation of Orientale, the youngest impact basin, and an upper boundary that is defined on the basis of D_L values (Table II). The Upper Imbrian rock units are distinct from older ones: Basin deposits are lacking, and two-thirds of the mare volcanic plains are in the Upper Imbrian (Figure 8). The Upper Imbrian was emplaced over a much longer time period than the Lower Imbrian . The extensive mare lavas forming Maria Serenitatis, Tranquillitatis, Crisium, Nectaris, Fecunditatis, Humorum, Nubium, Cognitum, eastern Imbrium, and western Oceanus Procellarum, and several other areas including all the farside mare plains, are part of the Upper

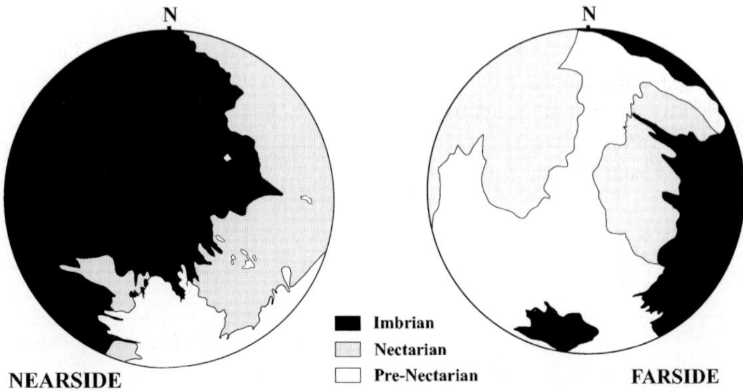

Figure 6. Geologic map of the nearside and farside of the pre-Eratosthenian moon showing the Imbrian, Nectarian, and pre-Nectarian Systems (compiled from Plates 3 A,B of Wilhelms, 1987).

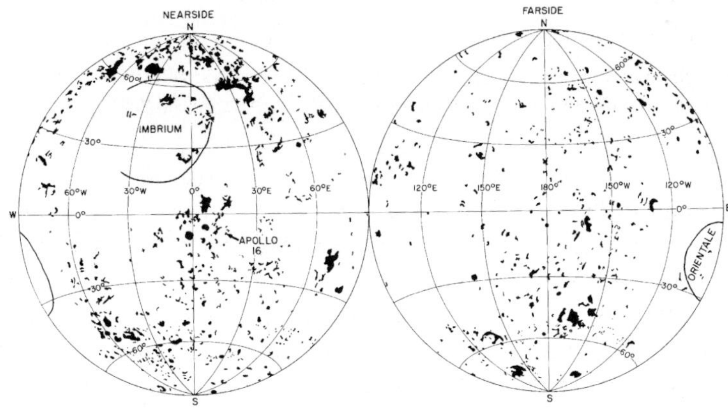

Figure 7. Geologic map of the distribution of "light plains" on the nearside and farside of the moon (from Wilhelms, 1987).

Imbrian Series (Figure 8). Their relative ages have been established on the basis of crater frequencies and by superposition, and their stratigraphic relationships have been elucidated with mineral-chemical data derived from earth-based spectral reflectance observations and from the Clementine and Prospector orbital data (e.g., Staid *et al.*, 1996; Jolliff *et al.*, 2000b). The Upper Imbrian also contains "dark mantling deposits" that have been correlated with volcanic glass of fire-fountain origin, examples of which have been sampled by the Apollo missions. With the exception of the Apollo 12 mission, all mare plains sampled by Apollo and Luna were Upper Imbrian, and no samples were from the oldest stratigraphic group.

3.5. ERATOSTHENIAN SYSTEM

The *Eratosthenian System* is less clearly defined than other systems (Table II). Its upper boundary is even more ambiguous than the lower. Operationally the distinction between the Eratosthenian and the subsequent Copernican systems was

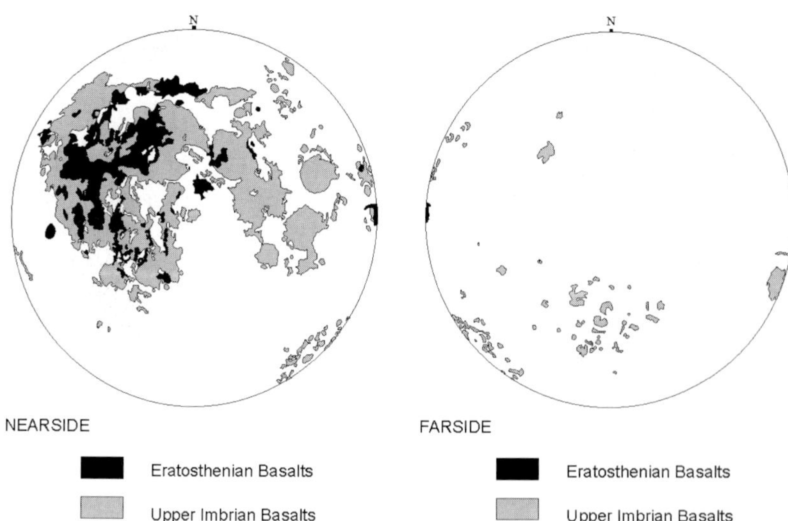

Figure 8. Geologic map of mare basalts of the Upper Imbrian Epoch and the Eratosthenian Period on the nearside and farside of the moon; compiled from Plates 9 A,B and 10 A,B of Wilhelms (1987).

made according to whether a crater was non-rayed (Eratosthenian) or bright-rayed (Copernican). However, the presence of rays depends not only on age, that is degree of impact erosion, but also on crater size and on compositional differences. Thus, a variety of criteria have been used, according to circumstance. Crater counts on ejecta blankets of Copernican craters are of limited use, because most craters are too small for good statistics (Table II; Figure 5). The same limited-area constraint applies to D_L ($\sim$140 m) as well. A new parameter of optical maturity derived from global orbital spectral reflectance measurements from the Clementine spacecraft is promising to enable a more rigorous establishment of the relative ages of Copernican and upper Eratosthenian craters.

The *Eratosthenian System* includes mare plains that are much less extensive than Upper Imbrian plains. They are absent from the lunar farside (other than Mare Smythii on the limb; Figure 8). The plains include those sampled at the Apollo 12 site in Oceanus Procellarum.

3.6. COPERNICAN SYSTEM

The *Copernican System* was first recognized by the rays of its craters, which were shown to be the youngest of lunar features because the are superposed on all other terrains (Table II; Figure 5). Despite the difficulty of using rays in defining a lower boundary, most rayed craters are indeed Copernican, and such craters are scattered all over the Moon. The upper boundary of the Copernican is the present day. Only a very small proportion of the Moon's face is Copernican.

4. Geologic Setting of the Apollo and Luna Landing Sites and Probable Meteorite Provenances

Six manned Apollo missions and three robotic Luna missions returned samples from different geologic settings on the Moon (Figure 1, Table I). 20 distinct meteorites are recognized as being of lunar origin; the types of provenance from which they must be derived can be constrained. We review the settings of the samples on which absolute time calibration of the relative stratigraphy is based, in numerical order of Apollo missions, Luna missions, and meteorites.

4.1. APOLLO 11 LANDING SITE, MARE TRANQUILLITATIS

The landing site, 40 km north-northeast of the nearest promontory of highlands material at the Kant Plateau (Figure 9a), is on intermediate-age-group basalts of the Upper Imbrian Series, the southern of two belts separated by the youngest-age-group. That the lavas are thin is suggested both by the lack of a mascon (gravity anomaly) in Tranquillitatis and the existence of structures over probable pre-existing features such as ridges.

Three patchy units of mare basalt are in the area within at least several tens of kilometers of this site (Grolier, 1970, b). The actual landing was on the most densely cratered of these units. The second oldest unit is exposed in substantial areas within a kilometer of the landing site. The youngest unit occurs in small rare patches, the nearest of which is a few kilometers away. The sampling site is approximately 400 m west of a sharp-rimmed, rayed crater approximately 180 m in diameter and 30 m deep (West Crater) (Figure 9b) in an area where the regolith is 3 to 6 m thick. Beaty and Albee (1978) suggested that most of the samples collected were ejected from West Crater. The samples include a diverse set of mare basalts and regolith breccias. Fragments within the soils and soil breccias show the presence of highland material, much in the form of feldspathic granulites and green glass.

4.2. THE APOLLO 12 LANDING SITE, MARE INSULARUM, SOUTHERN OCEANUS PROCELLARUM

The site is in a region of mare basalts of a younger age (Eratosthenian) and spectral type different from those at the Apollo 11 site (Figure 9c). Highland islands within about 15 km show that the basalts are quite thin, and the area has a complex topography. The nearby highlands are mainly Fra Mauro Formation, the ejecta blanket of the Imbrium basin. Crater frequencies and D_L show that the basalts at this site form a terrain distinguishably older than that about 1 km away to both east and west.

Nearly all of the sampled terrain is dominated by ejecta of several craters larger than 100 m (Figure 9d). The site is close to the rim of Surveyor crater (300 m diameter). Overlapping blankets contribute uncertainty in where particular samples were excavated. A ray from Copernicus crosses the site, and the astronauts observed high-albedo material in several locations. Since regolith appears to be only about half the thickness of the Apollo 11 regolith, craters only 3 m deep penetrate into bedrock. The samples are mainly mare basalts, with some regolith breccias. Among the smaller samples are ropy glasses, and impact melt fragments similar to those collected at the Apollo 14 landing site nearby. Some may represent Copernicus ejecta.

4.3. THE APOLLO 14 LANDING SITE, NEAR FRA MAURO CRATER

The landing site (Figure 9e) is about 1230 km south of the Imbrium basin center and 550 km south of its southern rim crest, near the outer edge of the Fra Mauro Formation. This Formation forms a broad continuous belt of ridges and grooves surrounding and morphologically sculptured by the Imbrium basin event and therefore interpreted as its continuous ejecta blanket (Figure 5). The exact emplacement mechanism and the proportion of Imbrium ejecta to local reworked material at the landing site remains debatable (Oberbeck, 1975; Wilhelms, 1987). Although it was first assumed that all of the material was primary Imbrium ejecta, later studies provided evidence that much of the material is locally derived.

The landing area forms a smooth terrain about 1100 m west of Cone Crater (Figure 9f), which is 340 m in diameter and about 75 m deep, with ejected blocks up to 15 m across. The landing area is densely populated with subdued craters up to several hundred meters across and a regolith 10 to 20 m thick (Swann *et al.*, 1971). Sampling was from both Cone Crater ejecta, which almost certainly represent the Fra Mauro Formation, and the smooth terrain. Most samples are fine-grained, clast-rich impact melt breccias; some are coarser, clast-free impact melt rocks, regolith breccias, and aluminous mare basalts. At Cone crater rim, feldspathic fragmental breccias were sampled. These polymict breccias contain a variety of lithic clasts such as aluminous mare basalts and diverse crustal rocks.

4.4. THE APOLLO 15 LANDING SITE, PALUS PUTREDINIS, AND HADLEY-APENNINES

The site is located on a mare plain of the youngest group of the Upper Imbrian Series (Table II; Figure 5), about 2 km from Hadley Rille, whose walls expose a layered mare basalt sequence (Figure 9g). The maria flood an embayment in the Apennine front, a scarp that is the Imbrium basin's main rim crest, rising abruptly to 3.5 km above the mare plains at Hadley Delta, only 4 km south of Apollo 15.

Extensive lava plains occur to the west of the landing site, and hummocky ejecta of the Imbrium basin to the east. The Lower Imbrian Apennine Bench Formation, a "light plains" unit (Figures 7 and 10) inside the Imbrium rim, is exposed within a few tens of kilometers of the site and probably underlies the Upper Imbrian lavas at the landing site. Rays from craters Autolycus or Aristillus cross the landing site. The regolith varies widely in thickness, according to terrain. It is only about 5 m deep near Apollo 15 and absent close to the Hadley rille.

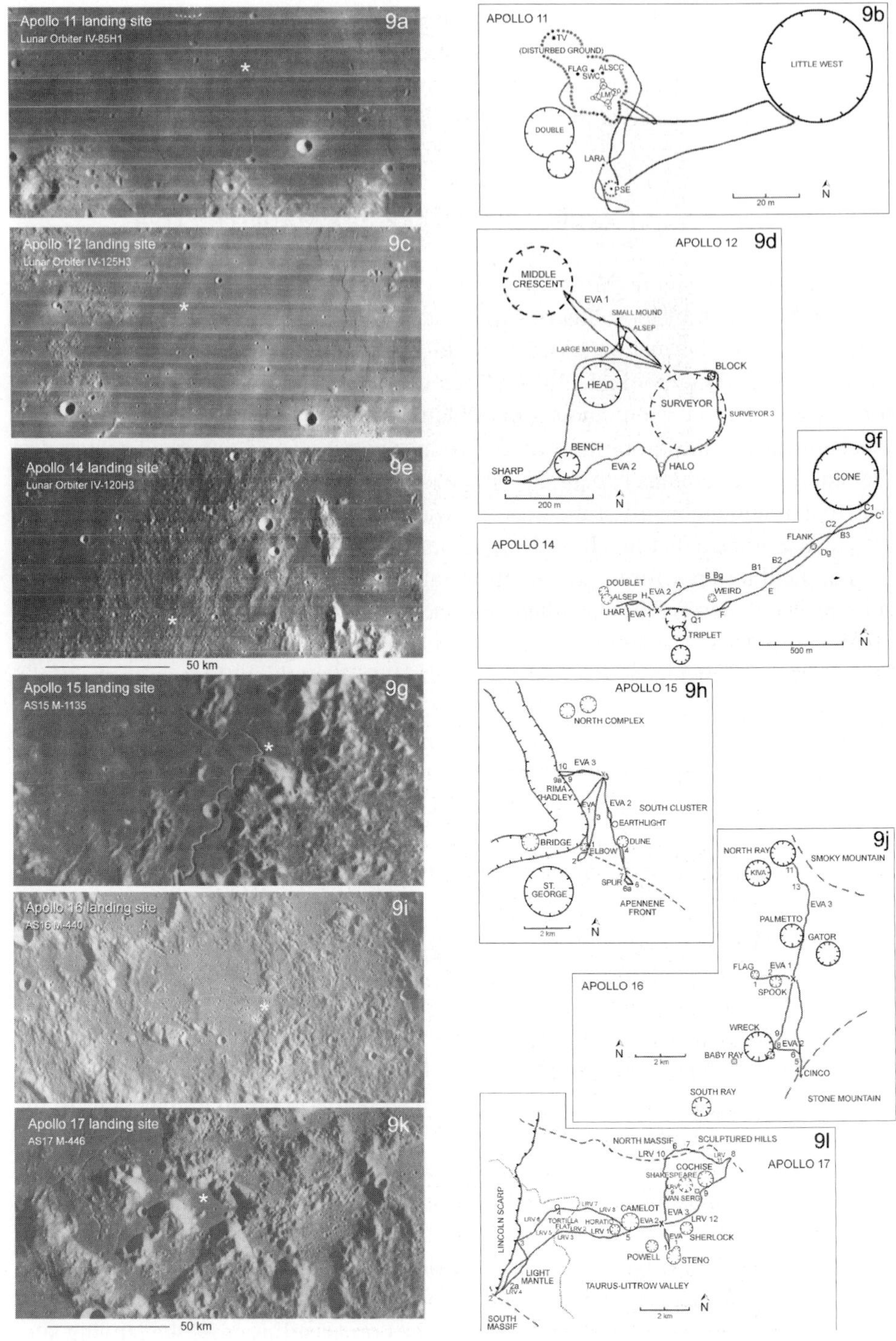

Apollo 11 landing site
Lunar Orbiter IV-85H1
9a
APOLLO 11
9b
TV
(DISTURBED GROUND)
FLAG
ALSCC
SWC
LM
LITTLE WEST
DOUBLE
LARA
PSE
20 m
N
Apollo 12 landing site
Lunar Orbiter IV-125H3
9c
APOLLO 12
9d
MIDDLE CRESCENT
EVA 1
SMALL MOUND
ALSEP
LARGE MOUND
BLOCK
HEAD
SURVEYOR
SURVEYOR 3
BENCH
EVA 2
HALO
SHARP
200 m
N
Apollo 14 landing site
Lunar Orbiter IV-120H3
9e
9f
CONE
C1
C2
C'
FLANK
B3
B2
Dg
B1
DOUBLET
A
B Bg
ALSEP H
EVA 2
WEIRD
E
LHAR
EVA 1
Q1
F
TRIPLET
APOLLO 14
500 m
N
50 km
Apollo 15 landing site
AS15 M-1135
9g
APOLLO 15
9h
NORTH COMPLEX
10
EVA 3
9a 9
RIMA HADLEY
EVA
3
EVA 2
SOUTH CLUSTER
EARTHLIGHT
BRIDGE
DUNE
ELBOW
2
SPUR
ST. GEORGE
APENNINE FRONT
2 km
N
Apollo 16 landing site
AS16 M-440
9i
9j
NORTH RAY
KIVA
11
SMOKY MOUNTAIN
13
EVA 3
PALMETTO
GATOR
FLAG
EVA 1
1
SPOOK
WRECK
9
EVA 2
BABY RAY
CINCO
SOUTH RAY
STONE MOUNTAIN
APOLLO 16
N
2 km
Apollo 17 landing site
AS17 M-446
9k
NORTH MASSIF
6 7
SCULPTURED HILLS
LRV 10
8
COCHISE
SHAKESPEARE
APOLLO 17
9l
VAN SERG
LRV 7
CAMELOT
LRV 6
TORTILLA
EVA 3
LRV 12
FLAT
HORATIO
LRV 2
EVA 2
EVA
LRV 5
LRV 1
SHERLOCK
LRV 3
5
POWELL
STENO
LIGHT MANTLE
TAURUS-LITTROW VALLEY
LRV 4
SOUTH MASSIF
2 km
N
50 km

Figure 9. (a) Apollo 11 landing area, Mare Tranquillitatis, (b) Map and sampling traverses at the Apollo 11 site, (c) Apollo 12 landing area, Oceanus Procellarum, (d) Map and sampling traverses at the Apollo 12 site, (e) Apollo 14 landing area, Fra Mauro Formation, (f) Map and sampling traverses at the Apollo 14 site, (g) Apollo 15 landing area, Palus Putrenis, Mare Imbrium, and Hadley Delta, (h) Map and sampling traverses at the Apollo 15 site, (i) Apollo 16 landing area, Descartes region, (j) Map and sampling traverses at the Apollo 16 site, (k) Apollo 17 landing area, Taurus-Littrow region, Mare Serenitatis, (l) Map and sampling traverses at the Apollo 17 landing site, (m) Landing areas of the Luna 16, 20, and 24 missions, Mare Crisium and Mare Fecunditatis.

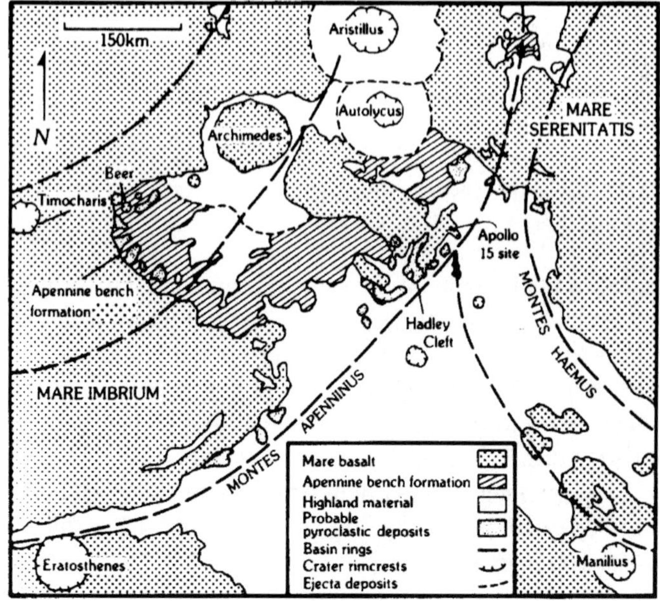

Figure 10. Geological map of the Apennine Bench Formation and the Apollo 15 landing area, Mare Imbrium (from Taylor, 1982).

Samples from the mare plains near the Hadley rille edge and in the so-called South Cluster are mainly mare basalts and regolith breccias, whereas samples found as high as 130 m on the Apennine Front (Figure 9h) are mainly varied impact breccias, some anorthosites, and clods of volcanic mare green glass. Small fragments of volcanic KREEP basalt were found in all sampling areas.

4.5. APOLLO 16 LANDING SITE, DESCARTES, CENTRAL LUNAR HIGHLANDS

The landing site was on the Cayley Formation, which is subdued smooth "light plains", Lower Imbrian Series, sculptured by the Imbrium event (Figures 5 and 7), and most probably part of its discontinuous ejecta. The site is 60 km west of the Kant Plateau, which is part of the Nectaris basin rim deposits about 100 km wide (Figure 9i). It is near to the hilly and furrowed Descartes Formation (Muehlberger *et al.*, 1980), which is most probably related to the Nectaris ejecta blanket. Light plains similar to the Cayley Formation are common around the Imbrium basin outside of the Fra Mauro Formation (Figure 7).

The landing area contains numerous overlapping craters in the 500 m size range, mostly subdued. Two young fresh craters, North Ray (1 km wide, 230 m deep) and South Ray (680 m wide, 135 m deep), as well as Stone Mountain and the subdued plains enabled to sample materials from both major Formations (Figure 9j). The samples from the rim of North Ray are almost certainly derived from the Descartes Formation. The regolith on both the Cayley and Descartes Formations is on average about 6 to 10 m thick (Freeman, 1981) although it varies from 3 to 15 m.

Dominantly friable feldspathic fragmental breccias and impact melt lithologies were collected. The latter have a wide range of textures and compositions from very feldspathic to mafic (aluminous basaltic). Anorthosites, mostly cataclastically brecciated, are common as both individual rocks and as clasts in polymict breccias; feldspathic granulites are common, mainly as clasts within breccias.

4.6. APOLLO 17 LANDING SITE, TAURUS-LITTROW VALLEY, SOUTHEAST SERENITATIS RIM

Apollo 17 landed on mare plains of the same intermediate-age-group of the Upper Imbrian Series that occupies northern Mare Tranquillitatis near Apollo 11 (Table II; Figure 5). The site is located in a mare-flooded valley, a radial graben in the massifs that form a main topographic rim of the Serenitatis basin (Figure 9k). While the massifs, rising to 2 km above the floor, are dominantly of Serenitatis origin (and therefore Nectarian), and consist of autochthonous and/or allochthonous pre-Serenitatis material, later events, including Imbrium, also influenced the topography. The subfloor basalt at the landing site is about 1.4 km thick. Much of the surface of massifs and mare in the area is covered with a "dark mantling material", correlated with volcanic orange glass sampled at the site.

Sampling was done in the mare valley floor near numerous fresh clustered craters, at the foot of the North Massif, and in a bright landslide of the South

Massif (Figure 9l). The latter and some of the clustered craters are presumably caused by ejecta from the ~ 0.1 Gyr old crater Tycho, whose rays extend through the area. The regolith on the subfloor basalts is up to 15 m thick. On the valley floor, dominantly mare basalts and some regolith breccias were collected, and dark mantle material was sampled as orange glass deposits. Most of the sampled boulders and large rocks that are derived from the Massifs are impact melt breccias, most commonly a mafic poikilitic variety. One boulder is composed of an aphanitic, chemically more diverse melt breccia. Fragments of old igneous rocks (dunites, norites etc.) are present as clasts in these melt breccias. One boulder is a norite. Conspicuously absent are anorthosites. Abundant small fragments suggest that much of the massif material is composed of feldspathic granulite.

4.7. LUNA 16 LANDING SITE, NORTHEASTERN MARE FECUNDITATIS

Luna 16 landed on mare lavas that are thin (slightly more than 1 km in the center and about 300 m at the landing site, De Hon and Waskom, 1976), but flood a large area of the extremely degraded, 690 km diameter Fecunditatis basin of pre-Nectarian age, about 400 km south of highlands formed by the ejecta blanket of Crisium basin (Figure 9m). The plains are in the middle to upper part of the Upper Imbrian System (Table II; Figure 5).The landing site is midway between the Eratosthenian crater Langrenus (132 km) and the Copernican crater Taruntius (56 km), whose ejecta and ray material as well as those of the more distal Theophilus and Tycho brighten the surface in the vicinity (McCauley and Scott, 1972).

Drilling to a depth of 35 cm (Vinogradov, 1971) provided 101 g of dark gray regolith with preserved stratigraphy. There was no visible layering, but 5 zones of mildly increasing grain size with depth were recognized (70 to 120 μm for the <1 mm fraction). Most of the few particles >3 mm are of felspathic mare basalt or minerals derived from them; others are glassy agglutinates and regolith breccias. A small amount of feldspathic highland material is present (e.g., Keil *et al.*, 1972).

4.8. LUNA 20 LANDING SITE, APOLLONIUS HIGHLANDS, SOUTHERN CRISIUM BASIN EJECTA

Luna 20 landed on the southern rim deposits of the Nectarian Crisium basin (Figure 9m), about 35 km north of the mare plains of Fecunditatis. The region consists of smooth rounded hills and shallow linear valleys, giving the area a hummocky appearance (Heiken and McEwen, 1972). The landing site, about 1 km higher than the surface of Mare Fecunditatis (Vinogradov, 1973), is on relatively smooth material characteristic of depressions in the area. Apollonius C (10 km) is a fresh Copernican crater only a few kilometers to the east and may have contributed to the site, but is in any case in similar highlands.

A core of $\sim$50 g of fine-grained light gray regolith (with a median grain size of about 70 microns) was collected by drilling analogous to Luna 16, but was returned only half full (note in Vinogradov, 1973). The core appeared to have no stratification, but it may have been mixed during transport. Most of the fragments are feldspathic granulites, although the bulk soil is somewhat less aluminous than Apollo 16 soil or the highlands meteorites. A few fragments are of mafic impact melt similar to those collected at the Apollo 16 and 17 landing sites. Mare basalt is a very minor component. Glasses and agglutinates are also present.

4.9. LUNA 24 LANDING SITE, SOUTHEASTERN MARE CRISIUM

Luna 24 landed on the Upper Imbrian mare plains that flood the Crisium basin (Figure 9m), about 40 km north of the basin rim which reaches 3.5 to 4 km above the mare level (Florensky *et al.*, 1977; Butler and Morrison, 1977). The landing site lies in the inner mare, inside an inner ring of Crisium that is the cause of mare ridges. Mare Crisium is fairly uniform but three successive main units have been mapped by Head *et al.* (1978). Luna 24 landed on the upper part of the middle-age-group that is common in the northern part of the basin but forms exposed patches in the south as well. Various lines of evidence suggest that the mare at the site is at least a kilometer and perhaps 2 kilometers thick. Numerous rays cross the basin, suggesting several possible sources of small amounts of highlands material (Maxwell and El-Baz, 1978). There are several bright patches in the vicinity, and crater Giordano Bruno is a likely cause (Florensky *et al.*, 1977).

The core device was different from that of Luna 16 and 24, and resulted in a mass of 170 g of regolith. The core was 160 cm long with a diameter of 12 mm, with a nominal penetration of $\sim$ 225 cm to a depth of $\sim$ 200 cm (intentionally off vertical). Stratification was well-preserved, and four depth zones were defined based on color and fragment sizes (Florensky *et al.*, 1977; Barsukov, 1977). Most of the core is clearly fine-grained regolith. Some larger particles, up to 10 mm in size and coarser than any found in the Luna 16 or 20 cores, are basaltic rocks ranging from very low Ti-basalt to olivine basalt. Other fragments include glasses, breccias, and agglutinates.

4.10. LUNAR METEORITE PROVENANCES

Twenty three meteorite fragments of lunar origin (Section 2.3) have been recognized by their petrographic, geochemical (e.g., FeO/MnO), and oxygen isotopic characteristics (Grossman, 2000). They probably represent 20 meteorite falls because some samples are paired. While the Apollo and Luna samples are from a limited though purposefully diverse region of the lunar frontside, meteorite samples are expected to be from randomly distributed locations. Eleven of the meteorites are regolith breccias of feldspathic highlands derivation; 1 is a feldspathic fragmental breccia of highland composition; 1 is a melt breccia from a KREEP-rich terrain; 6 are non-brecciated mare basalts; and 1 is a fragmental breccia almost entirely of mare derivation. Although their components and chronology in general ways constrain the history of the Moon, the meteorites cannot be assigned to specific geological units on the Moon, and thus do not help to calibrate the impact flux.

5. Radiogenic Isotope Ages of Lunar Rocks

Radiogenic isotope ages have been determined on many lunar rocks collected at the Apollo and Luna landing sites as individual fragments or as clasts within polymict breccias, and for lithic and mineral fragments extracted from lunar meteorites. The data are from several different methods, especially Rb-Sr and Sm-Nd isochron and ^{40}Ar-^{39}Ar stepwise-heating methods. Here, the data which directly date, or indirectly constrain, the age of morphological units are relevant (Section 6). In particular, these are crystallization ages for igneous rocks and impact melts, and metamorphic ages of recrystallized rocks. For complete data compilations, see the reviews by, e.g., Heiken *et al.* (1991), Nyquist and Shih (1992), Dalrymple (1991), Papike *et al.* (1998), Snyder *et al.* (2000), Nyquist *et al.* (2001a, 2001b).

Table III lists ages, mostly inferred to be crystallization ages, of specific plutonic and volcanic ancient highlands igneous rocks, including ancient mare basalts. They rarely date a unit directly, but provide a lower limit in some cases (Section 2).

Table IV, although not complete, represents crystallization and recrystallization age data of clasts in specific polymict highlands rocks, mainly impact melts and granulitic breccias. Most of these samples are fine-grained, so that ^{40}Ar-^{39}Ar ages dominate. Few of these ages directly date specific geologic units (Section 2). Some impact melt rocks, e.g. at Apollo 17, may, on geological grounds, directly date impact basins such as Serenitatis, as enclosed clasts in a breccia unit constrain its age. Some of these units have been inferred to be ejecta from large craters such as Copernicus, and reheated by those events, providing a means of dating them.

Unlike Tables III and IV, listing ages of specific samples, Table V presents age estimates for groups of multiple mare basalt samples, identified by their chemistry, isotopes, and petrography to belong to a single event and single unit. The different basalt group ages vary at each landing site, so that geological arguments are needed to determine the best age of the surface for which crater counts are available.

Exposure ages, dating when the surface of a sample was exposed to the vagaries of cosmic and solar influences, offer additional criteria, but are limited to rim deposits of young Copernican craters at the Apollo sites, such as North Ray and South Ray craters (Apollo 16), Cone Crater (Apollo 14), or the landslide and secondary cratering at the Apollo 17 site inferred to be from Tycho (Table VI). Others reflect purely local events of no great stratigraphic significance for this paper.

6. Absolute Ages of Cratered Lunar Surface Formations

6.1. PRE-NECTARIAN PERIOD

The pre-Nectarian Period as a time unit is the time span between the origin of the moon and the formation of the Nectaris basin, which is most plausibly ~3.92 Gyr old (next section). Since the oldest plausible age of solid lunar surface material

is 4.52 Gyr (Lee *et al.*, 1997; Halliday, 2000), a duration of the pre-Nectarian
Period of ~ 600 Myr is suggested. The pre-Nectarian system is recorded by (1) the
impact formations of some 30 multi-ring basins and their ejecta deposits identified photogeologically, and (2) returned samples of rocks whose absolute ages are
older than Nectaris. The suite of "plutonic" pre-Nectarian rocks comprises ferroan
anorthosites, alkali anorthosites and rocks of the so-called Mg-suite (troctolites,
norites, dunites, and gabbronorites). Clasts of aluminous mare basalts, rare clasts
of impact melt rocks, and granulitic lithologies also display pre-Nectarian ages
(Table III). All these rock types document the existence of magmatic, thermometamorphic, and impact processes throughout the pre-Nectarian Period. None of the
dated pre-Nectarian rock clasts can be directly related to the geologic unit (formation) in which they formed or to any specific pre-Nectarian surface unit because
they were all displaced after their formation by multiple impacts.

The relative ages of most of the pre-Nectarian multi-ring basins are documented
on the basis of crater counts on their ejecta formations. Wilhelms (1987) distin-
guishes 9 age groups in which the density of craters >20 km per 10^6 km^2 (Hartmann and Wood, 1971) increases from 79 (Nectaris) to 197 (Al-Khwarizimi/King)
(Figure 5). In Wilhelms' (1987) scenario, no multi-ring basins older than 4.2 Gyr
are unequivocally recorded, thus implying that the oldest basins, South Pole-Aitken
and Procellarum, and some 14 obliterated basins formed between 4.2 and 4.1 Gyr.

6.2. NECTARIAN PERIOD

Twelve multi-ring basins of Nectarian age have been identified (Wilhelms, 1987;
Spudis, 1993). The superimposed crater densities (craters $>$ 20 km per 10^6 km^2)
on the ejecta formations of these basins range from 31 for Bailly to 79 for Nectaris
(Figure 5). Ejecta are inferred to have been sampled at Apollo and Luna landing
sites for Nectaris, Crisium, and Serenitatis (Apollo 16, Luna 20, and Apollo 15 and
17, respectively). The attempts to assign absolute ages to these basins are based on
samples from these landing sites (Tables IV and VI).

6.2.1. *Age of the Nectaris Impact Basin*

The age of the Nectaris basin is mainly derived from radiometric ages of Apollo 16
samples (Table IV). The local stratigraphy of the Apollo 16 landing site defines
two major superimposed formations (Ulrich *et al.*, 1981): The older Descartes
Formation, most probably exposed by ejecta from the 50 Myr old North Ray crater
and the younger surficial Cayley Formation exposed as reworked regolith at the
whole landing site. North Ray crater is inferred to have excavated rocks that are
interpreted to be part of the continuous ejecta blanket of Nectaris (Stöffler *et al.*,
1981, 1985; Wilhelms, 1987). In the ejecta of North Ray crater highly feldspathic
fragmental breccias are common. Lithic clasts, both individual rock fragments of
the regolith and clasts within feldspathic fragmental breccias, provide the most
reliable age constraints for the Descartes Formation and hence for the age of the

TABLE III

Radiogenic crystallization ages (Gyr) for igneous lunar highlands rocks (decay constants from Steiger and Jäger, 1977; for references see Papike *et al.*, 1998).

Sample			^{40}Ar–^{39}Ar*	Rb–Sr	Sm-Nd	U-Pb, Pb-Pb
Ferroan anorthosites		22013,9002	4.51±?			
		60025			4.44 ± 0.02	4.51 ± 0.01
		67016 cl			4.56 ± 0.07	
		62236			4.36 ± 0.03	
		67435;33a cl	4.35 ± 0.05			
		67435;33b cl	4.33 ± 0.04			
Mg-rich plutonic rocks	*Troctolites*	76535	4.19 ± 0.02 4.16 ± 0.04 4.27 ± 0.08	4.51 ± 0.07	4.26 ± 0.06	4.27±?
		14306,150 (?)				4.245 ± 0.075
	Dunites	72417		4.47 ± 0.10		
	Norites	14305;91 (?)				4.211 ± 0.005
		15445;17			4.46 ± 0.07	
		15445;247			4.28 ± 0.03	
		15455;228		4.49 ± 0.13	4.53 ± 0.29	
		72255		4.08 ± 0.05		
		73215;46,25	4.19 ± 0.01			
		77215		4.33 ± 0.04	4.37 ± 0.07	
		78235				4.426 ± 0.065
		78236	4.39±? 4.11 ± 0.02	4.29 ± 0.02	4.43 ± 0.05 4.34 ± 0.04	
	Gabbronorites	67667			4.18 ± 0.07	
		73255c			4.23 ± 0.05	
Alkali rocks		14066;47 (?)				4.141 ± 0.005
		14304 cl b			4.34 ± 0.08[a]	4.108 ± 0.053
		14306;60 (?)				4.20 ± 0.03
		14321;16 c				4.028 ± 0.006
		67975;131				4.339 ± 0.005
KREEP basalt (Kb) and Quartz monzodiorite rocks (Qmr)	*A15 Kb*	15382	3.84 ± 0.05 3.85 ± 0.04	3.82 ± 0.02		
		15386		3.86 ± 0.04	3.85 ± 0.08	
		15434 particle		3.83 ± 0.05		
	A17 Kb	72275		3.93 ± 0.04 4.04 ± 0.08	4.08 ± 0.07	
	Qmr	15405,57				4.297 ± 0.035
		15405,145				4.309 ± 0.120
Granite and felsite rocks		12013		> 4.08		
		12033,507				3.883 ± 0.003
		12034,106				> 3.916 ± 0.17
		14082,49				4.216 ± 0.007
		14303 cl				4.308 ± 0.003
		14311,90				4.250 ± 0.002
		14321 B1(=cl?)	**K–Ca:**			4.010 ± 0.002
		14321 cl	4.060 ± 0.071	4.04 ± 0.03	4.11 ± 0.20	3.965 ± 0.025
		72215 mix		3.95 ± 0.03		
		73215,43		3.82 ± 0.05		
		73235,60				4.218 ± 0.004
		73235,63				4.320 ± 0.002
		73235,73				> 4.156 ± 0.003

* only given if suggestive of original crystallization age

[a] disturbed and suspect

cl = clast

TABLE IV

Representative radiogenic crystallization ages (Gyr) for polymict lunar highlands rocks (decay constants from Steiger and Jäger, 1977; for references see Papike *et al.*, 1998).

Sample*		Description	^{40}Ar–^{39}Ar	Rb–Sr
Fragmental	14064,31	KREEP melt clast	3.81 ± 0.04	
breccias	67015,320	feldspathic melt blobs	3.90 ± 0.01	
	67015,321	VHA melt blob	3.93 ± 0.01 (K–Ar)	
Glassy	61015,90	coat	1.00 ± 0.01	
breccias	63503 particle lm	glass fragment	2.26 ± 0.03	
and glass	67567,4	slaggy bomb	0.84 ± 0.03	
	67627,11	slaggy bomb	0.46 ± 0.03	
	67946,17	slaggy bomb	0.37 ± 0.04	
Crystalline	14063,215	poikilitic impact melt	3.89 ± 0.01	
melt breccias	14063,233	aphanitic impact melt	3.87 ± 0.01	
	14167,6,3	melt	3.82 ± 0.06	
	14167,6,7	melt	3.81 ± 0.01	
	15294,6	poikilitic, Gp.Y	3.87 ± 0.01	
	15304,7	ophitic, Gp. B	3.87 ± 0.01	
	15356,9	poikilitic, Gp. C	3.84 ± 0.01	
	15356,12	poikilitic, Gp. C	3.87 ± 0.01	
	60315,6	poikilitic	3.88 ± 0.05	
	63503 particle 1c	VHA?	3.93 ± 0.04	
	65015	poikilitic	3.87 ± 0.04	3.84 ± 0.02
	65785	ophitic	3.91 ± 0.02	
	72215,144	aphanite; felsite melts	3.83 ± 0.03	
	72255	aphanite; felsite melts	3.85 ± 0.04	
	72215,238b	aphanite	3.87 ± 0.02	
	73215	aphanite; felsite melts		3.84 ± 0.05
	77075,18	veinlet (Serenitatis)	3.93 ± 0.03	
	72395,96	poikilitic (Serenitatis)	3.89 ± 0.02	
	72535,7	poikilitic (Serenitatis)	3.89 ± 0.02	
	76055	magnesian, poikilitic	3.92 ± 0.05	
	76055,6	magnesian, poikilitic	3.78 ± 0.04	
	76055,6	magnesian, poikilitic		3.78 ± 0.04
Clast-poor	14073	subophitic 14310-group		3.80 ± 0.04
impact melts	14074	subophitic 14310-group	3.80 ± 0.04	
	14276	subophitic 14310-group		3.80 ± 0.04
	14310	subophitic	3.88 ± 0.05	3.79 ± 0.04
	14310	subophitic; plag	3.82 ± 0.04	
	65795	subophitic, very feldspathic		3.81 ± 0.04
	60635	subophitic, 68415-group		3.75 ± 0.03
	65055	subophitic, 68415-group	3.89 ± 0.02	
	67559	subophitic, 68415-group		3.76 ± 0.04
	68415	subophitic	3.80 ± 0.06	3.76 ± 0.04
	68416	subophitic, 68415-group		3.71 ± 0.02
Granulitic	14063,207		3.90 ± 0.02	
breccias and	14179,11	clast	3.97 ± 0.01	
granulites	15418,50		3.98 ± 0.06	
	67215,8		3.75 ± 0.11	
	67415		3.96 ± 0.04	
	67483,13,8		4.20 ± 0.05	
	72255,235b	clast	3.85 ± 0.02	
	77017,46		3.91 ± 0.02	
	78155		4.16 ± 0.04	
	78527		4.15 ± 0.02	
	79215		3.91±?	

* including split number if given by authors

TABLE V

Best estimates of crystallization ages of mare basalt flows at the Apollo and Luna landing sites. Data compiled from various sources; see especially Snyder *et al.* (2000), Burgess and Turner (1998), Nyquist and Shih (1992), Dalrymple (1991), Spangler *et al.* (1984), and references therein. Proposed ages for surface flows (crater retention ages) are given in bold (see Table VI).

Landing Site	Basalt group	Absolute Age (Gyr)
Apollo 11	High-K basalts	**3.58 ± 0.01**
	High-Ti basalts, groups B1-3	3.70 ± 0.02
	High-Ti basalts, group B2	**3.80 ± 0.02**
	High-Ti basalts, group D	3.85 ± 0.01
Apollo 12	Olivine basalt	3.22 ± 0.04
	Pigeonite basalt	3.15 ± 0.04
	Ilmenite basalt	3.17 ± 0.02
	Feldspathic basalt	3.20 ± 0.08
Apollo 15	Ol-normative basalt	**3.30 ± 0.02**
	Qz-normative basalt	3.35 ± 0.01
	Picritic basalt	3.25 ± 0.05
	Ilmenite basalt (15388)	3.35 ± 0.04
	Green glass	$\sim 3.4 - 3.3$
	Yellow glass	3.62 ± 0.07
Apollo 16	Feldspathic basalt	3.74 ± 0.05
Apollo 17	High-Ti basalt, group A	**3.75 ± 0.01**
	High-Ti basalt, group B1/2	3.70 ± 0.02
	High-Ti basalt, group C	**3.75 ± 0.07**
	High-Ti basalt, group D	3.85 ± 0.04
	Orange glass	$\sim 3.5 - 3.6$
Luna 16	Aluminous basalt	3.41 ± 0.04
Luna 24	Very-low-Ti-basalt (VLT)	**3.22 ± 0.02**
Lunar meteorite Asuka 881757	Basalt (gabbroic)	3.87 ± 0.06

Nectaris basin (Maurer *et al.*, 1978; Wacker *et al.*, 1983; Jessberger, 1983; Stöffler *et al.*, 1985). Their ages range from 3.84 Gyr to 4.14 Gyr. Since the youngest clast determines the age of the polymict impact breccia forming the basement of North Ray crater, an age as young as 3.85 ± 0.05 Gyr has been proposed for the Nectaris basin (Stöffler *et al.*, 1985; Table VI). This age may also be supported by the age distribution of lithic clasts of the Fra Mauro Formation excavated by Cone crater at the Apollo 14 site (Stadermann *et al.*, 1991). Other proposed ages (Table VI)

34   STÖFFLER AND RYDER

TABLE VI

Cumulative crater frequencies, crater degradation values D_L, and absolute ages of lunar surface units derived from isotope ages of lunar rocks; data are taken from the literature except for the absolute ages (this paper, Section 6).

Formation	D_L (m) (1,2)	Crater density, normalized to av. mare[a] (2)	$N=10^{-3}$ craters >4 km/km² (2)	$N=10^{-4}$ craters >1 km/km² (1)	$N=10^{-4}$ craters >1 km/km² (3)	$N=10^{-6}$ craters >10 km/km² (3)	Age (Gyr) (2,3)	Age (Gyr) New set a (4)	Age (Gyr) New set b (5,6)
Ancient highlands (oldest crust)	1150 ±200	10-36	564-677		3600 ±1100	920	4.3-4.55 4.35±0.10	?	?
Uplands		7-30	132-564				4.0-4.4	?	?
Nectaris basin	?	16			1200 ±400	310	4.10±0.10	3.92±0.03	3.92±0.03 3.85±0.05
A16/Descartes Formation					340±70	87	3.90±0.10	3.92±0.03	3.92±0.03 3.85±0.05
Crisium basin	?	?			570[#]	145[#]		3.89±0.02	3.84±0.04
Serenitatis basin	?	?			?	?	3.98±0.05	3.89±0.01	3.87±0.03
A16/Cayley F.	550±50	4.0	34.7					3.85±0.02	3.77±0.02
Imb. Apennines	350±30	3.0		250-480	?	89	3.91±0.10	3.85±0.02	
A14/Fra Mauro F.	350±30	2.8-3.0	47.7	250-480	370±70	94[#]	3.91±0.10	3.85±0.02	3.77±0.02
Imbrium basin							3.91±0.10	3.85±0.02	3.77±0.02
Orientale ejecta blanket	550 ±100	2.50	ND	220	220±?			ND	ND
Orientale basin								3.72-3.85?	3.72-3.77?
Oldest M. (Nubium)	315	2.5	ND	ND	ND	ND		ND	ND
Mare Nectaris	ND	ND	ND	ND	ND	ND		3.74	3.74
M. Tr., old (A11)	390	ND?	26.2?	200	90±18	23	3.72±0.10	3.80±0.02	3.80±0.02
M.Serenitatis (A17)	330-390	1.20		90	100±30	26[#]		3.75±0.01	3.75±0.01
M.Tr., young (A11)	280-390	1.39	15?	34	64±20	16	3.53±0.05	3.58±0.01	3.58±0.01
M.Fecunditatis (L16)	240-300	0.93	15.3		33±10	8.4	3.40±0.04	3.41±0.04	3.41±0.04
Mare Imbrium (A15)	255-285	0.43	8.01	26	32±11	8.2	3.28±0.10	3.30±0.02	3.30±0.02
Mare Crisium (L24)		0.43	8.17	26	30±10	7.6	3.30±0.10	3.22±0.02	3.22±0.02
O.Procellarum (A12)	210-215	0.72	13.6	24	36±11	9.2	3.18±0.10	3.15±0.04	3.15±0.04
Autolycus	160-200	ND	ND	ND	ND	ND		2.1±?	2.1±?
Copernicus	88-112	0.30	0.06		13±3	3.3	0.85±0.20	0.8±0.015	0.8±0.015
Tycho, A17	ND?	0.10	0.019	ND	0.9 ±0.18	0.23	0.109 ±0.004	0.109 ±0.004	0.109 ±0.004
Tycho	10-20							0.109 ±0.004	0.109 ±0.004
North Ray crater	4-5	ND	ND	ND	0.44 ±0.11	0.11	0.05 ±0.0014	0.053 ±0.008	0.053 ±0.008
Cone crater	ND	ND	ND	ND	0.21 ±0.05	0.05	0.026 ±0.0008	0.025 ±0.012	0.025 ±0.012
South Ray crater								0.002 ±0.0002	0.002 ±0.0002
Terrestrial craters (Phanerozoic)					3.6±1.1	9.2	0.375 ±0.075	0.375 ±0.075	0.375 ±0.075

(1) = Wilhelms (1987); (2) Hartmann et al. (1981); (3) Neukum and Ivanov (1994); (4) this paper, Ryder and Spudis (1987), Wilhelms (1987); (5) this paper; Deutsch and Stöffler (1987), Stadermann et al. (1991); (6) previous proposals, 3.85: Stöffler et al. (1985), 3.87: Jessberger et al. (1977); # from Neukum (1983); av. = average; ND = not determined; A = Apollo; F. = Formation; Imb. = Imbrium; L = Luna; M. = Mare; O. = Oceanus; Tr. = Tranquillitatis; a) average mare: 1.88×10^{-4} craters >4 km/km².

are 3.92 ± 0.03 Gyr (Wilhelms, 1987; Deutsch and Stöffler, 1987) and 3.95 Gyr (James, 1981). In part, the arguments for these older ages reflect the 3.85 Gyr age of Imbrium, which is younger than Nectaris. Wilhelms (1987) suggests 3.92 Gyr for Nectaris only because this age is most compatible with his assumption of a constant cratering rate in the pre-Nectarian and Nectarian time since about 4.2 Gyr (with 30 multi-ring basins formed between 4.2 and 3.92 Gyr and 12 basins formed between 3.92 and 3.85 Gyr, his inferred age of Imbrium). However, an assumed constant cratering rate is not a valid age constraint.

6.2.2. *Age of the Crisium Impact Basin*

The absolute age of the Crisium basin is tentatively inferred from radiometric ages of a few small particles from the Luna 20 regolith (Wilhelms, 1987; Spudis, 1993), collected from ejecta deposits of Crisium. Most of the fragments are feldspathic, KREEP-poor impact melt rocks not unlike some of the characteristic melt rocks at the Apollo 16 site although sample 22007,1 (3.87 Gyr, Podosek *et al.*, 1973) is similar to the more KREEP-rich, Apollo 17 crystalline melt rocks, interpreted as Serenitatis impact melt. One sample of the KREEP-poor impact melt lithology (22023,3,F) was dated at 3.895 ± 0.017 Gyr (Swindle *et al.*, 1991) which is proposed as a consistent age for the Crisium basin. Wilhelms (1987) suggested an age of 3.84 ± 0.04 Gyr for Crisium (Table VI). It remains uncertain whether any of the dated lithic clasts represent Crisium melt or even the youngest clasts of the continuous deposits of Crisium. Its actual age could be younger than 3.89 Gyr (Table VI) and nearly as young as the next younger dated basin (Serenitatis).

The relative ages of the Crisium and Serenitatis basins are not definitely clear. The crater density value for superimposed craters > 20 km per 10^6 km^2 is higher for Serenitatis (83?) than for Crisium (53) although it is based on very poor statistics, and Serenitatis has been extremely modified by Imbrium. Wilhelms (1987) argues on the basis of superposition and morphology characteristics that Serenitatis is younger than Crisium. This would set an age of 3.89 ± 0.01 Gyr, the proposed age for Serenitatis (see below), as the lower limit for the age of Crisium.

6.2.3. *Age of the Serenitatis Impact Basin*

The Apollo 15 and 17 landing sites are close to (though just outside of) the main rim of the Serenitatis basin, thus, samples from both sites were suitable for dating the Serenitatis event (Table IV; see Section 3). Apollo 17 samples were collected from massifs of the Taurus-Littrow region which are part of the eastern main rim of Serenitatis. This region is fairly undisturbed and only slightly modified by deposits of younger basins (Wilhelms, 1987; Spudis, 1993). In contrast, the Imbrium basin-forming event destroyed and buried the western rim formations of Serenitatis.

Most of the boulders and smaller rock fragments from the Taurus mountains' North and South Massifs represent a widespread unit of poikilitic, fragment-laden impact melt of uniform composition, inferred to be Serenitatis melt, spilled out of the growing cavity (e.g., Spudis and Ryder, 1981), with a tightly constrained age of 3.893 ± 0.009 Gyr (e.g., Dalrymple and Ryder, 1996). One boulder and few smaller fragments from the South Massif are aphanitic fragment-laden impact melts, chemically more varied and distinct from the poikilitic melts, and also more varied in fragment population. Their fine-grained, clast-rich nature is less conducive to radiometric dating. Inferred ages are in the range 3.86 − 3.93 Gyr, but on average younger than those of the poikilitic rocks. These rocks might be a variant of Serenitatis melt, or even from the Imbrium event. Previously, Serenitatis was dated as 3.86 ± 0.04 or 3.87 ± 0.03 Gyr (Jessberger *et al.*, 1974, 1977, 1978; Staudacher *et al.*, 1979; Wilhelms, 1987; Deutsch and Stöffler, 1987; Table VI).

6.3. EARLY IMBRIAN EPOCH

6.3.1. *Age of the Imbrium Impact Basin*
Imbrium basin deposits have been sampled at the Apollo 14, 15, and 16 sites, where different facies of Imbrium ejecta were deposited as indicated by photogeological interpretations (Wilhelms, 1987; Spudis, 1993) and by cratering models (Oberbeck, 1975; Schultz and Merrill, 1981; Melosh, 1989). *Apollo 15* sampled ejecta deposits (probably including impact melt) at the main rim of the Imbrium basin, *Apollo 14* lithic clasts of the continuous ejecta blanket (Fra Mauro Formation), and *Apollo 16* a zone of distal discontinuous ejecta (Cayley Formation).

Two major proposals for the age of Imbrium have been published in recent years after an age of 3.85 − 3.90 Gyr had been generally accepted before 1980. The ages proposed more recently are 3.85 ± 0.02 Gyr (Wilhelms, 1987; Ryder, 1990a, 1994; Spudis, 1993; Hartmann *et al.*, 2000) and 3.77 ± 0.02 Gyr (Deutsch and Stöffler, 1987; Stadermann *et al.*, 1991). The originally accepted age was mainly based on the measured ages of lithologies which were in some way supposedly reset by Imbrium event, and which displayed a peak of their frequency distribution within the 3.85 − 3.90 Gyr age range (e.g., Taylor, 1975). This approach is incorrect in view of our foregoing statements, yet continues to exist, e.g., Wilhelms' (1987) statement: "The time of the Imbrium impact seems to be well constrained at from 3.82 to 3.87 Gyr; the average and well represented age of 3.85 ± 0.03 Gyr is tentatively adopted here". The more recent proposals are not based on "histogram" approaches but on age constraints that apply to relevant geological units.

Arguments for a 3.85 ± 0.02 *Gyr age of the Imbrium impact basin (G. Ryder):* The age of Imbrium is established by the age of its ejecta blanket, because definite Imbrium impact melt has not been identified. Even the samples most likely on petrological grounds to be such melt (15445 and 15455) have disturbed ^{40}Ar-^{39}Ar ages, merely suggesting an age greater than about 3.82 Gyr. The age of the ejecta is established using lithic fragments within it: They must be contemporaneous with or be older than the Imbrium event, thus providing an upper age limit (oldest possible age). The age of the youngest lithic fragment is likely to be close to that of the impact. The ejecta is overlain and embayed by the Apennine Bench Formation (Hackmann, 1966) which is exposed near the Apollo 15 landing site. This flow-melt unit is younger than the Imbrium impact (or is contemporaneous with it in the unlikely event that it is its impact melt sheet), and thus provides a lower age limit.

The Imbrium ejecta were sampled at the Apollo 14 and 15 sites, and in a more diluted and complex form at the Apollo 16 site (Table IV). Cone Crater at Apollo 14 excavated material that is definitely Fra Mauro Formation. Melt fragments within these samples, including fragments in the friable light breccias, have a range of ages from ∼3.95 down to 3.85 Gyr (a few older fragments are not impact melts, e.g. Stadermann *et al.*, 1991). Samples which were collected outside of the Cone Crater ejecta blanket include melt samples with younger ages, down to nearly 3.7 Gyr (especially belonging to a single chemical group exemplified by 14310). However, these are not necessarily from the Fra Mauro Formation. The Cone Crater samples strongly suggest an age for the Imbrium ejecta blanket of 3.85 ± 0.02 Gyr.

Melt samples at the Apennine Front must be chiefly pre-Imbrian or contemporaneous with it, as no major impact events later affected the site. Samples collected are dominantly from Imbrium ejecta, continually exposed by mass wasting. Only a few are likely to be exotic. Dalrymple and Ryder (1993) obtained chronological data on the range of Apollo 15 impact melts defined by Ryder and Spudis (1987). All but one of the dated samples gave ages around $3.86 - 3.88$ Gyr, the other one gave an age of 3.84 ± 0.02 Gyr. The Apennine Front data strongly suggest an age for the ejecta blanket of 3.85 ± 0.02 Gyr. The Cayley Plains at Apollo 16 are less definitive but nearly all of the impact melts must pre-date Imbrium and nearly all have ages ≥ 3.86 Gyr. The main exception is a significant group with a composition similar to local regolith, to be described in the next section. Thus the Apollo 16 data are consistent with that from Apollo 15 as an upper limit on the age of Imbrium.

The Apennine Bench Formation (Figure 10) has the physical features of a volcanic unit. Gamma-ray orbital data (Apollo 15 and Prospector missions) show that the unit has the same thorium abundances as the KREEP basalts found as small fragments and a common regolith constituent at the Apollo 15 landing site (Spudis, 1978; Hawke and Head, 1978; Ryder, 1987). These volcanic rocks have a well-defined age of 3.85 ± 0.02 Gyr, indistinguishable from the upper limit for Imbrium defined by its ejecta. Thus both the upper and lower absolute age limits for Imbrium are the same, establishing the Imbrium basin as 3.85 ± 0.02 Gyr (Table VI).

Arguments for a 3.77 ± 0.02 Gyr age of the Imbrium impact basin (D. Stöffler): This age has been derived from detailed Consortium studies of the Apollo 14 and 16 highland breccia samples (e.g., Stöffler *et al.*, 1981, 1985, 1989; Stadermann *et al.*, 1991). The main arguments for the 3.77 Gyr age are given in Deutsch and Stöffler (1987) and supplemented by Stadermann *et al.* (1991). The youngest lithic clast of the basement breccias of the Apollo 14 and 16 sites, representing the Imbrium related Fra Mauro and Cayley Formations, respectively, must provide the age of the parent basin. At both sites, there are "young crystalline impact melt rocks" ranging in age from 3.71 ± 0.03 Gyr to 3.81 ± 0.01 Gyr. The 3.77 ± 0.02 Gyr age is mainly based on the group of anorthositic-noritic melt rocks (3 Apollo 16 melt rocks clustering at 3.75 ± 0.01 Gyr) and on the group of youngest Apollo 14 melt rocks, which are chemically distinct from them. The age of 3.77 ± 0.02 is covered by the age uncertainties of the two groups of subophitic melt rocks.

The samples younger than 3.82 Gyr belong to different textural and chemical groups and range in size from the cm- to the m-scale (e.g., boulder 68415/416). The subophitic samples (e.g., 14310 and 68415/416) represent clast-free, relatively coarse-grained and therefore slowly cooled impact melt rocks, particularly critical for the arguments against a post-Imbrian origin of these rocks. Deutsch and Stöffler (1987) argue that these rocks originate from large pre-Imbrian impact crater formations (melt sheets and polymict breccia deposits); they cannot be derived from erratic clasts ejected from local or distant post-Imbrium craters.

Deutsch and Stöffler (1987) questioned that the Apennine Bench Formation (Figure 10) is younger than Imbrium and that it is composed of the same type of KREEP basalts which occur as clasts at the Apollo 15 site dated at 3.85 ± 0.05 Gyr (e.g., Carlson and Lugmair, 1979). There is no direct geologic evidence that the volcanic "light plains" of the Apennine Bench extend to the Apollo 15 site forming the substratum of the mare basalts and covering Imbrium ejecta (Spudis, 1993, Figure 7.13) because these assumed relationships are not exposed at the Apollo 15 site. Therefore, the Apennine Bench Formation must be pre-Imbrian in age and formed on top of an older terra unit which assumed its present position between the inner ring and the main rim of the Imbrium multi-ring basin as a parautochthonous megablock of the pre-impact target not completely flooded by mare basalt flows.

6.3.2. *Age of the Orientale Impact Basin*

Orientale is the youngest of the multi-ring basins on the Moon (Wilhelms, 1987; Spudis, 1993), but its absolute age cannot be determined directly from measured ages because samples related to Orientale have not been identified at any of the landing sites. This can be hardly expected since only ray material could be present at the sites, which is difficult to identify in the sample collections. Wilhelms (1987) contends that Orientale must have formed at $\sim 3.85 - 3.72$ Gyr assuming that 3.85 Gyr is the age of Imbrium and 3.72 Gyr is a lower limit set by oldest age of nearby exposed mare basalts of Upper Imbrian age. Based on relative crater densities of these basins, he proposed a tentative age of 3.8 Gyr for Orientale. However, it could be almost as old as Imbrium, i.e., 3.84 Gyr. Based on the above proposed age of 3.77 ± 0.02 Gyr for Imbrium and on the relative crater densities, Orientale should be ≤ 3.75 Gyr old and could be as young as 3.72 Gyr (Table VI).

6.4. LATE IMBRIAN EPOCH

6.4.1. *Age of Apollo 17 Basalt Surfaces (3.70 − 3.75 Gyr)*

The Apollo 17 mare basalt samples collected over a wide area of several kilometers are high-titanium basalt. They fall into distinct chemical groups (Table V) that represent at least four distinct extrusions (Warner *et al.*, 1979; Neal *et al.*, 1990; Ryder, 1990b). Most of the samples are group A (3.75 Gyr) or the more complex group B (3.70 Gyr). Group C (~3.75 Gyr) samples have been identified only among the few Shorty Crater samples, and group D (~3.85 Gyr ?) only by one sample from the Van Serg regolith core. Samples from boulders at the rim of 650 m diameter Camelot (Station 5) presumably represent the deepest excavated basalt, perhaps 100 m; and all belong to group A basalts which occur at all mare sampling locations. Samples of boulders 150 m from the rim of 600 m diameter Steno (Station 1) presumably represent a shallower level and belong to group B basalts. Group B basalts are found throughout the mare sampling locations except Shorty crater. These relationships suggest that group A basalts underlie group B basalts, consistent with their radiometrically determined ages (Table V). Wolfe *et al.* (1981) suggested that group C basalts, dominating the ejecta of the small Shorty crater, were the youngest, but radiometric ages show that they are older than group B and similar in age to group A. Thus the youngest basalts which flood at least the eastern end of the Taurus-Littrow valley are the group B basalts (3.70 Gyr). However, to the west the covering by group B basalts may be patchy leaving group C and A basalts as the topmost bedrock (3.75 Gyr).

Both the regional and local area have dark mantle deposits, stratigraphically the youngest volcanic deposits which appear to be correlated with the sampled orange volcanic glass; its preferred age is ~3.5 Gyr (Tera and Wasserburg, 1976). The presence of dark mantle, the possible patchy distribution of the lava flows, and the considerable obscuration of the older cratering history by the production of the central cluster of craters (Lucchitta and Sanchez, 1975) at about 110 Myr, make relating a radiometric age to a crater density or crater degradation parameter an uncertain task at the Apollo 17 site. However, it seems likely that the mare plain at least to the immediate east of the landing site consists of lava flows with an age of 3.70 Gyr, while those extending out into Mare Serenitatis and Mare Tranquillitatis might be slightly older. It is unlikely that the oldest sampled basalts, group D, form any extensive surface in the region.

We infer that an age of 3.75 Gyr probably best represents the crater densities measured in basalts just inside the southeast rim of Serenitatis (Table VI).

6.4.2. *Age of Apollo 11 Basalt Surfaces (3.58 Gyr and 3.80 Gyr)*

The mare basalt samples collected from the very small area investigated on the Apollo 11 mission are all high-titanium varieties, but have a range of compositions and ages that represent at least four separately extruded basalt types (Table V). Group A (3.58 Gyr), the high-K basalt, is most abundant (10 out of 20 of the large basalt rocks, including the three largest, and 65% of the mass; and 14 out of 24 "peanut" samples). Group B1-B3 (3.70 Gyr), a complex group, comprises most of the rest of the samples, while the two oldest groups B2 (3.80 Gyr) and D (3.85 Gyr) are comparatively minor. Exposure data (Geiss *et al.*, 1977) indicate that the group A samples came from a surface exposure, and that the low-K basalts (groups B and D) came from a shielded site, most excavated in a single impact (possibly the only 30 m deep West Crater, but possibly from much further away).

Galileo and Clementine spectral reflectance data (Staid *et al.*, 1996) indicate that the landing site lies in but close to the edge of a western unit that is both the youngest and the highest in TiO_2 in Tranquillitatis, which they correlate with the group A basalts. A much more extensive nearby unit identified spectrally is older and extends a coherent surface as far north as the Apollo 17 landing site, consistent with this unit being the group B1-3 basalts which are similar in both age and composition to Apollo 17 basalts. Even older basalts identified spectrally as a little lower in TiO_2 may well correspond with group B2 or D (or both). Of the crater density units referred to by Wilhelms (1987) and Neukum and co-workers (e.g., Neukum and Ivanov, 1994) we infer that the young one is the 3.58 Gyr group A basalts, and the older one is group B2 or D which is about 3.80 Gyr.

The much smaller regolith thickness at the Apollo 11 landing site (3 to 6 m) than at the Apollo 17 landing site (<10 m) is consistent with the young basalts (3.58 Gyr) being a dominant surficial unit in the area, as is its chemistry. If the surface unit were 3.70 Gyr old we would expect a regolith thickness more similar that at Apollo 17. The thickness of the group A basalt unit is speculative, but at least some samples of it are coarse-grained and there is evidence that more than one flow was sampled and that differentiation was taking place, and thus constitutes a fairly thick unit. The evidence is thus suggestive that the local crater retention reflects a 3.58 Gyr surface (Table VI), although that might not be the case if the 3.58 Gyr old group A basalts form only a thin (30 m) sequence (if it is thin the crater densities might represent an older surface). It seems unlikely that it could be a generally thin sequence if it is laterally so extensive as inferred by Staid *et al.* (1996) though the Apollo 11 site must be close to its boundary.

6.4.3. *Age of Luna 16 Basalt Surface (3.41 Gyr)*

The tiny mare basalt fragments available from the Luna 16 regolith appear to be mainly a coherent chemical group that is more aluminous than typical mare basalts and with intermediate titanium contents (4-5% TiO_2) (Grieve *et al.*, 1972; Keil *et al.*, 1972; Kurat *et al.*, 1976; Ma *et al.*, 1979); they probably represent a single flow or related flows. The basalt fragments are all fine-grained, suggesting either a thin flow or a series of similar, overlapping thin flows. A Rb-Sr isochron and a ^{40}Ar-^{39}Ar age on a single fragment are consistent with an age of 3.41 Gyr for this basalt group (Table V; Papanastassiou and Wasserburg, 1972; Huneke *et al.*, 1972). Two separate fragments (3.45 ± 0.06 Gyr and 3.30 ± 0.15 Gyr) are consistent with this age (Cadogan and Turner, 1977).

The Luna 16 regolith, collected to a depth of ~30 cm, appears to be well-mixed (Reid *et al.*, 1972). Its chemical composition is a little less iron- and titanium-rich and a little more alumina-rich than the basalt particles. Although a few fragments suggested to be distinct from the main group of basalts are present, the regolith composition is consistent with being a mixture dominated by the main group of basalts and a more aluminous, incompatible trace-element-poor highland component. Thus the regolith composition indicates that the age of this group of basalts is representative of the mare surface at the Luna 16 landing site (Table VI). The sampling site is on a dark unit whose spectral class indicates a higher titanium content than most of Mare Fecunditatis. Presumably this unit has the crater density included in the table of Neukum and Ivanov (1994).

6.4.4. *Age of Apollo 15 Basalt Surface (3.30 Gyr)*

The great number of mare basalt samples collected from the mare plains on the Apollo 15 missions is dominated by two low-titanium varieties, the olivine-normative basalts (3.30 Gyr) and the quartz-normative basalts (3.35 Gyr) (Table V). The other rare basalt fragments are of a similar age, but were found as exotic fragments on the Apennine Front. Various types of volcanic glasses (~3.3 – 3.6 Gyr, Spangler *et al.*, 1984) occur only locally or dispersed in the regolith. Stratigraphically the olivine-normative mare basalts appear to be the highest and are dominant among small rock samples everywhere (Ryder, 1985). This is consistent with the younger radiometric ages. The rille walls show layers, consistent with a sequence of fairly thin flows of a single olivine-normative basalt magma that is suggested by the chemical variation and petrography of samples, and underlain at some level by a sequence of the quartz-normative flows (Ryder and Schuryatz, 2001).

The chemical composition of the Apollo 15 mare regolith samples demonstrates the domination by the olivine-normative mare basalt (Korotev, 1987), even at Dune Crater. The difference in ages of all the Apollo 15 mare basalt types (and probably the glass as well) is in any case so small and the total thickness so great that the crater density measured for this part of Palus Putredinis can be ascribed to an age of 3.30 Gyr with confidence (Table VI).

6.4.5. *Age of Luna 24 Basalt Surface (3.22 Gyr)*

Most of the basaltic fragments and at least a large proportion of the coarser mineral fragments from all levels of the Luna 24 regolith core represent a distinct very-low-titanium aluminous mare basalt type (Ryder and Marvin, 1978; Taylor*et al.*, 1978; Graham and Hutchison, 1980). Metabasalts, impact melts, and glasses have the same composition, indicating that it is a dominant component of the regolith, although other lithic types are present. The nature of the sample allows only fine-grained basalts to be recognized as such, but the mineral chemistry of the monomineralic fragments suggest that coarse-grained equivalents are present.

The available ages, which appear to all be on low-titanium mare basalts and metabasalts, show a rather narrow range around 3.22 Gyr (Burgess and Turner, 1998). That the metabasalts have the same ages suggests that they are metamorphosed flow margins and that the basalts consist of a sequence of overlapping flows of similar composition. One sample has a somewhat older ^{40}Ar-^{39}Ar-age, but might be compromised by the large amount of trapped argon. It is possible that a slightly younger age of 2.93 Gyr for one particle should be considered more reliable. Nonetheless it would appear that the basalt particles are dominated by a single component with an age of 3.22 Gyr (Table V).

The bulk regolith is very similar in chemical composition for both major and minor elements to that of the very-low-titanium basalts, identifying these basalts as the surface unit at this site. This is inconsistent with the remote-sensing data that show that surfaces with such low TiO_2 do not exist within tens of km of the nominal landing site (Blewett *et al.*, 1997). Either Luna 24 did not land where it was reported to have landed, or the basalts collected are representative of only a very small area surrounded by basalts with higher titanium that were not collected and thus not dated. However, according to Wilhelms (1987), Mare Crisium is stratigraphically amongst the most uniform, and therefore we consider the 3.22 Gyr age to be correlated with the typical crater density of southern Mare Crisium (Table VI).

6.5. ERATOSTHENIAN PERIOD

6.5.1. *Age of Apollo 12 Basalt Surface*

On the basis of chemical and isotopic characteristics, the collection of more than 40 mare basalt rocks from the Apollo 12 landing site represent three numerically subequal groups (olivine basalts, pigeonite basalts, and ilmenite basalts, and a single fragment of a fourth group (feldspathic basalt) (Neal *et al.*, 1994). The ilmenite and pigeonite basalt groups have very similar ages (3.15 − 3.17 Gyr), with the olivine basalts and the feldspathic basalt being perhaps slightly older (3.22 Gyr; Table V). This is consistent with stratigraphic relationships, where the ilmenite basalts are the only type found around the smaller craters, and the pigeonite and olivine basalts required excavation from larger craters (Surveyor, 200 m diameter and Middle Crescent, 400 m diameter). This would indicate that the ilmenite basalt is about 40 m thick (Rhodes *et al.*, 1977).

The overlapping ejecta blankets at the landing site make sample provenance and the relationships with the remotely-sensed data somewhat uncertain. On the basis of D_L, Soderblom and Lebofsky (1972) suggested that there were two surface units in the area, the older of which is at the landing site and the younger exposed about a kilometer away. The crater density measurements are presumably an average of these two units. However, Wilhelms confusingly states also that two units were included in crater density counts and that the younger one is at the landing site. There is some inconsistency in the use of crater density in plots of this site, even by the same author group; for instance, most of the Neukum papers show that the Apollo 12 landing area measured has a higher crater density than the Apollo 15 landing site, but in Neukum and Wise (1976) and Neukum (1977) and in Wilhelms (1987) which reports to use Neukum data, a lower density is used for the Apollo 12 landing site in crater density/age diagrams. The data for Apollo 12 (e.g., Neukum *et al.*, 1975) show an unusual kink at the critical point around 1-2 km sizes, deviating from a standard calibration. Possibly some secondary craters have not been identified and the actual count is indeed lower for Apollo 12 than for Apollo 15. We suggest that the interpolated lower count correlates with the surface basalt age of 3.15 Gyr (Table VI).

6.6. COPERNICAN PERIOD

6.6.1. *Age of Autolycus and Aristillus*

Early geologic analysis showed that a ray from either of the craters Aristillus or the older but nearer Autolycus crossed the Apollo 15 site and deposited exotic material. KREEP basalt fragments with an original crystallization age of $\sim$3.84 Gyr were shocked and thermally heated, and in one case shock-melted, at 2.1 Gyr (Ryder *et al.*, 1991). Autolycus lies in the Apennine Bench Formation, correlated with Apollo 15 KREEP basalts (Spudis, 1978) and is expected to contribute more material to this site than Aristillus, which would mainly supply mare basalt fragments. Seemingly, Autolycus formed at 2.1 Gyr (Table VI). If so, and assuming that Autolycus is Copernican, then that Period commenced earlier than Copernicus itself, which is less than 1 Gyr old. However, Autolycus is one of the most degraded Copernican craters due to the later Aristillus ejecta, and crater density measurements have not been made, although D_L measurements (180 $\pm$ 20) suggest that it is very close to the Copernican-Eratosthenian boundary (Wilhelms, 1987).

Although an age of $\sim$1.3 Gyr has been found for the stratigraphically younger Aristillus crater on the basis of the age of a 1m-block of KREEP impact melt on the Apennine Front (sample 15405; Bernatowicz *et al.*, 1978), this correlation is unreliable. More likely, sample 15405 is of a more local origin.

6.6.2. *Age of Copernicus*

The Apollo 12 mare site is heavily contaminated with KREEP materials, although the nearest non-mare outcrops are about 25 km away. Rays from Copernicus cross the landing site, and Meyer *et al.* (1971) suggested that KREEP glass in the samples was produced and ejected by the Copernicus event and thus could be used to date it. Subsequent ^{40}Ar-^{39}Ar dating of such materials suggested appreciable degassing at about 800 Myr (Eberhardt *et al.*, 1973; Alexander *et al.*, 1976). U,Th-Pb data also yielded an age of 850 $\pm$ 100 Myr for regolith disturbance (Silver, 1971). Bogard *et al.* (1994) found that a granite fragment encased in KREEP glass had been almost completely degassed at 800 $\pm$ 15 Myr. These ages are all from samples 12032 and 12033, the most immature and most KREEP rich regolith samples, which were probably both collected at Head Crater (Korotev *et al.*, 2000).

The 800 $\pm$ 15 Myr age is widely accepted as that of Copernicus (Table VI). If the dated samples are from Copernicus' rays, then this age is correct. However, not all of the KREEP at the site can be from Copernicus, even in a concentrate in a ray, and most of it may have arrived by other means (Korotev *et al.*, 2000; Jolliff *et al.*, 2000a). The dated samples are all from a restricted site (Head Crater) and the ropy glass is not found elsewhere, whereas a ray as seen from orbit might distribute materials more widely. In addition, Copernicus itself does not seem to have excavated mainly KREEP materials, although KREEP might have been an early, shallow ejected spray phase. With these caveats, either the age of Copernicus is well-defined at 800 $\pm$ 15 Myr, or it is known only to be younger than $\sim$2 Gyr.

6.6.3. *Age of Tycho*

The dating of the crater Tycho (diameter: 98 km) rests partly on the inference that a landslide on the slope of the South Massif (Apollo 17) was triggered by ejecta of Tycho, which is about 2200 km away. The exposure age near 0.1 Gyr of landslide material then represents the age of Tycho (Wolfe *et al.*, 1975; Arvidson *et al.*, 1976; Lucchitta, 1977; Drozd *et al.*, 1977). The "Central Cluster" craters at the Apollo 17 site also show an exposure age of about 0.1 Gyr, and were interpreted as secondary craters of Tycho (Wolfe *et al.*, 1975; Lucchitta, 1977). Thus, Drozd *et al.* (1977) proposed an age for Tycho of 109 ± 4 Myr (Table VI). However, the geological evidences for the South Massif landslide and the Central Cluster craters being formed by distal ejecta from Tycho are equivocal to some degree.

6.6.4. *Ages of Cone, North Ray, and South Ray Craters*

These young Copernican craters are of prime interest among all other young craters which have been dated on the basis of cosmic ray exposure ages because (1) samples collected from their ejecta deposits provide the basis for the age determination of the Nectaris and the Imbrium basins (see Sections 6.2 and 6.3) and (2) crater frequency data measured on their ejecta blankets are available (Moore *et al.*, 1980; Table VI). According to the exposure age data (Drozd *et al.*, 1977; Stadermann *et al.*, 1991; Drozd *et al.*, 1974; Eugster, 1999) the ages of Cone crater (Apollo 14 landing site) and of North Ray and South Ray craters (Apollo 16 landing site) are 25.1 ± 1.2 Myr, 50.3 ± 0.8 Myr and 2.0 ± 0.2 Myr, respectively (Table VI).

7. Cratering Rates in Lunar History: Implications for the Terrestrial Planets

We have presented the most recent determinations of the absolute ages of datable lunar surface formations, and measurements of the cumulative frequency of superimposed impact craters are available for them. These age data, the crater frequencies, and the widely used parameter of crater degradation (D_L) are summarized in Table VI. We derive revised calibration curves for the crater retention ages of lunar surfaces of varied age ranging from about 4 Gyr to the present (Figure 11 and 12; see also Neukum *et al.*, 2001). As recognized very early in the Apollo lunar science program (e.g., Hartmann, 1970; Soderblom and Lebofsky, 1972) such calibration curves are of fundamental importance for (1) determining the cratering rate in the Earth-moon system as a function of time, (2) establishing an absolute lunar stratigraphy, and (3) providing a standard reference curve for stratigraphic time applicable for other planetary bodies of the inner solar system.

Previous calibration curves for the lunar cratering rate and the absolute crater retention ages (Hartmann, 1972; Soderblom and Lebofsky, 1972; Neukum *et al.*, 1975; Neukum and König, 1976; Hartmann *et al.*, 1981; Neukum and Ivanov, 1994) are based on a similar set of ages derived for specific surface areas from isotope ages of lunar samples and were reproduced in many reference books such

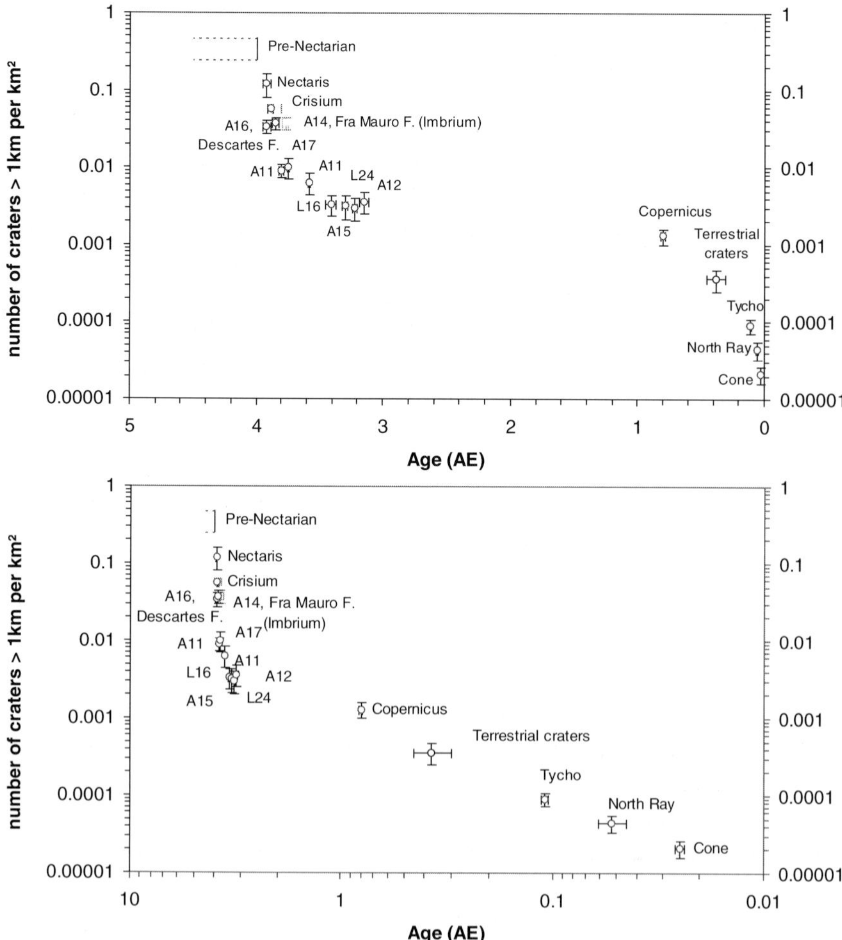

Figure 11. Cumulative crater frequencies for craters >1 km/km^2 as a function of lunar surface ages from Table VI in linear (*upper panel*) and logarithmic (*lower panel*) scale; for some impact basins alternative ages are given according to sets a and b of Table VI.

as Taylor (1982), Wilhelms (1987), and Heiken *et al.* (1991) and have been widely and sometimes uncritically accepted by the planetary science community.

During the evaluation of the presently available data base (see Sections 2, and 4 to 6) it became evident that there are several problems with previously used age calibration curves of the lunar crater frequency data (relative crater retention ages) from which absolute crater retention ages have been derived for lunar surface units of unknown age. The problems relate to the following:

1. The definition of coeval surface units for a specific set of crater counts.
2. Incorrect derivation of mare surface ages from ranges of mare basalt ages.
3. The use of outdated or even incorrect absolute ages (including incorrect uncertainties) of surface units based on wrong interpretations of lunar rock ages.

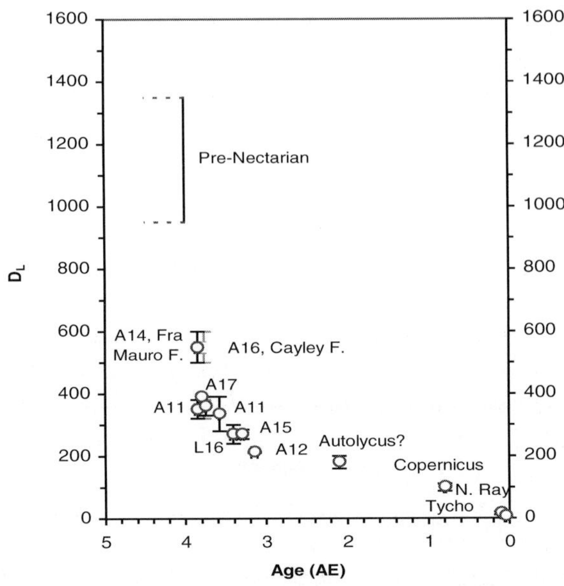

Figure 12. Crater degradation parameter D_L, defined in Section 2.2, as a function of the age of dated lunar surfaces (see Table VI; alternative ages for some impact basins from sets a and b of Table VI).

4. Incorrect ages of multi-ring basins including incorrect uncertainties.
5. The unsubstantiated assignment of an absolute age of > 4.3 to terrains of oldest lunar crust ("ancient highland", "lunar uplands") displaying the highest values for measured crater frequencies.

For some of these issues or open questions we present solutions or suggestions; some others remain open or at least disputable. We propose new best estimates for *ages of mare surfaces* at the Apollo and Luna landing sites and uncertainties for these ages which are lower than the 1-σ errors used in previous calibration curves by the Basaltic Volcanism Study Project (Hartmann *et al.*, 1981) and by Neukum and Ivanov (1994): 3.75 ± 0.01 Gyr (Apollo 17), 3.80 ± 0.02 Gyr (Apollo 11 older surface unit), 3.58 ± 0.01 Gyr (Apollo 11 younger surface unit), 3.41 ± 0.04 Gyr (Luna 16), 3.30 ± 0.02 Gyr (Apollo 15), 3.22 ± 0.02 Gyr (Luna 24), and 3.15 ± 0.04 Gyr (Apollo 12). These data and the corresponding values used previously are given in Tables V and VI and used for the Figures 11 and 12.

For the *ages of multi-ring basins* of the Nectarian and Imbrian Systems, Section 6 gives arguments for two differing data sets that may be used in parallel for the calibration curve until better data become available (Table VI, Figures 11 and 12). The different ages proposed for Imbrium (3.85 Gyr vs. 3.77 Gyr), for Nectaris (ranging from 3.92 Gyr to 3.85 Gyr), as well as for Crisium and Serenitatis do in fact not dramatically influence the shape of the calibration curves. However, the deletion of old outdated ages and of non-justified errors (Hartmann *et al.*, 1981; Neukum and Ivanov, 1994) for the Nectaris basins (4.1 ± 0.1 Gyr), the Descartes Formation (3.90 ± 0.1 Gyr), the Imbrium basin and the Fra Mauro

Formations (3.91 $\pm$ 0.1 Gyr) effects the curves, as do in addition discarding very old ages for the pre-Nectarian highlands such as the specific age of 4.35 $\pm$ 0.1 Gyr for the "ancient highlands" (Neukum and Ivanov, 1994) and the age ranges of 4.0 − 4.4 Gyr and 4.35 − 4.55 Gyr for the "most densely cratered province" and the "uplands" (Hartmann *et al.*, 1981). For these old ages no firm geologic evidence combined with any clear isotope data basis exists.

For the *ages of Eratosthenian and Copernican craters* and the related ejecta blankets appreciable uncertainties remain. Although some of the youngest ages are well constrained (Cone, North Ray and South Ray craters), in other cases geological interpretation is uncertain or equivocal (Autolycus, Copernicus, and Tycho).

We suggest that the improved data base presented in this paper should be implemented into the absolute age calibration for the lunar cratering rate as shown in Figures 11 and 12. This affects also all calibration curves for other terrestrial planets (see original curves in Hartmann *et al.*, 1981, Figures 8.6.1 to 8.6.5 as reprinted in Taylor, 1982, p. 105, and in Neukum and Ivanov, 1994, Figure 16). The calibration curve published in the Lunar Sourcebook (Heiken *et al.*, 1991; Figure 4.15) contains large errors (e.g., the data points for L24, A14 and A16 are wrong) and, hence, misleads by suggesting large uncertainties for the determination of absolute crater retention ages of lunar surfaces. For example, for a surface with 10^{-4} craters >4 km/km^2 the minimum and maximum values for its age differ by some 1.7 Gyr compared to about 0.9 Gyr read from our revised calibration. For an area with 5×10^{-4} craters >4 km/km^2 the corresponding value drops from about 0.55 Gyr to some 0.15 Gyr. This problem holds similarly with the lunar standard curve of Hartmann *et al.* (1981, Figure 8.6.1) but somewhat less with Neukum and Ivanov's (1994) curve because it contains the correct value for Luna 24 and smaller uncertainties for Apollo 16 and 14 than Heiken *et al.* (1991).

Our recommended *new calibration curve (Figures 11 and 12)* for the lunar cratering rate as a function of time is better constrained, with small uncertainties in the age range from about 4.0 to 3.0 Gyr corresponding to cumulative crater frequencies of $\sim1.5\times10^{-1}$ craters >1 km/km^2 to $\sim2\times10^{-3}$ craters >1 km/km^2. However, major uncertainties still exist for the pre-Nectarian Period ($\gtrsim4$ Gyr) and for the Eratosthenian and Copernican Periods ($\lesssim3$ Gyr). The steepness of the calibration curve older than ~3.75 Gyr, the possibility that the pre-Nectarian surfaces for which crater counts exist (Table IV; Figure 5), may not be older than 4.2 Gyr (Wilhelms, 1987), and the fact that impact melt lithologies older than 4.15 Gyr are lacking, indicate that the cratering rate may not smoothly increase according to the present calibration curve from 3.75 Gyr up to the time of the formation of the moon. This would be incompatible with the accretion rates required for the size of the moon (Ryder, 1990a). These observations at least mean that the cratering rate between 4.5 and 4.0 Gyr is not known and that there is still room for speculations about a possible late lunar cataclysm (Tera *et al.*, 1974; Ryder, 1990a; Hartmann *et al.*, 2000; Cohen *et al.*, 2000).

The cratering rates for the Eratosthenian and Copernican Periods are also not sufficiently well constrained because reliable absolute ages for surfaces formed between 3 Gyr and 1 Gyr are conspicuously lacking. In spite of the uncertain ages of the craters Autolycus and Copernicus, the tentative figures for their ages are largely compatible with a steady state and constant cratering flux since about 3 Gyr although the data for Copernicus (Figures 11 and 12) may indicate a slightly increased flux in the past 1 Gyr. Culler *et al.* (2000) have confirmed this recently by the non-uniform distribution of ^{40}Ar-^{39}Ar-ages of 155 glass spherules collected from the Apollo 14 regolith (Culler *et al.*, 2000). They suggest that the cratering rate decreased since about 3.5 Gyr by a factor of 2 to 3 to a minimum value at about $0.5 - 0.6$ Gyr and increased by a factor of 3.7 ± 1.2 in the past 0.4 Gyr in accordance with data for terrestrial craters (Grieve and Shoemaker, 1994) and astronomical constraints (Shoemaker *et al.*, 1994). Such changes in the post-Imbrian cratering flux of the Earth-moon system (Ryder, 2000; Hörz, 2000; Muller *et al.*, 2000) have important implications for the chronostratigraphy of Mars and other terrestrial planets. However, in terms of the fundamental task to improve the lunar standard reference for the cratering flux in the inner solar system, new sample return missions to Eratosthenian and Copernican regions of the moon are needed. They may be given a priority as high as sample return missions to Mars.

Acknowledgements

We thank Tim Swindle and Elmar Jessberger for detailed and constructive reviews, ISSI in Bern, and NASA for financial support, and Uwe Damerow, Diana Krüger, and Jackie Lyon for technical help with the manuscript, tables and figures. The Lunar and Planetary Institute operates under contract NASW-4066 with the National Aeronautics and Space Administration. This paper is LPI Contribution 1076.

References

Alexander, E.C., Jr., Bates, A., Coscio, M.R., Jr., Dragon, J.C., Murthy, V.R., Pepin, R.O., and Venkatesan, T.R.: 1976, 'K/Ar Ddating of Lunar Soils II', *Proc. 7th Lunar Sci. Conf.* **7**, 625–648.

Arvidson, R., Drozd, R.J., Guinness, E., Hohenberg, C.M., Morgan, C.J., Morrison, R.H., and Oberbeck, V.R.: 1976, 'Cosmic Ray Exposure Ages of Apollo 17 Samples and the Age of Tycho', *Proc. 7th Lunar Sci. Conf.*, 2817–2832.

Baldwin, R.B.: 1949, 'The Face of the Moon', University of Chicago Press, Chicago, p. 239.

Barsukov, V.L.: 1977, 'Preliminary Data for the Regolith Core Brought to Earth by Automatic Lunar Station Luna 24', *Proc. 8th Lunar Sci. Conf.*, 3303–3318.

Beaty, D.W., and Albee, A.L.: 1978, 'Comparative Petrology and Possible Genetic Relations Among the Apollo 11 Basalts', *Proc. 9th Lunar Sci. Conf.*, 359–463.

Bernatowicz, T.J., Hohenberg, C.M., Hudson, B., Kennedy, B.M., and Podosek, F.A.: 1978, 'Argon Ages for Lunar Breccias 14064 and 15405', *Proc. 9th Lunar Sci. Conf.*, 905–919.

Binder, A.: 1998, 'Lunar Prospector: Overview', *Science* **281**, 1475–1476.

Blewett, D.T., Lucey, P.G., Hawke, B.R., and Jolliff, B.L.: 1997, 'Clementine Images of the Lunar Sample-return Stations: Refinement of FeO and TiO_2 Mapping Techniques', *J. Geophys. Res.* **102**, 16,319–16,325.

Bogard, D.D., Garrison, D.H., Shih, C.-Y., and Nyquist, L.E.: 1994, '^{40}Ar-^{39}Ar Dating of Two Lunar Granites: The Age of Copernicus', *Geochim. Cosmochim. Acta* **58**, 3093–3100.

Boyce, J.M., and Dial, A.L., Jr.: 1975, 'Relative Ages of Flow Units in Mare Imbrium and Sinus Iridum', *Proc. 6th Lunar Sci. Conf.*, 2585–2595.

Burgess, R., and Turner, G.: 1998, 'Laser ^{40}Ar-^{39}Ar Age Determinations of Luna 24 Mare Basalts', *Met. Plan. Sci.* **33**, 921–935.

Butler, P., and Morrison, D.A.: 1977, 'Geology of the Luna 24 Landing Site', *Proc. 8th Lunar Sci. Conf.*, 3281–3301.

Cadogan, P.H., and Turner, G.: 1977, '^{40}Ar-^{39}Ar Dating of Luna 16 and Luna 20 Samples', *Phil. Trans. Royal Soc. London* **284**, 167–177.

Cameron, A.G.W.: 1984, 'The Impact Theory for Origin of the Moon', *in* W.K. Hartmann, R.J. Phillips, and G.J. Taylor (eds.), *Origin of the Moon*, LPI, Houston, pp. 609–616.

Cameron, A.G.W., and Ward, W.R.: 1976, 'The Origin of the Moon', *Lunar Science* **VII**, Lunar Science Institute, Houston, 120–122.

Canup, R.M., and Agnor, C.B.: 2000, 'Accretion of the Terrestrial Planets and the Earth-Moon System', *in* R.M. Righter and R. Canup (eds.), *Origin of the Earth and Moon*, Univ. Arizona Press, Tucson, pp. 113–129.

Carlson, R.W., and Lugmair, G.W.: 1979, 'Sm-Nd Constraints on Early Lunar Differentiation and the Evolution of KREEP', *Earth Planet. Sci. Lett.* **45**, 123–132.

Clayton, R.N., and Mayeda, T.K.: 1975, 'Genetic Relations Between the Moon and Meteorites', *Proc. 6th Lunar Sci. Conf.*, 1761–1769.

Cohen, B.A., Swindle, T.D., and Kring, D.A.: 2000, 'Support for the Lunar Cataclysm Hypothesis from Lunar Meteorite Impact Melt Ages', *Science* **290**, 1754–1756.

Culler, T.S., Becker, T.A., Muller, R.A., and Renne, P.R.: 2000, 'Lunar Impact History from ^{40}Ar/^{39}Ar Dating of Glass Spherules', *Science* **287**, 1785–1788.

Dalrymple, G.B.: 1991, *The Age of the Earth*, Stanford University Press, 474 pp.

Dalrymple, G.B., and Ryder, G.: 1993, '^{40}Ar/^{39}Ar Ages of Apollo 15 Impact Melt Rocks by Laser Step Heating and Their Bearing on the History of Lunar Basin Formation', *J. Geophys. Res.* **98**, 13,085–13,095.

Dalrymple, G.B., and Ryder, G.: 1996, '^{40}Ar/^{39}Ar Age Spectra of Apollo 17 Highlands Breccia Samples by Laser Step-heating and the Age of the Serenitatis Basin', *J. Geophys. Res.* **101**, 26,069–26,084.

De Hon, R.A., and Waskom, J.D.: 1976, 'Geologic Structure of the Eastern Mare Basins', *Proc. 7th Lunar Sci. Conf.*, 2729–2746.

Deutsch, A., and Stöffler, D.: 1987, 'Rb-Sr Analyses of Apollo 16 Melt Rocks and a New Age Estimate for the Imbrium Basin: Lunar Basin Chronology and the Early Heavy Bombardment of the Moon', *Geochim. Cosmochim. Acta* **51**, 1951–1964.

Drozd, R.J., Hohenberg, C.M., Morgan, C.J., and Ralston, C.E.: 1974, 'Cosmic-ray Exposure History at the Apollo 16 and Other Lunar Sites: Lunar Surface Dynamics', *Geochim. Cosmochim. Acta* **38**, 1625–1642.

Drozd, R.J., Hohenberg, C.M., Morgan, C.J., Podosek, F.A., Wroge, M.L.: 1977, 'Cosmic-ray Exposure History at Taurus-Littrow', *Proc. Lunar Sci. Conf.*, 3027–3043.

Eberhardt, P., Geiss, J., Grögler, N., and Stettler, A.: 1973, 'How Old is the Crater Copernicus?', *The Moon* **8**, 104–114.

Eugster, O.: 1999, 'Chronology of Dimict Breccias and the Age of South Ray Crater at the Apollo 16 Site', *Met. Planet. Sci.* **34**, 385–391.

Faure, G., 1986: *Principles of Isotope Geology*, (2nd. Ed.) Wiley, New York.

Florensky, C.P., Basilevsky, A.T., Ivanov, A.V., Pronin, A.A., and Rode, O.D.: 1977, 'Luna 24: Geological Setting of Landing Site and Characteristics of Sample Core (Preliminary Data)', *Proc. 8th Lunar Sci. Conf.*, 3257–3279.

Freeman, V.F.: 1981, 'Regolith of the Apollo 16 Site', *in* G.E. Ulrich, C.A. Hodges, and W.R. Muehlberger (eds.), *Geology of the Apollo 16 Area, Central Lunar Highlands, USGS Professional Paper* **1048**, 147–159.

Geiss, J., Eberhardt, P., Grögler, N., Guggisberg, S., Maurer, P., and Stettler, A.: 1977, 'Absolute Time Scale of Lunar Mare Formation and Filling', *The Moon - a new Appraisal from Space Missions and Laboratory Analyses, Royal Society of London Philosophical Transactions* **285**, 151–158.

Gilbert, G. K.: 1893, 'The Moon's Face, a Study of the Origin of its Features', *Philosophical Society of Washington Bulletin* **12**, 241–292.

Graham, A.L., and Hutchison, R.: 1980, 'Mineralogy and Petrology of Fragments from the Luna 24 Core', *Phil. Trans. Royal Soc. London* **A297**, 15–22.

Grieve, R.A.F., and Shoemaker, E.M.: 1994, 'The Record of Past Impacts on Earth', *in* T. Gehrels (ed.), *Hazards Due to Comets and Asteroids* Univ. Arizona Press, Tuscon, pp. 417–462.

Grieve, R.A.F., McKay G.A., and Weill, D .F.: 1972, 'Microprobe Studies of Three Luna 16 Basalt Fragments', *Earth Planet. Sci. Lett.* **13**, 233–242.

Grolier, M.J.: 1970, 'Geologic Map of Apollo Site 2 (Apollo 11); Part of Sabine D Region, Southwestern Mare Tranquillitatis', *USGS Map* **I-619** [ORB II-6 (25)], scale 1:25,000; 'Geologic Map of the Sabine Region on the Moon, Lunar Orbiter Site II P-6, Southwestern Mare Tranquillitatis, Including Landing Site 2', *USGS Map* **I-618** [ORB II-6 (100)], scale 1:100,000.

Grossman, J.N.: 2000, 'The Meteoritical Bulletin, No. 84', *Met. Planet. Sci.* **35**, A199–A225.

Hackmann, R.J.: 1966, 'Geologic Map of the Montes Apenninus Quadrangle of the Moon', *USGS Map* **I-463** (LAC-41), scale 1:1,000,000.

Halliday, A.N.: 2000, 'Terrestrial Accretion Rates and the Origin of the Moon', *Earth Planet. Sci. Lett.* **176**, 17–30.

Hartmann, W.K.: 1970, 'Lunar Cratering Chronology', *Icarus* **13**, 299–301.

Hartmann, W.K.: 1972, 'Paleocratering of the Moon: Review of Post-Apollo Data', *Astrophys. Space Sci.* **17**, 48–64.

Hartmann, W.K.: 1973, 'Ancient Lunar Mega-Regolith and Subsurface Structure', *Icarus* **18**, 634.

Hartmann, W.K., and Wood, C.A.: 1971, 'Moon: Origin and Evolution of Multi-ring Basins', *Moon* **3**, 3.

Hartmann, W.K., and Davis, D.R.: 1975, 'Satellite-sized Planetesimals and Lunar Origin', *Icarus* **24**, 504.

Hartmann, W.K., *et al.*: 1981, 'Chronology of Planetary Volcanism by Comparative Studies of Planetary Cratering', *Basaltic Volcanism on the Terrestrial Planets*, Pergamon Press, New York, pp. 1049–1128.

Hartmann, W.K., Phillips, R.J., and Taylor, G.J.: 1984, *Origin of the Moon*, LPI, Houston, pp. 781.

Hartmann, W.K., Ryder, G., Dones, L., and Grinspoon, D.H.: 2000, 'The Time-dependent Intense Bombardment of the Primordial Earth-Moon System', *in* R.M. Righter and R. Canup (eds.), *Origin of the Earth and Moon*, Univ. Arizona Press, Tucson, pp. 493–512.

Hawke, B.R., and Head, J.W.: 1978, 'Lunar KREEP Volcanism: Geologic Evidence for History and Mode of Emplacement', *Proc. 9th Lunar Planet. Sci. Conf.*, 3285–3309.

Head, J.W., Adams, J.B., McCord, T.B., Pieters, C.M., and Zisk, S.H.: 1978, 'Regional Stratigraphy and Geologic History of Mare Crisium', *in* R.B. Merrill and J.J. Papike (eds.), *Mare Crisium: The View from Luna 24*, Pergamon Press, New York, pp. 43–74.

Heiken, G., and McEwen, M.C.: 1972, 'The Geologic Setting of the Luna 20 Site', *Earth Planet. Sci. Lett.* **17**, 3–6.

Heiken, G., Vaniman, D., and French, B.M. (eds.): 1991, *The Lunar Sourcebook. A User's Guide to the Moon*, LPI and Cambridge Univ. Press, New York, pp. 736.

Hörz, F.: 2000, 'Time-variable Cratering Rates?', *Science* **288**, 2095a.

Huneke, J.C., Podosek, F.A., and Wasserburg, G.J.: 1972, 'Gas Retention and Cosmic-ray Exposure Ages of a Basalt Fragment from Mare Fecunditatis', *Earth Planet. Sci. Lett.* **13**, 375–383.

James, O.B.: 1980, 'Rocks of the Early Lunar Crust', *LPSCP* **11**, 365–393.

James, O.B.: 1981, 'Petrologic and Age Relations of Apollo 16 Rocks: Implications for the Sub-surface Geology and Age of the Nectaris Basin', *Proc. 12th Lunar Planet. Sci. Conf.*, 209–233.

Jessberger, E.K.: 1983, '^{40}Ar-^{39}Ar Dating of North Ray Crater Ejecta. I.', *Proc. 14th Lunar Planet. Sci. Conf.*, 349–350.

Jessberger, E.K., Huncke, J.C., Podosek, F.A., and Wasserburg, G.J.: 1974, 'High Resolution Argon Analysis of Neutron-irradiated Apollo 16 Rocks and Separated Minerals', *Proc. 5th Lunar Sci. Conf.*, 1419–1449.

Jessberger, E.K., Kirsten, T., and Staudacher, T.: 1977, 'One Rock and Many Ages – Further Data on Consortium Breccia 73215', *Proc. 8th Lunar Sci. Conf.*, 2567–2580.

Jessberger, E.K., Staudacher, T., Dominik, B., and Kirsten, T.: 1978, 'Argon-argon Ages of Aphanitic Samples from Consortium Breccia 73255', *Proc. 9th Lunar Planet. Sci. Conf.*, 841–854.

Jolliff, B.L., Gillis, J.J., Korotev, R.L., and Haskin, L.A.: 2000a, 'On the Origin of Nonmare Materials at the Apollo 12 Landing Site', *31st Lunar Planet. Sci. Conf.*, abstract #1671.

Jolliff, B.L., Gillis, J.J., Haskin, L.A., Korotev, R.L., and Wieczorek, M.A.: 2000b, 'Major Lunar Crustal Terranes: Surface Expressions and Crust-mantle Origins', *J. Geophys. Res.* **105**, 4197–4216.

Keil, K., Kurat, G., Prinz, M., and Green, J.A.: 1972, 'Lithic Fragments, Glasses and Chondrules from Luna 16 Fines', *Earth Planet. Sci. Lett.* **13**, 243–256.

Koblitz, J.: 1999, 'MetBase 4.0', Meteorite Data Retrieval Software, Fischerhude.

Korotev, R.L.: 1987, 'Mixing Levels, the Apennine Front Soil Component, and Compositional Trends in the Apollo 15 Soils', *Proc. 17th Lunar Planet. Sci. Conf.*, E411–E431.

Korotev, R.L., Jolliff, B.L., and Zeigler, R.A.: 2000, 'The KREEP Components of the Apollo 12 Regolith', *31st Lunar Planet. Sci. Conf.*, abstract #1363.

Kurat, G., Kracher, A., Keil, K., Warner, R., and Prinz, M.: 1976, 'Composition and Origin of Luna 16 Aluminous Mare Basalts', *Proc. 7th Lunar Sci. Conf.*, 1301–1321.

Lee, D.C., Halliday, A.N., Snyder, G.A., and Taylor, L.A.: 1997, 'Age and Origin of the Moon', *Science* **278**, 1098–1103.

Lucchitta, B.K.: 1977, 'Crater Clusters and Light Mantle at the Apollo 17 Site: A Result of Secondary Impact from Tycho', *Icarus* **30**, 80–96.

Lucchitta, B.K., and Sanchez, A.G.: 1975, 'Crater Studies in the Apollo 17 Region', *Proc. 6th Lunar Sci. Conf.*, 2427–2441.

Ma, M.S., Schmitt, R.A., Nielsen, R.L., Taylor, G.J., Warner, R.D., and Keil, K.: 1979, 'Petrogenesis of Luna 16 Aluminous Mare Basalts', *Geophys. Res. Lett.* **6**, 909–912.

Maurer, P., Eberhardt, P., Geiss, J., Grögler, N., Stettler, A., Brown, G.M., Peckett, A., and Krähen-bühl, U.: 1978, 'Pre-Imbrian Craters and Basins: Ages, Compositions and Excavation Depths of Apollo 16 Breccias', *Geochim. Cosmochim. Acta* **42**, 1687–1720.

Maxwell, T.A., and El-Baz F., 1978: 'The Nature of Rays and Sources of Highland Material in Mare Crisium', *in* R.B. Merrill and J.J. Papike (eds.), *Mare Crisium: The View from Luna 24*, Pergamon, New York, 89–103.

McCauley, J.F., and Scott, D.H.: 1972, 'The Geologic Setting of the Luna 16 Landing Site', *Earth Planet. Sci. Lett.* **13**, 225–232.

Melosh, H.J.: 1989, *Impact Cratering: A Geologic Process*, Oxford Univ. Press, New York, p. 245.

Meyer, C., Jr., Brett, R., Hubbard, N.J., Morrison, D.A., McKay, D.S., Aitken, F.K., Takeda, H., and Schonfeld, E.: 1971, 'Mineralogy, Chemistry and Origin of the KREEP Component in Soil Samples from the Ocean of Storms', *Proc. 2nd Lunar Sci. Conf.*, 393–411.

Moore, H.J., Boyce, J.M., and Hahn, D.A.: 1980, 'Small Impact Craters in the Lunar Regolith – Their Morphologies, Relative Ages, and Rates of Formation', *The Moon and the Planets* **23**, 231–252.

Muehlberger, W.R., Hörz, F., Sevier, J.R., and Ulrich, G.E.: 1980, 'Mission Objectives for Geological Exploration of the Apollo 16 Landing Site', *in* J.J. Papike and R.B. Merrill (eds.), *Proc. Conf. on the Lunar Highland Crust, Houston, Texas, 1979*, Pergamon Press, New York, pp. 1–49.

Muller, R.A., Becker, T.A., Culler, T.S., Karner, D.B., and Renne, P:R.: 2000, 'Time-variable Cratering Rates?', *Science* **288**, 2095a.

Neal, C.R., Taylor, L.A., Hughes, S.S., and Schmitt, R.A.: 1990, 'The Significance of Fractional Crystallization in the Petrogenesis of Apollo 17 Type A and B High-Ti Basalts', *Geochim. Cosmochim. Acta* **54**, 1817–1833.

Neal, C.R., Hacker, M.D., Snyder, G.A., Taylor, L.A., Liu, Y.-G., and Schmitt, R.A., 1994: 'Basalt Generation at the Apollo 12 Site I', *Meteoritics*, **29**, 334–348.

Neukum, G.: 1977, 'Different Ages of Lunar Light Plains', *The Moon* **17**, 383–393.

Neukum, G.: 1983, *Meteoritenbombardement und Datierung planetarer Oberflächen*, Habilitation Dissertation for Faculty Membership, Ludwig-Maximilians-University Munich. 186 pp.

Neukum, G., and König, B.: 1976, 'Dating of Individual Lunar Craters', *Proc. 7th Lunar Sci. Conf.*, 2867–2881.

Neukum, G., and Wise, D.U.: 1976, 'Mars: A Standard Crater Curve and Possible New Time Scale', *Science* **194**, 1381–1387.

Neukum, G., and Ivanov, B. A.: 1994, 'Crater Size Distributions and Impact Probabilities on Earth from Lunar, Terrestrial-planet, and Asteroid Cratering Data', *in* T. Gehrels (ed.), *Hazards due to Comets and Asteroids*, Univ. of Arizona Press, Tucson and London, pp. 359–416.

Neukum, G., König, B., Fechtig, H., and Storzer, D.: 1975, 'Cratering in the Earth-moon System: Consequences for Age Determination by Crater Counting', *Proc. 6th Lunar Sci. Conf.*, 2597–2620.

Neukum, G., Ivanov, B., and Hartmann, W.K.: 2001, 'Cratering Records in the Inner Solar System in Relation to the Lunar Reference System', *Space Sci. Rev.*, this volume.

Nozette, S., and the Clementine team: 1994, 'The Clementine Mission to the Moon: Scientific Overview', *Science* **266**, 1835–1839.

Nyquist, L.E., and Shih, C.-Y.: 1992, 'The Isotopic Record of Lunar Volcanism', *Geochim. Cosmochim. Acta* **56**, 2,213–2,234.

Nyquist, L.E., Bogard, D.D., Shih, C.-Y., Greshake, A., Stöffler, D., and Eugster, O.: 2001a, 'Ages and Geologic Histories of Martian Meteorites', *Space Sci. Rev.*, this volume.

Nyquist, L.E., Bogard, D.D., and Shih, C.-Y.: 2001b, 'Radiometric Chronology of Moon and Mars', *The Century of Space Science*, Kluwer Publishers, in press.

Oberbeck, V.R.: 1975, 'The Role of Ballistic Erosion and Sedimentation in Lunar Stratigraphy', *Reviews of Geophysics and Space Physics* **13**, 337–362.

Papanastassiou, D.A., and Wasserburg, G.J.: 1972, 'The Rb-Sr Age of a Crystalline Rock from Apollo 16', *Earth Planet. Sci. Lett.* **16**, 289–298.

Papike, J.J., Ryder, G., and Shearer, C.K.: 1998, 'Lunar Samples', *in* P.H. Ribbe (ed.), *Rev. Miner.* **36**, Planetary Material (Chapt. 5), Min. Soc. Amer., Washington DC, pp. 5-1–5-234.

Podosek, F.A., Huneke, J.C., Gancarz, A.J., and Wasserburg, G.J.: 1973, 'The Age and Petrography of Two Luna 20 Fragments and Inferences for Widespread Lunar Metamorphism', *Geochim. Cosmochim. Acta* **37**, 887–904.

Reid, A.M., Ridley, W.I., Harmon, R.S., Warner, J.L., Brett, R., Jakes, P., and Brown, R.W.: 1972, 'Highly Aluminous Glasses in Lunar Soils and the Nature of the Lunar Highlands', *Geochim. Cosmochim. Acta* **36**, 903–912.

Rhodes, J.M., Blanchard, D.P., Dungan, M.A., Brannon, J.C., and Rodgers, K.V.: 1977, 'Chemistry of Apollo 12 Mare Basalts: Magma Types and Fractionation Processes', *Proc. 8th Lunar Sci. Conf.*, 1305–1338.

Ryder, G.: 1985, 'Catalog of Apollo 15 Rocks', JSC Publ. No. 20787, Curatorial Branch Publ. 72, NASA Johnson Space Center, Houston, p. 1296.

Ryder, G.: 1987, 'Petrographic Evidence for Nonlinear Cooling Rates and a Volcanic Origin for Apollo 15 KREEP Basalts', *Proc. 17th Lunar Planet. Sci. Conf., J. Geophys. Res.* **92**, 331–339.

Ryder, G.: 1990a, 'Lunar Samples, Lunar Accretion, and the Early Bombardment of the Moon', *Eos, Trans. Amer. Geophys. Union* **71**, 313–333.

Ryder, G.: 1990b, 'A Distinct Variant of High-titanium Mare Basalt from the Van Serg Core, Apollo 17 Landing Site', *Meteoritics* **25**, 249–258.

Ryder, G.: 1994, 'Coincidence in Time of the Imbrium Basin Impact and the Apollo 15 Volcanic Flows: The Case for Impact-induced Melting', *in* B.O. Dressler, R.A.F. Grieve, and V.L. Sharpton (eds.), *Large Meteorite Impacts and Planetary Evolution*, Geol. Soc. Amer. **293**, 11–18.

Ryder, G.: 2000, 'Glass Beads Tell a Tale of Lunar Bombardment', *Science* **287**, 1768–1769.

Ryder, G., and Marvin, U.B.: 1978, 'On the Origin of Luna 24 Basalts and Soils', *in* R.B. Merrill and J.J. Papike (eds.), *Mare Crisium: The View from Luna 24*, Pergamon, New York, 339–355.

Ryder, G., and Spudis, P.D.: 1987, 'Chemical Composition and Origin of Apollo 15 Impact Melts', *J. Geophys. Res.* **92**, 432–446.

Ryder, G., and Schuryatz, B.C.: 2001, 'The Chemical Variation of the Large Apollo 15 Olivine-normative Mare Basalt Rock Samples', *J. Geophys. Res.*, in press.

Ryder, G., Bogard, D.D., and Garrison, D.: 1991, 'Probable age of Autolycus and Calibration of Lunar Stratigraphy', *Geology* **19**, 143–146.

Shearer, C.K., and Papike, J.J.: 1999, 'Invited Review: Magmatic Evolution of the Moon', *American Mineralogy* **84**, 1469–1494.

Schultz, P.H., and Merrill, R.B. (eds.): 1981, *Multi-Ring-Basins, Proc. 12th Lunar Planet. Sci.*, Pergamon Press, New York, p. 295.

Shoemaker, E.M., and Hackman, R.J.: 1962, 'Stratigraphic Basis for a Lunar Time Scale', *in* Z. Kopal and Z.K. Mikhailov (eds.), *The Moon*, Academic Press, London, pp. 289–300.

Shoemaker, E.M., Weissman, P.R., and Shoemaker, C.S.: 1994, 'The Flux of Periodic Comets Near Earth', *in* T. Gehrels (ed.), *Hazards Due to Comets and Asteroids* Univ. Arizona Press, Tuscon, pp. 313–336.

Silver, L.T.: 1971, 'U-Th-Pb Isotope Systems in Apollo 11 and 12 Regolithic Materials and a Possible Age for the Copernican Impact', *Eos, Trans. Amer. Geophys. Union* **52**, p. 534.

Snyder, G.A., Borg, L.E., Nyquist, L.E., and Taylor, L.A.: 2000, 'Chronology and Isotopic Constraints on Lunar Evolution', *The Origin of the Earth and the Moon*, Univ. Arizona Press, Tucson, pp. 361–395.

Soderblom, L.A., and Lebofsky, L.A.: 1972, 'Technique for Rapid Determination of Relative Ages of Lunar Areas from Orbital Photography', *J. Geophys. Res.* **77**, 279–296.

Spangler, R.R., Warasila, R., and Delano, L.W.: 1984, '^{40}Ar-^{39}Ar Ages for the Apollo 15 Green and Yellow Volcanic Glasses', *Proc. 14th Lunar Planet. Sci. Conf.*, B487–B497.

Spudis, P.D.: 1978, 'Composition and Origin of the Apennine Bench Formation', *Proc. 9th Lunar Planet. Sci. Conf.*, 3379–3394.

Spudis, P.D.: 1993, *The Geology of Multi-ring Impact Basins: The Moon and Other Planets*, Cambridge University Press, Cambridge, pp. 263.

Spudis, P.D., and Ryder, G.: 1981, 'Apollo 17 Impact Melts and Their Relation to the Serenitatis Basin, Multi-ring Basins', *Proc. 12th Lunar Planet. Sci. Conf.*, 133–148.

Stadermann, F.J., Heusser, E., Jessberger, E.K., Lingner, S., and Stöffler, D.: 1991, 'The Case for a Younger Imbrium Basin: New ^{40}Ar-^{39}Ar Ages of Apollo 14 Rocks', *Geochim. Cosmochim. Acta* **55**, 2339–2349.

Staid, M.I., Pieters, C.M., and Head, J.W., III: 1996, 'Mare Tranquillitatis: Basalt Emplacement History and Relation to Lunar Samples', *J. Geophys. Res.* **101**, 23,213–23,228.

Staudacher, T., Jessberger, E.K., Flohs, I., and Kirsten, T.: 1979, '^{40}Ar-^{39}Ar Age Systematics of Consortium Breccia 73255', *Proc. 10th Lunar Planet. Sci. Conf.*, 745–762.

Steiger and Jäger: 1977, Subcommission on Geochronology: Convention on the Use of Decay Constants in Geo- and Cosmochronology', *Earth Planet. Sci. Lett.* **36**, 359–362.

Stöffler, D., Knöll, H.D., Marvin, U.B., Simonds, C.H., and Warren, P.H.: 1980, 'Recommended Classification and Nomenclature of Lunar Highland Rocks - a Committee Report', *in* J.J. Papike and R.B. Merrill (eds.), *Proc. Conf. on the Lunar Highlands Crust, Houston, 1979*, Pergamon, New York, pp. 51–70.

Stöffler, D., Ostertag, R., Reimold, W.U:, Borchardt, R., Malley, J., and Rehfeldt, A.: 1981, 'Distribution and Provenance of Lunar Highland Rock Types at North Ray Crater, Apollo 16', *Proc. 12th Lunar Planet. Sci. Conf.*, 185–207.

Stöffler, D., *et al.*: 1985, 'Composition and Evolution of the Lunar Crust in the Descartes Highlands, Apollo 16', *J. Geophys. Res.* **89**, 449–506.

Stöffler, D., Bobe, K.D., Jessberger, E.K., Lingner, S., Palme, H., Spettel, B., Stadermann, F., and Wänke, H.: 1989, 'Fra Mauro Formation, Apollo 14: IV. Synopsis and Synthesis of Consortium Studies', *in* G.J. Taylor and P.H. Warren (eds.), *Workshop on Moon in Transition: Apollo 14 KREEP, and Evolved Lunar Rocks, LPI Tech. Rpt.* **89-03**, LPI, Houston, pp. 145–148.

Swann, G.A., Trask, N.J., Hait, M.H., and Sutton, R.L.: 1971, 'Geologic Setting of the Apollo 14 Samples', *Science* **173**, 3998, 716–719.

Swindle, T.D., Spudis P.D., Taylor G.J., Korotev, R.L., Nichols R.H., Jr., and Olinger, C.T.: 1991, 'Searching for Crisium Basin Ejecta: Chemistry and Ages of Luna 20 Impact Melts', *Proc. 21st Lunar Planet. Sci Conf.* **21**, 167–181.

Taylor, S.R.: 1975, *Lunar Science: A Post-Apollo View*, Pergamon, New York, p. 372.

Taylor, S.R.: 1982, *Planetary Science: A Lunar Perspective*, LPI, Houston, p. 481.

Taylor, L.A., Onorato, P.I.K., Uhlmann, D.R., and Coish, R.A.: 1978, 'Subophitic Basalts from Mare Crisium: Cooling Rates', *in* R.B. Merrill and J.J. Papike (eds.), *Mare Crisium: The View from Luna 24*, Pergamon, New York, pp. 473–482.

Tera, F., and Wasserburg, G.J.: 1976, 'Lunar Ball Games and Other Sports', *Lunar Sci.* **7**, 858–860.

Tera, F., Papanastassiou, D.A., and Wasserburg, G.J.: 1974, 'Isotopic Evidence for a Terminal Lunar Cataclysm', *Earth Planet. Sci. Lett.* **22**, 1–21.

Turner, G.: 1977, 'Potassium-argon Chronology of the Moon', *Phys. Chem. Earth* **10**, 145–195.

Ulrich, G.E., Hodges, C.A., and Muehlberger, W.R. (eds.): 1981, *Geology of the Apollo 16 Area, Central Lunar Highlands, USGS Professional Paper* **1048**, p. 539.

Vinogradov, A.P.: 1971, 'Preliminary Data on Lunar Ground Brought to Earth by Automatic Probe "Luna 16",' *Proc. 2nd Lunar Sci. Conf.*, 1–16.

Vinogradov, A.P.: 1973, 'Preliminary Data on Lunar Soil Collected by the Luna 20 Unmanned Spacecraft', *Geochim Cosmochim. Acta* **37**, 721–729.

Wacker, K., Müller, N., and Jessberger, E.K.: 1983, '^{40}Ar-^{39}Ar Dating of North Ray Crater Ejecta. II.', *Meteoritics* **18**, 201.

Warner, R.D., Taylor, G.J., Conrad, G.H., Northrup, H.R., Barker, S., Keil, K., Ma, M.S., and Schmitt, R.: 1979, 'Apollo 17 High-Ti Mare Basalts: New Bulk Compositional Data, Magma Types, and Petrogenesis', *Proc. 10th Lunar Planet. Sci. Conf.*, 225–247.

Wilhelms, D.E.: 1980, 'Stratigraphy of Part of the Lunar Near Side', *USGS Professional Paper* **1046-A**, A1–A71.

Wilhelms, D.E.: 1984, 'The Moon', *in* M. Carr *et al.* (eds.), *The Geology of the Terrestrial Planets, NASA SP* **469**, 107–205.

Wilhelms, D.E.: 1987, 'The Geologic History of the Moon', *U.S. Geol. Surv. Prof. Pap.* **1348**, pp. 302.

Wolfe, E.W., Lucchitta, B.K., Reed, V.S., Ulrich, G.E., and Sanchez, A.G.: 1975, 'Geology of the Taurus-Littrow Valley Floor', *Proc. 6th Lunar Sci. Conf.*, 2463–2482.

Wolfe, E.W., Bailey, N.G., Lucchitta, B.K., Muehlberger, W.R., Scott, D.H., Sutton, R.L., and Wilshire, H.G.: 1981, 'The Geologic Investigation of the Taurus-Littrow Valley: Apollo 17 Landing Site', *USGS Professional Paper* **1080**, p. 280.

Address for correspondence: Dieter Stöffler, Institut für Mineralogie, Museum für Naturkunde, Humboldt-Universität zu Berlin, Invalidenstrasse 43, D-10099 Berlin, Germany

CRATERING RECORDS IN THE INNER SOLAR SYSTEM IN RELATION TO THE LUNAR REFERENCE SYSTEM

G. NEUKUM[1], B.A. IVANOV[2] and W.K. HARTMANN[3]

[1] *DLR Institute of Space Sensor Technology and Planetary Exploration, 12484 Berlin, Germany*
[2] *Institute for Dynamics of Geospheres, Russian Academy of Sciences, Moscow, Russia 117939*
[3] *Planetary Science Institute, Tucson, AZ 85705, USA*

Received: 12 November 2000; accepted: 12 March 2001

Abstract. The well investigated size-frequency distributions (SFD) for lunar craters is used to estimate the SFD for projectiles which formed craters on terrestrial planets and on asteroids. The result shows the relative stability of these distributions during the past 4 Gyr. The derived projectile size-frequency distribution is found to be very close to the size-frequency distribution of Main-Belt asteroids as compared with the recent Spacewatch asteroid data and astronomical observations (Palomar-Leiden survey, IRAS data) as well as data from close-up imagery by space missions. It means that asteroids (or, more generally, collisionally evolved bodies) are the main component of the impactor family. Lunar crater chronology models of the authors published elsewhere are reviewed and refined by making use of refinements in the interpretation of radiometric ages and the improved lunar SFD. In this way, a unified cratering chronology model is established which can be used as a safe basis for modeling the impact chronology of other terrestrial planets, especially Mars.

1. Introduction

Several decades of lunar exploration allowed accumulation of enough data to present an approximate lunar chronology based on the ages of returned samples. The study of the size-frequency distribution (SFD) of lunar impact craters forms a solid basis to show the relative stability of the SFD shape from the time of the late heavy bombardment. The process of crater formation is still going on. Impact craters yield footprints of small body evolution and Solar System chronology.

Planetary cratering records show a picture of bombardment integrated through the whole geologic lifetime of the surfaces studied. Astronomical observations give a snapshot of the small body population. Planetary geologists can compare surfaces of various ages on different planets to reveal spatial and temporal variations of the crater-forming projectile flux.

To compare impact crater SFDs on different planets one needs to take into account many factors such as gravity, atmosphere, crustal strength, density and structure of these bodies, as well as differences in the projectile flux, size-frequency distribution and impact velocity spectrum. This problem has been under investigation for a long time. Comprehensive discussions may be found in, e.g., Hartmann *et al.* (1981), Hartmann (1977), Neukum (1983), and Strom and Neukum (1988).

The goals of the present paper are:
- to review the lunar impact crater production function (PF);
- to refine the lunar cratering chronology and establish a safe basis for conversion to other terrestrial planets, especially Mars;
- to estimate the size-frequency distribution (SFD) for the projectiles corresponding to the lunar PF;
- to compare the impact crater SFD of the terrestrial planets;
- to compare the crater-forming body ("projectile") SFD with the asteroid SFD for the Main Belt and planet-crossing asteroids.

The discussion of the topics listed above is important for the following discussion of the Mars/moon cratering rate comparison. Our basic approach is to establish a refined lunar cratering chronology and to determine the production function, or size distribution, of craters on Mars by using data from other planetary bodies.

2. Lunar Production Function

Despite several decades of studies of the Solar System cratering record, the discussion of the "exact" form of the size-frequency distribution of impact craters created on a fresh geologic unite (i.e. totally rejuvenated at the beginning of cratering with no further obliteration of craters) is far from a final answer. Partially it is the consequence of the simple fact that it is very hard to find enough large "fresh" surfaces — in most situations planetary surfaces have a complex geologic history with simultaneous crater accumulation and degradation (e.g., Hartmann, 1995). However, some general conclusions have been drawn. Below we compare two independent studies of the production SFD named "Hartmann's SFD" and "Neukum's SFD" along with names and results of the main proponents of important contributions to the issue of extracting the lunar SFD from various studies.

2.1. EARLY HISTORY

Crater counts. In the earliest research on the lunar crater size-frequency distributions, statistics were available only for the largest lunar craters resolvable from Earth, i.e., $D > 2$ km. The best statistics were for the range $4\,\text{km} < D < 100\,\text{km}$, which give relatively straight lines on plots of $\log N$ (no. of craters per km^2) vs. $\log D$. The earliest numerical study appears to have been that of Young (1940) who gave a value for the slope of this line as -2.5. Similarly, Brown (1960) analyzed the earliest crude data on log-log asteroid and meteorite size distribution, and fit straight lines to his data sets. These straight lines on plots of $\log N$ vs. $\log D$ are power laws, and thus the earliest literature introduced the idea that power laws gave good fits to the cratering data.

Hartmann (1964) used newly measured catalogs of lunar crater diameters to analyze the significance of the size distribution. Citing the work of Young (1940) and Brown (1960), he fit a slope of -2.1 to the $\log N$ vs. $\log D$ plot. Hartmann has consistently used log incremental plots where N = incremental number of craters in diameter bins of constant $\delta \log D$, and has emphasized the mathematical convenience that this plot gives the same slope as a plot of the cumulative number of craters larger than D. Hartmann (1964) also pointed out that Brown's asteroid/meteorite size distribution curve could be used to predict a power-law lunar crater diameter distribution with cumulative or log incremental slope of -2.4 on the $\log N$ vs. $\log D$ cumulative or log incremental plot, in fair agreement with the lunar data. Hartmann used this agreement to make an early argument that the lunar craters were created by asteroid/meteorite impact.

At the same time, Hartmann (1965) was able to use newly discovered impact craters on the Canadian shield to estimate the terrestrial/lunar crater production rate and use this to predict that the average lunar mare age would be 3 Gyr. This turned out to be almost exactly correct, giving some confidence to the overall procedure.

To a large extent, it was historical precedent and the limited D range of the early data sets that led to the choice of fitting early data to power law diameter distributions (Young, 1940; Hawkins, 1960; Hartmann, 1964). Note that the power law size distribution has the form

$$N = kD^{-b}.$$

N can be understood as either the cumulative number of craters of diameter $\geq D$, or Hartmann's log incremental number of craters in a logarithmic diameter bin (such as $1 - 1.4$ km, $1.4 - 2$ km, $2 - 2.8$ km, etc.). In this formulation, k is a constant depending on which definition of N is used, and b is the power law exponent, or slope of the $\log N$ vs. $\log D$ plot, and is the same for either definition of N. The plot of $\log N$ (either cumulative or log-incremental) vs $\log D$ gives a straight line

$$\log N = -b \log D + \log k$$

Note that all the papers discussed above were published before the discovery of the steep (so-called "secondary crater") branch in the distribution at smaller size (by the Ranger VII lunar impact probe in the summer of 1964), or the discovery of crater populations outside the Earth-moon system (by the Mariner IV Mars fly-by probe in the summer of 1965). Thus, at the time this convention was adopted, the sharp upward turn into a steep branch at $D < 2$ km was not known.

Catastrophic Collision Physics. The use of power laws in cratering and asteroid fragmentation work was extended into a more theoretical realm when Hartmann (1969) showed experimentally and empirically that power law segments gave good fits, often over two to three orders of magnitude in diameter, to size distributions of fragments in systems such as fractured basalt blocks, gravel in a dry streambed, rocks blown out of volcanic craters, and lunar rocks. Hawkins (1960) as well as earlier literature on mineral grinding cited by Hawkins (Gaudin, 1944) also suggested that the power law slope, $-b$, was lower (shallower on the plots) when fragments were produced from bodies broken in low energy-density events, and higher (steeper slope) for systems produced by repeated grinding or high energy density. The quantitative match between the steep slope of debris from high energy cratering events and secondary debris blown out of lunar craters supported Shoemaker's conclusion that the "secondary" steep branch of lunar craters ($D < 2$ km) was produced by rocks blown out of craters, which Shoemaker assumed were the large lunar craters themselves. However, early counts (Neukum et al., 1975; Neukum, 1983; König, 1977) and more recent work of Neukum and Ivanov (1994) suggest that many or most of the projectiles causing the presumed lunar secondary craters stem from collisional processes and the resulting products (fragments) in the asteroid belt, as part of the primary production function of the impactors from outside the moon, rather than debris ejected from lunar craters.

The early work of König et al. (1977) and more recent laboratory experimental work suggest that the fragments from collisional breakup and cratering events can have more complex size distributions than single power laws, being fitted by several power law segments or by a more complex curve, especially if the diameter distribution is traced over several orders of magnitude (e.g., Durda et al., 1998).

Non-power Law Approach. Therefore, starting in the 1970s and refined in the 1980s, Neukum and coworkers took two new steps. Given the three proposed "segments" of the production function, Neukum abandoned the fitting of straight-line power law segments on the log-log plots, and simply fitted a polynomial curve to the data, giving a somewhat S-shaped curve that fit the data. In some ways this procedure is more scientifically neutral or defensible than using power laws, since a polynomial can fit any curve, while there is no a priori guarantee that a power law gives the best fit. Second, applying the polynomial fits, Neukum compared branches of the size distribution from different sources, and constructed a universal polynomial production function size distribution that appeared to represent the initial or "input" size distribution of craters on a wide variety of surfaces, from the moon to asteroids and other bodies.

The shape of this SFD was also studied with respect to its stability through time and found to be invariant within the error limits of measurements (Neukum et al., 1975; Neukum, 1983; Neukum and Ivanov, 1994).

Given this background, our goal in this and the accompanying papers is to find a satisfactory degree of consistency between the hitherto independent approaches in order to derive isochrons and measure the approximate ages of Martian surface features, and compare them with radiometric ages of Martian meteorites.

2.2. HARTMANN'S PRODUCTION FUNCTION — HPF

Hartmann uses an log-incremental SFD representation with a standard diameter bin size. The number of craters per km^{-2} here is calculated for craters in the diameter

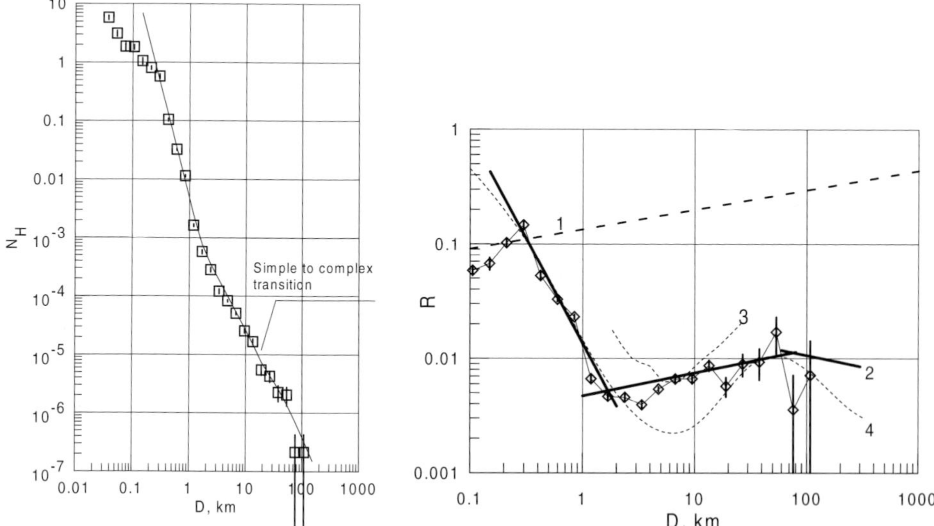

Figure 1. Hartmann's production function (HPF). N_H is the number of craters on an "average" mare surface per a $\sqrt{2}$ diameter bin. b) The same HPF data in R-plot (*diamonds with error bars*), Hartmann's fit to crater saturation (1) and to the production function (2). For comparison *dashed curves* represent Neukum's Orientale basin large crater count (3) and the analytical fit to the steep branch of the HPF for $D < 2$ km (4). See the discussion in the text and in Figure 8.

bin $D_L < D < D_R$, where D_L and D_R are the left and right bin boundary and the standard bin width is $D_R/D_L = \sqrt{2}$.

The tabulated HPF is an assemblage of data selected by Hartmann to present the production function for one specific moment of time — the average time of lunar mare surface formation. Here the condition to have a fresh surface is satisfied with the fact that most lunar mare basalt samples have a narrow band of ages grouping around 3.2 to 3.5 Gyr (Stöffler and Ryder, 2001), which restricts the lunar maria age to a narrow range of ages of a factor of 1.1.

As the tabulated HPF is the result of some averaging of individual crater counts in different areas, it should be treated as a relatively reliable model approach to the PF construction.

Hartmann (1999) propose to approximate the tabulated HPF in the form of a piece-wise three-segment power law:

$$\log N_H = -2.616 - 3.82 \log D_L, \quad D_L < 1.41 \text{ km}$$
$$\log N_H = -2.920 - 1.80 \log D_L, \quad 1.41 \text{ km} < D_L < 64 \text{ km} \tag{1}$$
$$\log N_H = -2.198 - 2.20 \log D_L, \quad D_L > 64 \text{ km}$$

This production function (named below HPF = Hartmann's production function) is compared in Figure 1 with the "best" averaged estimates of the crater density at an "average" mare. Note that in this set of equations the independent variable D_L is the left boundary of the diameter interval, while the data are plotted against the diameter interval geometric mean $D = \sqrt{D_L D_R}$). N_H gives the number of craters in each $\sqrt{2}D$ diameter bin.

2.3. NEUKUM'S PRODUCTION FUNCTION — NPF

Neukum in a series of publications — see the best summary by Neukum (1983), and a reworked version by Neukum and Ivanov (1994) — proposed an analytical function to describe the cumulative number of craters with diameters larger than a given diameter D per unit area. The function was constructed from pieces of impact crater SFD data in different areas of various ages.

Neukum showed that in the diameter accessible for counting craters and for all epochs of lunar history manifested in different geologic units on the moon, the PF had been stable from Nectarian to Copernican times, i.e. practically from more than 4 Gyr ago until now. However, not all diameter ranges for different ages on the moon are accessible, e.g. craters larger than 10 km in diameter do not exist on young areas in statistically significant numbers or craters < 1 km cannot be counted for PF determination on old areas because of erosion and saturation. On the other hand if there is no meaningful application for the moon in the non-accessible diameter ranges and since the smooth piece-wise fit suggested a continuation of the curve characteristics, Neukum's production function assumes a constant shape of the production SFD during all lunar history.

In contrast to piece-wise exponential equations for an incremental SFD, Neukum used a polynomial fit to the cumulative number of craters, N, per km^2 with diameters larger than a given value D.

In Figure 2 the NPF is given in comparison with power-law distributions found in the literature. It can be seen that the power-law representations of the lunar PF are average descriptions of the real behaviour as represented in much closer approximation by the NPF.

For the time period of 1 Gyr, $N(D)$ may be expressed (Neukum, 1983) as:

$$\log_{10}(N) = \sum_{j=0}^{11} a_j \times [\log_{10}(D)]^j \tag{2}$$

where D is in km, N is the number of craters per km^2 per Gyr, and the coefficients a_j are given in Table I. Equation (2) is valid for D from 0.01 km to 300 km.

More recently NPF was slightly reworked in the largest-crater part by careful remeasuring in the size range (Ivanov et al., 1999). "Old" (Neukum, 1983) and "new" (Ivanov et al., 1999, 2000) coefficients are discussed below. The time dependence of the a_0-coefficient is discussed in the following subsection.

A similar equation is used here to present the projectile SFD derived below. Coefficients for this projectile SFD are also listed in Table I. In the projectile SFD column the first coefficient a_0 has been set to zero for simplicity. This coefficient determines the absolute number of projectiles. The absolute value of a_0 for projectiles may be found by fitting to observational data (see Figures 17 and 18).

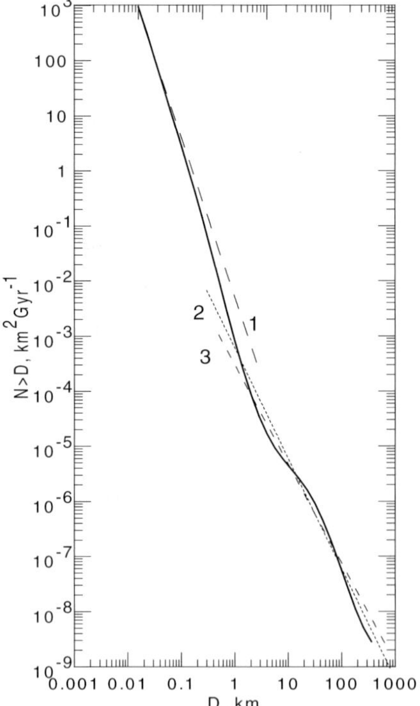

Figure 2. Comparison of the NPF with power-law distributions published in the literature: 1 – $N \sim D^{-2.9}$ (Shoemaker *et al.*, 1970); 2 – $N \sim D^{-2.0}$ (Hartmann, 1971); 3 – $N \sim D^{-1.8}$ (Baldwin, 1971; Hartmann *et al.*, 1981).

Other representations of the size-frequency distribution will be used below:
- the differential distribution: dN/dD;
- the R-distribution: $R = D^3 \times dN/dD$.

The differential distribution is, in practice, the number of craters δN in the diameter range from D to $D + \delta D$, divide by the bin width, δD. As we have an analytical representation of $N(D)$, we can simply differentiate Equation (2):

$$\frac{dN}{dD} = \frac{N}{D} \times \sum_{j=1}^{11} a_j \times [\log_{10}(D)]^{j-1} \tag{3}$$

where N is expressed by Equation (2).

Note: mathematically the derivative dN/dD is negative, since $N(D)$ is a decreasing function. However, for practical purposes, dN/dD should be the number of craters per unit interval of diameter, so it should have a positive value. Below we use the convention that dN/dD is positive and equal to the absolute value of that given by Equation (2). The function $N(D)$ is close to a power-law D^b where b is in the range from -4 to -1.5. Consequently dN/dD is also close to a power-law with the index $(b - 1)$. It is not very convenient for graphical presentation,

TABLE I

Coefficients in Equation (2)

Coefficient	"Old" $N(D)$ (1983) (Neukum, 1983)	"New" $N(D)$ (1999) (Ivanov *et al.*, 2000)	$R(D)$ for projectiles (Ivanov *et al.*, 2000)
a_0	-3.0768	-3.0876	0
a_1	-3.6269	-3.557528	$+1.375458$
a_2	$+0.4366$	$+0.781027$	$+1.272521 \times 10^{-1}$
a_3	$+0.7935$	$+1.021521$	-1.282166
a_4	$+0.0865$	-0.156012	-3.074558×10^{-1}
a_5	-0.2649	-0.444058	$+4.149280 \times 10^{-1}$
a_6	-0.0664	$+0.019977$	$+1.910668 \times 10^{-1}$
a_7	$+0.0379$	$+0.086850$	-4.260980×10^{-2}
a_8	$+0.0106$	-0.005874	-3.976305×10^{-2}
a_9	-0.0022	-0.006809	-3.180179×10^{-3}
a_{10}	-5.18×10^{-4}	$+8.25 \times 10^{-4}$	$+2.799369 \times 10^{-3}$
a_{11}	$+3.97 \times 10^{-5}$	$+5.54 \times 10^{-5}$	$+6.892223 \times 10^{-4}$
a_{12}	$-$	$-$	$+2.614385 \times 10^{-6}$
a_{13}	$-$	$-$	-1.416178×10^{-5}
a_{14}	$-$	$-$	-1.191124×10^{-6}

as N may vary over many orders of magnitude for the range of crater diameters of interest. To avoid this inconvenience one may use the so-called R-distribution, which is simply dN/dD multiplied by D^3 (cf. Arvidson *et al.*, 1978):

$$R(D) = D^3 \times \frac{dN}{dD} \tag{4}$$

If dN/dD is proportional to D^{-3}, then $R(D)$ is a constant; if dN/dD is proportional to D^{-2}, then $R(D)$ is proportional to D; if dN/dD is proportional to D^{-4}, then $R(D)$ is proportional to D^{-1}, etc.

The same equations may be used to describe the size-frequency distribution of projectiles. Below we express projectile dimensions via the projectile diameter D_P. One may use Equations (2) – (4) for projectiles, changing D into D_P.

2.4. NPF UPDATING

The new count of 6700 lunar impact craters in an area of 7.8×10^5 km^2 rejuvenated by the Orientale impact allows us to re-estimate the size-frequency distribution for the large-crater part (Ivanov *et al.*, 1999, 2000). The curve again is approximated by an 11th degree polynomial for 100 m $< D <$ 200 km (Figure 3). "New" coef-

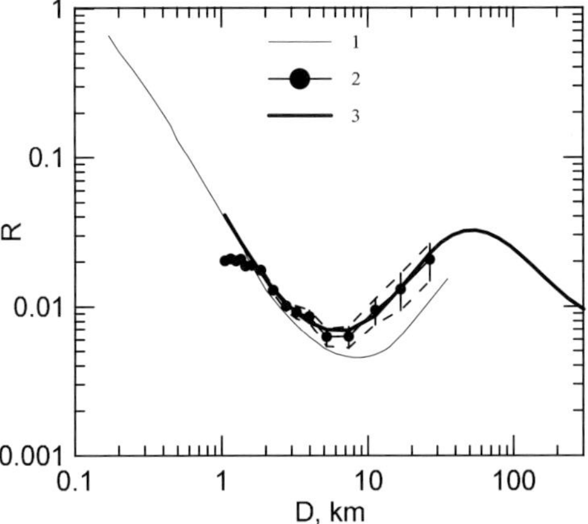

Figure 3. The "old" lunar calibration curve (1), the new crater count for the Orientale basin with error bars (2), and the "corrected" ("new") calibration curve (3). For $D < 2$ km and $D > 20$ km the "new curve" has the same shape as the "old" one.

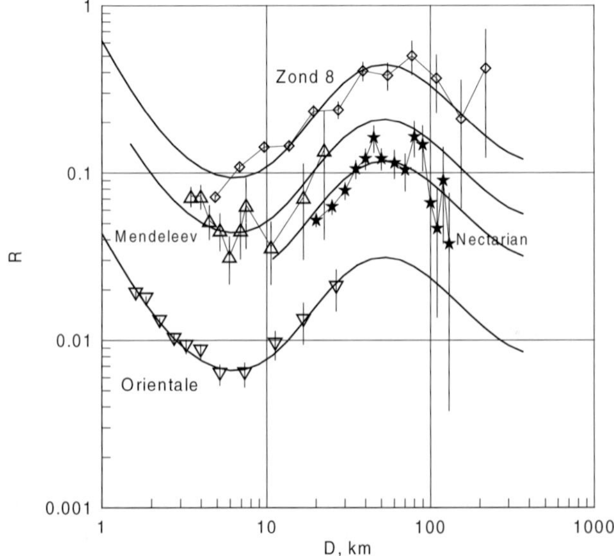

Figure 4. SFD analytical fit to counts on Imbrian, Nectarian and pre-Nectarian (highlands) geological units.

ficients in Equation (2) are listed in Table I in comparison with "old" coefficients from Neukum (1983).

Figure 4 illustrates the new Orientale counts and the crater SFD fit to these data and to the typical highland counts near the Tsiolkovsky area, on the farside, published by Ronca *et al.* (1981) and other highland counts by Neukum (1983) and for a Nectarian aged geological unit (Mendeleev) by Neukum (1977). Most of the craters were formed > 3 Gyr ago, so the question remains how the projectile

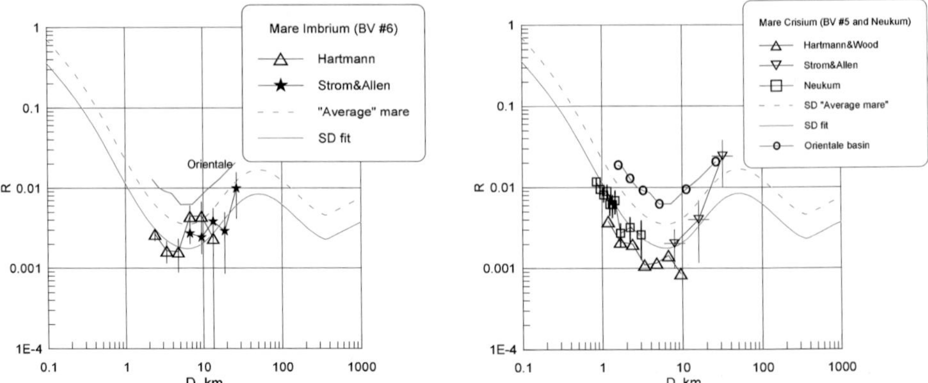

Figure 5. Left: Mare Imbrium counts. *Right:* Mare Crisium crater counts. (Data from Hartmann *et al.*, 1981).

population has evolved since the Orientale formation time. The problem of the stability in time of the production SFD for lunar craters has been under discussion for many years (e.g., Strom, 1977; Hartmann, 1984; Strom and Neukum, 1988; Neukum and Ivanov, 1994). Here we show the selected examples of lunar SFD distributions of various ages to illustrate the main features of the non-power-law SFD. Figure 5 illustrates published crater counts for Mare Imbrium and Mare Crisium in comparison with the proposed standard distribution.

While the obvious misfits of some counts should be discussed separately (see, for example, the discussion of a possible resurfacing by Hartmann, 1995), the important finding is the presence of a definitive R-minimum in the R-plots. Therefore, the general deviation of the mare crater SFD from a simple power law is well illustrated by individual crater counts. It seems that the suggested "flattening" of the "average mare" crater SFD is the result of the averaging of individual crater counts for separate areas. It is difficult to find a contiguous large geological unit much younger than the mare on the Moon. The recent Galileo and Clementine imaging data, however, allow us to estimate the SFD of rayed (Copernican) craters (McEwen *et al.*, 1993; 1997). Figure 6 *left* compares the SFD for farside rayed craters (McEwen *et al.*, 1997) with crater counts for Copernicus itself — see "old" counts by Hartmann *et al.* (1981), and "new" counts by Neukum cited by McEwen *et al.* (1993).

The non-power-law curve well approximates the size-frequency distribution for craters with an age of 1.5 Gyr (Neukum's dating in McEwen *et al.*, 1993), confirming the presence of the R-minimum at a crater diameter of 6 km (projectile diameter of 0.5 km). The presence of a steep branch of the lunar crater SFD for younger surfaces is confirmed by the crater count for Aristarchus (Figure 6 *right*).

To summarize the lunar production function examples Figure 7 compares the cumulative $N(D)$ curves (1) for the old Orientale basin, (2) – Erathosthenes crater and Eratostenian (and younger) craters (dated according to Wilhelms *et al.*, 1987),

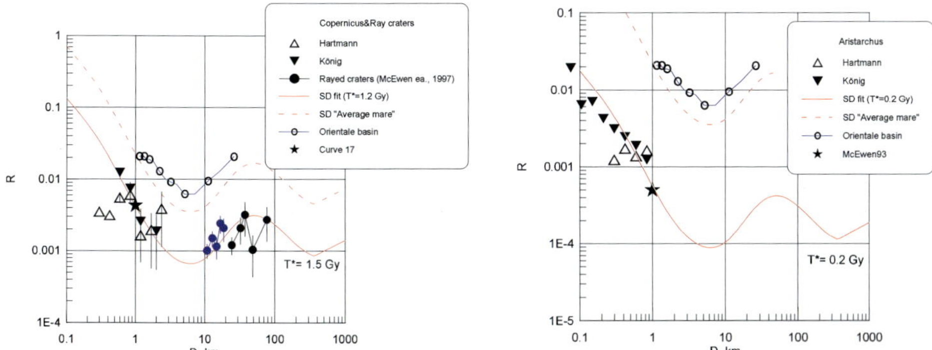

Figure 6. Left: The SFD for farside rayed craters (McEwen *et al.*, 1997) with crater counts for Copernicus itself in comparison with data for older areas. *Right:* The crater count for Aristarchus.

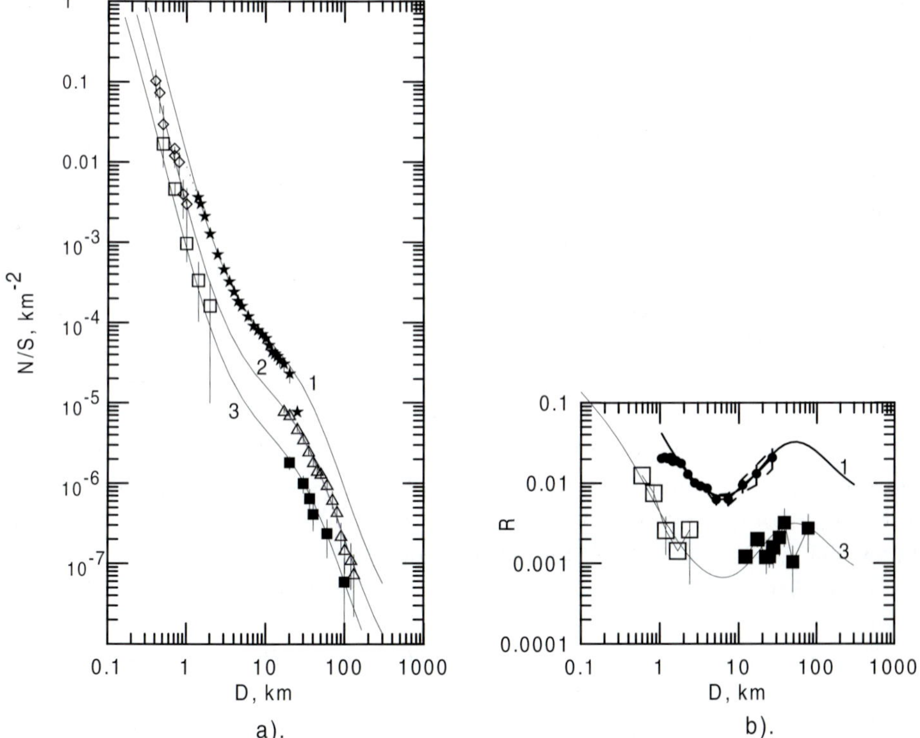

Figure 7. a) The cumulative $N(D)$ curves (1) for the Orientale basin, (2) for Erathosthenes and Eratostenian (and younger) craters (dated according to Wilhelms *et al.*, 1987), and (3) the population of rayed (Copernican) craters (after Moore and McEwen, 1996 and McEwen *et al.*, 1997) and Copernicus itself. b) Comparison of Orientale and Copernican craters in R-plot.

and (3) – the younger population of rayed (Copernican) craters (after Moore and McEwen, 1996, and McEwen *et al.*, 1997) and Copernicus itself. In Figure 6 *left* counts of Orientale and Copernican craters are compared in R-plot. All these three populations are rather representative for "time slices" of the cratering flux:

- the impact-created Orientale basin erased a large area at the time near the base of the Imbrian stratigraphic horizon;
- Eratosthenian-aged craters are well dated stratigraphically by Wilhelms;
- rays around craters are believed to have some limited life time, and thus also mark an approximate "time horizon".

For all 3 cases the NPF gives a good fit to observed crater counts, provided that Erathosthenes and Copernicus present the upper range of the correspondingly named subpopulations. At the time of the Orientale formation the cratering rate on the moon was roughly 100 times above the present level (Figure 11). Most Eratosthenian craters were formed with a flux 3 to 5 times above the modern one. Copernican craters were formed in the last 1 to 1.5 Gyr and reflect the same cratering flux as we have now. Having the good fit of the NPF to all 3 examined populations one can safely state that in the accuracy limits the projectile SFD has not changed dramatically.

2.5. HPF AND NPF COMPARISON

To compare the HPF and NPF functions it is suitable to present both of them in the R-representation (Equation 4) which allows to reduce the vertical scale and to clear fine details.

Figure 8 illustrates the HPF and NPF for the well dated Apollo 15 and Luna 24 (Mare Crisium) landing sites (the mare basalt sample's age is 3.2 Gyr) in comparison with the SFD for craters accumulated in the area rejuvenated by the Orientale basin formation. The NPF gives the fit to the crater counts with a proper assumed age. The HPF below $D = 1$ km also fits the observational data. However, above $D > 1$ km the HPF goes well above the NPF, meeting again the NPF at crater diameters $D \geq 40$ km. A maximum discrepancy of a factor of 3 between HPF and NPF is observed in the diameter bins around $D = 6$ km. Below $D = 1$ km and in the diameter range of 30 to 100 km HPF and NPF give the same or similar results.

The R-plot (Figure 8) shows, that there is moderately good agreement in the SFD shape between different researches. However the factor of 3 maximum discrepancy needs further investigation. One should be cautions interpreting data on craters in the diameter range of 2 to 20 km.

2.6. LUNAR CHRONOLOGY AND CRATERING RATE DECAY IN TIME

Most investigators believe that during the last 3 Gyr the cratering projectile flux was relatively constant with possible variations within a factor of 2. Before 3 Gyr ago the Planet-Crossing Asteroids (PCA) flux was much higher rapidly decaying

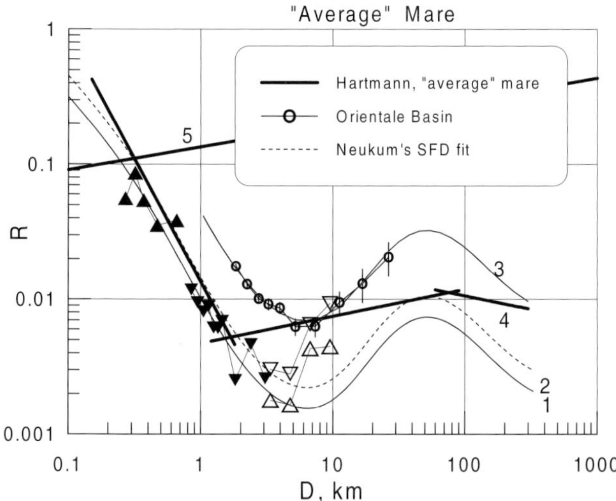

Figure 8. Comparison of production functions derived by Hartmann (HPF) and Neukum (NPF) in the *R*-plot representation. The maximum discrepancy with a factor of 3 between HPF and NPF is observed in the diameter bins around $D = 6$ km. Below $D = 1$ km and in the diameter range of 30 to 100 km HPF and NPF give the same or similar results.
1 – the NPF fit to crater counts for the Apollo 15 (*triangles*). *Filled triangles* — crater counts by Neukum (1983), *open triangles* — crater count data from Hartmann *et al.* (1981).
2 – the NPF fit to the steep brunch of HPF. Equation (5) gives the model time estimate of 3.6 Gyr for the "average mare" Hartmann's approximation.
3 – the NPF fit to the new wide range count of impact craters in the Orientale basin. The model age from Equation (5) is around 3.7 Gyr.

in time. This ancient period is named "late heavy bombardment". The dating of lunar rocks in parallel with the crater counts reveals the general character of the bombardment flux decay (Stöffler and Ryder, 2001).

There have been previously a number of attempts to combine crater frequency data and radiometric ages determined for the lunar landing sites and for other units on the moon for which a radiometric age could be derived indirectly from rock samples at the Apollo landing sites (e.g. for Tycho, sampling at the Apollo 17 landing site). The empirical relationship resulting from this kind of plotting crater frequency vs. radiometric age is called cratering chronology. The best established chronology models put forward are those by Hartmann (Basaltic Volcanism Project – Hartmann *et al.*, 1981, and Neukum, 1983, both displayed in Figure 9). The discrepancy between the two models has been known to be due partly to different data sets used, but mainly due to differences in PFs used for the data fit. In this paper, making use of the best currently available data (Stöffler and Ryder, 2001), and confident interpretations, especially by applying one and the same PF, the models are unified as shown in Figure 10. It gives the currently best interpretation for the cratering rate decay and the lunar cratering chronology.

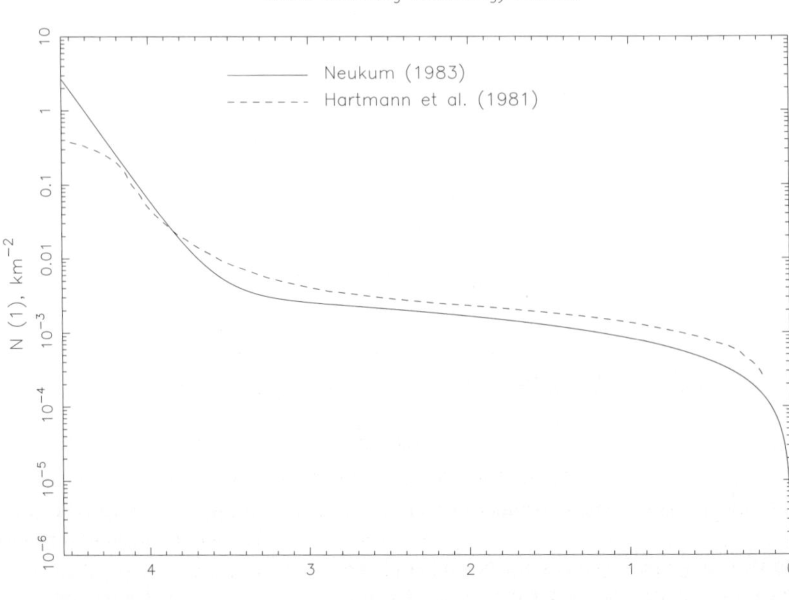

Figure 9. Comparison of two lunar cratering chronology models by Hartmann *et al.* (1981) and by Neukum (1983).

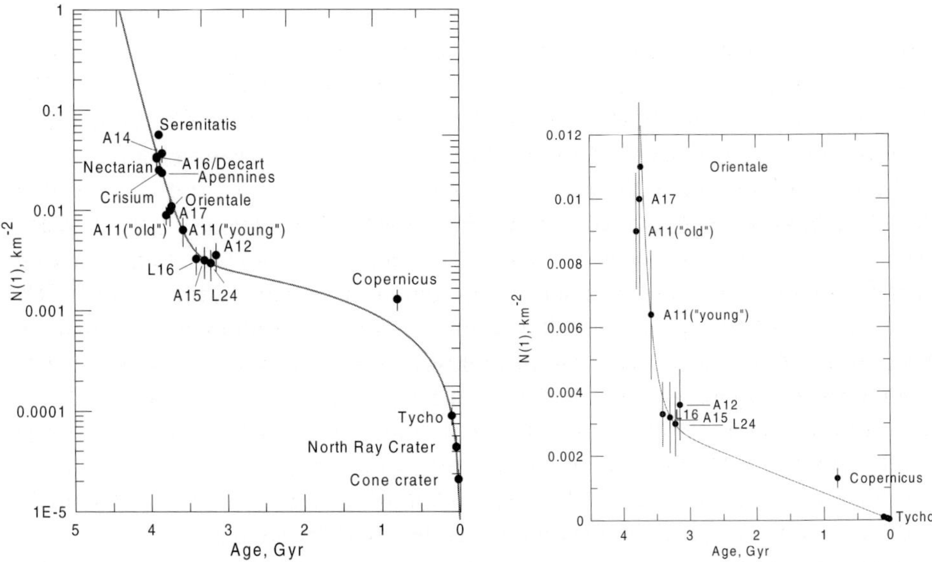

Figure 10. Left: Graphical representation of Equation (5) (lunar cratering chronology) in log-log format (see Table V of Stöffler and Ryder, 2001). *Right:* The part of the lunar cratering chronology in linear scale.

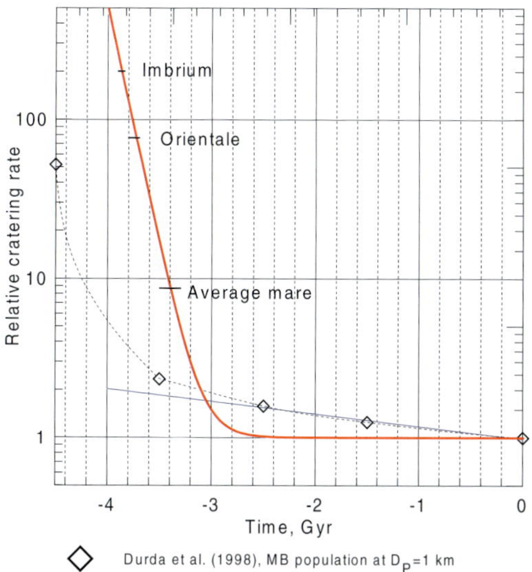

Figure 11. Estimate of the cratering rate as a function of time. The curve is a time derivative of the Equation (5), normalized to the modern impact rate. *Diamonds* present the model by Durda *et al.* (1998) where the gradual decrease of the number of the MB asteroids is due to the collision evolution only (no losses to planet crossing orbits).

Neukum (1983) expressed it analytically in the form (also used here)

$$N(1) = 5.44 \cdot 10^{-14}[\exp(6.93T) - 1] + 8.38 \cdot 10^{-4}T \quad , \tag{5}$$

which relates the number of craters equal to and larger than 1 km in diameter per km^2 in an area with the crater accumulation time (crater retention age) T in Gyr. Assuming the constant shape in time of the SFD for projectiles, Equation (5) is valid for any crater diameter with the proper numerical coefficients.

Figure 10 shows the graphical representation of Equation (5) and demonstrates that at approximately 4 Gyr age surfaces 95% of all craters were formed between 3 and 4 Gyr ago and only 5% of craters are younger than 3 Gyr. The time-derivative of the $N(T)$ relationship gives an expression for the cratering rate, dN/dt. The results is shown in Figure 11. One can see that the cratering rate 4 Gyr ago was 500 times higher than the constant rate during the last 3 Gyr in accord with earlier models by Hartmann (1970, 1984) and Neukum (1983). This high impact rate leads to a question about the nature of projectiles responsible for the late heavy bombardment: asteroids, comets, or left-over planetesimals?

This issue is discussed in the recent review by Hartmann (1999). Here we briefly outline some points regarding several possible options of interpretation.

The **cometary** impact flux is believed to have been stable at the same level or may have smoothly decayed during the last 4 Gyr. The Oort cloud and the Kuiper belt were populated by comets during the giant planet formation (e.g. Safronov, 1972; Weissman, 1999). The flux of long period (Oort cloud) comets may be modulated by galactic tides or a close star passage. For Jupiter family (Kuiper belt) comets one may assume a flux decay proportional to T^{-1} (Zahnle *et al.*, 1999) during the last 4 Gyr. As 95% of craters were formed between 3 and 4 Gyr ago, where the cratering rate was up to more than a hundred times greater, one might assume that during the heavy bombardment period a distinct "heavy bombardment projectile population" existed. The consequence a possible difference in the size frequency distributions of older and younger craters, created by projectiles with a different share of comets (Strom *et al.*, 1981; Strom and Neukum, 1988).

If, for example, a contribution of cometary impacts is at a comparable flux level as the asteroidal one three options can be discussed:
- comets have the same SFD as asteroids;
- the cometary flux changes in time exactly as the flux of asteroids;
- the cometary input is within the errorbars of our crater counts.

Otherwise one should expect the difference in SFD for craters produced at the end of the heavy bombardment period (Orientale) and in the last 1 Gyr (rayed craters). The first option is not compatible with cometary SFD estimates by Shoemaker and Wolfe (1982). The second option is not compatible with the current view of the cometary dynamics (Weissman, 1999). For these reasons we exploit the third option here. The real input of comets is an issue for more detailed future studies.

The question is still under discussion and investigation. However recent comparative studies of the impact crater SFDs on terrestrial planets show that some specific features exist (like on the moon) — characteristic deviations from a simple power law — which may be identified in cratering records for 3.8 Gyr-old and 0.5 Gyr-old cratering records respectively (Ivanov *et al.*, 1999, 2000). The characteristic deviations from a simple power law corresponds well with similar deviations in the Main Belt asteroid SFD, where the off-power-law deviations seem to be controlled by the strength-to-gravity transition in the collisional evolution of asteroids (e.g., Durda *et al.*, 1998). Hence, the "old" and "young" craters seem to have been produced by a collisionally well evolved population. This agrees with the discussion by Hartmann (1995) who presented evidence that the putative "old" size distribution on the moon is produced by obliteration processes.

The question now is: Are comets collisionally evolved bodies? For Kuiper belt objects the collisional evolution is possible under specific assumptions for the number of comets (Davies and Farinella, 1997). Were Oort cloud comets collisionally evolved before the ejection to the Solar System periphery (Stern and Weissman, 2000)? If comets are collisionally evolved bodies, is their SFD distinct in comparison with the SFD for asteroids? To answer these question one needs much more physical and dynamical studies in the future. In the 20 year old estimates by Shoemaker and Wolfe (1982) comets were assumed to have a simple power law SFD dramatically different from what we know about the impact crater SFD on terrestrial planets. Weidenschilling (1994) predicted that comets formed as aggregates of loosely bound 100 m scale bodies, so that even if they are collisionally evolved by fragmentation, they might have a different SFD than the asteroid population.

Unfortunately, it is impossible to be sure of the comet SFD or to estimate the real share of cometary craters in the cratering records on terrestrial planets.

If **asteroids** where the main kind of projectiles, one should discuss the mechanism of the re-supply of the PCA population. Under the assumption of catastrophic fragmentation and "direct delivery" to resonant orbits, the rate of the PCA supply is proportional to the collision rate of main belt asteroids. In this case the PCA production rate is proportional to the derivative of the Main Belt (MB) asteroid SFD. Consequently, the PCA SFD is steeper than the MB SFD, and the balance between PCA extinction and production (and the steady state number) is controlled by the PCA timelife and the collision rate in the main belt (Rabinowitz, 1997). The cratering rate in this case may varies if the Main Belt SFD varies in time.

If the chaotic migration is the main delivery mechanism for PCA, the number of MB asteroids converted to PCA in a given time period is proportional to the MB population. The gravity-driven chaotic migration acts equally for asteroids of any size so that the steady-state number of PCA is proportional to the number of bodies in the "feeding zones" of the Main Belt. Thus, the cratering rate is proportional to the number of bodies in the main belt at the respective time period.

Most models of the collisional evolution of the Main Belt (Davis *et al.*, 1985; Durda *et al.*, 1998) predict more or less steady state for the size-frequency distribution in the Main Belt. The bodies of a given size are periodically destroyed by catastrophic collisions (and removed from a given size bin) but the collision of larger bodies produce new fragments (and a part of them adds to the number of bodies in each size bin). Finally the population approximately is in a steady state. An example of the time variation of the MB population according to the model by Durda *et al.* (1998) is shown in Figure 11. For bodies of 1 km in diameter (the diameter bin $1 \text{km} < D_P < 1.41 \text{km}$) the total number in the asteroid belt 3 Gyr ago was only twice larger than now. Consequently one can assume that the PCA steady state number and the planetary cratering rate also changed no more than a factor 2. One should note that in the model published by Durda *et al.* only collisional evolution is taken into account while the measurable share of bodies at each time step should be removed from each size bin via "fast track" and "slow track" sinks. The non-collision body remove may change the collision rate and, consequently, the time scale for the evolution model.

If one assume the proportionality of the PCA and MB asteroid numbers, the hundreds time larger cratering rate before 3 Gyr automatically predicts the hundred times more populated Main Belt provided the currently operating MB-PCA delivery rout. In principle the modern theories predict the surface density of material in the protoplanet disk 10^3 to 10^4 times larger than observed now (Wetherill, 1989; Ruzmaikina *et al.*, 1989). Moreover, the main mass of the main belt is concentrated in several largest bodies. Alternatives to the "heavily populated" ancient MB are a different, than now, mechanism of the MB depopulation between 3 and 4 Gyr ago and/or left-over Ruzmaikina's planetesimals on long-living orbits (Hartmann *et al.*, 2001).

Planetesimals (cometesimals) ejected from the zone of giant planet growth are discussed as possible "killers" which removed the extra mass from the modern asteroid belt (Wetherill, 1989; Ruzmaikina *et al.*, 1989; Gil-Hutton and Brunini, 1999). The catastrophic collisions with asteroids in the MB seem to be a good candidate for explaining the main belt mass decrease. However, this leads to the assumption of some specific synchronization between the main belt depopulation, giant planet growth and the formation of the solid crusts on terrestrial planets. An open question, however, is the respective share of craters on terrestrial planets formed by fragmented asteroids and by direct impacts of the presumed planetesimals. Planetesimals (cometesimals) seem to have played their role at the very early stage of the cratering record accumulation, which essentially is not seen on the surfaces of the terrestrial planets anymore. In the "standard" scenario the late heavy bombardment projectiles are assumed to be the long-living "left-over" planetesimals of the terrestrial planet formation zone. The recent review by Hartmann *et al.* (2001) discuss the planetesimal scenario in greater details.

3. Inter-planetary Comparison

In this section we produce the "theoretical" planetary SFD from the lunar data by scaling it to other terrestrial planets, and compare it with published measurements for selected areas on Mercury, Venus, Earth, and Mars. As discussed in more details by Hartmann (1977) this requires that we derive the average impact velocity for each planetary body. The scaling law issue is discussed in Ivanov (2001).

3.1. THE PROJECTILE SIZE-FREQUENCY DISTRIBUTION

Average Impact Velocity.   Several approaches allow us to estimate the average impact velocity on Earth.
1. Astronomical catalogs of Earth-crossers (Shoemaker, 1977; Rabinowitz *et al.*, 1994). These data need to be recalculated from orbital parameters to impact probabilities and velocities. Shoemaker did this on the basis of Öpik's theory (Öpik, 1966; Shoemaker and Wolfe, 1982). A larger compendium of observational data may be found in the "astorb.dat" file updated at Lowell Observatory: `http://naic.edu/~nolan/astorb.html`.
   This file contains orbital osculating elements for $>30,000$ small bodies including a relatively large number of planet-crossing asteroids. These data seem to have a large observational bias; however we use them to make some statistically meaningful estimates.
2. A theoretical model by Rabinowitz (1993) who tried to estimate the total ("debiased") population of Earth crossers from observational data. The Öpik model may be used to convert the theoretical prediction of Rabinowitz to impact velocity and probability data.

The astronomical approach has a specific disadvantage: the lifetime of observed orbits is much smaller than the collisional time scale. Within their lifetimes orbits may evolve significantly (e.g., Milani *et al.*, 1989). So we are forced to apply the idea of some kind of equilibrium: the exchange of different "cells" of the phase space of planet crossers (semimajor axis a, eccentricity e and orbit inclination i) is in a steady state: if a body with current parameters a, e, i changes the orbit due to some perturbation or resonant interaction, statistically, another body will change its orbital parameters into the same a, e, i.

Typically the average velocities are similar for all sources of data, giving somewhat larger velocities for observed meteors and the debiased population of Earth crossers.

For other planets we can only use the observed projectile population, and most estimates here are done with the "astorb.dat" data file (Ivanov *et al.*, 2000, 2001).

Size of Cratering Projectiles. Using scaling laws (Ivanov, 2001) and estimated impact velocities, one can find the projectile SFD, dN/dD_{P}, for a given impact-crater SFD, dN/dD. To simplify the problem in this paper, we test the end-member hypothesis of a purely asteroidal projectile flux onto the terrestrial planets. Having such an estimate, we keep open the problem of the cometary impact fraction in the observed crater population. For the same reason we assume the same projectile density ($2.7 \ \mathrm{Mg\,m^{-3}}$) for all estimates. To facilitate the inter-planetary estimates, the projectile SFD is approximated in the same form as Equations (2) and (3) with a polynomial of 14th degree valid for projectile diameters from 1 m to 100 km. The estimates for the largest projectile sizes should be used with caution, as the source crater SFD in this range is based on the largest lunar basins, all of which are very old and do not appear in younger crater populations. Moreover, the basin assignment of a diameter D to a given basin involves interpretation of the origin of multi-ring structures.

Polynomial coefficients for the R-plot are listed in Table I. The estimated SFD is used below to produce a model ("lunar analog") for Mercury, Venus, Earth, and Mars. The projectile SFD is also compared with the recent data on the main belt SFD. The obtained projectile SFD is shown graphically in Figures 17– 19.

3.2. CRATERING RECORD ON THE TERRESTRIAL PLANETS

The lunar-based projectile SFD is applied below for terrestrial planets to produce the "lunar analog" of the crater SFD: the model population created with the same projectile SFD as for the moon, but with an account for specific impact velocity and gravity for each planetary body. For an assumed velocity of impact the gravitational focusing factor must also be taken into account:

$$g_{\mathrm{enh}} = 1 + (v_{\mathrm{esc}}/U_{\mathrm{inf}})^2$$

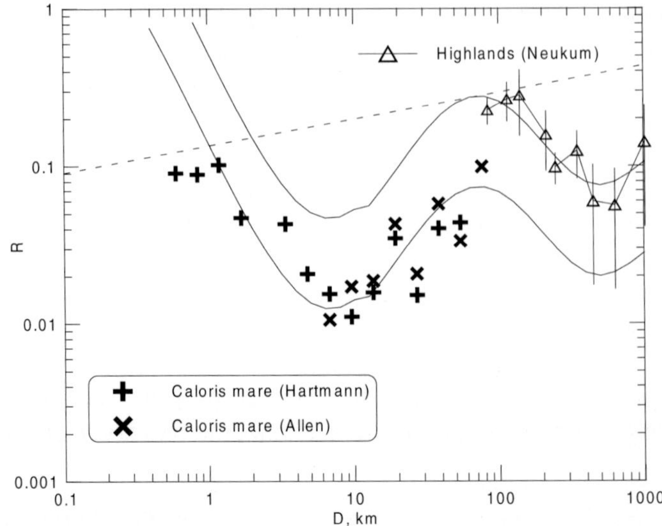

Figure 12. Crater counts for highlands and Caloris mare basin on Mercury in comparison with "lunar analog" curves. Dashed line show an approximate saturation level after Hartmann (1995).

where g_{enh} is the projectile flux enhancement factor due to the gravitational attraction to the target body (here v_{esc} and U_{inf} are the escape velocity and the average velocity of projectiles "at infinity").

Mercury. The mare surface in the Caloris basin is one of few areas suitable for production function measurements of small to intermediate-sized craters on Mercury. Figure 12 compares direct measurements and calculated SFD (the "lunar analog"). The good coincidence of these data shows a definite similarity of projectile SFDs on the Moon and Mercury in the projectile diameter range from 1 to $\sim$100 km with a steep part for smaller craters and the "$\mathcal{R}$-minimum" for craters with $D \sim 8$ km. However, the age of the Caloris basin is comparable to the age of the Orientale basin.

Venus. Magellan data allow us to compare the lunar data averaged over the last 3 Gyr with a planetary surface of $\sim$0.5 Gyr age. The presence of the atmosphere may be taken into account using a model of projectile atmospheric passage. A model and model results for Venus and Titan were presented by Ivanov *et al.* (1997). The resulting comparison (Figure 13) shows that Venusian craters were formed by a projectile population with a similar SFD for $D > 10$ km (projectile diameters $D_P > 2$ km). The R-maximum at $D \sim 50 - 70$ km exists both on the Moon (3 to 4 Gyr) and on Venus ($\sim$0.5 Gyr). One can conclude that the corresponding R-maximum in the projectile distribution in the range of $D_P \sim 5$ km is thus stable in time.

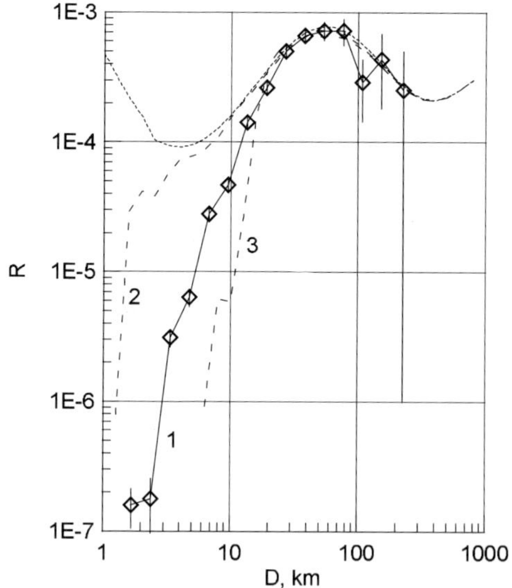

Figure 13. *R*-plot for the size-frequency distribution of Venusian craters (1) in comparison with the lunar curve recalculated for Venusian conditions with the Schmidt and Housen (1987) scaling law and Croft's (1985) model of the crater collapse. Dashed curves 2 and 3 present two models of atmospheric disintegration of projectiles after Ivanov *et al.* (1997).

Earth — Impact Craters. Hartmann (1965, 1966) pointed out that large terrestrial craters reflect an older population, while smaller craters are continually removed by erosion, producing an observed SFD that differ from the production function. The inspection of data for terrestrial North American and European cratons (Grieve and Shoemaker, 1994) shows that it is possible to distinguish two populations of craters:

– 8 craters with diameters from 24 to 39 km, the oldest of which is 115 Myr old, and

– 8 craters with diameters from 55 to 100 km and ages ranging from 214 to 370 Myr.

Adding the average time interval between impacts to the oldest age in each set, one can estimate the time of accumulation as ~135 Myr and 380 Myr, respectively, for the younger and older populations. For a proper balance between crater diameter bin width and the number of craters per bin, only two bins for each age sub-population may be used to represent the crater production rate.

We assume that craters smaller than ~20 km in the younger set and smaller than ~45 km in the older set are depleted by erosional processes. Figure 16 shows the incremental plot for these data sets in comparison with their "lunar analogs" calculated for assumed values of accumulation time and a constant flux (possible flux variability is discussed by Grieve and Shoemaker (1994), and McEwen*et al.,* 1997). As an illustrative estimate, the "lunar analog" for the Sudbury age is also

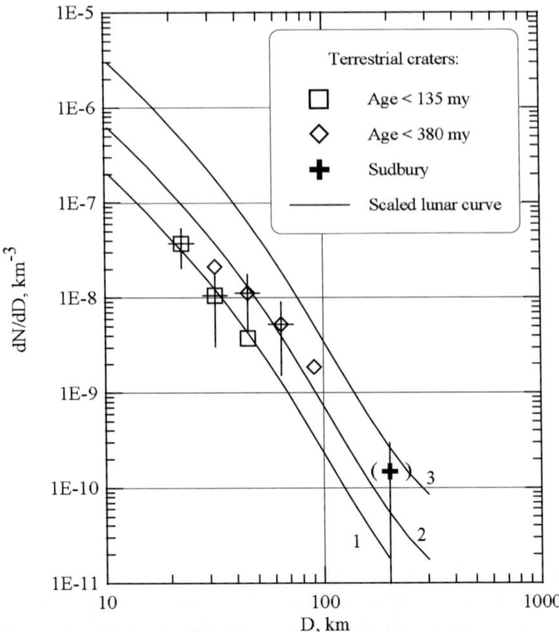

Figure 14. The differential SFD for terrestrial craters in comparison with data for cratons (North American + European).

plotted here to show that, statistically, Sudbury may be the largest impact crater formed in the area of 17.6×10^6 km^2 — the total area of measured cratons.

Figure 17 shows the R-plot of terrestrial data recalculated to the reference age of 1 Gyr assuming a constant crater production rate. This assumption does not contradict the lunar crater SFD recalculated to the terrestrial conditions. The recent terrestrial production rate estimate by Hughes (2000) gives the similar results to our simplified analysis (*dark circles* in Figure 17).

Earth — Bolides. Ivanov *et al.* (2000) use data of the satellite monitoring of light flashes in the atmosphere to estimate the energy of several tens of bright bolides (Nemtchinov *et al.*, 1997). Assuming that all bolide projectiles have the average NEA velocity and probability of impact, one can estimate the modern population of small NEAs. The data are in a good fit to the estimated NEA population.

Mars. The new data from Mars Global Surveyor are discussed in Hartmann and Neukum (2001; see also Hartmann *et al.*, 1999a, b). Here we present for consistency of the chapter only 2 examples of Martian cratering records which show the same trends as for other terrestrial planets (Figures 16, 17 and 18). The small craters on Mars have a large rate of obliteration, so one should add some model of crater "equilibrium" to analyze the data. However, the presence of the same R-maximum and R-minimum in the SFD is clear.

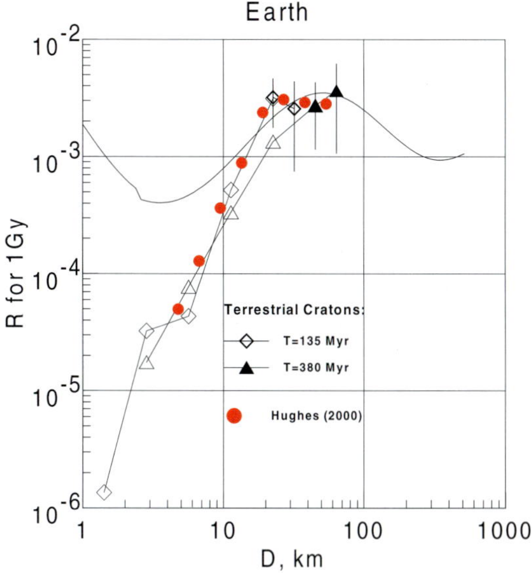

Figure 15. The same data as in Figure 16 in a *R*-plot. Data are divided by 0.135 and 0.380 to put them at the 1 Gyr position for checking the flux variation. Black dots are for estimates by Hughes (2000) made with a technique of "nearest neighbours".

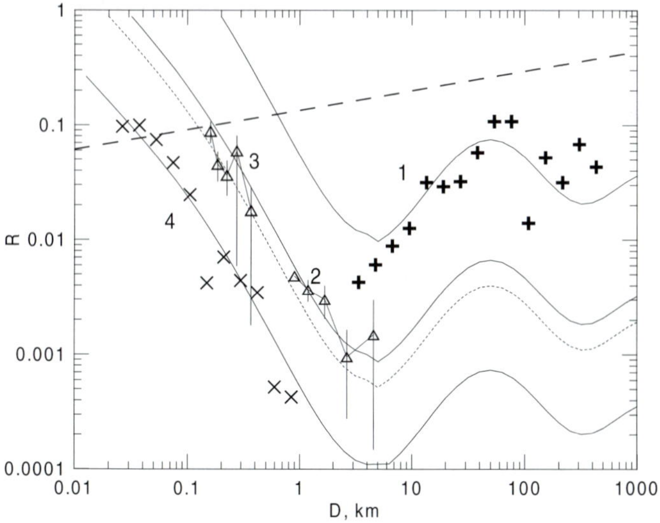

Figure 16. Mars: the crater SFD for heavily cratered terrain (1 — after Hartmann *et al.* (1981), Elysium Planitia and Alba Patera lava flow (2, 3 — after Neukum and Hiller, 1981) and a relatively young volcanic caldera floor (Hartmann *et al.*, 1999a). *Dashed line* shows an approximate saturation level after Hartmann.

3.3. Craters on Asteroids

The Galileo and NEAR spacecraft returned images of four asteroids: Gaspra, Ida, Mathilde, and Eros (Belton *et al.*, 1992, 1994; Chapman *et al.*, 1996a, b; Veverka *et al.*, 1997, 2000). All four bodies are covered by impact craters. Assuming an average velocity of impact of 5.5 km/s in the Main Belt, it is possible to estimate the small-projectile asteroid SFD (Ivanov *et al.*, 2000).

We can assume here that on Gaspra the impact crater SFD represents the production function, while Ida's craters are believed to be close to saturation (equilibrium) at small diameters, and the largest craters may be below the saturation limit (Chapman *et al.*, 1996a). Large craters on Ida may be formed in the gravity regime (Asphaug *et al.*, 1996).

For Mathilde and Eros we use the published impact crater SFD (Veverka *et al.*, 1997, 2000). The scaling of craters on Mathilde is not well defined due to the unusually low density (high porosity) of the target. As a first order approximation, here we use scaling parameters presented by Schmidt and Housen (1987) for the loosest soil.

With the assumptions discussed before a model projectile distribution for all imaged asteroids may be constructed (Figure 19). To simplify the perception, we use R-values for craters plotted vs. a projectile diameter (Figure 19a). In Figure 19b we made an attempt to fit all asteroid cratering data with a single curve. Using the Eros data as a reference level one concludes that in respect to "old" surface of Eros its "young" area (Veverka *et al.*, 2000) is cratered a factor 0.01 less. The same ratio is $\sim$0.15 for Gaspra $\sim$0.6 for Ida and Mathilde in accuracy limits of available scaling laws.

For comparison the lunar-derived projectile SFD is also shown. The model results for craters on asteroids demonstrate the presence of the R-minimum of the projectile SFD curve approximately in the same range of projectile diameters as for NEAs.

4. Size-frequency Distribution of Asteroids

4.1. Asteroid Count

Earth-based astronomical observations and the satellite infrared survey (IRAS) have revealed all Main Belt asteroids with diameters larger than about 40 km (van Houten *et al.*, 1970; Gradie *et al.*, 1989; Cellino *et al.*, 1991). For smaller diameters one usually supposes a power-law SFD:

$$dN/dD_{\mathrm{P}} \propto D_{\mathrm{P}}^{-k} \tag{6}$$

where the value of k may vary from 2.95 ("PLS-slope", after the Palomar-Leiden Survey — van Houten *et al.*, 1970) up to 3.5 ("Dohnanyi slope" – Dohnanyi, 1969).

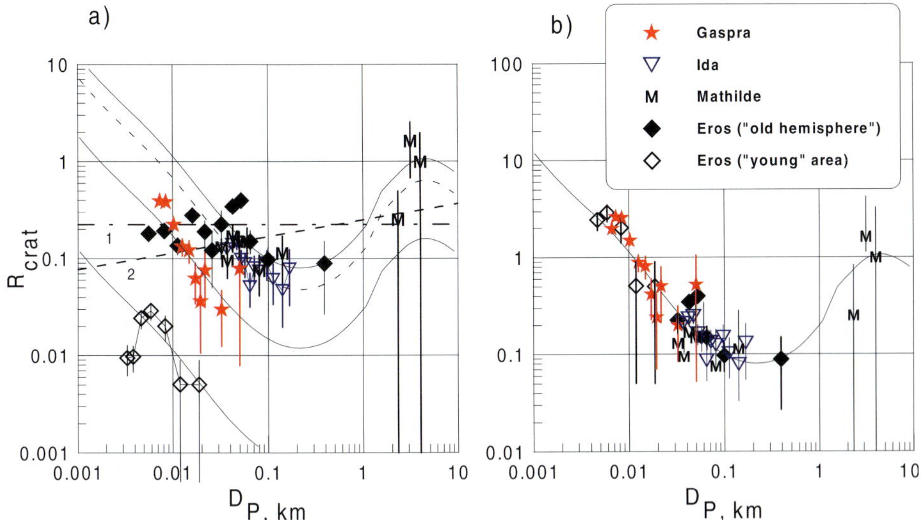

Figure 17. The *R*-plot for the derived projectile population for Gaspra, Ida, Mathilde, and Eros in comparison with the SFD derived from lunar cratering records. a) *R*-values calculated for craters plotted versus estimated projectile diameter; 1 — geometrical saturation limit; 2 — empirical saturation limit according to Hartmann. b) The same data shifted vertically to fit the one curve.

$k = 3.5$ is a typical value for a self-similar cascade of fragments. Davis *et al.* (1994) used a geometrical average of two possible power-law distributions — the PLS distribution and estimations by Cellino *et al.* (1991), to analyze the IRAS data (Figure 18).

Deviations from a simple power-law SFD of impact craters, considered above, suggest that the asteroid SFD also deviates from a simple power law at diameters smaller than the limit of completeness of detection. For large bodies ($\sim$100 km in diameter) the non-power-law SFD is usually discussed as an intrinsic feature of the initial distribution of small bodies before the Main Belt accumulation (Davis *et al.*, 1985). A possible mechanism for such deviations is based on results of modeling of impact evolution in the Main Belt — some models give a SFD (Campo Bagatin *et al.*, 1994a, b; Durda *et al.*, 1998).

Spacewatch Data for Main Belt.　Jedicke and Metcalfe (1998) have published an analysis of the Main Belt SFD based on absolute magnitudes measured as a part of the Spacewatch program. The basic purpose of the program is the automated search and study of near-Earth asteroids (NEA); however, during the survey of areas near the ecliptic about 60,000 Main Belt asteroids were also imaged. The duration of each observation was too small to determine the orbit, and statistical methods were used to estimate the distribution of orbital parameters from that of the magnitudes. The application of standard methods of the observation incompleteness modeling have allowed to estimate the distribution of absolute magnitudes in the Main Belt. The SFD is estimated from magnitudes using relative numbers of asteroids of

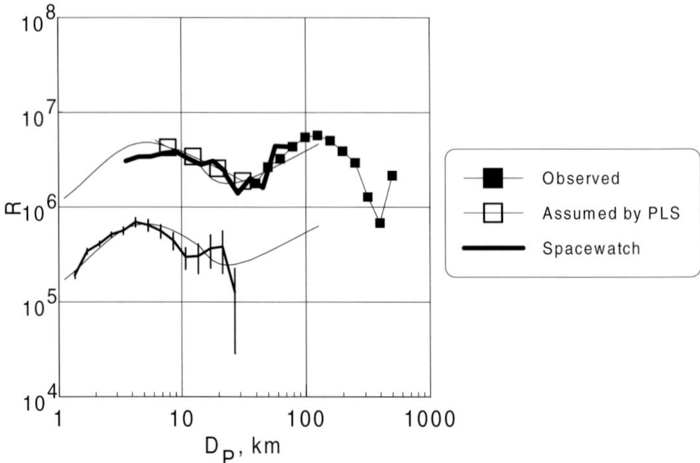

Figure 18. R-plot for Main Belt asteroids according to Davis *et al.* (1994) (*filled square* — observed asteroids, *open square* — assumed by PLS) and Spacewatch data by Jedicke and Metcalfe (1998) for all the Main Belt (*thick curve*) and the inner belt (*thin curve with error bars*) in comparison with the SFD for projectiles formed lunar craters.

various types and corresponding typical albedos in the Main Belt (Gradie*et al.*, 1989). The total sensitivity permits SFD estimates for asteroid diameters above 2 km. For comparison, the more direct IRAS SFD estimates are given for diameters exceeding $\sim$20 km (Cellino *et al.*, 1991).

Spacewatch Data for Main Belt. In Figure 18 data from direct observations (following Davis *et al.*, 1989) and estimations by Jedicke and Metcalfe (1998) are shown. The SFD of projectiles, received above from the SFD for impact craters, is used to approximate the data. Comparing these data one can conclude:

1. The direct astronomical data show a relative R-minimum at asteroid diameters $D \sim 30 - 40$ km; the depth of this minimum may vary for different semimajor axes a. This minimum would correspond to an impact crater size on the moon $\sim$300 km diameter. The lunar SFD does show such a minimum. Beyond this size of $\sim$300 km crater diameter the lunar SFD up to a size of$\sim$1000 km basin diameters show the same rising R-characteristics as the asteroidal projectile SFD as it approaches its R-maximum at $\sim$100 km asteroid diameter.

2. Estimations for the diameter range < 10 km demonstrate the presence of a second R-maximum at $D \sim 4 - 5$ km. This maximum is well visible in the inner and mid Main Belt ($2.0 < a < 2.6$ AU) and may be assumed for the exterior zone, being on the limit of detectability. The asteroidal R-maximum at $D \sim 4 - 5$ km corresponds to a lunar crater diameter of$\sim$50 km. The lunar SFD does show this maximum.

3. For asteroid diameters of 2 to 20 km the general shape of the SFD of the inner asteroid belt and the SFD for lunar projectiles calculated for impact craters looks identical within the error limits of available data.

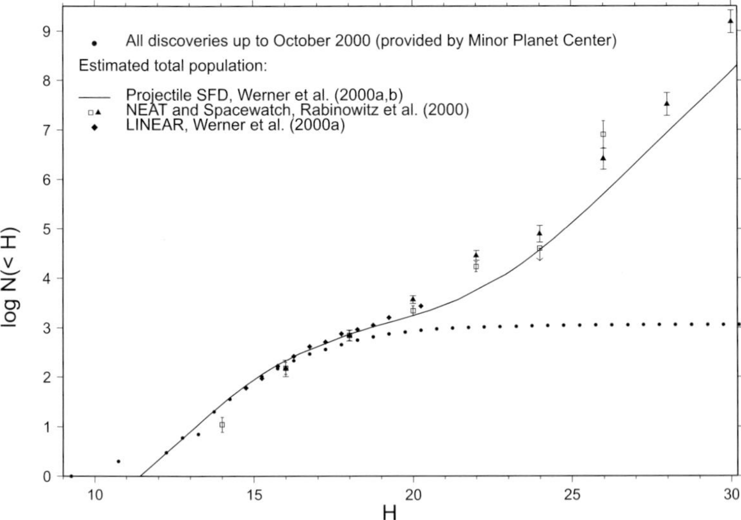

Figure 19. Comparison of the projectile SFD with the recent inventory of Near Earth Asteroids (NEA) in cumulative count vs. visible magnitude ($H = 18$ roughly corresponds to the body of 1 km in diameter (Weissman, 2000a, b). *Filled circles* represent the cumulative number ($\log N(< H)$) of known NEAs up to mid October 2000 as provided by the Minor Planet Center. *Squares and filled triangles* show the complete population estimates from bias corrections to the NEAT and Spacewatch surveys, according to Weissman (2000a, b). *Filled diamonds* represent the estimates of the total number of the NEA population up to $H = 20$ calculated via re–detection rates from the LINEAR survey for 1999 on the basis of observational data provided by the MPC, according to Rabinowitz *et al.* (2000). The curve is a lunar projectile SFD based on the polynomial expression of the lunar crater SFD formulated by Neukum and Ivanov (1994) with improved coefficients (Ivanov *et al.*, 2000).

The SFD deviations from a power law for impact craters, considered above, suggests that the SFD for asteroids in the Main Belt — the main source of projectiles for cratering on terrestrial planets — also deviates from a simple power law at diameters smaller than the limit of completeness of detection.

NEA. A similar SFD is found for Near Earth Asteroids (NEA). Figure 19 gives a recent summary of astronomical observations and debias modeling in comparison with the projectile SFD derived from lunar cratering records.

4.2. NEW RESULTS IN CATASTROPHIC COLLISION PHYSICS

Progress in catastrophic collision studies give some insight into the nature of the SFD deviation from a simple power law. We can use new ideas in the same way Hartmann previously used the physics of fragmentation to explain his early results (see Subsection 2.1). The numerical modeling of asteroid collisions by Love and Ahrens (1996) demonstrated the strength to gravity transition in the catastrophic events at unexpectedly low asteroid diameters ($\sim$ hundreds of meters). The estimates by Melosh and Ryan (1997) also show the importance of self-gravity for the reaccumulation of fragments after catastrophic disruption into a "rubble pile".

The revised impact strength model was used by Durda *et al.* (1998) in a standard numerical model of collisional evolution. The modeling demonstrated a very simple idea:

- Just above the transitional target diameter, larger critical projectiles ("killers") are required;
- The number of "killers" is less than for the range of pure strength confinement; hence at larger diameters a surplus of bodies exists; this surplus increases the probability of destruction of larger bodies, etc.
- As the result, an equilibrium is established where the SFD of asteroids deviates in a wavy manner from a simple power law.

Modeling results by Durda *et al.* (1998) are compared to the projectile SFD, described above by Ivanov *et al.* (2000). The idea of the transition from strength to gravity confinement of asteroids among catastrophic collisions allows us to reproduce the wavy deviation from a power law with the positions of R-maxima and R-minima corresponding to observational data. An important result for the interpretation of cratering on terrestrial planets is that according to the model the specific SFD shape is reached for less than 1 Gyr and remains approximately stable during more than 4 Gy. Without considering the depletion due to ejection into NEA orbits, the number of bodies in the Main Belt according to the model by Durda *et al.* (1998) decreased by less than three times during the last 3.5 Gyr.

5. Conclusions and Discussion

1. For the last 4 Gyr the shape of production function on the moon has not changed within the limits of observational accuracy.
2. Application of cratering scaling laws allows us to estimate the size- frequency distribution (SFD) of projectiles from the measured SFD of lunar impact craters. The estimated SFD of projectiles has a complex form with wavy deviations from a simple power law. Deficiencies of small craters, relative to production SFD, are primarily due to endogenic processes such as erosion, subaerial deposition, and resurfacing by sequential localized lava flows.
3. The estimated SFD of projectiles allows us to reproduce the impact crater SFD on all terrestrial planets. One can conclude that the majority of impact craters were formed by one population of projectiles, and the shape of the projectile SFD has not changed dramatically over the last 4 Gyr.
4. The SFD of the crater-forming bodies within the limits of accuracy of available data is similar to the SFD of asteroids in the Main Belt. Within the same limits of accuracy the contribution of comets to the crater formation is relatively insignificant or else the cometary SFD replicates the SFD for asteroids.

We suspect that asteroids, ejected from the Main Belt into NEAs, seem to present the main source of crater-forming objects on the terrestrial planets. One can conclude that these projectiles belong to a collisionally evolved family of objects.

We conclude that craters on the moon and Mars were created by the same family of projectiles. This family most probably are PCA which originated from the Main Belt due to gravitational interaction with planets. Under these assumptions we can transfer the lunar cratering chronology to Mars, what will be done in the next two chapters.

References

Arvidson, R., *et al.*: 1978, 'Standard Techniques for Presentation and Analysis of Crater Size-frequency Data', *Icarus* **37**, 467–474.

Asphaug, E., Moore, J.M., Morrison, D., Benz, W., Nolan, M.C., and Sullivan, R.J.: 1996, 'Mechanical and Geological Effects of Impact Cratering on Ida', *Icarus* **120**, 158–184.

Baldwin, R.B.: 1971, 'On the History of Lunar Impact Cratering: The Absolute Time Scale and the Origin of Planetesimals', *Icarus* **14**, 36–52.

Belton, M.J.S. *et al.*: 1992, 'Galileo Encounter with 951 Gaspra – First Pictures of an Asteroid', *Science* **257**, 1647–1652.

Belton, M.J.S. *et al.*: 1994, 'First Images of Asteroid 243 Ida', *Science* **265**, 1543.

Brown, H.: 1960, 'The Density and Mass Distribution of Meteoritic Bodies in the Neighborhood of the Earth's Orbit', *J. Geophys. Res.* **65**, 1679–1683.

Campo Bagatin, A., Cellino A., Davis, D.R., Farinella, P., and Paolicchi, P.: 1994a, 'Wavy Size Distribution for Collisional Systems with a Small-size Cutoff', *Planet. Space Sci.* **42**, 1049–1092.

Campo Bagatin, A., Farinella, P., and Petit, J.-M.: 1994b, 'Fragment Ejection Velocities and the Collisional Evolution of Asteroids', *Planet. Space Sci.* **42**, 1099–1107.

Cellino, A., Zappalà, V., and Farinella, P.: 1991, 'The Asteroid Size Distribution from IRAS Data', *Mon. Not. R. Astr. Soc.* **253**, 561–574.

Chapman, C., *et al.*: 1996a, 'Cratering on Ida', *Icarus* **120**, 77–86.

Chapman, C.R., Veverka. J. , Belton, M., Neukum, G. and Morrison, D.: 1996b, 'Cratering on Gaspra', *Icarus* **120**, 231–245.

Croft, S.K.: 1985, 'The Scaling of Complex Craters', *J. Geophys. Res.* **90**, 828–842.

Davis, D.R., Chapman, C.R., Weidenschilling, S.J., and Greenberg, R.: 1985, 'Collisional History of Asteroids: Evidence from Vesta and the Hirayama Families', *Icarus* **62**, 30–35.

Davis, D., Weidenshilling, S.J., Farinella, P., Paolicchi, P. and Binzel, R.P.: 1989, 'Asteroid Collisional History: Effects on Sizes and Spins', *in* R. Binzel, T. Gehrels, and M.S. Matthews (eds.), *Asteroids II*, Univ. Arizona Press, Tucson, pp. 805–826.

Davis, D.R., Ryan, E.V., and Farinella, P.: 1994, 'Asteroid Collisional Evolution: Results from Current Scaling Algorithm', *Planet. Space. Sci.* **43**, 599–610.

Dohnanyi, J.W.: 1969, 'Collisional Model of Asteroids and their Debris', *J. Geophys. Res.* **74**, 2531–2554.

Durda, D., Greenberg, R., and Jedicke, R.: 1998, 'Collisional Models and Scaling Laws: A new Interpretation of the Shape of the Main-belt Asteroid Distribution', *Icarus* **135**, 431–440.

Gaudin, A.: 1944, *Principles of Mineral Dressing*, McGraw-Hill Book Co., Inc., New York.

Gil-Hutton, R., and Brunini, A.: 1999, 'Collisional Evolution of the Early Asteroid Belt', *Planet. Space Sci.* **47**, 331–338.

Gradie, J.C., Chapman, C.R., and Tedesco, E.W.: 1989, 'Distribution of Taxonomic Classes and the Compositional Structure of the Asteroid Belt', *in* R. Binzel *et al.* (eds.), *Asteroids II*, Univ. Arizona Press, Tucson, pp. 316–335.

Grieve, R.A.F., and Shoemaker E.M.: 1994, 'The Record of the Past Impacts on Earth', *in* T. Gehrels (ed.), *Hazards due to Comets and Asteroids*, Univ. Arizona Press, Tucson, pp. 417–462.

Hartmann, W.K.: 1964, 'On the Distribution of Lunar Crater Diameters', *Comm. Lunar Planet. Lab.* **2**, 197–203.

Hartmann, W.K.: 1965, 'Terrestrial and Lunar Flux of Meteorites in the Last two Billion Years', *Icarus* **4**, 157–165.

Hartmann, W.K.: 1966, 'Martian Cratering', *Icarus* **5**, 565–576.

Hartmann, W.K.: 1969, 'Terrestrial, Lunar, and Interplanetary Rock Fragmentation', *Icarus* **10**, 201.

Hartmann, W.K.: 1970, 'Lunar Cratering Chronology', *Icarus* **13**, 299–301.

Hartmann, W.K.: 1971, 'Martian Cratering 3: Theory of Crater Obliteration', *Icarus*, **15**, 410–428.

Hartmann, W.K.: 1977, 'Relative Crater Production Rates on Planets', *Icarus* **31**, 260–276.

Hartmann, W.K.: 1984, 'Does Crater "Saturation Equilibrium" Occur in the Solar System?', *Icarus* **60**, 56–74.

Hartmann, W.K.: 1995, 'Planetary Cratering I: Lunar Highlands and Tests of Hypotheses on Crater Populations', *Meteoritics* **30**, 451–467.

Hartmann, W.K.: 1999, 'Martian Cratering VI: Crater Count Isochrons and Evidence for Recent Volcanism from Mars Global Surveyor', *Met. Planet. Sci.* **34**, 167–177.

Hartmann, W.K., and Neukum, G.: 2001, 'Cratering Chronology and the Evolution of Mars', *Space Sci. Rev.*, this volume.

Hartmann, W.K., *et al.*: 1981, 'Chronology of Planetary Volcanism by Comparative Studies of Planetary Craters', *Basaltic Volcanism on the Terrestrial Planets*, Pergamon Press, Elmsford, NY, pp. 1050–1127.

Hartmann, W.K., Berman, D., Esquerdo, G.A., and McEwen, A.: 1999a, 'Recent Martian Volcanism: New Evidence from Mars Global Surveyor' *Proc. 30th Lunar Planet. Sci. Conf.*, abstract #1270 (CD-ROM).

Hartmann, W.K., Malin, M.M., McEwen, A., Carr, M., Soderblom, L., Thomas, P., Danielson, E., James, P. and Veverka, J.: 1999b, 'Evidence for Recent Volcanism on Mars from Crater Counts', *Nature* **397**, 586–589.

Hartmann, W.K., Ryder, G., Dones, L., and Grinspoon, D: 2001, 'The Time-dependent Intense Bombardment of the Primordial Earth/moon System', *in* R. Canup and K. Righter (eds.), *Origin of the Earth and Moon*, Univ. Arizona Press, in press.

Hawkins, G.S.: 1960, 'Asteroid Fragments', *Astronom. J.* **65**, 318–322.

Hughes, D.W.: 2000, 'A new Approach to the Calculation of the Cratering Rate of the Earth over the Last 125 ± 20 Myr', *Mon. Not. Roy. Astron. Soc.* **317**, 429–437.

Ivanov, B.A.: 2001, 'Mars/Moon Cratering Rate Ratio Estimates, *Space Sci. Rev.*, this volume.

Ivanov, B.A., Basilevsky, A.T., and Neukum, G.: 1997, 'Atmospheric Entry of Large Meteoroids: Implication to Titan', *Planet. Space Sci.* **45**, 993–1007.

Ivanov, B.A., Neukum, G., and Wagner, R.: 1999, 'Impact Craters, NEA, and Main Belt Asteroids: Size-frequency Distribution', *Proc. 30th Lunar Planet. Sci. Conf.*, abstract # 1583 (CD-ROM).

Ivanov, B.A., Neukum, G., and Wagner, R.: 2000, 'Size–Frequency Distributions of Planetary Impact Craters and Asteroids', *in* H. Rickman, and M. Marov (eds.), *Collisional Processes in the Solar System*, ASSL, Kluwer Academic Publishers, Dordrecht, in press.

Jedicke, R., and Metcalfe, T.S.: 1998, 'The Orbital Absolute Magnitude Distributions of Main Belt Asteroids', *Icarus* **131**, 245–260.

König, B.: 1977, 'Investigations of Primary and Secondary Impact Structure on the Moon and Laboratory Experiments to Study the Ejecta of Secondary Particles', Ph.D. Thesis, Ruprecht Karl Univ., Heidelberg, 88 pp.

König, B., Neukum, G., and Fechtig, H.: 1977, 'Recent Lunar Cratering: Absolute Ages of Kepler, Aristarchus, Tycho', *Proc. 8th Lunar Planet. Sci. Conf.*, 555–557 (abstract) .

Love, S., and Ahrens, T.J.: 1996, 'Catastrophic Impacts on Gravity Dominated Asteroids', *Icarus* **124**, 141–155.

McEwen, A.S., Gaddis, L.R., Neukum, G., Hoffman, H., Pieters, C.M., and Head, J.W.: 1993, 'Galileo Observations of Post-Imbrium Lunar Craters During the First Earth-Moon Flyby', *J. Geophys. Res.* **98**, 17,207–17,231.

McEwen, A.S., Moore, J.M., and Shoemaker, E.M.: 1997, 'The Phanerozoic Impact Cratering Rate: Evidence from the Farside of the Moon', *J. Geophys. Res.* **102**, 9231–9242.

Melosh, H.J., and Ryan, E.V.: 1997, 'Note: Asteroids Shattered but not Dispersed', *Icarus* **129**, 562–564.

Milani, A., Carpino, M., Hahn, G., and Nobili, A.M.: 1989, 'Dynamics of Planet-crossing Asteroids: Classes of Orbital Behavior', *Icarus* **78**, 212–269.

Moore, J.M., and McEwen, A.S.: 1996, 'The Abundance of Large, Copernican-age Craters on the Moon', *Proc. 27th Lunar Planet. Sci.*, 899–900.

Nemtchinov, I.V., Svetsov, V.V., Kosarev, I.B., Golub, A.P., Popova, O.P., Shuvalov, V.V., Spalding, R.E., Jacobs, C., and Tagliaferri, E.: 1997, 'Assessement of Kinetic Energy of Meteoroids Detected by Satellite-based Light Sensors', *Icarus* **130**, 259–274.

Neukum, G.: 1977, 'Different Ages of Lunar Light Plains', *The Moon* **17**, 383–393.

Neukum, G.: 1983, *Meteoritenbombardement and Datierung Planetarer Oberflächen*, Habilitation Dissertation for Faculty Membership, Univ. of Munich, 186 pp.

Neukum, G., and Hiller, K.: 1981, 'Martian Ages', *J. Geophys. Res.* **86**, 3097–3121.

Neukum, G., and Ivanov, B.A.: 1994, 'Crater Size Distribution and Impact Probabilities on Earth from Lunar, Terrestrial-planet, and Asteroid Cratering Data', *in* T. Gehrels (ed.), *Hazards due to Comets and Asteroids*, Univ. Arizona Press, Tucson, pp. 359–416.

Neukum, G., König, B., and Arkani-Hamed, J.: 1975, 'A Study of Lunar Impact Crater Size Distributions', *The Moon* **12**, 201–229.

Öpik, E.J.: 1966, 'The Martian Surface', *Science* **153**, 255.

Rabinowitz, D.L.: 1993, 'The Size-distribution of the Earth-approaching Asteroids', *Astrophys. J.* **407**, 412–427.

Rabinowitz, D.L.: 1997, 'Are Main-belt Asteroids a Sufficient Source for the Earth-approaching Asteroids? Part II. Predicted vs. Observed Size Distribution', *Icarus* **130**, 287–295.

Rabinowitz, D.L., Bowell, E., Shoemaker, E. and Muinonen, K.: 1994, 'The population of Earth-crossing asteroids', *in* T. Gehrels (ed.), *Hazards due to Comets and Asteroids*, Univ. Arizona Press, Tucson, pp. 285–312.

Rabinowitz, D.L., Helin, E., Lawrence, K., and Pravdo, S.: 2000, 'A Reduced Estimate of the Number of Kilometer-sized Near-Earth Asteroids', *Nature* **403**, 165–156

Ronca, L.B., Basilevsky, A.T., Kryuchkov, V.P., and Ivanov, B.A.: 1981, 'Lunar Craters Evolution and Meteoroidal Flux in Pre-mare and Post-mare Times', *The Moon and the Planets* **245**, 209–229.

Ruzmaikina, T.V., Safronov, V.S., and Weidenschilling, S.J.: 1989, 'Radial Mixing of Material in the Asteroidal Zone', *in* R. Binzel, T. Gehrels, and M.S. Matthews (eds.), *Asteroids II*, Univ. Arizona Press, Tucson, PP. 681–700.

Safronov, V.S.: 1972, 'Ejection of Bodies from the Solar System in the Course of the Accumulation of the Giant Planets and the Formation of the Cometary Cloud', *in* G.A., Chebotarev, E.I. Kazimirchak-Polonskaia, and B.G. Marsden (eds.), *The Motion, Evolution of Orbits, and Origin of Comets*, Proc. IAU Symp. **45**, Leningrad, Reidel, Dordrecht, p. 329.

Schmidt, R.M., and Housen, K.R.: 1987, 'Some Recent Advances in the Scaling of Impact and Explosion Cratering', *Int. J. Impact Engng.* **5**, 543–560.

Shoemaker, E.M.: 1977, 'Astronomically Observable Crater-forming Projectiles', *in* D.J. Roddy, R.O. Pepin, and R.B. Merrill (eds.), *Impact and Explosion Cratering*, Pergamon Press, New York, pp. 639–656.

Shoemaker, E.M., and Wolfe, R.: 1982, 'Cratering Time Scales for the Galilean Satellites, *in* D. Morrison (ed.), *Satellites of Jupiter*, Univ. of Arizona Press, Tucson, pp. 277–339.

Shoemaker, E.M., Batson, R.M., Bean, A.L. *et al.*: 1970, 'Preliminary Geologic Investigation of the Apollo 12 Landing Site, Part A', *Geology of the Apollo 12 Landing Site*, Apollo 12 Prel. Sci. Rep., *NASA* **SP-235**.

Stern, S.A., and Weissman, P.R.: 2000, 'Collisional Processing of Proto-comets in the Primordial Solar Nebula', *Proc. 31st Lunar Planet. Sci. Conf.*, 1830.

Stöffler, D., and Ryder, G.: 2001, 'Stratigraphy and Isotope Ages of Lunar Geologic Units: Chronological Standard for the Inner Solar System', *Space Sci. Rev.*, this volume.

Strom, R.: 1977, 'Origin and Relative Age of Lunar and Mercurian Inter-crater Plains', *Phys. Earth Planet. Interiors* **15**, 156–172.

Strom, R.G., and Neukum, G.: 1988, 'The Cratering Record on Mercury and the Origin of Impacting Objects', *in* F. Vilas, C.R. Chapman, and M.S. Matthews (eds.), *Mercury*, Univ. Arizona Press, Tucson, pp. 336–373.

Strom, R.G., Woronow, A., and Gurnis, M.: 1981, 'Crater Populations on Ganymede and Callisto', *J. Geophys. Res.* **86**, 8659–8674

van Houten, C.J., van Houten-Groeneveld, I., Herget, P. and Gehrels, T.: 1970, 'The Palomar–Leiden Survey of Faint Minor Planets', *Astron. Astrophys. Suppl.* **2**, 339–448.

Veverka, J. *et al.*: 1997, 'NEAR's Flyby of 253 Mathilde: Images of a C Asteroid', *Science* **278**, 2109–2114.

Veverka, J., *et al.*: 2000, 'NEAR at Eros: Imaging and Spectral Results', *Science* **289**, 2088–2097.

Weidenschilling, S.J.: 1994, 'Origin of Cometary Nuclei as "Rubble Piles"', *Nature*, **368**, 721–723.

Weissman, P.R.: 1999, 'Diversity of Comets: Formation Zones and Dynamical Paths', *Space Sci. Rev.* **90**, 301–311.

Werner, S.C., Harris, A.W., Neukum, G., Ivanov, B.A., and Harris, A.W.: 2000a, 'A Comparison of Lunar Crater Size Frequency Distribution and Near-Earth Asteroid Population Characteristics: Strong Evidence for the Stability of their Size Frequency Distribution', *AAS Div. Planet. Sci. Meeting* **32**, p. 1409.

Werner, S.C., Harris, A.W., Neukum, G., and Ivanov, B.A.: 2000b, 'The Near-Earth Asteroid Size Frequency Distribution: A Snapshot of the Lunar Crater Size Frequency Distribution, *Icarus*, submitted.

Wilhelms, D.E., McCauley, J.F., and Trask, N.J.: 1987, 'The Geologic History of the Moon', U.S. Geol. Survey Prof. Paper 1348, Washington, U.S.G.P.O., and Denver, CO, U.S. Geol. Survey, 302 pp.

Wetherill, G.W.: 1989, 'Origin of the Asteroid Belt', *in* R. Binzel, T. Gehrels, and M.S. Matthews (eds.), *Asteroids II*, Univ. Arizona Press, Tucson, pp. 661–680.

Young, J.: 1940, 'A Statistical Investigation of Diameter and Distribution of Lunar Craters', *J. Brit. Astron. Assoc.* **50**, 309–326.

Address for correspondence: DLR Institute of Space Sensor Technology and Planetary Exploration, Rutherfordstrasse 2, 12484 Berlin - Adlershof, Germany; (gerhard.neukum@dlr.de)

MARS/MOON CRATERING RATE RATIO ESTIMATES

BORIS A. IVANOV

Institute for Dynamics of Geospheres, Russian Academy of Sciences, Moscow, Russia 117939

Received: 1 November 2000; accepted: 26 February 2001

Abstract. This article presents a method to adapt the lunar production function, i.e. the frequency of impacts with a given size of a formed crater as discussed by Neukum *et al.* (2001), to Mars. This requires to study the nature of crater-forming projectiles, the impact rate difference, and the scaling laws for the impact crater formation. These old-standing questions are reviewed, and examples for the re-calculation of the production function from the moon to Mars are given.

1. Crater Forming Projectiles

The modern crater forming projectiles are believed to be presented by three main populations: asteroids, Jupiter-Family comets (JFC), and long period comets (LPC). These bodies have different kinds of orbits, and physical and mechanical properties. According to Shoemaker and Wolfe (1982), for Earth and for all terrestrial planets the JFC impacts play a minor role in the formation of impact craters. Asteroids and long period comets may give comparable contributions to the modern cratering rate. However, the long period comets' flux in terms of mass of the projectiles is currently poorly known. Shoemaker and Wolfe (1982) used measurements of the LPC nuclear size by Roemer (1965, 1966), Roemer and Lloyd (1966) and Roemer *et al.* (1966), and the astronomically estimated frequency of fly-by through the solar system. As the estimated cratering rate from these assumptions was too high, Shoemaker and Wolfe (1982) ascribed the overestimate to the unresolved comae of measured LPCs, and drastically decreased the published nuclear diameter estimates by a factor of 3. Consequently, knowing the average probability and velocity of LPC impacts very well, there remains a severe problem of not knowing both the size distribution and the formation rate of impact craters by LPCs.

According to Neukum *et al.* (2001), the size-frequency distribution (SFD) of craters at best corresponds to the one of asteroids. At the same time, known estimates for the SFD of comet nuclei contradict the planetary cratering records (Shoemaker and Wolfe, 1982). Thus, as a first approximation this article compares the moon and Mars only for asteroid impacts.

Most asteroids occupy the area between the orbits of Mars and Jupiter, named the Main Belt (MB; see Binzel *et al.*, 1989, for a review and database). The relatively small sub-population of asteroids that currently cross the orbits of terrestrial planets are called *Planet Crossing Asteroids (PCA)*.

Chronology and Evolution of Mars **96** 87–104, 2001.
© 2001 *Kluwer Academic Publishers. Printed in the Netherlands.*

Recently, Morbidelli (1999) has reviewed the evolution of understanding the PCA origin. The first scenario developed by Wetherill (1979, 1988) and interpreted by Greenberg and Nolan (1989, 1993) assumes the catastrophic collision's fragment injection to resonance phase space area. Being in a resonance (ν_6 or 3/1), "new" fragmental asteroids change eccentricities and become PCAs. They further change orbits due to close encounters with terrestrial planets. The time scale for the transition from MB to PCA orbits is $\sim$1 Myr; the life time at the PCA's orbits is tens of Myr. If Mars alone controls the orbit evolution, an asteroid begins to cross the Earth orbit in 100 Myr ("slow track"). If Earth removes an asteroid from a resonance, the orbit evolves 10 to 100 times faster (1 to 10 Myr, "fast track").

Later, the "solar sink" was found to be an important mechanism to limit the time scale of the orbital evolution due to resonances. A chain of papers after Farinella*et al.* (1994), reviewed by Morbidelli (1999), demonstrated that resonances "pump" eccentricities of PCAs up to a sun-grazing orbit. The recent dynamic model by Gladman *et al.* (1997) predicts a median lifetime of 2 Myr for MB asteroids in resonant orbits, while 90% will impact Sun or planets, or will be ejected out of the inner solar system within 10 Myr. The short lifetime implies a very intensive re-supply of PCAs from the rest of the MB asteroids. The pure resonant orbital evolution thus looks like an effective "fast track" to create PCAs from MB asteroids.

The "slow track" interpretation was renovated by Migliorini*et al.* (1998) and Morbidelli (1999): Multiple weak resonances lead to chaotic evolution of the MB asteroids' orbits that are far from main mean motion and secular resonances with Jupiter and Saturn. These orbits "migrate" in phase space with a good chance to become Mars-crossing orbits on a timescale of 25 Myr. Then, Mars-crossers will evolve to PCAs by close encounters with Mars and other terrestrial planets.

Regarding planetary cratering rate estimates, three comments are important:

1. Before numerical models have accumulated better statistics in the behavior of small bodies in the solar system, the natural axiom assumed that the currently observed PCAs present a sensible "time slice" of a stationary distribution of orbital elements, so that impact probabilities and impact velocities of asteroids can be estimated from currently observed PCAs. With help of scaling laws the cratering rate is estimated. Most of the currently observed asteroids never hit any planet. The time to impact for a typical orbit is about $0.1 - 1$ Gyr. Thus, 4 to 10 "generations" of asteroids should replace one another on the same PCA orbits to create a new crater on one of the terrestrial planets or their satellites.

2. The "fast track" with ejection due to catastrophic collision differs from the "slow track", which operates equally for bodies of any size. The fragmentation ejection into resonant orbits may change the SFD of fragments. Smaller fragments are ejected faster, so that the "fast track" population of PCAs should be more abundant in small bodies compared to the MB population. In terms of a close to power law SFD, the PCA's SFD is "steeper" than the MB SFD.

3. Non-gravitational forces (such as the Yarkovsky effect) also can result in a steeper SFD for PCAs.

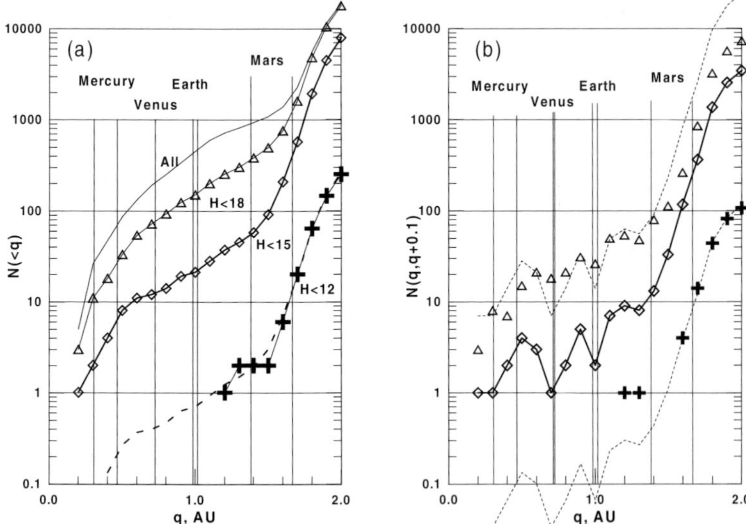

Figure 1. Cumulative (a) and incremental (b) perihelion distribution of planetary crossing asteroids (PCAs) of different size (characterized by magnitude *H*). The *N*(*q*) dependence of PCAs with $H < 15$ compromises the completeness of observation and the numerosity at near-Earth orbits.

Below we assume that asteroids from the MB may be treated as best candidates for the cratering rate comparison. Cometary input is not included in our current estimates. We also assume that the currently observed population of PCA is stationary (at least for the last 3 Gyr). For more ancient periods, e.g. the end of the heavy bombardment period, we assume that the relative cratering rates on terrestrial planets in respect to the moon was the same as relative cratering rates for the currently observed PCA. Each of these assumptions should be taken with care.

2. Projectile Flux Estimate

In the approach of Shoemaker and Wolfe (1982) for short period comets, the set of observed comet orbits is sorted in perihelion distance q, from which the average impact probability and velocity are estimated. This way, observed orbits are assumed to represent an ensemble of a statistically "averaged" steady-state population of PCAs. Here, the list of osculating orbits "astorb.dat", presented on *http://asteroid.lowell.edu* (February 2000), is used.

To remove bias in the observed asteroid population, several techniques have been developed (e.g., Rabinowitz, 1993, 1997; Rabinowitz *et al.*, 1994; Jedicke and Metcalfe, 1998; Bottke *et al.*, 2000a, b). Here, PCAs are sorted in their absolute magnitude and the obvious bias in the q-distribution is removed. Figure 1 plots "incremental" and "cumulative" distributions of the number of PCAs of different size, presented by magnitude. The incremental number of asteroids is calculated as their number per q-bin of width 0.1 AU. Comparing q-distributions of asteroids with $H < 15$ and $H < 12$ reveals similar behavior close to the Mars orbit.

The observations for Mars-crossers with $H < 15$ appear to be relatively complete, agreeing with theory (Bottke *et al.*, 2000b). A scaled curve derived from $N(q)$ distributions for $H < 15$ (Figure 1) and larger bodies fits to the data of $N(q)$ for larger bodies with $H < 12$. For Mars-crossers $N(q)$ functions look similar. For small q the number of $H < 12$ asteroids drops below 1. This means that the steady-state number of large NEAs is less than 1 – the delivery rate is smaller than the lifetime of large bodies, and they only sporadically appear in the Earth's vicinity. Here, it is assumed that the q-distribution of asteroids with $H < 15$ may be used to estimate the $N(q)$ function for smaller bodies.

The completeness of observation of small bodies on near Earth orbits (NEA) is now estimated as 40% for $H < 18$ (Bottke *et al.*, 2000b). At the same time, the trial fit of $N_{18}(q)$ with the $N_{15}(q)$ curve shows a more or less similar behavior for PCAs. A first order correction for PCAs is obtained by multiplication by some factor close to 2.5, according to Bottke *et al.* (2000b). For Mars-crossers with $H < 18$ the observational bias obviously increases for larger q (Figure 1).

With the procedure described above the impact rates on the moon and Mars are compared. Impact velocities and probabilities for all Earth crossers with $H < 18$, and for all Mars-crossers with $H < 15$ are calculated. To compare absolute impact rates on Mars and the moon, Mars estimates for each q are multiplied with the ratio $N_{H<15}/N_{H<18}$. This implies that unobserved Earth and Mars-crossers should have orbits similar to currently observed bodies. Figure 2 illustrates the impact velocity distribution for Mars and the moon binned in intervals of 1 km/s.

The Öpik formulas, refined by Wetherill (1967) for the general case of elliptic orbits for both target and projectile, are applied to all bodies in the "astorb.dat" file. A random orientation of the nodes latitude is assumed, so that all mutual positions of inclined elliptical orbits have the same probability. For each target (Mars or the moon) and projectile (PCA) the impact probability and velocity are calculated. The *total* impact probability, corrected for observational bias for small Mars-crossers, and the *average* impact velocity are needed to compute the cratering rate.

Before determining size-frequency distributions of projectiles, it helps to compare the estimated impact rate of asteroids of equal size on the moon and Mars.

For the moon the *average* probability of impact of *observed* NEA with $H < 18$ is 1.9×10^{-10} yr^{-1}. For the total number of projectiles of 155 and the moon radius of 1732 km the average impact rate (AIR) per year per km^2 is $AIR_{\mathrm{moon}} = 0.77 \times 10^{-15}$ yr^{-1} km^{-2} The average impact velocity is $\langle v_{\mathrm{moon}} \rangle = 16.2$ km/s.

The AIR estimate is difficult for Mars, because its orbit's eccentricity varies in the range of $\sim 0.01 - 0.1$ with a period of 2 Myr (Ward, 1992). Currently, this eccentricity is close to the upper limit $e = 0.094$, and Mars' distance to the Sun varies in the range $\sim 1.4 - 1.7$ AU. Figure 1 shows, how the number of Mars crossers grows with q within limits of the current Mars aphelia and perihelia: The number of potential impactors is 20 times larger in the aphelia than in the perihelia.

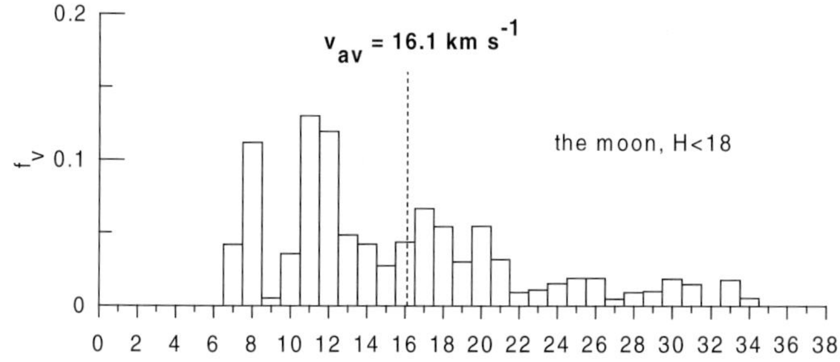

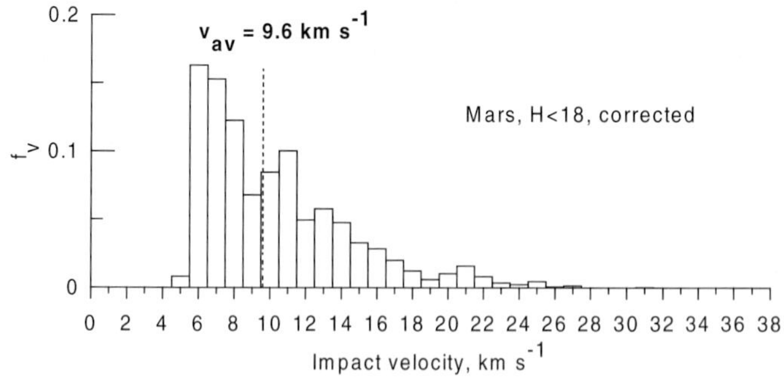

Figure 2. The impact velocity distribution for the moon and Mars (fraction of impacts per 1 km/s).

The calculations with "astorb.dat" Mars crossers show that the impact rate changes 7 times each Martian year, defining **"impact seasons"**.

For this reason, the AIR for Mars may either be calculated for shorter periods assuming a specific eccentricity, or may be averaged for one Martian year or even for the cycle of the Martian orbit eccentricity variation of $\sim$2 Myr. The long-time average, corrected for the unobserved $H < 18$ Mars crossers, is estimated as

$$AIR_{\mathrm{Mars}} = 1.57 \times 10^{-15}\mathrm{yr}^{-1}\mathrm{km}^{-2} \; .$$

The average impact velocity is $\langle v_{\mathrm{Mars}} \rangle = 8.62$ km s^{-1}. This number corresponds to the observed NEA with $H < 18$, and may present $\sim$40% of the real impact rate, as discussed above.

The Mars/moon impact rate ratio, averaged in time, called "R_{bolide}" or R_{b} following Hartmann and Neukum (2001), has a value

$$R_{\mathrm{b}} = AIR_{\mathrm{Mars}}/AIR_{\mathrm{moon}} = 1.57/0.77 = 2.04 \tag{1}$$

for asteroids *of the same size*. This is the ratio of frequencies of impacts per unit area. Due to the larger total surface of Mars, the ratio of the absolute numbers of impacts is larger than the value of Equation (1).

To convert this ratio, R_{bolide}, to the average cratering rate ratio, R, one needs to know the scaling law for impact cratering due to the different impact velocities and surface gravity on the moon and Mars, and the size-frequency distribution of projectiles to estimate the R value for *equal crater diameters*.

3. Scaling Laws

Crater diameters for the impact of the same projectile on the moon and on Mars depend on the efficiency of the cratering at different impact velocity and surface gravity. Nowadays, the scaling law by Schmidt and Housen (1987) is widely used, which conveniently is written as

$$D_t = 1.16(\delta/\rho)^{1/3} D_P^{0.78} (v \sin \alpha)^{0.43} g^{-0.22} \tag{2}$$

where D_t is the transient crater diameter, D_P is the projectile diameter, ρ and δ are densities of target and projectile materials, v is the impact velocity, α is the impact angle, and g is the gravity acceleration (e.g., Pierazzo *et al.*, 1997). For simple craters formed in the gravity regime their final diameter is about D_t.

The power law (Equation 2) may be more complex in extreme cases of small strength craters ($D < 50$ to 300 m (see Neukum and Ivanov, 1994), and large modified (collapsed) craters (see the review by Melosh and Ivanov, 1999).

The strength-to-gravity transition may be incorporated into the scaling law following Schmidt and Housen (1987), Neukum and Ivanov (1994), and Ivanov *et al.* (2000). Equation (2) may be rewritten in a form, where for sufficiently small events the crater diameter is directly proportional to the projectile diameter:

$$\frac{D_t}{D_P[(\delta/\rho)^{1/3}(v \sin \alpha)^{0.43}]^{0.78}} = \frac{(1.16)^{1/0.78}}{[(D_{sg} + D_t)g]^{0.22/0.78}} ,$$

where D_{sg} is the characteristic strength to gravity transition crater diameter; craters with $D_t \ll D_{sg}$ are formed in a strength regime, while craters with $D_t \gg D_{sg}$ are formed in a gravity regime. After this reduction the generalized form of the scaling law is

$$\frac{D_t}{D_P(\delta/\rho)^{0.26}(v \sin \alpha)^{0.55}} = \frac{1.28}{[(D_{sg} + D_t)g]^{0.28}} . \tag{3}$$

The scaling law (Equation 3) gives a smooth transition from the gravity cratering regime (Equation 2) to the assumed strength cratering regime where (for $D_t \ll D_{sg}$) the crater diameter is proportional to the projectile size:

$$D_t = \frac{1.28}{(D_{sg}g)^{0.28}} D_P(\delta/\rho)^{0.26}(v \sin \alpha)^{0.55} .$$

Such a model assumes that the formation of a small crater on a planetary surface may be described as the cratering process in a target with a constant strength. In reality, the strength of near surface rocks may be highly variable due to the presence of a regolith and a megaregolith. However, Equation (2) gives an upper limit for the crater size for a given projectile, implicitly assuming a low cohesion nature of shuttered and weathered near-surface rocks. Equation (3) gives a lower limit of the crater size, provided the near surface rocks have some finite cohesion.

The gravity collapse of a transient cavity makes the diameter of a final modified crater larger in comparison with a hypothetical simple crater which would be created by the same impact in the absence of the gravity collapse. For simple estimates one can use the semiempirical model derived by S. Croft (1985; see also Chapman and McKinnon, 1986) to find the transient cavity diameter, D_t, for an observed crater with a rim diameter D:

$$D_t = D_*^{0.15} \times D^{0.85} \tag{4}$$

for $D > D_*$. The value of a critical diameter D_*, which defines the boundary crater diameter when the collapse begins, depends on the target material strength and gravity. On Earth, the value of D_* is $\sim$4 km for crystalline rocks. For other terrestrial planets D_* varies approximately inversely proportional to the surface gravity acceleration (Pike, 1980). Using Equation (4) to estimate D_t, and Equation (3) for an assumed impact velocity v and impact angle α, one can estimate a projectile diameter needed to form the crater with a given final diameter D.

The best test for the model is to compare the theoretical and observed volume of the impact melt in well studied terrestrial impact craters. Numerical models of the impact events enable to estimate the impact melt volume. The results are confirmed by real underground nuclear tests. The reasonably good coincidence (Figure 3) between theory and observations gives confidence in the validity of scaling laws presented here (Ivanov, 1981; Pierazzo *et al.*, 1997).

The general review of the simple-to-complex transition has been published by Pike (1980). Most of these data reflect boundary diameters between ranges of different morphology styles of impact craters. In general, it is hard to say what is the best value of D_* in Equation (4) to estimate the crater diameter increase due to modification on a given planetary body. As a first approximation for terrestrial planets one may use the inverse proportionality of D_{sg} and D_* to gravity g (Pike, 1980). The recent Mars Global Surveyor data seem to significantly improve the D_* estimates (Garvin *et al.*, 2000).

4. Cratering Rate Comparison

The rigorous procedure of the interplanetary comparison of crater records consists of several natural steps.

1. Calculate the shape of the size-frequency distribution of projectiles using one of the bodies as a reference.
2. Calculate the production curve for new impact velocities and surface gravity, assuming the SFD of projectiles for the other planet is the same as for the reference planetary body.

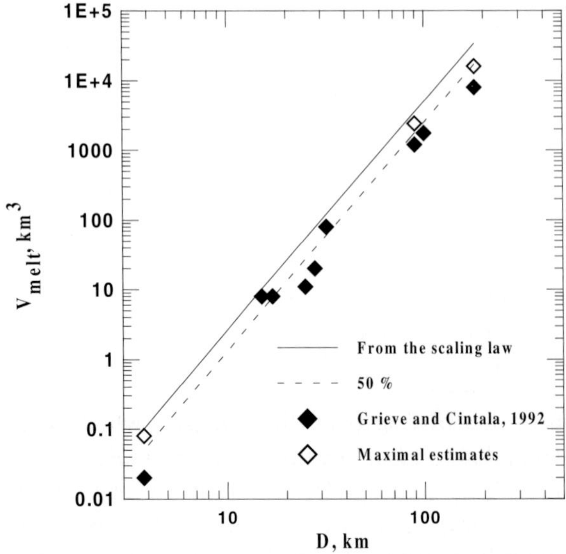

Figure 3. Comparison of calculated (Ivanov, 1981; Pierazzo *et al.*, 1997) and observed (Grieve and Cintala, 1992) impact melt volumes in terrestrial craters. The solid line shows the total theoretical melt production, the dashed line corresponds to an assumption that 50% of the produced impact melt is ejected beyond the crater rim.

3. Normalize the production function to the average impact rate ratio.
The known lunar production function is discussed by Neukum *et al.* (2001). The re-calculations of the Hartmann Production Function (HPF) and the one of Neukum (NPF) for the case of Mars are presented below. However, before discussing the relatively accurate procedure, some order-of-magnitude estimates are made.

Model 0 uses power law functions for both the cratering scaling law and the projectile SFD, and assumes one (average) impact angle and one (average) impact velocity. This was the first successful technique to compare impact cratering rates throughout the solar system (Hartmann, 1977; Hartmann *et al.*, 1981). In this model, the SFD as function of the projectile's diameter D_P is written as:

$$N(> D_P) \sim D_P^{-n} \tag{5}$$

The crater diameter is expressed as

$$D \sim D_P^{\alpha} v^{\beta} g^{-\gamma}.$$

The crater's SFD is also a power law

$$N(> D) \sim D^{-b} \tag{6}$$

where $b = n/\alpha$. For a known ratio of numbers of impacts of projectiles of a given size at a given area during a given time (named R_b in Equation 1), one can derive the ratio of cratering rates for a given crater diameter. Due to differences in the impact velocity and surface gravity, craters of the *same diameter* are formed by projectiles of *different size*. On planet #1 (let it be Mars) the crater of a diameter D is created by a projectile with a diameter

$$D_{PM} = D^{1/\alpha} v_M^{-\beta/\alpha} g_M^{\gamma/\alpha}$$

On planet #2 (the moon) the same crater is created by a projectile with the diameter

$$D_{Pm} = D^{1/\alpha} v_m^{-\beta/\alpha} g_m^{\gamma/\alpha}$$

Mars gravity is larger, and the average impact velocity is smaller than on the moon. Consequently, a larger projectile needs to strike Mars to create the same crater. Larger bodies are less numerous (Equation 5), and the ratio of the cratering rate, R, is smaller than the ratio of impact rates, R_b, by a "planetary" factor, f_P, derived from the scaling law and the size-frequency distributions:

$$f_P = (v_M/v_m)^{b\beta/\alpha} \cdot (g_M/g_m)^{-b\gamma/\alpha}$$

The final ratio of cratering rates (Mars/moon ratio, R) is given by the expression

$$R = f_P R_b$$

To make a bridge to previous estimates for interplanetary comparison of crater population, we test first "old" exponents, used by Hartmann (1977) and Hartmann *et al.* (1981): $\beta = 2/3.3 = 0.606$, $\gamma = 0.2$. For these parameters the ratio of crater diameters on Mars and the moon for velocities estimated above is

$$D_M/D_m = (v_M/v_m)^\beta (g_M/g_m)^{-\gamma} = 0.694 \tag{7}$$

The "planetary" factor, f_P, depends on the SFD's steepness (Equations 5 or 6).

For typical values of $b = 2$ to 3 the Mars/moon cratering rate ratio R varies in the range

$$R = (D_M/D_m)^b \times R_b = (0.33 \text{ to } 0.48) \times 4.8 = 1.6 \text{ to } 2.3 \tag{8}$$

"Modern" values of exponents in the scaling law ($\beta = 0.43$, $\gamma = -0.22$ (Equation 2) give close numbers for simple gravity craters:

$$D_M/D_m = (v_M/v_m)^\beta (g_M/g_m)^{-\gamma} = 0.64 \tag{9}$$

$$R = (D_M/D_m)^b \times R_b = (0.26 \text{ to } 0.41) \times 4.8 = 1.28 \text{ to } 1.99 \tag{10}$$

This comparison leads to the conclusion that the "old" scaling by Hartmann (1977) gives just the same D_M/D_m as more recent scaling laws for *simple* craters. The variations in scaling exponents and average impact velocities derived here are close to compensate one another. The "typical" impact on Mars creates the crater 1.5 times smaller than on the moon. Consequently, the cratering rate ratio is less than the average impact rate ratio.

TABLE I

Cratering rate comparison for simple craters for different values of the Mars orbit eccentricity, e, and time averaged estimates for the period of 2 Myr. M=Mars; m=moon; R_b is the Mars/moon impact rate ratio; $D_{D_{\mathrm{P}}=1}$ is the crater diameter for vertical impact of an asteroid of 1 km diameter; b is the exponent in the cumulative size distribution (Equation 6).

| Planet | v | R_b | $D_{D_{\mathrm{P}}=1}$ | $b = 1.8$ | | $b = 2.2$ | | $b = 3.82$ | |
	kms^{-1}		km	$\left(\frac{D_{\mathrm{M}}}{D_{\mathrm{m}}}\right)^b$	R	$\left(\frac{D_{\mathrm{M}}}{D_{\mathrm{m}}}\right)^b$	R	$\left(\frac{D_{\mathrm{M}}}{D_{\mathrm{m}}}\right)^b$	R
moon	16.09		15.32						
Mars									
$e = 0.093$	8.61	4.8	9.87	0.45	2.18	0.38	1.83	0.19	0.90
$e = 0.05$	10.32	2.02	10.65	0.52	1.05	0.45	0.91	0.25	0.51
$e = 0.01$	10.86	1.73	10.88	0.54	0.93	0.47	0.81	0.27	0.47
Time av.	**9.59**	**2.04**	**10.32**	**0.49**	**1.00**	**0.42**	**0.86**	**0.22**	**0.45**

Taking into account the variation of the Mars orbit eccentricity, one can derive R-values for different slopes of the cumulative SFD providing the *same* steepness of the cratering curve. The data in Table I well represent possible ranges of R-values for different Mars orbits and the crater cumulative distribution steepness. For the current Mars orbit (high e) the R value is in the range $0.9 - 2.2$ for different ranges of the $N(> D)$ distribution. During time periods, when the Mars orbit is almost circular, R decreases to values in the range $0.5 - 0.9$.

The last section of Table I shows the time averaged cratering rate ratio R for sinusoidal variation of e with time and for varying $N(> D)$ steepness: R is in the range $0.45 - 1.0$, providing one compares SFD brunches with the *same steepness* m. One cannot compare the number of craters on Mars and the moon for diameter ranges where $N(> D)$ curves have different steepness.

The possible celestial source of uncertainty in these estimates is connected with possible modulation of the Mars crossers orbital evolution with the evolution of the eccentricity of the Martian orbit. This modulation in principle may shift the orbital distribution of Mars crossers, which may shift time averaged impact velocity and probability for Mars. However, the time scale of the Mars orbit variation is about 2 Myr (Ward, 1992), while the typical time scale for Mars crossers evolution is about 30 Myr (Migliorini *et al.*, 1998; Morbidelli, 1999).

Model 0 gives a proper general estimate for the values of the Mars/moon cratering rate ratio R due to the bombardment of asteroids. If comets have not the significant input into the cratering rate (see Neukum *et al.*, 2001), it can be concluded that the cratering rates on Mars and the moon are close within a factor of

1.5. For steep $N(D)$ distributions $R \approx 0.5$, in a shallow $N(D)$ diapason, R may be close to 1.0, if averaged over the time variation of the Martian eccentricity.

The general information given above enables to construct the predictive $N(D)$ functions for Martian surfaces of varying ages, as presented in the following.

Model 1 is a simple approach (e.g., Neukum and Ivanov, 1994) to assume that all impact craters are created with projectiles having one (average) impact velocity and one (average) impact angle of 45°.

In this case, the production functions, HPF and NPF, can be easily recalculated from the moon to Mars. The procedure includes the estimate of the projectile SFD using Equations (2 – 3). The derived projectile population is assumed to strike Mars with the proper larger intensity. Using the average impact velocity for Mars, the correspondent crater population and the Martian production curve are estimated. It is important to take into account that the simple-to-complex transition on Mars occurs at smaller critical diameter $D_* = 7$ km (Garvin *et al.*, 2000), while on the moon we assume the D_* value of 15 km (Pike, 1980).

Model 2 is more complex but more exact. It automatically takes the change of crater scaling into account, which happens in different diapasons of crater diameters at planetary bodies with different gravity and surface mechanical properties. The most straightforward way is to derive the model SFD for one body and then include the variation of impact and target parameters to compute the crater SFD on the other body. The technical problem arises from the fact that craters of the same diameter may be created by impacts of projectiles of varying size depending on their impact angle and velocity.

With an analytical expression for the SFD the problem is solved in a more rigorous style. Ivanov (1999) and Ivanov *et al.* (1999, 2000) have estimated the relationship between projectile and crater SFDs using the real observed velocity distribution of asteroids. The model assumes that:

- The crater diameter is some function, i.e. the scaling law, $D = D(D_P, v, \alpha)$ of the projectile and impact parameters D_P (projectile diameter), v (impact velocity) and α (impact angle, $\alpha = 90°$ means vertical impact).

- The probability for a projectile to have the impact velocity in the range from v to $v + dv$ is $f_v(v)dv$. The minimum impact velocity is equal to the escape velocity of a planetary body, v_{esc}. The maximum velocity depends on the orbital parameters of a target body and projectiles.

- The share of impacts in the range of angles from α to $\alpha + d\alpha$ is a function $f_\alpha(\alpha)$. The maximum impact angle is naturally 90° (vertical impact). The minimum angle of impact α_{min} is defined as an angle, below which craters become elongated. From the restricted amount of experimental data (Gault and Wedekind, 1978) $\alpha_{min} = 15°$ can be estimated. Bottke *et al.* (2000c) recently have confirmed an angle of order 12° as the upper boundary for a significant ellipticity of impact craters.

Any crater of a diameter D may be formed by impact of projectiles of various diameters, depending on the impact velocity and impact angle. In differential form this condition is expressed as

$$\frac{dN}{dD} = \int\limits_{v_{min}}^{v_{max}} dv \int\limits_{\alpha_{min}}^{\alpha_{max}} d\alpha \left[\frac{dN}{dD_P} \frac{dD_P}{dD} \right] f_v(v) f_\alpha(\alpha) \quad , \tag{11}$$

where both values in square brackets should be calculated for the projectile diameter $D_P(D, v, \alpha)$, i.e. for the projectile of diameter D_P, which creates the crater of diameter D, impacting with velocity v at angle α.

The integral equation (11) enables to calculate the projectile SFD for a known SFD of craters, $D_P(D, v, \alpha)$, and velocity and angular distributions of impact events. Then, Equation (11) gives the model impact crater SFD (a production function) for a given planetary surface from the projectile SFD and the impact velocity spectra.

Scaling laws (Equations 2, 4) simplify the procedure. The analytical fit to the lunar data derived by Neukum (1983) and recently improved by Ivanov (1999) and Ivanov et al. (2000) is one example; however, the general equation (11) may be applied for any input crater distribution. The resulting projectile SFD is also presented as a polynomial function similar to Neukum's lunar production function. From the derived projectile (asteroid) SFD the model SFD for Martian craters is calculated. Note however, that the same asteroid SFD for Earth-crossers and Mars-crossers is assumed. This may not be completely true, if non-gravity effects change, depending on asteroid size, the populations of fast and slow track delivery routes.

Results. With all the assumptions and simplifications described above, the model results in a Mars-to-moon re-calculation of the production functions.

The *Hartman Production Function (HPF)*, recalculated to Mars using **Model 2**, is shown in Figure 4 in comparison with the original lunar HPF. Figure 5 plots the Mars-to-moon ratio for the same crater diameter bins. In the spirit of the lunar HPF, the production function for Mars may be split into 3 power law branches (for an "average" lunar mare surface of $\sim$3.4 Gyr on Neukum's time scale):

$$\log N_H = -2.894 - 3.82 \log D_L, \quad D < 1.0 \, \text{km}$$
$$\log N_H = -2.938 - 1.72 \log D_L, \quad 1.0 \, \text{km} < D < 32 \, \text{km}$$
$$\log N_H = -2.146 - 2.20 \log D_L, \quad D > 32 \, \text{km} \tag{12}$$

Note the slight change in the exponent (Equation 12) with respect to the lunar HPF (Neukum et al., 2001). As Martian craters created with the same projectile are smaller than on the moon, the boundary diameters of different power law branches are shifted to smaller sizes compared to the lunar HPF (Equation 1).

As the HPF is presented as the 3 power relationships (Neukum et al., 2001), the Mars-to-moon ratio has specific values for each power branch (Figure 5). The smaller crater dimensions on Mars for the same projectile diameter and the higher Martian impact rate barely compensate each other for $D > 2$ km. Thus, the Mars-to-moon crater number ratio is close to unity with the accuracy of $\pm 20\%$.

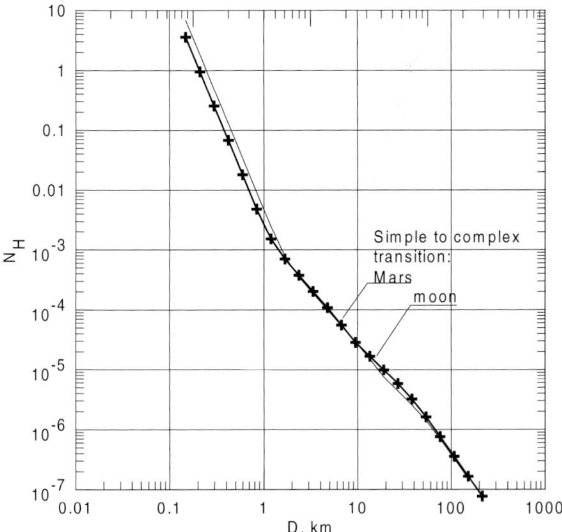

Figure 4. The Hartman production function for Mars in comparison with the lunar prototype. Crosses are plotted against the middle of $2^{1/2}$ diameter bins.

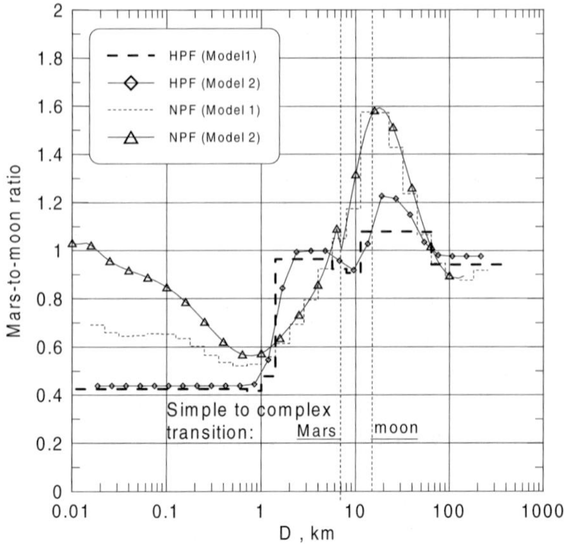

Figure 5. The Mars-to-moon ratio of a number of impact craters with equal diameter. Hartmann Production Function (HPF) and Neukum Production Function (NPF) are recalculated from the moon to Mars using **Model 1** (one average impact velocity and one average impact angle of $45°$) and **Model 2** (the full ensemble of impact velocities and impact angles). All re-calculations except NPF (**Model 2**) use the gravity crater scaling law with the Croft's collapse model. The NPF (**Model 2**) is recalculated assuming the strength-to-gravity cratering regime transition with D_{sg} at 300 m on the moon and 100 m on Mars. For details see text.

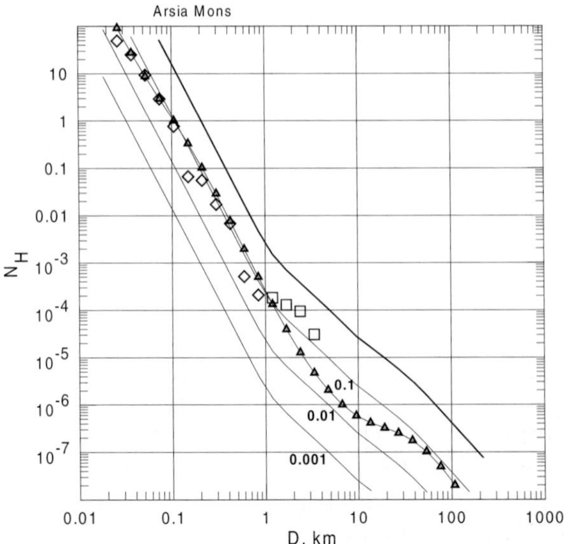

Figure 6. Martian "isochrones" based on the Hartmann Production Function (HPF) compared to the crater counts on the Arsia Mons volcano (*square*) and at its summit caldera floor (*diamonds*) by Hartmann (1999). The upper curve corresponds to the same age as the "average lunar mare" (~3.4 Gyr). Also shown are "isochrones" for surfaces that accumulated 1/10, 1/100, and 1/1000 less craters. The Arsia Mons caldera floor accumulated roughly 10 times less craters than surfaces of lunar mare "average" age. The NPF, derived below, is shown for comparison (*small triangles*).

The lunar HPF is constructed for the "average mare" surface age, estimated above as ~3.4 Gyr. If no excessive accuracy is required, the Martian HPF for various crater retention ages may be given as fractions of the "average mare" crater density. Figure 6 illustrates these HPF "isochrones" for Martian surfaces with 1/10 and 1/100 of the "average mare" crater counts. The recent MGS data for the caldera of Arsia Mons (Hartmann, 1999) corresponds to a crater areal density of 1/10 of the "average lunar mare". Martian crater counts are discussed by Hartmann and Neukum (2001), and Table II presents numerical values of the HPF for Mars.

The *Neukum Production Function (NPF)* originally is presented in the cumulative form, from which the incremental and *R*-plot distributions are easily produced. Figure 7 compares the Martian NPF for a 1 Gyr-old surface with the lunar NPF of the same age. Figure 5 shows this comparison in more detail, as the ratio of incremental number of craters in the same diameter intervals. The non-power law NPF has a steepness that differs from the HPF power law. Consequently, the Mars-to-moon ratio varies in a wider range.

The NPF may be approximated as a polynomial similar to the lunar NPF:

$$\log_{10}(N) = a_0 + \sum_{n=1}^{11} a_n \left[\log_{10}(D)\right]^n .$$

Coefficients a_i are listed in Table III; the fit is valid for ~15 m < D < 362 km.

TABLE II

Calculated crater number in each crater diameter bin for Mars and the moon according to Hartmann's (HPF) and Neukum's Production Function (NPF) for an age of the "average lunar mare" ($\sim$3.4 Gyr for NPF) and the ratio of NPF/HPF for Mars and moon.

D_L (km)	$N_H(Mars)$	$N_H(moon)$	$(NPF/HPF)_{Mars}$	$(NPF/HPF)_{moon}$
0.0156	8154	19210	0.21	0.15
0.0221	2170	5113	0.26	0.20
0.0313	577.4	1360	0.32	0.25
0.0442	153.6	362	0.41	0.31
0.0625	40.88	96.32	0.51	0.39
0.0884	10.88	25.63	0.65	0.49
0.1250	2.895	6.82	0.79	0.62
0.1768	0.7702	1.815	0.91	0.77
0.2500	0.205	0.4829	0.99	0.89
0.3536	0.05454	0.1285	1.02	0.98
0.5000	0.01451	0.03419	1.03	1.02
0.7071	3.787×10^{-3}	9.098×10^{-3}	1.09	1.03
1.0000	1.158×10^{-3}	2.421×10^{-3}	1.04	1.06
1.414	6.207×10^{-4}	6.443×10^{-4}	0.62	1.14
2.000	3.326×10^{-4}	3.453×10^{-4}	0.42	0.67
2.828	1.782×10^{-4}	1.850×10^{-4}	0.33	0.45
4.000	9.552×10^{-5}	9.915×10^{-5}	0.29	0.34
5.657	4.902×10^{-5}	5.313×10^{-5}	0.31	0.30
8.000	2.580×10^{-5}	2.847×10^{-5}	0.39	0.31
11.310	1.646×10^{-5}	1.526×10^{-5}	0.53	0.38
16.000	8.821×10^{-6}	8.177×10^{-6}	0.74	0.54
22.630	4.727×10^{-6}	4.382×10^{-6}	0.90	0.74
32.000	2.533×10^{-6}	2.348×10^{-6}	0.92	0.90
45.250	1.357×10^{-6}	1.258×10^{-6}	0.78	0.91
64.000	6.336×10^{-7}	6.736×10^{-7}	0.66	0.78
90.510	2.956×10^{-7}	3.142×10^{-7}	0.52	0.66
128.000	1.379×10^{-7}	1.466×10^{-7}	0.41	0.52
181.000	6.433×10^{-8}	6.839×10^{-8}	0.33	0.41
256.000	3.001×10^{-8}	3.191×10^{-8}	0.35	0.32

TABLE III

Coefficients for the Martian NPF

n	a_n	n	a_n	n	a_n	n	a_n
0	-3.384	3	0.7915	6	0.1016	9	-4.753×10^{-3}
1	-3.197	4	-0.4861	7	6.756×10^{-2}	10	6.233×10^{-4}
2	1.257	5	-0.3630	8	$-1.181\cdot10^{-2}$	11	5.805×10^{-5}

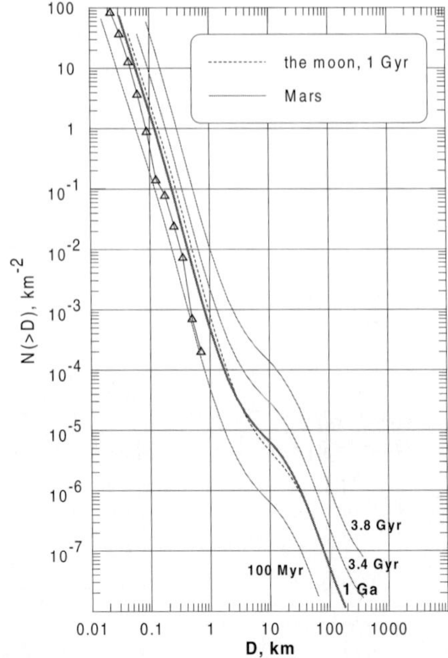

Figure 7. The Martian "isochrones" based on the Neukum Production Function (**Model 1**) with the gravity/collapse (no strength/gravity transition) scaling of cratering. The dashed line presents the 1 Gyr isochron for the moon. The detailed Mars-to-moon comparison is shown in Figure 5. The same data as in Figure 5 for the caldera floor of Arsia Mons are shown for comparison.

The coefficient a_0 in Table III is calculated for a crater retention age of 1 Gyr. It corresponds to the cumulative number of craters larger 1 km

$$N(1) = 10^{-3.312} = 4.13 \times 10^{-4} \text{ km}^{-2} .$$

Using the same time dependence as for the moon (Neukum, 1983; Neukum *et al.*, 2001) one can propose the similar time dependence for Mars

$$N(1) = 2.68 \cdot 10^{-14}(\exp(6.93T) - 1) + 4.13 \times 10^{-4}T \quad . \tag{13}$$

The introduction of a possible strength/gravity transition into the cratering scaling law adds some difference in the Mars-to-moon ratio of crater numbers. Figure 5 shows that in this case the Mars-to-moon ratio should increase to the one for crater diameters near $D = 10$ m.

Conclusions. According to the models described above, the impact crater number on Mars does not differ more than $\pm 50\%$ from the lunar crater number in the same diameter bins. The more frequent asteroid impacts on Mars do not result in a much larger cratering rate: the larger surface gravity and the smaller impact velocity decrease the crater size compared to the impact of the same body on the moon. Finally, the cratering rate Mars-moon ratio varies in the range $0.6 - 1.2$, depending on the steepness of the $N(D)$ distribution.

References

R.P. Binzel, T. Gehrels, and M.S.Matthews (eds).: 1989, *Asteroids II*, Univ. Arizona Press, Tucson.

Bottke, W.F., Jr., Morbidelli, A., Petit, J.M., Gladman, B., and Jedicke, R.: 2000a, 'Tracking NEAs from Their Source Regions to Their Observed Orbits', *Proc. 31st Lunar Planet Sci. Conf.*, abstract #1634.

Bottke, W.F., Jr., Jedicke, R., Morbidelli, A., Petit, J.M., and Gladman, B.,: 2000b, 'Understanding the Distribution of Near-Earth Asteroids', *Science* **288**, 2190–2194.

Bottke, W.F., Love, S.G., Tytell, D., and Glotch, T.: 2000c, 'Interpreting the Elliptical Crater Populations on Mars, Venus, and the Moon', *Icarus* **145**, 108–121.

Chapman, C.R., and McKinnon, W.B.: 1986, 'Cratering of Planetary Satellites', *Satellites*, Univ. Arizona Press, Tucson, pp. 492–580.

Croft, S.K.: 1985, 'The Scaling of Complex Craters', *J. Geophys. Res.* **90**, 828–842.

Farinella, P., Froeschle, Ch., Froeschle, C., Gonczi, R., Hahn, G., Morbidelli, A., and Valesechhi, G.B.: 1994, 'Asteroids Falling Into the Sun', *Nature* **371**, 314–317.

Garvin, J.B., Frawley, J.J., Sakimoto, S.E.H., and Schnetzler, C.: 2000, 'Global Geometric Properties of Martian Impact Craters: An Assessment from Mars Orbiter Laser Altimeter (MOLA) Digital Elevation Models', *Proc. 31st Lunar Planet Sci. Conf.*, abstract #1619.

Gault, D.E., and Wedekind, J.A.: 1978, 'Experimental Studies of Oblique Impact', *Proc. 9th Lunar Planet. Set. Conf.*, Pergamon Press, New York, pp. 3843–3875.

Gladman, B.J., *et al.*: 1997, 'Dynamical Lifetimes of Objects Injected Into Asteroid Main Belt Resonances', *Science* **277**, 197–201.

Greenberg, R., and Nolan, M.: 1989, 'Delivery of Asteroids and Meteorites to the Inner Solar System', *in* R.P. Binzel *et al.* (eds.), *Asteroids II*, Univ. Arizona Press, Tucson, pp. 778–804.

Greenberg, R., and Nolan, M.: 1993, 'Dynamical Relationships of Near-Earth Asteroids to Main-belt Asteroids', *Resources of Near-Earth Space*, Univ. Arizona Press, Tucson, pp. 473–492.

Grieve, R.A.F., and Cintala, M.J.: 1992, 'An Analysis of Differential Impact Melt-crater Scaling and Implications for the Terrestrial Impact Record', *Meteoritics* **27**, 526–538.

Hartmann, W.K.: 1977, 'Relative Crater Production Rates on Planets', *Icarus* **31**, 260–276.

Hartmann, W.K.: 1999, 'Martian Cratering VI. Crater Count Isochrons and Evidence for Recent Volcanism from Mars Global Surveyor', *Met. Planet. Sci.* **34**, 167–177.

Hartmann, W.K., and Neukum, G.: 2001, 'Cratering Chronology and the Evolution of Mars', *Space Sci. Rev.*, this volume.

Hartmann, W.K., *et al.*: 1981, 'Chronology of Planetary Volcanism by Comparative Studies of Planetary Craters', *in* W.M. Kaula, R.B. Merrill, and R. Ridings (eds.), *Basaltic Volcanism at the Terrestrial Planets*, Pergamon Press, New York, pp. 1050–1127.

Ivanov, B.A.: 1981, 'Cratering Mechanics', *Adv. Sci. Techn. VINITI, Mechanics of Deformable Solids* **14**, Moscow, VINITI Press, pp. 60–128, in Russian; English translation: NASA Tech. Memorandum 88477/N87-15662, 1986.

Ivanov, B.A.: 1999, 'Target – Earth. Size Frequency Distribution of Asteroids and Impact Craters', *Proc. Institute for Dynamic of Geospheres*, Moscow, Russia (in Russian), pp. 33–44.

Ivanov, B.A., Neukum, G., and Wagner, R.: 1999, 'Impact Craters, NEA, and Main Belt Asteroids: Size-frequency Distribution', *Proc. 30th Lunar Planet. Sci. Conf.*, abstract #1583 (CD-ROM).

Ivanov, B.A., Neukum, G., and Wagner, R.: 2000, 'Size–Frequency Distributions of Planetary Impact Craters and Asteroids', *in* H. Rickman, and M. Marov (eds.), *Collisional Processes in the Solar System*, ASSL, Kluwer Academic Publishers, Dordrecht, in press.

Jedicke, R., and Metcalfe, T.S.: 1998, 'The Orbital Absolute Magnitude Distributions of Main Belt Asteroids', *Icarus* **131**, 245–260.

Melosh, H.J., and Ivanov, B.A.: 1999, 'Impact Crater Collapse', *Ann. Rev. Earth Planet. Sci.* **27**, 385–415.

Migliorini, F.P., Michel., P., Morbidelli, A., Nesvorny, D., and Zappala, V.: 1998, 'Origin of Earth-crossing Asteroids: The New Scenario', *Science* **281**, 2022–2024.

Morbidelli, A.: 1999, 'Origin and Evolution of Near-Earth Asteroids', *Celestial Mech. and Dynamical Astron.* **73**, 39–50.

Neukum, G.: 1983, *Meteoritenbombardement and Datierung planetarer Oberflächen*, Habilitation dissertation for faculty membership, Univ. of Munich, 1983, 186 pp.

Neukum, G., and Ivanov, B.A.: 1994, 'Crater Size Distributions and Impact Probabilities on Earth from Lunar, Terrestrial Planet, and Asteroid Cratering Data', *in* T. Gehrels (ed.), *Hazards due to Comets and Asterois*, Univ. Arizona Press, Tucson, pp. 359–416.

Neukum, G., Ivanov, B., and Hartmann, W.K.: 2001, 'Cratering Records in the Inner Solar System in Relation to the Lunar Reference System', *Space Sci. Rev.*, this volume.

Pierazzo, E., Vickery, A.M., and Melosh, H.J.: 1997, 'A Re-evaluation of Impact Melt Production', *Icarus* **127**, 498–423.

Pike, R.J.: 1980, 'Control of Crater Morphology by Gravity and Target Type: Mars, Earth, moon', *Proc. 11th Lunar. Planet Sci. Conf.*, Pergamon Press, New York, pp. 2159–2189

Rabinowitz, D.: 1993, 'The Size-distribution of the Earth-approaching Asteroids', *Astrophys. J.* **407**, 412–427.

Rabinowitz, D.L.: 1997, 'Are Main-belt Asteroids a Sufficient Source for the Earth-approaching Asteroids? Part II. Predicted vs. Observed Size Distribution', *Icarus* **130**, 287–295.

Rabinowitz, D., Bowell, E., Shoemaker, E., and Muinonen, K.: 1994, 'The Population of Earth-crossing Asteroids', *in* T. Gehrels (ed.), *Hazards due to Comets and Asteroids*, Univ. Arizona Press, Tucson, pp. 285–312.

Roemer, E.: 1965, 'Observations of Comets and Minor Planets', *Astron. J.* **70**, 397–402.

Roemer, E.: 1966, 'The Dimensions of Cometary Nuclei', *Les Congrès et Colloques de Université de Liège* **37**, Université de Liège, pp. 23–28.

Roemer, E., and Lloyd, R.E.: 1966 'Observations of Comets, Minor Planets and Satellites', *Astron. J.* **71**, 443–457.

Roemer, E., Thomas, M., and Lloyd, R.E.: 1966 'Observations of Comets, Minor Planets, and Jupiter VIII', *Astron. J.* **71**, 591–601.

Schmidt, R.M, and Housen, K.R.,: 1987, 'Some Recent Advances in the Scaling of Impact and Explosion Cratering', *Int. J. Impact Engng.* **5**, 543–560.

Shoemaker, E.M., and Wolfe, R.F.: 1982, 'Cratering Time Scales for the Galilean Satellites', *in* D. Morrison (ed.), *Satellites of Jupiter*, Univ. Arizona Press, Tucson, pp. 277–339.

Ward, W.R.: 1992, 'Long-term Orbital and Spin Dynamics of Mars', *in* H.H. Kieffer, B.M. Jakovsky, C.W. Snyder, and M.S. Matthews (eds.), *Mars*, Univ. Arizona Press, Tucson, pp. 298–320.

Wetherill, G.W.: 1967 'Collisions in the Asteroid Belt', *J. Geophys. Res.* **72**, 2429–2444.

Wetherill, G.W.: 1979 'Steady-state Population of Apollo-Amor Object', *Icarus* **37**, 96–112.

Wetherill, G.W.: 1988, 'Where do the Apollo Objects Come From?', *Icarus]* **76**, 1–18.

Wetherill, G.W.: 1989, 'Origin of the Asteroid Belt', *in* R.P. Binzel *et al.* (eds.), *Asteroids II*, pp. 661–680.

Address for correspondence: Institute for Dynamics of Geospheres, Russian Academy of Sciences, Leninsky Prospect 38/6, Moscow, Russia 117939;

(baivanov@idg.chph.ras.ru)

AGES AND GEOLOGIC HISTORIES OF MARTIAN METEORITES

L.E. NYQUIST[1], D.D. BOGARD[1], C.-Y. SHIH[2], A. GRESHAKE[3], D. STÖFFLER[3]
and O. EUGSTER[4]

[1] *SN/Planetary Sciences, NASA Johnson Space Center, 2101 NASA Road 1,
Houston, TX 77058-3696, USA*
[2] *Basic and Applied Research Department, Lockheed-Martin Space Operations, 2400 NASA Road 1,
Houston, TX 77058-3696, USA*
[3] *Institut für Mineralogie, Museum für Naturkunde, Invalidenstr. 43, D-10115 Berlin, Germany*
[4] *Physikalisches Institut, University of Bern, Sidlerstr. 5, CH-3012 Bern, Switzerland*

Submitted: 28 July 2000; accepted: 16 February 2001

Abstract. We review the radiometric ages of the 16 currently known Martian meteorites, classified as 11 shergottites (8 basaltic and 3 lherzolitic), 3 nakhlites (clinopyroxenites), Chassigny (a dunite), and the orthopyroxenite ALH84001. The basaltic shergottites represent surface lava flows, the others magmas that solidified at depth. Shock effects correlate with these compositional types, and, in each case, they can be attributed to a single shock event, most likely the meteorite's ejection from Mars. Peak pressures in the range $15 - 45$ GPa appear to be a "launch window": shergottites experienced $\sim30 - 45$ GPa, nakhlites $\sim20 \pm 5$ GPa, Chassigny ~35 GPa, and ALH84001 $\sim35 - 40$ GPa. Two meteorites, lherzolitic shergottite Y-793605 and orthopyroxenite ALH84001, are monomict breccias, indicating a two-phase shock history *in toto*: monomict brecciation at depth in a first impact and later shock metamorphism in a second impact, probably the ejection event.

Crystallization ages of shergottites show only two pronounced groups designated S_1 (~175 Myr), including 4 of 6 dated basalts and all 3 lherzolites, and S_2 ($330 - 475$ Myr), including two basaltic shergottites and probably a third according to preliminary data. Ejection ages of shergottites, defined as the sum of their cosmic ray exposure ages and their terrestrial residence ages, range from the oldest (~20 Myr) to the youngest (~0.7 Myr) values for Martian meteorites. Five groups are distinguished and designated S_{Dho} (one basalt, ~20 Myr), S_L (two lherzolites of overlapping ejection ages, 3.94 ± 0.40 Myr and 4.70 ± 0.50 Myr), S (four basalts and one lherzolite, $\sim2.7 - 3.1$ Myr), S_{DaG} (two basalts, ~1.25 Myr), and S_E (the youngest basalt, 0.73 ± 0.15 Myr). Consequently, crystallization age group S_1 includes ejection age groups S_L, S_E and 4 of the 5 members of S, whereas S_2 includes the remaining member of S and one of the two members of S_{DaG}. Shock effects are different for basalts and lherzolites in group S/S_1. Similarities to the dated meteorite DaG476 suggest that the two shergottites that are not dated yet belong to group S_2. Whether or not S_2 is a single group is unclear at present. If crystallization age group S_1 represents a single ejection event, pre-exposure on the Martian surface is required to account for ejection ages of S_L that are greater than ejection ages of S, whereas secondary breakup in space is required to account for ejection ages of S_E less than those of S. Because one member of crystallization age group S_2 belongs to ejection group S, the maximum number of shergottite ejection events is 6, whereas the minimum number is 2.

Crystallization ages of nakhlites and Chassigny are concordant at ~1.3 Gyr. These meteorites also have concordant ejection ages, i.e., they were ejected together in a single event (NC). Shock effects vary within group NC between the nakhlites and Chassigny.

The orthopyroxenite ALH84001 is characterized by the oldest crystallization age of ~4.5 Gyr. Its secondary carbonates are ~3.9 Gyr old, an age corresponding to the time of Ar-outgassing from silicates. Carbonate formation appears to have coincided with impact metamorphism, either directly, or indirectly, perhaps via precipitation from a transient impact crater lake.

Chronology and Evolution of Mars **96** 105–164, 2001.
© 2001 *Kluwer Academic Publishers. Printed in the Netherlands.*

The crystallization age and the ejection age of ALH84001, the second oldest ejection age at 15.0 ± 0.8 Myr, give evidence for another ejection event (O). Consequently, the total number of ejection events for the 16 Martian meteorites lies in the range 4 − 8.

The Martian meteorites indicate that Martian magmatism has been active over most of Martian geologic history, in agreement with the inferred very young ages of flood basalt flows observed in Elysium and Amazonis Planitia with the Mars Orbital Camera (MOC) on the Mars Global Surveyor (MGS). The provenance of the youngest meteorites must be found among the youngest volcanic surfaces on Mars, i.e., in the Tharsis, Amazonis, and Elysium regions.

Keywords: shock effects, crystallization ages, cosmic ray exposure ages, ejection ages, provenance

1. Introduction

The clan of Martian meteorites, formerly called SNCs after *Shergotty*, *Nakhla* and *Chassigny*, now consists of 16 unpaired meteorites of magmatic origin (basalts and ultramafic cumulates). Generally young crystallization ages (with the exception of one pyroxenite), characteristic isotopic compositions of C, N, O, and noble gases, as well as distinct major and trace element concentrations, distinguish them from all other differentiated meteorites (McSween, 1994; Clayton and Mayeda, 1996; Dreibus and Wänke, 1987; Wänke and Dreibus, 1988; Wänke, 1991). The most convincing evidence for the Martian origin of these rocks is given by isotopic measurements of trapped gases in shock-melted glass of shergottites. It was found that the isotopic composition of these gases is indistinguishable from Martian atmosphere within the measurement errors of the mass spectrometer on board the Viking lander (e.g., Bogard and Johnson, 1983; Becker and Pepin, 1984; Swindle *et al.*, 1986; Marti *et al.*, 1995).

Those who have followed the development of the hypothesis of the Martian origin of these meteorites will recognize that much of the early information about them was obtained in attempts to verify that hypothesis. Once the determination of probable Martian origin had been made, observations about the meteorites could be generalized to become probable observations about Mars. Efforts to do so are hampered to variable degrees by lack of knowledge of the geologic settings from whence the meteorites came. This is also true of the radiometric age data.

The young radiometric ages of shergottites, first observed for the type example, Shergotty, by Geiss and Hess (1958), were among the first lines of evidence cited in support of their origin on a planetary-sized body, probably Mars (Nyquist *et al.*, 1979b; Wasson and Wetherill, 1979). However, some characteristics of the analytical data seemed to violate criteria developed for unambiguous interpretation of radiometric ages as igneous crystallizaton ages. In particular, the Rb-Sr and ^{39}Ar-^{40}Ar ages were discordant (Nyquist *et al.*, 1979a; Bogard *et al.*, 1979). Moreover, the Rb-Sr data showed considerable "scatter" about the best fit isochron. Thus, it did not initially appear possible to interpret the ~165 Myr Rb-Sr age of Shergotty, for example, as the crystallization age, particularly, since the ^{39}Ar-^{40}Ar age of a plagioclase separate was older at $\sim$250 Myr. Considerable experience with dating lunar samples had shown that although ^{39}Ar-^{40}Ar ages of bulk samples could be biased low due to diffusive loss of ^{40}Ar, ^{39}Ar-^{40}Ar ages of plagioclase separates generally gave reliable crystallization ages. Thus, the Rb-Sr age was initially interpreted as likely reflecting the time of a thermal metamorphism. Because the shergottites were more highly shocked than almost all other meteorites or lunar samples, initial attention was given to post-shock thermal metamorphism as the agent for resetting both types of ages (Nyquist *et al.*, 1979a; Bogard *et al.*, 1979). It has since been established that neither post-shock thermal metamorphism nor shock transformations of mineral phases are adequate to reset the Rb-Sr isotopic system. Thus, the *possibility* of shock resetting of the radiometric ages appears to have been a false lead.

Radiometric ages determined by other methods often were discordant also, and contributed to the confusion about the ages of shergottites. Here, it is sufficient to note that isotopic heterogeneities occurring within the rocks over distances of centimeters appear to complicate the isotopic data. For example, the existence of heterogeneity in initial ^{87}Sr/^{86}Sr between different samples of Zagami was shown by Nyquist *et al.* (1995). Papanastassiou and Wasserburg (1974) had observed a similar heterogeneity in initial ^{87}Sr/^{86}Sr for Nakhla much earlier. They took their observation as evidence that the Rb-Sr age of 1.30 ± 0.02 Gyr that they determined separately for two samples of Nakhla was an age of metamorphism rather than an igneous crystallization age. (Here, as elsewhere in this paper, we use the value of the ^{87}Rb decay constant recommended by Minster *et al.* (1982); i.e., $\lambda_{87} = 1.402 \times 10^{-11} \mathrm{yr}^{-1}$). Gale *et al.* (1975) obtained apparently well-defined, concordant, isochrons for two additional samples of Nakhla. They interpreted the isochron age for their total data set, 1.23 ± 0.01 Gyr, as the age of igneous crystallization. The present authors agree with that interpretation, but the issue of possible sample heterogeneity exists. The two samples studied by Papanastassiou and Wasserburg (1974) were obtained from separate sources, and weighed 0.7 g and 2.3 g, respectively (D. Papanastassiou, personal communication). The samples studied by Gale *et al.* (1975), also from two different sources, weighed 13 g and 18 g, respectively. It is possible that the larger samples used by Gale *et al.* (1975) effectively averaged out isotopic heterogeneity that could exist in Nakhla over small distances.

In spite of uncertainties in interpreting the isochron data, the early isotopic data of both Nakhla and Shergotty showed unambiguously that they were "young" by meteorite standards. For Nakhla, the evidence was Rb-Sr model ages in the range of $2.5 - 3.6$ Gyr, calculated relative to the initial ^{87}Sr/^{86}Sr of the solar system (Papanastassiou and Wasserburg, 1974; Gale *et al.*, 1975). These model ages provide a strict upper limit to the formation age of Nakhla. The Rb-Sr model ages of shergottites are $\sim$4.5 Gyr, indicative of the time of chemical differentiation of their parent body. Nyquist *et al.* (1979b) found the Sm-Nd model ages of three shergottites relative to initial ^{143}Nd/^{144}Nd of the solar system to be only $\sim$2.8 $-$ 3.6 Gyr, however. These Sm-Nd model ages also provide a strict upper limit to the formation ages of the meteorites, and they are especially remarkable in that they imply significant fractionation of the Sm/Nd ratios of the meteorites from chondritic values, which normally is not achieved in "simple" magmatic processes. Thus, these young model ages clearly pointed to "late" and complex magmatic activity on the parent body of the shergottites.

Another feature of the shergottites that makes them somewhat unusual among stony meteorites is their relatively young cosmic ray exposure (CRE) ages. Such ages are determined from the accumulation of nuclides spalled from target nuclides in the meteorites by high energy cosmic ray interactions. Spallation nuclides that are stable against radioactive decay accumulate continually at production rates that are a function of the chemical composition of the meteorite and of changes in the energy spectrum of primary and secondary cosmic rays as a function of the size of the meteoroid and the depth of the meteorite within it. Spallation radionuclides that undergo radioactive decay accumulate only to equilibrium levels at which they decay at the same rate at which they are produced. The production rates of stable nuclides can be determined either theoretically using nuclear cross section data, or empirically from the equilibrium activities of closely related radionuclides for which the ratio of the production rate to that of the stable nuclide is known. Determination of stable noble gas CRE ages, for example, require mass spectrometric techniques similar to those required for ^{39}Ar-^{40}Ar age dating, so stable noble gas CRE ages often are determined in the same investigations as those in which ^{39}Ar-^{40}Ar ages are determined. An early study of four shergottites, Shergotty, Zagami, ALH77005, and EET79001 (Bogard et $al.$, 1984) showed that, whereas three of them had very similar apparent exposure ages of $\sim$2–3 Myr, one (EET79001) had a much lower apparent exposure age, $<$ 1 Myr. One possible explanation was that the actual exposure age of EET79001 was the same as that of the others, but that the production rate of stable nuclides was lower in EET79001 because it was located deeper in a large meteoroid, where the effective flux of primary cosmic rays was attenuated by about a factor of 3–4. A variation of this scenario is that EET79001 was initially part of a very large meteoroid, and was essentially totally shielded from cosmic rays for most of its lifetime in space. With acquisition of additional data for both stable and radionuclides, both of these explanations have fallen into disfavor. The currently accepted interpretation of the spallogenic nuclide data is that exposure to cosmic rays was initiated by excavation of the meteorites from depths on Mars that were completely shielded from the effects of cosmic rays, and thus that the CRE ages plus the time the meteorites have been on Earth give directly the time since they were ejected from Mars ($cf.$ Eugster et $al.$, 1997b). However, the young CRE age of EET79001 presents a problem when viewed in the context of its crystallization age, and its apparent relationship to the other shergottites.

Thus, the earlier interpretation of the Rb-Sr and ^{39}Ar-^{40}Ar ages for shergottites of $\sim$180 Myr and $\sim$250 Myr, respectively, as giving the time of shock metamorphism during their ejection from Mars as large blocks to be broken up in later secondary collisions ($cf.$ Shih et $al.$, 1982) is no longer favored by the majority of of meteoriticists. Also, a steadily increasing number of Sm-Nd ages have been obtained that are concordant with the Rb-Sr age for the same meteorite. Furthermore, ^{39}Ar-^{40}Ar ages can be explained by the presence of excess, non-radiogenic ^{40}Ar from a variety of sources. These problems have been much less acute for the other types of Martian meteorites, for which the ages are comparatively well defined.

The desire to find possible mechanisms for launching Martian meteorites and for metamorphic resetting of the radiometric ages of shergottites stimulated thorough investigations of their shock-metamorphic features. These shock features of the meteorites provide insight into the environment of the rocks when they were ejected from Mars. Furthermore, the original igneous textures of their minerals provide insight into the environments in which the rocks originally crystallized. For example, mineral textures are influenced by the cooling rate of the rock during crystallization, which in turn is influenced by the thickness of a magma flow. Thus, observable features in the meteorites themselves tell us some things about their geologic setting at key points in their histories.

Because the Martian meteorites are the only samples from Mars currently available for laboratory studies, their properties are of great importance to understand the formation and evolution of our neighboring planet. In this review we summarize their crystallization and cosmic ray exposure ages, compare the shock levels to which they have been exposed, and briefly consider the implications of those data for the meteorites' provenance and for Martian evolution. But, first we describe the mineralogical and geochemical characteristics of the meteorites themselves.

2. Mineralogy, Petrography, and Geochemistry

According to their mineralogical composition and textural characteristics, the Martian meteorites represent igneous rocks of basaltic and ultramafic provenance. They appear to have crystallized either in lava flows as volcanic rocks or in mafic, probably shallow, intrusions as plutonic ultramafic rocks. They are divided into shergottites, consisting of a basaltic and a lherzolitic subgroup, nakhlites (clinopyroxenites), chassignites (dunites), and orthopyroxenites. Chassigny and ALH84001 are the only dunite and orthopyroxenite in the latter two groups (Figure 1). In the following, we describe the mineralogy, petrography, and geochemistry of the various Martian meteorites. For more details, the reader is referred to the review article by McSween (1994) and to the Mars Meteorite Compendium (Meyer, 1998).

2.1. BASALTIC SHERGOTTITES

The meteorites Shergotty, Zagami, EETA79001, QUE94201, Dar al Gani 476, and Los Angeles, as well as the recently found Dhofar 019 and Sayh al Uhaymir 005, belong to the group of basaltic shergottites. These rocks predominantly consist of augite and pigeonite, typically showing a strong irregular chemical zoning towards an Fe-rich rim, and of plagioclase in the form of shock-induced diaplectic

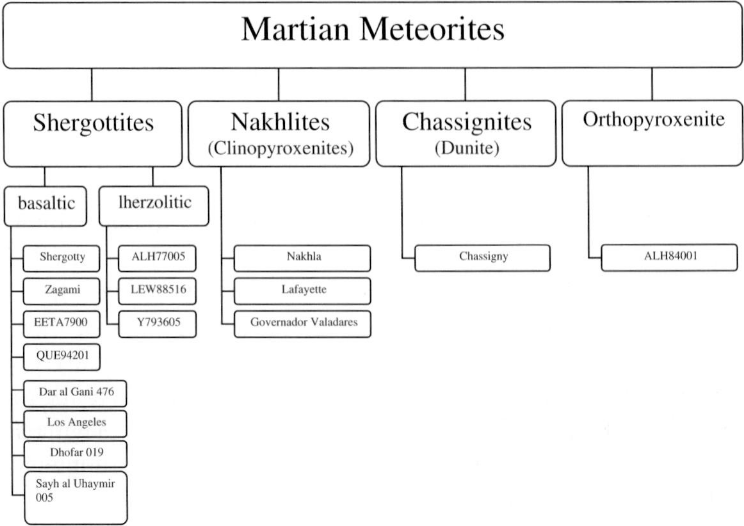

Figure 1. Classification of Martian meteorites (modified after Stephan *et al.*, 1999).

glass (maskelynite). Minor components are pyrrhotite, whitlockite, ilmenite and titanomagnetite (McSween, 1994). Pyroxene often contains small rounded to sub-rounded melt inclusions of kaersutite, spinel, and sulfides in a Si-rich glassy or microcrystalline groundmass, interpreted as trapped original melt (e.g., Treiman, 1985). EETA79001 contains two basaltic lithologies, termed A and B. The main differences between these units are the small grain size and the occurrence of olivine xenocrysts, orthopyroxene, and chromite in lithology A. Besides formation by simple mixing of basaltic liquids, it was suggested that lithology A repre-sents an impact melt (Mittlefehldt *et al.*, 1997). Among the basaltic shergottites, only Dar al Gani 476 and Sayh al Uhaymir 005 have some similarities to lithol-ogy A of EETA79001, containing large xenocrysts of olivine set into a matrix of clinopyroxene and maskelynite (Zipfel *et al.*, 2000; Grossman, 2000). The basaltic shergottites, except QUE94201 and Los Angeles, show a cumulate texture with mostly preferred orientations of the pyroxenes. However, the relatively small grain size of the pyroxenes and the petrographic similarities among the meteorites sug-gest that the alignment may have occurred by lava flow rather than by accumulation in a subsurface magma chamber (McCoy *et al.*, 1992; McSween, 1994).

Compared to terrestrial basalts, the basaltic shergottites are characterized by a high Fe/(Fe+Mg) ratio and low Al_2O_3 concentrations. All meteorites of this group have complex rare Earth element (REE) patterns with distinct depletions of light REE (LREE) but without a clear Eu anomaly. Except for water, all volatile elements are enriched (Wänke and Dreibus, 1988; Lodders, 1998; Zipfel *et al.*, 2000; Rubin *et al.*, 2000). Initial Sr, Nd, and Pb isotopic compositions are variable among the basaltic shergottites, possibly reflecting different stages of mixing Martian crust with an isotopically homogeneous magma.

2.2. LHERZOLITIC SHERGOTTITES

Three shergottites (ALH77005, LEW88516, and Y-793605) are lherzolitic as they contain <10 vol.% plagioclase. They are composed of relatively coarse-grained anhedral to euhedral olivine and chromite enclosed by large orthopyroxene crystals (Harvey *et al.*, 1993; McSween, 1994; Mikouchi and Miyamoto, 1997). Interstices are filled with accessory phases, i.e. maskelynite, pigeonite, augite, and whitlockite. The Fe-Mg-silicates of this suite of ultramafic rocks are much more magnesian than those of the basaltic shergottites, and an observed chemical disequilibrium between coexisting olivine and pyroxene gives evidence for non-linear cooling (Harvey *et al.*, 1993). High concentrations of Fe^{3+} in chromites suggest high oxygen fugacity during crystallization. Similarly to the basaltic shergottites, the lherzolitic ones are depleted in LREE (Dreibus *et al.*, 1982). However, the composition of radiogenic Sr isotopes in ALH77005 and LEW88516 shows that the two rocks crystallized from different magma sources (Borg *et al.*, 1998a, 1998b).

2.3. NAKHLITES

The group of nakhlites contains the three clinopyroxenites Nakhla, Lafayette, and Governador Valadares. They consist of Mg-rich augite and Fe-rich olivine set into a microcrystalline groundmass of mostly radiating crystalline plagioclase, which has not been transformed into maskelynite by shock, pigeonite, ferroaugite, titano-magnetite, pyrite, troilite, chlorapatite, and sometimes SiO_2-rich glass. In addition, they contain phyllosilicates (iddingsite) and evaporite mineral assemblages of secondary, but Martian, origin confirming the presence of liquid water on Mars (Gooding *et al.*, 1991; Bridges and Grady, 2000; Swindle *et al.*, 2000).

All three meteorites are cumulates, and lamellar inclusions of augite and magnetite in olivine from Nakhla and Governador Valadares confirm slow cooling under highly oxidizing conditions (Mikouchi and Miyamoto, 1998). In addition, olivine frequently contains small melt inclusions compositionally representing a major and distinct type of Martian magma.

All nakhlites have moderately high contents of volatiles and are enriched in LREE. The almost identical initial Sr and Nd isotopic composition of nakhlites is distinct from that of the shergottites, indicating that the two types of rocks formed from different parent magmas (Nakamura *et al.*, 1982b; McSween, 1994).

2.4. CHASSIGNY

Chassigny is the only dunite among the Martian meteorites, and consists of 90% Fe-rich olivine ($Fa_{\sim32}$), 5% pyroxene, 2% feldspar ($An_{\sim20}$; maskelynite), and 3% accessory phases (Floran *et al.*, 1978). High concentrations of Fe^{3+} in chromite as well as lamellar exsolutions in olivine indicate crystallization at high oxygen fugacity and low cooling rates (Floran *et al.*, 1978; Greshake *et al.*, 1998). Its cumulus fabric indicates a fractional crystallization of a mafic magma body.

Chassigny is enriched in LREE but shows no Eu anomaly. While this pattern is distinct from those of nakhlites, excluding a formation from the same magma, the initial Sr isotope compositions of Chassigny and nakhlites are identical (Nakamura *et al.*, 1982a; McSween, 1994).

2.5. ALH84001

ALH84001 is a coarse-grained brecciated orthopyroxenite with a modal composition of 96% orthopyroxene, 2% chromite, 1% plagioclase (maskelynite), and 0.15% phospate. Accessory phases are augite, olivine, pyrite and Fe-Mg-Ca-carbonates (Mittlefehldt, 1994). Texturally, ALH84001 is dominated by up to 6 mm long orthopyroxene crystals joined at $120°$ triple junctions and poikilitically enclosing Fe^{3+}-rich euhedral chromites (Berkley and Boynton, 1992; Mittlefehldt, 1994; McSween, 1994). Maskelynite and rarely chromite occur interstitially between orthopyroxene. Predominantly along fractures and in cataclastic areas, compositionally strongly zoned carbonates are found, often forming characteristic globules that appear either as concentric spherules or as flat "pancakes" (e.g., McKay *et al.*, 1996). Many carbonates resemble a "bull's-eye" with a center of dolomite-ankerite surrounded by concentric bands of siderite, magnesite, and sulfide, but they come in a multitude of varieties (Scott *et al.*, 1998). High-resolution Scanning Electron Microscope (SEM) images revealed worm-like features in the carbonates. Additionally, the morphologies of some magnetite grains in the carbonates resemble those formed by magnetotactic bacteria, and relatively high concentrations of polycyclic aromatic hydrocarbons (PAHs) have been found. From these observations, McKay *et al.* (1996) concluded early biologic activity was present on Mars. Meanwhile, various other non-biogenic formation mechanisms of the carbonates and magnetite assemblage have been proposed, including impact origin (Harvey and McSween, 1996; Scott *et al.*, 1998; Scott, 1999) and flood-evaporite formation (McSween and Harvey, 1998; Warren, 1998). The high concentrations of PAHs, present in all textural units of ALH84001 (Stephan *et al.*, 1998, 1999), could also be due to terrestrial contamination (Becker *et al.*, 1997; Jull *et al.*, 1998).

ALH84001 is depleted in the LREE and has a negative Eu anomaly. Its very low concentrations of siderophile elements led to the development of a model for the Martian mantle depleted in siderophile elements (Dreibus *et al.*, 1994).

2.6. ENVIRONMENTS OF IGNEOUS CRYSTALLIZATION

2.6.1. *Basaltic Shergottites*

The textures of the basaltic shergottites are consistent with those expected for surface flows of basaltic lava. McCoy *et al.* (1992) suggested that Zagami was the product of a two-stage magmatic history. The first stage occurred in a slowly cooling magma chamber. The presence of amphibole in the cores of pyroxene crystals requires pressures equivalent to depths >7.5 km on Mars. During the second stage, pyroxene crystals were entrained into a magma that either intruded to the near surface and cooled in a relatively thin dike or sill, or extruded to the surface and crystallized in a lava flow >10 m thick. This two-stage scenario is consistent with observations of volcanic constructs and flows in the Tharsis region of Mars.

2.6.2. *Lherzolitic Shergottites*

Detailed investigations of lherzolitic textures revealed a preferred crystallographic orientation of olivine, proving that the lherzolitic shergottites are real cumulates, formed in a plutonic sub-surface environment (Berkley and Keil, 1981; McSween, 1994). Ikeda (1994) suggested that the compositional discontinuities among the four zoning types of chromite in ALH77005 arise from magma mixing in shallow magma reservoirs on Mars. Their crystallization histories, as reconstructed by Harvey *et al.* (1993) and McSween (1994), require varying degrees of prolonged cooling to allow olivine to reequilibrate at comparatively low temperature. On Earth, lherzolites crystallize either at depths of >8 km (mantle rocks) or as cumulates in large magma chambers. Harvey *et al.* (1993) concluded that the trace element and minor element patterns of LEW88516 and ALH77005 minerals were essentially identical and consistent with large-volume, closed-system fractional crystallization followed by localized crystallization of isolated melt pockets.

2.6.3. *Nakhlites (Clinopyroxenites)*

The cumulate textures of the nakhlites, combined with the presence of lamellar inclusions of augite and magnetite, require slow cooling under highly oxidizing conditions. These textures, especially those of Nakhla and Governador Valadares, are analogous to those of terrestrial augite-rich igneous cumulate rocks of the Abitibi greenstone belt of northern Ontario (Treiman, 1987), where augite cumulates comprise the lower half of a 125 m thick flow. Augite cumulates in the middle third of a 300 m thick sill have little mesostasis, giving them textures more comparable to those of Lafayette. Treiman (1987) concluded from these comparisons that the nakhlites crystallized in thick flows, >125 m thick, or in shallow intrusions, probably less than 1 km deep, of basaltic or picritic magmas. He noted that volcanoes with thick lava flows and evidence of shallow intrusions were common in the Tharsis region of Mars. Furthermore, greenstone belt volcanism may be related to mantle hot spots, another potential analogy to the Tharsis region.

2.6.4. *Chassigny (Dunite)*

The texture and high modal abundance of olivine suggest that Chassigny is a cumulate, also. Floran *et al.* (1978) described the crystallization history as similar to that of nakhlites, except that olivine is much more abundant, and chromite, absent from naklites, crystallizes early. These characteristics suggest that Chassigny and the nakhlites might represent different parts of the same or similar layered igneous complexes. From their REE abundances, Wadhwa and Crozaz (1994) concluded that they could not have crystallized from the same magma, however.

2.6.5. *ALH84001 (Orthopyroxenite)*

Mittlefehldt (1994) interpreted the orthopyroxene and chromite in ALH84001 as cumulus phases. Their textural features indicate slow cooling either during magmatic crystallization, or metamorphic recrystallization, or both. Mineral compositions in ALH84001 are similar to those of lherzolitic shergottites or nakhlites. Mittlefehldt (1994) cited the uniform pyroxene compositions, unusual for Martian meteorites, as indicating that ALH84001 cooled more slowly than did the shergottites, nakhlites, or Chassigny; i.e., it formed at greater depth than they did. Kring and Gleason (1997) argued that the orthopyroxene-silica assemblage present in ALH84001 corresponded to magmatic temperatures of $\sim 1400 - 1470°C$, and to a static pressure of ~ 0.5 GPa, equivalent to a depth of ~ 40 km on Mars. Gleason *et al.* (1997) noted that its texture was reminiscent of those of cataclastic anorthosites from the ancient, heavily-cratered, lunar highlands. They, like Treiman (1995b), suggested that at least some of the secondary carbonate formed by replacement of plagioclase, and cite textural evidence as showing this occurred after plagioclase had been converted to maskelynite. Kring *et al.* (1998) concurred in that suggestion, noting that it implied formation of the carbonate after 3.92 ± 0.04 Gyr ago, the time of Ar-degassing of plagioclase according to Turner *et al.* (1997). However, Scott (1999) alternatively suggested that the original carbonates formed as evaporite deposits, probably prior to impact heating ~ 4 Gyr ago, when episodic floods were more common. He suggests that preservation of the carbonates for ~ 4 Gyr was aided by the impact, which sealed up the carbonate-bearing fractures and pores, making the rock less pervious to later infiltration of fluids.

3. Shock Metamorphism

3.1. EVIDENCE OF SHOCK

It is generally agreed that the Martian meteorites have been ejected from the planet's surface by large-scale impacts. The ejection velocity must have exceeded the escape velocity of Mars, which is about 5 km/s. Material accelerated by a shock wave to >5 km/s should be in a molten state according to basic shock wave physics. The fact that all known Martian meteorites are solid though strongly shocked rock fragments prompted Melosh (1984) to develop a model of the ejection process in which a special spallation mechanism provides most of the required ejection velocity for rock fragments ejected from a thin, uppermost layer of the impacted target without melting them. Although alternative mechanisms have been proposed (Nyquist, 1983; O'Keefe and Ahrens, 1986), this model has been widely accepted. The originally proposed spallation mechanism (Melosh, 1984) required the parent craters of the meteorites to be larger than ~ 10 km in diameter. Recent refinements of the model have reduced the size limit to >3 km (Head and Melosh, 2000).

As expected from the impact and ejection model for the origin of the Martian meteorites, the imposed extreme physical conditions caused significant changes in the textures, mineralogy, and possibly even the isotopic compositions of constituent mineral phases. It even led to a shock- induced implantation of Martian atmospheric gases into the meteorites (Duke, 1968; Stöffler *et al.*, 1986; Bogard *et al.*, 1986; Wiens and Pepin, 1988; McSween, 1994). *The important obervation is that all Martian meteorites are moderately to strongly shock metamorphosed by shock pressures ranging between about 15 and 45 GPa.* The understanding of the type and intensity of shock metamorphism of Martian meteorites is thus essential for the interpretation of the ejection and possible impact-induced relocation processes, which relate to some extent to the problem of the geological provenance, and to the interpretation of analyzed isotope systems, which may be disturbed by shock.

It has been recognized since the pioneering studies of Tschermak (1872) that shergottites and some other achondrites are severely shocked (Binns, 1967; Duke, 1968). Although shock effects in meteorites were known before the Martian origin of the SNC meteorites was suspected (e.g., Wood and Ashwal, 1981), the Martian origin hypothesis gave new impetus to their study. The degree of shock metamorphism in shergottites was first studied quantitatively on Shergotty (Lambert and Grieve, 1984; Stöffler *et al.*, 1986). In all shergottites the constituent minerals display specific, more or less similar shock effects, well known from naturally and experimentally shocked terrestrial, lunar, and meteoritic rocks (Stöffler, 1972; Stöffler *et al.*, 1988; Bischoff and Stöffler, 1992). Pyroxene shows strong mosaicism, mechanical twinning, shear fractures and various lattice defects such as high dislocation densities revealed in the Transmission Electron Microscope (TEM; e.g., Müller, 1993). Olivine, if present, is affected by strong mosaicism, deformation bands, planar fractures, planar deformation features and high dislocation densities (e.g., Greshake and Stöffler, 1999, 2000). Ostertag *et al.* (1984) attributed the brown staining of olivine in ALH77005 to a shock-induced oxidation of iron to Fe^{3+}. Plagioclase is transformed to diaplectic glass (maskelynite) and retains its primary crystal shape. Based on the experimentally calibrated refractive index of maskelynite, the peak shock pressure (final equilibrium shock pressure) of the host meteorite can be deduced (Stöffler *et al.*, 1986).

A typical feature of the shergottites is the presence of shock-produced veins and melt pockets caused by local pressure and temperature excursions of presumably up to $60 - 80$ GPa and 2000 °C (Stöffler *et al.*, 1986). These pressure estimates are based on experimental data (Kieffer *et al.*, 1976; Schaal and Hörz, 1977; Schmitt, 2000). In addition, high pressure phases such as very dense post-stishovite polymorphs of SiO_2 have been discovered in Shergotty (Sharp *et al.*, 1999; El Goresy *et al.*, 2000). It has to be pointed out that these high pressures were only produced very locally and do not represent the equilibration pressure as previously suggested (Sharp *et al.*, 1999). Recently, an assemblage of omphacite, stishovite and $KAlSi_3O_8$-hollandite was found in a shock vein of Zagami, indicating crystallization of these phases during decompression between 25 and 50 GPa (Langenhorst and Poirier, 2000). Also, the high-pressure polymorphs of olivine and pyroxene, ringwoodite and majorite, were tentatively reported from a melt vein in the basaltic shergottite EETA79001 (Steele and Smith, 1982). However, unambiguous identification of these two phases has failed so far (Boctor *et al.*, 1998). Moreover, the melt pockets in some Martian meteorites are obviously the host regions of gases of the Martian atmosphere that were first detected in EETA79001 (Bogard and Johnson, 1983; Becker and Pepin, 1984). These gases must have been implanted during shock metamorphism of the meteorite precursor rocks near the Martian surface. Experimental studies on shock implantation show that shock can relatively easily incorporate an ambient gas phase into solid material, even at temperatures well below melting (Bogard *et al.*, 1986; Wiens and Pepin, 1988).

Refractive index measurements of maskelynite gave quantitative estimates of the peak shock pressure for some shergottites: Shergotty: 29 ± 1 GPa (Stöffler *et al.*, 1986), Zagami: 31 ± 2 GPa (Stöffler *et al.*, 1986; Langenhorst *et al.*, 1991), EET79001: 34 ± 2 GPa (Lambert, 1985), ALH77005: 43 ± 2 GPa (McSween and Stöffler, 1980). For other shergottites the values, based on the overall shock effects in plagioclase, olivine, and pyroxene, and on the presence and abundance of localized melts, are less accurate: Dar al Gani 476: probably $30-35$ GPa (Greshake and Stöffler, 1999, 2000), QUE94201: $30-35$ GPa, Los Angeles: $35-40$ GPa, Dhofar 019: $35-40$ GPa, Sayh al Uhaymir: $35-40$ GPa, LEW88516: ca. $40-45$ GPa ("strongly shocked", Keller *et al.*, 1992), Y793605: ca. $40-45$ GPa. All estimated peak shock pressures of shergottites are summarized in Table I, along with the estimated post-shock temperatures.

Considering the range of shock pressures observed in shergottites, it is conspicuous that the basalts were all affected by similar shock pressures in the range of $\sim 30-35$ GPa, whereas the lherzolites reveal somewhat higher shock pressure ($\sim 40-45$ GPa). Also, Ott and Löhr (1992) noted that the ^{4}He content of lherzolite LEW88516 indicates complete loss of radiogenic ^{4}He acquired prior to its ejection from Mars ~ 3 Myr ago, consistent with its high post-shock temperature of $\sim 600\,^\circ$C (Table I). The type and homogeneity of shock damage observed in the constituent minerals of the basaltic shergottites indicates that each of them was affected by only one impact event (Stöffler *et al.*, 1986; Müller, 1993). This seems to be different for the lherzolitic shergottites, as observed for Y793605, which has been brecciated by a first impact and shock metamorphosed by a second impact. The three nakhlites are less intensely affected by shock metamorphism than the other Martian meteorites. Only weak undulatory extinction and a rather low dislocation density in olivine as well as entirely birefringent plagioclase suggest a peak shock pressure of $\sim \leq 20 \pm 5$ GPa (Bunch and Reid, 1975; Greshake, 1998).

Shock metamorphism in Chassigny was investigated in detail by optical and transmission electron microscope (Langenhorst and Greshake, 1999). Conversion of feldspars to diaplectic glass (maskelynite), the clino-/orthoenstatite inversion, strong mosaicism of olivine, and the activation of numerous planar fractures and c-dislocations in olivine are among the shock effects. High-resolution TEM revealed additionally the coexistence of planar fractures with discontinuous fractures in olivine. These findings point to a shock pressure of about 35 GPa.

TABLE I

Estimates of the peak shock pressure (final equilibration shock pressure) and the overall post-shock temperature increase in Martian meteorites. Data from Stöffler *et al.* (1986) and Stöffler (2000) except for Sayh al Uhaymir 005, Los Angeles, and Dhofar (this paper).

Meteorite	Shock pressure (GPa)	Post-shock temperature*
Shergotty	29 ± 1	200 ± 20
Zagami	31 ± 2	220 ± 50
EETA 79001	34 ± 2	250 ± 50
QUE94201	$\sim$30–35	$\sim$200 − 350
Dar al Gani 467	$\sim$35–40	$\sim$350 − 450
Los Angeles	$\sim$35–40	$\sim$350 − 450
Dhofar 019	$\sim$35–40	$\sim$350 − 450
Sayh al Uhaymir 005	$\sim$35–40	$\sim$350 − 450
ALHA77005	43 ± 2	$\sim$450 − 600
LEW88516	$\sim$45	$\sim$600
Y793605	$\sim$45	$\sim$600
ALH84001	$\sim$35–40	$\sim$300 − 400
Nakhlites	$\sim$20 ($\pm$5)	$\sim$100
Chassigny	$\sim$35	$\sim$300

*Relative to ambient pre-shock temperature.

In the orthopyroxenite ALH84001, shock metamorphism is documented by complex textures, such as localized brecciation in fine-grained shear zones, strong mosaicism and numerous irregular fractures in orthopyroxene, and by the conversion of all plagioclase to maskelynite. While Mittlefehldt (1994) explained these effects by a single impact event, Treiman (1998) invoked up to five impacts. We believe that the presence of maskelynite in both brecciated and non-brecciated regions indicates that at least two impact events are required: A first weak shock event producing the brecciation and a subsequent stronger shock event which transformed plagioclase to maskelynite throughout the whole rock.

3.2. ENVIRONMENTS AND IMPLICATIONS OF SHOCK METAMORPHISM

The observed shock metamorphism of Martian meteorites has important implications for their impact and ejection history and for their geologic provenance, if the geologic settings of their magmatic formation processes are taken into account. Summarizing the essential observations leads us to some general conclusions.

All Martian meteorites are moderately to severely shocked (Table I, Figure 7), with effects being homogeneously distributed throughout the rocks. These shock effects can be attributed to one specific event in each case, most probably the ejection event. A single stage shock history is implied for all basaltic shergottites and most likely for the nakhlites and for Chassigny. However, some of the ultramafic "plutonic" rocks such as lherzolitic shergottite Y-793605 and orthopyroxenite ALH84001 are shocked monomict breccias indicating a two-stage shock history: In a first impact, the rocks are brecciated at very low shock pressure at depth and relocated to the surface during the same event as commonly observed in terrestrial impact craters such as the Ries (e.g., Pohl *et al.*, 1977). The transformation of plagioclase to maskelynite indicates strong shock metamorphism in a second impact, most probably the ejection event. Although clear evidence for the two remaining lherzolitic basalts and for the nakhlites/chassigny group is unavailable, an impact-induced relocation of the"plutonic" ultramafic Martian meteorites from their primary deep-seated magmatic setting is highly plausible.

The second fundamental observation relates to the ranges of observed shock pressures for all the Martian meteorites and for particular groups of them (Table I, Figure 7). Although exact values are not yet available, we recognize 1) that the observed shock pressures are restricted to a range of about 15 to 45 GPa, and 2) that the basaltic shergottites range from about 30 to 35 GPa, the lherzolitic shergottites from about 40 to 45 GPa, and the nakhlites from about 15 to 25 GPa. This means that unshocked meteorites as well as shock-fused meteorites are lacking and that the observed 15–45 GPa range may be viewed as a typical "launch window" for Mars. The lower limit may indicate that unshocked rocks and rocks shocked to pressures lower than the Hugoniot Elastic Limit cannot be ejected, in contrast to what has been proposed by Melosh (1995), Mileikowsky *et al.* (2000), and Weiss *et al.* (2000). The upper limit indicates that melt ejecta are too much dispersed and, hence, too small to survive as meteoroids. Comparing meteorites from Mars, the Moon, and the eucrite parent body, it seems that the observed range of shock metamorphism related to the ejection event is a function of the size of the parent planetary body and therefore of the magnitude of the escape velocity: The present data indicate that lunar meteorites are shocked below about 20 GPa (Bischoff and Stöffler, 1992; Greshake *et al.*, 2001), and meteorites of the eucrite-howardite-diogenite group, possibly originating from the 550 km diameter asteroid Vesta, are at most mildly shocked, i.e. below ∼5–10 GPa (Metzler *et al.*, 1995).

A third implication of the observed shock metamorphism of Martian meteorites relates to the size of the precursor meteoroids and to their ejection ages. As known from terrestrial craters such as the Ries crater (Pohl *et al.*, 1977; Stöffler and Ostertag, 1983; von Engelhardt and Graup, 1984), the size of displaced shocked rock fragments is inversely proportional to the shock intensity. Crystalline rock fragments in polymict breccias of the Ries shocked to the range of the Martian shergottites (∼30−45 GPa) do not exceed 0.5 m or so, and most sizes are ∼0.1−10 cm. Typical shock stage III rocks (45−60 GPa; Stöffler, 1984) are consistently <50 cm in size. Additionally, distal ejecta (solid clasts) in the Ries (Reutter blocks: Upper Jurassic limestone fragments, Pohl *et al.*, 1977) are not only small, <∼20 cm, but are also derived from the uppermost layer of the target in agreement with the conditions invoked by the spallation model (Melosh, 1984). Consequently, the lherzolitic shergottites (40−45 GPa) must have originated from $\approx$0.1 m-sized rocks and cannot have been ejected in one block together with those basaltic shergottites that have the same ejection age (Figure 7a). Rather, the basalts and lherzolites may be derived from different surface regions of the same parent crater notwithstanding the fact that the lherzolites had to be relocated first to the surface by a previous impact. A similar case could be made for the nakhlites and Chassigny, which also have identical ejection and crystallization ages but different shock pressures.

The peak shock pressures experienced by the Martian meteorites are likely to be a consequence of their geometrical relationship to "ground zero" at the moment of the impact that is destined to launch them from the planet. The near-surface spall model, for example, outlines rather definite relationships between the impactor, the transient crater cavity, and the target region near ground zero (Melosh, 1984; Figure 11). Those fragments destined for ejection might be considered to constitute the "lid" of the transient cavity; a lid destined to be blown off. In the spallation model, the thickness of the "lid" is given by the depth of the spall zone, and was estimated by Warren (1994) to be 0.2−0.4 times the diameter of the projectile at a distance of 1−3 projectile radii from the the impact. Thus, the "lid" would be on the order of 50 to 100 m thick for a 10 km diameter crater. (See "Potential source terrains" later in the paper). In the lid, peak shock pressure increases in the downward direction from the surface, and decreases in the radial direction from ground zero. The nakhlites, being most lightly shocked, are thus implied to have been ejected from nearest the Martian surface, in spite of having probably crystallized near the center of a thick flow, $>\sim$100 m thick, or in a subsurface intrusion. Chassigny, being more severely shocked, is implied to have been ejected from a deeper region of the lid, if the nakhlites and Chassigny were ejected simultaneously. The lherzolitic shergottites are most highly shocked of all the Martian meteorites, and thus are expected to come from deep within the lid, close to the melt zone. If, for example, they and the basaltic shergottites were ejected simultaneously, the latter would have come from nearer to the surface, consistent with being recent lava flows. The orthopyroxenite ALH84001, which likely crystallized at the greatest depth of the Martian meteorites, experienced peak shock pressure equivalent to those of the shergottites, implying prior excavation from depth to the launch site. This is consistent with an ancient age and origin in the Martian highlands, which probably were "gardened" to depths on the order of a kilometer, or more (W. Hartmann, personal communication). Gardening of the surfaces of $\sim$180 Myr old basaltic shergottite lava should have been minimal, however, and shergottites are likely to have been ejected from their place of emplacement as lava flows.

Finally, we note that rocks of distinctly different shock pressures, e.g. nakhlites and Chassigny, or basaltic and lherzolitic shergottites, cannot have been ejected from Mars in one large rock unit. Such scenarios have been proposed in order to explain the different exposure ages within the shergottite group as due to later break-up in space. Indeed, the limited size of strongly shocked rocks ejected from

NYQUIST ET AL.

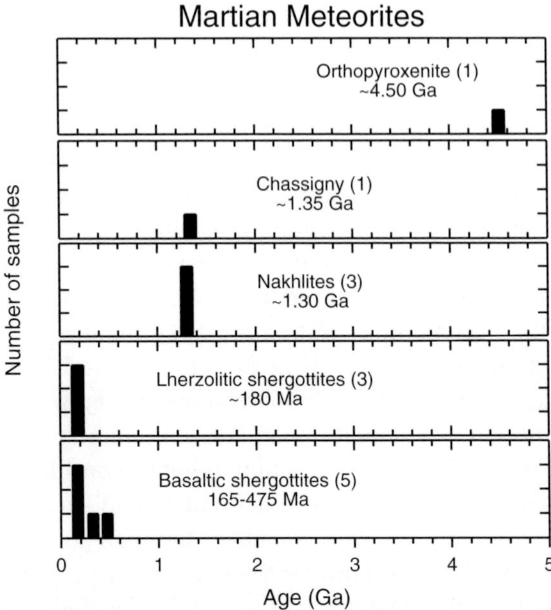

Figure 2. The crystallization ages of Martian meteorites separated by compositional group. The values plotted are the "preferred ages" from Tables II and III. The oldest meteorite in the Martian clan is ~4.5 Gyr old, and the youngest ~180 Myr old. Thus, Martian magmatism appears to have extended over most of solar system history, a conclusion that agrees with the time span of crater retention ages (Hartmann and Berman, 2000). The thirteen meteorites fall into only five age groups, leaving large gaps in Martian chronology as recorded by the meteorites.

the parent crater (see above) also argues against this possibility. In the case of the basaltic shergottites, however, the peak shock pressures are nearly equivalent, leaving the comparatively large size required of the initial ejecta fragments as the primary physical limitation on secondary break-up scenarios. In later sections, we will further discuss the issue of the number of required ejection events, in conjunction with the problem of the geological provenance of the Martian meteorites.

4. Radiometric Ages

Radiometric ages of Martian meteorites as reported in the literature are given in Tables II and III. We discuss the ages of individual meteorites within each of the meteorite classes separately. For the nakhlites and Chassigny (Table II), the ages determined by the various radiometric methods are in close agreement and present a coherent picture of when those rocks crystallized as thick magma flows or subsurface sills. For the other classes of Martian meteorites, notably the shergottites (Table III), the picture is more complicated.

Figure 2 gives an overview of those data that most reliably give the crystallization ages of the meteorites. The 13 meteorites define only 5 separate ages, covering

TABLE II

Summary of Radiometric Ages of Martian Meteorites: Nakhlites, Chassigny, ALH84001

Meteorite	K-Ar (Gyr)	^{39}Ar-^{40}Ar (Gyr)	Rb-Sr (Gyr)	Sm-Nd (Gyr)	U-Th-Pb (Gyr)	Preferred Age (Gyr)
Clinopyroxenites (Nakhlites):						
Nakhla	1.30 ± 0.03[a]	1.3[b]	1.23 ± 0.01[c] 1.30 ± 0.02[e] 1.36 ± 0.02[e]	1.26 ± 0.07[d]	1.28 ± 0.05[d] 1.24 ± 0.11[d]	**1.27 ± 0.01**
Governador Valadares		1.32 ± 0.04[f]	1.32 ± 0.01[g] 1.19 ± 0.02[h]	1.37 ± 0.02[h]		**1.33 ± 0.01**
Lafayette		1.33 ± 0.03[b]	1.25 ± 0.08[i]	1.32 ± 0.05[l]		**1.32 ± 0.02**
Dunite:						
Chassigny	1.39 ± 0.17[j]	1.32 ± 0.07[k]	1.22 ± 0.01[l]	1.36 ± 0.06[m]		**1.34 ± 0.05**
Orthopyroxenite :						
ALH 84001						
Silicates		3.92 ± 0.10[n] 4.07 ± 0.04[o] 4.10 ± 0.20[p]	4.55 ± 0.30[q] 3.89 ± 0.05[r]	~ 4.56[s] 4.50 ± 0.12[q]		**4.51 ± 0.11**
Carbonates		~ 3.6[t]	3.90 ± 0.04[u] 1.41 ± 0.10[r]		4.04 ± 0.10[u]	**3.92 ± 0.04**

References: [a]Stauffer (1962); [b]Podosek (1973); [c]Gale *et al.* (1975); [d]Nakamura *et al.* (1982a); [e]Papanastassiou and Wasserburg (1974); [f]Bogard and Husain (1977); [g]Wooden *et al.* (1979); [h]Shih *et al.* (1999); [i]Shih *et al.* (1998); [j]Lancet and Lancet (1971); [k]Bogard and Garrison (1999); [l]Nakamura *et al.* (1982b); [m]Jagoutz (1996); [n]Turner *et al.* (1997); [o]Ilg *et al.* (1997); [p]Bogard and Garrison (1999); [q]Nyquist *et al.* (1995); [r]Wadhwa and Lugmair (1996); [s]Jagoutz *et al.* (1994); [t]Knott *et al.* (1996); [u]Borg *et al.* (1999).

an age span from the formation of the planet extending nearly to the present day, and there is only a single Martian rock older than 1.3 Gyr. This is a rather incomplete sample of the ages of Martian surface rocks, and these ages only give a record of Martian evolution, if their geologic context is known. Isotopic data of Martian meteorites are most useful to study the Martian global geochemical evolution.

Radiometric ages, though, provide absolute calibration marks for relative ages from cratering records. By extrapolating the lunar cratering rate to Mars, the relative ages of Martian surface units can be estimated from the density of craters on them (Neukum *et al.*, 2001; Ivanov, 2001; Hartmann and Neukum, 2001). These ages are divided into three major chronostratigraphic units, or epochs: Noachian, Hesperian, and Amazonian. The Noachian and Amazonian are further subdivided into Early, Middle, and Late periods; whereas the Hesperian is simply divided into Early and Late periods (Tanaka, 1986; Tanaka *et al.*, 1992). We begin our discussion with early Mars, working towards the present day.

122   NYQUIST ET AL.

TABLE III

Summary of Radiometric Ages of Martian Meteorites: Shergottites

Meteorite	K-Ar (Myr)	^{39}Ar-^{40}Ar (Myr)	Rb-Sr (Myr)	Sm-Nd (Myr)	U-Th-Pb (Myr)	Preferred Age (Myr)
Shergottittes (Basalts) :						
Shergotty	580 ± 50[a1]	254 ± 10[b]	163 ± 12[d]	147 ± 20[e]	200 ± 4[g]	**165 ± 4**
	196 ± 40[w]	167[c]	165 ± 4[e]	360 ± 16[e]	437 ± 36[g]	
				620 ± 171[f]	600 ± 20[g]	
					217 ± 110[h]	
					189 ± 83[h]	
Zagami		242[c]	178 ± 3[f]	163 ± 7[i]	230 ± 5[g]	**177 ± 3**
			174 ± 14[i]		229 ± 8[g]	
			163 ± 19[i]			
Los Angeles			165 ± 11[j]	172 ± 8[j]		**170 ± 8**
EETA79001A		2035[c]	172 ± 18[k]		150 ± 15[g]	**173 ± 3**
					170 ± 36[g]	
EETA79001B			177 ± 12[k]	165 ± 43[l]		
			173 ± 3[l]			
QUE94201		730[c]	327 ± 12[m]	327 ± 19[m]		**327 ± 10**
DaG476				474 ± 11[n]		**474 ± 11**
				~800[o]		
Shergottites (Lherzolites) : :						
ALHA77005	1330 ± 130[p]	3500[c]	156 ± 6[q]	173 ± 7[r]		**179 ± 5**
			188 ± 11[f]			
			185 ± 11[r]			
LEW88516	2600[c]		183 ± 10[s]	166 ± 16[t]	~170[u]	**178 ± 8**
Y793605		1595[c]			212 ± 62[v]	**212 ± 62**

References: [a1]Geiss and Hess (1958), recalculated to the K-decay constants by Steiger and Jäger (1977); [a]Eugster *et al.* (1997a); [b]Bogard *et al.* (1979); [c]Bogard and Garrison (1999); [d]Nyquist *et al.* (1979a); [e]Jagoutz and Wänke (1986); [f]Shih *et al.* (1982); [g]Chen and Wasserburg (1986); [h]Sano *et al.* (2000); [i]Nyquist *et al.* (1995); [j]Nyquist *et al.* (2000); [k]Nyquist *et al.* (1986); [l]Nyquist *et al.* (2001); [m]Borg *et al.* (1997); [n]Borg *et al.* (2000); [o]Jagoutz *et al.* (1999), Jagoutz and Jotter (2000); [p]Miura *et al.* (1995); [q]Jagoutz (1989); [r]Borg *et al.* (2001b); [s,t]Borg *et al.* (1998a, 1998b); [u]Chen and Wasserburg (1993); [v]Misawa *et al.* (1997); [w]Terribilini *et al.* (1998).

4.1. ORTHPYROXENITE ALH84001: A CARBONATE-BEARING FRAGMENT OF THE NOACHIAN CRUST

ALH84001 is the only meteorite from the ancient Martian crust. We infer its crustal origin from its ancient age, and less directly from its composition. The old crystallization age of ALH84001 is direct evidence that portions of the Martian crust formed quickly after the planet accreted. There have been some variations in the ages reported for ALH84001 (Figure 3). Jagoutz *et al.* (1994) argued that the Sm-

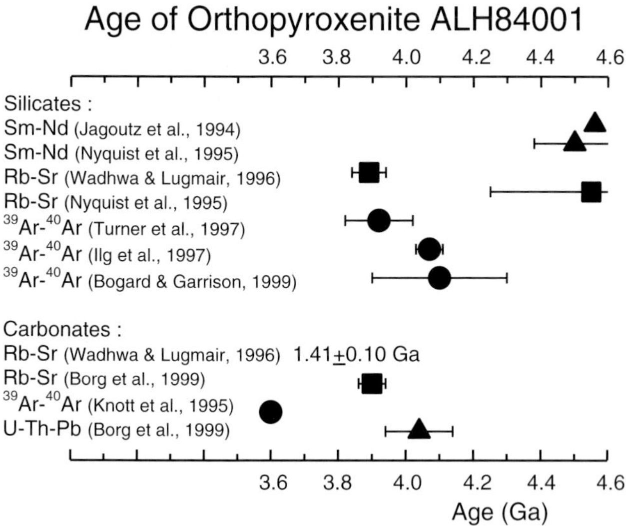

Figure 3. Radiometric ages of orthopyroxenite ALH84001, as determined by a variety of techniques. Sm-Nd isochron ages for silicate minerals indicate primary crystallization prior to 4.4 Gyr ago. ^{39}Ar-^{40}Ar ages for the silicates indicate Ar-outgassing between ∼3.9–4.1 Gyr ago in a secondary heating event, presumably related to impact cratering. Rb-Sr ages of different subsamples have given different results, apparently related to both primary crystallization and secondary reheating. Attempts to date secondary carbonates within ALH84001 have yielded ages as low as ∼1.4 Gyr and as high as ∼4.0 Gyr. We prefer a carbonate age of ∼3.9 Gyr, which would make carbonate formation directly or indirectly related to the cratering event that reset the ^{39}Ar-^{40}Ar age.

Nd isotopic data plotted along a 4.56 Gyr reference isochron. However, they neither reported an isochron regression nor estimated an uncertainty on the age. Their data were for bulk samples and acid leach/residue pairs, and thus any "isochron" also may be interpreted as an "unmixing" line. This is because phosphate minerals, major contributors to the REE budget, would be dissolved into the "leach" solutions, leaving the residue as a complementary end member. Nyquist *et al.* (1995) reported both Sm-Nd and Rb-Sr data for ALH84001. They obtained a ^{147}Sm-^{144}Nd isochron age of 4.50 ± 0.13 Gyr from a suite of bulk samples and a pyroxene mineral separate. Their isochron also could be interpreted as a mixing line between orthopyroxene and a second component of low Sm-Nd ratio, like phosphate, if that component dominated the REE budget of the rock. However, both phosphate and plagioclase have low Sm/Nd ratios, and contribute to the REE budget. Because both are in low abundance, both could be randomly distributed among the different bulk samples analysed, decreasing the likelihood that the isochron is simply a two-component mixing line. Nyquist *et al.* (1995) also determined initial $(^{146}Sm/^{144}Sm)_I = 0.0022 \pm 0.0010$ for ALH84001 from variations in $^{142}Nd/^{144}Nd$ caused by decay of ^{146}Sm $(T_{1/2} = 103$ Myr), initially present in the rock, to ^{142}Nd. This result suggests that ALH84001 formed more than one half-life of ^{146}Sm after the angrite meteorite LEW86010, for which $(^{146}Sm/^{144}Sm)_I$ was ∼0.0070–

0.0076 (Lugmair and Galer, 1992; Nyquist *et al.*, 1994). The $^{146}Sm/^{144}Sm$ ratio requires closure of the Sm-Nd system in ALH84001 no earlier than $\sim$115 Myr after formation of the angrite. Although $^{142}Nd/^{144}Nd$ measurements are analytically challenging, the long- and short-lived chronometers can be considered concordant for an age of $\sim$4.4 Gyr, the lower limit on the conventional ^{147}Sm-^{144}Nd age. Thus, the great antiquity of ALH84001 appears to be established, in spite of generally lower ^{39}Ar-^{40}Ar ages, and a lower Rb-Sr age (Wadhwa and Lugmair, 1996).

The ^{39}Ar-^{40}Ar ages of ALH84001 are in the range $\sim$3.8–4.2 Gyr (Figure 3). Thus, Ar-outgassing appears to have occurred after the parental rock solidified. Turner *et al.* (1997) derived an ^{39}Ar-^{40}Ar age of 3.92 $\pm$ 0.10 Gyr, whereas Ilg *et al.* (1997) reported an older age of 4.07 $\pm$ 0.04 Gyr. Martian Ar components in ALH84001 are difficult to characterize, and correcting for the uncertainty in trapped Martian ^{40}Ar allows ages in the broader interval 4.10 $\pm$ 0.20 Gyr (Bogard and Garrison, 1999). The time when ALH84001 was outgassed corresponds to the hypothesized period of the "terminal cataclysm" on the moon and HED parent body (Bogard, 1995). Perhaps Mars also experienced such a cataclysmic bombardment, but this remains a tentative conclusion, based on a single sample.

The Rb-Sr age of 3.84$\pm$0.05 Gyr for ALH84001 (Wadhwa and Lugmair, 1996) is significantly younger than the Sm-Nd age and also than the Rb-Sr age of Nyquist *et al.* (1995). Apparently, the secondary reheating event that reset the ^{39}Ar-^{40}Ar age affected different portions of the rock to different degrees. The rock contains crushed zones (Treiman, 1995b) that may have been more severely affected by the impact event than were intact orthopyroxenite areas. Furthermore, secondary carbonate mineralization is preferentially found within these crushed zones. Thus, we concur with the interpretation of Wadhwa and Lugmair (1996) that the young Rb-Sr age of $\sim$3.9 Gyr (^{87}Sr decay constant $\lambda_{87} = 1.402 \times 10^{-11}$ yr^{-1}, Minster *et al.*, 1982) represents a time of intense shock and post-shock thermal annealing. This interpretation implies that the Rb-Sr ages of some portions of the rock also were reset by the impact of a large meteoroid on Mars. Both interpretations may apply, as ALH84001 bears evidence of several major meteoroid impacts (*cf.* Treiman, 1998), not surprising for a rock from the Martian highlands, which have been "gardened" by meteoroid impact to a depth of $\sim$1 km (Hartmann *et al.*, 2000).

ALH84001 contains $\sim$1 vol.% of secondary carbonates. The secondary Sr or Nd in these carbonates may have disturbed the isotopic systems. The work of Borg *et al.* (1999), discussed below, shows that the leaching procedure used by Jagoutz *et al.* (1994) to determine the Sm-Nd age of ALH84001 would have dissolved both igneous phosphates and secondary carbonates. Also, the isochron of Nyquist *et al.* (1995), determined by bulk samples plus orthopyroxene, would be subject to variations in the relative proportions of phosphates and carbonates. The REE abundances in the primary phosphates probably are much higher than in the secondary carbonates, however, and it is likely that the presence of carbonates has not significantly affected the Sm-Nd isochrons.

Three attempts to date the carbonates have been reported (Figure 3). Knott *et al.* (1996) reported an age of $\sim$3.6 Gyr by laser-probe ^{39}Ar-^{40}Ar dating. Turner *et al.* (1997), however, interpreted those data as heavily influenced by outgassing from the plagioclase substrate beneath the carbonate grain they analysed. Wadhwa and Lugmair (1996), adopting the model of Treiman (1995b) for formation of the carbonates by replacement of plagioclase, proposed an age of 1.41 ± 0.10 Gyr for the carbonates from a two-point carbonate-plagioclase "isochron". However, the Sr-isotopic composition of plagioclase is variable, making pairing of carbonate and plagioclase for dating ambiguous. In the third investigation of the carbonate age, Borg *et al.* (1999) exploited the compositional zoning of the carbonate minerals in ALH84001 to selectively dissolve phases having different parent/daughter ratios for Rb-Sr, U-Pb, and Sm-Nd dating. Although REE concentrations in the resultant solutions were too low for Nd isotopic analysis, the Rb-Sr and U-Pb isotopic analyses yielded concordant ages of $\sim$3.9–4.0 Gyr, close to those originally obtained by laser probe ^{39}Ar-^{40}Ar dating (Knott *et al.*, 1996). These results, combined with the ^{39}Ar-^{40}Ar studies of ALH84001 silicates, suggest that plagioclase outgassing and carbonate formation were contemporaneous, and possibly even simultaneous. If so, the lower ^{39}Ar-^{40}Ar age reported for "carbonate" by Knott *et al.* (1996) may reflect some ^{40}Ar loss from this low-temperature secondary mineral phase.

Differences in interpretation of the radiometric age data for the carbonates may be related to the fact that there are several types, and possibly two or more generations, of carbonates present in ALH84001. Mittlefehldt (1994) identified two generations, "early" (pre-shock) carbonates, and "late" (post-shock) carbonates. Treiman (1995b) suggested the carbonates formed via replacement of plagioclase, a suggestion that strongly influenced the Rb-Sr study of Wadhwa and Lugmair (1996), as well as interpretation of the ^{39}Ar-^{40}Ar study of Knott *et al.* (1996). Gleason *et al.* (1997) and Kring *et al.* (1998) also favored carbonate formation via dissolution-replacement reactions between CO_2-charged fluids and maskelynite. They present as evidence carbonates filling small pockets in pyroxene previously occupied by maskelynite, as seen in photomicrographs of a thin section of the meteorite (Gleason *et al.*, 1997; Figure 4). Kring *et al.* (1998) argue from an electron microprobe study of K and Ca in six different complexly zoned carbonate patches in a single thin section that the laser probe ^{39}Ar-^{40}Ar study of Knott *et al.* (1996), as reported by Turner *et al.* (1997), does not give the age of the carbonates. They reached this conclusion because most of the data for which carbonate was identified as the target in the study of Turner *et al.* (1997) showed the presence of more K than could be accounted for by carbonates alone in the electron probe study. However, carbonates $\sim$100 μm in diameter also are easily visible with a binocular microscope along fractured surfaces of macroscopic pieces of the meteorite. There usually is no visible association with comparatively rare maskelynite, although sometimes such an association does exist. (The proportions of maskelynite and carbonate are subequal at $\sim$1%). These latter carbonates are of the globular variety, a photomicrograph of which was shown by McKay *et al.* (1996). These fracture-filling carbonates are described very completely in the paper by Scott *et al.* (1998), and were the intended objects of the investigation by Borg *et al.* (1999) of carbonate fragments picked from ~1 g of the meteorite.

It seems probable that some of the confusion concerning interpretation of the carbonate ages stems from occasionally inappropriate application of observations made on limited samples of the meteorite. Here, we follow most closely the discussion of Scott *et al.* (1998), who examined nine polished thin sections of the meteorite. Quoting: "Carbonates in ALH84001 occur in three distinct locations: in pyroxene fractures, in crushed zones (also called granular bands; Treiman, 1995b), and as massive grains and globules on pyroxene grain boundaries (e.g., Mittlefehldt, 1994; Treiman, 1995b) .... Carbonates in pyroxene fractures can be divided conveniently into three types according to their shape and nature of the fractures in which they formed: disks, dike-shaped veins, and irregularly shaped grains." These authors (and others) document that all types of carbonates are similarly compositionally zoned. The carbonates nucleated with Ca-rich cores and became richer in Fe and then Mg as they grew outward. Last to form were the magnesite rims. It is this zonation that the experiment of Borg *et al.* (1999) was designed to exploit: Enrichment of Sr and Pb over Rb and U in the Ca-rich cores, leaving enhanced Rb/Sr and U/Pb ratios in the last-formed magnesites. Similar zoning profiles in all types of carbonates imply that carbonate formation took place as a single event. Again quoting Scott *et al.*: "...there is much evidence that carbonates in fractures did not form by replacement of plagioclase glass."

The simplest interpretation of the observations appears to be:

1. Some carbonates formed by replacement reactions with crystalline plagioclase. A probable example is seen in Figure 4d of Gleason *et al.* (1997). If, as argued by the authors and Kring *et al.* (1998), replacement was of maskelynite, a prior shock event is required.

2. Not all carbonates formed by replacement reactions (Scott *et al.*, 1998). The majority of carbonates probably formed without need of plagioclase or maskelynite, but if some were present, reactions could occur. CO_2-enriched aqueous fluids apparently circulated through the rock, implying the prior existence of a fracture network. Thus, the rock had been brecciated prior to that time, probably by excavation from great depth to a surface or near-surface location. The compositional zoning of the carbonates was established at that time, and dated at ~4.0 Gyr ago by Borg *et al.* (1999).

3. A second shock fractured some carbonates and formed maskelynite and plagioclase glass, some of which can now be found in fractures. This shock also opened up some pre- existing fractures in which carbonates already had formed (Scott *et al.*, 1998, Figure 3a), and resealed others. This last shock is most likely the ejection event. The other Martian meteorites invariably show shock levels of 15–45 GPa, implying that such shock levels are required for their ejection from the planet. Thus, this second shock happened ~15 Myr ago, the ejection age of ALH84001, as discussed in a later section.

Treiman (1998) suggested a more complex scenario involving 4 "compositional", 6 to 8 "deformational", and 4 "impact" events. A critical difference to the scenario above is that Treiman's last impact event occurs without major shock metamorphism. Attempts to refine the inferred history of ALH84001, including the radiometric age of the carbonates, are likely to continue, but the concordant Rb-Sr and U-Pb ages of Borg *et al.* (1999) seem presently to be preferred.

The mechanism of carbonate formation has been debated in the context of "impact metasomatism" (Harvey and McSween, 1996) and "playa lake" models. Playa lake models have gained favor, seeming to be more consistent with various types of data (*cf.*, Warren, 1998). Also, in an experimental study, Golden *et al.* (2000a, 2000b) were able to reproduce the carbonate zonation profiles in ALH84001. They used a multi-step, sequential aqueous precipitation from fluids of changing composition, followed by a final reheating to 470°C. Whether this process mimics what might happen on Mars, perhaps in a Martian playa lake, has not been addressed in detail. However, one can easily envision a scenario in which Ar-outgassing from ALH84001 accompanied a crater-forming event that left ALH84001 either as part of the crater ejecta blanket, or, in the playa lake model, at the bottom of a crater. Scott *et al.* (1998) have suggested that ALH84001 may have been located beneath the central region of a large impact crater or basin that formed ~4 Gyr ago. The playa lake model would definitely be favored over impact metasomatism if Ar outgassing were earlier than carbonate formation, occuring at most 4.1 Gyr ago. The carbonate age data would then be consistent with precipitation from a crater lake filled slightly later by surface runoff into the crater. Alternatively, formation of a crater lake might be triggered by formation of the crater itself, filled by melted groundwater released by the heat of the impact. Newsom *et al.* (1996) have argued that formation of large (> 65 km diameter) impact craters on Mars may have been accompanied by the creation of ice-covered impact crater lakes, which would not freeze totally over a lifetime of $\sim 10^4$ years. Supply of water to them from deep aquifirs might provide a connection to possible life residing in the aquifirs (Boston *et al.*, 1992). Thus, although the suggestion by McKay *et al.* (1996) that certain worm-like morphological features in ALH84001 might be relics of Martian life has proven controversial, that interpretation is consistent with the apparent age of the carbonates, the time of formation of equivalent lifeforms on earth, and possible access to potential subsurface habitats via crater formation. The crater lake scenario remains speculative, but is consistent with suggested modes of carbonate formation, and is made more plausible by the apparent near coincidence of the outgassing and carbonate formation ages. The apparent presence of liquid water at later times, perhaps even up to the present day (Malin and Edgett, 2000), appears to be permissive of later carbonate formation also. Currently, however, there appears to be little rationale to consider alternate scenarios.

Although ALH84001 is a sample of the Noachian crust, it should not be considered a "typical" Martian crustal rock. It is an orthopyroxenite cumulate with much higher MgO and FeO, and lower Al_2O_3, SiO_2, and K_2O than typical Martian crustal rocks such as the Pathfinder "sulfur free rock" (Rieder *et al.*, 1997; Bell *et al.*, 2000). Although rocks at the Pathfinder site are thought to be derivative from the southern Martian highlands, the Al_2O_3 content of the "sulfur free rock" is much lower than that of lunar highland soils and even lower than in lunar mare soils. On average, the Martian crust appears to be rather Al_2O_3-poor. McSween and Keil (2000) conclude that if the global Martian dust is representative of the Martian upper crustal composition, the planet's surface geology is dominated by contributions from evaporitic salts and a basaltic protolith chemically similar to basaltic shergottites. Thus, fractionated MgO- and FeO-rich magmas may have been common within the crust, and orthopyroxenite ALH84001 may have formed as a mafic cumulate in a layered igneous province. The mode and timing of its formation may have been analogous to those of rocks of the lunar "Mg-suite", except that the composition of the surrounding crust was basaltic rather than anorthositic.

Finally, viewed from the context of the lunar samples, it seems fortuitous that even one out of 16 dated Martian rocks would have preserved such an old age. Only a few lunar crustal rocks have been reliably dated to have ages in the range of $\sim$4.3–4.5 Gyr. Furthermore, the dated lunar ferroan anorthosites were small clasts extracted from lunar highland breccias, in which they often were surrounded by impact melt glass. The rarity of unadulterated "original" crustal material among the lunar highlands rocks poses some questions relative to ALH84001 and the Martian crust: Did Mars and Moon both experience the same heavy meteor bombardment early in their history? Did Mars experience a "terminal cataclysm" of bombardment? Was ALH84001 excavated from a considerable depth where it was shielded from bombardment? If 4.5 Gyr-old crustal rocks are fairly common on Mars, but not on the Moon, it may imply that the moon had a much heavier terminal bombardment that physically destroyed its crust. ALH84001 has given us a few clues, but answers to these questions await combination of spacecraft orbital imaging and absolute dating of samples returned from heavily cratered areas on Mars.

4.2. THE NAKHLITES AND CHASSIGNY: AMAZONIAN OR HESPERIAN CUMULATES?

So far, we have no Martian meteorites that are clearly Hesperian in age; i.e., $\sim$3.5–1.8 Gyr old, according to the Hartmann-Tanaka (HT) cratering model, or even older according to Hartmann and Neukum (2001). The $\sim$1.3 Gyr radiometric ages of four cumulate rocks, three clinopyroxenite nakhlites (Nakhla, Lafayette, and Governador Valadares), and the dunite Chassigny (Figure 2) are Early Amazonian in the HT model, but close to the lower age limit of the Hesperian. However, they are squarely in the Middle Amazonian in the Neukum-Wise (NW) model. Close agreement of the ages of these four meteorites by four dating techniques, Rb-Sr, Sm-Nd, ^{39}Ar-^{40}Ar, and U-Pb, apparently unambiguously define their crystallization ages at $\sim$1.3 Gyr (Figure 4). Type localities for the Early and Middle Amazonian are Amazonis Planitia (EA) and Acidalia Planitia (MA), respectively (Tanaka, 1986).

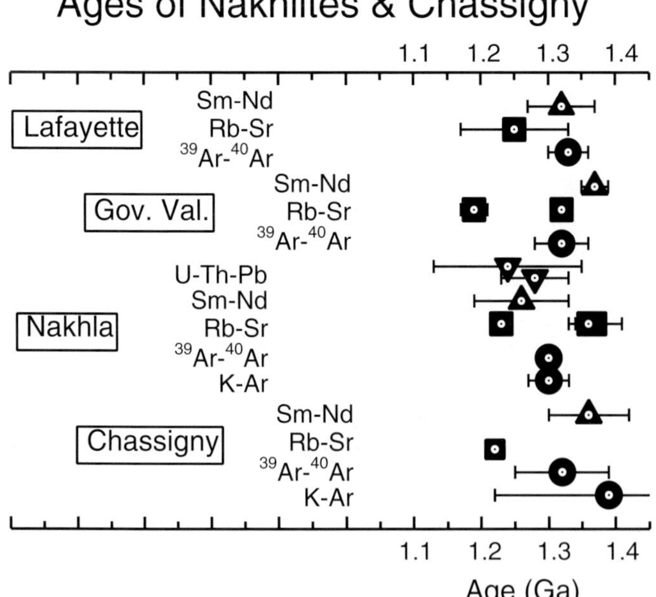

Figure 4. Radiometric ages of the Nakhlites and Chassigny. Nearly all the ages are compatible with an average age of ~1.3 Gyr.

Although the radiometric ages of the nakhlites are in general well-defined, they do show some isotopic disturbances, first evident in the Sr-isotopic heterogeneity noted by Papanastassiou and Wasserburg (1974). Some of the disturbance may be attributed to iddingsite, an apparent Martian weathering product formed during alteration by water. Some of the isotopic heterogeneity may perhaps be magmatic in origin. Attempts to date the formation time of iddingsite in the Lafayette nakhlite by the K-Ar and Rb-Sr techniques have yielded apparent ages of 600–700 Myr (Swindle *et al.*, 1999; Shih *et al.*, 1998), suggesting liquid water activity on Mars ~ 650 Myr ago. Malin and Edgett (2000) have cited a number of lines of evidence, such as the observations of gullies within the walls of a small number of impact craters, as indicating groundwater seepage and surface run-off on even younger Martian landforms. Examples are shown in this book (Hartmann, 2001).

4.3. THE SHERGOTTITES: LATE AMAZONIAN VOLCANISM

The preferred radiometric ages of basaltic and lherzolitic shergottites lie in the range ~165–475 Myr. Individual ages from the literature and unpublished data from the JSC lab are given in Table III and are shown with error limits in Figure 5. Confusion about the ages of the shergottites is slowly being dispelled. The initial apparent age discordance of the shergottites has been shown to arise from three sources: a) The presence of trapped ^{40}Ar from the Martian atmosphere and mantle, and possibly excess, inherited, radiogenic ^{40}Ar as well, in sufficient quan-

Age of Shergottites

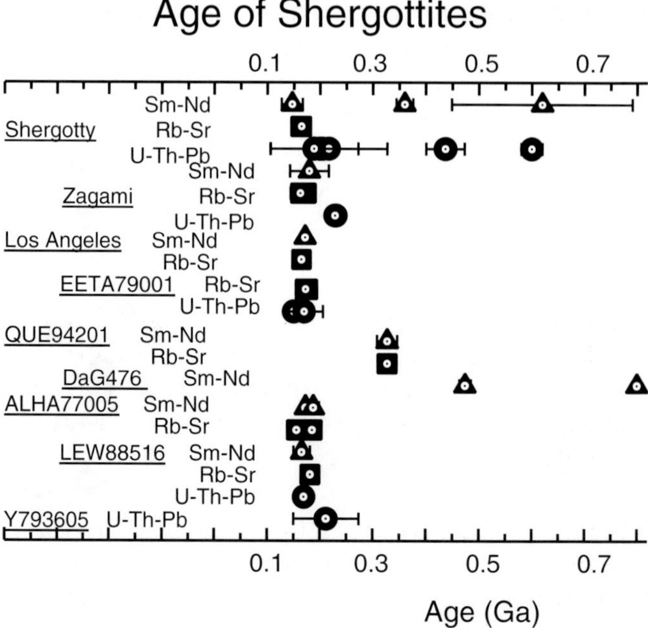

Figure 5. Radiometric ages of the shergottites. The Shergotty ages >0.3 Gyr seem to be in error due to unexplained analytical effects. Concordant Rb-Sr and Sm-Nd ages of ~327 Myr indicate that age is the true crystallization of QUE94201. The ubiquitous presence of terrestrial contamination has prevented determination of a Rb-Sr age for DaG476. The ubiquity of such contamination causes us to favor the 475 ± 11 Myr Sm-Nd age of Borg *et al.* (2000) to the older ~800 Myr age reported by Jagoutz *et al.* (1999) and Jagoutz and Jotter (1999) for this meteorite. See Table II for references.

tities to significantly affect measured ^{39}Ar-^{40}Ar ages. b) Analytical difficulties accompanying isotopic analyses of young samples with low abundances of the trace elements being analysed. c) The apparent presence of isotopic heterogeneity probably preserved in the basalts in the cores of pyroxene and olivine phenocrysts. These difficulties are being worked out, and preferred ages can be given with a degree of confidence (Table III).

It is also worth noting that a major contributor to initial confusion about the ages of the shergottites, and indeed all the SNC meteorites, was simply an early reluctance of meteoriticists to accept radiometric ages significantly less than ~4.5 Gyr as giving the time of igneous crystallization of any meteorite. The first radiometric age for a meteorite now considered to be of Martian origin was the K-Ar age of 580 ±50 Myr (recalculated with the decay parameters by Steiger and Jäger, 1977) determined for Shergotty by Geiss and Hess (1958) in an early study of the K-Ar ages of stony meteorites. Geiss and Hess (1958) considered their age to be "too young". They excluded the possibilities of K contamination or heterogeneity in the K content of the meteorite. Not knowing of Shergotty's Martian origin and, hence, assuming that Ar extracted from the meteorite could only consist of the radiogenic, spallogenic, and (terrestrial) atmospheric components, they considered Ar loss as being unlikely to explain its young age. A loss of $\sim$95% of the Ar would be required, and they concluded that diffusive loss of that magnitude would be unlikely either via solar heating at the earth's orbit or beyond, or via frictional heating during passage through earth's atmosphere. We now know that correction for the "atmospheric" component using ^{40}Ar/^{36}Ar $\sim$2000 for the Martian atmosphere, instead of 296 for the terrestrial atmosphere, gives an even younger age, more in agreement with currently accepted values for the Martian meteorites. Of course, the possible presence of a second atmospheric component, and the possibility that these young rocks might contain a mantle Ar component as well, introduces ambiguity in correcting for non-radiogenic ^{40}Ar. Later work showed that K-Ar ages <1 Gyr were relatively common among shocked chondritic meteorites and are produced by impact heating on meteorite parent bodies.

The question of the age of Shergotty was revisited by Bogard *et al.* (1979), who redetermined the K-Ar age using the ^{39}Ar-^{40}Ar technique, and also by Nyquist *et al.* (1979b), who determined the Rb-Sr age as well. The ^{39}Ar-^{40}Ar ages of stepped extractions of Ar from a whole rock were variable, but similar to the value of Geiss and Hess (1958). The ^{39}Ar-^{40}Ar ages of stepped extractions from a plagioclase separate were approximately constant and gave a good age plateau at 254$\pm$10 Myr. The Rb-Sr age was younger still at 165$\pm$11 Myr. Nyquist *et al.* (1979b) interpreted the $\sim$165 Myr Rb-Sr age of Shergotty as due to metamorphic resetting because it was the lower of the Rb-Sr and ^{39}Ar-^{40}Ar ages. Subsequent work showed that not only was the Rb-Sr age younger than the ^{39}Ar-^{40}Ar age of 254 $\pm$ 10 Myr, but also younger than the still older Sm-Nd and U-Pb ages of Shih *et al.* (1982), Jagoutz and Wänke (1986), and Chen and Wasserburg (1986) (Table III; Figure 5). These latter measurements mostly have not been repeated, but it now seems likely that they were influenced by sample contamination, or other sources of analytical errors.

A recent *in situ* isotopic analysis by ion probe of U-Th-Pb in phosphates in Shergotty (Sano *et al.*, 2000) yields the time of closure of the U-Pb system in Shergotty phosphates as 204$\pm$68 Myr ago. Whether this result excludes those Sm-Nd and U-Pb "ages" in excess of 300 Myr in Figure 5 as real crystallization ages depends on the actual mode of petrogenesis of these rocks. Jagoutz and Wänke (1986) preferred the older Sm-Nd age of 360 $\pm$ 16 Myr obtained from a pyroxene-leachate isochron to the younger Sm-Nd age of 147 $\pm$ 20 Myr they obtained from an isochron including the whole rock data. However, the lower value of $\sim$147 Myr normally would be favored by the isotopic systematics, since data for the bulk ("whole rock") sample must lie on the isochron for closed-system evolution of the Sm-Nd system. Jagoutz and Wänke (1986) argued for an open Sm-Nd system, and that the phosphates crystallized from a metasomatic contaminant infiltrating a pyroxene cumulate. This possibility continues to be allowed by the results of Sano *et al.* (2000). Thus, the interpretation of Jagoutz and Wänke (1986) probably cannot be totally excluded, but it is nevertheless weakened by several observations. First, it requires Sr in plagioclase, as well as Nd in phosphates, to be derived from the metasomatic contaminant. Second, the Sm-Nd data for Shergotty also can be explained by recent terrestrial contamination, which may have been a greater problem than previously recognized. Third, Bogard and Garrison (1999) decomposed the Ar released from an irradiated sample of Shergotty into a radiogenic component that would be produced in 165 Myr of decay, and a trapped component with an $^{40}Ar/^{36}Ar$ ratio of 1780. This $^{40}Ar/^{36}Ar$ ratio is within the range of values believed representative of the Martian atmosphere, supporting the validity of this approach. Because most of the K and radiogenic $^{40}Ar^*$ is contained in plagioclase, this result would require that most of the K, as well as Nd and Sr, be attributed to the hypothesized metasomatic fluid. These more recent observations substantially support the earlier arguments of Jones (1986) against metamorphic resetting of the Rb-Sr ages of Shergotty and other shergottites by either thermal or hydrothermal events. Finally, a Sm-Nd age of 163 ± 7 Myr was found for Zagami, a "twin" of Shergotty (Nyquist et al., 1995), concordant with three determinations of the Rb-Sr age averaging 177 ± 3 Myr. These arguments suggest that an age of crystallization as old as ~360 Myr for Shergotty is unlikely. Our preferred age of 165 ± 4 Myr is the weighted average of the two Rb-Sr ages (Nyquist et al., 1979b; Jagoutz and Wänke, 1986) converted to a ^{87}Sr decay constant $\lambda_{87} = 0.01402$ Gyr^{-1}, as used throughout this paper (Minster et al., 1982).

That the young radiometric ages of the SNCs are indeed crystallization ages seems incontrovertible in light of the recent data summarized in Tables II and III. Three other basaltic shergottites, Zagami, EET79001, and Los Angeles have similar preferred ages as Shergotty; i.e., 177 ± 3 Myr, 173 ± 3 Myr, and 170 ± 7 Myr, respectively. For Zagami, the value is based on three Rb-Sr ages and one Sm-Nd age (Table III). For EET79001, the preferred age is based on three Rb-Sr ages, one Sm-Nd age, and two U-Th-Pb ages. In this case the weighted mean is greatly influenced by a precise Rb-Sr age of 173 ± 3 Myr recently determined in the JSC lab (Nyquist et al., 2001). For Los Angeles, the age is the weighted average of Rb-Sr and Sm-Nd ages. Although the preferred age for Shergotty appears to be slightly younger, its resolution from the other ages is problematic. The calculated uncertainty of ±4 Myr for the Shergotty age may be unrealistically low in light of apparent cm-scale isotopic heterogeneity in Zagami (Nyquist et al., 1995).

Two of the basaltic shergottites, QUE94201 and DaG 476, have older crystallization ages. Concordant Rb-Sr and Sm-Nd ages of 327 ± 12 and 327 ± 19 Myr were obtained for QUE94201 by Borg et al. (1997). QUE94201 contained easily-leachable components which may have formed as Martian weathering products. The presence of these phases and also of impact-produced glass veining throughout the rock led to significant complications of the isotopic systematics, but, nevertheless, the age appears to be robustly determined.

The age of DaG476, another basaltic shergottite with compositional similarities to QUE94201, is currently debated. Jagoutz et al. (1999), and Jagoutz and Jotter (2000) have presented Sm-Nd data leading to an apparent age of ~800 Myr. Borg et al. (2000) found instead an age of 474 ± 11 Myr. This shergottite is heavily weathered, making the Rb-Sr data useless for age determination. Terrestrial contamination accompanies the weathering throughout the meteorite and is evident in elevated K and LREE abundances in some mineral phases (Crozaz and Wadhwa, 1999). Because the effect of terrestrial contamination would be to displace the apparent age to higher values, we favor the lower value of $\sim$474 Myr as most likely to be the true crystallization age.

Concordant Rb-Sr and Sm-Nd ages have been determined for two of the lherzolitic shergottites, ALH77005, and LEW88516. Our preferred ages for these two meteorites are 178 $\pm$ 6 Myr and 179 $\pm$ 6 Myr, respectively (Table III). An U-Pb age of $\sim$170 Myr for LEW88516 (Chen and Wasserburg, 1993) is also concordant with these values, as is an U-Pb age of 212 $\pm$ 62 Myr for the third lherzolitic shergottite, Yamato 793605 (Misawa *et al.*, 1997). These crystallization ages are the same, within uncertainties of a few percent, as the crystallization ages of several of the basaltic shergottites. Historically, these shergottites have been referred to as having crystallization ages of $\sim$180 Myr as suggested by Jones (1986), based on the 178 $\pm$ 3 Myr Rb-Sr age reported for Zagami by Shih *et al.* (1982).

The interpretation of the radiometric data for shergottites has been controversial in part because of the complexities that often exist both within and between radiometric systems. One of the most puzzling problems has been the observation that the apparent ^{39}Ar-^{40}Ar ages of the shergottites are systematically older than the Rb-Sr ages, as we have already mentioned for Shergotty. Although ^{39}Ar-^{40}Ar ages in terrestrial basalts can sometimes be "too old" because of inherited radiogenic ^{40}Ar, such situations are rare among meteorites. Bogard and Johnson (1983) first found that melt glass in the EET79001 shergottite contained trapped Martian atmospheric gases, including substantial amounts of ^{40}Ar, which had been shock-implanted by impacts on the Martian surface. Martian atmospheric gases have also been found in Zagami, ALH77005, and Y793605. In addition, an elementally fractionated component of the Martian atmosphere has been measured in some samples of the nakhlites and ALH84001 (see references in Bogard and Garrison, 1998). The ^{40}Ar/^{36}Ar ratio of the Martian atmosphere has a relatively high value of $\sim$1800 (Bogard and Garrison, 1999), which makes it difficult to correct for atmospheric ^{40}Ar using ^{36}Ar. Further, it is now recognized that some Martian meteorites contain a trapped volatile component from the Martian interior, which appreciably differs in elemental and isotopic composition from the atmospheric component (Ott, 1988; Bogard and Garrison, 1998; Marti and Matthew, 2000). The ^{40}Ar/^{36}Ar ratio of this interior component is not known and is probably variable. In many phases of Martian meteorites the Martian atmospheric and interior volatile components occur as mixtures in variable proportions.

Because of the presence of multiple Ar components, shergottites do not generally give reliable ^{39}Ar-^{40}Ar ages (Bogard and Garrison, 1999), as can be seen from the ^{39}Ar-^{40}Ar ages given for them in Table III. Most of the listed ^{39}Ar-^{40}Ar ages are from the compilation of Bogard and Garrison (1999), and are "total ^{40}Ar ages" for stepped ^{39}Ar-^{40}Ar analyses. Those ^{39}Ar-^{40}Ar ages that are closest to the preferred crystallization ages of the samples are for plagioclase separates of basaltic shergottites with relatively high modal abundance of plagioclase, and thus relatively high K-contents. Relatively good ^{39}Ar-^{40}Ar plateau ages of $\sim$254 Myr and $\sim$242 Myr were determined for Shergotty and Zagami feldspar. Also, as already mentioned, it was possible to decompose the $\sim$387 Myr total ^{40}Ar age for a bulk sample of Shergotty into trapped and radiogenic ^{40}Ar components. Data for Zagami feldspar are consistent with a similar decomposition of Ar components for a crystallization age of 180 Myr. However, Bogard and Garrison (1999) were unable to reliably determine the amount of radiogenic ^{40}Ar for other shergottite samples because of the multiplicity of Ar components that might have been present in the analyses. Terribilini *et al.* (1998) determined a conventional ^{40}K-^{40}Ar age of 196 $\pm$ 40 Myr from an isochron plot of ^{40}Ar/^{36}Ar versus K/^{36}Ar in a bulk sample and separated minerals of Shergotty. The corresponding ratio for trapped ^{40}Ar/^{36}Ar was found to be $\sim$1100, however, a value significantly lower than ^{40}Ar/^{36}Ar $\sim$1800 in the Martian atmosphere, showing that the shergottites contain both Martian atmospheric Ar and a mantle Ar component of significantly lower ^{40}Ar/^{36}Ar ratio. These two components mix in variable proportions in Martian meteorites, making it necessary to independently determine the ^{40}Ar/^{36}Ar ratio for each sample.

Subtle isotopic inconsistencies are present in the other isotopic systems as well. Some of these are manifest in the U-Pb ages summarized in Table III. Blichert-Toft *et al.* (1999) note also that Lu-Hf data for bulk shergottites do not show isochron relationships. Rb-Sr isochrons of different samples of the shergottites can give identical ages for different initial ^{87}Sr/^{86}Sr ratios (e.g., Nyquist *et al.*, 1995). Some isotopic inconsistencies for nakhlites probably are due to the presence of Martian weathering products. The differences in initial ^{87}Sr/^{86}Sr ratios among subsamples of shergottites and nakhlites require unusual petrogenetic processes.

As already mentioned, the Rb-Sr ages were originally interpreted as dating the time of shock metamorphism accompanying their ejection from Mars. An attraction of that explanation was that it accounted for the simultaneity of ages near 180 Myr. Nyquist *et al.* (1979b) argued that subsolidus *isotopic* equilibration might be achieved by heating at low temperatures ($\sim$300 $-$ 400 °C) for long times ($\sim$10^4 yr) while allowing elemental zoning to be preserved. Jones (1986) criticized that interpretation on the grounds that preservation of elemental zoning patterns in major mineral phases of the shergottites, in spite of shock-induced transformation of plagioclase to maskelynite, precluded identification of the Rb-Sr ages with the time of shock metamorphism.

Because of the discordant ages obtained for Shergotty and other shergottites by the various methods, Shih *et al.* (1982) sought an approach that would "see through" secondary events, if such were the explanation of the young, $\sim$180 Myr ages. They noted that the whole rock Sm-Nd data for the basaltic shergottites, Shergotty and Zagami, combined with that of the lherzolitic shergottite, ALH77005, defined an apparent "isochron" of slope corresponding to an age of $\sim$1.34 Gyr, in remarkable agreement with the ages of the nakhlites. This coincidence was

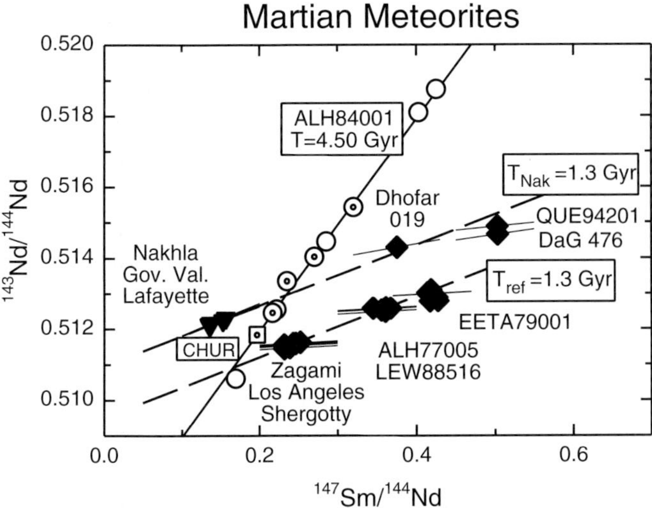

Figure 6. Whole rock Sm-Nd data for Martian meteorites. A 4.50 Gyr reference isochron is shown for bulk samples (*dotted circles*) and mineral separates (*open circles*) for ALH84001 orthopyroxenite. Dhofar019: preliminary data, Borg *et al.* (2001a). DaG476: constructed from the mineral isochron (Borg *et al.*, 2000) and the bulk ^{147}Sm/^{144}Nd ratio (Jagoutz *et al.*, 1999). The other data are from the literature. A reference 1.3 Gyr isochron (T_{ref}) has been drawn through the data for Shergotty, Zagami, and Los Angeles. For reasons that are unclear, the data of the shergottites Dhofar019, QUE94201, and DaG476 appear to lie along T_{nak}, a nakhlite isochron for the average age of the nakhlites Nakhla, Governador Valadares, and Lafayette. The linear alignment along T_{ref} from the traditional shergottites to EETA79001 has been interpreted as a mixing line between a mantle component to the right of EETA79001 and a "crustal" component to the left of the intersection of this line with the $\sim$4.5 Gyr isochron. The isotopic data for a Chondritic Uniform Reservoir (CHUR) fall on this isochron (*squares*). Short lines through the individual data points show the slopes of mineral isochrons determined for these rocks. Rocks satisfying a simple two-stage isotopic evolution history would be derived from Martian mantle source regions having ^{147}Sm/^{144}Nd ratios determined by the intersection of the mineral isochrons with the primary $\sim$4.5 Gyr mantle differentiation isochron. Such an intersection would occur at ^{147}Sm/^{144}Nd $\sim$0.3 for QUE94201, DaG476, and Dhofar 019. QUE94201, DaG476, and the nakhlites also have measured excesses of ^{142}Nd from decay of 103 Myr ^{146}Sm (Harper *et al.*, 1995; Borg *et al.*, 1997; Jagoutz and Jotter, 2000), showing that Martian differentiation occurred very early. The likely addition of a crustal component to the parental magmas of the younger, $\sim$175 Myr old shergottites would have displaced the apparent mantle ^{147}Sm/^{144}Nd ratios to the lower left along the primary mantle isochron.

reinforced by later data for basaltic shergottite EET79001. It seemed to suggest that the shergottites and the nakhlites might be related, if only indirectly, via mantle processes. It also suggested that the ~1.3 Gyr age might have significance for the shergottites. However, later interpretations have favored the view that this "isochron" is a mixing line, representing mixing among "crustal" and "mantle" end-members as suggested by Jones (1989) and Longhi (1991).

Figure 6 shows the currently available whole rock Sm-Nd data for Martian meteorites. Whole rock analyses and mineral separates for ALH84001 define an

~4.5 Gyr reference isochron, used here as a reference isochron giving the approximate age of the planet. The "traditional" basaltic shergottites (Shergotty, Zagami, and Los Angeles), plus the lherzolitic shergottites (ALH77005 and LEW88516) and basaltic shergottite EET79001 plot along an ~1.3 Gyr reference isochron similar to the one originally defined for Shergotty, Zagami, and ALH77005 by Shih *et al.* (1982). The line in the figure is constrained to pass through the data for the traditional basaltic shergottites. According to the isotopic mixing models of Jones (1989) and Longhi (1991), the linear alignment of data is due to mixing of more radiogenic Nd (higher ^{147}Sm/^{144}Nd and ^{143}Nd/^{144}Nd ratios) from the Martian mantle with less radiogenic Nd (lower ^{147}Sm/^{144}Nd and ^{143}Nd/^{144}Nd ratios) from the Martian crust. End member compositions are to be found along the line to the right or left of the data array. This interpretation attributes no time significance to the linear data array, and suggests that the correspondence of the slope of the line to an apparent Sm-Nd age of ~1.3 Gyr is a coincidence. This approach has much to recommend it. Norman (1999) found self-consistent results for the Nd isotopic compositions and REE abundances in Shergotty as a mixture between a LREE-depleted mantle-derived magma similar in composition to EET79001A and a LREE-enriched "crustal" component with >10 ppm Nd. The success of such models depends in large part on identification of at least one of the end member components. A number of meteorites have been suggested as representative of mantle-derived Martian magmas, including the nakhlites (Jones, 1989; Longhi, 1991), QUE94201 (Borg *et al.*, 1997), and EET79001 (Norman, 1999).

Recent data for newly found Martian meteorites have reopened the issue of whether Sm-Nd whole rock "isochrons" may have time significance, however. Jagoutz and Jotter (2000) note that leachates (~phosphates) from Nakhla, DaG476, and QUE94201 lie along a line of slope corresponding to an isochron age of 1.2 Gyr, and that these samples all have significant excess ^{142}Nd anomalies, showing that they all were derived from an early-formed mantle source with significant depletion in LREE relative to HREE (H for *H*eavy; *cf.* Harper *et al.*, 1995). Figure 6 shows that whole rock data for the nakhlites (Shih *et al.*, 1998, 1999), QUE94201 (Borg *et al.*, 1997), and newly found Dhofar019 (Borg *et al.*, 2001a) all plot close to the extension of the 1.3 Gyr nakhlite isochron. Because whole rock data for DaG476 are unreliable due to terrestrial contamination, we show a value calculated to lie on the mineral isochron of Borg *et al.* (2000) for the ^{147}Sm/^{144}Nd ratio measured by Jagoutz *et al.* (1999). If the ~1.3 Gyr alignments have any time significance, it may be due to "mantle" end members generated by successive episodes of partial melting and magma extraction over an extended interval from before ~1.3 Gyr ago until the crystallization ages of the basalts. Melt extraction on a massive scale may account for the large depletion of LREE abundances in QUE94201 and DaG476, and may have left surface evidence for large expanses of volcanic flows ~1.3 Gyr ago. Models for the Nd and Sr isotopic evolution of QUE94201 (Borg *et al.*, 1997), however, require successive melt extractions from the source to be restricted to times near the basalt crystallization age.

In contrast to the Sm-Nd data, the whole rock Rb-Sr age for most Martian meteorites is $\sim$4.5 Gyr (Shih *et al.*, 1982). Thus, the Rb-Sr data require planet-wide differentiation at $\sim$4.5 Gyr, accompanied by establishment of mantle reservoirs that remained essentially closed systems to Rb/Sr fractionation thereafter. That major fractionation in Sm-Nd could occur $<\sim$1.3 Gyr ago without some additional fractionation of Rb/Sr is a strong constraint on isotopic models for crust/mantle differentiation. The models of Jones (1989) and Longhi (1991) for the petrogenesis of SNC meteorites invoked crustal assimilation following partial melting of the Martian mantle for nakhlite genesis at 1.3 Gyr ago and shergottite genesis at 180 Myr ago. Borg *et al.* (1997) presented a related multi-stage model of mantle melting and crustal assimilation for petrogenesis of QUE94201 at 330 Myr ago.

5. Ejection Ages and Events

Assuming a meteorite is ejected as a small object from the Martian surface and comes directly to Earth without secondary breakup in space, its "ejection age", i.e. the time since its ejection, equals the sum of its cosmic ray exposure (CRE) age and its terrestrial age (Eugster *et al.*, 1997b). Table IV gives all   presently available CRE and terrestrial ages of Martian meteorites, and its notes explain how these ages were calculated. The CRE ages based on the stable noble gas isotopes ^{3}He, ^{21}Ne, and ^{38}Ar and appropriate production rates, T_s, may be subject to several sources of systematic bias, such as diffusive loss of spallogenic noble gases, which is possibly related to particular orbital parameters, errors in correction of production rates for variations in chemical composition, variations in shielding from cosmic ray particles, and contributions of solar particles to ^{21}Ne production. However, the ^{81}Kr-Kr ages, T_{81}, and, to a lesser extent, the ^{10}Be-^{21}Ne ages, T_{10}, are less subject to all these sources of bias (Marti, 1967; Eugster *et al.*, 1967). Terribilini *et al.* (2000) obtained T_{81} ages for 7 Martian meteorites. Their averages are 3% lower than the average T_s ages. The T_{10} ages, calculated for 5 Martian meteorites by several authors, are on an average also 3% lower than the T_s ages. It appears that these T_s ages have little or none of the biases mentioned above.

Figure 7a plots peak shock pressures from Table I versus ejection ages from Table IV. Both of these parameters are expected to be related to ejection of the meteorites from Mars. Most of the data, with exception of that for the nakhlites, plot within a range of shock pressures from $\sim$30 GPa to $\sim$80 GPa corresponding, respectively, to the pressures at which plagioclase is converted to maskelynite (P_{mask}), and at which shock melting occurs (P_{melt}). Melosh (1985) notes that in the spallation model for meteorite ejection, the ejection velocity is proportional to the pressure *gradient*, not the pressure. The implication of Figure 7 is that just prior to ejection, the shock pressure due to a compressive pulse at the shergottite location had reached nearly the maximum sustainable without shock melting. Furthermore, in the context of the Melosh model, the pressure gradient due to interfering com-

NYQUIST ET AL.

TABLE IV

Cosmic-ray exposure ages, terrestrial ages, and ejection ages of Martian meteorites in Myr.

	T_3	T_{21}	T_{38}	Meth.	$T_{s,av}$	T_{81}	T_{10}	T_{pref}	T_{terr}	T_{ej}
Shergottites (basalts)										
Dhofar019	(13.7[δ])	18.1[δ]	21.4[δ]	A	19.8 ±2.3			19.8 ±2.3		19.8 ±2.3
Los Angeles	(1.9[a])	3.0[a]	2.8[a]	A	3.02 ±0.30	3.10[b] ±0.70	3.10[β] ±0.20	3.10 ±0.20		3.10 ±0.20
	(1.3[γ])	3.2[γ]	3.1[γ]							
QUE94201	2.17[c]	3.12[c]	2.56[c]	A	2.50 ±0.25	2.10[b] ±0.25	2.6[o] ±0.5	2.42 ±0.20	0.29[o] ±0.05	2.71 ±0.20
	2.20[d]	2.60[d]	2.50[d]	A						
	1.91[e]	*	2.75[e]	A						
	2.0[f]	3.4[f]	2.3[f]	A						
Sayh al Uhaymir		1.5[ε]		A	1.5 ±0.3			1.5 ±0.3		1.5 ±0.3
Shergotty	2.56[g]	3.56[g]	2.54[g]	A	2.91 ±0.25	2.71[b] ±0.30	2.1[p] ±0.2	2.73 ±0.2		2.73 ±0.20
px mask	3.31[g]	3.17[g]	2.71[g]	A						
		3.9[g]	3.00[g]	A						
	2.4[h]	3.2[h]	2.3[h]	B		2.7[q] ±0.9				
	2.7[i]	3.9[i]	2.1[i]	B						
	2.5[f]	3.3[f]	2.2[f]	A						
	2.9[j]			B						
	3.2[k]	4.0[k]	2.7[k]	B						
	2.0[l]	3.4[l]	2.4[l]	B						
Zagami	2.84[c]	3.31[c]	2.56[c]	A	2.85 ±0.45	3.05[b] ±0.30		2.92 ±0.15		2.92 ±0.15
	3.2[k]	3.8[k]	2.6[k]	B						
	3.4[m]	4.3[m]	1.8[m]	B						
	2.1[n]	2.1[n]	2.2[n]	B						
Lherzolites										
ALHA 77005	3.64[t]	2.45[t]	2.50[t]	A	2.98 ±0.45		2.5[p] ±0.3	2.87 ±0.20	0.19[r] ±0.07	3.06 ±0.20
	3.8[m]	2.6[m]	2.9[m]	B			2.8[q] ±0.6		0.19[s] ±0.07	
									0.21[A] ±0.08	
LEW 88516	4.52[c]	4.90[c]	3.92[c]	A	4.01 ±0.40		3.0[w]	3.92 ±0.40	0.021[u] ±0.001	3.94 ±0.40
	4.42[x]	4.22[x]		B					0.021[v] ±0.001	
	3.96[z]	3.0[z]		B						
	4.4[w]	3.8[w]	3.0[w]	B						
Y793605	4.72[g]	3.98[g]	3.13[g]	A	4.67 ±0.50			4.67 ±0.50	0.035[A] ±0.035	4.70 ±0.50
	4.9[f]	5.2[f]	4.6[f]	A						
	5.36[B]	5.46[B]		A						

TABLE IV

(continued)

	T_3	T_{21}	T_{38}	M.	$T_{s,av}$	T_{81}	T_{10}	T_{pref}	T_{terr}	T_{ej}
Shergottite/Lherzolites										
DaG476	1.08^H	1.26^H	1.14^H	A	1.16 ±0.11			1.16 ±0.11	0.085^α ±0.050	1.24 ±0.12
DaG489	1.05^T	1.36^T	1.09^T	A	1.17 ±0.19			1.17 ±0.19	0.085^α ±0.050	1.25 ±0.20
EETA 79001 A	0.68^m	0.52^m	0.74^m	B	0.54 ±0.09		0.78^L ±0.14	0.60 ±0.09	0.012^K ±0.002	0.73 ±0.15
	0.51^M	0.43^M	0.30^M	B			0.5^P ±0.1		0.320^L ±0.170	
	0.61^L	0.54^L	0.45^L	B			0.73^q ±0.19		$<0.06^P$	
		0.45^N		B					(av.0.13) ±0.12	
EETA 79001 B	0.45^m	0.69^m	1.02^m	B			0.90^q ±0.17			
EETA 79001 C	0.46^M	0.45^M		B			$T_{10,av}=0.73$			
		0.42^N		B						
Nakhlites										
Governador Valadares	12.2^P	12.3^P	6.7^P	B	10.0 ±2.1			10.0 ±2.1		10.0 ±2.1
		9.5^Q	9.2^Q	B						
Lafayette	10.1^n	12.4^n	8.8^n	B	11.9 ±2.2			11.9 ±2.2	0.0089^U ±0.0013	11.9 ±2.2
	13.7^R	16.0^R	10.3^R	B						
Nakhla	11.4^i	12.4^i	8.4^i	B	12.2 ±1.5	10.75^b ±0.40		10.75 ±0.40		10.75 ±0.40
	12.1^J	12.0^J	10.6^J	B						
	14.8^R	16.0^R	12.2^n	B						
Dunite										
Chassigny	13.3^g	11.1^g	10.5^g	A	11.6 ±1.6	10.7^b ±1.8		11.3 ±0.6		11.3 ±0.6
	14.7^F	13.9^F	7.1^F	B						
	15.1^E	13.7^E	8.8^E	B						
	12.3^i	11.8^i	7.1^i	B						
Orthopyroxenite										
ALH 84001	15.4^ζ	17.0^ζ	12.3^ζ	A	14.7 ±0.9	15.8^b ±3.3		15.0 ±0.8	0.0065^v ±0.0010	15.0 ±0.8
	15.6^c	13.3^c	12.2^c	A						
opx	15.4^c	15.4^c	12.9^c	A					$\sim0.013^Y$	
	14.9^G	17.7^G	17.5^G	A						
	15.3^t	15.4^t	12.8^t	A						
	14.6^f	14.1^f	12.2^f	A						

TABLE IV

(continued)

Notes

T_3, T_{21}, T_{38}: Cosmic-ray exposure ages based on ^{3}He, ^{21}Ne, ^{38}Ar and appropriate production rates.

Meth., M.: Method used for calculating T_3, T_{21}, T_{38}:

    A – CRE ages as given by author(s). Applies to papers published after 1990.

    B – CRE age calculated using production rates according to Eugster and Michel (1995). Applies to papers published before 1990 and to work where authors do not give CRE ages.

$T_{s,av}$: Mean value of all T_3, T_{21}, and T_{38} ages for a particular meteorite. Errors are $2\sigma_{mean}$.

T_{81}: ^{81}Kr-Kr CRE age (Terribilini *et al.*, 2000).

T_{10}: ^{10}Be-^{21}Ne CRE age as given by authors.

T_{pref}: Preferred CRE age, $T_{pref} = 0.5 \times [T_{s,av} + 0.5 \times (T_{81} + T_{10,av})]$.

    Error of T_{pref}, $\Delta T_{pref} = 2\sqrt{\sum \left(2\Delta_{s,av}^2 + \Delta_{81}^2 + \Delta_{10,av}^2\right)/12}$.

T_{terr}: Terrestrial age as given by authors.

T_{ej}: Mars ejection age, $T_{ej} = T_{pref} + T_{terr}$. Error of T_{ej}, $\Delta T_{ej} = \sqrt{(\Delta T_{pref})^2 + (\Delta T_{ej})^2}$.

References

[a] Garrison and Bogard (2000); [b] Terribilini *et al.* (2000); [c] Eugster *et al.* (1997b); [d] Dreibus *et al.* (1996); [e] Swindle *et al.* (1996); [f] Garrison and Bogard (1998); [g] Terribilini *et al.* (1998); [h] Becker and Pepin (1986); [i] Ott (1988); [j] Eberhardt and Hess (1960); [k] Heymann *et al.* (1968); [l] Müller and Zähringer (1969); [m] Bogard *et al.* (1984); [n] Ott (1989), unpubl. data (see Schultz and Franke, 2000); [o] Nishiizumi and Caffee (1996); [p] Nishiizumi *et al.* (1986); [q] Pal *et al.* (1986); [r] Schultz and Freundel (1984); [s] Evans *et al.* (1992); [t] Miura *et al.* (1995); [u] Nishiizumi *et al.* (1992); [v] Jull *et al.* (1994); [w] Treiman *et al.* (1994); [x] Becker and Pepin (1993); [z] Ott and Löhr (1992); [A] Nishiizumi and Caffee (1997); [B] Nagao *et al.* (1997); [D] Bogard (1995); [E] Lancet and Lancet (1971); [F] Schultz and Signer (1973), unpublished; [G] Swindle *et al.* (1995); [H] Zipfel *et al.* (2000); [J] Stauffer (1962); [K] Jull and Donahue (1988); [L] Sarafin *et al.* (1985); [M] Becker and Pepin (1984), for R=35 to >1000 and d=6–80 g/cm^2; [N] Swindle *et al.* (1986); [P] Bogard and Husain (1977); [Q] Swindle *et al.* (1989); [R] Ganapathy and Anders (1969); [T] Folco *et al.* (1999), same production rates used as for DaG476; [U] Jull *et al.* (1997); [Y] Jull *et al.* (1995); [α] Nishiizumi *et al.* (1999; same T_{terr} for the paired meteorites DaG476/489); [β] ^{21}Ne and ^{10}Be data from Garrison and Bogard (2000) and Nishiizumi and Masarik (2000), respectively; [γ] Lorenzetti and Eugster (2001); [δ] Shukolyukov *et al.* (2000); [ε] Paetsch *et al.* (2000); [ζ] Eugster (1994).

pressive and tensile pulses, the latter reflected from the free surface of the planet, is sufficient to accelerate ejecta to escape velocity. These conditions define an annulus around the impact site from which material is ejected (*cf.* Warren, 1994).

From Figure 7a there appear to be 7 ejection events: at $\sim$20, 15, 12, 4.5, 3, 1.3 and 0.7 Myr, respectively. Five of the total are for either basaltic or lherzolitic shergottites, suggesting that shergottites must be widespread on the Martian surface, or that the ejection mechanism preferentially selects basaltic compositions. Interestingly, both the oldest and the youngest events are for shergottites: Dhofar 019 at $\sim$20 Myr ago, and EET79001 at $\sim$0.7 Myr ago.

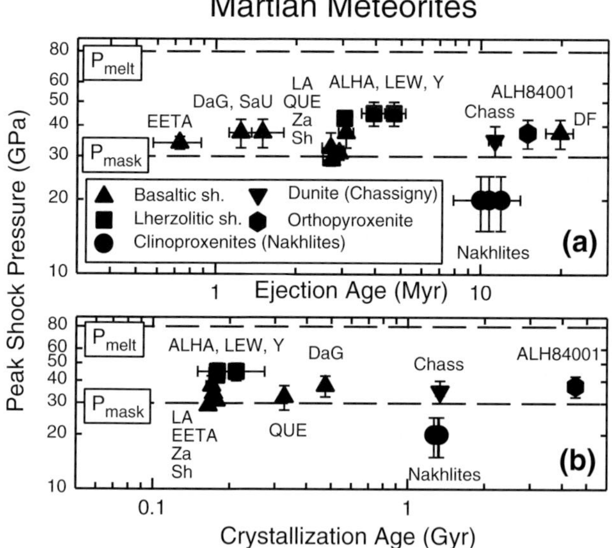

Figure 7. Peak shock pressures in Martian meteorites vs. their a) ejection ages and b) crystallization ages (EETA: EETA79001; DaG: Dar al Gani476; SaU: SaU005; LA: Los Angeles; QUE: QUE94201; Za: Zagami; Sh: Shergotty; ALHA: ALHA77005; LEW: LEW88516; Y: Y793605; Chass: Chassigny); DF: Dhofar019). Shock pressure data from Stöffler *et al.* (1986) and Stöffler (2000) except for SaU, LA, and DF (this paper; Table I); age data from Tables II, III, and IV (see references there). P_{max} and P_{melt} are the approximate peak shock pressures at which plagioclase is converted to maskelynite and basalts are shock-melted, respectively.

The picture changes somewhat when shock pressures are plotted against the crystallization ages of the meteorites (Figure 7b). To the extent that different Martian surface units are composed of rocks of distinct crystallization ages, both crystallization ages and ejection ages might be viewed as "event discriminators". Nearly the entire Martian surface has been classified according to relative age as determined from the density of meteorite impact craters per unit area. These "crater retention ages" reflect ca. the upper 1 km of near-surface layers, so that lava flows of a variety of absolute ages may be present in a given area (Hartmann, 1999). Nevertheless, the crater retention ages provide a Martian context in which to view the potential number of ejection events. In this context, there appear to be only 4–5 events: one on old terrain, one on terrain of intermediate-to-young 1.3 Gyr age, 1 or 2 on young-to-intermediate, 0.3–0.5 Gyr terrain, and one on very young, ~0.18 Gyr terrain. The crystallization ages of Dhofar019 and SaU005 are not available yet, and might define additional events, possibly bringing the total to 6–7 separate events. More realistically, the total is apt to increase by no more than one, since SaU005 appears to be paired ejection age with DaG476. Nevertheless, 7 meteorites, more than half of those for which crystallization ages have been determined, derive from ~0.18 Gyr terrain. This presents an apparent "paradox" of too many young meteorites from too little young Martian terrain (Nyquist *et al.*, 1998).

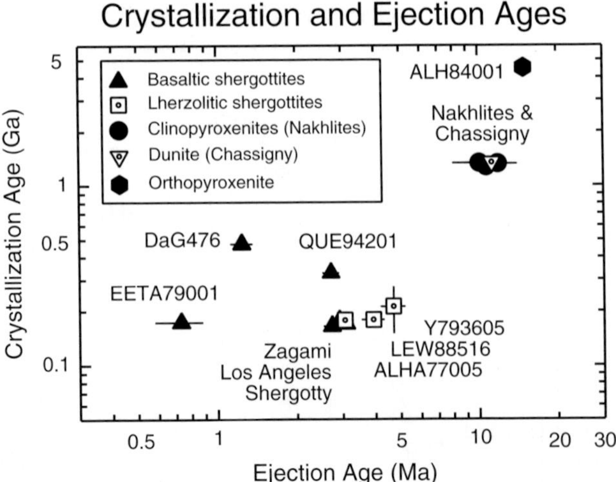

Figure 8. Preferred values of the crystallization ages of the Martian meteorites (Tables II and III) plotted vs. the ejection ages of the meteorites (Table IV).

Figure 7b seems to show that meteorites ejected in a given event may experience a variety of peak shock pressures, as shown for the "nakhlite–chassignite" and "young shergottite" events, respectively.

Figure 8 plots the crystallization ages of the Martian meteorites directly versus their ejection ages. SaU005 and Dhofar019 are two important omissions, particularly because the Dhofar019 event 19.8 ± 2.3 Myr ago on basaltic shergottite terrain is the oldest event of which we have record. The next event, which ejected ALH84001 from old (Noachian) terrain 15.0 ± 0.8 Myr ago, is clearly identified. ALH84001 apparently experienced a peak shock pressure of $\sim35 - 40$ GPa on ejection, typical for Martian meteorites. This rock of initially deep-seated origin must have been previously moved to its place of ejection, probably a few meters to tens of meters beneath the Martian surface, assuming it came from just beneath the spall zone of an impactor ~1–10 km in diameter (Melosh, 1984; 1985). This is consistent with its complicated shock history, consisting of two or more stages.

The next ejection event was the nakhlite event. The nakhlites, Governador Valadares, Lafayette, and Nakhla, were recovered at three widely separated localities: Egypt, Brazil, and Indiana. They are clinopyroxenites, have the same radiometric crystallization ages, and the same ejection age of ~11.4 Myr. The dunite, Chassigny, is a fourth meteorite with a ~1.3 Gyr crystallization age and an ejection age of 11.3 ± 0.6 Myr. Its radiogenic isotope composition and the abundances of several key incompatible trace elements suggest a close relationship to the nakhlites. Although Chassigny is mineralogically distinct from the nakhlites, it probably was ejected simultaneously with them. If so, it and the nakhlites probably were previously moved to their place of ejection. If, however, they crystallized near the center of a ~100 m thick flow (Treiman, 1987), they may have been ejected directly by a large event. Alternatively, they may have experienced a two-stage shock history, with the earlier, lighter shock overprinted by the shock of the ejecting impact. In any case, early Amazonian Martian terrain, on the HT timescale, appears to have been sampled only once. We call this the NC event following Treiman (1995a).

The shergottites present the greastest puzzle. Taken at face value, the CRE ages indicate 4 ejection events within 5 Myr on terrain of three different, young, crystallization ages. Four basaltic shergottites, Shergotty, Zagami, Los Angeles, and QUE94201, have ejection ages in a narrow range of 2.7–3.1 Myr, and one lherzolitic shergottite, ALH77005, has a similar ejection age of 3.06 ± 0.20 Myr. The ejection ages of two other lherzolites (3.94 ± 0.40 Myr for LEW88516 and 4.70 ± 0.50 Myr for Y793605), however, do not overlap with those of the basaltic shergottites. The exposure ages of the three lherzolites are somewhat uncertain because of uncertainty in production rates, and because no ^{81}Kr ages yet exist. Further, ALH77005 and LEW88516 are similar in several properties (Treiman *et al.*, 1994) and might be expected to have the same ejection age. Thus, the differences in ejection ages among these three lherzolites are not easily interpreted.

Let us consider a scenario in which these 4 shergottites and the 3 lherzolites were ejected in a single event, and assume that the higher ages of LEW88516 and Y793605 are due to an exposure to cosmic rays prior to ejection from Mars (Nagao *et al.*, 1997). Nishiizumi and Caffee (1997) also concluded that for Y793605 it is not possible to completely eliminate, based on radionuclide activities, the possibility of a pre-irradiation on the surface of the parent body. On the other hand, Treiman *et al.* (1994) found from the radionuclide activities of LEW88516 that most of its cosmic-ray exposure probably occurred as an individual meteoroid. Furthermore, the cosmogenic ^{22}Ne/^{21}Ne ratios are 1.227 ± 0.035 for LEW88516 (Eugster *et al.*, 1997a) and 1.207 ± 0.020 for Y793605 (Terribilini *et al.*, 1998), indicating irradiation as small bodies for the total duration of exposure. A complicated exposure history beginning with excavation of LEW88516 and Y793605, followed by ejection within 1-2 Myr, or reburial of these two rocks between their initial excavation and ejection, seems improbable. Nevertheless, the shock features of Y793605 do suggest a two-phase shock history. These plutonic rocks probably crystallized at considerable depth, and were relocated to the ejection site. Peak shock pressures for all three are the highest among the Martian meteorites, suggesting that they were excavated from deeper depths than the other shergottites, and thus that they were more heavily shielded against cosmic-ray irradiation immediately prior to ejection. Nevertheless, the more brecciated texture of Y793605 suggests that it may have spent some time in a relatively shallow surface location. Thus, it may have acquired some cosmogenic noble gases between its first and second shock, the latter being coincident with its ejection from Mars. Nearly identical crystallization ages of 179 ± 5 Myr and 178 ± 8 Myr, respectively, for ALH77005 and LEW88516 (Table III), raise the possibility that all the lherzolites were ejected simultaneously with ALH77005, Shergotty, Zagami, and Los Angeles, even though the "ejection ages" of LEW88516 and Y793605 are analytically resolved from the others.

Additional events seem to be required for the shergottites/lherzolites DaG476 and 489, and EET79001. The ejection ages for EET79001, DaG476, and DaG489, are 0.73 ± 0.15 Myr, 1.24 ± 0.12 Myr, and 1.25 ± 0.20 Myr, respectively. DaG476 and 489 are likely to be part of a single fall, so the coincidence of their ejection ages is not surprising. The ages of DaG476/489 do not appear to overlap that of EET79001, and are clearly younger than those for the shergottites and the lherzolites. Zipfel *et al.* (2000) derive a slightly older ejection age of 1.35 ± 0.10 Myr for DaG476. In any event, its old crystallization age of $\sim$474 Myr indicates a separate ejection event for it. The ejection age of EET79001 has some ambiguity. Most CRE ages for EET79001 have been determined for lithology A, leading to a preferred value $T_{pref} = 0.60 \pm 0.09$ Myr (Table IV). This value, when combined with the averaged terrestrial age of 0.13 ± 0.12 Myr, yields $T_{ej} = 0.73 \pm 0.15$ Myr (Table IV). A significantly higher value of 1.22 ± 0.24 Myr would be obtained, however, by combining the ^{10}Be-^{21}Ne CRE age of 0.90 ± 0.17 Myr for EET79001 lithology B (Pal *et al.*, 1986) with the terrestrial age of 0.32 ± 0.17 Myr reported for EET79001 (Sarafin *et al.*, 1985). This is the highest value derivable from the EET79001 data and is in good agreement with the preferred ejection age of DaG476 (Table IV). Thus, the resolution of the EET79001 and DaG476 ages may not be completely established. The disparity in apparent ejection ages of lithologies A and B of EET79001 may be related to the chemical differences between them, which affect the spallation nuclide production rates used to calculate CRE ages.

6. Provenance of the Martian Meteorites

6.1. ALTERNATIVE EJECTION SCENARIOS

Inconsistent numbers of Martian ejection events are inferred from crystallization and CRE ages, respectively. One may argue, that either some of the CRE ages are due to secondary collisions after ejection from Mars, or that there are vast expanses on Mars that not only are "young", but are the same age, within current experimental uncertainty. In the following, we characterize these two approaches.

6.1.1. *Cosmic-Ray Exposure Events: Without Secondary Collisions*

This assumption takes the ejection ages at face value (Eugster *et al.*, 1997b). Figures 7a and 8 show the following events: S_{Dho} (Dhofar019 shergottite), O (orthopyroxenite), NC (nakhlites and Chassigny), S_L (lherzolitic shergottites, especially LEW88516 and Y793605), S (traditional basaltic shergottites), S_{DaG} (DaG476/489 and SaU005), and S_E (EET79001). These 7 events include 5 related to shergottites. Three of these are on terrain <500 Myr old. A modification of this scenario might allow an additional event, S_Q, for QUE94201 because of its unique crystallization age. Thus, in this scenario there are 7-8 ejection events, 5-6 of which are for shergottites <$\sim$500 Myr old, and three include shergottites $\sim$175 Myr old.

6.1.2. *Events Based On Crystallization Age: Cosmic-Ray Exposure With Secondary Collisions*

Nyquist *et al.* (1998) preferred the term "small body ages" to "ejection ages" to emphasize that the event directly dated was initiation of exposure of "small" bodies to cosmic radiation. "Small" was defined by Bogard *et al.* (1984) as $<\sim6$ m as a safe upper limit. A practical definition of "small" is that size below which a two-stage exposure becomes detectable in the total radionuclide and stable spallation nuclide data. This size is difficult to quantify, but is apt to be somewhat smaller, perhaps ~3-4 m in diameter. In a "strong" version of this model, some CRE ages may be unrelated to the actual ejection events. In this case, the minimum number of events suggested by the crystallization age data are: O, NC, and S (Nyquist *et al.*, 1998). An additional two events S_Q, and S_{DaG}, might be allowed because the crystallization ages of QUE94201 and DaG476 differ from ~175 Myr, and another two events may be permissible because precise crystallization ages of Dhofar019 and SaU005 are not yet available. The whole rock Sm-Nd data (Figure 6), and initial Sr- and Nd-isotopic data (not shown) suggest that DaG476, QUE94201, and Dhofar 019 constitute a suite of samples separate from the ~175 Myr shergottites. Treiman (1995a) considered a larger array of petrologic, chemical, and chronological data, and suggested two distinct sites of origin for the then known SNC meteorites, which did not include QUE94201, DaG476/489, or Dhofar019. Thus, the probable number of events in this scenario is $\geq$ 4: O, NC, S_1, and S_2, where subscripts 1 and 2 separate the ~175 Myr shergottites from $\sim300-500$ Myr shergottites. Subgroup S_2 may be further sub-divided into as many as three. The ~3 Myr exposure age for QUE94201 perhaps should exclude it from S_2. Also, SaU 005 is almost certainly paired with DaG476/489 on the basis of their CRE age. Note that for 4-6 ejection events the "secondary collision" hypothesis is required to hold only in a "weak" form for the young exposure age of EET79001. In that case, ejection events would have occurred at ~20 Myr, ~15 Myr, ~12 Myr, ~3 Myr, and ~1.3 Myr, respectively, assuming that some of the lherzolitic shergottites had some exposure on the Martian surface.

That half or more of the total number of Martian meteorite ejection events yielded shergottites $<\sim500$ Myr old continues to be a paradox. According to new spacecraft observations, very late lava flows appear to cover an area of $\sim10^7$ km^2 (Keszthelyi *et al.*, 2000), $\sim7\%$ of the total area of Mars, or $\sim12\%$ of the volcanic surface area (Tanaka *et al.*, 1992). The previously recognized Late Amazonian volcanic surface was only $\sim3.3 \times 10^6$ km^2 (Tanaka *et al.*, 1992), so the apparent over-representation of shergottite-ejection events compared to the total number of events and to the number of orthopyroxenite and nakhlite events persists. This shergottite age paradox (Nyquist *et al.*, 1998) is most pronounced for scenarios involving no secondary collisions in meteorite transit from Mars. That is, the probability that of 3–4 random events, at $\sim4.3(?)$, ~3.0, ~1.3, and ~0.7 Myr ago, 2–3 ejected meteorites having indistinguishable crystallization ages within an analytical precision of a few percent, would seem to be very low. If these are truly separate, random, events, there must be great expanses of $\sim$175 Myr lavas on the Martian surface. Keszthelyi *et al.* (2000) noted that the $\sim$10^7 km^2 of newly recognized flood basalts in Elysium and Amazonis Planitia is an area roughly the size of Canada. The age paradox appears to require an even larger area; i.e., $\sim$50% of the 84×10^6 km^2 total volcanic surface area of Mars (Tanaka *et al.*, 1992) appears to be required to be $<\sim$500 Myr old, and $\sim$2/3–3/4 of that to be $\sim$175 Myr basalt if the events at $\sim$4.3, $\sim$3.0, and $\sim$0.7 are due to separate, random impacts on the Martian surface. The fewer the number of shergottite ejection events, the more easily reconcilable is their number with the Martian cratering record, and the number of nakhlite and orthopyroxenite ejection events (one each).

Finally, it should be noted that Martian surface ages are derived by scaling an assumed Martian cratering rate as a function of time to the lunar cratering rate as a function of the absolute ages of lunar surfaces as determined on returned lunar samples. Hartmann (1999) estimates the uncertainty in this process to be on the order of a factor of three. If the Martian cratering rate were higher than assumed in Tanaka *et al.* (1992), the apparent ages of Martian surface units would be shifted to lower values. Surface terrain of age $\sim$1.3 Gyr, assumed to be present in the Early Amazonian on the HT model, and representing $\sim$9% of the volcanic surface of Mars, would actually be found in the Late Hesperian, representing $\sim$14% of the volcanic surface. Surface terrain of $\sim$0.5 Gyr or less, found in the Middle and late Amazonian in the HT model, would be shifted to Early Amazonian, representing $\sim$15% of the volcanic surface. Together they would represent a sizeable fraction of the total surface ($\sim$29%), and a significant probability that Martian meteorites would be $\leq$1.3 Gyr old. Even in this case, however, the relative frequency of shergottite events to nakhlite events would be expected to be only $\sim$1:1.

Mars may have been more active in recent times than previously thought, and late, thin lava flows may be relatively ubiquitous on the Martian surface, increasing the probability of young meteorites. Clarification of whether the ages of the meteorites are in proportion to the exposed area of young surfaces, however, will require a) continued and refined chronological studies of the Martian meteorites; b) absolute calibration of the Martian crater-frequency curve via returned Martian samples; c) more high resolution imagery of the Martian surface; and d) evaluation of the compositional biases, if any, that may exist in the yield of meteorites from surfaces of different compositional types.

6.2. POTENTIAL SOURCE TERRAINS

Several authors have sought to identify potential Martian source terrains and candidate source craters for the Martian meteorites (Nyquist, 1983; Mouginis-Mark *et al.*, 1992; Treiman, 1995a; Barlow, 1997; Nyquist *et al.*, 1998). Such attempts have relied primarily on the crystallization ages of the meteorites and the perceived characteristics of the resulting impact craters. Mouginis-Mark *et al.* (1992) identified 9 candidate source craters for the SNC meteorites on the Tharsis plains of

Mars, the only area of the planet seen to have lava flows ≤ 1.3 Gyr old on the HT model using Viking imagery. The higher resolution imagery from the Mars Global Surveyor (MGS) has shown the presence of thin, nearly craterless, lava flows on areas that are more heavily cratered at the Viking resolution, however (Hartmann, 1999; Hartmann and Berman, 2000). These new observations significantly extend the Martian surface area known to be covered by young basalts, as well as lower some cratering ages into the range of the ages of SNCs by either model. Crater densities on these young lavas are 10^{-3} to 10^{-2} times the densities on lunar maria of ages ~ 3.5 Gyr, so that some Martian basalts appear to be as young as ~ 10 Myr or less (Hartmann and Berman, 2000).

6.3. IMPLICATIONS OF EJECTION MODELS

Three recent developments loosen the restrictions on possible source terrains for the shergottites: a) The MGS data show young, thin, lava flows covering older flows in some localities, particularly in Elysium and Amazonis Planitia (*cf.* Keszthelyi *et al.*, 2000). b) The older crystallization ages of ~ 330 Myr, and ~ 475 Myr found for QUE94201 and DaG476, respectively, further extend the potential source terrains. c) The lower theoretical limiting crater size for meteorite ejection of ≥ 3 km (Head and Melosh, 2000) increases the number of candidate source craters, as compared to those ≥ 10 km (Mouginis-Mark *et al.*, 1992).

The decrease in crater size required for meteorite ejection according to the Melosh (1984) spallation model greatly increases the number of fragments potentially ejected from Mars. According to Mileikowsky *et al.* (2000) the number of fragments ejected from craters ~ 13 km diameter on Mars with shock pressures <1 GPa over 4 Gyr is 9.5×10^{10}. They estimate the number ejected at <1 GPa is about 2% of the total. For craters of this size, the mean ejecta fragment diameter is ~ 0.3 m. These values may be compared to $\sim 9.4 \times 10^9$ fragments of mean diameter ~ 0.9 m and shock pressure <1 GPa for a 30 km diameter crater, and $\sim 2.0 \times 10^8$ fragments of mean diameter ~ 3 m and shock pressure <1 GPa for a ~ 100 km diameter crater. The maximum fragment diameter is estimated to be about four times the mean diameter. The spallation model thus favors a high proportion of small meteorites, with the relative abundance falling off by more than 2 orders of magnitude for an order of magnitude increase in size. Although there is a high proportion of small meteorites among the lunar meteorites, small meteorites do not seem to be preferred among Martian meteorites (*cf.* Warren, 1994).

The largest Martian meteorite, Nakhla, has a recovered mass of 40 kg, corresponding to a spherical meteoroid ~ 0.3 m in diameter. This, of course, is a lower limit on the pre-atmospheric size of Nakhla. Although 0.3 m is not large in an absolute sense, it is ~ 60 times larger in mass, or ~ 4 times larger in radius, than the largest lunar meteorite (Warren, 1994), implying proportionally larger impactors. For the spallation model, Mileikowsky *et al.* (2000) give fragment size $l = 3 \times 10^{-4} \, L$, where L is the impactor diameter. Since this scales as $V_{ej}^{-2/3}$, where V_{ej} is the ejection velocity, we expect $l_{Mars}/l_{Moon} \sim 1.6$, implying moderately larger Martian launch craters than lunar launch craters. A much more significant difference would be implied if a "proto-EET79001" ~ 3 m in diameter is required for later involvement in a secondary collision. A fragment this large would be at the upper size limit for a 30 km crater. A larger crater, ~ 100 km in diameter, is inferred from the spallation model if the mean fragment size is to be ~ 3 m.

From Gladman (1997; Figure 11), we estimate that a Martian meteorite ejected 4 Myr ago would on average spend ~ 0.15 Myr in the main asteroid belt, where the collision rate is relatively high. Wetherill (1988) gave the half-life, τ (Myr), against collision in the main belt as $\tau \sim 1.2r^{1/2}$ for meteoroids of radius r (cm). Thus, for $r = 150$ cm, τ is 15 Myr, and on average $\sim 1\%$ of objects launched 4 Myr ago would undergo secondary collisions in the main belt. If a large number of fragments were launched simultaneously, as implied by recovery of 3–7 individuals from the shergottite (S) event, then the probability that one of those individuals would have undergone a secondary collision would be ~ 3–7%. Thus, the possibility of secondary collisions in space after ejection of meterorites from Mars cannot be discounted *a priori*. As noted earlier in the paper, crystalline rock fragments from the ~ 25 km diameter Ries crater shocked to the range of the Martian shergottites do not exceed a size of ~ 0.5 m. Fragment size should scale as the product of dynamic tensile strength and projectile diameter (Melosh, 1985). An increase in this product of a factor of ~ 6 or more for the Martian source crater of the shergottites compared to the Ries would appear to be required to yield proto-EET79001 fragments.

Mileikowsky *et al.* (2000) have estimated the minimum crater size theoretically consistent with proto-EET79001s of ~ 3 m at ~ 30 km. For ~ 0.3 m fragments, their estimate of an ~ 13 km final crater diameter is in close agreement with ~ 15 km estimated by Warren (1994) for Nakhla. This is ~ 15 times larger than the largest source craters Warren estimates for the lunar meteorites on the same basis, an unexpected difference considering $l_{Mars}/l_{Moon} \sim 1.6$. Warren (1994) also estimates that a crater ~ 11 km in diameter is required to launch the largest shergottite, Zagami. Thus, the relatively large size of the Martian meteorites may indicate derivation from relatively large craters, in spite of the lowering of the theoretical limit of the Melosh (1984, 1985) model for the smallest crater required for launch.

6.3.1. *Implications of the Apparent Shock Pressure Launch Window*

The observation of an apparent "launch window" of peak shock pressures for the Martian meteorites may have implications for their launch mechanism and provenance. Early attempts to account for the existence of Martian meteorites sought the explanation in unusual, relatively rare, events because conventional cratering theory did not account for them. It was argued, for example, that ejection velocities could not exceed twice the "particle velocity" produced in a compressive wave at the peak shock pressures the meteorites had experienced. This maximum velocity could be achieved only at a free surface. Stöffler *et al.* (1986) gave the particle velocities according to the Hugoniot equation of state for the then known shergottites as $\sim$1.5 $-$ 2.0 km/s for peak shock pressures of 29 $-$ 43 GPa, corresponding to ejection velocities of 3 $-$ 4 km/s. The additional meteorites included here widen the "launch window" to $\sim$20 $-$ 45 GPa, corresponding to particle velocities of 0.8 $-$ 2.2 km/s for the range of materials involved. Thus, their free surface launch velocities are $\sim$1.6 $-$ 4.4 km/s, and are below Martian escape velocities.

Nyquist (1983) suggested that vapor drag might provide the needed acceleration, especially in "richocheting", very oblique angle, impacts. Numerical simulations supported the idea that oblique impacts could launch meteorites from planets (O'Keefe and Ahrens, 1986). Vickery (1986) considered a range of possibilities involving vapor phase acceleration, and concluded that if the proto-SNCs had diameters of a few centimeters or less, gas acceleration from a 30-km-diameter crater would be consistent with their ejection from Mars. Interestingly, the equivalent spherical radii of Martian meteorites corresponding to their recovered masses are in the range $\sim$2-30 cm. Larger craters would give higher final velocities. For example, at a launch position, x, one crater radius, r_p, from the point of impact, 0.5 m rocks would be accelerated by factors of $\sim$1.7 for a 30 km diameter crater and of $\sim$2.9 for a 130-km crater. Vickery concluded that these factors, combined with the distribution of spall velocities at $x/r_p = 1$ according to the Melosh (1984) model, would not result in velocities exceeding the Martian escape velocity. In fact, the effect of gas drag at $x/r_p \leq 0.4$ was to decelerate "ejecta" from the spall model. Thus, vapor drag did not appear to augment the number of meteorites ejected compared to those ejected by spallation alone. Nevertheless, when applied to possible free surface launch velocities of $\sim$3-4 km/s for the shergottites, these factors result in velocities $\geq$5 km/s, as required for launch from Mars. Thus, vapor drag should not be ignored as a potential acceleration mechanism.

The oblique impact hypothesis has the attractive feature that craters produced by this process have a distinctive "butterfly" morphology that allows them to be identified. Nyquist (1983) and Nyquist *et al.* (1998) identified some candidate oblique impact source craters for the SNC meteorites, whereas Barlow (1997) identified a candidate oblique impact source crater for the orthopyroxenite ALH84001. Recently, attention has fallen on the Chicxulub "K-T impact" crater as of probable oblique impact origin. Schultz and D'Hondt (1996) show a time-lapse photographic sequence of the evolution of an impact-generated vapor cloud as well as summarizing the evidence for Chicxulub as an oblique impact. Hydrocode modeling of Chicxulub (Pierazzo and Melosh, 1999) and of oblique impacts more generally (Pierazzo and Melosh, 2000) describe the development of downstream melt and vapor plumes in such impacts. If surface fragments of Mars were launched into such oblique impact vapor plumes at velocities close to the escape velocity, perhaps some would acquire the "boost" needed for launch. Kadano and Fugiwara (1996) experimentally verified that for nylon projectiles impacting Cu targets, solid ejecta fragments are accelerated to velocities approaching that of the expanding vapor cloud. Continued studies should show whether oblique impacts could have a significant role in launching Martian meteorites.

However, the spallation model of Melosh (1984) is most widely considered. Warren (1994) applied it in detail to the launch of both lunar and Martian meteorites. In general the model worked well for lunar meteorites, with some modification to account for the high abundance of regolith breccias and the shallow launch depths of the lunar meteorites. There are no regolith breccias among the Martian meteorites, however. In other respects, also, including a preponderance of small meteorites relative to larger ones, and a majority of unshocked meteorites relative to moderately to highly shocked ones, the spallation model matches the lunar meteorites better than the Martian meteorites.

In the spallation model, interference of a reflected rarefaction wave from the free planetary surface with the direct compressive wave reduces shock pressures in a zone of interference, where near-surface material can be ejected at high velocity without experiencing high compressive shock pressures. Spalls form at the boundary of the interference zone with a lower "free-field" zone. In the "hydrodynamic ejection model" (Melosh, 1984, 1985, 1989), the ejection velocity is given as

$$V_{ej} \approx 2V_p(r) \left[1 + (s/d)^2\right]^{-1/2} \tag{1}$$

which reproduces Equation 5.5.3 of Melosh (1989). Here, $V_p(r)$ is the particle velocity in the compressive shock at a distance $r = (s^2 + d^2)^{1/2}$ from the "equivalent center" of the impact. The horizontal distance s is measured across the surface from the impact site to the ejection site. In analogy to nuclear explosions, d is the "equivalent depth of burst." The particle velocity varies with distance according to

$$V_p(r) \approx C_v \, (U/2) \, (a/r)^{1.87} \tag{2}$$

Here, U is the velocity of an impactor of mean radius a, and $C_v \sim 1$ is a "coupling constant." The equivalent depth of burst is given approximately by

$$d \approx 2a \, (\rho_P/\rho_t) \tag{3}$$

Here, ρ_P and ρ_t are the projectile and target densities.

For $s = 0$, Equation (1) appears to give the "velocity doubling rule" for free surfaces, i.e., ejection velocity equal to twice the particle velocity. However, particle velocities corresponding to the observed peak shock pressures would lead to ejection velocities of only $\sim$1.6$-$4.4 km/s for the Martian meteorites, as mentioned earlier. Equation (1) can only lead to ejection velocities in excess of 5.0 km/s if the theoretical particle velocity of Equation (2) is decoupled from the empirical particle velocity inferred from the peak shock pressures experienced by the meteorites. In this model, this "decoupling" of theoretical and empirical values of particle velocity is due to the interference of the compression and rarefaction waves.

Launch acceleration is proportional to the upward component of the pressure gradient acting on a volume of material (*cf.*, Melosh, 1984, 1989). Let the peak shock pressure in the compressive wave be P_{cw}, that in the rarefaction wave P_{rw}, and the resultant pressure in the zone of interference be P_{res}. Because the compressive wave arrives first at a given point in the target, P_{res} rises from zero as the compressive wave approaches, attains a maximum value, $P_{max} < P_{cw}$ as the tensile wave arrives, falls to a value P_{ten}, limited by the tensile strength of the material, as the maximum of the of the compressive wave passes, and finally returns to zero as the tensile wave passes (*cf.*, Melosh, 1985, Figure 3). The empirical particle velocities corresponding to peak shock pressures registered in the meteorites are those corresponding to P_{res}, whereas the particle velocity V_p in Equation (1) corresponds to P_{cw} of the compressive wave. Thus, in the interference zone very near the free surface, the velocity-doubling rule no longer applies, and further, as noted by Melosh (1985), the ejection velocity "is exactly twice the particle velocity in the compression wave only at the (unphysical) point on the surface lying directly above the equivalent center."

The complete decoupling of theoretical and empirical particle velocities appears to apply for lunar, but not for the Martian meteorites, in which the observed shock levels are at least partially coupled to the ejection phenomenon (Figure 7). Only a boost in acceleration above that recorded in the Martian meteorites is needed for their ejection, accomplishable as part of the spallation process, or by another ejection mechanism instead of, or in addition to, the spallation. Melosh (1989, Figure 5.8) presents contours of peak shock pressure and ejection velocity for the spallation model as a function of depth in units of the projectitle diameter. Ejection velocity contours are given in units of impact velocity, U, and pressure contours in units of $\rho_t U^2$. For a typical impact velocity $U = 15$ km/s and target density $\rho_t = 3.3$ g/cm^3, peak shock pressures of 15 GPa $\leq P_{max} \leq 45$ GPa for Martian meteorites give $\sim 0.2 \leq (P_{max}/\rho_t U^2) \leq 0.6$, corresponding to depth to diameter ratios, d/a, in the range $\sim 0.1 - 0.3$ very near the impact site and just below the near surface interference zone. Ejection velocities lie in the range $0.2 \leq (V_{ej}/U) \leq 0.5$, i.e., 3.5 km/s $\leq V_{ej} \leq 7.5$ km/s for $U = 15$ km/s. Thus, for impactors ~ 1 km in diameter, the depth of origin of the Martian meteorites is implied to be $\sim 100 - 300$ m. Such impactors would produce final craters ~ 13 km in diameter (Mileikowsky *et al.*, 2000).

One apparent discrepancy between the predictions of this model and observations of the Martian meteorites concerns the apparent absence of lightly shocked Martian meteorites from our collections. According to the model (e.g., Melosh, 1989, Equation 6.4.3), the mass m_{ej} of material ejected at velocities greater than V_{ej} and shocked to pressures less than P_{max} is

$$m_{ej}/m = 1.2\,(P_{max}/\rho_t c_L U)\left[1 - \left(2V_{ej}/U\right)^{1/3}\right] \tag{4}$$

where m is the projectile mass, and c_L is the seismic wave velocity of the target. The other quantities are as previously defined. For $c_L = 6$ km/s, $P_{max} = 15$ GPa, and other quantities as above, this equation predicts $m_{ej}/m = 0.08$. Likewise, $m_{ej}/m = 0.23$ is predicted for $P_{max} = 45$ GPa. Thus, we would expect that approximately 1/3 of the meteorites would have been shocked to pressures <15 GPa, whereas none are observed. The seriousness of this discrepancy is unclear because the number of recovered meteorites may be too few to give reliable statistics. Indeed, agreement between observation and theory is significantly improved if the lower value of P_{max} is shifted to ~20 GPa, so that the nakhlites populate a low-shock bin. In this case, the observed ratio of "low-shock" to "high-shock" meteorites becomes 1: 6 in comparison to the predicted value of 1: 2.3. Nevertheless, the apparent lack of lightly shocked Martian meteorites is puzzling.

Because the functional relation between P_{cw} and particle velocity V_p varies with composition, there could be a compositional bias in the material ejected. Also, because both maximum pressure and particle velocity in the compressive wave decrease approximately as the inverse square of the distance from the impact (Equation 2; Melosh, 1985, Figure 1), and the area of an annulus of interference increases with the square of the distance, the majority of the material ejected will be near the largest value of r and the smallest value of V_p consistent with acceleration to escape velocity. This effect may lead to relatively long delays between arrival of the compression and rarefaction waves, and, thus, to relatively high values of P_{res} for the Martian meteorites.

Strong coupling of compressive shock into ejected Martian meteorites implies that they come from very near the surface contour of the spall zone. The most likely place of origin of the meteorites is from within the zone of "Grady-Kipp fragments" lying just below the spall zone illustrated in Figure 11 of Melosh (1984). The locus of the spall contour is given by z_p in Equation (3) of Melosh (1985). Warren (1994) estimates the ratio of z_p to projectile radius r_p as $z_p/r_p \sim 0.2 - 0.4$, and $z_p \sim 50 - 100$ m for a ~ 10 km diameter crater produced by an impactor of radius $r_p \sim 250$ m. These depths are consistent with those estimated above by comparing meteorite shock pressures to the predicted shock contours of Melosh (1989), and allow excavation of the basaltic shergottites and possibly even the nakhlites from their places of igneous crystallization. The textures of the lherzolitic shergottites show that they crystallized in a large volume (Harvey *et al.*, 1993), and thus at greater depths, and suggests they were relocated to the place where they were ejected. Because coherent, homogeneous material is required for build-up of maximal pressure gradients, highly brecciated material from the Martian highlands probably is discriminated against in the launch process. Thus, ALH84001, although previously shocked, probably was part of a larger coherent block prior to its ejection from Mars, and not part of the Martian regolith.

7. Concluding Remarks

Further study of the shock metamorphic histories of the Martian meteorites, combined with an improved quantitative understanding of the ejection mechanism(s), can make an important contribution to determining launch conditions and pre-launch sample locations. As additional meteorites are discovered, it will be particularly important to establish whether the "launch window" of peak shock pressures is maintained. Head and Melosh (2000) conclude that most Martian meteorites come from relatively small events of crater diameter ~ 3 km, just large enough to eject candidate Martian meteorite material. Nevertheless, several characteristics of the meteorites, including a relatively high percentage of "large" meteorites, strong launch pairing, sampling of a limited number of events, pre-launch shielding from cosmic rays, and, possibly, secondary break-ups in space, favor larger events. Warren (1994) noted that in the context of the spallation model, Nakhla and Zagami, representatives of two main ejection groups of Martian meteorites, probabably came from craters ~ 15 km and ~ 11 km in diameter, respectively. The launch mechanism for the Martian meteorites is sufficiently uncertain that a number of possible mechanisms should continue to be evaluated, however.

Because the places of origin of the Martian meteorites are unknown, use of their ages for calibrating the cratering rate is distinctly limited. Nevertheless, the observation of young igneous crystallization ages among the meteorites, down to ~ 165 Myr, shows that Martian volcanism continues essentially until the present day. Moreover, the observation of a high proportion of young ages suggests that Mars has been volcanically relatively active at recent times. The grouping of Martian meteorite ages around certain preferred values emphasizes the need to correct for "launch-pairing" among them, in contrast to the lunar case, where most individual meteorites appear to represent individual ejection events (Warren, 1994). We note that efforts to determine reliable launch-pairing of the meteorites will enable better interpretation of a variety of mineralogical, geochemical, and isotope geochemical data obtained for these rocks, currently our only samples of Mars. Finally, it must be noted that the degree of reliability currently achieved for radiometric dating of the Martian samples results from two decades of experience and improvements in laboratory techniques. Not all of the problems encountered in dating these samples have been analytical. Martian rocks appear to bear the record of a complex series of igneous and secondary processes, and the return of Martian samples to terrestrial laboratories will be required to answer the many remaining questions of Martian chronology. Those samples, too, will doubtless hold surprises for unwary analysts, but, if adequate samples are returned, dating them absolutely should be possible.

Acknowledgements

We thank the International Space Science Institute for hosting the Workshop on the Evolution of Mars, held at Bern, Switzerland, 10–14 April, 2000, and especially Johannes Geiss and Reinald Kallenbach for its organization. W. Hartmann and an anonymous reviewer tactfully prodded us into making this a more easily readable and comprehensive contribution. Technical support was provided by D. Garrison, H. Wiesmann, and Y. Reese (Lockheed-Martin, Houston), and A. Chaoui (Physikalisches Institut, Bern). Financial support by NASA's Cosmochemistry Program, RTOP 344-31-30, to L. Nyquist, D. Bogard, C.-Y. Shih; the Swiss

National Science Foundation to O. Eugster; and the Deutsche Forschungsgemeinschaft (DFG) to A. Greshake and D. Stöffler is gratefully acknowledged.

References

Barlow, N.G.: 1997, 'The Search for Possible Craters for Martian Meteorite ALH84001', *Proc. 28th Lunar Planet. Sci. Conf.*, 65–66.

Becker, R.H., and Pepin, R.O.: 1984, 'The Case for a Martian Origin of the Shergottites: Nitrogen and Noble Gases in EET 79001', *Earth Planet. Sci. Lett.* **69**, 225–242.

Becker, R.H., and Pepin, R.O.: 1986, 'Nitrogen and Light Noble Gases in Shergotty', *Geochim. Cosmochim. Acta* **50**, 993–1000.

Becker, R.H., and Pepin, R.O.: 1993, 'Nitrogen and Noble Gases in a Glass Sample from the LEW88516 Shergottite', *Meteoritics* **28**, 637–640.

Becker, L., Glavin, D.P., and Bada, J.L.: 1997, 'Polycyclic Aromatic Hydrocarbons (PAHs) in Antarctic Martian Meteorites, Carbonaceous Chondrites, and Polar Ice', *Geochim. Cosmochim. Acta* **61**, 475–481.

Berkley, J.L., and Keil, K.: 1981, 'Olivine Orientation in the ALHA77005 Achondrite', *Am. Mineral.* **66**, 1233–1236.

Berkley, J.L., and Boynton, N.J.: 1992, 'Minor/major Element Variation Within and Among Diogenite and Howardite Orthopyroxenite Groups', *Meteoritics* **27**, 387–394.

Binns R.W.: 1967, 'Stony Meteorites Bearing Maskelynite', *Nature* **214**, 1111–1112.

Bischoff, A., and Stöffler, D.: 1992, 'Shock Metamorphism as a Fundamental Process in the Evolution of Planetary Bodies: Information from Meteorites', *Eur. J. Minerals* **4**, 707–755.

Bell, J.F. III, McSween, H.Y., Jr., Crisp, J.A., Morris, R.V., Murchie, S.L., Bridges, N.T., Johnson, J.R., Britt, D.T., Golombek, M.P., Moore, H.J., Ghosh, A., Bishop, J.L., Anderson, R.C., Brückner, J., Economou, T., Greenwood, J.P., Gunnlaugsson, H.P., Hargraves, R.M., Hviid, S., Knudsen, J.M., Madsen, M.B., Reid, R., Rieder, R., and Soderblom, L.: 2000, 'Mineralogic and Compositional Properties of Martian Soil and Dust: Results from Mars Pathfinder', *J. Geophys. Res.* **105**, 1721–1755.

Blichert-Toft, J., Gleason J.D., Telouk, P., and Albarède, F.: 1999, 'The Lu-Hf Isotope Geochemistry of Shergottites and the Evolution of the Martian Mantle-crust System', *Earth Plan. Sci. Lett.* **173**, 25–39.

Boctor, N.Z., Fei, Y., Bertka, C.M., O'D Alexander, C.M., and Hauri, E.: 1998, 'Vitrification and High Pressure Phase Transition in Olivine Megacrysts from Lithology A in Martian Meteorite EETA79001', *Proc. 29th Lunar Planet. Sci. Conf.*, abstract #1492.

Bogard, D.D.: 1995, 'Impact Ages of Meteorites: A Synthesis', *Meteoritics* **30**, 244–268.

Bogard, D.D., and Husain, L.: 1977, 'A new 1.3 Aeon-young Achondrite', *Geophys. Res. Lett.* **4**, 69–71.

Bogard, D.D., and Johnson, P.: 1983, 'Martian Gases in an Antarctic Meteorite?', *Science* **221**, 651–654.

Bogard, D.D., and Garrison, D.H.: 1998, 'Relative Abundances of Argon, Krypton, and Xenon in the Martian Atmosphere as Measured in Martian Meteorites', *Geochim. Cosmochim. Acta* **62**, 1829–1835.

Bogard, D.D., and Garrison, D.H.: 1999, 'Argon-39-argon-40 "Ages" and Trapped Argon in Martian Shergottites, Chassigny, and Allan Hills 84001', *Met. Planet. Sci.* **34**, 451–473.

Bogard, D.D., Husain, L., Nyquist, L.E.: 1979, '^{40}Ar-^{39}Ar Age of the Shergotty Achondrite and Implications for its Post-shock Thermal History', *Geochim. Cosmochim. Acta* **43**, 1047–1056.

Bogard, D.D., Nyquist, L.E., and Johnson, P.: 1984, 'Noble Gas Contents of Shergottites and Implications for the Martian Origin of SNC Meteorites', *Geochim. Cosmochim. Acta* **48**, 1723–1739.

Bogard, D.D., Hörz, F., and Johnson, P.: 1986, 'Shock-implanted Noble Gases: An Experimental Study with Implications for the Origin of Martian Gases in Shergottite Meteorites', *Proc. 17th Lunar Planet. Sci. Conf., J. Geophys Res.* **91**, E99–E114.

Borg, L.E., Nyquist, L.E., Taylor, L.A., Wiesmann, H., Shih, C.-Y.: 1997, 'Constraints on Martian Differentiation Processes from Rb-Sr and Sm-Nd Isotopic Analyses of the Basaltic Shergottite QUE 94201', *Geochim. Cosmochim. Acta* **61**, 4915–4931.

Borg, L.E., Nyquist, L.E., Wiesmann, H.: 1998a, 'Rb-Sr Isotopic Systematics of the Lherzolitic Shergottite LEW88516', *Proc. 29th Lunar Planet. Sci. Conf.*, abstract #1233 (CD-ROM).

Borg, L.E., Nyquist, L.E., Wiesmann, H., Reese, Y.: 1998b, 'Samarium-neodymium Isotopic Systematics of the Lherzolitic Shergottite Lewis Cliff 88516' *Met. Planet. Sci.* **33**, A20.

Borg, L.E., Connelly, J.N., Nyquist, L.E., Shih, C.-Y., Wiesmann, H., Reese, Y.: 1999, 'The Age of the Carbonates in Martian Meteorite ALH84001', *Science* **286**, 90–94.

Borg, L.E., Nyquist, L.E., Wiesmann, H., Reese Y., Papike, J.J.: 2000, 'Sr-Nd Isotopic Systematics of Martian Meteorite DaG476', *Proc. 31st Lunar Planet. Sci.*, abstract #1036 (CD-ROM).

Borg, L.E., Nyquist, L.E., Reese, Y., Wiesmann, H., Shih C.Y., Taylor, L.A., and Ivanova, M.: 2001a, 'The Age of Dhofar 019 and its Relationship to the Other Martian Meteorites', *Proc. 32nd Lunar Planet. Sci. Conf.*, abstract #1144 (CD-ROM).

Borg, L.E., Nyquist, L.E., Wiesmann, H., and Reese, Y.: 2001b, 'Constraints on Secondary Alteration, Shock Metamorphism, and Petrogenetic Relationships of the Martian Meteorites LEW 88516 and ALHA77005 from Their Rb-Sr and Sm-Nd Isotopic Systematics', *Geochim. Cosmochim. Acta*, submitted.

Boston, P.J., Ivanov, M.V., and McKay, C.P.: 1992, 'On the Possibility of Chemosynthetic Ecosystems in Subsurface Habitats on Mars', *Icarus* **95**, 300–308.

Bunch, T.E., and Reid, A.M.: 1975, 'The Nakhlites. Part I. Petrography and Mineral Chemistry', *Meteoritics* **10**, 303–315.

Bridges, J.C., and Grady, M.M.: 2000, 'Evaporite Mineral Assemblages in the Nakhlite (Martian Meteorites)', *Earth Planet. Sci. Lett.* **176**, 267–279.

Chen, J.H., and Wasserburg, G.J.: 1986, 'Formation Ages and Evolution of Shergotty and its Parent Planet from U-Th-Pb Systemics', *Geochim. Cosmochim. Acta* **50**, 955–968.

Chen, J.H., and Wasserburg, G.J.: 1993, 'LEW88516 and SNC Meteorites', *Proc. 24th Lunar Planet. Sci. Conf.*, 275–276, (abstract).

Clayton, R.N., and Mayeda T.K.: 1996, 'Oxygen Isotope Studies of Achondrites', *Geochim. Cosmochim. Acta* **60**, 1999–2017.

Crozaz, G., and Wadhwa, M.: 1999, 'Chemical Alteration of hot Desert Meteorites: The Case of Shergottite Dar al Gani 476', *in: Workshop on Extraterrestrial Materials from Cold and Hot Deserts*. Kwa Maritane, Pilanesberg, South Africa, July 6–8.

Dreibus, G., Palme, H., Rammensee, W., Spettel, B., Weckwerth, G., and Wänke, H.: 1982, 'Composition of the Shergotty Parent Body: Further Evidence of a two Component Model for Planet Formation', *Proc. 13th Lunar Planet. Sci. Conf.*, 186–187.

Dreibus, G., and Wänke, H.: 1987, 'Volatiles on Earth and Mars: A Comparison', *Icarus* **71**, 225–240.

Dreibus, G., Burghele, A., Jochum, K.P., Spettel, B., Wlotzka, F., and Wänke, H.: 1994, 'Chemical and Mineral Composition of ALH84001: A Martian Orthopyroxenite', *Meteoritics* **29**, 461.

Dreibus, G., Spettel, B., Wlotzka, F., Schultz, L., Weber, H.W., Jochum, K.P., and Wänke, H.: 1996, 'QUE94001: An Unusual Martian Basalt', *Met. Planet. Sci.* **31 Suppl.**, A39–A40 (abstract).

Duke, M.B.: 1968, 'The Shergotty Meteorite: Magmatic and Shock Metamorphic Features', *in* B.M. French and N.M. Short (eds.), *Shock Metamorphism of Natural Materials*, Mono Book Corp., Baltimore, pp. 613–621.

Eberhardt, P., and Hess, D.C.: 1960, 'Helium in Stone Meteorites', *Astrophys. J.* **131**, 38–46.

El Goresy, A., Dubrovinsky, L., Sharp, T.G., Saxena, S.K., and Chen, M.: 2000, 'A Monoclinic Post-shishovite Polymorph of Silica in the Shergotty Meteorite', *Science* **288**, 1632–1634.

Engelhardt, W., von, and Graup, G.: 1984, 'Suevite of the Ries Crater, Germany: Source Rocks and Implications for Cratering Mechanics', *Geol. Rundschau* **73**, 447–481.

Eugster, O.: 1994, 'Orthopyroxenite ALH84001: Ejection from Mars (?) 15 Ma Ago', *Meteoritics* **29**, 464 (abstract).

Eugster, O., and Michel, T.: 1995, 'Common Asteroid Break-up Events of Eucrites, Diogenites, and Howardites and Cosmic-ray Production Rates for Noble Gases in Achondrites', *Geochim. Cosmochim. Acta* **59**, 177–199.

Eugster, O., Eberhardt, P., and Geiss, J.: 1967, '^{81}Kr in Meteorites and ^{81}Kr Radiation Ages', *Earth Planet. Sci. Lett.* **2**, 77–82.

Eugster, O., Polnau, E., and Terribilini, D.: 1997a, 'Ejection Age of Martian Lherzolite Yamato 793605, Chassigny, and Shergotty and Formation Age of Shergotty Maskelynite', *Met. Planet. Sci.* **32**, A40.

Eugster, O., Weigel, A., and Polnau, E.: 1997b, 'Ejection Times of Martian Meteorites', *Geochim. Cosmochim. Acta.* **61**, 2749–2757.

Evans, J.C., Wacker, J., and Reeves, J.H.: 1992, 'Terrestrial Ages of Victoria Land Meteorites Collected by the United States Expeditions 1985–1987', *in* Marvin and Mc Pherson (eds.), *Smithson. Contrib. Earth Sci.* **30**, Washington DC, 45–56.

Floran, R.J., Prinz, M., Hlava P.F., Keil, K., Nehru C.E., and Hinthorne, J.R.: 1978, 'The Chassigny Meteorite: A Cumulate Dunite with Hydrous Amphibole-bearing Melt Inclusions', *Geochim. Cosmochim. Acta* **42**, 1213–1229.

Folco, L., Franchi, I.A., Scherer, P., Schultz, L., and Pillinger, C.T.: 1999, 'Dar al Gani 489 Basaltic Shergottite: A new Find from the Sahara Likely Paired with Dar al Gani 476', *Meteorit. Planet. Sci.* **34** (Suppl.), A36–A37 (abstract).

Gale, N.H., Arden. J.W., Hutchison. R.: 1975, 'The Chronology of the Nakhla Achondritic Meteorite', *Earth Planet. Sci. Lett.* **26**, 195–206.

Ganapathy, R., and Anders, E.: 1969, 'Ages of Calcium-rich Achondrites - II. Howardites, Nakhlites, and the Angra dos Reis Angrite', *Geochim. Cosmochim. Acta* **33**, 775–787.

Garrison, D.H., and Bogard, D.D.: 1998, 'Isotopic Composition of Trapped and Cosmogenic Noble Gases in Several Martian Meteorites', *Met. Planet. Sci.* **33**, 721–736.

Garrison, D.H., and Bogard, D.D.: 2000, 'Cosmogenic and Trapped Noble Gases in the Los Angeles Martian Meteorite', *63rd Annual Met. Soc. Mtg. Aug. 28–Sept. 1, 2000, Chicago*, (abstract).

Geiss, J., and Hess, D.C.: 1958, 'Argon-potassium Ages and the Isotopic Composition of Argon from Meteorites', *Astrophys. J.* **127**, 224–236.

Gladman, B.: 1997, 'Destination: Earth. Martian Meteorite Delivery', *Icarus* **130**, 228–246.

Gleason, J.D., Kring, D.A., Hill, D.H., and Boynton, W.V.: 1997, 'Petrography and Bulk Chemistry of Martian Orthopyroxenite ALH84001: Implications for the Origin of Secondary Carbonates', *Geochim. Cosmochim. Acta* **61**, 3503–3512.

Golden, D.C., Ming, D.W., Schwandt, C.S., Lauer, H.V., Socki, R.A., Morris, R.V., Lofgren, G.E., and McKay, G.: 2000a, 'Inorganic Formation of Zoned Mg-Fe-Ca Carbonate Globules with Magnetite and Sulfide Rims Similar to those in Martian Meteorite ALH84001', *Proc. 31st Lunar Planet. Sci. Conf.*, LPI, Houston, abstract #1799 (CD-ROM).

Golden, D.C., Ming, D.W., Schwandt, C.S., Morris, R.V., Yang S.V., and Lofgren, G.E.: 2000b, 'An Experimental Study on Kinetically-driven Precipitation of Calcium-magnesium-iron Carbonates from Solution: Implications for the Low-temperature Formation of Carbonates in Martian Allan Hills 84001', *Met. Planet. Sci.* **35**, 457–465.

Gooding, J.L., Wentworth, S.J., and Zolensky, M.E.: 1991, 'Aqueous Alteration of the Nakhla Meteorite', *Meteoritics* **26**, 135–143.

Greshake, A.: 1998, 'Transmission Electron Microscope Charactrization of Shock Defects in Minerals from the Nakhla SNC Meteorite', *Meteorit. Planet. Sci.* **33**, A63.

Greshake, A., and Stöffler, D.: 1999, 'Shock Metamorphic Features in the SNC Meteorite Dar al Gani 476', *Proc. 30th Lunar Planet. Sci.*, LPI, Houston abstract #1377 (CD-ROM).

Greshake, A., and Stöffler, D.: 2000, 'Shock Related Melting Phenomena in the SNC Meteorite Dar al Gani 476', *Proc. 31st Lunar Planet. Sci. Conf.*, LPI, Houston abstract #1043 (CD-ROM).

Greshake, A., Stephan, T., and Rost, D.: 1998, 'Symplectic Exsolutions in Olivine from the Martian Meteorite Chassigny: Evidence for Slow Cooling Under Highly Oxidizing Conditions', *Proc. 29th Lunar Planet. Sci. Conf.*, abstract #1069.

Greshake, A., Schmitt, R.T., Stöffler, D., Pätsch, M., and Schultz, L.: 2001, 'Dhofar 081: A new Lunar Highland Meteorite', *Meteoritics and Planet. Sci.*, in press.

Grossman, J.N.: 2000, The Meteoritical Bulletin, No. 84, *Meteorit. Planet. Sci.* **35**, A199–A225.

Harper, C.L., Jr., Nyquist, L.E., Bansal, B.M., Wiesmann, H., and Shih, C.-Y.: 1995, 'Rapid Accretion and Early Differentiation of Mars Indicated by $^{142}Nd/^{144}Nd$ in SNC Meteorites', *Science* **267**, 213–216.

Hartmann, W.K.: 1999, 'Martian Cratering VI: Crater Count Isochrons and Evidence for Recent Volcanis from Mars Global Surveyor', *Met. Planet. Sci.* **34**, 167–177.

Hartmann, W.K.: 2001, 'Martian Seeps and Their Relation to Youthful Geothermal Activity', *Space Sci. Rev.*, this volume.

Hartmann, W.K., and Berman, D.C.: 2000, 'Elysium Planitia Lava Flows: Crater Count Chronology and Geological Implications', *J. Geophys. Res.* **105**, 15,011–15,025.

Hartmann, W.K., and Neukum, G.: 2001, 'Cratering Chronology and Evolution of Mars', *Space Sci. Rev.*, this volume.

Hartmann, W.K., Anguita, J., de la Casa, M., Berman, D.C., and Ryan, E.V.: 2000, 'Martian Cratering 7: The Role of Impact Gardening', *Icarus* **148**, in press.

Harvey, R., and McSween, H.Y., Jr.: 1996, 'A Possible High-temperature Origin for the Carbonates in the Martian Meteorite ALH84001', *Nature* **382**, 49–51.

Harvey, R.P., Wadhwa, M., McSween, H.Y., Jr., and Crozaz, G.: 1993, 'Petrography, Mineral Chemistry and Petrogenesis of Antarctic Shergottite LEW88516', *Geochim. Cosmochim. Acta* **57**, 4769–4783.

Head, J.N., Melosh, H.J.: 2000, 'Launch Velocity Distribution of the Martian Clan Meteorites', *Proc. 31st Lunar Planet. Sci.*, LPI, Houston, abstract #1937 (CD-ROM).

Heymann, D., Mazor, E., and Anders, E.: 1968, 'Ages of Ca-rich achondrites - I. Eucrites', *Geochim. Cosmochim. Acta* **32**, 1241–1268.

Ikeda, Y.: 1994, 'Petrography and Petrology of the ALH-77005 Shergottite', *Proc. NIPR Symp. Antarct. Meteorites* **7**, 9–29.

Ilg, S., Jessberger, E.K., El Goresy, A.: 1997, 'Argon-40/argon-39 Laser Extraction Dating of Individual Maskelynites in SNC Pyroxenite Allan Hills 84001', *Met. and Planet. Sci.* **32**, A65.

Ivanov, B.: 2001, 'Mars/Moon Cratering Rate Ratio Estimates', *Space Sci. Rev.*, this volume.

Jagoutz, E.: 1989, 'Sr and Nd Isotopic Systematics in ALHA77005: Age of Shock Metamorphism in Shergottites and Magmatic Differentiation on Mars', *Geochim. Cosmochim. Acta* **53**, 2429–2441.

Jagoutz, E.: 1996, 'Nd Isotopic Systematics of Chassigny', *Proc. 27th Lunar Planet. Sci.*, 597–598 (abstract).

Jagoutz, E., and Wänke H.: 1986, 'Sr and Nd Isotopic Systematics of Shergotty Meteorite', *Geochim. Cosmochim. Acta* **50**, 939–953.

Jagoutz, E., and Jotter, R.: 1999, 'SNC Meteorites: Relatives Finally Finding Each Other', *Met. Planet. Sci.* **34**, A59.

Jagoutz, E., and Jotter, R.: 2000, 'New Sm-Nd Isotope Data on Nakhla Minerals', *Proc. 31st Lunar Planet. Sci.*, abstract #1609 (CD-ROM).

Jagoutz, E., Sorowka, A., Vogel, J.D., Wänke, H.: 1994, 'ALH84001: Alien or Progenitor of the SNC Family?', *Meteoritics* **29**, 478–479 (abstract).

Jagoutz, E., Bogdanovski, O., Krestina, N., Jotter, R.,: 1999, 'DAG: A new age in the SNC Family, or the First Gathering of Relatives', *Proc. 30th Lunar Planet. Sci.*, abstract #1808 (CD-ROM).

Jones, J.H.: 1986, 'A Discussion of Isotopic Systematics and Mineral Zoning in the Shergottites: Evidence for a 180 Myr Igneous Crystallization Age', *Geochim. Cosmochim. Acta* **50**, 969–977.

Jones, J.H.: 1989, 'Isotopic Relationships Among the Shergottites, Nakhlites, and Chassigny', *Proc. 19th Lunar Planet. Sci. Conf.* , 465–474.

Jull, A.J.T., and Donahue, D.J.: 1988, 'Terrestrial ^{14}C Age of the Antarctic Shergottite EETA79001', *Geochim. Cosmochim. Acta* **52**, 1309–1311.

Jull, A.J.T., Cielaszyk, E., Brown, S.T., and Donahue, D.J.: 1994, '^{14}C Terrestrial Ages of Achondrites from Victoria Land Antarctica', *Proc. 25th Lun. Planet. Sci.*, 647–648.

Jull, A.J.T., Eastoe, C.J., Xue, S., and Herzog, G.F.: 1995, 'Isotopic Composition of Carbonate in the SNC Meteorites ALH84001 and Nakhla', *Meteoritics* **30**, 311–318.

Jull, A.J.T., Eastoe, C.J., and Clout, S.: 1997, 'Terrestrial Age of the Lafayette Meteorite and Stable-isotopic Composition of Weathering Products', *Proc. 28th Lun. Planet. Sci.*, 685–686.

Jull, A.J.T., Courtney, C., Jeffrey, D.A., and Beck, J.W.: 1998, 'Isotopic Evidence for a Terrestrial Source of Organic Compounds Found in Martian Meteorites Allan Hills 84001 and Elephant Moraine 79001', *Science* **279**, 366–369.

Kadano, T., and Fugiwara, A.: 1996, 'Observation of Expanding Vapor Cloud Generated by Hypervelocity Impact', *J. Geophys. Res.* **101**, 26,097–26,109.

Keller, L.P., Treiman A.H., and Wentworth S.J.: 1992, 'Shock Effects in the Shergottite LEW88516: Optical and Electron Microscope Observations', *Meteoritics* **27**, 242.

Keszthelyi, L., McEwen, A.S., and Thordarson, T.;, 2000, 'Terrestrial Analogs and Thermal Models for Martian Flood Lavas', *J. Geophys. Res.* **105**, 15,027–15,049.

Kieffer, S.W., Schaal, R., Gibbons, R., Hörz, F., Milton, D., and Duba, A.: 1976, 'Shocked Basalt from Lonar Crater, India, and Experimental Analogues', *Proc. 7th Lunar Planet. Sci. Conf.*, 1391–1412.

Knott, S.F., Ash, R.D., and Turner, G: 1996, '^{40}Ar-^{39}Ar Dating of ALH84001: Evidence for the Early Bombardment of Mars', *Proc. 26th Lunar Planet. Sci. Conf.*, 765–766 (abstract).

Kring, D.A., and Gleason, J.D.: 1997, 'Magmatic Temperatures and Compositions on Early Mars as Inferred from the Orthopyroxene-silica Assemblage in Allan Hills 84001', *Meteoritics* **32**, A74.

Kring, D.A., Swindle, T.D., Gleason, J.D., and Grier, J.A.: 1998, 'Formation and Relative Ages of Maskelynite and Carbonate in ALH84001', *Geochim. Cosmochim. Acta* **62**, 2155–2166.

Lambert, P.: 1985, 'Metamorphic Record in Shergottites', *Meteoritics* **20**, 690–691.

Lambert, P., and Grieve, R.A.F.: 1984, 'Shock-experiments on Maskelynite-bearing Anorthosite', *Earth Planet. Sci. Lett.* **68**, 159–171.

Lancet, M.S., and Lancet, K.: 1971, 'Cosmic-ray and Gas Retention Ages of the Chassigny Meteorite', *Meteoritics* **6**, 81–86.

Langenhorst, F., and Greshake, A.: 1999, 'A TEM Study of Chassigny: Evidence for Strong Shock Metamorphism', *Met. Planet. Sci.* **34**, 43–48.

Langenhorst, F., and Poirier, J.-P.: 2000, '"Eclogitic" Minerals in a Shocked Basaltic Meteorite', *Earth Planet. Sci. Lett.* **176**, 259–265.

Langenhorst, F., Stöffler, D., and Klein, D.: 1991, 'Shock Metamorphism of the Zagami Achondrite', *Proc. 22nd Lunar Planet. Sci. Conf.*, 779–780.

Longhi, J.: 1991, 'Complex Magmatic Processes on Mars: Inferences from the SNC Meteorites', *Proc. Lunar Planet. Sci. Conf.* **21**, 695–709.

Lodders, K.: 1998, 'A Survey of Shergottite, Nakhlite and Chassigny Meteorites Whole-rock Compositions', *Met. Planet. Sci.* **33**, A183–A190.

Lorenzetti, S., and Eugster, O.: 2001, 'Ejection Ages of the Shergottites Los Angeles and Say al Uhaymir', to be published.

Lugmair, G.W., and Galer, S.J.: 1992, 'Age and Isotopic Relationships Among the Angrites Lewis Cliff 86010 and Angra dos Reis', *Geochim. Cosmochim. Acta* **56**, 1673–1694.

Malin, M.C., and Edgett, K.S.: 2000, 'Evidence for Recent Groundwater Seepage and Surface Runoff on Mars', *Science* **288**, 2330–2335.

Marti, K.: 1967, 'Mass-spectrometric Detection of Cosmic-ray-produced Kr-81 in Meteorites and the Possibility of Kr-Kr Dating', *Phys. Rev. Lett.* **18**, 264–266.

Marti, K., and Matthew, K. J.: 2000, 'Ancient Martian Nitrogen', *Geophys. Res. Lett.* **27**, 1463–1466.

Marti, K., Kim, J.S., Thakur, A.N., McCoy, T.J., and Keil, K.: 1995, 'Signatures of the Martian Atmosphere in Glass of the Zagami Meteorite', *Science* **267**, 1981–1984.

McCoy, T.J., Taylor, G.J., and Keil, K.: 1992, 'Zagami: Product of a Two-stage Magmatic History', *Geochim. Cosmochim. Acta* **56**, 3571–3582.

McKay, D.S., Gibson, E.K., Jr., Thomas-Keprta, K.L., Vali, H., Romanek, C.S., Clemett, S.J., Chillier, X.D.F., Maechling, C.R., and Zare, R.N.: 1996, 'Search for Life on Mars: Possible Relic Biogenic Activity in Martian Meteorite ALH84001', *Science* **273**, 924–930.

McSween, H.Y., Jr.: 1994, 'What Have we Learned about Mars from SCN Meteorites', *Meteoritics* **29**, 757–779.

McSween, H.Y., Jr., and Stöffler, D.: 1980, 'Shock Metamorphic Features in ALHA77005 Meteorite', *Proc. 11th Lunar Planet. Sci. Conf.*, 717–719.

McSween, H.Y., Jr., and Harvey, R.P.: 1998, 'An Evaporation Model for Formation of Carbonates in the ALH84001 Martian Meteorite', *Intern. Geol. Rev.* **40**, 774–783.

McSween, H.Y., Jr., and Keil, K.: 2000, 'Mixing Relationships in the Martian Regolith and the Composition of Globally Homogeneous Dust', *Geochim. Cosmochim. Acta* **64**, 2155–2166.

Melosh, H.J.: 1984, 'Impact Ejection, Spallation, and the Origin of Meteorites', *Icarus* **59**, 234–260.

Melosh, H.J.: 1985, 'Ejection of Rock Fragments from Planetary Bodies', *Geology* **13**, 144–148.

Melosh, H.J.: 1989, 'Impact Cratering: A Geologic Process', *Sky and Telescope* **78**, 382.

Melosh, H.J.: 1995, 'Cratering Dynamics and the Delivery of Meteorites to the Earth', *Meteoritics* **30**, 545–546.

Metzler, K., Bobe, K.D., Palme, H., Spettel, B. and Stöffler, D.: 1995, 'Thermal and Impact Metamorphism of the HED-asteroid', *Planet. Space Sci.* **43**, 499–525.

Meyer, C.: 1998, *Mars Meteorite Compendium - 1998*, NASA, Houston, Texas, 237 pp.

Mikouchi, T., and Miyamoto, M.: 1997, 'Yamato-793605: A new Lherzolitic Shergottite from the Japanese Antarctic Meteorite Collection', *Antarct. Meteorite Res.* **10**, 41–60.

Mikouchi, T., and Miyamoto, M.: 1998, 'Pyroxene and Olivine Microstructures in Nakhlite Martian Meteorites: Implication for Their Thermal History', *Proc. 29th Lunar Planet. Sci.*, abstract #1574.

Mileikowsky, C., Cucinotta, F.A., Wilson, J.W., Gladman, B., Horneck, G., Lindegren, L., Melosh, J., Rickman, H., Valtonen, M., and Zheng, J.Q.: 2000, 'Natural Transfer of Viable Microbes in Space', *Icarus* **145**, 391–427.

Minster, J.-F., Birck, J.-L., Allègre, C.-J.: 1982, 'Absolute Age of Formation of Chondrites by the ^{87}Sr-^{87}Sr Method', *Nature* **300**, 414–419.

Misawa, K., Nakamura, N., Premo, W.R., Tatsumoto, M.: 1997, 'U-Th-Pb Isotopic Systematics of Lherzolitic Shergottite Yamato 793605', *Antarctic Meteorites* **XXII**, NIPR, 115–117.

Mittlefehldt, D.W.: 1994, 'ALH84001, A Cumulate Orthopyroxenite Member of the Martian Meteorite Clan', *Meteoritics* **29**, 214–221.

Mittlefehldt, D.W., Lindstrom, D.J., Lindstrom, M.M., and Martinez, R.R.: 1997, 'Lithology A in EETA79001 – Product of Impact Melting on Mars', *Proc. 28th Lunar Planet. Sci. Conf.*, 961–962.

Miura, Y.N., Nagao, K., Sugiura, N., Sagawa, H., and Matsubara, K.: 1995, 'Orthopyroxenite ALHA84001 and Shergottite ALHA77005: Additional Evidence for a Martian Origin from Noble Gases', *Geochim. Cosmochim. Acta* **59**, 2105–2113.

Mouginis-Mark, P.J., McCoy, T.J., Taylor, G.J., Keil, K.: 1992, 'Martian Parent Craters for the SNC Meteorites', *J. Geophys. Res.* **97**, 10,213–10,225.

Müller, W.F.: 1993, 'Thermal and Deformational History of the Shergotty Meteorite Deduced from Clinopyroxene Microstructure', *Geochim. Cosmochim. Acta* **57**, 4311–4322.

Müller, H.W., and Zähringer, J.: 1969, 'Rare Gases in Stony Meteorites', *in* P.M. Millmann (ed.) *Meteorite Research*, D. Reidel, Dordrecht, pp. 845–856.

Nagao, K., Nakamura, T., Miura, Y., and Takaoka, N.: 1997, 'Noble Gases and Mineralogy of Primary Igneous Materials of the Yamato-793605 Shergottite', Natl Inst. Polar Res., Tokyo, *Anarct. Meteorit. Res.* **10**, 125–142.

Nakamura, N., Komi, H., and Kagami, H.: 1982a, 'Rb-Sr Isotopic and REE Abundances in the Chassigny Meteorite', *Meteoritics* **17**, 257–258.

Nakamura, N., Unruh, D.M., Tatsumoto, M., and Hutchison, R.: 1982b, 'Origin and Evolution of the Nakhla Meteorite Inferred from the Sm-Nd and U-Pb Systematics and REE, Ba, Sr, Rb Abundances', *Geochim. Cosmochim. Acta* **46**, 1555–1573.

Neukum, G., Ivanov, B., and Hartmann, W.K.: 2001, 'Cratering Records in the Inner Solar System in Relation to the Lunar Reference System', *Space Sci. Rev.*, this volume.

Newsom, H.E., Brittelle, G.E., Hibbitts, C.A., Crossey, L.J., and Kudo, A.M.: 1996, 'Impact Crater Lakes on Mars', *J. Geophys. Res.* **101**, 14,951–14,955.

Nishiizumi, K., and Caffee, W.: 1996, 'Exposure History of Shergottite Queen Alexandra Range 94201', *Proc. 27th Lun. Planet. Sci. Conf.*, 961–962.

Nishiizumi, K., and Caffee, M.W.: 1997, 'Exposure History of Shergottite Yamato' - 793605 (abstract). *Antarctic Meteorites* **XXII**, Natl Inst. Polar Res., Tokyo, 149–151.

Nishiizumi, K., Klein, J., Middleton, R., Elmore, D., Kubik, P.W., and Arnold, J.R.: 1986, 'Exposure History of Shergottites', *Geochim. Cosmochim. Acta* **50**, 1017–1021.

Nishiizumi, K., Arnold J.R., Caffee, M.W., Finkel, R.C., and Southon, J.: 1992, 'Exposure Histories of Calcalong Creek and LEW88516 Meteorites', *Meteoritics* **27**, 270 (abstract).

Nishiizumi, K., Masarik, J., Welten, K.C., Caffee, M.W., Jull A.J.T., and Klandrud, S.E.: 1999, 'Exposure History of new Martian Meteorite Dar al Gani 476', *Proc. 30th Lun. Planet. Sci. Conf.*, abstract #1966 (CD-ROM).

Nishiizumi, M.W., and Masarik, J.: 2000, 'Cosmogenic Radionuclides in Los Angeles Martian Meteorite', *63rd Annual Met. Soc. Mtg. Aug. 28–Sept. 1, 2000, Chicago* (abstract).

Norman, M.D.: 1999, 'The Composition and Thickness of the Crust of Mars Estimated from Rare Earth Elements and Neodymium-isotopic Compositions of Martian Meteorites', *Met. Planet. Sci.* **34**, 439–449.

Nyquist, L.E.: 1983, 'Do Oblique Impacts Produce Martian Meteorites?' *Proc. 13th Lunar Planet. Sci. Conf., J. Geophys. Res.* **88 Suppl.**, A785–A798.

Nyquist, L.E., Wooden, J., Bansal, B., Wiesmann, H., McKay, G., Bogard, D.D.: 1979a, 'Rb-Sr age of the Shergotty Achondrite and Implications for Metamorphic Resetting of Isochron Ages', *Geochim. Cosmochim. Acta* **43**, 1057–1074.

Nyquist, L.E., Bogard, D.D., Wooden, J.L., Wiesmann, H., Shih, C.-Y., Bansal, B.M., and McKay, G.: 1979b, 'Early Differentiation, Late Magmatism, and Recent Bombardment on the Shergottite Parent Planet', *Meteoritics* **14**, 502.

Nyquist, L.E., Wiesmann, H., Shih, C.-Y., Bansal, B.M.: 1986, 'Sr Isotopic Systematics of EETA79001', *Proc. 17th Lunar Planet. Sci. Conf.*, 624–625 (abstract).

Nyquist, L.E., Bansal, B., Wiesmann, H., Shih, C.-Y.: 1994, 'Neodymium, Strontium and Chromium Isotopic Studies of the LEW86010 and Angra dos Reis Meteorites and the Chronology of the Angrite Parent Body', *Meteoritics* **29**, 882–885.

Nyquist, L.E., Bansal, B.M., Wiesmann, H., Shih, C.-Y.: 1995, '"Martians" Young and old: Zagami and ALH84001', *Proc. 26th Lunar Planet. Sci. Conf.*, 1065–1066 (abstract).

Nyquist, L.E., Borg, L.E., Shih, C.-Y.: 1998, 'The Shergottite Age Paradox and the Relative Probabilities for Martian Meteorites of Differing Ages', *J. Geophys. Res.* **103**, 31,445–31,455.

Nyquist, L.E., Reese, Y.D., Wiesmann, H., Shih, C.-Y., and Schwandt, C.: 2000, 'Rubidium-Strontium Age of the Los Angeles Shergottite', *Met. Planet. Sci.* **35**, A121–A122.

Nyquist, L.E., Reese, Y., Wiesmann, H., and Shih, C.-Y.: 2001, 'Age of EET79001B and Implications for Shergottite Origins', *Proc. 32nd Lunar Planet. Sci. Conf.*, abstract #1407 (CD-ROM).

O'Keefe, J.D., and Ahrens, T.J.: 1986, 'Oblique Impact: A Process for Obtaining Meteorite Samples from Other Planets', *Science* **234**, 346–349.

Ostertag, R., Amthauer, G., Rager, H., and McSween, H.Y., Jr.: 1984, 'Fe^{3+} in Shocked Olivine Crystals of the ALHA77005 Meteorite', *Earth Planet. Sci. Lett.* **67**, 162–166.

Ott, U.: 1988, 'Noble Gases in SNC Meteorites Shergotty, Nakhla, Chassigny', *Geochim. Cosmochim. Acta* **52**, 1937–1948.

Ott, U., and Löhr, H.P.: 1992, 'Noble Gases in the new Shergottite LEW88516', *Meteoritics* **27**, 271 (abstract).

Paetsch, M., Altmaier, M., Herpers, U., Kosuch, H., Michel, R., and Schultz, L.: 2000, 'Exposure Age of the New SNC Meteorite Sayh al Uhaymir 005', *Met. Planet. Sci.* **35**, A124–A125 (abstract).

Pal, D.K., Tuniz, C., Moniot, R.K., Savin,W., Kruse, T., and Herzog, G.F.: 1986, 'Beryllium-10 Contents of Shergottites, Nakhlites and Chassigny', *Geochim. Cosmochim. Acta* **50**, 2405–2409.

Papanastassiou, D.A., Wasserburg, G.J.: 1974, 'Evidence for Late Formation and Young Metamorphism in the Achondrite Nakhla', *Geophy. Res. Lett.* **1**, 23–26.

Pierazzo, E., and Melosh, H.J.: 1999, 'Hydocode Modeling of Chicxulub as an Oblique Impact Event', *Earth Planet. Sci. Lett.* **165**, 163–176.

Pierazzo, E., and Melosh, H.J.: 2000, 'Melt Production in Oblique Impacts', *Icarus* **145**, 252–261.

Podosek, F.A.: 1973, 'Thermal History of the Nakhlites by the ^{40}Ar-^{39}Ar Method', *Earth Planet. Sci. Lett.* **19**, 135–144.

Pohl, J., Stöffler, D., Gall, H., and Ernstson, K.: 1977, 'The Ries Impact Crater', *in* D.H. Roddy, R.O. Pepin, and R.B. Merrill (eds.), *Impact and Explosion Cratering*, Pergamon Press, New York, pp. 343–404.

Rieder, R., Economou, T., Wänke, H., Turkevich, A., Crisp, J., Brückner, J., Dreibus, G., McSween, H.Y., Jr.: 1997, 'The Chemical Composition of Martian Soil and Rocks Returned by the Mobile Alpha Proton X-ray Spectrometer: Preliminary Results from the X-ray Mode', *Science* **278**, 1771–1774.

Rubin, A.E., Warren, P.H., Greenwood, J.P., Verish, R.S., Leshin, L.A., Hervig, R.L., Clayton, R.N., and Mayeda, T.K.: 2000, 'Los Angeles: the Most Differentiated Basaltic Martian Meteorite', *Geology*, submitted.

Sano, Y., Terada, K., Takeno, S., Taylor, L.A., McSween, H.Y.: 2000, 'Ion Microprobe Uranium-thorium-lead Dating of Shergotty Phosphates', *Met. Planet. Sci.* **35**, 341–346.

Sarafin, R., Herpers, U., Signer, P., Wieler, R., Bonani, G., Hofmann, H.J., Morenzoni, E., Nessi, M., Suter, M., and Wölfli, W.: 1985, '^{10}Be, ^{26}Al, ^{53}Mn, and Light Noble Gases in the Antarctic Shergottite EETA79001(A)', *Earth Planet. Sci. Lett.* **75**, 72–76.

Schaal, R.B., and Hörz, F.: 1977, 'Shock Metamorphism of Lunar and Terrestrial Basalts', *Proc. 8th Lunar Planet. Sci. Conf.*, 1679–1729.

Schmitt, R.T.: 2000, 'Shock Experiments with the H6 Chondrite Kernouvé: Pressure Calibration of Microscopic Shock Effects', *Met. Planet. Sci.* **35**, 545–560.

Schultz, L., and Freundel, M.: 1984, 'Terrestrial Ages of Antarctic Meteorites', *Meteoritics* **19**, 310 (abstract).

Schultz, P.H., and DŠHondt, S.: 1996, 'Cretaceous-Tertiary (Chicxulub) Impact Angle and its Consequences', *Geology* **24**, 963–967.

Schultz, L., and Franke, L.: 2000, 'Helium, Neon, and Argon in Meteorites, a Data Compilation', Max-Planck-Institut für Chemie, Mainz.

Scott, E.R.D.: 1999, 'Origin of Carbonate-magnetite-sulfide Assemblages in Martian Meteorite ALH84001', *J. Geophys. Res.* **104**, 3803–3813.

Scott, E.R.D., Krot, A.N., and Yamagouchi, A.: 1998, 'Carbonates in Fractures of Martian Meteorite ALH84001: Petrologic Evidence for Impact Origin', *Met. Planet. Sci.* **33**, 709–719.

Sharp, T.G., El Goresy, A., Wopenka, B., and Chen, M.: 1999, 'A Post-stihovite SiO$_2$ Polymorph in the Meteorite Shergotty: Implications for Impact Events', *Science* **284**, 1511–1513.

Shih, C.-Y., Nyquist, L.E., Bogard, D.D., McKay, G.A., Wooden, J.L., Bansal, B.M., and Wiesmann, H.: 1982, 'Chronology and Petrogenesis of Young Achondrites, Shergotty, Zagami, and ALHA 77005: Late Magmatism on a Geologically Active Planet', *Geochim. Cosmochim. Acta* **46**, 2323–2344.

Shih, C.-Y., Nyquist, L.E., Reese, Y., and Wiesmann, H.: 1998, 'The Chronology of the Nakhlite Lafayette: Rb-Sr and Sm-Nd Isotopic Ages', *Proc. 29th Lunar Planet. Sci.*, abstract #1145 (CD-ROM).

Shih, C.-Y., Nyquist, L.E., and Wiesmann, H.: 1999, 'Sm-Nd and Rb-Sr Systematics of Nakhlite Governador Valadares', *Met. Planet. Sci.* **34**, 647–655.

Shukolyukov, Yu.A., Nazarov, M.A., and Schultz, L.: 2000, 'Dhofar019: A Shergottite with an Approximately 20-million-year Exposure Age', *Met. Planet. Sci.* **35**, A (abstract).

Stauffer H.: 1962, 'On the Production Rates of Rare Gas Isotopes in Stone Meteorites', *J. Geophys. Res.* **67**, 2023–2028.

Steele, I.M. and Smith, J.V.: 1982, 'Petrography and Mineralogy of two Basalts and Olivine-pyroxene-spinel Fragments in Achondrite EETA79001', *J. Geophys. Res.* **87**, A375–A384.

Steiger, R.H., and Jäger, E.: 1977, 'Subcommission on Geochronology: Convention on the use of Decay Constants in Geo- and Cosmochronology', *Earth Planet. Sci. Lett.* **36**, 359–362.

Stephan, T., Rost ,D., Jessberger, E.K., and Greshake, A.: 1998, 'Polycyclic Aromatic Hydrocarbons are Everywhere in Allan Hills 84001', *Met. Planet. Sci.* **33**, A149–A150.

Stephan, T., Greshake, A., Herpers, U., Jessberger, E.K., Jochum, K.-P., Michel, R., Ott, U., and Stöffler, D.: 1999, 'Systematic and Interdisciplinary Analyses of all SNC Meteorites', Proposal, Münster, pp. 42.

Stöffler, D.: 1972, 'Deformation and Transformation of Rock-forming Minerals by Natural and Experimental Shock Processes: I. Behavior of Minerals Under Shock Compression', *Fortschr. Miner.* **49**, 50–113.

Stöffler, D.: 1984, 'Glasses Formed by Hypervelocity Impact', *J. Non-Cryst. Sol.* **67**, 465–502.

Stöffler, D.: 2000, 'Maskelynite Confirmed as Diaplectic Glass: Indication for Peak Shock Pressures Below 45 GPa in all Martian Meteorites', *Proc. 31st Lunar Planet. Sci.*, abstract #1170.

Stöffler, D., and Ostertag, R.: 1983, 'The Ries Impact Crater', *Fortschr. Miner.* **61**, Beiheft 2, 71–116.

Stöffler, D., Ostertag, R., Jammes, C., Pfannschmidt, G., Sen Gupta, P.R., Simon, S.B., Papike, J.J., and Beauchamp, R.H.: 1986, 'Shock Metamorphism and Petrography of the Shergotty Achondrite', *Geochim. Cosmochim. Acta* **50**, 889–913.

Stöffler, D., Bischoff, A., Buchwald, U., and Rubin, A.E.: 1988, 'Shock Effects in Meteorites', *in* J.F. Kerridge and M.S. Matthews (eds.), *Meteorites and the Early Solar System*, Univ. Arizona Press, Tucson, pp. 165–205.

Swindle, T.D., Caffee, M.W., and Hohenberg, C.M.: 1986, 'Xenon and Other Noble Gases in Shergottites', *Geochim. Cosmochim. Acta* **50**, 1001–1015.

Swindle, T.D., Nichols, R., and Olinger C.T.: 1989, 'Noble Gases in the Nakhlite Governador Valadares', *Proc. 20th Lunar Planet. Sci. Conf.*, 1097–1098 (abstract).

Swindle, T.D., Grier, J.A., Burkland, M.K.: 1995, 'Noble Gases in Orthopyroxenite ALH84001: A Different Kind of Martian Meteorite with an Atmospheric Signature', *Geochim. Cosmochim. Acta* **59**, 793–801.

Swindle, T.D., Li, B., and Kring, D.A.: 1996, 'Noble Gases in Martian Meteorite QUE94201', *Proc. 27th Lunar Planet. Sci. Conf.*, 1297–1298 (abstract).

Swindle, T.D., Treiman, A.H., Lindstrom, D.J., Burkland, M.K., Cohen, B.A., Grier, J.A., Li, B., Olson, E.K.: 1999, 'Nobel Gases in Iddingsite from the Lafayette Meteorite: Evidence of Liquid Water on Mars in the Last few Hundred Million Years', *Met. Planet. Sci.* **35**, 107–115.

Swindle T.D., Treiman, A.H., Lindstrom, D.J., Burkland, M.K., Cohen, B.A., Grier, J.A., Li B., and Olson E.K.: 2000, 'Noble Gases in Iddingsite from the Lafayette Meteorite', *Met. Planet. Sci.* **35**, 107–116.

Tanaka, K.L.: 1986, 'The Stratigraphy of Mars', *Proc. 17th Lunar Planet. Sci. Conf., J. Geophys. Res. Suppl.* **91**, E139–158.

Tanaka, K.L., Scott, D.H., and Greeley, R.: 1992, 'Global stratigraphy', *in* H.H. Kieffer *et al.*(ed.) *Mars*, Univ. of Ariz. Press, Tucson, pp. 345–382.

Terribilini, D., Eugster, O., Burger, M., Jakob, A., and Krähenbühl, U.: 1998, 'Noble Gases and Chemical Composition of Shergotty Mineral Fractions, Chassigny, and Yamato-793605', *Met. Planet. Sci.* **33**, 677–684.

Terribilini, D., Busemann, H., and Eugster O.: 2000, '^{81}Kr - Kr Cosmic-ray Exposure Ages of Martian Meteorites Including the new Shergottite Los Angeles', *Met. Planet. Sci.* **35**, A155.

Treiman, A.H.: 1985, 'Amphibole and Hercynite Spinel in Shergotty and Zagami: Magmatic Water, Depth of Crystallization, and Metasomatism', *Meteoritics* **20**, 229–243.

Treiman, A.H.: 1987, 'Geology of the Nakhlite Meteorites: Cumulate Rocks from Flows and Shallow Intrusions', *Proc. 18th Lunar Planet. Sci.*, 1022–1023.

Treiman, A.H.: 1995a, 'S$\neq$NC: Multiple Source Areas for Martian Meteorites', *J. Geophys. Res.* **100**, 5329–5340.

Treiman, A.H.: 1995b, 'A Petrographic History of Martian Meteorite ALH84001: Two Shocks and an Ancient Age', *Meteoritics* **30**, 294–302.

Treiman, A.H.: 1998, 'The History of ALH84001 Revised: Multiple Shock Events', *Meteorit. Planet. Sci.* **33**, 753–764.

Treiman, A.H., McKay, G.A., Bogard, D.D., Mittelfehldt, D.W., Wang, M.-S., Keller, L., Lipschutz, M.E., Lindstrom, M.M., and Garrison, D.: 1994, 'Comparison of the LEW88516 and ALH77005 Martian Meteorites: Similar but Distinct', *Meteoritics* **29**, 581–592.

Tschermak, G.: 1872, 'Die Meteoriten von Shergotty und Gopalpur', *Sitzungsberichte der mathematisch-naturwissenschaftlichen Classe der kaiserlichen Akademie der Wissenschaften* **65**, 122–146.

Turner, G., Knott, S.F., Ash, R.D., Gilmour, J.D.: 1997, 'Ar-Ar Chronology of the Martian Meteorite ALH84001: Evidence for the Timing of the Early Bombardment of Mars', *Geochim. Cosmochim. Acta* **61**, 3835–3850.

Vickery, A.: 1986, 'Effect of an Impact-generated Gas Cloud on the Acceleration of Solid Ejecta', *J. Geophys. Res.* **91**, 14,139–14,160.

Warren, P.H.: 1994, 'Lunar and Martian Meteorite Delivery Services', *Icarus* **111**, 338–363.

Warren, P.H.: 1998, 'Petrologic Evidence for Low-temperature, Possible Flood-evaporitic Origin of Carbonates in the ALH84001 Meteorite', *J. Geophys. Res.* **103**, 16,759–16,773.

Wänke, H.: 1991, 'Chemistry, Accretion and Evolution on Mars', *Space Sci. Rev.* **56**, 1–8.

Wänke, H., and Dreibus, G.: 1988, 'Chemical Composition and Accretion History of Terrestrial Planets', *Phil. Trans. Roy. Soc. London* **A325**, 545–557.

Wasson, J.T., and Wetherill, G.W.: 1979, 'Dynamical, Chemical, and Isotopic Evidence Regarding the Formation Locations of Asteroids and Meteorites', *in* T. Gehrels (ed.), *Asteroids*, Univ. Arizona Press, Tucson, pp. 926–974.

Wadhwa, M., and Crozaz, G.: 1994, 'Rare Earth Element Distributions in Chassigny: Clues to its Petrogenesis and Relation to the Nakhlites', *Proc. 25th Lunar Planet. Sci.*, 1451–1452.

Wadhwa, M., and Lugmair, G.W.: 1996, 'The Formation Age of Carbonates in ALH84001', *Met. Planet. Sci.* **31**, A145.

Weiss, B.P., Kirschvink, J.L., Baudenbacher, F.J., Vali, H., Peters, N.T., Macdonald, F.A., and Wikswo, J.P.: 2000, 'A Low Temperature Transfer of ALH84001 from Mars to Earth', *Science* **290**, 791–795.

Wetherill, G.W.: 1988, 'Where do the Apollo Objects Come from?' *Icarus* **111**, 1–18.

Wiens, R.C., and Pepin, R.O.: 1988, 'Laboratory Shock Emplacement of Noble Gases, Nitrogen, and Carbon Dioxide into Basalt, and Implications for Trapped Gases in Shergottite EET79001', *Geochim. Cosmochim. Acta* **52**, 295–307.

Wood, C.A., and Ashwal, L.D.: 1981, 'SNC Meteorites: Igneous Rocks from Mars', *Proc. 12th Lunar Planet. Sci. Conf.*, 1359–1375.

Wooden, J.L., Nyquist, L.E., Bogard, D.D., Bansal, B., Wiesmann, H., Shih C.-Y., and McKay, G.A: 1979, 'Radiometric Ages for the Achondrites Chervony Kut, Governador Valadares, and Allan Hills 77005', *Proc. 10th Lunar Planet. Sci.*, 1379–1381 (abstract).

Zipfel, J., Scherer, P., Spettel, B., Dreibus, G., and Schultz, L.: 2000, 'Petrology and Chemistry of the new Shergottite Dar al Gani 476', *Met. Planet. Sci.* **35**, 95–106.

Address for correspondence: SN/Planetary Sciences, NASA Johnson Space Center, 2101 NASA Road 1, Houston, TX 77058-3696, USA;
(laurence.e.nyquist1@jsc.nasa.gov)

CRATERING CHRONOLOGY AND THE EVOLUTION OF MARS

WILLIAM K. HARTMANN[1] and GERHARD NEUKUM[2]

[1]*Planetary Science Institute, 620 N. 6th Avenue, Tucson AZ 85705-8331, USA*
[2]*Deutsches Zentrum für Luft- und Raumfahrt, D-12489 Berlin, Germany*

Received: 16 November 2000; accepted: 27 February 2001

Abstract. Results by Neukum *et al.* (2001) and Ivanov (2001) are combined with crater counts to estimate ages of Martian surfaces. These results are combined with studies of Martian meteorites (Nyquist *et al.*, 2001) to establish a rough chronology of Martian history. High crater densities in some areas, together with the existence of a 4.5 Gyr rock from Mars (ALH84001), which was weathered at about 4.0 Gyr, affirm that some of the oldest surfaces involve primordial crustal materials, degraded by various processes including megaregolith formation and cementing of debris. Small craters have been lost by these processes, as shown by comparison with Phobos and with the production function, and by crater morphology distributions. Crater loss rates and survival lifetimes are estimated as a measure of average depositional/erosional rate of activity.

We use our results to date the Martian epochs defined by Tanaka (1986). The high crater densities of the Noachian confine the entire Noachian Period to before about 3.5 Gyr. The Hesperian/Amazonian boundary is estimated to be about 2.9 to 3.3 Gyr ago, but with less probability could range from 2.0 to 3.4 Gyr. Mid-age dates are less well constrained due to uncertainties in the Martian cratering rate. Comparison of our ages with resurfacing data of Tanaka *et al.* (1987) gives a strong indication that volcanic, fluvial, and periglacial resurfacing rates were all much higher in approximately the first third of Martian history. We estimate that the Late Amazonian Epoch began a few hundred Myr ago (formal solutions 300 to 600 Myr ago). Our work supports Mariner 9 era suggestions of very young lavas on Mars, and is consistent with meteorite evidence for Martian igneous rocks 1.3 and $0.2-0.3$ Gyr old. The youngest detected Martian lava flows give formal crater retention ages of the order 10 Myr or less. We note also that certain Martian meteorites indicate fluvial activity younger than the rocks themselves, 700 Myr in one case, and this is supported by evidence of youthful water seeps. The evidence of youthful volcanic and aqueous activity, from both crater-count and meteorite evidence, places important constraints on Martian geological evolution and suggests a more active, complex Mars than has been visualized by some researchers.

1. Background: Cratering Studies and the Relation to Martian Rocks

Through the process of impact cratering, Nature randomly stamps circular bowls of known shape on planetary surfaces. This fact offers us a tool for interpreting planetary surfaces. Though the accumulated numbers of impact craters, we can assess ages. Through the modification of the crater shapes by erosion, dust deposition, lava flow coverage, etc., we can assess geological processes of the planet.

In this volume Stöffler and Ryder (2001) summarized the basic radiometric dating that dates lunar surfaces, and correlates with impact crater density. Neukum *et al.* (2001) laid out evidence that the shape of the crater size distribution and

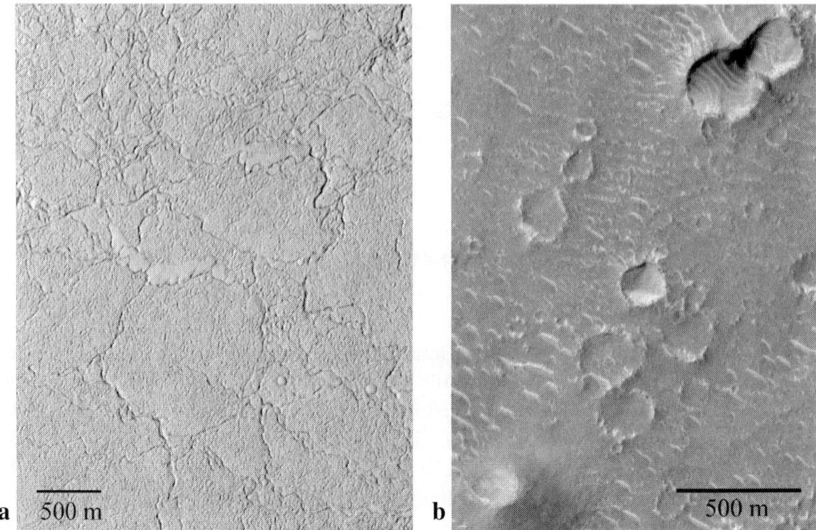

Figure 1. Comparison of young and old Martian surfaces. a) On a relatively fresh lava flow, such as this example in Amazonis Planitia at latitude 25.4 N, longitude 166.5 W, the lack of accumulated impact craters directly reflects the young age. b) On the older highland surface in Isidis Planitia, shown here, crater density is higher, and dust drifts cover small craters at latitude 8.4 N, longitude 276.5. The net effect of such dust drifts is to obliterate small craters preferentially, relative to larger craters in older areas. MGS images M00-00536 (a) and FHA-00521 (b). All MGS images in this chapter are courtesy Malin Space Science Systems and JPL Mars Global Surveyor Project.

time dependence of cratering is known in the inner solar system. They show that a relatively uniform shape of size distribution is observed on relatively undisturbed surfaces, such as lunar maria, lunar ejecta blankets, and some asteroids. This shape – more specifically the number of craters/km^2 produced on a surface in a given time as a function of diameter D – is called the "production function" for impact craters. Ivanov (2001) derives the cratering rate of Mars relative to the moon. Combining the results of those chapters, we know the rate of production of craters on Mars of any given size, and hence the absolute crater retention age of different stratigraphic units on Mars within an uncertainty factor of about two or three.

For a surface undisturbed by non-impact processes, e.g. a deep lava flow covering a large area, the analysis of accumulated craters to determine an age is conceptually straightforward (Figure 1). This situation is approached in the lunar maria, which formed mostly between about 2.9 and 3.9 Gyr ago – a factor of only 1.3 in age. However, the relatively active geology of Mars produces a more complex case than the moon. Younger lava flows may interfinger with older background surfaces with higher crater density. Martian rocks as well as crater counts show that Martian lava flows cover a much larger range of age, a factor of 100 or more.

"Crater retention age" (CRA) was defined by Hartmann (1966a) as the average time interval during which craters of diameter D are preserved on a given surface. For a young lava flow, where no craters have been lost, it should be the age of the flow itself. For older surfaces, it is D-dependent and refers to the preservation time of the craters. For example, on an idealized surface, with constant net accumulation of dust on crater floors, small craters would be obliterated faster, and the CRA would be small for small craters, and larger for larger craters. For the largest craters the CRA may indeed be the age of the formation of the surface.

Various stratigraphic units have been mapped on Mars and their relative ages have been determined by a combination of superposition relations and crater densities (Tanaka, 1986; Scott *et al.*, 1987). In principle, absolute ages can be estimated through impact crater densities. However, the absolute chronology and absolute ages of different Martian stratigraphic units have been known only crudely due to uncertainties, primarily in the Martian impact flux and methodologies used to scale the lunar cratering to Mars. Viking and Mariner 9 analysis produced a wide range of chronologic systems, with no clear consensus on absolute ages (Hartmann, 1973; Soderblom *et al.*, 1974; Neukum and Wise, 1976; Hartmann *et al.*, 1981; Neukum and Hiller, 1981; Strom *et al.*, 1992).

In recent years, as developed in detail throughout this book, the absolute Martian chronology has been loosely constrained by two complementary data sets.

First, as discussed by Nyquist *et al.* (2001), Martian meteorites give precise radiometric crystallization ages for rocks from a small number ($4 - 8$) of impact sites on Mars. The nakhlites and Chassigny appear to represent a mafic igneous intrusion 1.3 Gyr ago, and the basaltic shergottites appear to represent surface flows somewhere on Mars about $165 - 475$ Myr ago. In addition, one Martian meteorite, ALH84001, gives a crystallization age of 4.5 Gyr and with subsequent carbonate formation at about $3.9 - 4.0$ Gyr ago, and little subsequent. This sample suggests that in at least some regions, primordial crust is not only preserved but also exposed, relatively near the surface. Cratering studies (Melosh, 1989) indicate that the ejected rocks are likely to come from near surface layers, no more than a few hundred meters down, although this is not regarded as proven, because the impact models to not appear to have reached a state of sophisticated maturity.

The Martian impact sites not only reveal recent igneous activity, but also show evidence of liquid water-based activity after the rocks formed. Shih *et al.* (1998) and Swindle *et al.* (2000) dated weathered minerals in the 1.3 Gyr-old nakhlite, Lafayette, concluding that it had been exposed to liquid water around 670 Myr ago. In a similar vein, Sawyer *et al.* (2000) and Bridges and Grady (2000) find that the nakhlites Lafayette, Nakhla, and Governador Valadares all contain evaporite minerals, such as gypsum, anhydrite, and clays, caused by exposure to evaporating, seawater-like brines more recently than 1.3 Gyr ago. In addition, Malin and Edgett (2000a) identified water seeps and resultant gullies on crater-free hillsides that are relatively free of dust accumulation; these have unknown ages, but appear to have less dust cover than some of the young lavas (Hartmann, 2001).

In summary, the available Martian rocks establish a Martian chronology of igneous activity and water based weathering that stretches from the beginning of Martian history to the last few percent of Martian time.

The second constraint on Martian absolute chronology is the impact record, and it appears consistent with the rock data. Crater counts on different units, especially since Mars Global Surveyor (MGS) imagery was available, show a wide range of ages, including geologically young activity. The earliest Mariner data, from 1965 to 1971, revealed heavily cratered areas where the largest craters, $D > 64$ km, had crater densities similar to those in the lunar highlands, with inferred ages of the order $3.8 - 4.5$ Gyr (Leighton *et al.*, 1965). In these same regions, smaller craters (250 m $< D < 16$ km) have lower numbers than in the lunar highlands and a wide range of degradation states, suggesting losses of smaller craters by erosion and deposition, as first suggested by Öpik (1965, 1966). Similar losses of small craters occur on Earth. The numbers and losses of craters of various smaller diameters and depths offer a way to characterize "crater retention ages" and the rate of geologic activity in terms of the time scale needed to fill or obliterate the craters, as discussed by Hartmann (1966a). Much of the early, Mariner-era work was devoted to deciphering the history of obliteration processes, and generally suggested that the craters revealed strong obliteration, with a higher rate in early Martian history (Öpik, 1965, 1966; Hartmann, 1966a, Hartmann, 1971; Chapman *et al.*, 1969; Jones, 1974; Soderblom *et al.*, 1974). From Mariner 9 images, Hartmann (1973) derived $3 - 4$ Gyr ages in the uplands, along with enhanced erosion/deposition in early Martian time, but also proposed volcanic activity about 300 Myr ago in Tharsis, and this was supported this with later analysis of the impact rates and (Hartmann, 1978; Hartmann *et al.*, 1981). Several other early chronological studies of the 1970s (Soderblom *et al.*, 1974; Neukum and Wise, 1976) derived somewhat older ages for the upper Amazonian volcanic features, emphasizing that most Martian geologic activity was concentrated in an early period. Later efforts, such as Neukum and Hiller (1981), Hartmann (1998), and the Martian rock data (Nyquist *et al.*, 2001) allowed for of a tail of igneous and other activity extending into the present. MGS confirmed massive layering and mobility of dust and fine material on Mars (Malin *et al.*, 1998), which supports the idea that small craters are removed by erosional/depositional effects, as on Earth, and this must be taken into account when interpreting crater retention ages (see also Greeley *et al.*, 2001). One way of expressing the modern issue is to ask about the frequency distribution of igneous, erosional, and depositional activity as a function of time.

The crater-dating results and the Mars meteorite results are independent, and their present agreement gives some confidence that the two methods give accurate ages. The early suggestions of young igneous units on Mars in 1973 came before any meteoritic suggestion of young igneous activity. Papanastassiou and Wasserburg (1974) noted that Nakhla's properties implied late formation on a body that might be bigger than most asteroids, but that body was still unknown. Just before the wide recognition of young Mars meteorites, NASA convened a "Basaltic Volcanism Study Project," within which a consortium including Strom, Shoemaker, Weidenschilling, Chapman, Dence, Grieve, Hartmann and others used asteroid/comet data to estimate relative Mars/moon cratering rates, and then used crater counts by various researchers to infer Mars ages – an antecedent of the method used here. The consortium concluded that the Martian uplands are very old, but that some of the younger plains of Mars, such as the Tharsis volcanism, had ages in the range of a few hundred Myr to around 2.0 Gyr (Hartmann *et al.*, 1981). The recognition of similar Mars meteorite ages culminated a year or two after that work was done. Suggestions began to be floated as early as 1979-81 that Mars might be a source for such objects (Nyquist *et al.*, 1979; Wood and Ashwal, 1981) but acceptance of this idea came a few years later when Martian atmospheric gases were found in them (Bogard and Johnson, 1983).

Mars Global Surveyor, orbiting Mars since 1997, added a new twist to the understanding of the youngest volcanic units, by means of much higher resolution images (1.5 m/pixel). In certain restricted areas, such as Elysium Planitia, these revealed lava flows with fresh surface textures and very sparse numbers of well-preserved impact craters (Keszthelyi *et al.*, 2000). For some flows, initial counts suggested densities of small craters of the order of one percent those of the lunar maria, implying ages of <100 Myr, possibly <10 Myr (Hartmann and Berman, 2000). This work affirmed a conclusion by Plescia (1990) that Elysium Planitia has very young lavas. In addition, other early MGS analyses pointed to other broad areas of lava flows, such as the Tharsis volcanoes and Amazonis Planitia, where ages of at least some flows appeared to be <100 Myr. Thus, the cratering data independently support the conclusion from the Mars meteorites that igneous geologic activity on Mars has persisted from early times down to the present geologic era, though the rate may have decreased, since the youngest flows are rare.

The meteorite data and the cratering data are complementary in another way. Meteorites give good ages but poor statistical sampling of Mars; craters give poorly constrained ages but good statistical sampling of known units. These units may not be studied by sample return or in situ studies for decades to come. A major goal of this chapter and earlier chapters is to refine the methodology and tie these two complementary systems together, so that crater densities can give reasonably accurate absolute ages (potentially within 20%) for all stratigraphic units, and can be tested for consistency with the known Martian rock samples.

2. Issues with Regard to Crater Count Chronologies on Mars

The crater count methodology has several fundamental limitations. First, the total crater density is measured by statistically fitting the crater count data (size distribution) to the known production function over a wide range of D. The fit, by least squares techniques, is limited to about 10% accuracy, and curves fit to data from different workers on the same images have about the same level of repeatability (Hartmann *et al.*, 1981). Neukum and co-workers achieved a higher degree of accuracy and repeatability by using stereo coverage of many units, such as in their lunar work (Neukum, 1983; Neukum *et al.*, 1975).

A second limitation is that several slightly different definitions for the boundaries of a given stratigraphic unit, such as a particular set of lava flows or a putative paleolake deposit, may exist. Thirdly, the crater retention age for a given unit, such as a sequence of lava flow, represents only a new age, biased towards the younger surface subunits. Fourth, in order to extend the crater counts to larger diameters, it is necessary to cover larger areas in order to achieve good statistics, and this may be in conflict with the need to define a specific small geologic unit.

Hartmann *et al.* (1981) and Lissauer *et al.* (1988) reviewed these problems and concluded that crater counts generally give repeatable characterization of overall crater density and relative ages of various units to an uncertainty factor of about 1.2 to 1.3. These issues suggest an ultimate limit of perhaps 5% to 20% uncertainty in ages simply due to the process of defining a homogeneous geologic unit, and gathering good cratering statistics, not counting the (presently larger) uncertainties in crater production rate R_{crater} on Mars relative to the moon.

These issues are exacerbated by a peculiar circumstance of Martian spacecraft exploration to date. Mariner 9 and the two Viking orbiters carried low resolution cameras and MGS carried a high resolution camera. With the exception of a small number of late Viking and early MGS images, there is a "hole" between the two data sets, such that it is difficult to get good mid-size crater statistics at D ~250 to 500 m. Also, MGS does not give 100% areal coverage, so that the small crater populations in broad units must be characterized by "postage-stamp" samples. This leads to problems of characterizing the small crater populations of broad units.

3. The Importance of the R_{bolide} Value

Hartmann (1977) and Hartmann *et al.* (1981) stressed that the modern-day ratio of Mars/moon cratering (expressed in terms of bolides/km²-yr, which is convertible to craters/km²-yr at a fixed crater diameter, as shown by Ivanov, 2001) is a critical, measurable parameter for determining Martian surface ages, and hence the overall chronology of Mars. Any estimate of the Martian absolute chronology involves, implicitly or explicitly, an estimate of the Mars/moon cratering rate ratio, R. Hartmann (1977, 1999) adopted this approach and defined the ratio R, which is developed by Ivanov (2001), in terms of

$$R_{bolide} = [\text{bolides}/\text{km}^2-\text{yr on Mars}]/[\text{bolides}/\text{km}^2-\text{yr on the moon}] \qquad (1)$$

at a fixed bolide diameter. This can be derived from direct observations, as discussed in more detail by Ivanov (2001), and also from dynamical considerations involving asteroid and comet populations, as treated by Bottke (in preparation). This leads in turn to the definition of

$$R_{crater} = [\text{craters}/\text{km}^2-\text{yr on Mars}]/[\text{craters}/\text{km}^2-\text{yr on the moon}]. \qquad (2)$$

Neukum and Ivanov (1994) and, more explicitly, Ivanov (2001) show that R_{crater} is a function of crater diameter, because of factors involving the slope of the size distribution and also the transition from simple to complex craters.

The largest current uncertainty is in R_{bolide}. A review of recent estimates suggests that the uncertainty is as much as a factor 2 in either direction, caused partly by an uncertainty in the contribution of cratering by comets. As reviewed by Ivanov, our estimate of R_{bolide} is primarily based on asteroids. From tabulations of known Mars- and Earth-crossing asteroids, Ivanov (2001) derived $R_{\text{bolide}} = 2.0$, and Bottke (personal communication, 1999) also derives $R_{\text{bolide}} = 2.0$ from dynamical studies of asteroid feeding mechanism.

The treatment by Ivanov (2001) is currently the best available summary of R_{crater} and its application to Mars chronology. Figure 2 introduces the resulting crater count diagram, and plots the isochrons derived by Ivanov from the Neukum data and from the Hartmann data, along with the isochrons derived below for other ages. "Isochron" is defined as the number of craters that would be produced on a single surface of specified age, and would be still visible today in the absence of any obliterative effects. As seen in Figure 2, the Neukum and Hartmann isochron systems, each with its own independent history stretching over more than 20 years, lie within a few percent of each other 125 m $< D <$ 1 km and at 22 km $< D <$ 45 km, but differ by as much as a factor 2 to 3 at diameters around 2 to 11 km.

The difference at 2 to 11 km apparently stems from differences in our original data bases of post-mare impact craters. We independently averaged over different lunar mare units to characterize the mare crater density. We are still investigating this difference; it illustrates the danger of basing age determinations on too narrow a range of D. However, we are gratified at the close agreement over most of the D range, which shows that a good fit of crater data to isochrons should be possible if crater counts are used over a sufficiently wide range of D. Based on our entire discussion, we believe we can use crater count data to fit isochrons with an effective 1-sigma uncertainty of about a factor 2 in the age determination.

Figure 2 shows the isochrons in a format developed by Hartmann and co-workers for MGS data, but the remaining plots of data in this paper several different formats. The first two formats are "standard formats" recommended by a cratering consortium (Crater Analysis Techniques Working Group "CATWG," 1979). These formats have both advantages and disadvantages. The first is cumulative, which has an advantage of smoothing the data, but the disadvantage of suppressing any turndown in crater population toward smaller diameters. For example, in an incremental plot, removal of craters smaller than D due to flooding of an area by a lava flow would produce an dramatic turndown in an incremental curve, while the cumulative plot merely levels off. Also, the CATWG group recommended using equal scales for the ordinate and abscissa, but this makes the curve is graphically so steep that it is hard to see structure in the curves or differences between them. The second plot, known as the R-plot, plots the data in increments of log D relative to an arbitrary -2 power law, which approximates the size distribution of larger

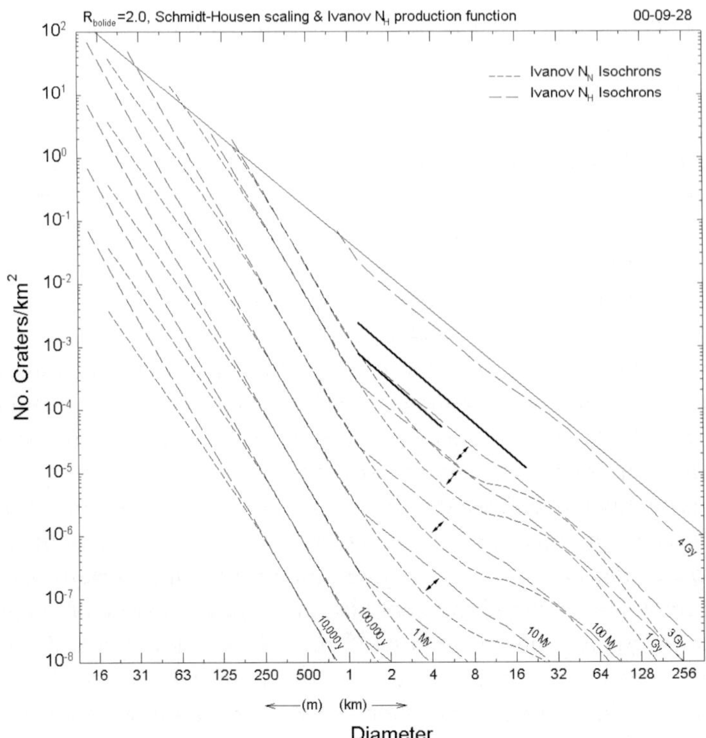

Figure 2. Comparison of isochrons derived by Ivanov (2001) from lunar crater count data by Neukum (it upper curve of each pair) and independent lunar crater count data by Hartmann (*lower curve*). The isochrons from the two independent sources mostly average within 30% of each other, but in the few-kilometer diameter range have discrepancies of a factor 2 to 3. The agreement, combined by the uncertainty in R_{crater} (see Ivanov, 2001) gives a measure of the uncertainty of derived ages. We estimate this uncertainty at a factor about 2. Top solid line shows saturation equilibrium level (see Figure 3). Short solid lines show definitions by Tanaka (1986) of boundaries between Amazonian (*bottom*), Hesperian (*middle*), and Noachian (*top*) Periods. See text for derivation of isochron positions. Figure a) shows standard cumulative format and b) shows the "*R*"-plot, both as defined by Crater Analysis Techniques Working Group (1979). Figure c) shows the log differential plot, as developed for Mars Global Surveyor (*see text*).

craters. The *R*-plot magnifies structure in the curves and discrepancies between various curves, which has both advantages and disadvantages. The concept is less obvious for readers not involved in the cratering field. It can lead to discussion of "structure" that may be noise, and the reference line of -2 slope is not especially significant, since it is not directly observed in any population. For these reasons, the third format, mentioned by CATWG (Appendix I), was chosen and developed by Hartmann and colleagues for the MGS data sets. This is designed to follow the design principle that a graph is easiest to read if the average apparent slope is near 45°. It plots numbers of craters in logarithmic increments in D, giving sensitivity to turndown, and it also gives numerically the same slope as the cumulative plot.

In addition to Ivanov's basic 3.4-Gyr isochron, Figure 2 shows isochrons for other surface ages on Mars. We will now derive those isochrons. The time behavior of the cratering rate, as measured in the Earth-moon system, is fairly well known (Stöffler and Ryder, 2001; Neukum *et al.*, 2001; Ivanov, 2001). Neukum (1983) gave a numerical solution for this time dependence, quoted by Neukum *et al.* (2001, Equation 5). The Neukum time dependence indicates that the isochron for 4.0 Gyr age should be about 16 times higher than that for 3.4 Gyr.

An upper limit exists near this point. Hartmann (1966b) showed empirically that the lunar upland crater density at $D > 2$ km is a factor of 32 higher than that for average mare. Hartmann (1984), Neukum and Ivanov (1994), and Hartmann and Gaskell (1997) have shown that craters reach a saturation equilibrium curve at this level, corresponding to ages greater than $\sim$4 Gyr. This result on saturation is also confirmed by Phobos (Figure 3a), which, orbiting above the Mars atmosphere, has no losses due to the Martian surface erosion regime, and displays cratering that accurately fits the saturation equilibrium curve defined by Hartmann (1984; Figure 3b). This saturation equilibrium curve is thus viewed as the empirical upper limit for crater density, and is shown by the heavy solid curve that bounds the top of Figure 2. Neukum and Ivanov (1994) noted that the shape of the saturation equilibrium curve depends in principle upon the input production function size distribution, although the general level, for the known size distributions, is expected from numerical simulations to lie at the level of the curve shown here (Hartmann and Gaskell, 1997). We believe that the saturation curve is not a precise fixed limit, but fluctuates around the curve shown here, as time goes on, depending on factors such as the age and location of the last largest impact basin (which can spread ejecta, obliterate craters, and reset the surface age at sites near the basin).

Precise isochrons between 4.0 and 3.4 Gyr are hard to map because the cratering rate declined rapidly during this period. On the positive side, any counts falling above the 3.4 Gyr isochrons are restricted to a very early era of Martian history and thus have value in constraining geophysical evolution. Isochrons for ages younger than 4 Gyr have the shape of the production function, except that they truncate where they hit the saturation level, because crater densities can't easily rise above this level. We adopt the argument by Neukum *et al.* (2001) that the shape of the isochrons, i.e. the production function, has been constant through time, though Strom *et al.* (1992) have suggested that it was different in the earliest history of Mars. Counter arguments are given elsewhere (Hartmann, 1984, 1999).

Starting with Ivanov's isochron for a 3.4 Gyr surface on Mars, we now derive isochrons for younger ages. We use the Neukum (1983) time dependence, which shows the cratering rate declining somewhat after 3.4 Gyr ago and leveling out by 3 Gyr ago. As reviewed by Stöffler and Ryder (2001), Neukum *et al.* (2001), and Ivanov (2001), evidence suggests that the crater production rate on Mars, averaged over 100 Myr timescales, has been constant within about a factor 2 since 3 Gyr ago (see also Neukum, 1983; Grieve and Shoemaker, 1994; Grier *et al.*, 2000). Thus, isochron crater density levels scale roughly with age after 3 Gyr ago.

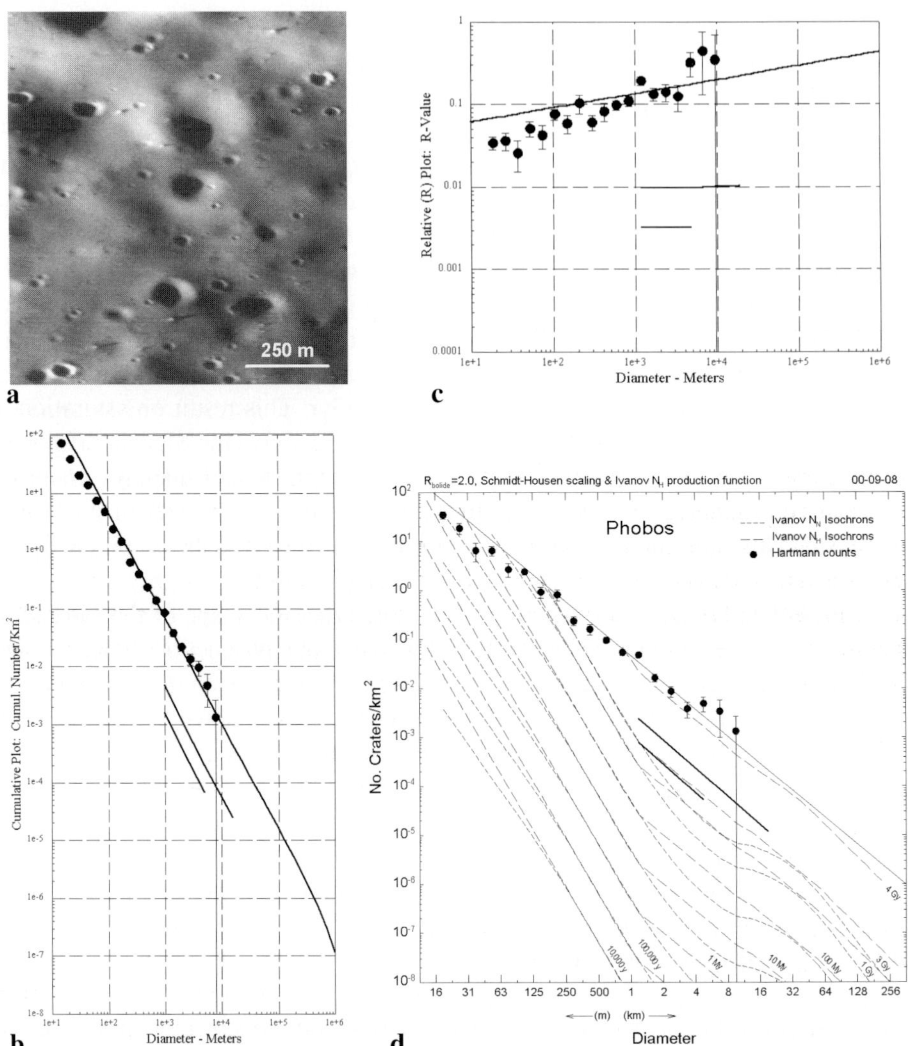

Figure 3. a) MGS image SP2-55103, showing Phobos. b-d) Crater densities on Phobos measured by Hartmann from Mariner 9, Viking, and MGS imagery, plotted in the three formats of Figure 2. The solid line is the least squares fit by Hartmann (1984) to the most densely cratered surfaces on the lunar far side, Phobos, and outer solar system satellites, and is regarded as the empirical upper limit due to saturation equilibrium. Note contrast with Mars data in other figures.

Note that because we start with the lunar mare crater densities (Ivanov, 2001), the Mars isochron positions depend fairly strongly on the age assigned to the lunar maria, because the cratering rate was changing during that time. Table I summarizes this issue by tabulating the total accumulated number of craters on a 3 Gyr surface, relative to various other ages, assuming Neukum's cratering time dependence and, for comparison, a constant cratering rate. The table shows how the

TABLE I

Crater Density at 3000 Myr Relative to that on Older Surfaces

Assumed age of lunar mare surface, T	[Crater density on 3000 Myr surface]/[Crater density on surface of age T]	
	Neukum time dependence	Constant cratering since T
3400 Myr	0.68	0.88
3500 Myr	0.54	0.86
3600 Myr	0.38	0.83

assumed mean mare age affects positioning of the 3 Gyr isochron (and hence all the isochrons for younger ages). We adopt Ivanov's mean age of 3.4 Gyr and multiply Ivanov's Mars crater densities at 3.4 Gyr by 0.68 to get the position of the 3 Gyr isochron. To be conservative, we estimate a 20% uncertainty in ages introduced in this stop. While Figure 2 shows the Ivanov/Hartmann/Neukum 3.4 Gyr isochron, because it is represents our initial input data, the remaining figures in this paper show the 3.0 Gyr isochron for ease of interpolation; it has a slightly lower position.

4. Young Surfaces: Evidence for Youthful Volcanism

As emphasized by the MGS Imaging Team, it is important to study the youngest Martian volcanic stratigraphic units for several reasons. First, their dates constrain models of geothermal evolution of the planet. Second, a deep, youthful lava flow, if unaltered by still more recent dunes or dust drifts, is a perfect surface for recording the production function diameter distribution of impact craters, and offers a test of our assumed production function shape (i.e. isochron shape). Third, as noted by Hartmann (2001), youthful volcanism might provide a key to understanding recent aqueous phenomena, such as young water deposition and erosion features (Shih *et al.*, 1998; Swindle *et al.*, 2000; Sawyer *et al.*, 2000; Bridges and Grady, 2000; Malin and Edgett, 2000a), by providing a geothermal source for melting permafrost. We now discuss several examples of young lavas.

a) Arsia Mons. This volcano was studied by Hartmann *et al.* (1999) in one of the first MGS reports on young volcanism, where ages of 40–200 Myr were suggested in the summit caldera. Figure 4a shows a sample MGS image of the young lava textures along the rim of the summit caldera. Figure 4b-d shows our current data set and the isochrons diagrams derived here. The data in the caldera give a best fit to a mean lava flow age in the central caldera of about 200–400 Myr, with a somewhat older, less constrained age of about 500–2000 Myr for the average of the whole volcano surface flows, based on larger craters.

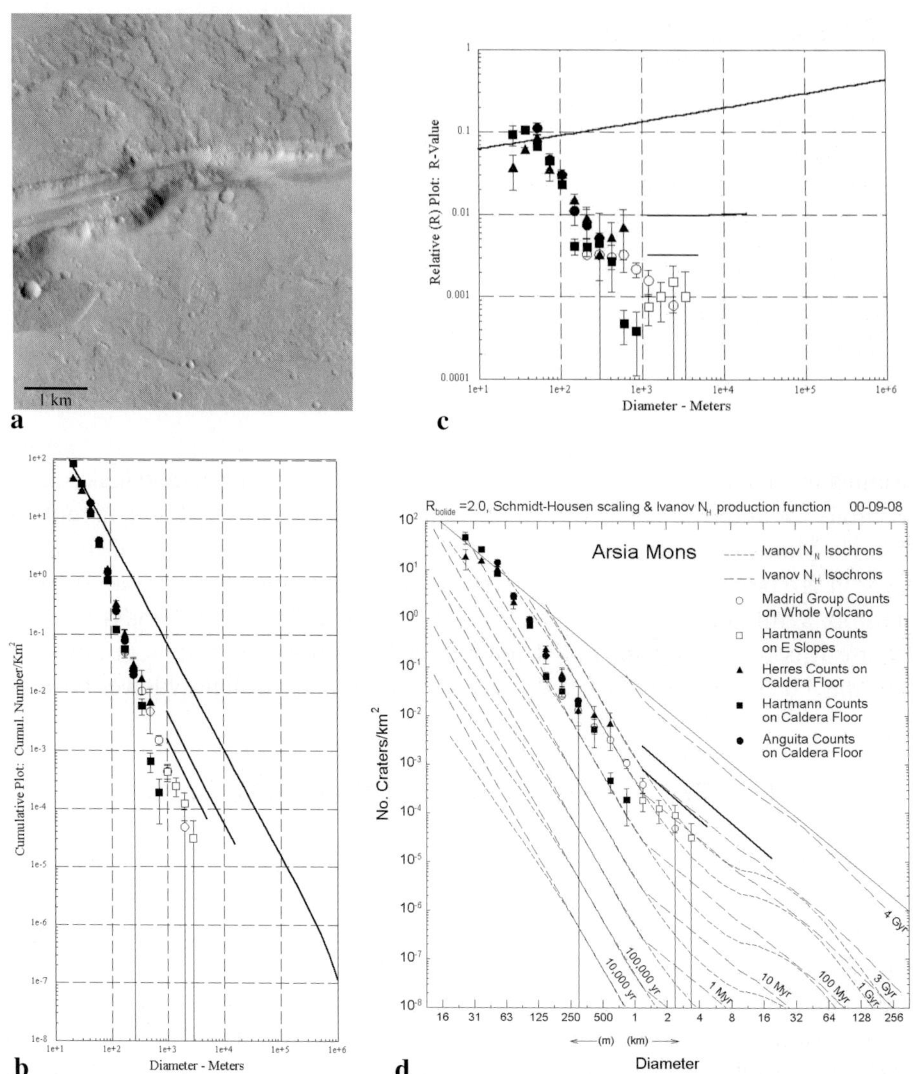

Figure 4. Young lavas on Arsia Mons. a) The outer rim of the Arsia Mons caldera (*top*) and the caldera floor (*bottom*) show overlapping young lava flows. (Lat. 8.7 S, long. 120.8 W, MGSAB1-03308). b-d) Comparison of crater counts and isochrons derived in this paper suggests a characteristic age of a few hundred Myr for most surface flows on the flanks and caldera floor of Arsia Mons.

b) Elysium Planitia. This area was correctly described as young lavas by Plescia (1990). Keszthelyi *et al.* (2000) confirmed unusually fresh-appearing lava textures, similar to examples of flood basalts found in Iceland. Hartmann and Berman (2000) derived an age of a few Myr to 100 Myr from crater counts. Figure 5a shows MGS images of young lava flows at different scales. Figures 5 b-d along with plots of the crater counts against the isochrons derived here. These data at $D < 500$ m suggest very young ages of a few to 30 Myr for some of the youngest flows.

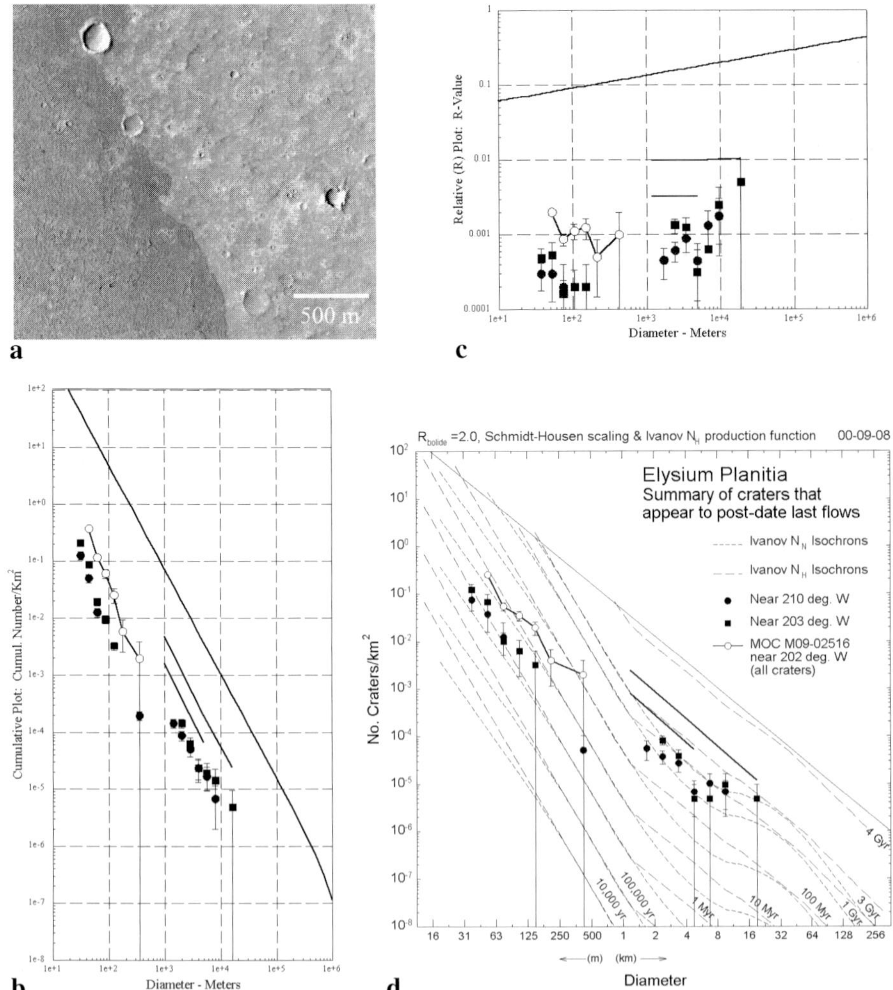

Figure 5. Young lavas in Elysium Planitia. a) Younger, darker flow (*bottom*) flows across an older, cratered surface and around the largest crater. (Lat. 5.5 N, long. 214.3 W, MGS MO3-03779). b-d) Crater counts reflect heterogeneity, with some older background flows having ages of order 900 Myr, while some of the youngest lava flows suggest ages of 10 Myr or less.

Hartmann and Berman (2000) raised an important point about the dating of the Martian lava plains. While the lunar mare lavas date almost entirely from 2.9 to 3.9 Gyr, or at most from around 2.0 to 4.0 Gyr (a factor of 1.3 to 2.0 in age), the Martian lavas, even within a restricted area such as Elysium Planitia, appear to cover a much larger range in age, because the youngest flows are so young. Thus, if the Martian flow ages in Elysium Planitia range from 10 Myr to 300 Myr, their range of crater densities spans a factor 30, rather than the factor mentioned for the moon. For this reason, individual high resolution MGS images which happen to fall in the youngest individual flows can fit much younger isochrons than the average crater densities at larger diameters, derived over larger areas from low-resolution Viking images. This effect can cause some difficulty in fitting crater density data to our isochrons, with some MGS frames giving much lower ages at low diameter ($D \lesssim 500$ m) than at high diameter ($D \gtrsim 1$ km). However, we note that individual flows shown on individual MGS frames fit our isochrons reasonably well, as seen by the connected dots in Figures 5b-d.

c) Amazonis Planitia. Amazonis Planitia is a region northeast of the young Elysium Planitia lavas, which appear to flow into the Amazonis area (Plescia, 1990; Keszthelyi *et al.*, 2000). Figures 1a and 6a show aspects of the area, including a very uncratered flow overlapping a young background. As shown in connected dots in Figure 6b-d, some of these sparsely cratered flows fit the Neukum and Hartmann isochrons for ages as young as 3 to 20 Myr. Our average of data over the older background flows (*solid symbols*) gives a fairly good fit to the Neukum and Hartmann isochrons for age $100-200$ Myr, all the way $D = 31$ m to $D = 1$ km. At larger sizes, $D > 1$ km, the isochrons suggest older ages, $\sim 500 - 900$ Myr on the Hartmann system to $0.6 - 2.0$ Gyr on the Neukum system. The 5 to 10 km craters counted on Viking frames could actually predate a few of the final, thin (4-m?), 10-Myr old flows that dot the Amazonian plains.

d) Olympus Mons. Olympus Mons is of special interest as the largest volcanic construct on Mars. Figure 7a shows an example of an individual recent lava flow running from top center to bottom center. Figures 7b-d show our various data sets. Using the Hartmann or Neukum isochrons, the data in the range 45 m $< D <$ 700 m suggest a characteristic age of the order 100 to 200 Myr, respectively, for the uppermost exposed lavas on the slopes of Olympus Mons. Data including lower resolution views at $D > 700$ m suggest an older age of the order $300 - 500$ Myr for flows in the upper few hundred meters. Some MGS frames such as MGS/MOC SPO1-41105 show individual flows with much lower crater densities giving ages of the order 10 Myr in either isochron system (*connected open circles*).

5. Comment on Mid Range Ages by Crater Count Methods

Figure 8 shows the region of the Viking 1 landing site in Chryse Planitia. This area and many other plains are older than the young lava flows we have been discussing. Both isochron systems indicate ages in the range of 3 to 4 Gyr, probably involving craters formed in underlying strata, whose rims and ejecta are still exposed. The Chryse Planitia plains appear to be cut at their western edge by massive flow features, where water apparently emptied into the area.

Although we can assign older absolute ages to such plains than to the young lavas, we wish to point out a fundamental limitation of the crater count method as applied to Mars at the present level of our knowledge. The uncertainty of a factor 2 in ages, arising primarily from the uncertainty in the factor R, presents an

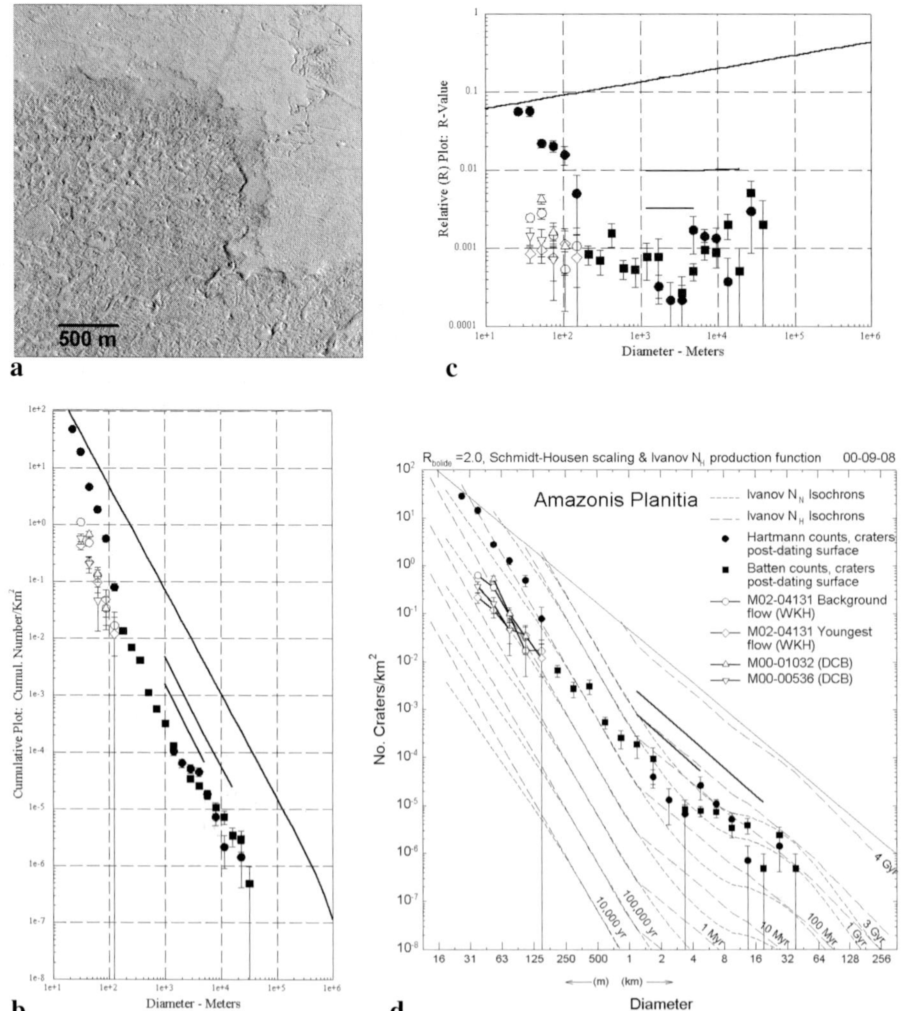

Figure 6. Young lavas in Amazonis Planitia. a) Fresh lava textures, similar to those of Figure 5a in Elysium Planitia. (Lat. 26.5 N, long. 167.5, MGS 02-04131). b-d) Crater counts suggest an average age of the order 200 Myr, with the youngest individual flows having ages as young as 10 Myr or less.

unfortunate situation for dating events in "mid-Martian" history. For example, if our best dating of a given feature is 2.0 Gyr, the actual 1-σ range of ages could be from 1 to 4 Gyr. Such an age has little value in placing constraints on the geological history or geophysical evolution of the planet. This is the reason we emphasized the youngest volcanism in Section 4. If we obtain an age of the youngest volcanic features of, say, 20 Myr, then even a 4-σ error would give an age of 80 Myr, and we would appear to have a robust constraint on geologically young volcanism.

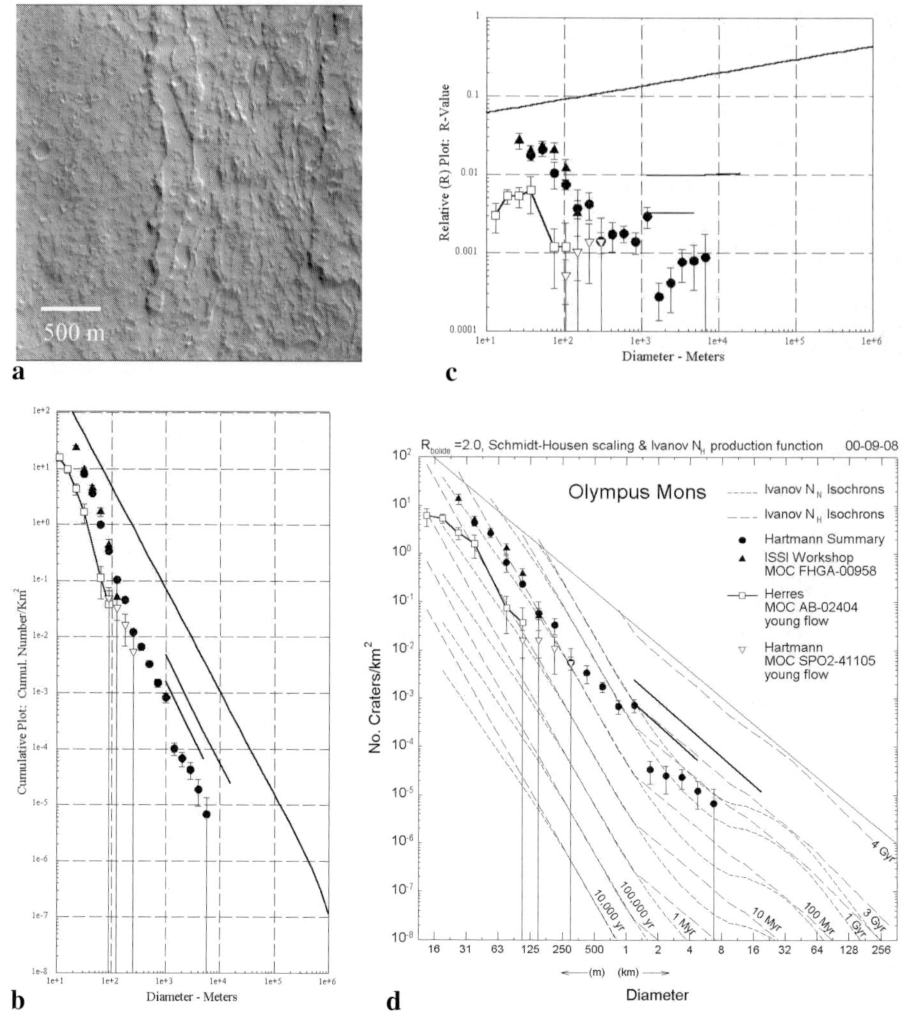

Figure 7. Young lavas on Olympus Mons. a) Tongue of lava the flank, leaving a negative relief channel with levees at the top, changing to a positive relief flow front at the bottom. Lat. 20.0 N, 133.3 W, MGS M09-05643. b-d) Crater counts suggest an average age of a few hundred Myr on the slopes, with the youngest flows having ages of the order 10 − 100 Myr.

6. Older Areas: Steady State Size Distribution and Long-term Crater Infill

Before 3.5 Gyr ago, the cratering rate was higher. Surfaces of that age approach the saturation equilibrium density in terms of accumulated impacts, but the observed numbers of craters has been reduced by cumulative effects of erosion and deposition during or since that era.

Mars Global Surveyor images affirm that mobile dust drifts and thin lava flows are a strong influence in obliterating smaller craters (Malin *et al.*, 1998; Keszthelyi

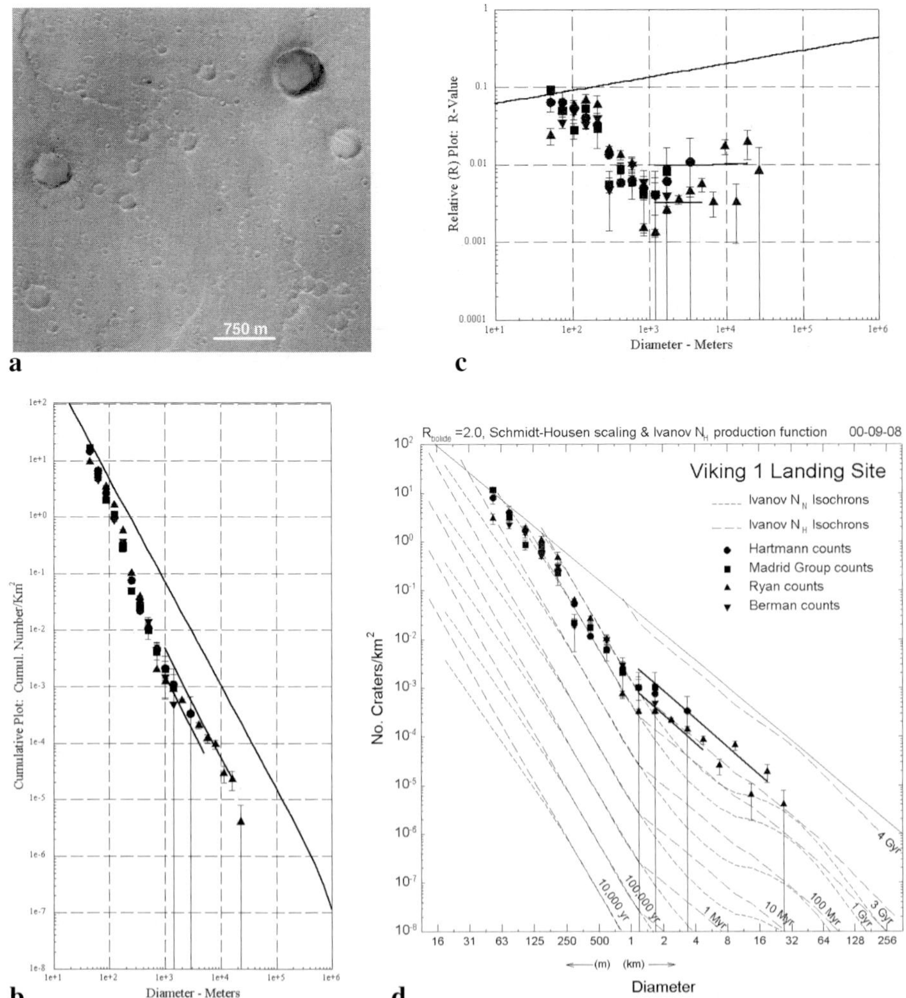

Figure 8. Viking 1 landing site area. a) Cratered plains of Chryse Planitia near the Viking 1 landing site. Lat. 22.5 N, long. 48.0, MGS SP1-23503. b-d) Crater counts suggest an age of the order 1–3.5 Gyr for the plains in this region. See text for discussion.

et al., 2000; Hartmann and Berman, 2000). Greeley *et al.* (2001) discuss the pervasive effects of aeolian deposition, and the possibility that some older surfaces have been covered and then exhumed, reducing the crater density. In principle, actual ages could thus be larger than ages derived from observed craters on such a flow.

Could such effects negate our conclusions about young volcanism? Probably not, for several reasons. 1) Even if all these areas had spent, on average, half the time buried, the derived CRA would be half the true age, which still would evidence geologically young volcanism on Mars. 2) To argue that the true ages are ~3 Gyr, one must argue that all these areas have spent 97−99% of their history buried with-

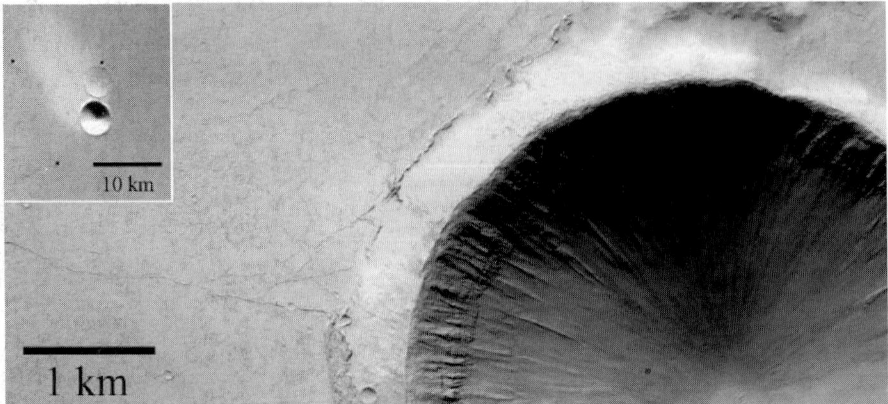

Figure 9. Young, sparsely cratered lava flows nearly covering a 4 km crater in Elysium Planitia. Thin flows have lapped up against the crater rim, and additional flows could have breached the rim and covered the interior, while leaving the rims an interiors of larger craters intact. Inset shows a Viking frame of the same crater, illustrating that crater counts on low-resolution Viking frames may include seemingly fresh craters that are actually postdated by thin lava flows that lap up against their rims or ejecta blankets, without covering them. Lat. 26.4 N, long 165.8 W, MGS M02-00364.

out accumulating craters, and that vast areas from Elysium Planitia to the summit of Arsia and Olympus Mons have been exhumed very recently. This would, in itself, require recent massive geologic activity, though not volcanism. 3) Additionally, one would need to argue that the measurements of Mars meteorite ages by different labs with different isotopic systems are seriously and systematically in error.

Averaging over large areas, mobile dust does gradually accumulate on crater floors, because they are potential wells. The net effect of dust migration and, in certain areas, continued lava flows, is to cover and obliterate smaller craters while leaving larger ones. How can we predict crater size distributions for the conditions in which mobile dust deposits and other cumulative infill processes tend to obliterate craters? The early modeling work of 1966-71, by Öpik (1966), Hartmann (1966a, 1971), Chapman *et al.* (1969), and Chapman (1974) treated crater floors as potential wells and assumed that during long term episodes of deposition and deflation, there would be a net deposition in low spots. In the first-order model, the crater was assumed to disappear when the dust infill or lava flows reached the top of the rim. MGS images also show thin lava flows lapping up against the otherwise sharply-defined rims or rampart ejecta blankets of fresh-looking bowl-shaped craters (Figure 9). If the lava reaches the top of the rim, lava would flow into the crater and partially or totally fill it, obliterating smaller craters while leaving larger craters relatively fresh-looking, at least at low resolution. If the average rate of dust deposition in crater floors, or the average rate of lava accumulation around rims, is assumed to be constant in a simple model, then the lifetime of a crater would be proportional to crater depth (or rim height, which is roughly proportional to crater depth), at least to first order.

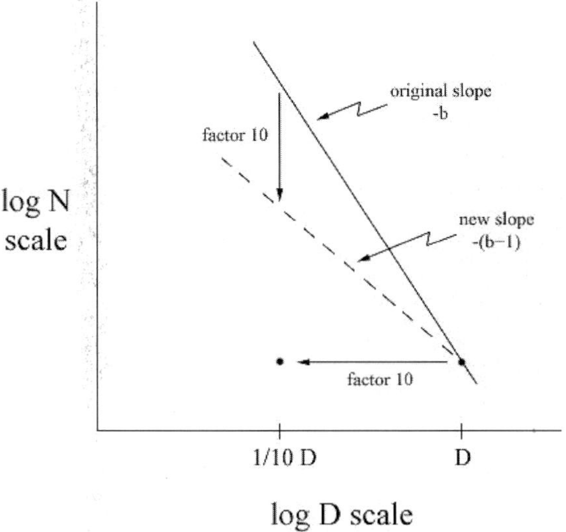

Figure 10. Schematic diagram illustrating the effect of gradual, constant crater infill in producing a steady state size distribution. If craters tend to fill in (due to processes such as net dust deposition and/or lava flows) the lifetime will be proportional to depth and rim height, which is roughly proportional to diameter *D*. This causes a loss proportional to *D*, which reduces the slope by approximately unity. See text for further discussion.

This in turn means that small crater lifetimes would be roughly proportional to their size, with small craters disappearing more rapidly, leading to a shallower slope in the crater size distribution on old surfaces. The basic idea can be understood graphically as in Figure 10. Suppose a crater of size *D* has depth *d*, and after time *T* it has been just filled in by dust, and the rim has been worn down and perhaps mantled by drifts, to the extent that it is not detected in crater counts. Now consider a crater of size 0.1 *D*. As a thought experiment, suppose depth is proportional to diameter, so that it has a depth 0.1 *d*. Then the oldest small craters would be 1/10 as old as the oldest big crater. If we assume constant crater production, we would see only 1/10 of the total number that had formed. This bends the *D* distribution down to a shallower slope, by unity in this example, since one decade in *N* is lost for one decade decrease in *D*. Chapman *et al.* (1969) and Hartmann (1971) made similar analyses, and Hartmann (1999) modified it with better data on the depth-diameter relation. The latter curve is used here. To be more realistic, if a proportionality exists between the a declining cratering rate and a declining infill/obliteration rate during the first 1 Gyr, then a similar shape of curve still results (Hartmann, in preparation). This predicted behavior fits surprisingly well with observed data, suggesting that the older regions of Mars have been shaped by measurable mean net infill of craters. The crater populations, in the oldest area of Mars, are dramatically different than those in younger areas or in the old, unflooded uplands of the moon

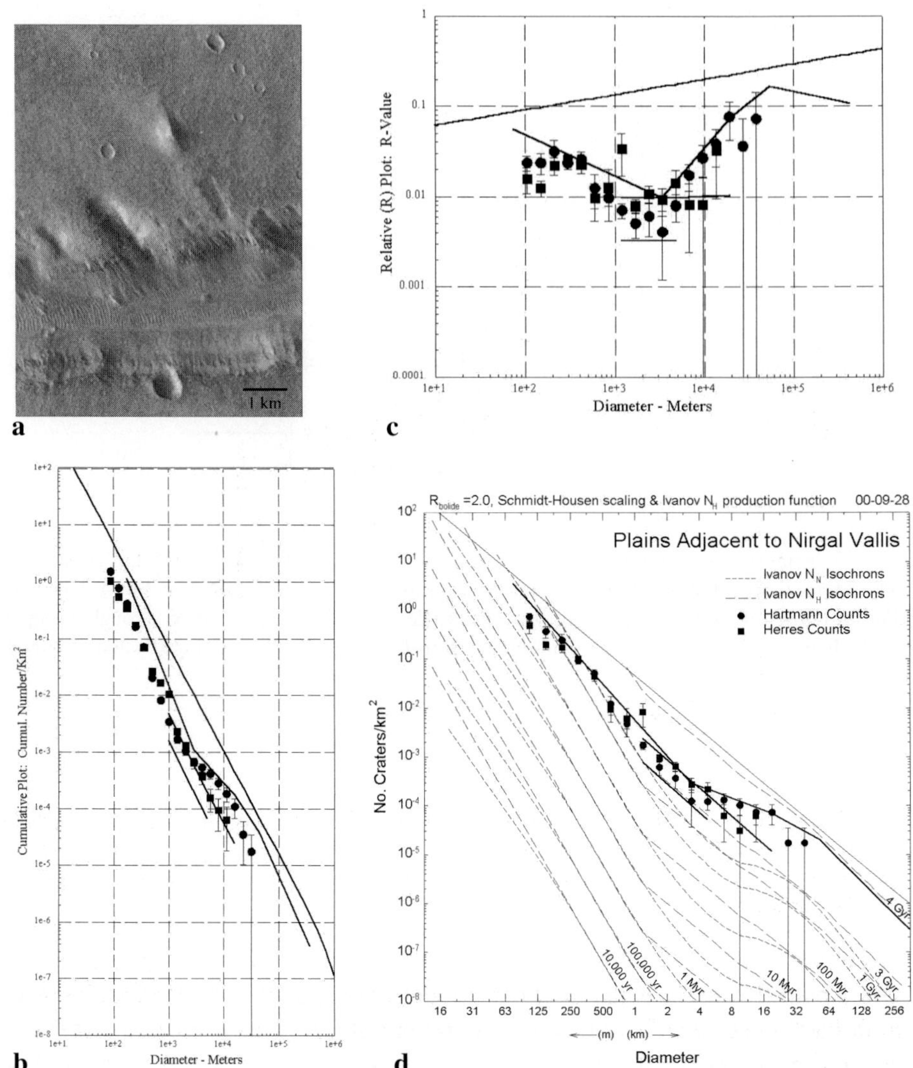

Figure 11. Plains Adjacent to Nirgal Vallis. a) Examples of degraded craters. Lat. 28.6 S, long. 41.6 W, MGS AB1-00605. b-d) Crater counts show good fit to the predicted steady state line.

(Hartmann, 1971; 1995). We now give examples of older, upland surfaces that support the principles discussed above. Hartmann (1999) gives other examples.

   a) Uplands adjacent to Nirgal Vallis. An example of a moderately old Martian upland is given in Figure 11, showing the surface and crater counts around Nirgal Vallis, in the southern uplands. The MGS image shows a range of degraded morphologies with intercrater flat areas that may be covered with dust sediments. The crater counts approach saturation at $D>16$ km, and follow the steady state deposition law derived above at smaller diameters.

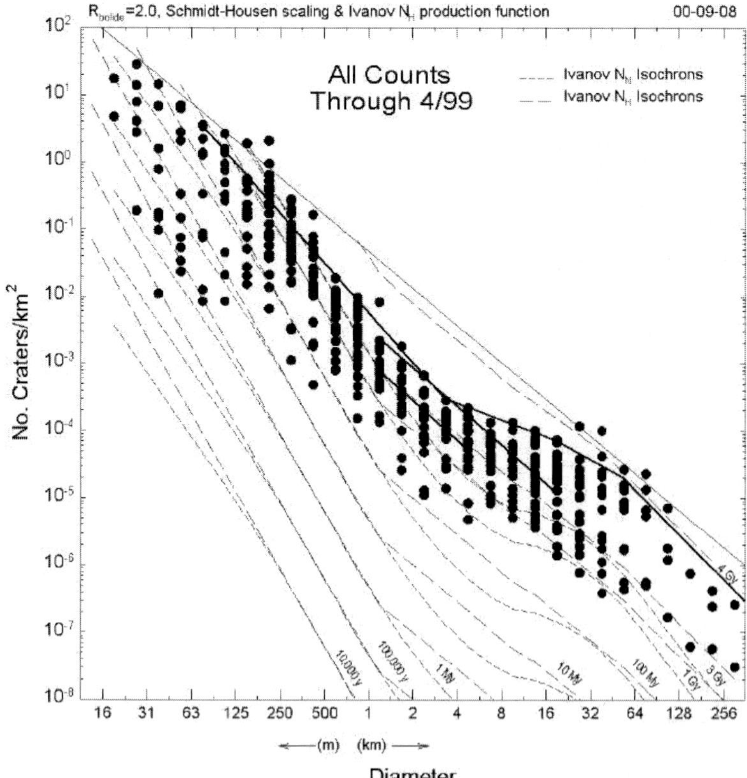

Figure 12. Potpourri of Martian crater counts on exposed Martian surfaces show a general fit to the predicted steady state line (*solid bent line*), with a scattering of many younger surface ages. A comparison is made to the Phobos counts (*top*), proving that losses must have been experienced on the Martian surface by erosion and deposition.

b) Potpourri. To demonstrate the behavior of the oldest regions, Figure 12 shows a "grab-bag" sampling of crater counts from the PSI group from many different terrains on Mars. Of great interest is the upper envelope on crater density in the oldest areas. It falls dramatically below the saturation line found for Phobos (Figure 3), but fits the profile suggested for long term infill and erasure of craters, described above. The interpretation is that the oldest craters of $D \gtrsim 22$ km are in saturation and date back to very early times. The oldest craters of 500 m $< D <$ 22 km have lifetimes less than the age of the planet and fall well below saturation. Still smaller craters have even shorter lifetimes, but form fast enough to maintain densities up to saturation on old surfaces. The data support the view that, under typical Martian conditions (just as in the more extreme case of Earth) erosion and deposition limit the crater densities to a steady state curve below that on the lunar highlands.

From these principles, one can estimate net mean infill rates of craters from Figures 10-12. For example, the lifetimes (maximum ages) of 350 m scale craters are of order 3.0 Gyr. These craters have depth of order 70 m (Pike, 1977; Strom *et al.*, 1992), suggesting a mean infill rate of order 20 nm/yr on crater floors, averaged over the last 3.0 Gyr. Larger craters of $D = 16$ km have lifetimes around 3.5 Gyr and depth 1300 m, giving a mean infill rate of order 400 nm/yr since that earlier time. The infill rate in the first Gyr must have been higher than the later rate, according to these numbers. A more systematic approach (Hartmann, in preparation) suggests infill rates were one to two orders of magnitude higher prior to ages around 3 Gyr ago, supporting a result in the next section. These numbers, for net deposition on crater floors, are not inconsistent with estimates of erosion. Golombek and Bridges (2000) list 100 to 10,000 nm/yr in the Noachian, 100 to 1000 in the Noachian to Hesperian, and 0.1 to 10 nm/yr from Hesperian to present.

7. Dating the Amazonian/Hesperian/Noachian Relative Stratigraphic System

A goal of Martian chronology studies is to derive the absolute dates of the relative stratigraphic periods defined by Tanaka (1986). In principle, such dating is now straightforward, because we can use Tanaka's defining crater densities to measure the ages from our isochron system. In practice there are several complications.

First, Tanaka defined the boundaries only in the diameter range $1 \text{ km} < D > 16 \text{ km}$, which leaves open the question of the boundaries that could be determined (for example) from MGS images at crater diameters of 11 to 500 m. Second, Tanaka assumed that the crater production function followed a -2 slope cumulative power law in this region, and calculated crater densities at $D = 1$ km and 4 to 10 km by extrapolating from densities at $D = 2$ km (Tanaka, 1986, Table 2 footnote). Current data show the production function is shallower than Tanaka's fit. This means the Tanaka assumed isochron shape does not exactly fit ours, producing a D dependence of inferred age. Worse yet, the D range of Tanaka's definitions, especially from 4 to 16 km diameter unfortunately overlaps the region where the Neukum and Hartmann systems, as reduced by Ivanov (2001), give the least consistent ages. Thus we can give only approximate ages for the boundaries. Finally, some of these approximate ages mostly fall in mid-Martian history (1 to 3 Gyr), and as explained in Section 5, these are the least valuable in constraining Martian geologic time, because of the uncertainty factor in our absolute ages, of about two.

While Tanaka's definitions of the beginnings of the epochs are precise, they are not completely internally consistent because of point 2 above, and there is a slight imprecision in defining the boundaries of the epochs. In view of this, and in view of the importance of the beginning of the Amazonian, we have re-examined the crater densities at the beginning of the Early Amazonian, fitting crater data to a wide size spectrum, not just to the diameters cited by Tanaka in the original definition. We use counts (from Neukum's group) for the Chryse/Arcadia Planitia type area of 57,500 km^2. Figure 13 shows a fit of 100 craters with $D > 1$ km to the Neukum production function derived by Ivanov (2001). The smaller craters,

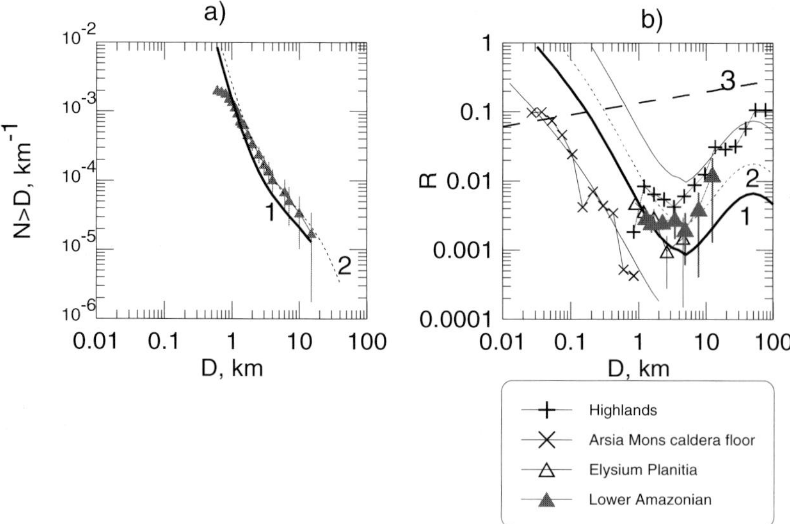

Figure 13. Crater densities in type areas related to the beginning of the Lower Amazonian, relative to Martian highlands and to Arsia Mons caldera. The fits of the Neukum production function shape to the smaller (*curve 1*) and larger (*curve 2*) craters in Chryse/Arcadia Planitia are shown. The *R*-plot in b) includes a surface of similar stratigraphic age in Elysium Planitia (*see text*). Curve 3 represents the saturation equilibrium level defined by Hartmann (1995) and found for Phobos (cf. Figure 3).

at $0.9 < D < 1.3$ km, fit a slightly lower isochron than the larger craters at $2.5 < D < 15$ km, as typical of our earlier results. An interpretation is that the larger-crater part of the distribution (Figure 13, curve 2) is related to a stratum which was subsequently eroded, lost craters of $D < 1.3$ km, and was later re-cratered to produce the population in curve 1, which would mark the base of the Early Amazonian ($\sim$3.1 Gyr in the Neukum system). It is not certain that this interpretation is correct, and whether curve 2 might be closer to the beginning of the Early Amazonian ($\sim$3.4 Gyr in the Neukum system). Neukum's group made additional counts for Elysium Planitia (Figure 13b) which give a more uniform CRA than the Chryse/Arcadia Planitia site, favoring the choice of 3.1 Gyr for the base of the Early Amazonian. This example illustrates the range of uncertainty for even a single system of isochrons, not counting the additional differences between the Hartmann and Neukum systems in the size range of $D \sim$ a few km.

Note that we make no effort here to redefine these boundaries or apply a correction to the -2 power law production function shape assumed by Tanaka. Although current data suggest a certain intrinsic "fuzziness" in the Tanaka definitions of the boundaries, we retain his definitions in terms of specific crater densities at different Ds, and then try to make the best possible estimate of the age at each boundary.

In spite of the range of uncertainties, we offer an overview of the Tanaka stratigraphy in Figure 14. As noted by Ivanov (2001) and this paper, the geologically recent (Amazonian) epochs give the most leverage on establishing the chronologic

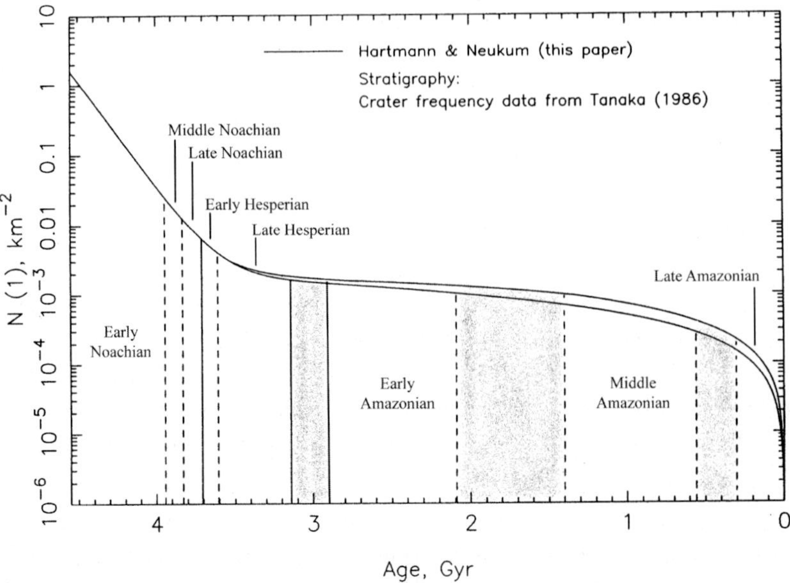

Figure 14. Mars cratering chronology model based on work in the present paper, using Tanaka's (1986) definition of stratigraphy based on crater densities at $D > 1$ km (plus our rediscussion of the definition of Lower Amazonian), and Ivanov's (2001) derivation of isochrons from Neukum and Hartmann data. The solid lines give model ages based primarily on the Ivanov-Neukum isochrons combined with the Neukum equation for time dependence of cratering (with essentially constant cratering rate after 3 Gyr ago). The left curve (older ages) is from Neukum data, the right curve (younger ages) from Hartmann. The diagram shows why uncertainties are greatest in mid-Martian histories. The model ages assume $R_{bolide} = 2.0$. Model ages younger than $\sim$3.0 Gyr are proportional to $1/R_{crater}$ (which is roughly proportional to $1/R_{bolide}$) and thus an additional uncertainty enters for those younger ages.

system because the present cratering rate is best known, and because earlier dates crowd around $3.5 - 4.1$ Gyr because of the high cratering rate at that time. To estimate ages in the Tanaka system we start with the Tanaka crater density definitions (taking into account the above discussion of the Early Amazonian beginning) and then combine these data with the Neukum equation for crater density as a function of time (Neukum *et al.*, 2001, Equation 5). We find the following results.

1. The entire Noachian Period lies before 3.5 to 3.7 Gyr ago according to both sets of isochrons. This appears to be a fairly robust result. Note that Stöffler and Ryder (2001) re-evaluated the ages of lunar basin impacts (placing all of them essentially between 3.7 and 3.9 Gyr ago. As a result their Figure 11 implies that the curve in our Figure 14 turns up much more steeply at about 4.0 Gyr than we show. We regard their age intepretations as intriguing but still unproven. In any case, they do not strongly affect our result, because the upturn is essentially within the Early Noachian. Indeed, a stronger upturn would even more tightly constrain early ages on Mars (Figure 14), because all $N(1)$ crater densities higher than $\sim$0.005 would be forced into the age range of $\sim$3.7 − 4.1 Gyr.

2. The boundary between Hesperian and Amazonian lies fairly early in Martian history, probably around 2.9 (Hartmann system) to 3.3 (Neukum system) Gyr ago. The position defined by Tanaka at $D = 1$ km lies very close to 3.1 Gyr ago in both systems. Figure 14 uses his definitions at $D = 1$ km. Tanaka's positions at 2 to 5 km show a somewhat greater range of age in the two systems, and this difference persists in examining younger epochs. Errors in R_{bolide} could conceivably reduce the boundary age to as little as 2.0 Gyr.

3. The beginning of the mid Amazonian lies around 1.4 (H system) to 2.1 Gyr (N system). This is the biggest discrepancy in absolute ages, occurring in mid-Martian history for reasons mentioned earlier. Errors in R_{bolide} could conceivably increase the uncertainty range to $\sim 1 - 3$ Gyr. Further reconciliation of the H and N systems, and sample return or in situ dating from this Epoch would be extremely valuable to reduce uncertainties in the system.

4. The beginning of the Late Amazonian is placed at about 0.3 Gyr (H) to 0.6 Gyr (N). Errors in R_{bolide} are unlikely to make the Late Amazonian older than 1.0 Gyr. The important result here is that Late Amazonian geology robustly is not confined to the ancient past but extends into the recent part of Martain history. (Note also that any argument for shifting Late Amazonian ages outside this range would have consequences in shifting all other ages accordingly, though the Noachian is generally constrained to before about 3.5 Gyr, due to the high crater densities, in any interpretation.)

Tanaka *et al.* (1987) used stratigraphic mapping to tabulate the total areas of Mars resurfaced by volcanism, and fluvial, periglacial, or other processes in each epoch. We divide the total area surfaced by the newly estimated duration of the epochs to calculate the rate of activity (km^2/yr) as a function of time. Regardless of whether the Neukum or Hartmann isochrons are used, the total resurfacing rates (km^2/yr) by volcanic, fluvial, periglacial, and cratering processes were all much greater in the Noachian and Hesperian before about 3 Gyr ago (Figures 15c-f). The data suggest that eolian resurfacing has continued at a more nearly constant rate. Tanaka *et al.* (1987) pioneered this analysis and obtained a similar result, but with a wider range of uncertainty in available chronological models. The modest recent upturn in reconstructed fluvial, periglacial, and cratering activity (Figure 15c-e) may result from errors in the assigned durations of recent epochs, or from the fact that the most recent units are better mapped. Measuring ages of the Martian epochs allows us to study not only geologic evolutionary processes but the nature of the Martian surface. Figure 15a shows that only modest percentage of the known volcanics (or of all units) are younger than 1.3 Gyr, raising the issue as to why 3 out of 4 (or 7 out of 8?) Martian impact sites have produced such young rocks. The problem is aggravated if one argues for older ages than we have suggested. The statistic may also mean that Martian uplands are covered by deep, gardened, loosely evaporite-cemented sediments that do not efficiently produce Martian meteorites.

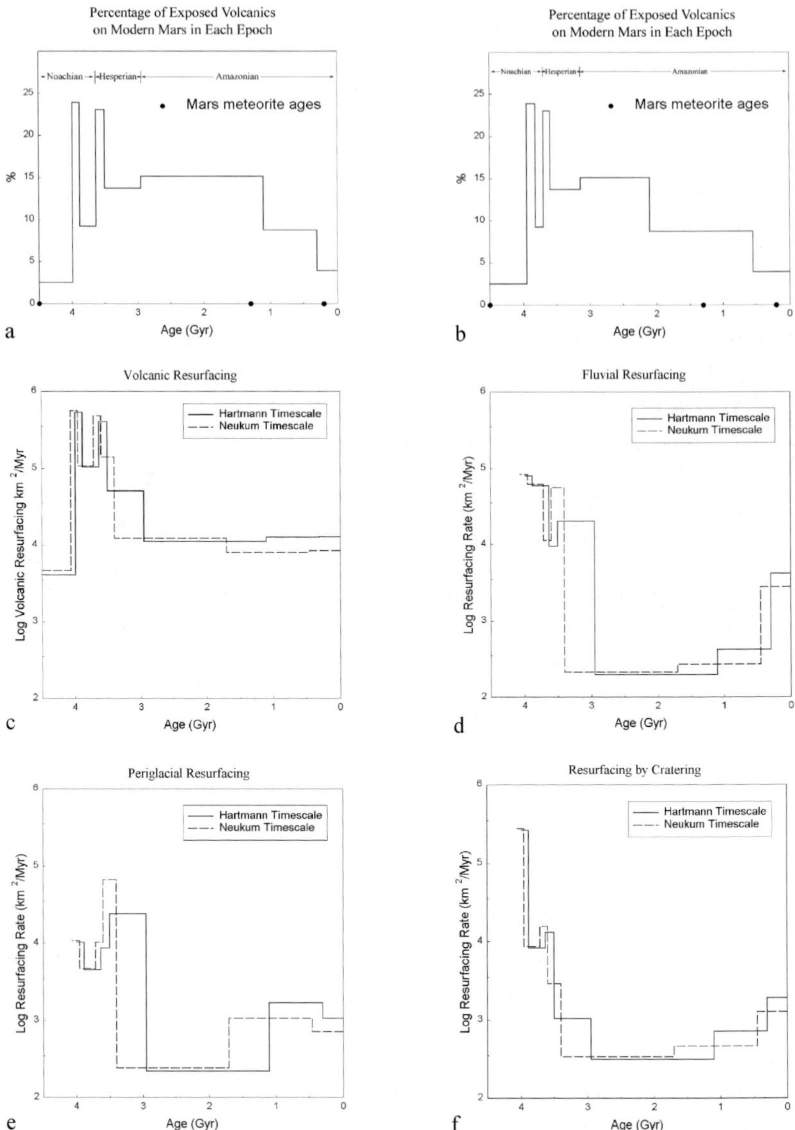

Figure 15. a-b) Age distributions for volcanic surface units on Mars, using epoch definitions and areas covered by volcanics from Tanaka *et al.* (1987), and dating systems of Hartmann/Ivanov (a) and Neukum/Ivanov (b). The two results are similar, and consistent with existence of young Martian SNC's. Larger numbers of Martian meteorites might allow a test of whether this age distribution applies, or whether ancient upland surfaces of 2 to 4.4 Gyr age are too weakly consolidated to produce meteorites. c-f) Time distributions of rate of volcanic, fluvial, periglacial, and impact resurfacing activity, based on Tanaka *et al.* (1987). Ages of epochs are drawn from Hartmann and Neukum systems (reduced by Ivanov, 2001), but slightly different from Figure 14, because of using broader diameter ranges to define the epochs. The data robustly show higher rates of activity by one or two orders of magnitude before ∼3 Gyr ago. With less certainty, the data raise the possibility of increased fluvial and perhaps other activity within the last few hundred Myr, though this may merely reflect easier identification of younger features.

8. Conclusions: Implications for Martian Geological History

We have shown that cratering data offer a valuable complement to Mars meteorites in understanding the absolute chronology of Mars. Meteorites give precise dates from a few (unknown) stratigraphic units, while crater data give rough dates from all stratigraphic units. With modern understanding of orbital dynamics and impact rates, crater counts provide dates with an total uncertainty that we estimate at a factor 2. Sample return or in situ dating would calibrate the crater dating and thus vastly improve planet-wide dating.

The combination of rock and crater data offers the following view of Martian history. Crustal rock units formed as long as 4.5 Gyr ago, as evidenced by ALH84001. The fact that one of the first dozen Mars rock samples is of this age, whereas such rocks are relatively rare from lunar samples, suggests that the Martian situation is very different from the that on the moon. Crater densities indicate that the old highlands should have been gardened to a depth of a kilometer or so, but the apparent aqueous weathering and carbonate deposits in ALH84001 at 4 Gyr ago, combined with the evidence for early fluvial resurfacing and for river and lake formation on Mars (Malin and Edgett, 2000b) suggests that any early megaregolith was subject to aqueous activity and probable cementing by carbonates and salts. The exposed megaregolith crustal units dating from about 4.4 to 3 or even 2 Gyr ago, being less consolidated, sedimentary-rich materials, may produce fewer meteorites, or fewer recognizable meteorites, than the primordial crust or the young volcanic units. This may explain the "missing meteorites" in the 1.3 Gyr to 4.5 Gyr age range. Both the Mars meteorite collection and the crater counts give strong, independent lines of evidence that volcanic and fluvial activity continued throughout Martian history into recent times. At least two impact sites on Mars have produced rocks with crystallization ages of 1300 and about 170-300 Myr ago. MGS images show extremely fresh-looking lava flows with crater count ages less than 100 Myr, and possibly as low as 3 to 10 Myr. Mars meteorites and MGS images also suggest sporadic ongoing aqueous activity. Aqueous alteration in nakhlites has been dated at 670 Myr ago. Virtually uncratered hillsides have apparent aqueous seep features that are much younger. Our understanding of Mars must allow for relatively recent volcanic activity and water mobility.

Acknowledgements

We thank Boris Ivanov, Roland Wagner, and the staff a Planetary Science Institute for helpful discussions. A number of research assistants and student workers helped us compile some of the crater counts, and these include Dr. Jennifer Grier, Dr. Eileen Ryan, Daniel Berman, Greg Herres, Julie Batten, Roland Wagner and, through the courtesy of an ISSI-sponsored workshop, group of Spanish student researchers from the Universidad Complutense, Madrid, headed by Jorge Anguita and Miguel de la Casa. We thank Prof. Johannes Geiss and the staff of ISSI for hospitality and encouragement. WKH acknowledges support from the NASA Mars Global Surveyor program through the Jet Propulsion Laboratory and the NASA Planetary Geology and Geophysics program and through the office of Dr. Carl Pilcher.

References

Bogard, D.D., and Johnson, P.: 1983, 'Martian Gases in an Antarctic Meteorite?', *Science* **221**, 651–654.

Bridges, J.C., and Grady, M.M.: 2000, 'Evaporite Mineral Assemblages in the Nakhlite (Martian) Meteorites', *Earth Planet. Sci. Lett.* **176**, 267–279.

Chapman, C.R.: 1974, 'Cratering on Mars. I Cratering and Obliteration History. II Implications for Future Cratering Studies from Mariner 4 Reanalysis', *Icarus* **22**, 272–300.

Chapman, C.R., Pollack, J., and Sagan, C.: 1969, 'An Analysis of the Mariner 4 Photography of Mars', *Astron. J.* **74**, 1039–1051.

Golombek, M.P., and Bridges, N.T.: 2000, 'Erosion Rates on Mars and Implications for Climate Change: Constraints from the Pathfinder Landing Site', *J. Geophys. Res.* **105**, 1841–1853.

Greeley, R., Kuzmin, R.O., and Haberle, R.M.: 2001, 'Aeolian Processes and Their Effects on Understanding the Chronology of Mars', *Space Sci. Rev.*, this volume.

Grier, J.A., Hartmann, W.K., Berman, D.C., Goldman, E.B., Esquerdo, G.A.: 2000, 'Constraining the Age of Martian Polar Strata by Crater Counts', *Amer. Astron. Soc., DPS Meeting*, abstracts #32, #58.03.

Grieve, R.A., and Shoemaker, E.: 1994, 'The Record of Past Impacts on Earth', *in* T. Gehrels (ed.), *Hazards Due to Comets and Asteroids*, Univ. Arizona Press, Tucson, Arizona, pp. 417–462.

Hartmann, W.K.: 1966a, 'Martian Cratering', *Icarus* **5**, 565–576.

Hartmann, W.K.: 1966b, 'Early Lunar Cratering', *Icarus* **5**, 406–418.

Hartmann, W.K.: 1971, 'Martian Cratering III: Theory of Crater Obliteration', *Icarus* **15**, 410–428.

Hartmann, W.K.: 1973, 'Martian Cratering 4: Mariner 9 Initial Analysis of Cratering Chronology', *J. Geophys. Res.* **78**, 4096–4116.

Hartmann, W.K.: 1977, 'Relative Crater Production Rates on Planets', *Icarus* **31**, 260–276.

Hartmann, W.K.: 1978, 'Martian Cratering V: Toward an Empirical Martian Chronology, and its Implications', *Geophys. Res. Lett.* **5**, 450–452.

Hartmann, W.K.: 1984, 'Does "saturation" Cratering Exist in the Solar System?', *Proc. 15th Lunar Planet. Sci. Conf.*, 348–349 (abstract).

Hartmann, W.K.: 1995, 'Planetary Cratering I: Lunar Highlands and Tests of Hypotheses on Crater Populations', *Meteoritics* **30**, 451.

Hartmann, W.K.: 1998, 'Martian Crater Populations and Obliteration Rates: First Results from Mars Global Surveyor', *Proc. 29th Lunar Planet. Sci. Conf.*, abstract #1115.

Hartmann, W.K.: 1999, 'Martian Cratering VI. Crater Count Isochrons and Evidence for Recent Volcanism from Mars Global Surveyor', *Meteoritics Planet. Sci.* **34**, 167–177.

Hartmann, W.K.: 2001, 'Martian Seeps and Their Relation to Youthful Geothermal Activity', *Space Sci. Rev.*, this volume.

Hartmann, W.K., and Gaskell, R.W.: 1997, 'Planetary Cratering 2: Studies of Saturation Equilibrium', *Meteoritics Planet. Sci.* **32**, 109–121.

Hartmann, W.K., and Berman, D.C.: 2000, 'Elysium Planitia Lava Flows: Crater Count Chronology and Geological Implications', *J. Geophys. Res.* **105**, 15,011–15,025.

Hartmann, W.K., Cruikshank, D.P., Degewij, J., Capps, R.W.: 1981, 'Surface materials on unusual planetary object Chiron', *Icarus* **47**, 333–341.

Hartmann, W.K., Malin, M.C., McEwen, A., Carr, M., Soderblom, L., Thomas, P., Danielson, E., James, P., and Veverka, J.: 1999, 'Recent Volcanism on Mars from Crater Counts', *Nature* **397**, 586–589.

Ivanov, B.A.: 2001, 'Mars/Moon Cratering Rate Ratio Estimates', this volume.

Jones, K.L.: 1974, 'Martian Obliterational History', Ph.D. Thesis Brown Univ., Providence, RI, USA.

Keszthelyi, K., McEwen, A.S., and Thordarson, T.: 2000. 'Terrestrial Analogs and Thermal Models for Martian Flood Lavas', *J. Geophys. Res.* **105**, 15,027–15,050.

Leighton, R.B., Murray, B., Sharp, R., Allen, J., and Sloan, R.: 1965, 'Mariner IV Photography of Mars: Initial Results', *Science* **149**, 627–630.

Lissauer, J.J., Squyres, S.W., and Hartmann, W.K.: 1988, 'Bombardment History of the Saturn System', *J. Geophys. Res.* **93**, 13,776–13,804.

Malin, M.C., and Edgett, K.S.: 2000a, 'Evidence for Recent Groundwater Seepage and Surface Runoff on Mars', *Science* **288**, 2330–2335.

Malin, M.C., and Edgett, K.S.: 2000b, 'Sedimentary Rocks of Early Mars', *Science* **290**, 1927–1937.

Malin, M.C., Carr, M.H., Danielson, G.E., Davies, M.E., Hartmann, W.K., Ingersoll, A.P., James, P.B., Masursky, H., McEwen, A.S., Soderblom, L.A., Thomas, P., Veverka, J., Caplinger, M.A., Ravine, M.A., Soulanille, T.A., and Warren, J.L.: 1998, 'Early Views of the Martian Surface from the Mars Orbiter Camera of Mars Global Surveyor', *Science* **279**, 1681.

Melosh, H.J.: 1989, 'Impact Cratering. A Geologic Process.', Oxford Univ. Press, New York.

Neukum, G.: 1983, *Meteoritenbombardement und Datierung planetarer Oberflächen*, Habilitation Dissertation for Faculty Membership, Ludwig-Maximilians Univ. München, Munich, Germany, 186 pp.

Neukum, G., and Wise, D.U.: 1976, 'Mars - A Standard Crater Curve and Possible new Time Scale', *Science* **194**, 1381–1387.

Neukum, G., and Hiller, K.: 1981, 'Martian Ages', *J. Geophys. Res.* **86**, 3097–3121.

Neukum, G., and Ivanov, B.A.: 1994, 'Crater Size Distributions and Impact Probabilities on Earth from Lunar, Terrestrial-Planet, and Asteroid Cratering Data', *in* T. Gehrels (ed.), *Hazards due to Comets and Asteroids*, Univ. Arizona Press, Tucson, pp. 359–416.

Neukum, G., König, B., Storzer, D., and Fechtig, H.: 1975, 'Chronology of Lunar Cratering', *Proc. 6th Lunar Planet. Sci. Conf.*, 598 (abstract).

Neukum, G., Ivanov, B., and Hartmann, W.K.: 2001, 'Cratering Records in the Inner Solar System in Relation to the Lunar Reference System', *Space Sci. Rev.*, this volume.

Nyquist, L.E., Wooden, J., Bansal, B., Wiesmann, H., McKay, G., and Bogard, D.D.: 1979, 'Rb-Sr Age of the Shergotty Achondrite and Implications for Metamorphic Resetting of Isochron Ages', *Geochim. Cosmochim. Acta* **43**, 1057–1074.

Nyquist, K., Bogard, D., Shih, C.-Y., Greshake, A., Stöffler, D., and Eugster, O.: 2001, 'Ages and Geologic Histories of Martian Meteorites', *Space Sci. Rev.*, this volume.

Öpik, E.J.: 1965, 'Mariner IV and Craters on Mars', *Irish Astron. J.* **7**, 92.

Öpik, E.J.: 1966, 'The Martian Surface', *Science* **153**, 255.

Papanastassiou, D.A., and Wasserburg, G.J.: 1974, 'Rb-Sr Age', *Geophys. Res. Lett* **1**, 23.

Pike, R.J.: 1977, 'Apparent Depth/Apparent Diameter Relation for Lunar Craters', *Proc. 8th Lunar Sci. Conf.* **3**, Pergamon Press, New York, USA, pp. 3427–3436.

Plescia, J.B.: 1990, 'Recent Flood Lavas in the Elysium Region of Mars', *Icarus* **88**, 465–490.

Sawyer, D.J., McGehee, M.D., Canepa, J., and Moore, C.B.: 2000, 'Water Soluble Ions in the Nakhla Martian Metorite', *Met. Planet. Sci.* **35**, 743–747.

Scott, D.H., Tanaka, K.L., Greeley, R., and Guest, J.E.: 1987, Geologic Maps of the Western Equatorial, Eastern Equatorial and Polar Regions of Mars, Maps. I-1802-A, B and C, Miscellaneous Investigation Series, 1986-1987, U.S. Geological Survey, Flagstaff.

Shih, C.-Y., Nyquist, L.E., Reese, Y., and Wiesmann, H.: 1998, 'The Chronology of the Nakhlite, Lafayette: Rb-Sr and Sm-Nd Isotopic Ages', *Proc. 29th Ann. Lunar Planet. Sci. Conf.*, abstract #1145.

Soderblom, L.A., Condit, C.D., West, R.A., Herman, B.M., and Kreidler, T.J.: 1974, 'Martian Planetwide Crater Distributions – Implications for Geologic History and Surface Processes', *Icarus* **22**, 239–263.

Stöffler, D., and Ryder, G.: 2001, 'Stratigraphy and Isotope Ages of Lunar Geologic Units: Chronological Standard for the Inner Solar System', *Space Sci. Rev.*, this volume.

Strom, R.G., Croft, S., and Barlow, N.: 1992, 'The Martian Impact Cratering Record', *in* H. Kieffer *et al.* (eds.), *Mars*, Univ. Arizona Press, Tucson, pp. 383–423.

Swindle, T.D., Treiman, A.H., Lindstrom, D.J., Burkland, M.K., Cohen, B.A., Grier, J.A., Li, B., and Olson, E.K.: 2000, 'Noble Gases in Iddingsite from the Lafayette Meteorite: Evidence for Liquid Water on Mars in the last few Hundred Million Years', *Met. Planet. Sci.* **35**, 107–115.

Tanaka, K.L.: 1986, 'The Stratigraphy of Mars', *Proc. 17th Lunar Planet. Sci. Conf.*, *J. Geophys. Res.* **91, suppl.**, 139–158.

Tanaka, K.L., Isbell, N.K., Scott, D.H., Greeley, R., and Guest, J.E.: 1987, 'The Resurfacing History of Mars – A Synthesis of Digitized, Viking-based Geology', *Proc. 18th Lunar Planet. Sci. Conf., 1987*, Cambridge University Press/LPI, 1988, Cambridge and New York/Houston, TX, USA, pp. 665–678.

Wood, C.A., and Ashwal, L.D.: 1981, 'Meteorites from Mars: Prospects, Problems and Implications', *Proc. 12th Lunar Planet. Sci. Conf.*, 1197–1199 (abstract).

Address for correspondence: William K. Hartmann, Planetary Science Institute, 620 N. 6th Avenue, Tucson AZ 85705-8331, USA; (hartmann@psi.edu)

II: EVOLUTION OF THE INTERIOR
AND SURFACE OF MARS

THE ACCRETION, COMPOSITION AND EARLY DIFFERENTIATION OF MARS

A.N. HALLIDAY[1], H. WÄNKE[2], J.-L. BIRCK[3] and R.N. CLAYTON[4]

[1]*Institute for Isotope Geology and Mineral Resources, Department of Earth Sciences, ETH Zentrum, NO C61, CH-8049 Zürich,*
[2]*Max-Planck-Institut für Chemie, Becher-Weg 27, D-55128, Mainz, Germany*
[3]*Laboratoire de Géochimie-Cosmochimie, IPGP, 4 Place Jussieu, 75252 Cedex 05, Paris, France*
[4]*Enrico Fermi Institute, University of Chicago, Chicago, IL 60637, USA*

Received: 1 November 2000; accepted: 27 February 2001

Abstract. The early development of Mars is of enormous interest, not just in its own right, but also because it provides unique insights into the earliest history of the Earth, a planet whose origins have been all but obliterated. Mars is not as depleted in moderately volatile elements as are other terrestrial planets. Judging by the data for Martian meteorites it has Rb/Sr $\approx$ 0.07 and K/U $\approx$ 19,000, both of which are roughly twice as high as the values for the Earth. The mantle of Mars is also twice as rich in Fe as the mantle of the Earth, the Martian core being small ($\sim$20% by mass). This is thought to be because conditions were more oxidizing during core formation. For the same reason a number of elements that are moderately siderophile on Earth such as P, Mn, Cr and W, are more lithophile on Mars. The very different apparent behavior of high field strength (HFS) elements in Martian magmas compared to terrestrial basalts and eucrites may be related to this higher phosphorus content. The highly siderophile element abundance patterns have been interpreted as reflecting strong partitioning during core formation in a magma ocean environment with little if any late veneer. Oxygen isotope data provide evidence for the relative proportions of chondritic components that were accreted to form Mars. However, the amount of volatile element depletion predicted from these models does not match that observed – Mars would be expected to be more depleted in volatiles than the Earth. The easiest way to reconcile these data is for the Earth to have lost a fraction of its moderately volatile elements during late accretionary events, such as giant impacts. This might also explain the non-chondritic Si/Mg ratio of the silicate portion of the Earth. The lower density of Mars is consistent with this interpretation, as are isotopic data. ^{87}Rb-^{87}Sr, ^{129}I-^{129}Xe, ^{146}Sm-^{142}Nd, ^{182}Hf-^{182}W, ^{187}Re-^{187}Os, ^{235}U-^{207}Pb and ^{238}U-^{206}Pb isotopic data for Martian meteorites all provide evidence that Mars accreted rapidly and at an early stage differentiated into atmosphere, mantle and core. Variations in heavy xenon isotopes have proved complicated to interpret in terms of ^{244}Pu decay and timing because of fractionation thought to be caused by hydrodynamic escape. There are, as yet, no resolvable isotopic heterogeneities identified in Martian meteorites resulting from ^{92}Nb decay to ^{92}Zr, consistent with the paucity of perovskite in the martian interior and its probable absence from any Martian magma ocean. Similarly the longer-lived ^{176}Lu-^{176}Hf system also preserves little record of early differentiation. In contrast W isotope data, Ba/W and time-integrated Re/Os ratios of Martian meteorites provide powerful evidence that the mantle retains remarkably early heterogeneities that are vestiges of core metal segregation processes that occurred within the first 20 Myr of the Solar System. Despite this evidence for rapid accretion and differentiation, there is no evidence that Mars grew more quickly than the Earth at an equivalent size. Mars appears to have just stopped growing earlier because it did not undergo late stage (>20 Myr) impacts on the scale of the Moon-forming Giant Impact that affected the Earth.

Space Science Reviews **96**: 197–230, 2001.
© 2001 *Kluwer Academic Publishers. Printed in the Netherlands.*

1. Introduction

Mars is about one-eighth the mass of the Earth and it may provide an analogue of what the Earth was like when it was at such an early stage of accretion. The further growth of the Earth was sustained by major collisions with planetesimals and planets such as that which resulted in the formation of the Earth's moon (Hartmann and Davis, 1975; Cameron and Ward, 1976; Wetherill, 1986; Cameron and Benz, 1991). This late accretionary history, which lasted more than 50 Myr in the case of the Earth (Halliday, 2000a, b), appears to have been shorter and less catastrophic in the case of Mars (Harper *et al.*, 1995; Lee and Halliday, 1997). In this article we review the basic differences between the bulk composition of Mars and the Earth and the manner in which this plays into our understanding of the timing and mechanisms of accretion and core formation. We highlight some of the evidence for early cessation of major collisional growth on Mars. Finally, we reevaluate the isotopic evidence that Mars differentiated quickly.

Fundamental differences between the composition of Mars and that of other terrestrial planets are apparent from the planet's slightly lower density and from the compositions of Martian meteorites. The low density is partially explicable if there is a greater proportion of more volatile elements. However, given that the terrestrial planets are chiefly (>80%) composed of O (mass ~16), Mg (24), Si (28) and Fe (56), the proportion of Fe is clearly another critical parameter. For example, the high density of Mercury has long been considered to be the result of a large metallic Fe core. In a similar vein one could argue that the somewhat low density of Mars is the result of a smaller proportion of Fe in the planet as a whole.

Two explanations have been advanced for these differences in density and (inferred) Fe content between the terrestrial planets. The first is that they relate to incomplete condensation of metal and silicate from a hot solar nebula of chondritic composition (Turekian and Clark, 1969; Lewis, 1972). In this model, the innermost planets formed directly from higher temperature condensates before all of the more volatile elements had condensed. These may have been blown away during the early T-Tauri stages of the Sun. This requires that the terrestrial planets and their cores formed very early, before dispersal of a still hot solar nebula. Although condensation theory explains certain features of the distribution of elements in refractory inclusions (Grossman, 1972; Grossman and Larimer, 1974) there is now significant evidence against planetary accretion models based simply on condensation from a hot solar nebula. For example, the preservation of presolar grains in all classes of chondrites (Huss, 1988) is difficult to explain if these objects simply condensed from a very hot gas. More recent models explain refractory inclusions in terms of very localized heating within the proto-planetary disk (Shu *et al.*, 1997).

An alternative explanation for silicate depletion and variations in planet density within the inner solar system is that these are the result of collisional stripping associated with the later stages of formation of the planets (Benz *et al.*, 1987). Dynamic simulations (Safronov, 1954; Wetherill, 1986) provide powerful evidence that accretion of the terrestrial planets was highly protracted. There is now strong evidence that such protracted accretion ultimately produces giant impacts (Cameron and Benz, 1991; Canup and Agnor, 1998). These processes would have been extraordinarily energetic and should have resulted in dramatic changes to the planet's constitution and possibly bulk composition.

2. Accretion Dynamics and the Composition of Mars

Accretion is nowadays thought to have involved three broadly overlapping stages (Canup and Agnor, 1998). In the first stage it is necessary to get the dust and ice of the circumstellar disc to co-adhere and build bodies of the size where gravitational interactions dominate. This first stage is by far the most poorly understood at present. Getting dust grains to stick is understood (Blum, 2000) but how rock-sized objects accrete to build objects that are big enough for gravitational attraction to hold them together is more difficult (Benz, 2000). Ultimately one has to achieve bodies of about a kilometer in size. The second stage involves runaway gravitational growth of kilometer-sized objects (Lin and Papaloizou, 1985). This builds the planetesimal to $\sim$1,000 km diameter from material sourced in a reasonably broad provenance. As the runaway stage consumes most of the smaller material in the inner solar system the subsequent accretion of the terrestrial planets becomes dominated by collisions between these planetesimals and, ultimately, with other planets, as proposed by Safronov (1954). As such, growth would be highly protracted. This is a stochastic process such that one can not readily predict growth histories for the terrestrial planets. So dynamic models for the accretion of Mars and other planets have largely been based on Monte Carlo simulations that generate some solutions that result in terrestrial planets with the correct (broadly speaking) size and distribution. By focusing on these solutions and tracking the growth of each such planet, Wetherill (1986) showed that all of the terrestrial planets would require a time scale approaching 10^8 years for accretion. He also noted that the terrestrial planets would accrete at roughly exponentially decreasing rates. The half-mass accretion time (time for half of the present mass to accumulate) was comparable (about 5 to 7 Myr), and in reality indistinguishable, for Mercury, Venus, Earth and Mars using such simulations. Of course these objects, being of different size, would have had very different absolute growth rates. At one time it had been assumed that the accretion rates of the planets would decrease with distance from the Sun because the time taken to achieve runaway accretional growth from localized "feeding zones" should increase markedly with heliocentric distance and more "space" from which accretion must occur. On this basis Earth would form more quickly than Mars. However, Wetherill (1994a) showed that this concept was wrong and that the material from which the terrestrial planets formed was derived from such a broad provenance that only a small difference in average feeding zone and accretion rate existed between the planets.

The rapid conversion of kinetic energy to heat in a planet growing by accretion of planetesimals and other planets means that it is inescapable that silicate and metal melting temperatures are achieved (Sasaki and Nakazawa, 1986; Benz and Cameron, 1990; Melosh, 1990). Core segregation should take place early, although the mechanisms are not well understood (Shaw, 1978; Stevenson, 1981, 1990; Minarik *et al.*, 1996). This being the case the possibility exists that "glancing blow" collisions between already differentiated planets ("giant impacts") will preferentially remove major portions of the outer silicate portions of the planet as it grows. It has been argued that proto-atmospheres (Melosh and Vickery, 1989; Ahrens, 1990; Benz and Cameron, 1990; Zahnle, 1993), moderately volatile elements (Lodders, 2000; Halliday and Porcelli, 2001) and even refractory silicates (Benz *et al.*, 1987) could be eroded by these processes. Mercury, with its high density, is a prime candidate for a body that lost a great deal of its silicate material by giant impacts (Benz *et al.*, 1987). Conversely Mars, with a density lower than that of the Earth, may actually be a closer approximation of the material from which Earth accreted than Earth itself is (Halliday and Porcelli, 2001). The proportional size of the Earth's core (like Mercury's) may well have been increased as a consequence of giant impacts. Similarly the non-chondritic Si/Mg ratio of the Earth's primitive mantle (Allègre *et al.*, 1995) may in part be the product of impact erosion of its outer, more differentiated portions.

3. Martian Meteorites and the Bulk Composition of Mars

Martian meteorites are mafic and ultramafic igneous rocks derived from the Martian crust and possibly also the mantle. They were produced by melting of the mantle and provide the most important direct information on the composition of the Martian interior. One group of Martian meteorites, the shergottites, although cumulate enriched, represent basaltic magmas. Hence, they directly reflect the chemical composition of the Martian interior. The first extensive study on the bulk composition of Mars was by Dreibus and Wänke (1984, 1985) based on data for SNC meteorites. There are several important papers that summarize and discuss the composition of Martian meteorites (Wänke, 1981; Dreibus and Wänke, 1985, 1987; Treiman *et al.*, 1986, 1987; Wänke and Dreibus, 1988), the most up to date compilation being that of Lodders (1998).

Before discussing the approach of these estimates we should give some cautionary advice. It is important to realize that a planet that is one-eighth the mass of the Earth will have distinct phase stability as a function of depth (Longhi *et al.*, 1992; Bertka and Fei, 1997). Furthermore, the very different budgets of volatile elements will dramatically affect a range of features including phase transitions (Bertka and Fei, 1997), core formation (Righter and Drake, 1996), the stability of minor accessory phases, the partitioning of trace elements during melting and the rheology of the mantle (Breuer *et al.*, 1997). The meteorites may not even

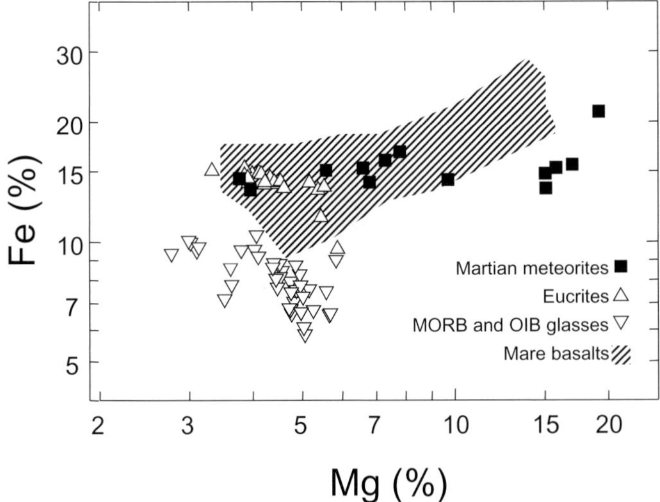

Figure 1. Plot showing the relationship between Fe and Mg concentrations for Martian meteorites, eucrites, mare basalts and terrestrial OIB and MORB glasses. Data from Kitts and Lodders (1998), Lodders (1998), Paul Warren (personal communication) and Yi *et al.* (2000). Note that the Earth's mantle appears to be more depleted in Fe than any other basalt sources sampled.

be representative. The crustal dichotomy of Mars possibly reflects fundamental differences in the geochemistry and history of the underlying mantle (Breuer *et al.*, 1993). There is little evidence for large-scale convective overturn on Mars (Breuer *et al.*, 1997). There are few if any plate-tectonic like features (Sleep, 1994) and shield volcanoes are sited on a static lithosphere (Carr, 1973) and fed by isolated plumes (Harder and Christensen, 1996). So the convective stirring that homogenizes the terrestrial mantle and results in uniform trace element compositions in basalts (Hofmann *et al.*, 1986) is lacking on Mars.

The ramifications of these differences are considerable and interpretations based on these meteorites with extrapolations from terrestrial geochemistry need to be considered with caution. Having said this, we also emphasize that certain aspects of Martian meteorite composition are so consistent that representation is not likely to be an issue. Furthermore, the lack of mantle homogenization has a very positive side. The isotope geochemistry of the Martian meteorites, even though these objects are relatively young, can be expected to display heterogeneity preserved from far earlier in Solar System history than is recoverable from terrestrial basalts. Indeed, the data from Martian meteorites provide a picture of the development of Mars that is very consistent with the information obtained from geophysics, geology and direct sampling.

Major elements: Martian meteorites lack the depletion in Fe and certain other siderophile elements found in terrestrial basalts (Figure 1). In this and in some other respects they are more like eucrites or lunar basalts. They do not display the striking Ti enrichment of lunar basalts however, which probably reflects ilmenite

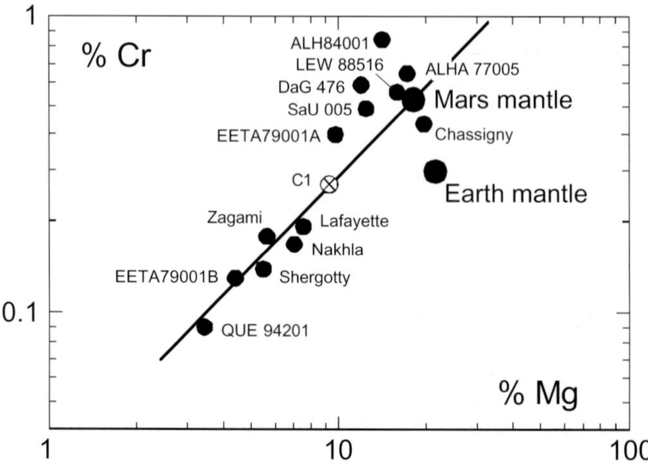

Figure 2. Cr versus Mg. Unlike in the Earth, Cr is not depleted in the Martian mantle. All data from the Mainz laboratory.

in the lunar mantle. Manganese is also relatively enriched in SNC meteorites. Under normal planetary conditions with olivine, orthopyroxene and clinopyroxene being the major FeO- and MnO-bearing mantle phases, the liquid-solid partition coefficients of FeO and MnO are very similar and only slightly above 1. The FeO concentrations in all shergottites are very similar with a mean value for FeO of 18.9% and 0.48% for MnO. These average values are similar to estimates of FeO and MnO in the shergottite parent liquids reported by Longhi and Pan (1989). Assuming a C1 abundance, a concentration of 0.46% MnO would be expected. Hence, it is quite clear that contrary to the Earth's mantle, manganese is not depleted in the Martian mantle, but like in the eucrite parent body (possibly asteroid Vesta) is present in C1 abundance, or, normalized to C1 and Si, MnO = 1.00. Comparing the FeO/MnO ratio of 39.5 ± 1.2 with the C1 FeO/MnO ratio of 100.6, we obtain a normalized FeO abundance of 0.39 or an absolute concentration of 17.9 ± 0.6 wt% in the martian mantle. In the Earth's upper mantle the FeO concentration is about 8 wt.% or only less than half. So while the density of Mars is explained by a lower proportion of Fe, SNC meteorites provide evidence that the Martian mantle is in fact far richer in Fe than that found in Earth.

As iron resides as FeO in the mantle of the planets, but as metal and FeS in the core, the FeO definitely deviates most from chondritic abundance compared to MgO, SiO_2, Al_2O_3, and CaO, for which C1 abundances in the Martian mantle can be assumed. As seen from Figure 2, chromium follows Mg in all Martian meteorites with a C1 Cr/Mg ratio. Hence, as with Mn there is a clear indication that Cr is not depleted in the Martian mantle contrary to its depletion in the terrestrial mantle. With the assumption of a C1- and Si-normalized Mg abundance of 1.00, we find also that the normalized Cr abundance is 1.00 in the Martian mantle. So Mn/Cr ratios were not fractionated by core formation on Mars.

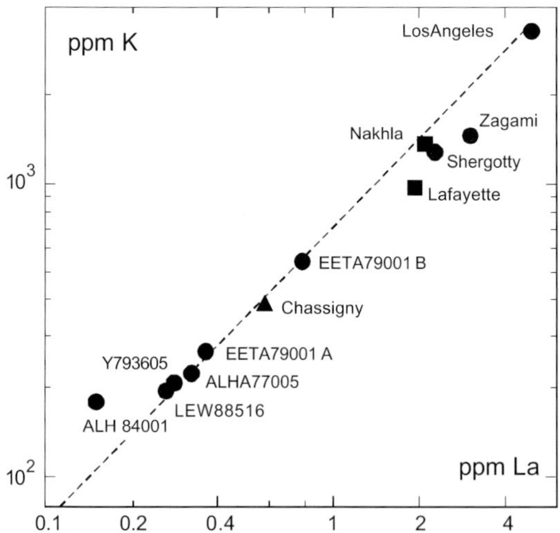

Figure 3. K versus La. Assuming a C1 abundance for the refractory element La, an abundance of 0.30 is obtained for the moderately volatile element K. All data from the Mainz laboratory except values for Los Angeles and Y793605 which are taken from Warren *et al.* (2000) and Warren and Kallemeyn (1997).

Potassium, rubidium and cesium: As with the Moon, the eucrite parent body and to a lesser extent the Earth, an excellent correlation of K with La is observed for all Martian meteorites (Figure 3) with a K/La ratio of 635 compared to the C1 ratio of 2110. If one assumes a normalized abundance of 1.00 for the refractory element La, a value of 0.30 is obtained for the abundance of K. Shergottites have a K/Rb ratio of 289 ± 39 and Rb/Cs ratio of 15.8 ± 0.8. Based on the K abundance of 0.30 one obtains abundances of 0.26 for Rb and 0.20 for Cs. As with the Earth, the depletion is greater in alkalis with higher theoretical condensation temperature and this may be because the mechanisms of some of the volatile losses were unrelated to "canonical" nebular condensation (Halliday and Porcelli, 2001).

Phosphorous and high field strength elements: For not yet fully understood reasons, phosphorus in the Earth's mantle is considerably depleted while in the Martian mantle phosphorus is normal. This is the reason why phosphates play an important role in the fractionation of LIL (large ion lithophile) elements. In the Martian rocks the REE (rare earth elements), but also U and Th, reside mainly in the phosphates. As phosphates dissolve easily in slightly acidified solutions, the bearing for Sm/Nd, Lu/Hf and U/Pb dating is obvious. Phosphates may also have a strong effect on partitioning during melting. The extreme fractionations of Zr/Nb found in Martian meteorites are unlike anything found on Earth and possibly reflect the increased role of phosphate in the Martian mantle (Figure 4). Similarly the large time-integrated fractionations in Lu/Hf and Sm/Nd (Blichert-Toft *et al.*, 1999) could be related to the increased importance of phosphate. It has been proposed that these are caused by ilmenite in the source region, as is the case

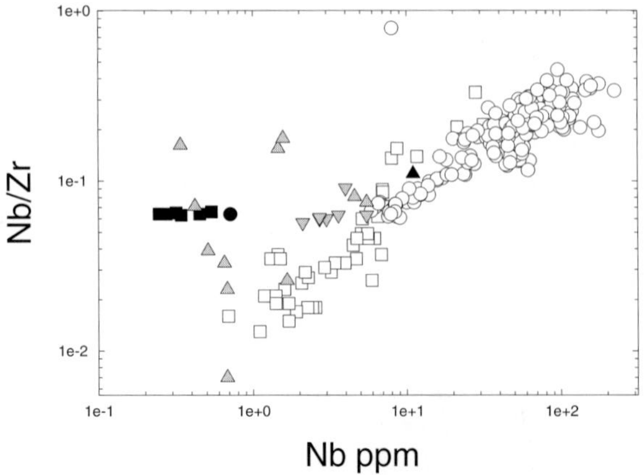

Figure 4. A plot of Nb/Zr vs. Nb concentration for terrestrial mid-ocean ridge basalts (*open squares*; Hofmann *et al.*, 1986; Yi *et al.*, 2000), ocean island basalts (*open circles*; Hofmann *et al.*, 1986; Halliday *et al.*, 1996, and references therein; Yi *et al.*, 2000), chondrite classes (*filled squares*; Newsom, 1995), Martian meteorites (*upright triangles*; Lodders, 1998), eucrites *inverted triangles*; Kitts and Lodders, 1998), the bulk silicate Earth (*filled circle*; McDonough and Sun, 1995) and the average continental crust of the Earth (*filled upright triangle*; Taylor and McLennan, 1985). Note the uniform Nb/Zr in chondrites and eucrites, and the systematic variations with Nb concentration in terrestrial basalts. In contrast Martian meteorites display considerable variability but no relationship with Nb. This probably relates to the increased importance of phosphorus in the Martian mantle (Figure 8) and the effect of phosphate on the partitioning of high field strength elements.

for lunar basalts. However, Ti contents of Martian basalts are not especially high, whereas mare basalts that formed by re-melting cumulates formed in a magma ocean have Ti contents that are an order of magnitude higher than is found on Earth or Mars. Indeed, the evidence for an early magma ocean stage on Mars (Righter and Drake, 1996), unlike for the Moon, is based almost entirely on siderophile element abundances.

Other lithophile trace elements: Using a number of element correlations found for the shergottites, together with other constraints from studying Martian meteorites, Wänke and Dreibus (1988) estimated the abundance of F, Cl, Br, I, Na, P, Co, Zn, Ga, Mo, In, Tl, and W. For all refractory trace lithophile major and trace elements they assumed C1 abundance values, including the radioelements U and Th. For the majority of elements the values from Dreibus and Wänke (1985, 1987) agree well with those from Treiman *et al.* (1986), who used a similar approach. However, the latter authors use for U, Th and Al, Ti the ratios observed in Martian meteorites, which differ by up to a factor of 2 from the C1 values. Later Longhi *et al.* (1992) adopted the Mainz model (Wänke and Dreibus, 1988, and references therein). They argued that it is more difficult to envision a series of nebular and/or accretion processes that would lead to planet-wide fractionations of refractory lithophile elements than to account for changes in the U/La and Al/Ti ratios in shergottites with a multi-stage petrogenetic history.

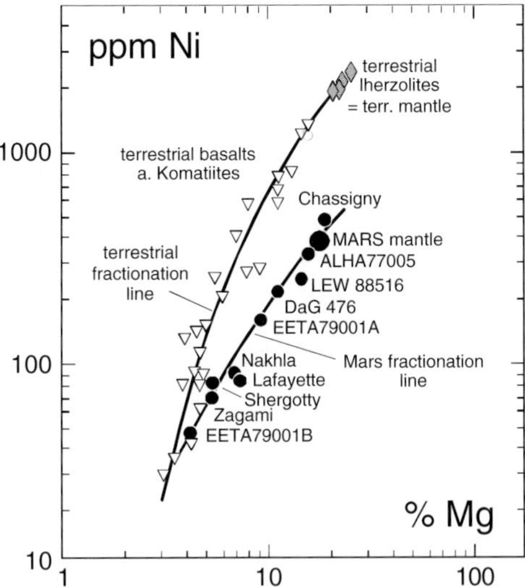

Figure 5. Ni versus Mg. The abundance of Ni is found from the Mars fractionation line of this diagram, assuming a C1 abundance for Mg. All data from the Mainz laboratory.

Cobalt and nickel: These two moderately siderophile elements are of special importance for understanding the accretion history of Mars. Cobalt shows a good correlation with the sum of $MgO + FeO$ in the terrestrial and Martian rocks. Hence, knowing the FeO abundance in the Martian mantle and assuming for MgO an abundance of 1.00, one obtains for Co an abundance of 0.070. Two distinct fractionation lines for terrestrial and Martian rocks are observed in the Ni versus Mg diagram (Figure 5), yielding a concentration of 400 ppm for Ni in the Martian mantle, corresponding to an abundance of Ni = 0.019.

Highly siderophile elements: The highly siderophile elements (HSE) have been difficult to determine precisely but new Re and Os measurements are now available and provide important constraints. Judging from ultramafic Martian meteorites the mantle has a composition that is close to that of the terrestrial mantle in terms of Re and Os concentrations. In the production of basalts by mantle melting, Os is always strongly compatible, Re displays variable behaviour in the different planetary bodies for which samples are available for laboratory analyses (Birck and Allègre, 1994). Osmium contents are highly variable but Re for a given planet or planetary body is constant within less than an order of magnitude. Figure 6 represents this property and shows that from an ultramafic reservoir similar in composition for the different planets, distinct planetary trends are found in basalts or related materials. For typical basalts containing 10 ppt Os, the Re/Os in Mars is close to two orders of magnitude higher than in chondrites, but this is still about a factor of 10 lower than typical terrestrial basalts. The interpretation of these trends has been discussed extensively. A possible contamination of basaltic melts having very low Os contents with impacting chondritic materials has been excluded (Birck and

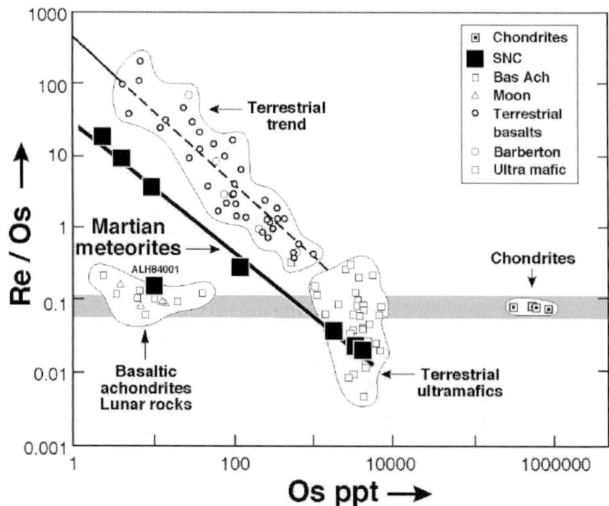

Figure 6. Comparison of Re/Os ratios and Os abundances in chondrites and various differentiated samples of the solar system (from Birck and Allègre, 1994). Os abundances refer to non radiogenic Os. For the Moon and Basaltic achondrites the Re/Os ratio is calculated from the $^{187}Os/^{186}Os$ ratio, the formation age, and assuming a chondritic evolution for the source material of these rocks. A buffered Re trend with variable Os would plot along a -1 slope in this diagram: the terrestrial trend scatters around such a line (mean value 475 ppt Re). The reverse would produce a vertical line as is illustrated by the terrestrial ultramafic rocks.

Allègre, 1994; Warren and Kallemeyn, 1996). The origin most probably relates to the oxygen fugacity in the mantle source of the basalts, which varies between the bodies (Figure 7). This stems from the different oxidation states possible for Re. In very reducing conditions as in the Moon, Re behaves as a compatible element like Os. Mars is more oxidised in its interior (Wänke and Dreibus, 1988) which results in higher partitioning of Re into the melt phase.

Water: Dreibus and Wänke (1985, 1987) estimated the water content of the Martian mantle in two different ways and obtained a value of 36 ppm. Hence, the Martian interior is very dry in spite of the clear evidence for abundant water on the surface of Mars. Without plate tectonics, the water from the surface never found its way into the interior of the planet. The planetary interior may have remained dry and all the water added during accretion may have been consumed by reaction with metallic iron. This would release large amounts of hydrogen that may have escaped from the planet, particular with loss of an early atmosphere (Hunten *et al.*, 1987). Part also have been dissolved into the core (Zharkov, 1996).

The estimates of the composition of the Martian mantle derived in the above manner are listed in Table I together with values for the Earth's mantle for comparison. As is obvious from Table I and Figure 8, all moderately volatile elements are higher on Mars by about a factor of 2 aside from Mn and Cr. The higher abundances of moderately volatile elements means that K/U and Rb/Sr are significantly higher on Mars than on Earth (Figure 9). Also P is much higher in the Martian

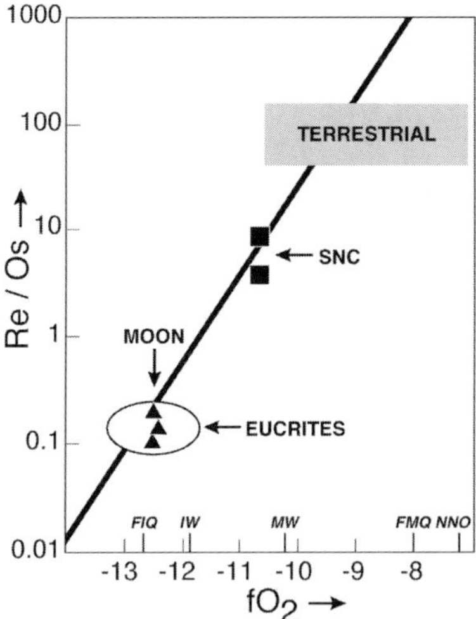

Figure 7. Re/Os fractionation versus oxygen fugacity. The Re/Os range for a planetary body is taken only for basaltic rocks containing ca. 10 ppt of total Os. Extraterrestrial oxygen fugacity values are mostly intrinsic oxygen fugacity measurements. The terrestrial box includes the whole range of intrinsic measurements and other estimates from sample chemistry. Fugacities are compared for a temperature of 1200 °C and literature data were adjusted to this temperature if not available. Data from Birck and Allègre (1994), Christie *et al.* (1986), Delano and Arculus (1980), Haggerty (1981), Hewins and Ulmer (1983), Higuchi and Morgan (1975), Sato (1976), Sato *et al.* (1973), Stolper (1977), Warren and Kallemeyn (1996) and Wolf *et al.* (1979). Symbols as in Figure 6.

TABLE I

Major and selected minor and trace elements in the mantles of Earth and Mars.

Element	Earth[1]	Mars[2]	Mars [3]	Element	Earth[1]	Mars[2]	Mars [3]
MgO [%]	36.9	30.2	1.00	Br [ppb]	45.6	145	0.029
Al_2O_3	4.19	3.02	1.00	I	13.3	32	0.029
SiO_2	46.0	44.0	1.00	Co [ppm]	105	68	0.070
CaO	3.54	2.45	1.00	Ni	2110	400	0.019
TiO_2	0.23	0.14	1.00	Cu	28	5.5	0.027
FeO	7.58	17.9	0.39	Zn	48	62	0.099
Na_2O	0.39	0.50	0.38	Ga	3.8	6.6	0.37
P_2O_5	0.015	0.16	0.36	Mo [ppb]	18	118	0.066
Cr_2O_3	0.44	0.76	1.00	In	–	14	0.090
MnO	0.13	0.46	1.00	La [ppm]	0.52	0.48	1.00
K [ppm]	231	305	0.30	Tl [ppb]	–	3.6	0.013
Rb	0.74	1.06	0.26	W	24	105	0.6
Cs	0.01	0.07	0.20	Th	–	56	1.00
F	19.4	32	0.31	U	21	16	1.00
Cl	11.8	38	0.029				

[1] Wänke *et al.* (1984); [2] Wänke and Dreibus (1988); [3] Rel. to Si and C1

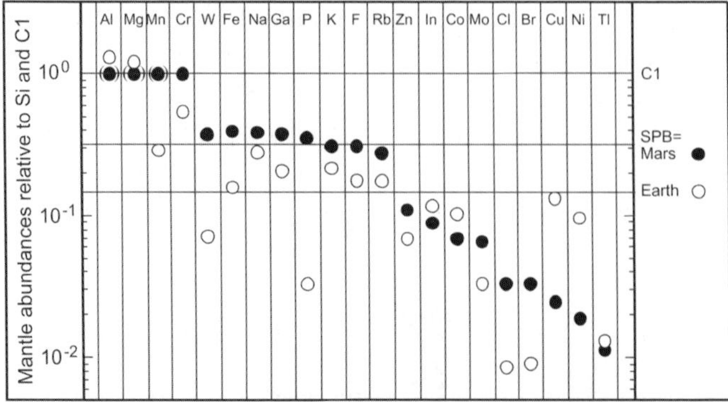

Figure 8. Composition of the Martian and terrestrial mantles relative to carbonaceous chondrites. Relative to the Earth, Mars shows less pronounced depletion in P, Mn, Cr, and W as a consequence of core formation.

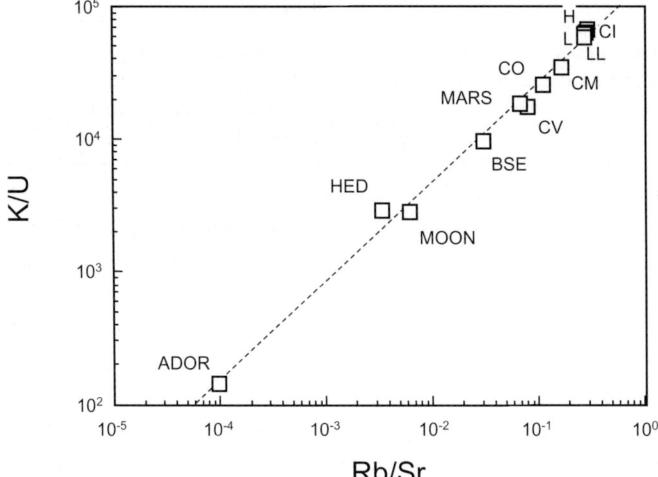

Figure 9. A plot of K/U vs. Rb/Sr for chondrites, Mars and the Earth (BSE) as well as smaller differentiated bodies such as the Eucrite (HED) Parent Body (believed to be Asteroid 4 Vesta) and the Moon. Data for the volatile element depleted meteorite Angra dos Reis are shown as well. Sources: Kitts and Lodders (1998), Lodders (1998), Lugmair and Galer (1992), Newsom (1995), Nyquist *et al.* (1994).

mantle presumably for the same reason. Contrary to the terrestrial situation, Mn and Cr are not at all depleted in the Martian mantle; the Mn/Cr ratio is roughly chondritic (Figure 8). This has a direct bearing on the interpretation of Cr isotopic data, as will be discussed below. Mn and Cr are probably depleted in the Earth's mantle because they partially reside in the core. Tungsten is also not as strongly depleted in the Martian mantle as it is in the Earth's (Figure 8) and this has a bearing on the interpretation of the Hf-W isotope data, also discussed below. So some elements that are moderately siderophile on Earth appear to have behaved more like lithophile elements on Mars – this probably relates to more volatile-rich oxidizing, S-rich conditions during core formation. How this relates to the behavior of Re/Os (Figure 7) is currently unclear. Other moderately siderophile elements especially Co and Ni are higher in the Earth's mantle and this formed the basis for the proposition of an inhomogeneous accretion of the Earth (Wänke, 1981). The low abundances of Co and Ni in Martian basalts provide evidence of homogeneous accretion and strong partitioning of these elements during core formation.

4. Mantle Phase Transitions and the Size of the Core

The composition of the Martian mantle derived from studies of Martian meteorites can be used to constrain the composition and size of the core, as well as the phase assemblages and transitions that would be likely in a homogeneous primitive Martian mantle. The solubility of S in silicate melts is a strong function of their Fe content (Wallace and Carmichael, 1992). As demonstrated above, the FeO content of the Martian mantle is thought to be twice that in the Earth's mantle so, not surprisingly, Martian rocks also contain large amounts of sulfur. In fact, it has been argued that sulfur determines the oxygen fugacity of Martian rocks (Wänke and Dreibus, 1994). The amount of S in the Martian mantle and core have been the subject of considerable debate. Most authors favor high ($>10\%$) S concentrations in the core consistent with the limited depletion in volatile elements (Longhi *et al.*, 1992). However, recent experimental partitioning studies have yielded far lower S concentrations (Gaetani and Grove, 1997).

The high FeO content of Martian meteorites and the Martian mantle provides evidence for major differences between Earth and Mars in terms of the conditions of core formation. A greater proportion of the Fe in Mars resides in its mantle. Whereas the core occupies $\sim31\%$ of the present Earth's mass, in the case of Mars the figure is closer to 21% (Wänke and Dreibus, 1988). Evidently core formation did not proceed to the same extent on Mars, perhaps because conditions were more oxidizing. Another possibility is that an early large core reacted with the later accreted more oxidized mantle and actually became smaller during the process of accretion as oxidized material was added. However, there are no well understood mechanisms for such a process. Assuming a S abundance equal to that of other moderately volatile elements, a composition of 77% Fe, 15% S, 7.6% Ni, and 0.4% Co was estimated for the core of Mars (Wänke and Dreibus, 1988).

The phases that should be stable as a function of depth within the Martian mantle exert a fundamental influence on the rheological behaviour of the planet. Likely phase assemblages have been reviewed by Longhi *et al.* (1992). A more recent experimental assessment of the phase transitions is given by Bertka and Fei (1997). These authors used the independently estimated composition of the Martian mantle established by Dreibus and Wänke (1985). Because of the lower pressure on Mars the upper mantle would be dominated by orthopyroxene, clinopyroxene, olivine assemblages with small amounts of garnet down to a depth of roughly

1,000 km where there should be transition to spinel peridotite with majorite garnet. It is thought that a thin (~200 km) perovskite-bearing layer may be present above the core-mantle boundary at 2,000 km depth. It should be no surprise if garnet was involved in trace element partitioning during Martian mantle melting. This is consistent with isotopic evidence from Martian meteorites (Blichert-Toft *et al.*, 1999). The geophysical effects of the phase transitions are of great interest (Breuer *et al.*, 1997) and bear upon the entire issue of heterogeneity. In particular, if convection was limited (Harder and Christensen, 1996; Breuer *et al.*, 1997) isotopic heterogeneity in the mantle could be great.

5. Chemical Evidence of Homogenous Accretion

Wänke and coworkers have used the composition of the Martian mantle and core derived from meteorite studies to deduce the nature of the components that accreted to form Mars. These components do not necessarily correspond to the composition of a particular Solar System material. Furthermore, describing the data in terms of such components does not necessarily imply a particular accretion history. However, the concept is useful for comparisons with the Earth. Indeed, this work is based on earlier ideas on Earth's accretionary components by Ringwood (1979). To explain the chemical composition of the Earth's mantle, he introduced a two-component model. In order to account for the high abundances of moderately siderophile elements in the Earth's mantle, Wänke (1981) took up this model with only slight changes. According to this model the Earth was formed from:

Component A: Highly reduced and free of all elements with equal or higher volatility than Na, but containing all other elements in C1 abundance ratios. Iron and all siderophile elements in metallic form, and even part of Si as metal.

Component B: Oxidized and containing all elements, including the volatiles, in C1 abundances. Iron and all siderophile and lithophile elements are present mainly as oxides.

Wänke (1981) favored an inhomogeneous accretion for the Earth with a mixing ratio of component A : component B of 85:15, according to which most of component B was only added after the Earth had reached about two thirds of its present mass. The obvious depletion of all chalcophile elements in the Martian mantle (Figure 8) provides the key observational evidence in favour of a homogeneous accretion of Mars. We have already pointed out (Table I and Figure 8) that the Martian mantle has relatively low abundances of certain siderophile elements (Co, Ni, Cu) that have strong chalcophile affinities, yet higher abundances of siderophile elements (W, Cr, Mn) that have weak chalcophile affinities. So with large amounts of S one could extract a major fraction of the Co, Ni and Cu into the core, without removing much W, Mn or Cr. Dreibus and Wänke (1984, 1985) used these observations to postulate a homogeneous accretion scenario for Mars during which sulfur reacted with metallic iron forming a sulfur-rich FeNi phase.

There is little evidence for a late oxidized "veneer" component as is generally considered necessary to produce the chondritic relative abundances of highly siderophile elements in the Earth. Indeed the highly siderophile elements appear to be strongly fractionated (Treiman *et al.*, 1986, 1987; Righter and Drake, 1996; Brandon *et al.*, 2000) as if residual from equilibrium with core-forming liquids. Note, however, that the Martian trend in Figure 6 also incorporates rocks that are cumulates (Nakhla, Chassigny). All of these agree with the non-cumulate samples with the exception of ALH84001. This is the only really early Martian meteorite so far identified and it displays unusually low Re. Its composition most probably relates to its origin in a mantle depleted in Re by core formation before subsequent addition of about 1% chondritic material. So this meteorite may provide evidence of very early mantle heterogeneity preserved on Mars that developed by internal differentiation before the completion of accretion and core formation.

It can be shown that Mars could have been formed with the same two model components as for the Earth, but with a mixing ratio of 60:40. In other words, the abundance of the elements from component B, i.e. the moderately volatile and volatile elements on Mars are about three times more abundant than on Earth. In the two-component model for the formation of the terrestrial planets (Ringwood, 1977, 1979; Wänke, 1981), sulfur is supplied by the oxidized, volatile-bearing component B, while metallic Fe is derived by the reduced, volatile free component A. Segregation of the sulfur-rich phase extracted chalcophile elements from the mantle of Mars according to their sulfide-silicate partition coefficients. The model of a homogeneous accretion of Mars and its sulfur-rich core was also advocated by Treiman *et al.* (1987) and is now widely accepted.

Water, added in rather large amounts from component B, may have reacted with the metallic iron from component A. The huge amounts of hydrogen generated by the reaction $Fe + H_2O \rightarrow FeO + H_2$ would be dissolved in the core or lost by hydrodynamic escape (Hunten *et al.*, 1987). So it could be a consequence of the reaction of metal with water, that Mars acquired an FeO-rich mantle and, compared to the Earth, a smaller core. Furthermore, because of homogeneous accretion and the presence of metal the Martian mantle kept rather dry. Water was restricted to the surface layers of Mars and due to the lack of plate tectonics never found its way to the interior of the planet (Carr and Wänke, 1992).

6. Oxygen Isotopic Evidence for the Components that Formed Mars

The isotopic compositions of several light elements (H, C, N, O, S) are variable in terrestrial materials, and provide the basis for a large field of stable isotope geochemistry (Clayton, 1986, 1993). Variations in these compositions in Martian samples indicate that there are processes and sources peculiar to Mars, which may place constraints on its materials of construction, and elucidate some specific Martian processes. Oxygen is especially useful, since a three-isotope system al-

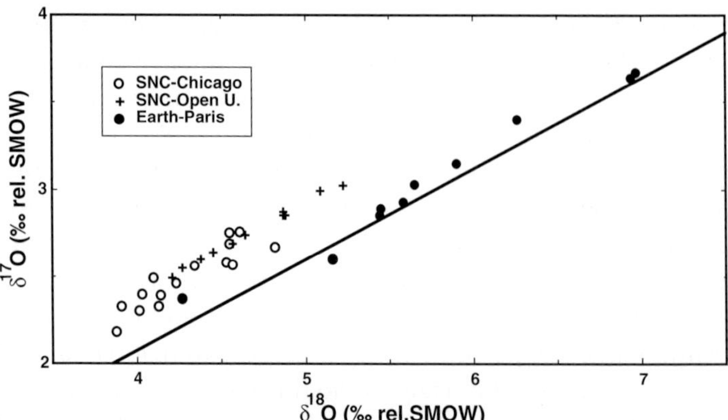

Figure 10. The oxygen three-isotope graph shows separate fractionation trends for Earth and Mars, reflecting different proportions of precursory planetesimals. The Mars samples are SNC whole-rocks, from Clayton and Mayeda (1996) and Franchi *et al.* (1999); the Earth samples are mafic igneous whole-rocks from Robert *et al.* (1992).

lows discrimination between mass-dependent planetary processes and primordial nebular heterogeneities inherited during planet formation (Clayton and Mayeda, 1996; Franchi *et al.*, 1999). The silicate rocks from Earth and Mars define two separate mass-fractionation trends (Figure 10), reflecting a whole-planet isotopic difference derived from a different mix of precursory planetesimals (Lodders and Fegley, 1997). The implied relative proportions of volatile-rich and volatile-poor constituents are in the opposite sense from those derived in the Wänke-Ringwood mixing models (Clayton and Mayeda, 1996).

If the planetesimal building blocks were constituted from the same range of materials now represented by chondritic meteorites, it might be possible to relate the differences in bulk chemical composition between Earth and Mars to the proportions of chondritic precursors. As outlined above, Wänke and Dreibus (1988) modeled these two planets as two-component mixtures, one component volatile-free and reduced, the other volatile-rich and oxidized. The former has no obvious meteoritic counterpart; the latter may resemble carbonaceous chondrites. Oxygen isotopic compositions provide an excellent tracer for these two types of material. The ordinary chondrites and enstatite chondrites (anhydrous but not greatly depleted in moderately volatile elements) occupy a restricted range in the oxygen three-isotope graph. Whereas volatile-rich carbonaceous chondrites of classes CI, CM and CR are considerably enriched in the heavy isotopes as a direct consequence of the low-temperature hydration processes that produced the phyllosilicates which characterize these meteorites. As explained above, in terms of moderately-volatile elements such as the alkalis and halogens, Mars is more volatile-rich than the Earth. However, the oxygen-isotope compositions of Mars (based on analysis of SNC meteorites) show a smaller proportion of carbonaceous-chondrite-like material in Mars than in Earth. It thus appears that, unless the Earth lost a significant proportion of its moderately volatile elements after it formed, the principal carrier of moderately volatile elements in formation of the terrestrial planets was not of carbonaceous chondrite composition. However, it is quite plausible, given the dynamics of planetary accretion discussed above that the Earth did indeed loose a major fraction of its moderately volatile elements. This has been advanced as an explanation by Lodders (2000) and is supported by Sr isotope data for early solar system objects (Halliday and Porcelli, 2001).

7. Manganese-53 and Early Heterogeneity in the Accretion Disk

It has often been proposed that extinct radionuclides provide evidence, not just of time-scales for the early solar system, but also of spatial heterogeneities in freshly synthesized nuclides within the disk, the origins of which are obscure. ^{26}Mg isotopic effects related to the decay of ^{26}Al (half-life = 0.73 Myr), have so far only been very clearly identified in either refractory inclusions or Al rich components of chondrites (Lee *et al.*, 1977, 1998; Hinton and Bischoff, 1984; Hinton *et al.*, 1988; Zinner and Göpel, 1992; Russell *et al.*, 1996; Galy *et al.*, 2000). At the bulk rock scale no such effects have been detected so far with present day analytical tools. For larger planetary bodies like the Earth, the Moon or Mars no detectable differences are indeed expected and for the SNC group of meteorites no measurement is available in the literature. One has to look at longer-lived systems.

The case for ^{53}Mn (half-life = 3.7 Myr) induced effects on Cr isotopes is significantly different as the average solar system parent/daughter ratio is more favorable by a factor of ten and the half-life is about five times longer. Furthermore, at the present time isotopic measurements of Cr have roughly a 10 times higher precision than Mg measurements (Lugmair and Shukolyukov, 1998). At these high levels of precision systematic differences in Cr isotopic composition are observed between samples originating in different planetary bodies or classes of bodies at the bulk rock scale. With regard to Mars, only three SNCs (Shergotty, EETA 79001 and ALH 84001) have been investigated so far with high precision techniques (Lugmair and Shukolyukov, 1998). All ^{53}Cr/^{52}Cr ratios, whether they represent separated minerals or bulk rocks, are different from the terrestrial composition and display an excess of ^{53}Cr of 0.23ϵ. As far as we can tell from meteorites, Mars is isotopically homogeneous for Cr, as would be expected from a large size planet with geologic activity that was ongoing over an extended period of time with respect to the short time scale of the 53Mn radioactivity. Nevertheless the Cr isotopic composition is clearly distinct from the Earth and from the different chondrite groups, as well as from the basaltic achondrite family (HED).

As the presence of live ^{53}Mn in the early solar system is now well established (Birck and Allègre, 1988), the small difference between large objects like the Earth and Mars is thought to be produced by this radioactivity. Regarding the ^{53}Mn-^{53}Cr system, there are two kinds of model for the evolution of the solar system, both taking into account the fact that the value of Mars (SNCs) is intermediate between the Earth and the asteroid belt samples (chondrites and HEDs).

The first explanation for this is by Lugmair and Shukolyukov (1998), who propose a radial gradient of ^{53}Mn atomic abundance in the early solar system with an increase of the ^{53}Mn/^{55}Mn ratio with heliocentric distances. This model has the advantage of decoupling the observed chromium isotopic differences between the major bodies, from time; this releases the time constraint for the formation of terrestrial planets within the life-time of ^{53}Mn, which is short when compared to the current accretion times discussed above. The ^{53}Mn heterogeneity model has a few difficulties however. Most observational and theoretical studies of star collapse involve extensive mixing in the early solar nebula (Shu et $al.$, 1987). Furthermore, as already noted, the studies of Wetherill indicate that the provenance of the planets is very broad. So it is hard to see how such an early feature related to heliocentric distance survives. Having said this, it is widely noted that the Asteroid Belt preserves radial-from-Sun compositional heterogeneities, carbonaceous chondrites beginning abruptly at 2.5–2.7 AU. There are however, "collateral consequences" of such Mn-Cr isotopic heterogeneities in the form of other isotopic variability that are not observed (Nichols, 2000). Finally, the Mn-Cr isotopic data for CAIs (Birck and Allègre, 1988) and chondrules (Nyquist et $al.$, 1997) do not support the low solar system initial ^{53}Mn/^{55}Mn that is implied from the Mn-Cr systematics of eucrites (Lugmair and Shukolyukov, 1998). These data provide evidence that an early process that changes the Mn/Cr of the material from which planets and planetesimals form may provide a better explanation for variations in Cr isotopic composition in the inner solar system.

An alternative series of models have been proposed that rely on the difference in volatility of the two elements to match the observed composition of the different planets or planetary bodies with respect to Mn/Cr (Halliday et $al.$, 1996; Cassen and Woolum, 1997; Birck et $al.$, 1999). Assuming that the solar nebula started with homogeneous ^{53}Mn/^{55}Mn, ^{53}Cr/^{52}Cr and chondritic Mn/Cr ratio, the bulk solar system should display an evolution similar to that of C1 carbonaceous chondrites. This puts very strong constraints on the time at which the Mn/Cr fractionated in the materials constituting the planets. The differences in Cr isotopic composition then relate to the degree of depletion in more volatile Mn relative to less volatile Cr. As differences are observed in the Cr isotopic ratios this has to happen during the lifetime of ^{53}Mn which cannot exceed 20 Myr. In the model of Halliday et $al.$ (1996) it was assumed that the ^{53}Mn/^{55}Mn at the start of the Solar System was that defined by an Allende CAI (Birck and Allègre, 1988) and on this basis the material from which the Earth accreted was depleted in more volatile Mn within no more than roughly 3.5 Myr after the start of the Solar System. Cassen and Woolum (1997) presented a more comprehensive model of volatile depletion taking place at the very start of the solar system as a consequence of early heating in the disk. The Cr isotopic compositions then reflects the degree and exact timing of depletion in Mn relative to Cr, the effects of which are greater closer in to the Sun.

Recently, Birck *et al.* (1999) have also assumed that the differences relate to the timing of volatile element depletion but with all bodies forming from precursor material that originally had chondritic Mn/Cr. This model implies that the materials constituting the Earth (the planetesimals) are separated from the solar nebula with a depleted Mn/Cr, no later than 2.2 Myr after the formation of C1 carbonaceous chondrites. Similarly for Mars the isolation of materials with lower than solar Mn/Cr occurs 3 Myr after the Earth. Although this model is inconsistent with chemical models for Mars that imply a chondritic Mn/Cr ratio for this planet (Wänke and Dreibus, 1988) it is in good agreement with a number of other isotopic chronometers for the development of solar system bodies (Birck *et al.*, 1999).

So the Mn-Cr results provide fascinating insights into early solar system processes. However, which exact processes occurred is unclear! The data could either reflect incomplete mixing of material in the circumstellar disk, or different degrees and timing of volatile element depletion. Although the latter model appears more likely, it is not yet consistent with independently estimated Mn/Cr ratios.

8. Early Isotopic Heterogeneity on Mars

The time required for a planet to heat up and differentiate into a core, mantle, crust and atmosphere/hydrosphere is strongly model dependent. Early models (e.g., Toksöz and Hsui, 1978; Solomon, 1979) assumed that the terrestrial planets were formed from cold material, which heated up slowly from the outside inward as a consequence of the conversion of gravitational energy from late accretion phases, radioactive decay and the density-driven process of core formation. In these models it took periods of $> 10^8$ or even $> 10^9$ years to form the core of Mars. Such models for a while superceded the concept that the material that accreted to form Mars was hot and freshly condensed from a nebula (Clark *et al.*, 1972; Grossman, 1972) heated by frictional and other mechanisms in the primitive disk (Boss, 1990). However, as it became clear that late stage growth from collisions was a fundamental part of planetary growth, the possibility arose that one might easily melt an entire planet at an early stage (Sasaki and Nakazawa, 1986; Melosh, 1990; Melosh *et al.*, 1993; Benz and Cameron, 1990). So the more recent models for core formation on Mars assume an early magma ocean environment (Righter and Drake, 1996). Similarly, outgassing and formation of an early atmosphere are viewed as very early processes (Melosh and Vickery, 1989; Carr and Wänke, 1992; Jakosky and Jones, 1997; Carr, 1999).

Models for the accretion and early differentiation of Mars can be tested with chronometers that have a longer half-life than ^{26}Al ($T_{1/2} = 0.73 \times 10^6$ yrs) or ^{53}Mn ($T_{1/2} = 3.7 \times 10^6$ yrs). Several such chronometers have provided evidence of very early heterogeneity preserved within Mars. These include ^{182}Hf ($T_{1/2} = 9 \times 10^6$ yrs), ^{129}I ($T_{1/2} = 16 \times 10^6$ yrs), ^{244}Pu ($T_{1/2} = 82 \times 10^6$ yrs) and ^{146}Sm ($T_{1/2} = 103 \times 10^6$ yrs). However, even the longer lived chronometers ^{235}U ($T_{1/2} = $

7.04×10^8 yrs), ^{238}U ($T_{1/2} = 4.47 \times 10^9$ yrs), and ^{87}Rb ($T_{1/2} = 4.88 \times 10^{10}$ yrs), reveal a remarkable early isotopic heterogeneity on Mars, not preserved on Earth.

Isotopic data for noble gases, in particular for xenon yield evidence of early segregation of the atmosphere from the mantle (Ott, 1988; Pepin, 1994; Marti *et al.*, 1995; Jakosky and Jones, 1997; Swindle and Jones, 1997; Bogard and Garrison, 1998; Gilmour *et al.*, 1998; Matthew and Marti, 2001). The history of these different reservoirs is a matter of some uncertainty. Martian atmospheric Xe is more enriched in ^{129}Xe than the "mantle" Xe as recorded e.g. in the Chassigny meteorite (Matthew and Marti, 2001). In a similar manner the atmospheric Ar is more enriched in ^{40}Ar than is the mantle Ar. In both cases these relationships are the opposite of that found on Earth and the explanation is unclear. It has been proposed that an early halogen-rich reservoir may have become a repository for ^{129}I, which subsequently degassed ^{129}Xe into the atmosphere. There are further complications with trying to define the ^{244}Pu fission xenon component given that mass fractionation by hydrodynamic escape (Hunten *et al.*, 1987) is thought to have affected the atmosphere (Swindle and Jones, 1997). It has been proposed that a component of the atmosphere was outgassed before much ^{244}Pu decay had occurred. Suffice it to say that there is clear evidence of a very early heterogeneity preserved in the noble gases.

The isotopic compositions of lithophile elements in Martian meteorites provide evidence for the preservation of such heterogeneity with the Martian interior. A particularly striking feature is found in the Sr isotope data. The bulk rock values for these define an $\sim$4.5 Gyr isochron implying a remarkable lack of remixing in the Martian mantle since its earliest differentiation (Shih *et al.*, 1982; Jagoutz *et al.*, 1994; Borg *et al.*, 1997). These data provide no direct evidence regarding the time scales for core formation. However, they do imply early melting and a lack of large scale mixing subsequently.

Chen and Wasserburg (1986) also argued that the differentiation of Mars was early, as the proximity of the Pb isotopic data of shergottites to the Geochron indicates that U/Pb fractionation was early. It is commonly assumed that the major fractionation of U/Pb takes place as a consequence of core formation. Evaluating the degree to which core formation as opposed to volatile loss resulted in depletion in volatile siderophile elements like Pb is complex.

The standard method for estimating the abundance of volatile elements in the bulk Earth makes use of the volatile lithophile element depletion trend, defined by their primitive mantle abundances as a function of half-mass condensation temperature (Newsom, 1990; McDonough and Sun, 1995; Yi *et al.*, 2000). However, this requires knowledge of the speciation, ambient pressures etc., which in turn requires an understanding of the exact mechanisms of volatile loss.

Alternatively, Allègre *et al.* (1995) have used the trends defined by the ratios of volatile elements to refractory lithophile elements in chondrites to estimate abundances in the bulk Earth. This latter method circumvents the uncertainty over the exact volatility of the elements that accreted to form the Earth. However, it assumes that volatile element depletion processes recorded in carbonaceous chondrites, which probably come from the outer reaches of the asteroid belt, are analogous to those that affected the inner solar system. Both approaches assume that (1) volatile elements were lost before core formation, (2) volatility is the only factor that affects the elemental abundances relative to chondrites and (3) lithophile elements do not enter the Earth's core in significant amounts. Whichever approach is used, it is necessary to make a reasonable estimate of the primitive mantle abundance of the element of interest in order to use the data to place constraints on models of core formation.

Cerium and Pb have similar bulk distribution coefficients during melting (Hofmann *et al.*, 1986). From the $Ce/^{204}Pb$ of basaltic Martian meteorites we can determine the $Ce/^{204}Pb$ and hence the degree of Pb depletion of the Martian mantle. The continental crust carries a major proportion ($>50\%$) of the silicate Earth's Pb because it has been selectively enriched by ocean floor hydrothermal and arc processes over billions of years. Assuming that this is not the case for Mars and no major silicate reservoir with fractionated Ce/Pb exists, one can also use the $Ce/^{204}Pb$ of the basaltic Martian meteorites to determine the degree of Pb depletion of the bulk silicate Mars. If one then uses the degree of depletion in alkalis as judged from Rb/Sr, Rb/Ba or K/U one can conclude that the degree of moderately volatile element depletion of Mars is overall comparable to that found in CV carbonaceous chondrites like Allende (Figure 9). From the $Ce/^{204}Pb$ as a function of the Rb/Sr of chondrite classes we can conclude that the bulk Mars should have $Ce/^{204}Pb$ that is about an order of magnitude higher than that deduced for the bulk silicate Mars. On this basis a sizeable fraction (perhaps 90%) of Pb was removed by core formation. The Pb in Martian meteorites is not as radiogenic as terrestrial Pb because the degree of volatile element depletion is much less. However core formation did produce a major fractionation and the Pb data would therefore appear to support the view that the core formed reasonably early, i.e. within 10^8 yrs.

9. Hafnium - Tungsten Chronometry and Rapid Accretion and Differentiation

The rates of accretion and core formation of Mars and other inner solar system objects can be more directly studied using the newly developed ^{182}Hf-^{182}W chronometer. The former decay of ^{182}Hf to ^{182}W ($T_{1/2} = 9$ Myr) has resulted in variations in W isotope composition that reflect early solar system time integrated Hf/W ratios (Halliday and Lee, 1999). This approach offers more leverage than other short-lived nuclide systems for several reasons. Firstly, the half-life of 9 Myr is ideal for the kinds accretionary time-scales proposed by Safronov and Wetherill. Secondly, both parent and daughter elements are highly refractory and in known, chondritic proportions in Mars. Third, the initial ^{182}Hf atomic abundance at the start of the solar system was higher than hitherto realized. Fourth, Hf and W are strongly fractionated by the early processes of melting and core formation.

Harper *et al.* (1991), were the first to provide a hint of a W isotopic difference between the iron meteorite Toluca and the silicate Earth. It has now been convincingly demonstrated that there is a ubiquitous clearly resolvable deficit in ^{182}W in iron meteorites and the metals of ordinary chondrites and enstatite relative to the atomic abundance found in the silicate Earth (Lee and Halliday, 1995, 1996, 2000a, b; Harper and Jacobsen, 1996; Jacobsen and Harper, 1996; Horan *et al.*, 1998).

If ^{182}Hf is sufficiently abundant then minerals, rocks and reservoirs with higher Hf/W ratio will develop W that is significantly more radiogenic (higher ^{182}W/^{184}W or ϵW) compared with the initial W isotopic composition of the solar system. Conversely metals with low Hf/W that segregate at an early stage from bodies with chondritic Hf/W (most early planets and planetesimals) will sample unradiogenic W, as found. Early segregated metals are deficient by roughly 3 to 4 ϵW units (300–400 ppm) relative to the silicate Earth. The only known explanation for this difference is that these metals sampled early solar system W before live ^{182}Hf had decayed. A similar picture emerges from the study of achondrites. Eucrites represent samples of the silicate (high Hf/W) portion of a body that differentiated early. They yield a range of W isotopic compositions from near chondritic to very radiogenic (Lee and Halliday, 1997; Quitté *et al.*, 2000). The data for eucrites implicate rapid accretion and core formation on the eucrite parent body within about 10 Myr of the start of the solar system (Lee and Halliday, 1997).

A very different picture is derived from studying the Earth however. The W isotopic difference between early metals and the silicate Earth reflects the time integrated Hf/W of the material that formed the Earth and its reservoirs during the lifetime of ^{182}Hf. The Hf/W ratio of the silicate Earth is determined to lie in the range $10 - 40$ (Newsom *et al.*, 1996). This is an order of magnitude higher than in carbonaceous and ordinary chondrites and is a consequence of terrestrial core formation. Although the bulk silicate Earth (BSE) has non-chondritic Hf/W because of core formation, it has a chondritic W isotopic composition (Lee and Halliday, 1995, 1996). This is inconsistent with models that involve the completion of terrestrial accretion and core formation within the first 10 Myr of solar system history, such as early heterogeneous accretion of silicate and metal from a fractionated partially condensed nebula. Protracted accretion (Wetherill, 1986) of material that has, on average, chondritic compositions with respect to Hf-W is more in accord with the chondritic W isotopic composition of the BSE. Most early-formed low Hf/W metal and high Hf/W silicate that was added to the Earth during accretion must have largely equilibrated isotopically with the growing BSE, otherwise its W isotopic composition would not be chondritic (Halliday, 2000a, b).

With the current exception of ALH84001, Martian meteorites provide us with samples of igneous rocks that formed by melting the interior of Mars over the past 1 Gyr or so, long after ^{182}Hf became extinct. Therefore, the W isotopic variations have nothing to do with the measured Hf/W. The Hf/W ratio of the Martian mantle is $\sim 2 - 3$, more than five times smaller than found for the Earth (Halliday and Lee, 1999). This is presumably related to the greater oxidation of the Martian interior. Yet Martian samples include some samples with radiogenic W, quite unlike the silicate Earth. Given the low Hf/W ratio of the Martian mantle this implicates the effects of very early differentiation on Mars. Evidently the accretion and differentiation of Mars was complete within ~ 20 Myr (Lee and Halliday, 1997).

There are other striking differences between Earth and Mars highlighted by the W isotope data. Mars clearly differentiated very early and this explains the W isotopic anomalies. However, it is also remarkable that such an anomaly has survived at all as a feature of the Martian mantle. In the Earth 4.5 Gyr of convection has stirred the mantle sufficiently vigorously to eliminate all trace of such early heterogeneity. The earliest isotopic heterogeneities sourced in the convecting mantle of the Earth are less than 3 billion years in age. In contrast the Martian mantle has heterogeneities that can only have been produced in the first 50 Myr of the solar system. So mantle mixing must be far less effective on Mars. This is consistent with fluid dynamic models for the Martian interior (Breuer *et al.*, 1997; Harder and Christensen, 1996). Large-scale convective overturn such as drives the Earth's plates and makes continents move must be absent from Mars as has long been suspected (e.g. Carr, 1973). Indeed the W data are difficult to reconcile with any kind of long-term, plate tectonic activity on Mars (cf. Sleep, 1994).

10. Integrated Models of Accretion and Early Differentiation based on Zr, Nd, Hf, W and Os Isotope Geochemistry

The W isotope data display a very significant relationship with ^{146}Sm-^{142}Nd data (Harper *et al.*, 1995) for the same SNC meteorites (Lee and Halliday, 1997; Figure 11a). Those with radiogenic Nd, reflecting time-integrated depletion in light rare earths and other incompatible elements, are also characterized by radiogenic W (Figure 11a). This striking agreement between two short-lived chronometers provides a powerful insight into the early development of Mars. Two relevant processes can fractionate Hf from W-silicate melting (Newsom *et al.*, 1986, 1996) and (metallic) core formation (Rammensee and Wänke, 1977; Figure 12). In contrast only silicate melting can produce a significant fractionation of Sm/Nd. The question arises as to whether the W data provide any constraints on the timing of core formation. In order to generate a correlation between the W and Nd isotopic data the two parent / daughter element ratios must have fractionated together at an extremely early stage, but was metal segregation involved?

Hf/W and Sm/Nd ratios are both fractionated by partial melting in the silicate Earth. Whereas Hf, Sm and Nd are all moderately incompatible during melting, W is extremely incompatible like Ba (Newsom *et al.*, 1986) leading to a very large fractionation in Hf/W. In Figure 12 we show the data for MORB and OIB from Newsom *et al.* (1986) as f(Hf/W) vs. f(Sm/Nd), where f is the ratio nor-

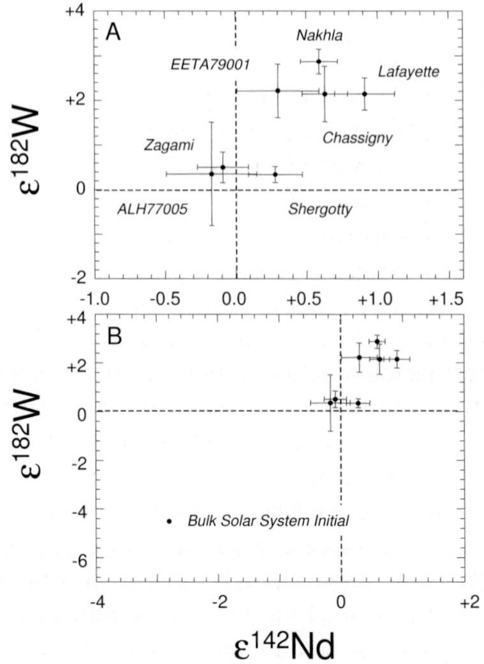

Figure 11. (a) Nd and W isotope data for Martian meteorites show correlated effects implying a close relationship between the W isotopic anomalies and the timing and extent of early silicate melt depletion in the Martian mantle. In (b) the same data shown in (a) are compared with the initial isotopic compositions of the solar system. Data from Harper *et al.* (1995) and Lee and Halliday (1997).

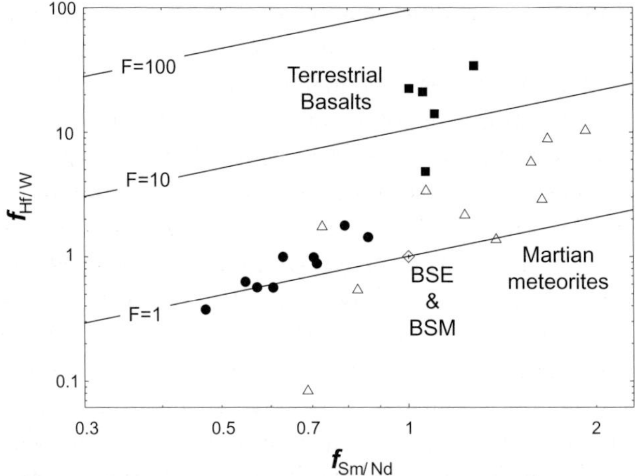

Figure 12. Fractionation of Sm / Nd and Hf / W are correlated in silicate melting and this could be the explanation for the data in Figure 11. Data from Newsom *et al.* (1986) and Lodders (1998). See text for discussion.

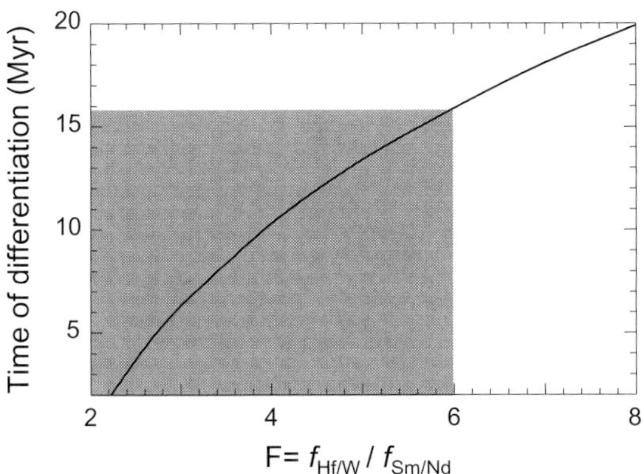

Figure 13. Systematics of the relationship between the fractionation of Hf/W relative to Sm/Nd, and the timing of fractionation implied by the slope of the data shown in Figure 11.

malized to a primitive mantle with chondritic Sm/Nd but Hf/W that is 15 times chondritic because of core formation. It can be seen that the data define a strong positive correlation between Hf/W and Sm/Nd. Note that for the silicate Earth the continental crust also exists as a major reservoir ($>$50% of the W). Data for lunar basalts and eucrites would be roughly collinear with the data for terrestrial samples (Halliday and Lee, 1999). In Figure 12 we also show the corresponding data for Martian meteorites but they are normalized to a primitive mantle with chondritic Sm/Nd and Hf/W. It can be seen that the trends are similar and the normalized data overlap.

If we assume that the Nd and W isotopic effects were produced as a result of coeval fractionation of Sm/Nd and Hf/W we can model the data together to yield the timing of differentiation they imply. The timing of differentiation is defined by the slopes of the lines connecting the initial Nd and W isotopic composition of the solar system with the compositions of the meteorites (Figure 11b). These are all approximately the same (1.19–1.65). The time required to generate a slope of about 1.5 is shown in Figure 13, where it is plotted as a function of the parameter $F = [f(\text{Hf/W})]/[f(\text{Sm/Nd})]$. The isotopic evolution depends on the Sm/Nd as well as the Hf/W ratio. To construct the curve shown in Figure 13 we have assumed that $f(\text{Sm/Nd})$ is 1.3, similar to that found in ALH84001. This is not a unique solution. However, given that F is $<$6 in all Martian meteorites the data are consistent with large-scale melting of Mars within the first $\sim$15 Myr of the Solar System.

So in principle the Nd and W isotopic data could be interpreted in terms of early silicate melt fractionation of Sm/Nd and Hf/W with no direct relationship to core formation. However, there are several lines of evidence that this is not the case. First, it is striking that some of the data are close to chondritic, implying no fractionation of Hf/W and Sm/Nd. The combined Nd and W data are collinear with both the present day chondritic compositions and the initial compositions for the solar system. If the Martian meteorite data really reflect varying degrees of depletion of a chondritic body up to 15 Myr after the start of the solar system the strong correlation is surprising given the strong possible fractionation of Hf/W relative to Sm/Nd. Explaining the combined near-chondritic compositions of both systems becomes a problem. One alternative explanation is that the Martian meteorite data reflect a basic heterogeneity in the material that accreted to form Mars. However, there is no evidence to support this.

Clearly in more recent times W has behaved as a lithophile trace element on Mars, as on Earth, and is fractionated from Hf purely by silicate melting. However, even in the early history of Mars there was not much W depletion from core formation. So was it significant at all? The only way to evaluate this is with the Ba/W ratios of Martian meteorites because the Ba and W should have about the same bulk distribution coefficients (Newsom et $al.$, 1986). The Ba/W ratios of chondrites are about 25, whereas all SNC meteorites have Ba/W between 35 and 95 with an average of roughly 50. So these data provide clear evidence in support of a depletion in W of a factor of $\sim$2 (Wänke and Dreibus, 1988; Figure 8).

If the core formation that brought about this W depletion on Mars were later than the silicate melt depletion responsible for the high Sm/Nd, the presence of a metal phase in the Martian mantle would have buffered W concentrations. This would have resulted in an inverse correlation between Sm/Nd and Hf/W in the melting products. The positive correlation between Nd and W isotopic compositions is inconsistent with this model.

If the core formed before the silicate melt depletion responsible for the high Sm/Nd, the enhanced Hf/W ratio of the silicate Mars would have driven the W isotopic composition to greater than chondritic values while the Martian mantle maintained chondritic Sm/Nd and Nd isotopic composition. Although the meteorites with chondritic Nd isotopic compositions also have near chondritic W isotopic compositions, there is a hint that the W isotopic compositions may be offset to slightly radiogenic figures (Figure 11a). This would be consistent with metal segregation starting before large-scale melting.

If different portions of the Martian mantle lost different amounts of W to core forming metallic liquids, this might be preserved in both the W isotopic compositions and the Ba/W ratios. The variations in Ba/W amongst Martian meteorites are in fact large (Figure 14). Clearly the meteorites with high Ba/W all have radiogenic W. Conversely, the samples with chondritic W all have low Ba/W. Perhaps there are processes other than early source metal segregation that have resulted in the fractionated Ba/W found in these samples. Certainly the data do not define a clear linear trend. However, the relationships raise the question of whether these samples of the Martian mantle preserve ancient chemical as well as isotopic heterogeneities that are residual from the process of core formation, and furthermore, provide strong support for the view that the W isotope data do not just reflect the effects of early silicate melting.

All the ^{142}Nd/^{144}Nd and ^{182}W/^{184}W isotopic compositions measured thus far have been chondritic or radiogenic implying varying degrees of melt and metal de-

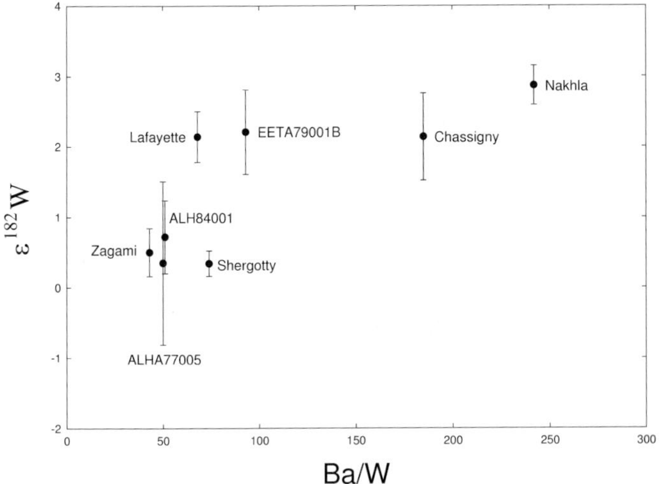

Figure 14. Relationship between W isotopic composition and Ba/W in Martian meteorites.

pletion of a chondritic parent (Figure 11). Therefore the source regions of the Martian meteorite parent magmas appear to have suffered early depletion from melting and core formation but yield no sign of the corresponding enriched reservoirs with low Hf/W and Sm/Nd that would have generated unradiogenic ^{142}Nd/^{144}Nd and ^{182}W/^{184}W. This would be consistent with the outer portions of Mars representing a depleted residue or floating cumulate layer from which early melts and metals segregated inward toward the center of the planet (Lee and Halliday, 1997). However, the existing dataset is based on a very small population of samples.

Recently, Brandon *et al.* (2000) have produced Os isotopic data for Martian meteorites and these display a very pronounced correlation with the ^{142}Nd and ^{182}W anomalies. The Martian mantle has clearly been heterogeneous in Re/Os for at least 2.5 Gyr with both suprachondritic (Zagami, ALH-77005) and sub-chondritic Re/Os regions (Nakhla, Chassigny). Despite the long-lived nature of ^{187}Re decay, the Os isotope systematics appear to record very early fractionation of Re from Os as may have occurred during segregation of metallic and S-bearing liquids during core formation in a magma ocean environment. The correlation of the initial Os isotopic composition in the source region of the samples with ^{142}Nd/^{144}Nd and ^{182}W/^{184}W indicates that Re/Os fractionation occurred very early when the two short-lived isotopes ^{146}Sm and ^{182}Hf were still alive. The fractionation appears to have involved melt extraction for one group and melt addition for the other.

Although this heterogeneity has not been erased other short- and long-lived lithophile systems show no early isotopic heterogeneity. Most notably Zr isotopic effects resulting from decay of ^{92}Nb to ^{92}Zr (half-life = 36 Myr) are absent within uncertainties (Sanloup *et al.*, 2000). This indicates that the high field strength elements were not fractionated in the early stages of the development of Mars, unless the effects were subsequently re-homogenized. Nb/Zr ratios are not strongly fractionated during shallow mantle melting so this is not too surprising. However, there are significant poorly understood fractionations in Nb/Zr among Martian meteorites (Figure 4) and major fractionations would be expected in an early magma ocean environment if perovskite was a liquidus phase (Kato *et al.*, 1988; Münker *et al.*, 2000). The absence of Zr isotopic anomalies in Martian meteorites is entirely consistent with the fact that perovskite is unlikely in a Martian magma ocean given the depth to the core / mantle boundary. It is also possible that Zr isotopes were subsequently selectively re-homogenized without affecting Nd and W. The latter seems unlikely, although the strange trace element fractionations in Nb and Zr (Figure 4) might relate to such a process. Furthermore, the long lived systems ^{147}Sm-^{143}Nd and ^{176}Lu-^{176}Hf show less of a clear record of an early differentiation history (Blichert-Toft *et al.*, 1999) and this would be consistent with some process of selective re-equilibration during the geochemical evolution of the Martian mantle.

Finally the Mars isotopic data can be compared with isotopic constraints for the accretion of the Earth. A body the size of Mars (1/8th Earth mass) would form in about 10 Myr with the absolute accretionary rates considered for the Earth (Halliday, 2000a, b). This is consistent with the above evidence from W and Nd data for the rate of accretion of Mars. So the two planets could have grown at the same rates. The isotopic evidence is certainly insufficient to rule this out. The striking difference between Mars and the Earth is probably not the accretion rate but that Mars did not experience such a devastating protracted history of late giant impacts as those that produced the Earth's Moon. This may be because the growth of Jupiter perturbed the trajectories for late accretion in the outer portions of the inner solar system (Wetherill, 1994b).

References

Ahrens, T.J.: 1990, 'Earth Accretion', *in* H.E. Newsom and J.H. Jones (eds.), *Origin of the Earth*, Oxford Univ. Press, Oxford, pp. 211–227.

Allègre, C.J., Poirier, J.-P., Humler, E., and Hofmann, A.W.: 1995, 'The Chemical Composition of the Earth', *Earth Planet. Sci. Lett.* **134**, 515–526.

Benz, W.: 2000, 'Low Velocity Collisions and the Growth of Planetesimals', *in* W. Benz, R. Kallenbach, and G.W. Lugmair (eds.), *From Dust to Terrestrial Planets, Space Sci. Rev.* **92**, 279–294.

Benz, W., and Cameron, A.G.W.: 1990, 'Terrestrial Effects of the Giant Impact', *in* H.E. Newsom and J.H. Jones (eds.), *Origin of the Earth*, Oxford Univ. Press, Oxford, 61–67.

Benz, W., Cameron, A.G.W., and Slattery, W.L.: 1987, 'Collisional Stripping of Mercury's Mantle', *Icarus* **74**, 516–528.

Bertka, C.M., and Fei, Y.J.: 1997, 'Mineralogy of the Martian Interior up to Core-mantle Boundary Pressures', *J. Geophys. Res.* **102**, 5251–5264.

Birck, J.-L., and Allègre, C.J.: 1988, 'Manganese-chromium Isotope Systematics and the Development of the Early Solar System', *Nature* **331**, 579–584.

Birck, J.-L., and Allègre, C.J.: 1994, 'Contrasting Re/Os Fractionation in Planetary Basalts', *Earth Planet. Sci. Lett.* **124**, 139–148.

Birck, J.L., Rotaru, M., and Allègre, C.J.: 1999, [53]Mn-[53]Cr Evolution of the Early Solar System', *Geochim. Cosmochim. Acta* **63**, 4111–4117.

Blichert-Toft, J., Gleason, J.D., Télouk, P., and Albarède, F.: 1999, 'The Lu-Hf Isotope Geochemistry of Shergottites and the Evolution of the Martian Mantle-crust System', *Earth Planet. Sci. Lett.* **173**, 25–39.

Blum, J.: 2000, 'Laboratory Experiments on Preplanetary Dust Aggregation', *in* W. Benz *et al.* (eds.), *From Dust to Terrestrial Planets*, *Space Sci. Rev.* **92**, 265–278.

Bogard, D.D., and Garrison, D.H.: 1998, 'Relative Abundances of Argon, Krypton, and Xenon in the Martian Atmosphere as Measured in Martian Meteorites', *Geochim. Cosmochim. Acta* **62**, 1829–1835.

Borg, L.E., Nyquist, L.E., Taylor, L.A., Wiesmann, H., and Shih, C.-Y.: 1997, 'Constraints on Martian Differentiation Processes from Rb-Sr and Sm-Nd Isotopic Analyses of the Basaltic Shergottite QUE94201', *Geochim. Cosmochim. Acta* **61**, 4915–4931.

Boss, A.P.: 1990, '3D Solar Nebula Models: Implications for Earth Origin', *in* H.E. Newsom and J.H. Jones (eds.), *Origin of the Earth*, Oxford Univ. Press, Oxford, pp. 3–15.

Brandon, A.D., Walker, R.J, Morgan, J.W., and Goles, G.G.: 2000, 'Re-Os Isotopic Evidence for Early Differentiation of the Martian Mantle', *Geochim. Cosmochim. Acta* **64**, 4083–4095.

Breuer, D., Spohn, T., and Wüllner, U.: 1993, 'Mantle Differentiation and the Crustal Dichotomy of Mars', *Planet. Space Sci.* **41**, 269–283.

Breuer, D., Yuen, D.A., and Spohn, T.: 1997, 'Phase Transitions in the Martian Mantle: Implications for Partially Layered Convection', *Earth Planet. Sci. Lett.* **148**, 457–469.

Cameron, A.G.W., and Ward: 1976, 'The Origin of the Moon', *Lunar Science* **7**, Lunar Science Institute, Houston.

Cameron, A.G.W., and Benz, W.: 1991, 'Origin of the Moon and the Single Impact Hypothesis IV.', *Icarus* **92**, 204–216.

Canup, R.M., and Agnor, C.: 1998, 'Accretion of Terrestrial Planets and the Earth-moon System', *Origin of the Earth and Moon*, LPI Contribution 597, LPI, Houston, pp. 4–7.

Carr, M.H.: 1973, 'Volcanism on Mars', *J. Geophys. Res.* **78**, 4049–4062.

Carr, M.H.: 1999, 'Retention of an Atmosphere on Early Mars', *J. Geophys. Res.* **104**, 21,897–21,909.

Carr, M.H., and Wänke, H.: 1992, 'Earth and Mars: Water Inventories as Clues to Accretional Histories', *Icarus* **98**, 61–71.

Cassen, P., and Woolum, D.S.: 1997, 'Nebular Fractionations and Mn-Cr Systematics', *Proc. 28th Lunar Planet. Sci. Conf.*, 211–212.

Chen, J.H., and Wasserburg, G.J.: 1986, 'Formation Ages and Evolution of Shergotty and its Parent Planet from U-Th-Pb Systematics', *Geochim. Cosmochim. Acta* **50**, 955–968.

Christie, D.M., Carmichael, I.S.E., and Langmuir, C.H.: 1986, 'Oxidation States of Mid-ocean Ridge Basalt Glasses', *Earth Planet. Sci. Lett.* **79**, 397–417.

Clark, S.P., Jr., Turekian, K.K., and Grossman, L.: 1972, 'Model for the Early History of the Earth', *in* E.C. Robertson (ed.), *The Nature of the Solid Earth*, McGraw Hill, pp. 3–18.

Clayton, R.N.: 1986, 'High Temperature Isotope Effects in the Early Solar System', *in* J.W. Valley, H.P. Taylor, and J.R. O'Neil (eds.), *Stable Isotopes in High Temperature Geological Processes*, Mineral. Soc. Am., Washington D.C., pp. 129–140.

Clayton, R.N.: 1993, 'Oxygen Isotopes in Meteorites', *Ann. Rev. Earth Planet. Sci.* **21**, 115–149.

Clayton, R.N., and Mayeda, T.K.: 1996, 'Oxygen Isotope Studies of Achondrites', *Geochim. Cosmochim. Acta* **60**, 1999–2017.

Delano, J.W., and Arculus, R.J.: 1980, 'Nakhla: Oxidation State and Other Constraints', *Proc. 11th Lunar Planet. Sci.*, 219–221.

Dreibus, G, and Wänke, H.: 1984, 'Accretion of the Earth and the Inner Planets', *Proc. 27th Int. Geol. Conf.* **11**, VNU Science Press, Utrecht, pp. 1–20.

Dreibus, G., and Wänke, H.: 1985, 'Mars: A Volatile Rich Planet', *Meteoritics* **20**, 367–382.

Dreibus, G., and Wänke, H.: 1987, 'Volatiles on Earth and Mars: A Comparison', *Icarus* **71**, 225–240.

Franchi, I.A., Wright, I.P., Sexton, A.S., and Pillinger, C.T.: 1999, 'The Oxygen Isotopic Composition of Earth and Mars', *Met. Planet. Sci.* **34**, 657–661.

Gaetani, G.A., and Grove, T.L.: 1997, 'Partitioning of Moderately Siderophile Elements Among Olivine, Silicate Melt and Sulfide Melt: Constraints on Core Formation in the Earth and Mars', *Geochim. Cosmochim. Acta* **61**, 1829–1846.

Galy, A., Young, E.D., Ash, R.D., and O'Nions, R.K.: 2000, 'High Precision Magnesium Isotopic Composition of Allende Material: A Multiple Collector Inductively Coupled Mass Spectrometry Study', *Lunar Planet. Sci.* **31**, 1193–1194.

Gilmour, J.D., Whitby, J.A., and Turner, G.: 1998, 'Xenon Isotopes in Irradiated ALH84001: Evidence for Shock-induced Trapping of Ancient Martian Atmosphere', *Geochim. Cosmochim. Acta* **62**, 2555–2571.

Grossman, L.: 1972, 'Condensation in the Primitive Solar Nebula', *Geochim. Cosmochim. Acta* **36**, 597–619.

Grossman, L., and Larrimer, J.W.: 1974, 'Early Chemical History of the Solar System', *Rev. Geophys. Space Phys.* **12**, 71–101.

Haggerty, S.R.: 1981, 'Opaque Mineral Oxides in Terrestrial Igneous Rocks', *Oxide Minerals*, Miner. Soc. Am., pp. Hg101–Hg278.

Halliday, A.N.: 2000a, 'Terrestrial Accretion Rates and the Origin of the Moon', *Earth Planet. Sci. Lett.* **176**, 17–30.

Halliday, A.N.: 2000b, 'Hf-W Chronometry and Inner Solar System Accretion Rates', *in* W. Benz *et al.* (eds.), *From Dust to Terrestrial Planets*, Space Sci. Rev. **92**, 355–370.

Halliday, A.N., and Lee, D.-C.: 1999, 'Tungsten Isotopes and the Early Development of the Earth and Moon', *Geochim. Cosmochim. Acta*, (C.J. Allègre 60th Birthday Volume) **63**, 4157–4179.

Halliday, A.N., and Porcelli, D.: 2001, 'In Search of Lost Planets – the Paleocosmochemistry of the Inner Solar System', *Earth Planet. Sci. Lett.*, in submission.

Halliday, A.N., Rehkämper, M., Lee, D.-C., and Yi, W.: 1996, 'Early Evolution of the Earth and Moon: New Constraints from Hf-W Isotope Geochemistry', *Earth Planet. Sci. Lett.* **142**, 75–90.

Harder, H., and Christensen, U.R.: 1996, 'A One-plume Model of Martian Mantle Convection', *Nature* **380**, 507–509.

Harper, C.L., and Jacobsen, S.B.: 1996, 'Evidence for ^{182}Hf in the Early Solar System and Constraints on the Timescale of Terrestrial Core Formation', *Geochim. Cosmochim. Acta* **60**, 1131–1153.

Harper, C.L., Völkening, J., Heumann, K.G., Shih, C.-Y., and Wiesmann, H.: 1991, ^{182}Hf-^{182}W: New Cosmochronometric Constraints on Terrestrial Accretion, Core Formation, the Astrophysical Site of the r-Process, and the Origin of the Solar System', *Proc. 22nd Lunar Planet. Sci. Conf.*, 515–516.

Harper, C.L., Nyquist L.E., Bansal, B., Wiesmann, H., and Shih, C.-Y.: 1995, 'Rapid Accretion and Early Differentiation of Mars Indicated by ^{142}Nd/^{144}Nd in SNC Meteorites', *Science* **267**, 213–217.

Hartmann, W.K., and Davis, D.R.: 1975, 'Satellite-sized Planetesimals and Lunar Origin', *Icarus* **24**, 504.

Hewins, R.H., and Ulmer, G.C.: 1983, 'Intrinsic Oxygen Fugacity Measurements for Clasts in Diogenites and Mesosiderites', *Proc. 14th Lunar Planet. Sci. Conf.*, 311–312.

Higuchi, H., and Morgan, J.W.: 1975, 'Ancient Meteoritic Component in Apollo 17 Boulder', *Proc. 6th Lunar Planet. Sci. Conf.*, 1625–1651.

Hinton, R.W., and Bischoff, A.: 1984, 'Ion Microprobe Magnesium Isotope Analysis of Plagioclase and Hibonite from Ordinary Chondrites', *Nature* **308**, 169–172.

Hinton, R.W., Davis, A.M., Scatena-Wachel, D.E., Grossman, L., and Draus, R.J.: 1988, 'A Chemical and Isotopic Study of Hibonite-rich Refractory Inclusions in Primitive Meteorites', *Geochim. Cosmochim. Acta* **52**, 2573–2598.

Hofmann, A.W., Jochum, K.P., Seufert, M., and White, W.M.: 1986, 'Nb and Pb in Oceanic Basalts: New Constraints on Mantle Evolution', *Earth Planet. Sci. Lett.* **79**, 33–45.

Horan, M.F., Smoliar, M.I., and Walker, R.J.: 1998, ^{182}W and ^{187}Re-^{187}Os Systematics of Iron Meteorites: Chronology for Melting, Differentiation, and Crystallization in Asteroids', *Geochim. Cosmochim. Acta* **62**, 545–554.

Hunten, D.M., Pepin, R.O., and Walker, J.G.C.: 1987, 'Mass Fractionation in Hydrodynamic Escape', *Icarus* **69**, 532–549.

Huss, G.R.: 1988, 'The Role of Presolar Dust in the Formation of the Solar System', *Earth, Moon, and Planets* **40**, 165–211.

Jacobsen, S.B., and Harper, Jr., C.L.: 1996, 'Accretion and Early Differentiation History of the Earth Based on Extinct Radionuclides', *in* A. Basu and S. Hart (eds.), *Earth Processes: Reading the Isotope Code*, AGU, Washington D.C., pp. 47–74.

Jagoutz, E., Sorowka, A., Vogel, J.D., and Wänke, H.: 1994, 'ALH84001: Alien or Progenitor of the SNC Family?', *Meteoritics* **28**, 548–479.

Jakosky, B.M., and Jones, J.H.: 1997, 'The History of Martian Volatiles', *Rev. Geophys.* **35**, 1–16.

Kato, T., Ringwood, A.E., and Irifune, T.: 1988, 'Experimental Determination of Element Partitioning Between Silicate Perovskites, Garnets and Liquids: Constraints on Early Differentiation of the Mantle', *Earth Planet. Sci. Lett.* **89**, 123–145.

Kitts, K., and Lodders, K.: 1998, 'Survey and Evaluation of Eucrite Bulk Compositions', *Met. Planet. Sci.* **33 Suppl.**, A197–A213.

Lee, D.-C., and Halliday, A.N.: 1995, 'Hafnium-tungsten Chronometry and the Timing of Terrestrial Core Formation', *Nature* **378**, 771–774.

Lee, D.-C., and Halliday, A.N.: 1996, 'Hf-W Isotopic Evidence for Rapid Accretion and Differentiation in the Early Solar System', *Science* **274**, 1876–1879.

Lee, D.-C., and Halliday, A.N.: 1997, 'Core Formation on Mars and Differentiated Asteroids', *Nature* **388**, 854–857.

Lee, D.-C., and Halliday, A.N.: 2000a, 'Hf-W Isotopic Systematics of Ordinary Chondrites and the Initial ^{182}Hf/^{180}Hf of the Solar System', *Chem. Geol.* (G.J. Wasserburg Spec. Iss.) **169**, 35–43.

Lee, D.-C., and Halliday, A.N.: 2000b, 'Accretion of Primitive Planetesimals: Hf-W Isotopic Evidence from Enstatite Chondrites', *Science* **288**, 1629–1631.

Lee, T., Papanastassiou, D.A., and Wasserburg, G.J.: 1977, 'Aluminum-26 in the Early Solar System: Fossil or Fuel?', *Astrophys. J.* **322**, L107–L110.

Lee, T., Shu, F.H., Shang, H., Glassgold, A.E., and Rehm, K.E.: 1998, 'Protostellar Cosmic Rays and Extinct Radioactivities in Meteorites', *Astrophys. J.* **506**, 898–912.

Lewis, J.S.: 1972, 'Metal/silicate Fractionation in the Solar System', *Earth Planet. Sci. Lett.* **15**, 286–290.

Lin, D.N.C., and Papaloizou, J.: 1985, 'On the Dynamical Origin of the Solar System', *in* D.C. Black and M.S. Matthews (eds.), *Protostars and Planets II*, 981-1072, Univ. Arizona Press, Tucson, pp. 981–1072.

Lodders, K.: 1998, 'A Survey of Shergottite, Nakhlite and Chassigny Meteorites Whole-rock Compositions', *Met. Planet. Sci.* **33 Suppl.**, A183–A190.

Lodders, K.: 2000, 'An Oxygen Isotope Mixing Model for the Accretion and Composition of Rocky Planets', *in* W. Benz *et al.* (eds.), *From Dust to Terrestrial Planets*, Space Sci. Rev. **92**, 341–354.

Lodders, K., and Fegley, B.: 1997, 'An Oxygen Isotope Model for the Composition of Mars', *Icarus* **126**, 373–394.

Longhi, J., and Pan, V.: 1989, 'The Parent Magmas of the SNC Meteorites', *Proc. 19th Lunar Planet. Sci. Conf.*, Cambridge Univ. Press, Cambridge, pp. 451–464.

Longhi, J., Knittle, E., Holloway, J.R., and Wänke, H.: 1992, 'The Bulk Composition, Mineralogy and Internal Structure of Mars', *in* H.H. Kieffer, B.M. Jakosky, C.W. Snyder, and M.S. Matthews (eds.), *Mars*, Univ. Arizona Press, Tucson, pp. 84–208.

Lugmair, G.W., and Galer, S.J.G.: 1992, 'Age and Isotopic Relationships Among the Angrites Lewis Cliff-86010 and Angra Dos Reis', *Geochim. Cosmochim. Acta* **56**, 1673–1694.

Lugmair, G.W., and Shukolyukov, A.: 1998, 'Early Solar System Timescales According to ^{53}Mn-^{53}Cr Systematics', *Geochim. Cosmochim. Acta* **62**, 2863–2886.

Marti, K., Kim, J.S., Thakur, A.N., McCoy, T.J., and Keil, K.: 1995, 'Signatures of the Martian Atmosphere in Glass of the Zagami Meteorite', *Science* **267**, 1981–1984.

Matthew, K.J., and Marti, K.: 2001, 'Early Evolution of Martian Volatiles: Nitrogen and Noble Gas Components in ALH84001 and Chassigny', *J. Geophys. Res.*, in press.

McDonough, W.F., and Sun, S.-S.: 1995, 'The Composition of the Earth', *Chem. Geol.* **120**, 223–253.

Melosh, H.J.: 1990, 'Giant Impacts and the Thermal State of the Early Earth, *in* H.E. Newsom and J.H. Jones (eds.), *Origin of the Earth*, Oxford Univ. Press, Oxford, pp. 69–83.

Melosh, H.J., and Vickery, A.M.: 1989, 'Impact Erosion of the Primordial Atmosphere of Mars', *Nature* **338**, 487–489.

Melosh, H.J., Vickery, A.M., and Tonks, W.B.: 1993, 'Impacts and the Early Environment and Evolution of the Terrestrial Planets', *in* E.H. Levy and J.I. Lunini (eds.), *Protostars and Planets III*, Univ. Arizona Press, Tucson, 1339–1370.

Minarik, W.G., Ryerson, F.J., and Watson, E.B.: 1996, 'Textural Entrapment of Core-forming Melts', *Science* **272**, 530–533.

Münker, C., Weyer, S., Mezger, K., Rehkämper, M., Wombacher, F., and Bischoff, A.: 2000, ^{92}Nb-^{92}Zr and the Early Differentiation History of Planetary Bodies', *Science* **289**, 1538–1542.

Newsom, H.E.: 1990, 'Accretion and Core Formation in the Earth: Evidence from Siderophile Elements', *in* H.E. Newsom and J.H. Jones (eds.), *Origin of the Earth*, Oxford Univ. Press, Oxford, pp. 273–288.

Newsom, H.E.: 1995, 'Composition of the Solar System, Planets, Meteorites, and Major Terrestrial Reservoirs', *Global Earth Physics, A Handbook of Physical Constants*, AGU Reference Shelf **1**, American Geophysical Union.

Newsom, H.E., White, W.M., Jochum, K.P., and Hofmann, A.W.: 1986, 'Siderophile and Chalcophile Element Abundances in Oceanic Basalts, Pb Isotope Evolution and Growth of the Earth's core', *Earth Planet. Sci. Lett.* **80**, 299–313.

Newsom, H.E., Sims, K.W.W., Noll, Jr., P.D., Jaeger, W.L., Maehr, S.A., and Bessera, T.B.: 1996, 'The Depletion of W in the Bulk Silicate Earth', *Geochm. Cosmochim. Acta* **60**, 1155–1169.

Nichols: 2000, 'Short-lived Radionuclides in Meteorites: Constraints of Nebular Timescales for the Production of Solids', *in* W. Benz *et al.* (eds.), *From Dust to Terrestrial Planets, Space Sci. Rev.* **92**, 113–122.

Nyquist, L.E., Bansal, B., Wiesmann, H., and Shih, C.-Y.: 1994, 'Neodymium, Strontium and Chromium Isotopic Studies of the LEW86010 and Angra-Dos-Reis Meteorites and the Chronology of the Angrite Parent Body', *Meteoritics* **29**, 872–885.

Nyquist, L.E., Lindstrom, D., Shih, C.-Y., Weismann, H., Mittlfelhdt, D., Wentworth, S., and Martinez, R.: 1997, 'Mn-Cr Systematics of Chondrules from the Bishunpur and Chainpur Meteorites', *Proc. 28th Lunar Planet. Sci.*, 1033–1034.

Ott, U.: 1988, 'Noble Gases in SNC Meteorites: Shergotty, Nakhla, Chassigny', *Geochim. Cosmochim. Acta* **52**, 1937–1948.

Pepin, R.O.: 1994, 'Evolution of the Martian Atmosphere', *Icarus* **111**, 289–304.

Quitté, G., Birck, J.-L., and Allègre, C.J.: 2000, '^{182}Hf-^{182}W Systematics in Eucrites: The Puzzle of Iron Segregation in the Early Solar System', *Earth Planet. Sci. Lett.* **184**, 83–94.

Rammensee, W., and Wänke, H.: 1977, 'On the Partition Coefficient of Tungsten Between Metal and Silicate and its Bearing on the Origin of the Moon', *Proc. 8th Lunar Sci. Conf.*, 399–409.

Righter, K., and Drake, M.J.: 1996, 'Core Formation in Earth's Moon, Mars, and Vesta', *Icarus* **124**, 513–529.

Ringwood, A.E.: 1977, 'Composition of the Core and Implications for Origin of the Earth', *Geochem. J.* **11**, 111–135.

Ringwood, A.E.: 1979, *On the Origin of Earth and Moon*, Springer, New York.

Robert, F., Rejou-Michel, A., and Javoy, M.: 1992, 'Oxygen Isotope Homogeneity of the Earth: New Evidence', *Earth Planet. Sci. Lett.* **108**, 1–10.

Russell, S.S., Srinivasan, G., Huss, G.R., Wasserburg, G.J., and MacPherson, G.J.: 1996, 'Evidence for Widespread ^{26}Al in the Solar Nebula and Constraints for Nebula Time Scales', *Science* **273**, 757–762.

Safronov, V.S.: 1954, 'On the Growth of Planets in the Protoplanetary Cloud', *Astron. Zh.* **31**, 499–510.

Sanloup, C., Blichert-Toft, J., Télouk, P., Gillet, P., and Albarède, F.: 2000, 'Zr Isotope Anomalies in Chondrites and the Presence of Live ^{92}Nb in the Early Solar System', *Earth Planet. Sci. Lett.* **184**, 75–81.

Sasaki, S., and Nakazawa, K.: 1986, 'Metal-silicate Fractionation in the Growing Earth: Energy Source for the Terrestrial Magma Ocean', *J. Geophys. Res.* **91**, 9231–9238.

Sato, M.: 1976, 'Oxygen Fugacity and Other Thermochemical Parameters of Apollo 17 High-Ti Basalts and Their Implications on the Reduction Mechanism', *Proc. 7th Lunar Planet. Sci. Conf.*, 1323–1344.

Sato, M., Hickling, N.L., and McLane, J.E.: 1973, 'Oxygen Fugacity Values of Apollo 12, 14 and 15 Lunar Samples and Reduced State of Lunar Magmas', *Proc. 4th Lunar Planet. Sci. Conf.*, 1061–1079.

Shaw, G.H.: 1978, 'Effects of Core Formation', *Phys. Earth Plan. Inter.* **16**, 361–369.

Shih, C.-Y., Nyquist, L.E., Bogard, D.D., Mckay, G.A., Wooden, J.L., Bansal, B.M., and Wiesmann, H.: 1982, 'Chronology and Petrogenesis of Young Achondrites, Shergotty, Zagami and ALHA77005: Late Magmatism on a Geologically Active Planet', *Geochim. Cosmochim. Acta* **46**, 2323–2344.

Shu, F.H., Adams, F.C., and Lizano, S.: 1987, 'Star Formation in Molecular Clouds – Observation and Theory', *Ann. Rev. Astron. Astrophys.* **25**, 23–81.

Shu, F.H., Shang, H., Glassgold, A.E., and Lee, T.: 1997, 'X-rays and Fluctuating X-winds from Protostars', *Science* **277**, 1475–1479.

Sleep, N.H.: 1994, 'Martian Plate Tectonics', *J. Geophys. Res.* **99**, 5639–5655.

Solomon, S.C.: 1979, 'Formation, History, and Energetics of Cores in the Terrestrial Planets', *Earth Planet. Sci. Lett.* **19**, 168–182.

Stevenson, D.J.: 1981, 'Models of the Earth's Core', *Science* **214**, 611–619.

Stevenson, D.J.: 1990, 'Fluid Dynamics of Core Formation', *in* H.E. Newsom and J.H. Jones (eds.), *Origin of the Earth*, Oxford University Press, Oxford, pp. 231–249.

Stolper, E.: 1977, 'Experimental Petrology of Eucritic Meteorites', *Geochim. Cosmochim. Acta* **41**, 587–611.

Swindle, T.D., and Jones, J.H.: 1997, 'The Xenon Isotopic Composition of the Primordial Martian Atmosphere: Contributions from Solar and Fission components', *J. Geophys. Res.* **102**, 1671–1678.

Taylor, S.R., and McLennan, S.M.: 1985, *The Continental Crust: Its Composition and Evolution*, Blackwell Sci.

Toksöz, M.N., and Hsui, A.T.: 1978, 'Thermal History and Evolution of Mars', *Icarus* **34**, 537–547.

Treiman, A.H., Drake, M.J., Janssens, M.-J., Wolf, R., and Ebihara, M.: 1986, 'Core Formation in the Earth and Shergottite Parent Body (SPB): Chemical Evidence from Basalts', *Geochim. Cosmochim. Acta* **50**, 1071–1091.

Treiman, A.H., Jones, J.H., and Drake, M.J.: 1987, 'Core Formation in the Shergottite Parent Body and Comparison with the Earth', *J. Geophys. Res.* **92**, 627–632.

Turekian, K.K., and Clark, S.P., Jr.: 1969, 'Inhomogeneous Accumulation of the Earth from the Primitive Solar Nebula', *Earth Planet. Sci. Lett.* **6**, 346–348.

Wänke, H.: 1981, 'Constitution of Terrestrial Planets', *Phil. Trans. R. Soc. Lond.* **A303**, 287–302.

Wänke, H., and Dreibus, G.: 1988, 'Chemical Composition and Accretion History of Terrestrial Planets', *Phil. Trans. R. Soc. Lond.* **A325**, 545–557.

Wänke, H., and Dreibus, G.: 1994, 'Chemistry and Accretion of Mars', *Phil. Trans. R. Soc. Lond* **A349**, 285–293.

Wänke, H., Dreibus, G., and Jagoutz, E.: 1984, 'Mantle Chemistry and Accretion History of the Earth', *in* A. Kroner, G.N. Hanson, and A.M. Goodwin (eds.), Springer, New York, pp. 1–24.

Wallace, P., and Carmichael, I.S.E.: 1992, 'Sulfur in Basaltic Magmas', *Geochim. Cosmochim. Acta* **56**, 1863–1874.

Warren, P.H., and Kallemeyn, G.W.: 1996, 'Siderophile Trace Elements in ALH84001, Other SNC Meteorites and Eucrites: Evidence for Heterogeneity, Possibly Time-linked, in the Mantle of Mars', *Met. Planet. Sci.* **31**, 97–105.

Warren, P.H., and Kallemeyn, G.W.: 1997, 'Yamato-793605, EET79001, and Other Presumed Martian Meteorites: Compositional Clues to Their Origin', *Antarct. Meteorite Res.* **10**, 61–81.

Warren, P.H., Greenwood, J.P., Richardson, J.W., Rubin, A.E., and Verish, R.S.: 2000, 'Geochemistry of Los Angeles, a Ferroan, La- and Th-rich Basalt from Mars', *Proc. 31st Lunar Planet. Sci. Conf.*, LPI, Houston, abstract #2001 (CD-ROM).

Wetherill, G.W.: 1986, 'Accumulation of the Terrestrial Planets and Implications Concerning Lunar Origin', *in* W.K. Hartmann, R.J. Phillips, and G.J. Taylor (eds.), *Origin of the Moon*, LPI, Houston, pp. 519–550.

Wetherill, G.W.: 1994a, 'Provenance of the Terrestrial Planets', *Geochim. Cosmochim. Acta* **58**, 4513–4520.

Wetherill, G.W.: 1994b, 'Possible Consequences of Absences of "Jupiters" in Planetary Systems', *Astrophys. Space Sci.* **212**, 23–32.

Wolf, R., Woodrow, A., and Anders, E.: 1979, 'Lunar Basalts and Pristine Highland Rocks: Comparison of Siderophile and Volatile Elements', *Proc. 10th Lunar Planet. Sci. Conf.*, 2107–2130.

Yi, W., Halliday, A.N., Alt, J.C., Lee, D.-C., Rehkämper, M., Garcia, M., Langmuir, C., and Su, Y.: 2000, 'Cadmium, Indium, Tin, Tellurium and Sulfur in Oceanic Basalts: Implications for Chalcophile Element Fractionation in the Earth', *J. Geophys. Res.* **105**, 18,927–18,948.

Zahnle, K.J.: 1993, 'Xenological Constraints on the Impact Erosion of the Early Martian Atmosphere', *J. Geophys. Res.* **98**, 10,899–10,913.

Zharkov, V.N.: 1996, 'The Internal Structure of Mars: A Key to Understanding the Origin of Terrestrial Planets', *Sol. Syst. Res.* **30**, 456–465.

Zinner, E., and Göpel, C.: 1992, 'Evidence for ^{26}Al in Feldspars from the H4 Chondrite Ste Marguerite', *Meteoritics* **27**, 311.

Address for correspondence: Institute for Isotope Geology and Mineral Resources, Department of Earth Sciences, ETH Zentrum, NO C61, CH-8049 Zürich, Switzerland; (alex.halliday@erdw.ethz.ch)

GEOPHYSICAL CONSTRAINTS ON THE EVOLUTION OF MARS

TILMAN SPOHN[1], MARIO H.ACUÑA[2], DORIS BREUER[1], MATTHEW
GOLOMBEK[3], RONALD GREELEY[4], ALEXANDER HALLIDAY[5], ERNST
HAUBER[6], RALF JAUMANN[6] and FRANK SOHL[1]

[1] *Institut für Planetologie, Westfälische Wilhelms-Universität,
W. Klemmstrasse 10, D-48149 Münster, Germany*
[2] *NASA Goddard Space Flight Center, Greenbelt, MD 20771; USA*
[3] *Jet Propulsion Laboratory, Pasadena, Ca 91109, USA*
[4] *Department of Geology, Arizona State University, Tempe, AZ 85287-1404, USA*
[5] *Department of Earth Sciences, ETH Zentrum, CH-8092 Zürich, Switzerland*
[6] *Deutsches Zentrum für Luft- und Raumfahrt DLR, D-12489 Berlin, Germany*

Received: 28 November 2000; accepted: 28 February 2001

Abstract. The evolution of Mars is discussed using results from the recent Mars Global Surveyor
(MGS) and Mars Pathfinder missions together with results from mantle convection and thermal
history models and the chemistry of Martian meteorites. The new MGS topography and gravity data
and the data on the rotation of Mars from Mars Pathfinder constrain models of the present interior
structure and allow estimates of present crust thickness and thickness variations. The data also allow
estimates of lithosphere thickness variation and heat flow assuming that the base of the lithosphere
is an isotherm. Although the interpretation is not unambiguous, it can be concluded that Mars has a
substantial crust. It may be about 50 km thick on average with thickness variations of another ±50
km. Alternatively, the crust may be substantially thicker with smaller thickness variations. The former
estimate of crust thickness can be shown to be in agreement with estimates of volcanic production
rates from geologic mapping using data from the camera on MGS and previous missions. According
to these estimates most of the crust was produced in the Noachian, roughly the first Gyr of evolution.
A substantial part of the lava generated during this time apparently poured onto the surface to produce
the Tharsis bulge, the largest tectonic unit in the solar system and the major volcanic center of Mars.
Models of crust growth that couple crust growth to mantle convection and thermal evolution are
consistent with an early 1 Gyr long phase of vigorous volcanic activity. The simplest explanation
for the remnant magnetization of crustal units of mostly the southern hemisphere calls for an active
dynamo in the Noachian, again consistent with thermal history calculations that predict the core
to become stably stratified after some hundred Myr of convective cooling and dynamo action. The
isotope record of the Martian meteorites suggest that the core formed early and rapidly within a
few tens of Myr. These data also suggest that the silicate rock component of the planet was partially
molten during that time. The isotope data suggest that heterogeneity resulted from core formation and
early differentiation and persisted to the recent past. This is often taken as evidence against vigorous
mantle convection and early plate tectonics on Mars although the latter assumption can most easily
explain the early magnetic field. The physics of mantle convection suggests that there may be a few
hundred km thick stagnant, near surface layer in the mantle that would have formed rapidly and may
have provided the reservoirs required to explain the isotope data. The relation between the planform
of mantle convection and the tectonic features on the surface is difficult to entangle. Models call for
long wavelength forms of flow and possibly a few strong plumes in the very early evolution. These
plumes may have dissolved with time as the core cooled and may have died off by the end of the
Noachian.

Space Science Reviews **96**: 231-262, 2001.

1. Introduction

Like that of any other planet, the evolution of Mars is to a large extent determined by its interior structure and composition and by its material properties. A significant part of these quantities can be measured applying geophysical methods at the planet, in orbit or *in situ*. Moreover, model calculations and laboratory measurements are instrumental.

Mars is among the best explored planets, second only to the Earth and Moon, thanks to the number of missions to the planet and thanks to the significant interest it has attracted, not the least because of its similarity in various respects with the Earth. Among the most important recent accomplishments in terms of planetary physics (borrowing to some extent from cosmochemistry) are the

- accurate measurement of the topography through laser ranging by Mars Global Surveyor (MGS) (Smith *et al.*, 1999a).
- measurement and mapping of the gravity field through two-way Doppler tracking of the MGS spacecraft (Smith *et al.*, 1999b).
- determination of the precession rate of the rotational axis by Mars Pathfinder (Folkner *et al.*, 1997).
- detection and measurement of remnant magnetisation in the southern hemisphere (Acuña *et al.*, 1998).
- identification of young lava flows (<100 Myr) in Mars Observer Camera (MOC) images (Hartmann *et al.*, 1999).
- *in situ* measurement of the composition of Martian rock and soil by Mars Pathfinder and detection of rock that is similar in composition to terrestrial andesites (Rieder *et al.*, 1997).
- detection and mapping of two crustal rock types, basaltic and andesitic, on the surface by MGS (Bandfield *et al.*, 2000).

The first three items above constrain interior structure. From the first two, models of crust and lithosphere thickness were derived (Zuber *et al.*, 2000). The third allowed an accurate measurement of the moment-of-inertia factor (MoI-factor) which, together with the mass of the planet is the most important geophysical piece of data to constrain global models of interior structure. The MoI-factor is defined as the dimensionless ratio between the moment of inertia of the planet about the rotation axis divided by the mass and the square of the planet radius.

The last four discoveries listed above constrain models of the evolution of the planet: The remnant magnetization of the southern hemisphere crust strongly suggests that the planet once had a self-sustained magnetic field which it is lacking at the present time. Crustal units devoid of magnetization must have formed when the dynamo was not active. The detection of young lava flows of less than 100 Myr of age suggests that the planet has been volcanically active up to the recent past. This observation ties in nicely with the young crystallization ages of the Shergottite meteorites from Mars (Nyquist *et al.*, 2001; Halliday *et al.*, 2001). The detection of two chemical components in Martian rock, one basaltic like the composition of the Martian meteorites and one more silica rich (Wänke *et al.*, 2001), constrains models of the mantle and its differentiation history. Apparently, differentiation may have employed crust recycling and remelting. Thermal Emission Spectrometer TES data from the MGS mission (Bandfield *et al.*, 2000) suggest that there is a dichotomy in the composition of the surface rock with the northern hemisphere being largely composed of the silica rich (or andesitic) component and the southern hemisphere dark regions largely basaltic. This observation indicates a striking albeit yet little understood difference in evolution between the two hemispheres that goes well beyond their difference in age (Hartmann and Neukum, 2001).

The geologic features of the planet can be briefly characterized as such: Mars is a one-plate planet, lacking any sign of present-day plate tectonics. Rather, its deep interior is likely covered by a one-plate lithosphere which in places is pierced by volcanic vents that may sample the sublithosphere layers. The most prominent topographic features of the surface (Figure 2 of Hartmann *et al.*, 2001) are the dichotomy between the northern and the southern hemispheres, the Tharsis bulge and the Hellas impact basin. While the southern hemisphere is heavily cratered, the northern hemisphere is cratered to a much lesser extent and, therefore, younger. The Tharsis bulge, a gigantic volcanic and tectonic region likely formed during the early evolution of the planet and persisted and has been active unto the recent past. The formation of the Tharsis bulge and the volcanic activity associated with it is a major challenge for planetary modeling. The Hellas basin is the largest, deepest and most prominent impact basin on Mars.

We begin this review by summarizing observational constraints on Mars evolution models and then we discuss the results of model calculations. These constraints are as presently understood for the volcanic and tectonic history, the crust and lithosphere thicknesses, the structure of the deep interior and the magnetic field. We conclude the article by pointing out major open questions that remain to be solved through further exploration of the planet including seismic exploration and modeling of its interior dynamics and evolution.

2. Observational Constraints on Evolution Models

2.1. TOPOGRAPHY AND GRAVITY

The topography of Mars and its gravity field have recently been measured with much improved accuracy over previous data by the MOLA and Radio Science teams on Mars Global Surveyor, (e.g., Smith *et al.*, 1999a, 1999b, 2000; Zuber *et al.*, 2000). The presently available new topography model has a spatial resolution of 1/32° by 1/64° ($\approx$ 2 by 4 km) and an absolute accuracy of 13 m with respect to the center of mass. The gravity field measured by X-band Doppler tracking of the MGS spacecraft has a spatial resolution of about 178 km and an accuracy of ± 10 mGal (10^{-5} m s^{-2}) at the poles, ± 20 mGal at the equator, and ± 100 mGal over the Tharsis Montes and Olympus Mons.

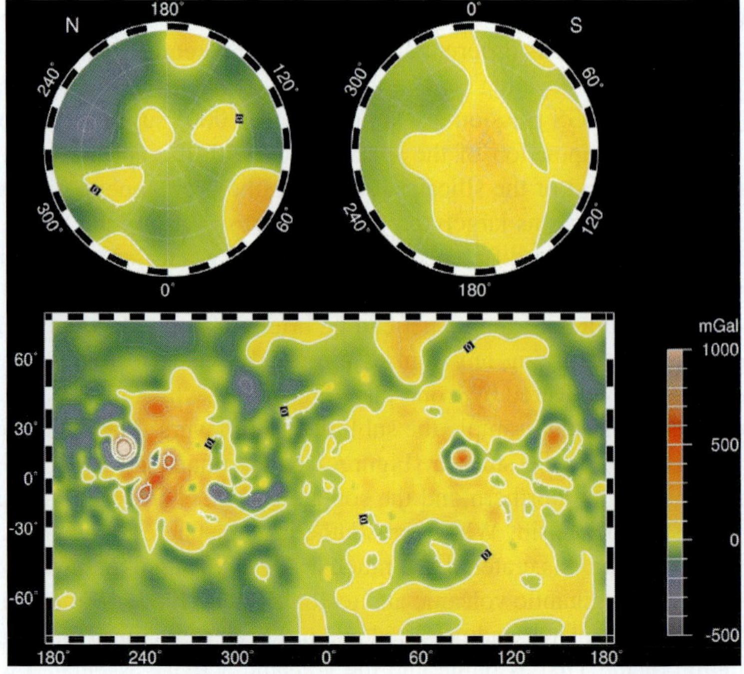

Figure 1. MGS (free air) gravity field from Smith *et al.* (1999a)

The topography of Mars exhibits a clear dichotomy which divides the surface into a southern highland hemisphere (about 60% of the surface) rising several thousands of meters above the zero level and a northern lowlands hemisphere well below the datum. The most prominent features besides the dichotomy are the Tharsis bulge and the Hellas impact basin (compare Figure 2 of Hartmann *et al.*, 2001). The Tharsis bulge is of enormous size covering the entire western hemisphere of the planet and rising 10 km above the datum. The topographic data reveal that Tharsis actually consists of two parts, a larger nearly circular feature that contains the Tharsis volcanoes in the south and an elongated rise in the north that contains Alba Patera. The southern rise extends onto the southern highlands while the northern rise is in the lowlands. The southern Tharsis province contains a scorpion-tail shaped ridge that is apparently considerably older, of Noachian age (see Section 2.3.1 below, for a brief definition of the Martian stratographic time system), and has not been flooded by the Tharsis volcanic material. A significant part of the Martian topography can be correlated with a roughly 3 km center of mass - center of figure offset southward along the polar axis (Smith *et al.*, 2000). When the COM-COF offset is subtracted from the topography, the Tharsis province stands even further out (Smith *et al.*, 1999a). In this representation, the topography shows a $\ell=2$ characteristics, a chain of highs and lows that circle Mars' equator.

The (free air) gravity field (Figure 1) is generally rather smooth although it is rougher in the northern hemisphere than in the southern hemisphere. Peaks in gravity at the Tharsis and Elysium volcanoes indicate that these are not compensated, i.e. not in isostatic equilibrium. Major features of the (free air) gravity field include a broad complex shaped high over Tharsis and a similar high over Isidis. There is no clear association with the dichotomy boundary. The Valles Marineris canyon system is clearly visible on the gravity map as a chain of deep lows on the Eastern flank of Tharsis. Large impact basins such as Hellas and Syrtis Major are characterized by relative highs surrounded by rings of lows.

The topography and gravity data from the MGS mission have been interpreted in terms of crust and elastic lithosphere thickness models (Zuber *et al.*, 2000). Moreover, the variation of the surface heat flow has been estimated from the lithosphere thickness map. Due to the inherent non-uniqueness in any interpretation of gravity data, these models rely on additional assumptions such as constant crust and mantle densities and average crustal thickness. Nevertheless, crust and lithosphere thickness estimates constrain models of cooling and thermo-chemical evolution.

A rough estimate of crustal thickness variation and mean crust thickness can be derived from the simple assumption of Airy isostasy and the amplitudes of the topography variations. (Airy isostasy assumes that the crust is of homogeneous density and that, in equilibrium, any variations of topography are compensated by equivalent variations of crust thickness. Pratt isostasy, an alternative model, assumes that the topography is caused by lateral variations in crust density). In an Airy model the crustal thickness variation is simply $(1-\rho_c/\rho_m)^{-1}$ times the amplitude of the topography variation where ρ_c is the crust density and ρ_m the mantle density. Assuming 2900 kg m^{-3} for ρ_c and 3500 kg m^{-3} for the density of the mantle on which the crust is assumed to float, $(1-\rho_c/\rho_m)^{-1}$ is about 6 and with a total variation of the topography of roughly 20 km a thickness variation of 120 km is obtained. Assuming further that the crust thickness is approximately zero underneath the deepest regions (the large impact basins such as Hellas) and that the zero topography level is representative of the average crust, an average crustal thickness of about 50 km is calculated from the roughly 8 km depth of the Hellas Basin. This crust comprises about 4.4 % of the planet's volume.

A more sophisticated approach (Wieczorek and Phillips, 1998; Zuber *et al.*, 2000) uses the (Bouguer) gravity map to calculate crustal thickness variations under the assumption of Airy isostasy. The model crust map of Zuber *et al.* (2000) is based on an assumed constant crustal density of 2900 kg m^{-3}, a constant mantle density of 3500 kg m^{-3} and an assumed average crust thickness of 50 km. This yields a maximum thickness of the crust of 92 km underneath Syria Planum and a minimum thickness of 3 km beneath the Isidis basin. The model crust thickness is roughly constant underneath the northern lowlands and Arabia Terra at a thickness of 40 km and increases underneath the southern highlands towards the south pole to reach a thickness of $\sim$70 km. The average crust thickness of 50 km is a minimum thickness, assuming that the crust covers the whole planet. A problem with the assumption of Airy isostasy is that in perfect equilibrium the (free air) gravity should be uncorrelated with topography, which in general is not the case on Mars.

Sohl and Spohn (1997) have derived average crustal thicknesses of more than 100 km in their models of the global structure which are based on the Martian meteorite chemistry, the MoI-factor and the mass using the same crust and upper mantle average densities. (Note that the Zuber *et al.* and the Sohl and Spohn models differ in the constraints used.) Zuber *et al.* (2000) argue that a crust of this thickness could not sustain the inferred low order crustal thickness variations over the planet's history. Rather, crustal thickness variations would relax unless the viscosity of the lower crust were larger than 10^{22} Pa s. A crust of 50 km thickness could be maintained for more than 10^8 years if the viscosity of the lower crust was more than 10^{20} Pa s. While this argument is reasonable, it is not entirely convincing in supporting a 50 km crust with Airy-model thickness variations of approximately ± 50 km. A possible alternative is the suggestion of lateral density variations as in the Pratt-model. Again we make an argument based on a simple model. With Pratt-isostasy the density variation $\Delta\rho$ to support a given topography variation ΔT_{op} is $\bar{\rho}(\bar{z}/\Delta T_{\mathrm{op}} + 1)^{-1}$ where $\bar{z}$ is the average crust thickness. Assuming 100 km and 3000 kg m^{-3} for the average crust thickness and density, respectively, and taking the topography variations to be ± 5 km thereby ignoring the largest variations associated with large impact basins and volcanoes, we require a $\Delta\rho$ of a modest ± 150 kg m^{-3}. This value would bring the density of the lowland crust close to (but still smaller than) the densities of the SNC meteorites of 3230 to 3350 kg m^{-3} and would require a density in the highland crust of 2850 kg m^{-3}. The crustal thickness variations in this model would be small, about equal to the topography variations. Therefore, the model avoids the difficulty of maintaining substantial crustal thickness variations over geologic history against viscous relaxation. We will not claim that this model should apply to the entire Martian crust, however. For instance, it is reasonable to assume crustal thinning caused by mantle rebound underneath large impact basins. Rather, we stress the potential importance of lateral crust density and chemistry variations and the non-uniqueness of gravity models. A seismic network on Mars like the one to be installed by the European Netlander mission presently under development (Harri *et al.*, 2000) will certainly help to provide more reliable crustal thickness models.

Using both topography and gravity data, Zuber *et al.* (2000) have calculated the Martian effective elastic lithosphere thickness, which can be understood as the thickness of the coldest outermost layer of the planet that supports stresses over geologically long time scales. This effective lithosphere thickness varies considerably over the Martian surface and not always as one might expect. The thickest lithosphere values of ~ 200 km are found at the largest volcanoes Olympus Mons, Pavonis Mons (a Tharsis volcano) and Tharsis in general; the smallest values are at Arabia Terra and Nochis Terra. Generally, the southern hemisphere shows small values of ~ 20 km while typical values for the Northern hemisphere are ~ 100 km. Since the calculation applies models of elastic lithosphere flexure it is argued that the lithosphere thicknesses are actually the thicknesses at the time of loading, an interpretation that has been successfully used on Earth. Consequently, the elastic lithosphere has increased from 20 km during the time of formation of the units on the southern hemisphere to $\sim$100 km at the time of formation of units on the northern hemisphere. Loading by the Tharsis volcanoes, according to this hypothesis, occurred at a time when the elastic lithosphere was already more than 100 km, $\sim$200 km, thick. The elastic lithosphere thickness is often roughly associated with an isotherm thereby acknowledging the temperature dependence of mantle rheology. From the definition of the lithosphere base as an isotherm, the surface heat flow again at the time of loading, can be estimated. The exact value of the basal temperature is a matter of debate but the assumed value of 925 K should be realistic. Zuber *et al.* (2000) arrive at heat flow values between $\sim$60 mW/m^2 for the southern hemisphere to values around a few tens of mW/m^2 at the time of loading of the Tharsis volcanoes. These values provide a coarse but interesting constraint for thermal evolution models.

2.2. COMPOSITION, MOMENT OF INERTIA FACTOR, AND STRUCTURE

The Martian or SNC meteorites (Nyquist *et al.*, 2001) possess distinctive chemical compositions that hold clues for the composition of the planet as a whole and for the compositions and masses of the crust, mantle and core (Halliday *et al.*, 2001). These basalts are not as depleted in Fe as terrestrial basalts and the Martian mantle appears to be twice as rich in FeO as the terrestrial mantle. On the basis of the SNC chemistry and assuming that Mars is overall chondritic in composition, Dreibus and Wänke (1985) estimated that the core comprises $\sim$20% of the planet's mass, and further suggested that the core contains about 15 wt-% sulfur.

The Martian mantle appears to have been more oxidizing than the Earth's mantle during core formation, because it is apparently not as depleted in moderately volatile elements as is the Earth. For example, the K/U ratio of Mars is approximately 19,000 as compared with the Earth's K/U ratio of about 10,000 (Lodders, 1998). The Th/U ratio should be about the same as on Earth, because both Th and U are refractory lithophile elements. U, Th, and K are the most important radioactive heat producing elements in a terrestrial planet's interior.

Despite the recent accurate measurement of the moment of inertia factor of Mars by Pathfinder (Folkner *et al.*, 1997), present geophysical data do not constrain the interior structure of Mars unambiguously. Although a number of models have been published in the past two decades (for a recent review see Spohn *et al.*, 1998), uncertainty persists mainly because of the small relative mass of the Martian core. The uncertainty can be illustrated by considering the simplest of models consisting just of an iron-rich core and a silicate shell. The three unknowns in this model, the core density and radius and the silicate shell density, are balanced by only two known quantities, the mass and the MoI-factor. The usual procedure is then to assume the core density and calculate the mantle density and the core radius. Even if we take the core composition to be constrained by the chemistry of the Martian meteorites and assign the core a concentration of 15 wt-% sulfur and a density of

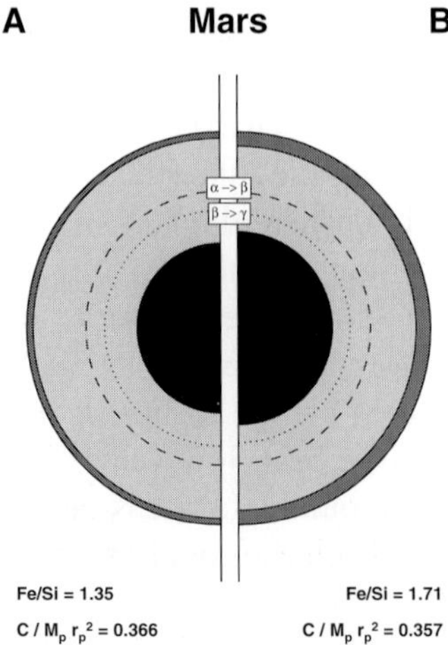

Figure 2. Comparison of the internal structure of Mars for a model **A** that satisfies the recent values of the MoI-factor and a model **B** that has a global solar Fe/Si value of 1.71. The global structures of the two models are similar with a basaltic crust, an upper and a lower mantle separated by the α-olivine to β-spinel transition, and a metallic core. The lower mantle is further subdivided into β-spinel and very thin γ-spinel layers. The models differ significantly in the thickness of the crusts and the radii of the liquid cores. Fe/Si and $C/M_p r_p^2$ denote the calculated global iron to silicon ratio and the dimensionless polar moment of inertia factor, respectively.

7000 kg/m^3 at standard pressure and temperature(10^5 Pa, 273 K), there will be an uncertainty in the latter value. If we assign an optimistic 5% error and correct for compression, we get a core radius of 1700 ± 200 km, half the planetary radius, the accuracy will be $\sim$12%, and the mass of the core is then 23 ± 9 wt-% of the planet's mass. The (average) silicate shell density of 3489 kg/m^3 is much better constrained, formally within $\sim$1%. This density allows for a FeO-rich mantle composition.

These uncertainties have to be kept in mind, when we discuss more detailed models in the following. In Figure 2, two models of the present interior structure of Mars are compared using identical sets of material parameter values for crust, mantle, and core derived from Martian meteorite data (see Sohl and Spohn, 1997, for a more complete discussion). The first model (A) is consistent with the geophysical constraint of a polar moment of inertia factor of 0.3663, while the second (B) satisfies the geochemical postulate of a chondritic bulk composition in terms of the bulk Fe/Si ratio of 1.71. Model (A) gives a global ratio Fe/Si=1.35 and model (B) produces a polar moment of inertia factor of 0.357. It is not possible on the basis of the present data to reconcile a MoI factor of 0.366 with the solar Fe/Si value of 1.71 (see also Bertka and Fei, 1998).

The two models have a basaltic crust, an upper olivine-rich mantle, a lower spinel mantle further subdivided into β-spinel and γ-spinel layers, and an Fe-rich core. The mineralogy is consistent with the sequence of phase assemblages stable in the Martian mantle at elevated temperatures and up to core-mantle boundary pressures obtained by Bertka and Fei (1997) from high-pressure and high-temperature experiments using synthetic mineral mixtures according to the mantle composition of Dreibus and Wänke (1985).

The locations of the olivine-β-spinel transition and the β-spinel to γ-spinel transition are similar in both models. (We will discuss below how phase transitions can have an effect on the evolution of the planet.) The pressure required for the phase transformations to take place is 13 GPa at the α-olivine–β-spinel transition and 18 GPa at the β-spinel–γ-spinel transition. The olivine to β-spinel transformation is located at a depth of about 1000 km, while the β-spinel to γ-spinel transition is situated several hundred kilometers above the core-mantle boundary. These transitions occur at a greater depth than in the Earth's mantle simply because of the smaller gravitational acceleration of Mars which results in lower pressures in the Martian mantle; the phase transitions are spread over a wider depth range because of the smaller pressure gradient.

The pressure at the core-mantle boundary is $\sim 22 - 23.5$ GPa. It is lower than the pressure of the γ-spinel to perovskite phase transition (Chopelas *et al.*, 1994) if the temperature is <2100 K. Since the γ-spinel to perovskite phase transition is strongly temperature dependent, a perovskite layer may have existed in early Mars when deep mantle temperatures were higher. Almost half of the bulk Fe by mass is in the core, which is found to be entirely liquid lacking a solid inner core and comprising 15% (model A) to 21% (model B) of the planet's mass. The core may be even smaller if it is solid or partly solid (although this may be inconsistent with the absence of a global magnetic field at present). For that case, recent data on the solid Fe-FeS system at high pressure and high temperature suggest a high-pressure, high-density phase change to a hexagonal NiAs superstructure (Fei *et al.*, 1995). The core size would then only be ~ 1400 km, assuming a S content in the Martian core of 12 to 16 wt-%, and a ~ 200 km thick perovskite layer would be possible.

Figure 3 shows the density ρ as a function of the radial distance from the planet's center. For both models the density varies similarly with depth despite differences in crust thicknesses and core radii. The density increases almost linearly with depth in the mantle and crust and parabolically in the core. There are prominent density discontinuities of 600 kg m^{-3} at the crust-mantle boundary and of 3000 kg m^{-3} at the core-mantle boundary. Phase transformations cause much smaller density discontinuities of ~ 250 kg m^{-3} across the olivine–β-spinel transition and ~ 130 kg m^{-3} across the β-spinel–γ-spinel transition.

2.3. GEOLOGIC HISTORY, VOLCANISM, AND TECTONICS

2.3.1. *Geologic History*

The geologic history of Mars is reviewed by Head *et al.* (2001) and will only be briefly outlined here. Three major stratigraphic time systems are used to characterize the geologic history, the Noachian, Hesperian and Amazonian (Tanaka *et al.*, 1992). The assignment of absolute ages to the epochs is based on crater density measurements and is dependent on modeled cratering rates. The new synthesis

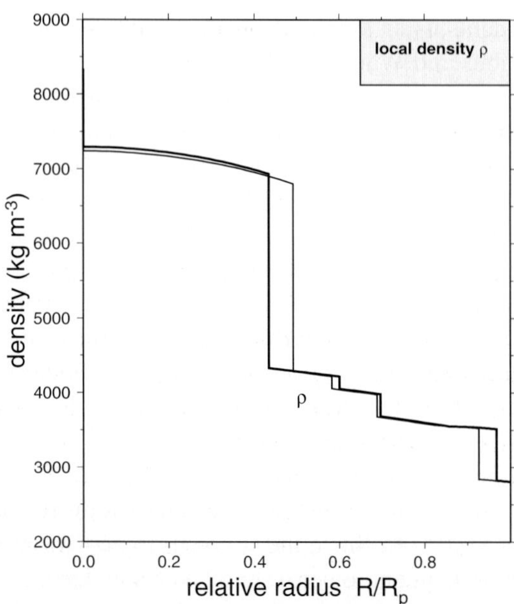

Figure 3. Radial distribution of density for model **A** (bold-lined) and model **B** (*thin-lined*). Note that there is little difference between crust and mantle densities in the two models despite of significant differences in crust thicknesses and core radii.

by Hartmann and Neukum (2001) places the Amazonian-Hesperian boundary at 2.9 − 3.3 Gyr ago and the Hesperian-Noachian boundary at 3.5 − 3.7 Gyr ago.

Most of the southern highland units are of Noachian and Early Hesperian age. However, younger units of Amazonian age are exposed in small areas. The northern plains are younger than the southern highlands and range in age from Upper Hesperian to Upper Amazonian (Tanaka *et al.*, 1992). The plains are diverse in origin: Most common around the volcanic centers of Tharsis and Elysium are lava plains with clearly developed flow fronts. However, the vast majority of the northern plains lack direct volcanic characteristics but are textured and fractured in a way that has been attributed to the interaction with ground ice and sedimentation as a result of large flood events, finally modified by wind (Carr, 1996). Nevertheless, it is believed that most of the Martian surface in particular the Northern lowlands are of volcanic origin (e.g., Greeley and Spudis, 1978; Tanaka *et al.*, 1988; Mouginis-Mark *et al.*, 1992; Greeley *et al.*, 2000).

2.3.2. *Volcanism*

It has been argued that the entire surface of Mars may be of volcanic origin (e.g., Tanaka *et al.*, 1988) but Tharsis and Elysium are undoubtedly the two most prominent volcano provinces on Mars. The three large Tharsis volcanoes Arsia, Pavonis and Ascraeus Montes are oriented along the southwest-northeast tending top of the bulge with their summits 27 km above the datum. Olympus Mons, the largest volcano in the solar system is located on the northwest flank of Tharsis. Each volcano has a large and complex summit caldera, and lava flows and channels are visible all over Tharsis. The large volcanoes seem to be formed by lava with only little pyroclastic activity (Pike, 1978; Plescia and Saunders, 1979; Tanaka *et al.*, 1992); the smaller volcanoes on Tharsis show the same characteristics as the bigger ones. The flanks of the volcanoes exhibit lava flows with only a few superimposed impact craters indicating a very young age, whereas extended flows between the large volcanoes are older. The huge dimension of the Tharsis complex and the high effusion rates of Martian volcanoes (Greeley and Spudis, 1978) demonstrate that volcanism was an ongoing process throughout the planet's history (Neukum and Hiller, 1981; Tanaka *et al.*, 1992) up to the recent (<100 Myr) past, as interpretation of data from the Mars Observer Camera (MOC) on MGS suggest (Hartmann *et al.*, 1999; Hartmann and Berman, 2000; Hartmann and Neukum, 2001). Alba Patera to the north of Tharsis has the largest areal extent but rises just a few km above the surrounding surface. The flanks of Alba Patera are dissected by young branching valleys, which are interpreted as evidence for easily erodable pyroclastic deposits such as ash (Wilson and Mouginis-Mark, 1987). Similar channeled deposits have been observed at Hecates Tholus in Elysium and Tyrrhena Patera in the highlands. Thus, the Martian volcanism seems to have experienced quiet effusion of lava and to some extent also violent pyroclastic eruptions. The Elysium volcanoes are smaller but similar to those on Tharsis.

Geological mapping of the areal extent of volcanic deposits through time (Greeley and Schneid, 1991), coupled with estimates of the thickness of the deposits, enabled an estimation of the volume of volcanic materials produced on Mars of 6.86×10^7 km^3. These authors assumed a ratio of intrusive to extrusive materials of 8.5 to 10 (based on values for the Earth) and estimated a total magma volume of 6.54×10^8 km^3. This volume is equivalent to a global layer of 4.5 km thickness, $\sim10\%$ of the likely minimum thickness of the crust estimated from the gravity data. For the 3.8 Gyr age-span (mostly Hesperian to Amazonian) of the volcanic materials analyzed, this gives a magma production rate of ~0.17 km^{3}yr^{-1}. However, there are uncertainties in these estimates due to poorly constrained values for thickness and uncertainties in the intrusive to extrusive ratio. The recent identification in MOC images of extensive thin layering in the walls of Valles Marineris suggests that volcanism in the Noachian has been much more extensive than previously recognized (McEwen *et al.*, 1999). If these layers are indeed flood lavas and are representative globally, their volume will increase the estimated magma volume by an order of magnitude and bring the thickness of the layer close to the minimum crust thickness estimate from gravity! Furthermore, if most of this volcanism is Noachian in age as suggested by the mapping, the volcanic output would have peaked in the Noachian with a general decrease with time.

2.3.3. *Tectonics*

Mars tectonics centers on Tharsis, the single largest tectonic entity in the solar system. Tharsis is surrounded by radial extensional rifts and grabens and concentric compressional wrinkle ridges that extend thousands of kilometers from its center. Most grabens on Mars are narrow (a few kilometers wide) and tens to hundreds of kilometers long structures bounded by inward dipping normal faults. Wider and deeper structures, more analogous to rifts on the Earth that rupture the entire lithosphere, can also be found in Tempe Terra, Valles Marineris and Thaumasia. Wrinkle ridges, linear to arcuate asymmetric topographic highs, are the most common compressional structures and form patterns of distributed compressional deformation. Large compressional ridges and buckles with greater strain in Noachian terrain to the south and southwest of Tharsis have also been suggested (Schultz and Tanaka, 1994). The total circumferential strain at a radius of 2500 km from the center of Tharsis is estimated to be ~0.4%, not accounting for buried structures, however (Banerdt and Golombek, 2000, and references therein).

Mapping of geologic and tectonic features (e.g., Wise *et al.*, 1979, Tanaka *et al.*, 1992, Banerdt *et al.*, 1992) within the stratigraphic framework of Mars has revealed that the area has been active throughout most of the history of the planet. Its complex structural history involves 5 stages of tectonic activity at Tharsis with changes in the derived centers of activity through time (e.g., Anderson *et al.*, 2000). More than half of the structures mapped have Noachian age (stage 1; >3.8 − 4.3 Gyr), and are exposed in Tempe Terra, Ceraunius Fossae, Syria Planum, Claritas Fossae, Thaumasia, and Sirenum. By Late Noachian-Early Hesperian (stage 2), activity was concentrated in Thaumasia and Valles Marineris. Middle Hesperian (stage 3) included the development of concentric wrinkle ridges concentrated along the edge of the topographic rise, although wrinkle ridges may have continued to form later due to global compression (e.g., Tanaka *et al.*, 1991). Normal faulting also occurred north of Alba, in Tempe Terra, in Ulysses Fossae, in Syria Planum and Valles Marineris, and in Claritas Fossae and Thaumasia. Stage 4 activity during the Late Hesperian-Early Amazonian was concentrated in and around Alba Patera. Middle to Late Amazonian activity (stage 5) was concentrated on and around the Tharsis Montes volcanoes. These events all produced radial grabens centered at slightly different locations (local centers of volcanic and tectonic activity) within the highest standing terrain of Tharsis. This indicates most importantly that the basic structure of Tharsis has changed little since the Middle to Late Noachian (Anderson *et al.*, 2000).

Lithospheric deformation models based on the Mars Global Surveyor gravity and MOLA topography appear to have simplified the stress states required to explain most of the tectonic features around Tharsis (Banerdt and Golombek, 2000). Flexural loading stresses based on present day gravity and topography appear to explain the location and orientation of most tectonic features. This observation coupled with the mapping results which suggest that more than half of the radial structures formed by the Middle to Late Noachian, argues strongly that the basic

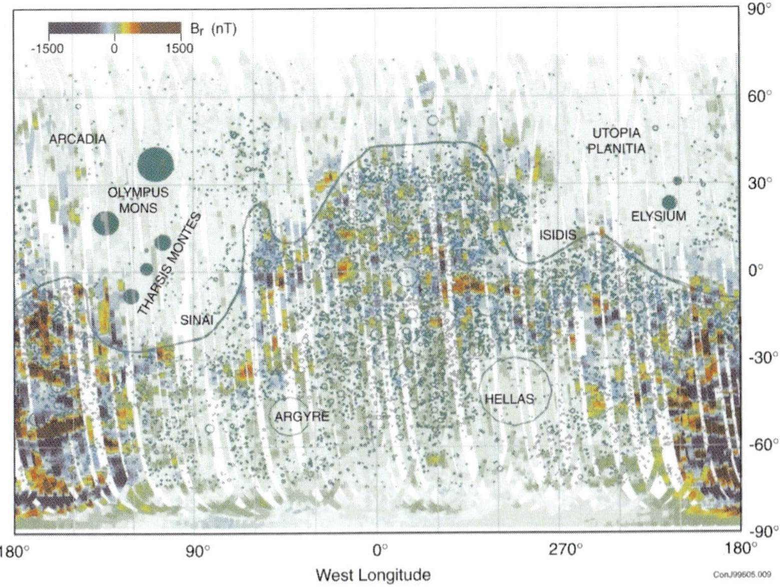

Figure 4. Magnetic field map of Mars compiled from MGS data (Acuña *et al.*, 1999).

structure of the bulge has probably changed little since the Middle/Late Noachian. The lithospheric deformation models and the mapping require a huge load, of the scale of Tharsis, to have been in place by the Middle Noachian. This load likely was formed by Noachian volcanic extrusions. The load appears to have produced a flexural moat, which shows up most dramatically as a negative gravity ring, and an antipodal dome that explains the first order topography and gravity of the planet (Phillips *et al.*, 2001). Many ancient fluvial valley networks, which likely formed during an early warmer and wetter period on Mars, flowed down the present large-scale topographic gradient, further arguing that Tharsis loading was very early. This agrees with Figure 15c of Hartmann and Neukum (2001), which shows an order of magnitude decrease in the volcanic resurfacing rate (km^2/yr) after the Noachian.

2.4. MAGNETISM

The MAG/ER investigation flown on Mars Global Surveyor has established conclusively that Mars does not currently possess a magnetic field of internal origin and reported the discovery of strongly magnetized regions in its crust (compare Figure 4), closely associated with the ancient, heavily cratered terrain of the highlands south of the dichotomy boundary (Acuña *et al.*, 1998, 1999). Evidence for a magnetic history of Mars did not come entirely unexpected, though. Models of Mars' interior and thermal evolution (Stevenson *et al.*, 1983; Schubert and Spohn, 1990; Spohn, 1991; Spohn *et al.*, 1998) predicted the existence of a thermal convection driven dynamo for at least the first billion years after accretion. Prior to the arrival of MGS at Mars the remnant magnetization of Martian meteorites had already suggested that the planet may have had in its past a surface magnetic field comparable in magnitude to that of the Earth (Curtis and Ness, 1988) and different regions of the Martian crust could have been magnetized with varying strengths and orientations representative of prior epochs of magnetic activity (Leweling and Spohn, 1997). Spacecraft missions prior to MGS, established that the current global magnetic field, if one existed at all, is weak with a surface field strength of less than 50 nT. A recently improved estimate of the upper limit to the magnitude of the current Mars dipole moment derived from the MGS data yields $M \leq 2 \times 10^{17}$ A m^{-2} which corresponds to a surface equatorial field strength of ≤ 0.5 nT.

Complementing the discovery of crustal magnetism at Mars, the MGS MAG/ER experiment observed linear patterns of strikingly intense magnetization in the southern hemisphere ($\sim$1600 nT at $h = 100$ km), particularly over Terra Sirenum and Terra Cimmeria (Connerney et al., 1999). These linear magnetic structures or "magnetic lineations" are clearly visible in Figure 4 and have been interpreted to result from processes similar to those associated with seafloor spreading at Earth. This association would imply that some form of plate tectonics and magnetic field reversals existed in Mars' early history. However, the sharp magnetization contrast required by the model of Connerney et al. (1999) suggests that alternative explanations involving the fracturing of a magnetized, thin crustal layer by tectonic stresses perhaps associated with the Tharsis rise may also be considered.

We cannot claim that the magnetization pattern of the Martian crust and its implications for the thermal evolution of the planet are fully understood. However, the MGS observations show that the majority of the crustal magnetic sources lie south of the dichotomy boundary on the ancient, densely cratered terrain of the highlands and extend $\sim$60° south of this boundary. The formation of the dichotomy boundary should simply be interpreted to postdate the cessation of dynamo action. The absence of detectable crustal magnetization north of the dichotomy boundary in spite of a widespread record of active volcanism and magmatic flows suggests that dynamo action had ceased at this stage of crustal formation.

The southernmost limit of the crustal magnetization region appears to be associated with the destruction or modification of the magnetized crust by the impacts that created the Argyre and Hellas basins. The magnetic field observations acquired over the Hellas, Argyre and other impact basins indicate the presence of weak fields most likely of external origin, suggesting that the dynamo had ceased to operate when these basins were formed. It is estimated (e.g., Hartmann and Neukum, 2001, and references therein), that the Hellas and Argyre impacts took place less than about 300 Myr after Mars accretion was completed and therefore Mars dynamo cessation should coincide or predate this epoch if this interpretation is correct. The distribution of magnetization suggests that processes that took place after the cessation of dynamo action only modified the ancient, magnetized, thin crust through deep impacts, magmatic flows, tectonics or reheating above the Curie point. Weaker crustal magnetic sources detected in the northern hemisphere (Acuña et al., 1999) are located in the younger, Amazonian plains and dipole models

(Acuña *et al.*, 1998; Ness *et al.*, 1999) tend to yield source depths in excess of 100km close to the estimated depth to the Curie isotherm in this region (Smith *et al.*, 2000; Zuber *et al.*, 2000).

The above interpretation of the magnetic record of Mars has recently been criticized by Schubert *et al.* (2000). These authors note that the punching of a homogeneously magnetized southern highlands crust should have left characteristic magnetic anomalies. While this is correct, it should be noted that the dissipation of impact energy may have resulted in heating and thermal demagnetization of a much extended region with the effect that the expected anomaly pattern there would have been lost. In any case, the authors are correct in pointing out that a high spatial resolution measurement of the magnetic field is required to fully solve the puzzle of the origin of its peculiar pattern.

3. Interior Evolution

3.1. ACCRETION, EARLY DIFFERENTIATION AND CORE FORMATION

Unraveling a planet's evolution is a difficult task that requires an integration of evidence from various fields into consistent models. Starting with the earliest times, it is now widely accepted that Mars formed by homogeneous accretion (Halliday *et al.*, 2001). The time required for a planet to differentiate into a core, mantle, crust and atmosphere/hydrosphere depends on the basically unknown conditions in the early planet. Most accretion scenarios result in hot planets possibly with magma oceans (e.g., Benz and Cameron, 1990). Several isotopic chronometers applied to Martian meteorites have provided evidence of very early isotopic heterogeneity caused by differentiation and preserved within Mars to the present day. Most importantly, Chen and Wasserburg (1986) showed using the Pb isotopic systematics of Shergottites, that Mars formed a core early.

The time span of accretion and the time of core formation can be more directly and accurately studied using the newly developed ^{182}Hf-^{182}W chronometer (e.g., Lee and Halliday, 1997; Halliday *et al.*, 2001). Hafnium and W are strongly fractionated by the processes of melting and core formation with W being much more incompatible than Hf and, consequently, with W being enriched in the melt and depleted in the residue. Some of the Martian meteorites have anomalously high concentrations of ^{182}W, the daughter product of ^{182}Hf (measured with respect to ^{184}W). These anomalies must have formed after the core had separated and W had been depleted. Evidently the accretion and differentiation of Mars was complete within a few Hf halflives or $\sim$20 Myr (Lee and Halliday, 1997). On the Earth, the isotopic record suggests that core formation took substantially longer.

There are other striking differences between the Earth and Mars in the W isotope data. Since Martian meteorites nearly all formed long after ^{182}Hf became extinct, the variations of W isotopic abundances in the meteorites observed today indicate that the meteorite source regions must have somehow preserved these ancient heterogeneities. In the Earth 4.5 Gyr of convection has stirred the mantle sufficiently to eliminate all trace of very early heterogeneity. In contrast, the Martian mantle has heterogeneities that can only have been produced in the first 50 Myr of the solar system. Thus, mantle mixing must have been far less effective on Mars or the source regions of some SNC meteorites must have escaped vigorous mixing, as has been concluded earlier from Sr isotope data (e.g., Jagoutz *et al.*, 1994)

The W data display a very significant relationship with the ^{146}Sm-^{142}Nd data (Harper *et al.*, 1995) for the same Martian meteorites. This striking agreement provides a powerful insight into the early development of Mars. In order to generate a correlation between the W and Nd isotopic data, the two parent-to-daughter element ratios must have fractionated together at an extremely early stage. Lee and Halliday (1997) have interpreted these data as evidence of coeval, co-genetic and very early silicate melting and metal segregation (core formation). Although the source regions of the Martian meteorite parent magmas appear to have suffered early depletion from melting and core formation they yield no sign of the corresponding enriched reservoirs. This would be consistent with the outer portions of Mars representing a depleted residue or floating cumulate layer from which early enriched melts and metals segregated inward toward the center of the planet. However, the existing dataset is based on a very small number of samples.

A strong relation between core formation and silicate melting is also suggested from the physics of core formation. It has been argued (e.g., Stevenson, 1990) that the separation of iron melt from a silicate matrix is difficult, if not impossible because the melt would form isolated pockets instead of connected networks. This is a consequence of the surface energy of iron melt in comparison with the surface energy of silicate rock. The ratio between the surface energies is more favorable of iron segregation if there is silicate melt. However, Bruhn *et al.* (2000) have shown that iron can form a connecting network and flow if the solid silicate matrix is subject to shear such as would be provided by large-scale convective overturn.

In conclusion, the evidence from Martian meteorite isotope variations strongly argue for a hot early evolution with rapid core formation accompanied by early mantle melting and, perhaps, differentiation. The mantle differentiation may have led to chemical mantle layering which has perhaps drawn too little attention to date by scientists who model the mantle dynamics.

3.2. EVOLUTION AFTER CORE FORMATION

Since planets can be regarded as heat engines that convert thermal energy to mechanical work and magnetic field energy, the evolution of the planet subsequent to formation and early differentiation is strongly coupled to its thermal evolution or cooling history. The thermal evolution, in turn, is strongly related to the dynamics of the planet's mantle. Despite the arguments against mantle mixing above, there is little doubt that at least a significant part of the Martian mantle should have been convecting for most of if not all of its history. The argument is based on a comparison between the radiogenic heat production rate in the silicate shell and the heat transfer rates provided by heat conduction and thermal convection (e.g., Tozer, 1967; Stevenson and Turner, 1979). Given that the thermal conductivity of silicates is small, a few W/mK, heat production by radioactive elements even at their present concentrations after 4.5 Gyr of decay should lead to large scale planetary melting. However, the strongly temperature dependent viscosity of the mantle rock will decrease as the temperature increases, and even more so as partial melting begins, to allow large scale thermal convection driven by the temperature difference between the deep interior and the surface or between the deep interior and the temperature at the base of rigid lithosphere. Heat transfer by convection is generally thought to be efficient enough to eventually remove the heat generated by radioactives and to cool the interior. Partial melting will lead to the growth of the crust thereby further differentiating the silicate shell and transferring heat sources to the crust. The latter occurs because the radiogenic elements are lithophile due to their large ionic radii and valence states and tend to be enriched in the melt.

3.3. PLANFORM OF MANTLE CONVECTION

The planform of convection, the pattern of flow in the Martian mantle, remains speculative. Intuitively, it appears natural that the surface distribution of major tectonic and volcanic units at present and over time should somehow reflect the flow at depth. For instance, Matyska *et al.* (1998) have analyzed the distribution of volcanoes on the Martian surface and have related the latter to possible planforms of mantle convection. Moreover, the origin and genesis of the dichotomy is ascribed variously either to long-wavelength mantle convection (Wise *et al.*, 1979) or to the flow associated with post-accretional core formation (Davies and Arvidson, 1981) sweeping up most of the crustal material into one large protocontinental mass. The MOLA topography (Smith *et al.*, 2000) shows no trace of a giant impact that could have caused the dichotomy (Wilhelms and Squyres, 1984). Most recently, Zhong and Zuber (2001) proposed that the dichotomy and the crust thickness variation was the result of a degree-1 mantle convection planform. A low order convection pattern may also be suggested by the $\ell = 2$ pattern of the topography (Section 2.1). In return, it has been speculated that a dichotomy in crustal thickness between the northern and southern hemispheres may have caused large scale differences in the mantle convection pattern underneath these hemispheres (Breuer *et al.*, 1993) with consequences for crustal evolution. The dominance and stability of Tharsis and its continued activity until the recent past suggests that it may have been formed by a super-plume or a family of large plumes. The recent findings on the origin and history of Tharsis (Section 2.3.3) are consistent with a superplume but do not require the plume to be existent under Tharsis throughout Martian history (W. Banerdt, personal communication). This superplume must have supplied lava in the Noachian with an enormous rate.

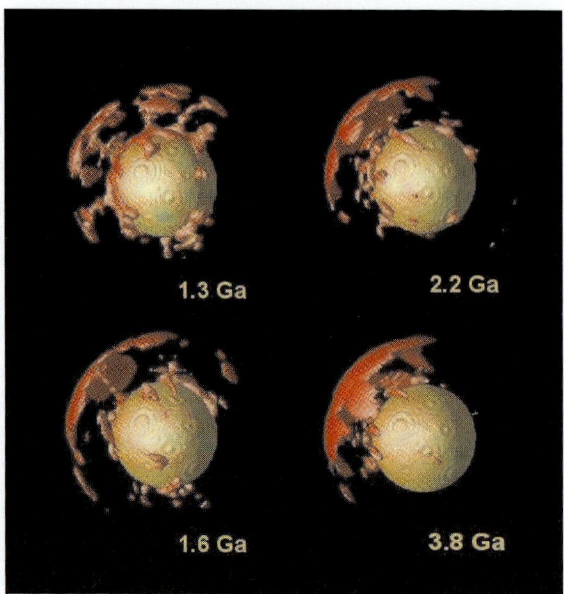

Figure 5. 3D mantle convection model showing plumes and their interaction with the ol-sp mantle phase transitions after Breuer *et al.* (1997). As can be seen from the figure, the model starts out with a number of small scale plumes. With the passage of time, the number of plumes is reduced and a super-plume survives. A similar evolution will be found as a consequence of the interaction of the flow with the spinel-perovskite transition.

Numerical convection modeling by Weinstein (1995), Zhou *et al.* (1995), Harder and Christensen (1996), Breuer *et al.* (1996, 1997, 1998) and Harder (2000) have suggested that the interaction of the olivine-spinel and, more strongly and more directly, the spinel-perovskite phase transformations with the mantle flow may suppress small, less energetic plumes and amplify large scale, strong plumes and thus force the upwelling flow into a small number of super-plumes. Figure 5 shows a series of snapshots from a 3D mantle convection calculation that illustrates the formation of a super-plume. The time scale of superplume formation in the calculation may be considered too long. However, this is mostly caused by an unrealistically low vigor of convection in the calculation forced by limitations in present day computer power. In early Mars the formation of the plume would probably have been about an order of magnitude faster.

The calculations of Martian mantle convection with phase transformations assumed that the temperature at the base of the mantle or the heat flow from the core was constant in time, assumptions often made for simplicity in numerical mantle convection calculations. However, recent 2D and 3D calculations, to be reported in detail elsewhere, that relax this assumption and allow the core to cool, cast some doubt on the applicability of these models to the entire thermal history. If the constant thermal boundary condition is relaxed and if the core is allowed to cool, it is found that the plumes decrease in strength on a time scale of a few 100 Myr depending on the initial superheat of the core. Eventually the plumes fade away in step with the decrease of the heat flow from the core. The flow in the mantle is then increasingly dominated by downwelling plumes and broad upwelling flow as it is commonly found in internally heated convection experiments and calculations. These results suggest that the thermal evolution of the Martian interior after an initial phase of plume convection may basically be independent of mantle phase transitions. Plumes may persist for longer times only if the core provides additional heating. The amount of this heating is, however, expected to be small compared to the heat produced by radioactive elements in the mantle.

3.4. THERMAL HISTORY MODELS USING PARAMETERIZED CONVECTION

Thermal evolution calculations with 2D or 3D mantle convection codes are very time-consuming on present-day computers. Therefore, to calculate thermal histories, convective heat transfer is often parameterized by relating the heat flow to the convective vigor through a suitable scaling law. The models are particularly useful if it can be assumed that the flow is simple in nature and that complications such as those that arise from interactions of the flow with phase transitions can be ignored. Early models have simply applied scaling laws derived from constant viscosity convection experiments. For the Earth with plate tectonics the parameterization seems to work quite well. These models were also applied to one-plate planets such as Mars (e.g., Stevenson et $al.$, 1983) and later modified to include the effects of a growing lithosphere (Schubert and Spohn, 1990; Schubert et $al.$, 1992; Spohn, 1991), using an equation for lithosphere growth first proposed by Schubert et $al.$ (1979). The base of the lithosphere in these models is an isotherm assumed to be characteristic for the transition from viscous deformation to rigid response to loads applied over geologic time scales. A representative value of this temperature is $\sim 10^3$ K. Heat transfer through the lithosphere is by heat conduction.

In recent years, experimental and numerical studies (e.g. Richter et $al.$, 1983; Davaille and Jaupart, 1993; Solomatov, 1995; Moresi and Solomatov, 1995; Grasset and Parmentier, 1998) have improved our understanding of convection in layers of fluids with strongly temperature dependent rheologies. In these layers, a so-called stagnant lid will form beneath the cold surface and above a convecting sub-layer. The convection in the sub-layer is driven by the temperature difference across it. The viscosity variation in the sub-layer is small, only one to two orders of magnitude. Most of the viscosity variation occurs in the stagnant lid. For such a lid to form, the rate of change of viscosity η with temperature T, $-d \ln \eta / dT$ times the temperature difference ΔT across the entire layer must be larger than ~ 10 (Solomatov, 1995). For commonly assumed mantle rheology parameters (e.g., Turcotte and Schubert, 1982) and Martian deep mantle temperatures the product is much larger than 10 placing the planet squarely in the field of stagnant lid convection.

The temperature at the base of the stagnant lid T_b (the top temperature is the surface temperature of the planet) can be expressed in terms of the temperature of the

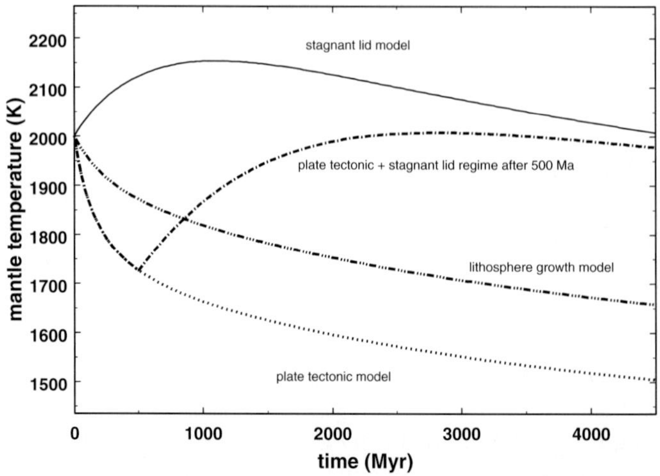

Figure 6. Thermal histories calculated for 4 different mantle heat transfer models

convecting sub-layer. If the latter is T_i, then T_b is approximately $T_i/(2.3RT_i/E+1)$ where R is the universal gas constant and E is the activation enthalpy for creep. Choosing representative values for $T_i \approx 1800$ K and $E \approx 400$ kJ/mole, the difference between T_i and T_b is small, of the order of 100 K. Thus, the stagnant lid contains the rheological lithosphere and is likely to be much thicker then the latter. Heat transfer through the lid again is by heat conduction.

A problem with the stagnant lid model, however, is that it may be ignoring some important aspect of mantle convection since the Earth also meets the criteria for stagnant lid convection but instead has plate-tectonics. Although it is universally accepted that at least present day Mars lacks plate-tectonics there may be other modes of efficient heat removal that we just have no way of appreciating at the present time. It has been speculated that Mars went through a phase of early plate tectonics: Sleep (1994) proposed that the Northern lowlands were formed in a phase of early Martian plate tectonics and that the Tharsis volcanoes may be the remnants of an island arc volcanic chain. Connerney *et al.* (1999) proposed that plate tectonics caused the stripe pattern of magnetic anomalies over Terra Sirenum and Terra Cimmeria, and Nimmo and Stevenson (2000) invoked plate tectonics to explain the early magnetic field of Mars altogether.

There are some important differences between models using these different parameterization schemes (Figures 6 and 7). Thermal history calculations applying the stagnant lid parameterization scheme show that the stagnant layer will thicken as the planet cools while the temperature of the convecting sub-layer will change comparatively little. In a series of evolution calculations with 2D and 3D convection models for the Moon, Spohn *et al.* (2001) have confirmed these principles by finding that the model lunar mantle cooled mostly by thickening its stagnant lid.

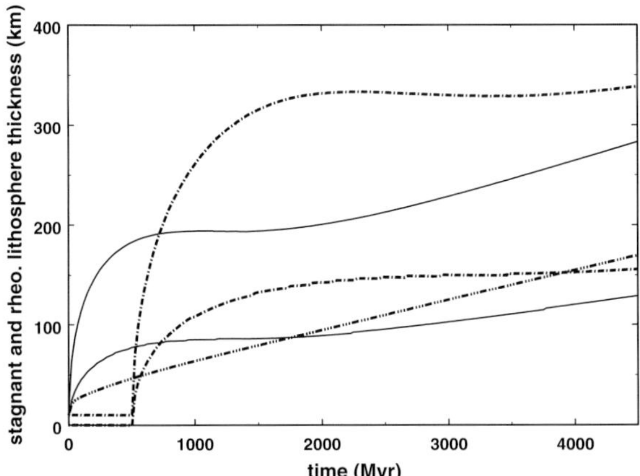

Figure 7. Lithosphere and stagnant lid thickness for 3 of the heat transfer models of Figure 6. The solid lines are for the stagnant lid model; the lower curve gives the thickness of the lithosphere characterized by a 1075 K isotherm, and the upper plots the thickness of the stagnant lid. The dash-dotted lines are for the plate tectonics model that transfers into a stagnant lid model after 500 Myr. The dash-triple dotted line refers to the lithosphere growth model with a basal temperature of 1075 K.

Figures 6 and 7 include a model with an early 500 Myr long phase of plate tectonics that transitions into the stagnant lid regime. The model is ad-hoc in the timing of the transition just as the Nimmo and Stevenson model. It is unclear how to model this transition and why the planet should undergo the transition in the first place. With plate tectonics there is no lid thickening and the cooling of the deep interior is most efficient. However, as the figures show, the model compensates the increased deep mantle cooling caused by plate tectonics through warming thereafter and the lack of lid thickening in the first 500 Myr through increased thickening. The lithosphere thickening model is inbetween the two extremes.

Figures 8 and 9 compare the surface heat flows and the accumulated heat removed from the models. It is interesting to note that these quantities fail to differ significantly between the models. Therefore, plate tectonics is not cooling the entire planet more effectively. The difference between the models is mostly a difference between cooling of the outer and the deeper layers. Plate tectonics cools the deep interior most effectively while stagnant lid convection cools the outer layer most effectively but cools the deep interior least effectively. It is further interesting to note that the surface heat flow shown in Figure 8 is consistent with the estimates of Zuber *et al.* (2000) based on gravity and topography modeling. Of the three models, the stagnant lid model is closest to the Zuber *et al.* estimate because the surface heat flow shows the smallest variation over the past 4 Gyr.

Geochemical heterogeneity may persist in the upper mantle due to stagnant lid convection because this mode removes a sizable fraction of the silicate shell from convective mixing. For lid thicknesses of up to 350 km, 100 km thickness is equivalent to approximately 10% of silicate shell volume. Figure 7 suggests that the

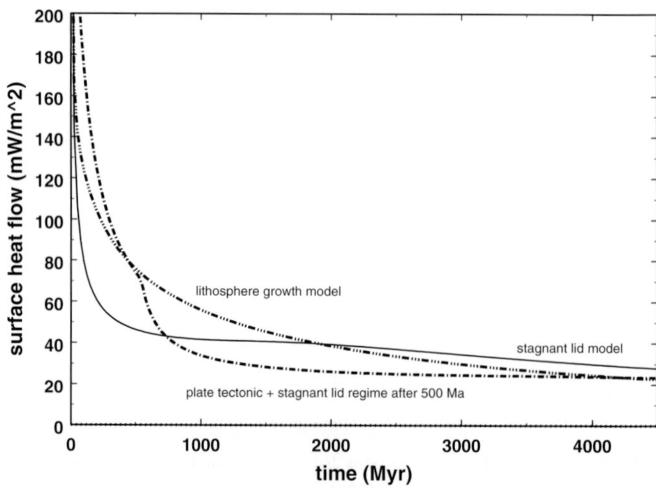

Figure 8. Surface heat flow calculated for the lithosphere growth model, the stagnant lid model, and the model with plate tectonics followed by stagnant lid convection.

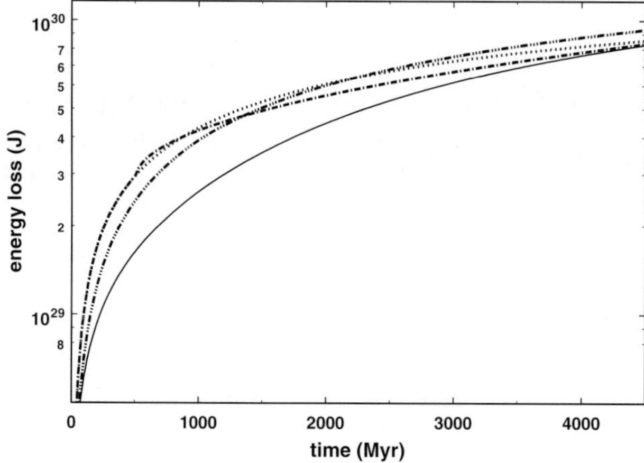

Figure 9. Cumulative heat flow calculated for the stagnant lid model (*solid line*), the lithosphere growth model (dash-triple dotted line), and the model with plate tectonics followed by stagnant lid convection (*dash dotted line*).

lid grows quickly initially. It may already be 100 km thick at the end of the overturn that accompanies core formation. The lid could contain the depleted reservoir and Martian meteorite source region required by the geochemical data.

3.4.1. *Crust Formation*

Parameterized convection models can easily be extended to include a model of crust growth (e.g., Spohn, 1991; Schubert *et al.*, 1992). The mechanism of crust growth in these models may differ to some extent but typically depend on the vigor of mantle convection and on the concentration of the crustal component in the

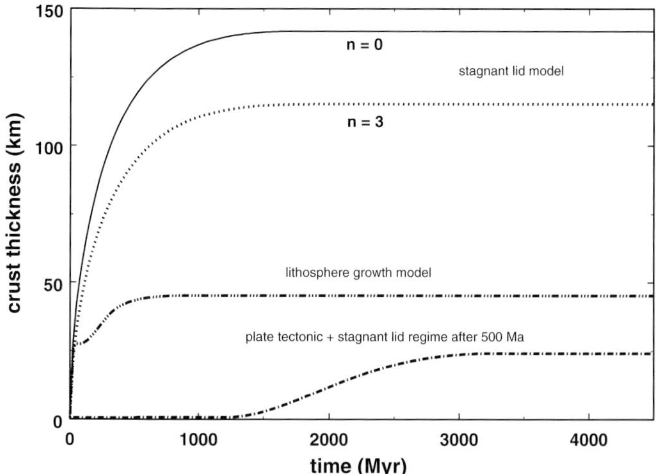

Figure 10. Crust thickness calculated for 3 models of Martian heat transfer through mantle convection, the lithosphere growth model (*dashed line*), the stagnant lid model (*solid line*), and model with plate tectonics followed by stagnant lid convection (*dotted line*).

mantle. In Figure 10 we show crustal thicknesses recently calculated by Breuer and Spohn (2001a, 2001b) for the stagnant lid, lithosphere growth, and plate tectonics models that additionally relate the growth rate to the degree of melting of the mantle and the thickness of the lid. The crust growth rate is given by

$$\frac{dD_c}{dt} = \frac{D_{pot} - D_c}{D_m} u\, m_a \left(1 - \frac{D_l}{D_{cr}}\right)^n ,\tag{1}$$

where D_c is the thickness of the crust, t is time, D_{pot} the maximum crustal thickness for complete separation between crustal reservoir and residue assumed to be 200 km, D_m the thickness of the mantle, u the mean mantle flow velocity calculated from the Rayleigh number to the power of 2/3, m_a the mean melt concentration in the mantle i.e. the degree of partial melting, D_l the thickness of the stagnant lid, D_{cr} a critical depth, below which the melt cannot rise to the surface, and n an exponent that accounts for the increasing frustration of melt transfer through the growing lid with increasing thickness. A value of D_{cr} of 600 km has been assumed. The pressure at this depth is equivalent to the pressure where the density of magma due to compression becomes equal to the mantle density (Stolper *et al.*, 1981). The use of n is certainly ad-hoc, albeit reasonable in its effect on the crustal growth rate. For $n = 3$, for example, only $\sim$60% of the melt will pass through a stagnant lid of 100 km thickness. The remainder will be intruded into the lid. The melt concentration m_a has been obtained by intersecting the solidus of the mantle material with the calculated mantle temperature profile. A linear increase of degree of melting between the solidus and the liquidius of the mantle has been assumed.

As Figure 10 shows, the crust grows fast to its final thickness within $\sim$1 Gyr for the stagnant lid and the lithosphere growth models. The growth rate of the crust decreases with time in these models mainly because m_a decreases with temperature and with time. The final crustal thickness is strongly model-dependent; the stagnant lid model provides the thickest crust of $\sim$140 km. If the crustal growth rate is assumed to decrease with increasing lid thickness (n>0), the final thickness will be smaller. The stagnant lid model produces the thickest crust because of the higher mantle temperatures predicted by this model (Figure 6).

The plate tectonics model includes crustal growth only after the cessation of plate tectonics thereby assuming that crust produced by plate tectonics will be recycled. Since plate tectonics cools the mantle such that it is subsolidus ($m_a = 0$) at the time of the transition to the stagnant lid regime, it takes $\sim$700 Myr for the mantle to warm again and partially melt. Then, the crustal production rate begins to increase and peaks at $\sim$2 Gyr. The present-day crust is a factor of 2 and 6 thinner than for the lithosphere growth and the stagnant lid models, respectively.

Crustal growth will act to additionally cool the mantle by moving heat sources from the mantle to the crust. In the stagnant lid and lithosphere growth models, this will lead to thicker lids and lithospheres with time. For the models presented in Figure 10, these layers are up to 100 km thicker as for the models in Figure 7. An important parameter that determines the final thickness of the crust, the lithosphere, and the lid is the initial temperature which is assumed to be 2000 K as in Figure 6. With an initial mantle temperature of 1800 K a present-day crust thickness of only 70 km is obtained for the stagnant lid model. In this case, the crust production rate does not decrease continuously with time as in Figure 10 but shows a weak peak after $\sim$1 Gyr, as Tanaka and Davis (1988) also have suggested on the basis of geologic data. Further parameters that determine the crustal evolution are the rheology parameters and the mantle thermal conductivity (Breuer and Spohn, 2001a, b).

The Martian meteorite compositions, in particular those of the Shergottites, the Pathfinder analyses, and the results of TES on MGS suggest at least two crustal types for Mars: Basalt and a more fractionated, more silica-rich andesitic type. According to the TES results, basalts appear to be mostly restricted to the southern hemisphere, while the andesites are prominent in the northern hemisphere but are also found in the south (Bandfield *et al.*, 2000). It seems possible that both types are derived by partial melting of mantle rock and fractional crystallization in magma chambers. However, it is also possible that the andesitic rocks are produced by remelting of basalt. This may have occurred through an earlier phase of plate tectonics on Mars or through crust (partial) remelting. For instance, if the Tharsis bulge is indeed a gigantic volcanic pile, as the MGS results suggest, then it appears possible that the lower layers of this pile were eventually remolten to produce a more fractionated magma.

The calculations presented above do not include effects of crust reservoir recycling. An attempt at modeling the growth of two reservoirs has been presented by Breuer *et al.* (1993). The present results show that the Martian mantle may produce a voluminous crust during approximately the Noachian. The crust growth rate actually may differ locally. Heterogeneities in crust and upper mantle structure may focus a substantial part of the melt in a region such as Tharsis. It has been speculated, for instance by Watts *et al.* (1991), that the Hellas impact may have caused fracturing and may have triggered or enhanced volcanic activity in the Tharsis region. The net crust growth rate may be reduced by crust recycling with the mantle as through a phase of plate tectonics, but other mechanisms of recycling such as crust delamination may work as well locally. A further growth limiting factor is the frustration of melt transfer by lid growth. Large scale crust growth will cease as the planet cools and as the mantle becomes subsolidus ($m_a = 0$) on average. However, this is not to say that volcanism should cease altogether. Even in an on average sub-solidus mantle, melt and volcanic activity may occur locally. The rate of melt production should, however, be small compared to the early activity.

3.4.2. *Core Convection and the Dynamo*

The observation of strong remnant crustal magnetization by the MGS-mission suggests a strong, self-generated magnetic field very early in Martian history. The much weaker magnetization of the SNC meteorites may indicate a weaker magnetic field at the time of the crystallization of their parent magmas $\sim$1 Gyr ago.

It is widely accepted that a planetary magnetic field is produced by regenerative dynamo action in a fluid or partly fluid iron-rich core. The dynamo is thought to be driven by vigorous thermal or chemical convection. Thermal convection in the core like thermal convection in the mantle is driven by a sufficiently large super-adiabatic temperature difference between the core and the mantle. Because the core fluid is of comparatively low viscosity ($\sim$1 Pa s), the required temperature differences are much smaller than those between the mantle and the surface required to drive mantle convection. The necessary temperature differences arise naturally when mantle convection removes heat from the core at a sufficiently large rate. This heat flow must be larger than the heat flow supplied by thermal conduction along the core adiabat. The latter heat flow therefore serves as a criterion for the existence of thermally driven convection in the core. The critical value of core heat flow depends on the adiabatic temperature gradient in the core and on the thermal conductivity. Nimmo and Stevenson (2000) have recently estimated the critical heat flow for Mars to be $\sim$5 $-$ 20 mW/m^2.

Figure 11 shows the heat flow from the Martian core as a function of time calculated from the above thermal history models. The heat flow is below the critical value for thermal convection and thermally driven dynamo action at the present time for all models. This result has been previously obtained from similar calculations (Stevenson *et al.*, 1983; Schubert and Spohn, 1990; Spohn, 1991). In early Martian history, the models predict dynamo action for periods of some 100 Myr to $\sim$1 Gyr depending on model details such as the initial temperatures and rheology of mantle and core, and the style of heat transfer. Thus, the models are consistent with estimates of the history of the Martian magnetic field that are based on the absence of magnetization in Hellas to suggest that the field had ceased to be produced by the time of formation of the Hellas and Isidis basins (Acuña *et al.*, 1999). Note that

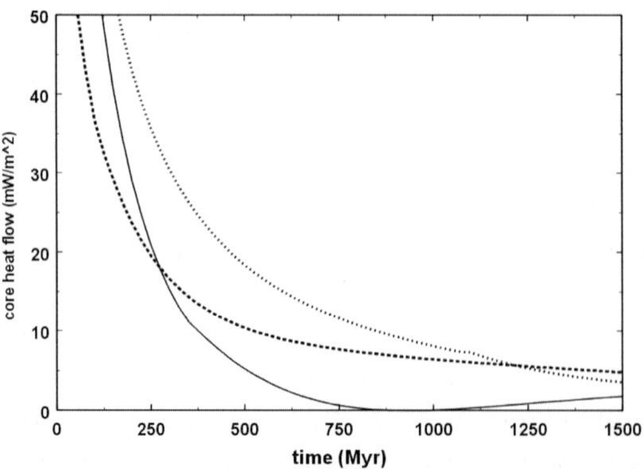

Figure 11. Core heat flow calculated for three models of heat transfer through Martian mantle convection, the lithosphere growth model, the stagnant lid model, and the model with plate tectonics followed by stagnant lid convection.

the stagnant lid model in particular and the lithosphere thickening model to a lesser degree require a substantial initial temperature difference between the core and the mantle because in these models cooling of the planet's deep interior is ineffective. This temperature difference may arise from a superheating of the core during core formation. Stevenson (1990) and Nimmo and Stevenson (2000) have argued that this superheating may be difficult to obtain because the Fe in the initial iron-silicate mixture may be finely distributed, and the Fe may be expected to be in thermal equilibrium with the silicates. Plate tectonics bring cold, near surface temperature material to the base of the mantle and cool the core more effectively than the other forms of heat transfer discussed above. The plate-tectonics model will have the least difficulty in explaining the strong early magnetic field of Mars. However, it is possible that the core contains some radiogenic heat sources as is thought to be possible for the Earth (e.g., Zindler and Hart, 1986; Breuer and Spohn, 1993)

Chemical convection can drive a dynamo more effectively than thermal convection (e.g., Braginsky, 1964; Stevenson *et al.*, 1983). While the conversion of thermal energy to magnetic field energy has a Carnot efficiency factor such as any other heat engine, a chemically driven dynamo has no similar restriction in efficiency. Chemical convection may arise when a solid core freezes in a core that contains some light alloying element. Sulfur is usually taken as the most likely candidate because it is cosmochemically abundant and because of its chemical affinity to Fe. If the core has a non-eutectic composition, buoyant fluid will be released as the inner core grows at the surface of the inner core and will drive convection even if the core is stably stratified against thermal convection. Schubert and Spohn (1990) have concluded from their thermal evolution models that the Martian core is likely to be entirely liquid if it contains $\sim$15 wt-% S, as data from the SNC meteorites suggest. A stagnant lid model predicts even greater temperatures in the lower mantle and core. A fluid core also results from the model with an early phase of plate tectonics. Thus a fluid core appears to be the simplest explanation for the present absence of a magnetic field at Mars and its early magnetization. Note, however, that inviscid solutions of an $\alpha\omega$ geodynamo (Jault, 1996) indicate that the dynamo action may be strongly impeded and strong-field solutions may be prohibited if an inner solid core grows larger than 0.35 of the core size.

Phase transitions in the Martian mantle and their interactions with the mantle flow may bring additional complexities to the evolution of the core and the magnetic field. Spohn *et al.* (1998) have discussed how a possible perovskite layer may affect the thermal history of the core. It is quite well established and has been substantiated by 2D and 3D convection calculations reported in Breuer *et al.* (1997) that the perovskite layer will actually reduce the heat flow from the core because the perovskite layer will form an additional (convective) layer between the mantle above and the core. Therefore, the hypothesis of an (early) perovskite layer will face the challenge of explaining the early strong dynamo if there was one. Core superheating during core formation and core heat sources are possibilities. The perovskite layer could provide an explanation for the magnetization of the Martian meteorites though. Their crystallization ages suggest that they were formed after the dynamo had been shut off. As discussed in Spohn *et al.* (1998) the perovskite layer may thin as the planet cools and eventually disappear because the transition is significantly temperature dependent. When the perovskite layer becomes so thin that it will no longer provide a barrier to the heat flow from the hot core, the heat flow may suddenly increase and may start or restart the dynamo thermally or perhaps even chemically. A thermally driven dynamo, as model calculations suggest, may operate for a few 100 Myr before it (again) becomes extinct. A chemically driven dynamo would, however, be likely to operate to the present day.

4. Concluding Remarks

We have discussed the evolution of Mars making use of the results of recent space missions, the chemistry of Martian meteorites and new insights into the physics of mantle convection. Some of the new results such as the gravity data and the magnetic field data cannot be unambiguously interpreted and certainly open new questions. Moreover, the confidence in our evolution models is still limited at present mostly because of our ignorance in some key areas such as dynamo theory, rheology and the interplay between convection and tectonics. Still an improved understanding of the evolution of Mars emerges: The history of Mars apparently proceeds from still little understood planet formation and rapid early differentiation by core formation and perhaps primordial crust formation to an early phase during which the planet was very active. Much of the major features on Mars such as, for instance, the Tharsis bulge was formed then. This period, basically the Noachian, was probably characterized by magnetic field generation, vigorous volcanic and tectonic activity, perhaps even plate tectonics, and a surface heat flow similar to the one of the present Earth's continents. Thereafter, the activity declined to reach the present day level. However, the recent past witnessed apparently more activity then was previously thought. This very likely includes recent volcanic activity.

Future missions to Mars have a high potential for providing missing information that will help to complete the story of the evolution of Mars. In addition to the outstandingly important task of providing an absolute scale for timing events in Martian history, completing the present day picture of Mars will be extremely helpful. Among the expected and much needed data are a determination of the core radius and the state of the core. Is the core entirely liquid or is there a solid inner core? Moreover, an absolute determination of crust thickness by seismology will help to anchor the crust thickness maps derived from gravity just as stratigraphy will be anchored by absolute age measurements. Similarly, a few heat flow measurements will anchor the heat flow estimates from gravity. The discovery of crustal magnetization on Mars, of course, motivates the desire for a further more detailed look with increased resolution. For modeling, we will need some increased effort to come to grips with the effects of rheology, partial melting, and chemical layering on the evolution of a planet. While our understanding of the effects of temperature dependent rheology on the heat transfer properties and the cooling history of the Martian interior are definite accomplishments it is disturbing that the only planet we know in detail does NOT behave the way the theory predicts. For the Earth, we have some, but not a complete understanding of why. For Mars, as well as for Mercury, Venus and any other of the earthlike planets and moons we do not even know whether or not these planets behave like the model predicts. A better knowledge of the interior structure will help.

Acknowledgements

We thank J. Geiss and R. Kallenbach of the International Space Science Institute for hosting the workshop, B. Hartmann and R. Kallenbach for their patient and thorough editing, and B. Jakosky and B. Hartmann for thoughtful reviews and constructive criticism. A significant part of this review was written while T. Spohn spent a sabbatical at UCLA. He thanks his colleagues at UCLA, in particular G. Schubert and M. Kivelson, for fruitful discussions.

References

Acuña, M.H., *et al.*:1998, 'Magnetic Field and Plasma Observations at Mars: Initial Results of the Mars Global Surveyor Mission', *Science* **279**, 1676–1680.

Acuña, M.H., *et al.*: 1999, 'Global Distribution of Crustal Magnetism Discovered by the Mars Global Surveyor MAG/ER Experiment', *Science* **284**, 790–793.

Anderson, R.C., *et al.*: 2000, 'Primary Centers and Secondary Concentrations of Tectonic Activity Through Time in the Western Hemisphere of Mars', *J. Geophys. Res.*, in press.

Bandfield, J.L., Hamilton, V.E., and Christensen, P.R.: 2000, 'A Global View of Martian Surface Compositions from MGS-TES', *Science* **287**, 1626–1630.

Banerdt, W.B., Golombek, M.P., and Tanaka, K.L.:1992, 'Stress and Tectonics on Mars', *in* H.H. Kieffer, B.M. Jakosky, C.W. Snyder, and M.S. Matthews (eds.), *Mars*, Univ. Arizona Press, Tucson, pp. 249–297.

Banerdt, W.B., and Golombek, M.P.: 2000, 'Tectonics of the Tharsis Region, Insights from MGS Topography and Gravity', *Proc. 31st Lunar Planet. Sci. Conf.*, 2038.

Bertka, C. M. and Fei, Y.: 1997, 'Mineralogy of the Martian Interior up to Core-Mantle Boundary Pressures', *J. Geophys. Res.* **102**, 5251–5264.

Bertka, C. M. and Fei, Y.: 1998, 'Density Profile of an SNC Model Martian Interior and the Moment of Inertia Factor of Mars', *Earth Planet. Sci. Lett.* **157**, 79–88.

Benz, W., and Cameron, A.G.W.: 1990, 'Terrestrial Effect of the Giant Impact', *in* H.E. Newsom and J.H. Jones (ed.), *Origin of the Earth*, Oxford Univ. Press, Oxford, 61–67.

Braginsky, S. I.: 1964, 'Magnetohydrodynamics of the Earth's Core', *Geomag. Aeron.* **4**, 698–712.

Breuer, D., and Spohn, T.: 1993, 'Cooling of the Earth, Urey Ratios, and the Problem of Potassium in the Core', *Geophys. Res. Lett.* **20**, 1655–1658.

Breuer, D., Spohn, T., and Wüllner, U.: 1993. 'Mantle Differentiation and the Crustal Dichotomy of Mars', *Planet. Space Sci.* **41**, 269–283.

Breuer, D., Zhou, H., Yuen, D.A., and Spohn, T.: 1996, 'Phase Transitions in the Martian Mantle: Implications for the Planet's Volcanic Evolution', *J. Geophys. Res.* **101**, 7531–7542.

Breuer, D., Yuen, D.A., and Spohn, T.: 1997, 'Phase Transitions in the Martian Mantle: Implications for Partially Layered Convection', *Earth Planet. Sci Lett.* **148**, 457–469.

Breuer, D., Yuen, D.A., Spohn, T., and Zhang, S.: 1998, 'Three Dimensional Models of Martian Mantle Convection with Phase Transitions', *Geophys. Res. Lett.* **25**, 229–232.

Breuer, D., and Spohn, T.: 2001a, 'Thermal, Volcanic, and Magnetic Field History of Mars', *Planet. Space Sci.*, submitted.

Breuer, D., and Spohn, T.: 2001b, 'Plate Tectonics Versus one-Plate Tectonics on Mars: Constraints From the Crustal Evolution', *J. Geophys. Res.*, submitted.

Bruhn, D., Groebner, N., and Kohlstedt, D. L.: 2000, 'An Interconnected Network of Core-forming Melts Produced by Shear Deformation', *Nature* **403**, 883–886.

Carr, M.H.: 1996, '*Water on Mars*', Oxford Univ. Press, New York.

Chen, J.H., and Wasserburg, G.J.: 1986, 'Formation Ages and Evolution of Shergotty and its Parent Planet From U-Th-Pb Systematics', *Geochim. Cosmochim. Acta* **50**, 955–968.

Chopelas, A., Boehler, R., and Ko, T.: 1994, 'Thermodynamics and Behavior of γ-Mg$_2$SiO$_4$ at High Pressure: Implications for Mg$_2$SiO$_4$ Phase Equilibrium', *Phys. Chem. Min.* **21**, 351–359.

Connerney, J.E.P., *et al.*: 1999, 'Magnetic Lineations in the Ancient Crust of Mars', *Science* **284**, 794–798.

Curtis, S.A., and Ness, N.F.: 1988, 'Remanent Magnetism at Mars', *Geophys. Res. Lett.* **15**, 737–739.

Davaille, A., and Jaupart, C.: 1993, 'Transient High-Rayleigh-Number Thermal Convection With Large Viscosity Variations', *J. Fluid Mech.* **253**, 141–166.

Davies, G.F., and Arvidson, R.E.: 1981, 'Martian Thermal History, Core Segregation, and Tectonics', *Icarus* **45**, 339–346.

Dreibus, G., and Wänke, H.: 1985, 'Mars: A Volatile Rich Planet', *Meteoritics* **20**, 367–382.

Fei, Y., Prewitt, C.T., Mao, H.K., and Bertka, C.M.: 1995, 'Structure and Density of FeS at High Pressure and High Temperature and the Internal Structure of Mars', *Science* **268**, 1892–1894.

Folkner, W.M., Yoder, C.F., Yuan, D.N., Standish, E.M., and Preston, R.A.: 1997, 'Interior Structure and Seasonal Mass Redistribution of Mars From Radio Tracking of Mars Pathfinder, *Science* **278**, 1749–1752.

Grasset, O., and Parmentier, E.M.: 1998, 'Thermal Convection in a Volumetrically Heated, Infinite Prandtl Number Fluid With Strongly Temperature-Dependent Viscosity: Implications for Planetary Thermal Evolution, *J. Geophys. Res.* **103**, 18,171–18,181.

Greeley, R., and Spudis, N.F.: 1978, 'Volcanism in the Cratered Terrain Hemisphere of Mars', *Geophys. Res. Lett.* **5**, 453–455.

Greeley, R., and Schneid, B.D.: 1991, 'Magma Generation on Mars: Amounts/Rates, and Comparisons With Earth, Moon, and Venus', *Science* **254**, 996–998.

Greeley, R., Bridges, N.T., Crown, D.A., Crumpler, L.S., Fagents, S.A., Mouginis-Mark, P.J., and Zimbleman, J.R.: 2000, 'Volcanism on the Red Planet: Mars', *in* J.R. Zimbelmann and T.K.P. Gregg (eds.), *Environmental Effects on Volcanic Eruptions: from Deep Oceans to Deep Space*, Kluwer Academic/Plenum Publ., New York, pp. 75–112.

Halliday, A.N., and Lee, D.-C.: 1999, 'Tungsten Isotopes and the Early Development of the Earth and Moon', *Geochim. Cosmochim. Acta* **63**, 4157–4179.

Halliday, A.N., Birck, J.L., Clayton, R.N., and Wänke, H.: 2001, 'The Accretion, Composition and Early Differentiation of Mars', *Space Sci. Rev.*, this volume.

Harder, H.: 2000, 'Mantle Convection and the Dynamic Geoid of Mars', *Geophys. Res. Lett.* **27**, 301–304.

Harder, H., and Christensen, U.: 1996, 'A One-Plume Model of Martian Mantle Convection, *Nature* **380**, 507–509.

Harper, C.L., Nyquist, L.E., Bansal, B., Wiesmann, H., and Shih C.-Y.: 1995, 'Rapid Accretion and Early Differentiation of Mars Indicated by ^{142}Nd/^{144}Nd in SNC Meteorites', *Science* **267**, 213–217.

Harri, A.M., *et al.*: 2000, 'Network Science Landers for Mars', *Adv. Space Res.*, in press.

Hartmann, W.K., and Berman, D.C.: 2000, 'Elysium Planitia Lava Flows: Crater Count Chronology and Geological Implications', *J. Geophys. Res.* **105**, 15,011–15,026.

Hartmann, W.K., Malin M.C., McEwen, A., Carr, M., Soderblom, L., Thomas, P., Danielson, E., James, P., and Veverka, J.: 1999, 'Recent Volcanism on Mars from Crater Counts', *Nature* **397**, 586–589.

Hartmann, W.K., and Neukum, G.: 2001, 'Cratering Chronoloy and the Evolution of Mars', *Space Sci. Rev.*, this volume.

Hartmann, W.K., Kallenbach, R., Geiss, J., and Turner, G.: 2001, 'Summary: New Views and New Directions in Mars Research', *Space Sci. Rev.*, this volume.

Head, J.W., Greeley, R., Golombek, M.P., Hartmann, W.K., Hauber, E., Jaumann, R., Masson, P., Neukum, G., Nyquist, L.E., and Carr, M.H.: 2001, 'Geological Processes and Evolution', *Space Sci. Rev.*, this volume.

Jagoutz, E., Sorowka, A., Vogel, J.D., and Wänke, H.: 1994, 'ALH84001: Alien or Progenitor of the SNC Family', *Meteoritics* **28**, 548–579.

Jault, D.: 1996, 'Sur l'Inhibition de la régénération du Champ Magnétique dans Certains Modèles de Dynamo Planétaire en Présence d'une Graine Solide', *C. R. Acad. Sci. Paris* **323**, 451–458.

Lee, D.-C., and Halliday, A.N.: 1997, 'Core Formation on Mars and Differentiated Asteroids', *Nature* **388**, 854–857.

Leweling, M., and Spohn, T.: 1997, 'Mars: a Magnetic Field due to Thermoremanence?', *Planet. Space Sci.* **45**, 1389–1400.

Lodders K.,: 1998, 'A Survey of Shergottite, Nakhlite and Chassigny Meteorites Whole-rock Compositions', *Met. Planet. Sci.* **33**, A183–A190.

Longhi, J., Knittle, E., Holloway, J.R., and Wänke, H.: 1992, 'The Bulk Composition, Mineralogy, and Internal Structure of Mars', *in* H.H. Kieffer *et al.* (eds.), *Mars*, University of Arizona Press, Tucson, pp. 184–208.

Matyska, C., Yuen, D., Breuer, D., and Spohn, T.: 1998, 'Symmetries of Volcanic Distributions on Mars and its Interior Dynamics', *J. Geophys. Res.*, **103**, 28,587–28,597.

McEwen, A.S., Malin, M.C., Carr, M.H., and Hartmann, W.K.: 1999, 'Voluminous Volcanism on Early Mars Revealed in Valles Marineris', *Nature* **397**, 584–586.

Moresi, L.N., and Solomatov, V.S.: 1995, 'Numerical Investigation of 2D Convection With Extremely Large Viscosity Variations', *Phys. Fluids* **7**, 2154–2162.

Mouginis-Mark, P.J., Wilson, L., and Zuber, M. T.: 1992, 'The Physical Volcanology of Mars', *in* H.H. Kieffer *et al.* (eds.), *Mars*, Univ. Arizona Press, Tucson, pp. 424–452.

Ness, N.F., *et al.*: 1999, 'MGS Magnetic Fields and Electron Reflectometer Investigation: Discovery of Paleomagnetic Fields due to Crustal Remanence', *Adv. Space Res.* **23**, 1879–1886.

Neukum, G., and Hiller, H.: 1981, 'Martian Ages', *J. Geophys. Res.* **86**, 3097–3121.

Nimmo, F., and Stevenson, D.J., 2000, 'Influence of Early Plate Tectonics on the Thermal Evolution and Magnetic Field of Mars', *J. Geophys. Res.* **105**, 11,969–11,979.

Nyquist, K., Bogard, D., Shih, C.-Y., Greshake, A., Stöffler, D., and Eugster, O.: 2001, 'Ages and Geologic Histories of Martian Meteorites', *Space Sci. Rev.*, this volume.

Phillips, R.J, *et al.*: 2001, 'Ancient Geodynamics and Global Scale Hydrology on Mars', *Science*, in press.

Pike, R.J.: 1978, 'Volcanoes on the Inner Planets: Some Preliminary Comparison of Gross Topography', *Proc. 9th Lunar Planet. Sci. Conf.*, 3239–3273.

Plescia, J.B., and Saunders, R.S.: 1979, 'The Chronology of the Martian Volcanoes', *Proc. 10th Lunar Planet. Sci. Conf.*, 2841–2859.

Reasenberg, R.D., Shapiro, I.I., and White, R.D.: 1975, 'The Gravity Field of Mars', *Geophys. Res. Lett.* **2**, 89–92.

Richter, F.M., Nataf, H.C., and Daly, S.F.: 1983, 'Heat Transfer and Horizontally Averaged Temperature of Convection With Large Viscosity Variations', *J. Fluid Mech.* **129**, 173–192.

Rieder, R., Economou, T., Wänke, H., Turkevich, A., Crisp, J., Brückner, J., Dreibus, G., McSween, H.Y., Jr.: 1997, 'The Chemical Composition of Martian Soil and Rocks Returned by the Mobile APXS: Preliminary Results from the X-ray Mode', *Science* **278**, 1771–1774.

Schubert, G., Cassen, P., and Young, R.E.: 1979, 'Subsolidus Convective Cooling Histories of Terrestrial Planets, *Icarus* **38**, 192–211.

Schubert, G., and Spohn, T.:1990, 'Thermal History of Mars and the Sulfur-Content of its Core', *J. Geophys. Res.* **95**, 14,095–14,104.

Schubert, G., Solomon, S.C., Turcotte, D.L., Drake, M.J., Sleep, N.H.: 1992, 'Origin and Thermal Evolution of Mars', *in* H.H. Kieffer *et al.* (eds.), *Mars*, Univ. Arizona Press, Tucson, pp. 147–183.

Schubert, G., Russel C.T., Moore, W.B.L.: 2000, 'Timing of the Martian Dynamo', *Science* **408**, 666–667.

Schultz, R.A., and Tanaka, K.L.: 1994, 'Lithospheric-Scale Buckling and Thrust Structures on Mars: The Coprates Rise and South Tharsis Ridge Belt', *J. Geophys. Res.* **99**, 8371–8385.

Sleep, N.H.: 1994, 'Martian Plate Tectonics', *J. Geophys. Res.* **99**, 5639–5655.

Smith, D.E., *et al.*: 1999a, 'The Global Topography of Mars and Implication for Surface Evolution', *Science* **284**, 1495–1503.

Smith, D.E., Sjogren, W.L., Tyler, G.L., Balmino, G., Lemoine, F.G., and Konopliv, A.S.: 1999b, 'The Gravity Field of Mars: Results from Mars Global Surveyor', *Science* **286**, 94–97.

Smith, D.E., *et al.*: 2000, 'Mars Orbiter Laser Altimeter (MOLA): Experiment Summary After the First Year of Global Mapping of Mars', *J. Geophys. Res.*, submitted.

Sohl, F., and Spohn, T.: 1997, 'The Interior Structure of Mars: Implications from SNC Meteorites', *J. Geophys. Res.* **102**, 1613–1635.

Solomatov, V. S.: 1995, 'Scaling of Temperature- and Stress-Dependent Viscosity ', *Phys. Fluids* **7**, 266–274.

Spohn, T.: 1991, 'Mantle Differentiation and Thermal Evolution of Mars, Mercury, and Venus', *Icarus* **90**, 222–236.

Spohn, T., Sohl, F., and Breuer, D.: 1998, 'Mars', *Astron. Astrophys. Rev.* **8**, 181–235.

Spohn, T., Konrad, W., Breuer, D., and Ziethe, R.: 2001, 'The Longevity of Lunar Volcanism: Implications of Thermal Evolution Calculations With 2D and 3D Mantle Convection Models', *Icarus* **149**, 54–65.

Stevenson, D.J.: 1990, 'Fluid Dynamics of Core Formation', *in* H.E. Newsom and J.H. Jones (eds.), *Origin of the Earth*, Oxford Univ. Press, New York, pp. 231–250.

Stevenson, D.J., and Turner, J.S.: 1979, 'Fluid Models of Mantle Convection', *in* M.W. McElhinny (ed.), *The Earth, Its Origin, Evolution, and Structure*, Wiley, New York, pp. 227–263.

Stevenson, D. J., Spohn, T., and Schubert, G.: 1983, 'Magnetism and Thermal Evolution of the Terrestrial Planets', *Icarus* **54**, 466–489.

Stolper, E., Walker, D., Hager, B. H., and Hays, J. F.: 1981, 'Melt Segregation From Partially Molten Source Regions: The Importance of Melt Density and Source Region Size', *J. Geophys. Res.* **91**, 6261–6271.

Tanaka, K.L., and Davis, P.A.: 1988, 'Tectonic History of the Syria Planum Province of Mars', *J. Geophys. Res.* **93**, 14,893–14,917.

Tanaka, K.L., Isbell, N.K., Scott, D.H., Greeley, R., and Guest, J.E.: 1988, 'The Resurfacing History of Mars', *Proc. 18th Lunar Planet. Sci. Conf.*, 665–678.

Tanaka, K. L., Golombek, M. P., and Banerdt, W. B.: 1991, 'Reconciliation of Stress and Structural Histories of the Tharsis Region of Mars', *J. Geophys. Res.* **96**, 15,617–15,633.

Tanaka, K.L., Scott, D.H., and Greeley, R.: 1992, 'Global Stratigraphy', *in* H.H. Kieffer *et al.* (eds.), *Mars*, Univ. Arizona Press, Tucson, pp. 345–382.

Tozer, D.C.: 1967, 'Towards a Theory of Thermal Convection in the Mantle', *in* T.F. Gaskell (ed.), *The Earths's Mantle*, Academic Press, London, pp. 325–353.

Turcotte, D.L., and Schubert, G.: 1982, *Geodynamics*, Wiley, New York.

Treiman, A.H., Drake, M.J., Janssens, M.-J., Wolf, R., and Ebihara, M.: 1986, 'Core Formation in the Earth and Shergottite Parent Body (SPB): Chemical Evidence From Basalts', *Geochim. Cosmochim. Acta* **50**, 1071–1091.

Wänke, H., and Dreibus, G.: 1988, 'Chemical Composition and Accretion History of Terrestrial Planets', *Phil. Trans. R. Soc. Lond.* **A325**, 545–557.

Wänke, H., Brückner, J., Dreibus, G., and Ryabchikov, I.: 2001, 'Chemical Composition of Rocks and Soils at the Pathfinder Site', *Space Sci. Rev.*, this volume.

Watts, A.W., Greeley, R., and Melosh, H.J.: 1991, 'The Formation of Terrains Antipodal to Major Impacts', *Icarus* **93**, 159–168.

Weinstein, S.A.: 1995, 'The Effects of a Deep Mantle Endothermic Phase Change on the Structure of Thermal Convection in Silicate Planets', *J. Geophys. Res.* **100**, 11,719–11,728.

Wieczorek, M.A., and Phillips, R.J.:1998, 'Potential Anomalies on a Sphere: Applications to the Thickness of the Lunar Crust', *J. Geophys. Res.* **103**, 1715–1724.

Wilhelms, D.E., and Squyres, S.W.: 1984, 'The Martian Hemispheric Dichotomy may be due to a Giant Impact', *Nature* **309**, 138–140.

Wilson, L., and Mouginis-Mark, P.J.: 1987, 'Volcanic Input to the Atmosphere from Alba Patera on Mars', *Nature* **330**, 354–357.

Wise, D.U., Golombek, M.P., and McGill, G.E.: 1979, 'Tectonic Evolution of Mars', *J. Geophys. Res.* **84**, 7934–7939.

Zhou, H., Breuer, D., Yuen, D.A., and Spohn, T.: 1995, 'Phase Transitions in the Martian Mantle and the Generation of Megaplumes', *Geophys. Res. Lett.* **15**, 1945-1948.

Zindler, A., and Hart, S.R.: 1986, 'Chemical Geodynamics', *Ann. Rev. Earth Planet. Sci.* **14**, 493–571.

Zuber, M.T., *et al.*: 2000, 'Internal Structure and Early Thermal Evolution of Mars from Mars Global Surveyor Topography and Gravity', *Science* **287**, 1788–1793.

Zhong, S., and Zuber, M.T.: 2001, 'Degree-1 Mantle Convection and Martian Crustal Dichotomy', *Nature*, submitted.

Address for correspondence: Institut für Planetologie, Westfälische Wilhelms-Universität, W. Klemmstrasse 10, D-48149 Münster, Germany

GEOLOGICAL PROCESSES AND EVOLUTION

J.W. HEAD[1], R. GREELEY[2], M.P. GOLOMBEK[3], W.K. HARTMANN[4], E. HAUBER[5],
R. JAUMANN[5], P. MASSON[6], G. NEUKUM[5], L.E. NYQUIST[7] and M.H. CARR[8]

[1] *Department of Geological Sciences, Brown University, Providence, RI 02912 USA*
[2] *Department of Geology, Arizona State University, Tempe, AZ 85287 USA*
[3] *Jet Propulsion Laboratory, 4800 Oak Grove Drive, Pasadena, CA 91109 USA*
[4] *Planetary Science Institute, Tucson, AZ 85705 USA*
[5] *DLR Institute of Space Sensor Technology and Planetary Exploration,
Rutherfordstrasse 2, 12484 Berlin-Aldershof, Germany*
[6] *University of Paris-Sud, 91405, Orsay Cedex France*
[7] *Johnson Space Center, Houston TX 77058 USA*
[8] *US Geological Survey, Branch of Astrogeological Studies,
345 Middlefield Road, Menlo Park, CA 94025 USA*

Received: 14 February 2001; accepted: 12 March 2001

Abstract. Geological mapping and establishment of stratigraphic relationships provides an overview
of geological processes operating on Mars and how they have varied in time and space. Impact craters
and basins shaped the crust in earliest history and as their importance declined, evidence of extensive
regional volcanism emerged during the Late Noachian. Regional volcanism characterized the Early
Hesperian and subsequent to that time, volcanism was largely centered at Tharsis and Elysium, continuing until the recent geological past. The Tharsis region appears to have been largely constructed
by the Late Noachian, and represents a series of tectonic and volcanic centers. Globally distributed
structural features representing contraction characterize the middle Hesperian. Water-related processes involve the formation of valley networks in the Late Noachian and into the Hesperian, an ice
sheet at the south pole in the middle Hesperian, and outflow channels and possible standing bodies
of water in the northern lowlands in the Late Hesperian and into the Amazonian. A significant part of
the present water budget occurs in the present geologically young polar layered terrains. In order to
establish more firmly rates of processes, we stress the need to improve the calibration of the absolute
timescale, which today is based on crater count systems with substantial uncertainties, along with a
sampling of rocks of unknown provenance. Sample return from carefully chosen stratigraphic units
could calibrate the existing timescale and vastly improve our knowledge of Martian evolution.

1. Introduction and Constraints on Geological Evolution

Geological processes represent the major dynamic forces shaping the surfaces,
crusts and lithospheres of planets. They may be linked to interaction with the atmosphere (e.g., eolian, polar), the hydrosphere (e.g., fluvial, lacustrine), the cryosphere
(e.g., glacial and periglacial), or with the crust, lithosphere and interior (e.g., tectonism and volcanism) (Figure 1). Geological processes may also be linked to
interaction with the external environment (e.g. impact cratering), and may vary in
relative importance with time or at any point on the surface of a planet at a given

Space Science Reviews **96**: 263–292, 2001.
© 2001 *Kluwer Academic Publishers. Printed in the Netherlands.*

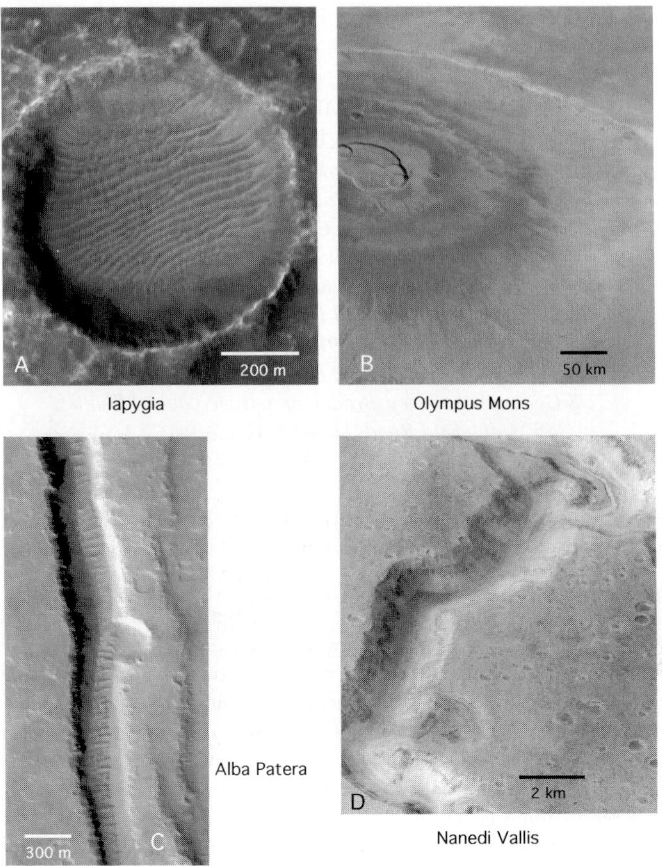

Figure 1. Examples of landforms produced by different geological processes and forming the units of the Martian crust, as seen in very high resolution Mars Orbiter Camera (MOC) images. a) Impact crater modified by dunes. MOC image M10-03652; 279.14W, 16.88S. b) Volcanic activity as exemplified by the 500-600 km diameter volcanic edifice Olympus Mons. MOC images 26301, 26302. c) Tectonic activity as exemplified by Alba Patera graben. MOC image M07-03562; 102.91W, 30.42N. d) Fluvial activity as exemplified by Nanedi Vallis channel in the Xanthe Terra region. MOC image 8704.

time. For example, impact cratering was a key process in forming and shaping planetary crusts in the first one-quarter of solar system history, but its global influence has waned considerably with time.

The geological history of a planet can be reconstructed from an understanding of the products or deposits of these geological processes and how they are arranged relative to one another. The geological history of Mars has been reconstructed through careful mapping using the global Viking image data set to delineate geological units (e.g., Tanaka, 1986; 1990; Tanaka *et al.*, 1992; Scott and Tanaka, 1986; Tanaka and Scott, 1987; Greeley and Guest, 1987), and superposition and cross-cutting relationships to establish their relative ages. Also important in estab-

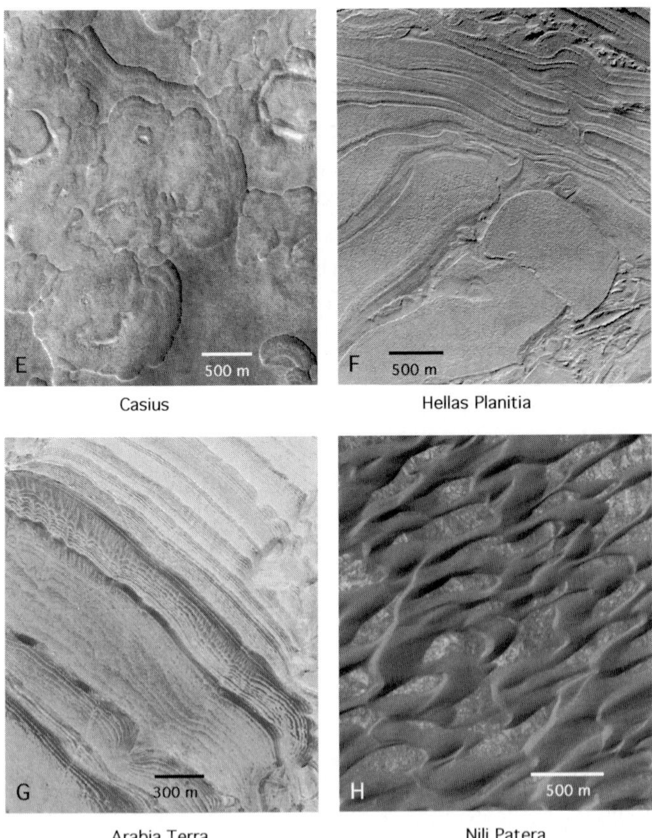

Figure 1. (continued) e) Periglacial features as exemplified by irregular depressions in the northern lowlands. MOC image M04-02077; 268.56W, 45.52N. f) Smooth deposits on the floor of the Hellas basin, interpreted as modified lake deposits. MOC image MOC2-277, 306.7W, 39.7S. g) Layering on a crater floor in western Arabia Terra. MOC image M09-01840; 7.33W, 8.34N. h) Eolian activity as exemplified by a major dune field in Nili Patera in Syrtis Major. MOC image FHA-00451; 292.93W, 8.83N.

lishing relative ages is the number of superposed impact craters on the surface of these units. Together these data have permitted the reconstruction of the relative geological history of Mars, establishment of the main themes in the evolution of Mars (Figure 2), and the relative importance of processes as a function of time. Three major time periods are defined during which the observed surface features and units have been formed: Noachian, Hesperian and Amazonian. The rocks emplaced during these periods form systems and each system is subdivided into series with corresponding time units (epochs).

But the evolution of a planet, the absolute ages of its events, the duration of the periods, and the phases of history cannot really be known without the pinning of the relative chronology to an absolute chronology. This step can be accomplished in several ways: first, if the flux of impact craters is known with certainty, as is generally the case in the lunar environment, then the number of superposed craters can be equated with an absolute age. Secondly, units can be directly dated by returning samples from broad geological units which act as marker horizons. The absolute ages of several of these units constrain the ages of other units, and the crater flux can be calibrated to date other similarly cratered units more confidently. Due to the lack of samples from Mars whose context and provenance are known, the assignment of absolute ages to the epochs based on crater densities depends on models to estimate cratering rates. Different earlier models have placed the Amazonian-Hesperian boundary between $1.8 - 3.5$ Gyr and the Hesperian-Noachian boundary between $3.5 - 3.8$ Gyr (Tanaka *et al.*, 1992; Neukum and Wise, 1976; Hartmann, 1978; Hartmann *et al.*, 1981; Neukum and Hiller, 1981). Ages associated with Hartmann tended to be younger, and those associated with Neukum, older. As discussed below, Hartmann and Neukum (2001; Figure 14) have revisited this problem, placing the divisions between the Periods and Epochs generally within the ranges suggested by the earlier literature, but tightening the age constraints.

2. Geological Processes and their Importance with Time

2.1. IMPACT CRATERING

Impact cratering dominates the character of Mars, especially the heavily cratered, old uplands (Figure 1; Strom *et al.*, 1992; Smith *et al.*, 1999a; Kreslavsky and Head, 1999, 2000; Aharonson *et al.*, 2001), and several large basins (Hellas and Argyre) dominate the topography of that part of the planet. Impact cratering causes the vertical excavation and lateral transport of crustal material, and the ejecta deposits in younger craters provide important information on the nature of the substrate and of the cratering process. Impact craters have also provided geothermal sites due to heating and impact melt emplacement, penetrated the cryosphere to release groundwater, and served as sinks for ponded surface water (e.g., Carr, 1996; Cabrol and Grin, 2001).

Impact cratering is also important in terms of chronometric information. Papers in Part I of this volume review the methodology. Using this system, Hartmann and Neukum (2001) show that appreciable areas of late Amazonian young lavas and other units have ages in the range of a few 100 Myr, agreeing with martian meteorite ages, while, at the other end of the time scale, most of the Noachian dates before 3.5 Gyr ago.

Hartmann and Neukum (2001) use the Tanaka *et al.* (1987) tabulation of areas (km^2) resurfaced by different geological processes in different epochs, to graph the *rate* of resurfacing by those processes as a function of time. A consistent result from those diagrams is that many endogenic processes, including volcanic and fluvial resurfacing, show much stronger activity before roughly 3 Gyr ago, and decline, perhaps sharply, to a lower level after that time. Those results support

Mars: Major events in geological history

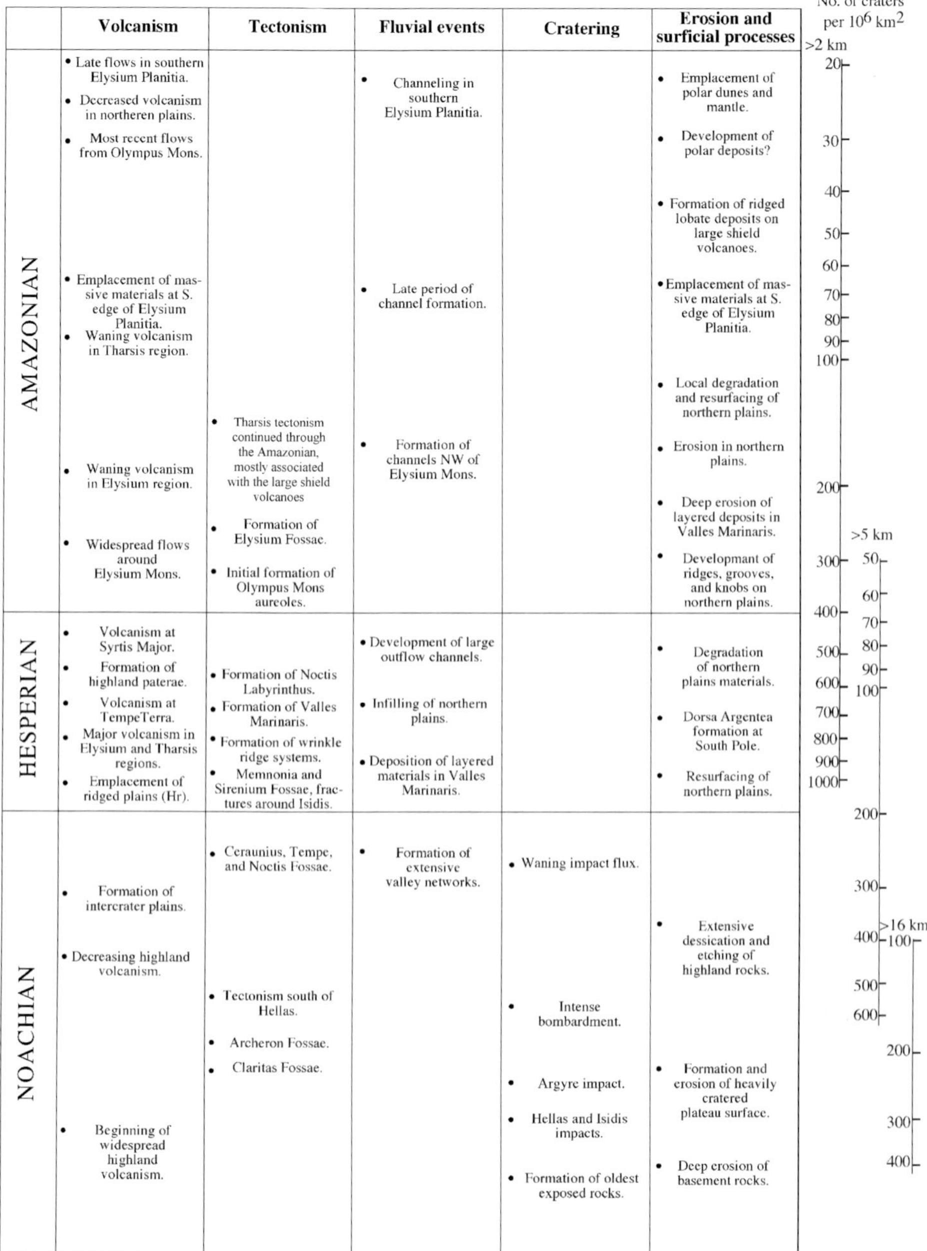

	Volcanism	Tectonism	Fluvial events	Cratering	Erosion and surficial processes
AMAZONIAN	• Late flows in southern Elysium Planitia. • Decreased volcanism in northeren plains. • Most recent flows from Olympus Mons. • Emplacement of massive materials at S. edge of Elysium Planitia. • Waning volcanism in Tharsis region. • Waning volcanism in Elysium region. • Widespread flows around Elysium Mons.	• Tharsis tectonism continued through the Amazonian, mostly associated with the large shield volcanoes • Formation of Elysium Fossae. • Initial formation of Olympus Mons aureoles.	• Channeling in southern Elysium Planitia. • Late period of channel formation. • Formation of channels NW of Elysium Mons.		• Emplacement of polar dunes and mantle. • Development of polar deposits? • Formation of ridged lobate deposits on large shield volcanoes. • Emplacement of massive materials at S. edge of Elysium Planitia. • Local degradation and resurfacing of northern plains. • Erosion in northern plains. • Deep erosion of layered deposits in Valles Marinaris. • Developmant of ridges, grooves, and knobs on northern plains.
HESPERIAN	• Volcanism at Syrtis Major. • Formation of highland paterae. • Volcanism at TempeTerra. • Major volcanism in Elysium and Tharsis regions. • Emplacement of ridged plains (Hr).	• Formation of Noctis Labyrinthus. • Formation of Valles Marinaris. • Formation of wrinkle ridge systems. • Memnonia and Sirenium Fossae, fractures around Isidis.	• Development of large outflow channels. • Infilling of northern plains. • Deposition of layered materials in Valles Marinaris.		• Degradation of northern plains materials. • Dorsa Argentea formation at South Pole. • Resurfacing of northern plains.
NOACHIAN	• Formation of intercrater plains. • Decreasing highland volcanism. • Beginning of widespread highland volcanism.	• Ceraunius, Tempe, and Noctis Fossae. • Tectonism south of Hellas. • Archeron Fossae. • Claritas Fossae.	• Formation of extensive valley networks.	• Waning impact flux. • Intense bombardment. • Argyre impact. • Hellas and Isidis impacts. • Formation of oldest exposed rocks.	• Extensive dessication and etching of highland rocks. • Formation and erosion of heavily cratered plateau surface. • Deep erosion of basement rocks.

Figure 2. Geologic features formed and processes operating during the major periods of Martian history, the Noachian, Hesperian, and Amazonian. Derived from Scott and Tanaka (1986), Tanaka and Scott (1987) and Greeley and Guest (1987).

many suggestions (Jones, 1974; Soderblom *et al.*, 1974; Chapman, 1974; Hartmann, 1971, 1973; Neukum and Hiller, 1981) that Martian geological processes were more active in the early part of Martian history, and declined, perhaps rapidly, at the end of that period. The results also show the great advance to be made if the dates of the stratigraphic epochs could be measured with lower uncertainties, since the detailed history of various processes (Figure 2) could then be deciphered.

2.2. VOLCANISM

Mariner 9 provided the first clear evidence for the importance of volcanic processes in the history of Mars (McCauley *et al.*, 1972; Carr, 1973). Images revealed shield and dome volcanoes of the Tharsis and Elysium regions, extensive lava plains, and numerous other features of volcanic origin, including low-profile contructs, called patera, characterized by central craters and radial channels. Subsequently, data from the Viking Orbiters (Carr *et al.*, 1977b) allowed mapping and characterization of the extent, timing, and styles of volcanism on Mars (Figure 1; Greeley and Spudis, 1978, 1981; Mouginis-Mark *et al.*, 1992; Greeley *et al.*, 2000). High resolution images (Malin *et al.*, 1998) (Figure 1), information on surface compositions (McSween *et al.*, 1998), and topographic data (Smith *et al.*, 1998, 1999a, b) from the Mars Global Surveyor (MGS) permit comparison of Martian volcanism with theoretical analysis of the ascent and eruption of magma on Mars (e.g., Wilson and Head, 1994).

Although much of the ancient crust of Mars is likely to be of volcanic origin, obvious morphological features, such as flows and structures which might be vents, are not seen in images of the oldest martian terrains. Nonetheless, by analogy with the early history of the Earth and the Moon (Stöffler and Ryder, 2001) and from models for Mars (Spohn *et al.*, 2001), magmatic activity was probably involved in the formation of the crust. Moreover, Thermal Emission Spectrometer (TES) data suggest primarily basaltic compositions for the martian highlands where most of the ancient crust is found (Christensen *et al.*, 2000). Subsequent geological processes, including those associated with wind and water, have modified the highlands so extensively that morphological traces related to early putative volcanism are not readily found with currently available data.

The first recognized volcanic features are the paterae in the Hellas region, e.g. Tyrhenna, Hadriaca, Amphitrites, and Peneus Paterae (Greeley and Spudis, 1978). The first two have been observed in sufficient detail to derive their volcanic history. Early eruptions seem to have involved magmas rising through water-rich megaregolith (Crown and Greeley, 1993), leading to extensive phreatic-magmatic activity and the emplacement of ash shields, the presence of which is suggested by the style of erosion on the flanks of the patera. These eruptions were apparently followed by effusive activity, emplacing complex sequences of flows which radiate from central calderas.

Alba Patera is a central vent structure covering more than 4.4×10^6 km^2, making it one of the largest volcanoes in the Solar System. It is composed of tube-and-channel fed flows and flows emplaced as massive sheets. It contains a caldera some 100 km across, the floor of which includes small cones of probable spatter and pyroclastic origin (Cattermole, 1987).

The most impressive volcanoes on Mars are in the Tharsis and Elysium regions, where more than a dozen major constructs exist, including classic shield volcanoes. High-resolution images show that these volcanoes were built from countless individual flows, many of which were emplaced through channels and lava tubes, signaling a style of volcanism analogous to Hawaiian eruptions (Greeley, 1973). Complex summit calderas suggest multiple stages of magma ascent and withdrawal (Crumpler *et al.*, 1996). The flanks of many of the volcanoes are marked by terrain which appears to have failed by gravitational collapse, leaving mass wasted deposits covering hundreds of square kilometers in some cases. Some of the deposits in the Elysium region have a morphology suggestive of lahars, i.e., to have been emplaced as water-rich slurries (Christensen, 1989). The subdued appearance of some of the Elysium volcanic summits has been proposed as pyroclastic material, suggesting Plinian styles of eruption (Mouginis-Mark *et al.*, 1988).

By far the greatest areal extent and inferred volume of volcanic materials on Mars are found in the various plains, the youngest of which show flow fronts and embayment into older terrains (Greeley and Schneid, 1991). These materials likely represent high-volume flood eruptions. The extensive ridged plains, typified by Hesperia Planitia, are more difficult to interpret. Characterized by "wrinkle" ridges, these units resemble mare basalts on the Moon. TES data suggest basaltic compositions in contrast to the younger northern plains which have more siliceous composition (Christensen *et al.*, 2000), suggesting magma evolution with time.

In addition to these major volcanoes and plains units, a wide variety of smaller volcanoes and volcanic features are recognized on Mars, including possible composite cones found in the highlands, small shield volcanoes (many of which have associated flowlike channels), and fields of small cones with summit craters, similar to pseudocraters in Iceland, which result from local phreatic eruptions as lavas flow over water-logged ground (Frey *et al.*, 1979).

In summary, volcanic processes appear to have operated throughout the history of Mars and could possibly be active today (Hartmann and Berman, 2000; Keszthelyi *et al.*, 2000). Although from the morphologies of the flows and most of the volcanoes basaltic compositions are inferred to be dominant, remote sensing data from orbit and measurements made on the surface by Mars Pathfinder suggest that slightly more evolved magmas might also have been present in the evolution of the surface (see discussion in Bibring and Erard, 2001).

The generation of magma and its extrusion onto the surface through volcanism is a direct indicator of interior activity. Age, location, and type of volcanic materials give insight into the evolution of the interior. Volcanism has dominated much of the history of Mars. Nearly half of its surface has materials inferred to be of volcanic origin (Greeley and Spudis, 1978; Mouginis-Mark *et al.*, 1992; Greeley *et al.*, 2000). These materials form either central volcanoes, such as the shields in the Tharsis region, or vast plateaus formed by the eruption of flood lava flows. From geological mapping of the areal extent of various volcanic deposits through time, coupled with estimates of the thicknesses of the deposits, the volumes of volcanic materials produced on Mars. Greeley and Schneid (1991) assumed a ratio of intrusive to extrusive materials of 8.5 to 1 (based on values for the Earth) and estimated a total magma volume of 6.54×10^8 km^3. For the 3.8 Gyr age-span of the volcanic materials analyzed, this gives a magma production rate of $\sim$0.17 km^3 yr^{-1}. Note that uncertainties in the derivation of these estimates include poorly constrained values for thicknesses and the fact that the intrusive to extrusive ratio might be different on Mars. The values based on photogeologic mapping can be refined through the application of MGS data, including topographic information.

The recent identification in Mars Observer Camera (MOC) images of extensive thin layering in the walls of Valles Marineris suggests that volcanism in the Noachian has been much more extensive than previously recognized (McEwen *et al.*, 1999). If these layers are flood lavas, as suggested by their morphology, the volume of material increases the magma volume discussed above by an order of magnitude. Furthermore, if most of this volcanism is Noachian in age as suggested by the mapping (McEwen *et al.*, 1999), a peak in the volcanic output occurred in the Noachian, with a general decrease with time.

2.3. TECTONISM

The morphology of the martian surface as observed in imaging data provides ample evidence for tectonic processes (e.g., Carr, 1981; Banerdt *et al.*, 1992; Figure 1). Brittle failure of the crust and the lithosphere is indicated by a variety of structural features, both extensional (simple graben, complex graben, rifts, tension cracks, troughs, and polygonal troughs) and compressional (wrinkle ridges, lobate scarps). While the relative ages of the features and, therefore, the processes responsible for their formation may be dated by structural mapping and crater counts, additional information on topography and gravity is required to model loads and to derive stresses in the lithosphere (Golombek and Banerdt, 1999).

Global-scale processes like thermal contraction cooling, despinning, or polar wandering cause stresses in the lithosphere of a planet. While such processes may have operated on Mars, tectonic evidence for them is weak or ambiguous (Banerdt *et al.*, 1992). The global dichotomy is the most fundamental physiographic feature on the planet and formed early in Martian history. Exogenic processes (i.e. one or several mega-impacts, Wilhelms and Squyres, 1984; Frey and Schultz, 1988) have been invoked to account for it, but recent Mars Orbital Laser Altimeter (MOLA) investigations (Smith *et al.*, 1999a, b) did not find any single or several large circular topographic depressions to confirm this hypothesis (except for the Utopia basin, which had been speculated to be an impact basin even in the pre-MGS era; McGill,

1989). Endogenic processes seem to offer attractive alternatives, and a variety of convection or subduction mechanisms has been proposed. A plate- tectonics scenario has been suggested to explain many of the surface features in and around the northern lowlands (Sleep, 1994). Unfortunately, little photogeologic evidence supports the suite of features that would be expected to be observed at the various plate tectonic boundaries defined in Sleep's model (Pruis and Tanaka, 1995; Tanaka, 1995). Detailed structural mapping of key locations (e.g. the dichotomy boundary) required to further test the hypothesis is underway with new MGS data (Turcotte, 1999). An ancient phase of plate tectonics has also been proposed to explain a group of magnetic anomalies in a portion of the cratered highlands (Connerney *et al.*, 1999), although many alternate hypotheses are being considered to explain these features.

Most graben on Mars are narrow (few km wide) and long ($\sim 10 - 100$ km) structures bounded by inward dipping normal faults (Banerdt *et al.*, 1992). Wider (up to 100 km) and deeper structures (many km), more analogous to rifts on the Earth that rupture the entire lithosphere, are found in Tempe Terra, Valles Marineris, and Thaumasia. Linear to arcuate asymmetric topographic highs ("wrinkle ridges") are the most common compressional structures and form patterns of distributed compressional deformation (e.g., Chicarro *et al.*, 1985; Watters and Maxwell, 1986; Watters, 1988, 1993; Plescia and Golombek, 1986; Golombek *et al.*, 1991). Schultz and Tanaka (1994) have reported a system of large compressional ridges and buckles with greater strain in Noachian terrain to the south and southwest of Tharsis, and topographic data of the northern plains show a system of ridges generally concentric to Tharsis (Smith *et al.*, 1999a; Head *et al.*, 2001).

Regional-scale deformation seems to be responsible for most of the observed tectonic features. While minor faulting patterns are associated with the Elysium volcanic province and some impact structures, by far the dominant element in martian tectonics is the Tharsis bulge (e.g. Carr, 1981, Banerdt *et al.*, 1982, 1992; Golombek and Banerdt, 1999). In particular, a plethora of extensional structures (simple and complex graben, rifts, and troughs) radiate outwards from Tharsis. Additionally, Tharsis is the center of a concentric pattern of compressional structures (wrinkle ridges) (Lucchitta and Anderson, 1980; Maxwell, 1982; Chicarro *et al.*, 1985). Several processes have been proposed to explain the formation of the huge topographic bulge: Domal uplifting (e.g. Phillips *et al.*, 1973; Carr, 1973; Hartmann, 1973), magmatic intrusion (Sleep and Phillips, 1979, 1985; Willeman and Turcotte, 1982), and volcanic loading (Solomon and Head, 1982). For reviews of Tharsis stress models see Carr (1974b), Banerdt *et al.* (1982; 1992), Sleep and Phillips (1979), Solomon and Head (1982), Mége and Masson (1996a), Golombek and Banerdt (1999), and Banerdt and Golombek (2000). Prior to the MGS mission the formation of the extremely large system of graben (radial graben occur both on the elevated flanks of Tharsis *and* in regions outside Tharsis; Tanaka *et al.*, 1991) was explained with models that require two distinct stress states. The then available topographic and gravity data implied isostatic conditions to generate the

TABLE I

Major stages and locations of magmatic and tectonic activity in Tharsis (after Anderson and Dohm, 2000).

	Stratigraphy[1]	Age[2] (Gyr before present)	Centers of local and regional activity
Stage 1	Noachian	4.65–3.7	**Claritas**, Tempe, Ascraeus
Stage 2	Late Noachian - Early Hesperian	3.8–3.6	**Valles Marineris, Thaumasia,** Warrego Valles
Stage 3	Early Hesperian	3.7–3.6	**Syria NW**, Tempe, Ulysses, Valles Marineris
Stage 4	Late Hesperian - Early Amazonian	3.6–2.1	**Alba**, Tempe NW, Tempe SE
Stage 5	Middle - Late Amazonian	2.1–0	**Pavonis**, Tharsis Montes

[1] after Tanaka (1986)
[2] after Hartmann and Neukum (2001) (this volume)

graben on the flanks, while flexure was invoked to explain those farther out. However, the considerable improvement in the topography (Smith and Zuber, 1999) and gravity (Zuber and Smith, 1999) fields provided by MGS changed that view: Recent models of the lithospheric support of Tharsis based on topography, gravity, and structural mapping suggest that flexure alone can explain the structural pattern (Banerdt and Golombek, 2000). Many radial graben may also be the surface manifestation of dikes propagating from magmatic centers in central Tharsis (Wilson and Head, 2000; 2001).

The overall scheme of Tharsis as being the center of volcano-tectonic activity on Mars throughout most of its history is complicated by the fact that there are several local and regional sub-centers within Tharsis (Anderson and Dohm, 2000). A regional variation of Tharsis-related deformation is also in agreement with new modeling results (Banerdt and Golombek, 2000) which attribute the extensional faults *within* Tharsis (which previously seemed to be incompatible with pure loading models) to such variations unresolved in the pre-MGS data sets. Such centers of extensional as well as compressional features seem to have changed in space and time. Several studies tried to decipher the tectonic record in and around Tharsis as provided by surface faults (e.g. Wise *et al.*, 1979a, b; Plescia and Saunders, 1982; Watters and Maxwell, 1983, 1986; Tanaka and Davis, 1988; Golombek, 1989; Tanaka, 1990; Tanaka *et al.*, 1991; Schultz, 1991; Scott and Dohm, 1990; Mége and Masson, 1996b; Schultz, 1998; Dohm and Tanaka, 1999). Statistical analyses of recent hemispheric-scale structural mapping (Anderson *et al.*, 1998) indicate five measurable successive stages of volcano-tectonic activity within Tharsis (Table I) (Anderson and Dohm, 2000).

To summarize Martian tectonism, mapping of geologic and tectonic activity (e.g., Carr, 1974a, 1974b; Wise *et al.*, 1979a, 1979b; Plescia and Saunders, 1982; Scott and Dohm, 1990; Tanaka *et al.*, 1991; Banerdt *et al.*, 1992; Frey, 1979) within the stratigraphic framework of Mars (Tanaka, 1986, 1990; Scott and Tanaka, 1986, Tanaka and Davis, 1988, Tanaka, 1990; Dohm and Tanaka, 1999) has revealed a complex structural history involving five stages of tectonic activity with changes in the derived centers of activity through time (Anderson *et al.*, 2000). More than half of the structures mapped on Mars are Noachian in age (stage 1; >3.8 − 4.3 Gyr), concentrated in exposures of Noachian age crust exposed in Tempe Terra, Ceraunius Fossae, Syria Planum, Claritas Fossae, Thaumasia, and Sirenum. By Late Noachian-Early Hesperian (stage 2) activity was concentrated in Thaumasia and Valles Marineris. Middle Hesperian (stage 3) included the development of concentric wrinkle ridges concentrated along the edge of the topographic rise, although wrinkle ridge development may have continued later due to global compression (e.g., Tanaka *et al.*, 1991). Normal faulting during this time also occurred north of Alba, in Tempe Terra, in Ulysses Fossae, in Syria Planum and Valles Marineris, and in Claritas Fossae and Thaumasia. Stage 4 activity during the Late Hesperian-Early Amazonian was concentrated in and around Alba Patera and Middle to Late Amazonian activity (stage 5) was concentrated on and around the Tharsis Montes volcanoes with additional activity in Elysium Planitia along Cerberus Rupes. All of these events produced radial graben centered at slightly different locations (local centers of volcanic and tectonic activity) within the highest standing terrain of Tharsis, indicating that the basic loading pattern of Tharsis has changed little since the Middle Noachian (Anderson *et al.*, 2000).

Lithospheric deformation models based on MGS gravity and topography appear to have simplified the stress states required to explain most of the tectonic features around Tharsis (Banerdt and Golombek, 2000). Flexural loading stresses based on present day gravity and topography appear to explain the type, location, orientation and strain of most tectonic features around Tharsis (Banerdt and Golombek, 2000). These models require the load to be huge (of the scale of Tharsis and thus large relative to the radius of the planet) and the mapping requires the load that caused the flexure to have been in place by the Middle to Late Noachian (>3.8 − 4.3 Gyr), with minor changes since that time. Fine layers within Valles Marineris revealed by the Mars Orbiter Camera have been interpreted as being lava flows that are Late Noachian or older (McEwen *et al.*, 1999). These volcanics are therefore likely the load that caused the flexure around Tharsis. This enormous load appears to have produced a flexural moat, which shows up most dramatically as a negative gravity ring, and an antipodal dome that explains the first order topography and gravity of the planet (Phillips *et al.*, 2001). New altimetric data provide evidence that the circum-Tharsis mid-Hesperian wrinkle-ridge system extends into the northern lowlands, a radius of 7000 km from the center of Tharsis (Head *et al.*, 2001). Many ancient fluvial valley networks, which may have formed during an early warmer and wetter period on Mars, flowed down the present large-scale topographic gradient, further arguing that Tharsis loading was very early (Phillips *et al.*, 2001). It is in fact possible that the formation of Tharsis actually produced this early warmer and wetter environment. If the load is composed of volcanics as suggested by fine layers within Valles Marineris (McEwen *et al.*, 1999), water released with the magma would be equivalent to a global layer up to 100 m thick, which could have had a significant impact on the martian climate (Phillips *et al.*, 2001).

2.4. FLUVIAL FEATURES AND PROCESSES

Mariner 9 images in 1972 first showed giant channels and smaller branching valley networks on Mars, and Viking provided more detail indicating many flow-like features (Figure 1), which seemed to be formed by running water (e.g., Masursky *et al.*, 1977; Baker and Kochel, 1979; Pieri, 1980; Carr, 1981). Present-day conditions on Mars preclude liquid water flowing on the surface, however, liquid water may have existed on the surface in the past (e.g., Sharp and Malin, 1975; Carr, 1981, Mars Channel Working Group, 1983; Baker *et al.*, 1992). Alternatives to water for the outflow channels have been proposed (lava flows, winds, debris flows and liquid hydrocarbons; Carr, 1974a, b; Schonfeld, 1976; Cutts and Blasius, 1981; Nummedal, 1978; Nummedal and Prior, 1981; Yung and Pinto, 1978), but none of these theories explains the formation of the associated flow-like features as readily as water (e.g., Baker *et al.*, 1992; Carr, 1996; Lucchitta and Anderson, 1980). Most workers account for the observed channel features as due to the catastrophic release of groundwater under stable, metastable or non-stable surface conditions for liquid water. Masson *et al.* (2001) give a more thorough discussion of fluvial features such as outflow channels and valley networks.

The valley networks are almost entirely restricted to the old uplands and the simplest explanation is that the valleys are old themselves and the climatic requirements for valley formation were met early in the planet's history and rapidly declined during the subsequent evolution. A warmer, wet Mars with a dense atmosphere at the time after the heavy bombardment is supposed to provide the conditions for valley formation by running water (Malin and Edgett, 1999). Based on the evaluation of high resolution MOC images Malin and Edgett (1999; Carr and Malin, 2000) concluded that the valleys were formed by fluid erosion, and in most cases the source was ground water.

A detailed review on the current knowledge about the ages of the fluvial features is also given by Masson *et al.* (2001). We only refer to a few results here: The relatively small size of Martian valleys, the modification by aeolian processes, and restrictions by the resolution and coverage of existing images makes it difficult to use crater counting for age determination. Thus, only a small number of channels and almost no valley networks are dated by this method. For instance, Neukum and Hiller (1981) have estimated crater retention ages of the floors of particular channels and their significantly older surroundings. The number of craters in the surroundings and the number representing a younger resurfacing period translates to crater model ages of a few Gyr. Neukum and Hiller (1981) also show a sequence of the circum-Chryse valleys with most of the Kasei Vallis floors being oldest followed by the mouth of Maya Vallis, Ares Vallis, Tiu Vallis and the mouth of Kasei as the youngest unit in this area. Channels on volcanoes are too small for crater counting on their floors but they dissect volcanic units (for example at the flanks of Alba Patera) and are considered to be as young as a few 100 Myr (Gulick and Baker, 1990). In general, although valley networks are found on units that range in age from Noachian to Amazonian (Scott and Dohm, 1992; Carr, 1995), they are predominantly in the oldest units (Noachian). Formation of outflow channels appears to have peaked in the late Hesperian.

The distribution of channels in time can also be analyzed using relative superposition and intersection relationships (Ivanov and Head, 2001). For example, Nelson and Greeley (1999) mapped the circum-Chryse channels in detail and found the youngest units of different channels, ranging from Mawrth and Ares Vallis oldest, followed by Tiu and Simud Vallis to Shalbatana Vallis and the youngest channel units exposed at the mouth of Kasei Vallis, indicating that the last flood events of each channel were younger from east to west around Chryse.

Relatively young small-scale alcove-like gullies combined with small channels and aprons in the walls of impact craters, south polar pits and two of the larger valleys indicate even more recent groundwater seepage and probably short-term surface runoff under almost current climatic conditions (Malin and Edgett, 2000). These are discussed in more detail by Hartmann (2001).

2.5. PERIGLACIAL AND GLACIAL LANDFORMS AND PROCESSES

Freezing and thawing of the ground and the presence of permanently frozen ground or permafrost (Tricart, 1968; 1969) are the characteristics of periglacial surfaces on Earth. A number of Martian surface features seen in Mariner 9 and Viking Orbiter imagery have been attributed to periglacial or permafrost processes (Figure 1). Although there are analogies between terrestrial periglacial features and Martian landforms it has to be kept in mind that these landforms are the result of freeze-thaw cycles in the active layer above the permanently frozen ground which is not possible under the present climatic conditions on Mars (e.g. Carr, 1996). In addition, the dimensions of most of the periglacial-like features are one or more orders of magnitude larger than those on Earth. All types of terrain softening are mostly seen in a latitudinal belt between 30° and 60°. A detailed description of these landforms as geomorphologic evidence for liquid water on Mars and many references are given in the accompanying paper by Masson *et al.* (2001).

One of the periglacial landforms are the *lineated valley fills* which to some extent resemble terrestrial median moraines on glaciers. However it is not clear to what level transverse versus longitudinal flow contributed to the generation of the lineations. They are found at the upland/lowland boundary on Mars, where lobate debris aprons with well defined flow fronts and convex-upward surfaces extend from the highlands and from isolated mesas into the low-lying plains (Mangold and Allemand, 2001; Carr, 2001). In valleys, opposing scarp walls confine material flows and the flow fronts converge, resulting in ridges and grooves, commonly called lineated valley fill. In some places, terrain softening appears also in craters where material has obviously moved down the inner crater wall and probably forms *concentric crater fill*. Another periglacial landform on Mars is the viscous flow of ice-lubricated debris, also compared to *rock glaciers*. These features are flow-like but the rheology and composition of the ductile material is not clear (e.g. ice versus CO_2 clathrate hydrate; Hamlin *et al.*, 2000). The ejecta of so-called *rampart craters* on Mars are marked by distinct lobes and provide strong evidence for subsurface ice or water (Strom *et al.*, 1992; Mouginis-Mark, 1979; Barlow, 1988; Schultz and Gault, 1979). The lobes seem to flow around pre-existing obstacles and are outlined by a low ridge or rampart. Viking imagery showed fractured plains marked by giant *polygons* 30 km across, thought to originate from the cooling and fracturing of ice rich sediments and/or self-compaction and accommodation to the underlying terrain, or the desiccation of wet sediments deposited by water from the giant outflow channels (see review in Hiesinger and Head, 2000). Polygons of much smaller sizes (diameters of $10 - 100$ m) are also observed on Mars. They are interpreted to be the result of thermal contraction in ice-rich soils based on the similar scales as compared to terrestrial ice-wedge polygons (Mellon and Jakosky, 1995).

Other periglacial or glacial landforms on Mars are *thermokarsts, frost mounds, eskers*, and *moraines* (Masson *et al.*, 2001, and references therein). Among these, the most convincing morphological evidence for ancient glaciation on Mars are the eskers (e.g. Carr *et al.*, 1980), which are long sinuous ridges mainly found in mid and high latitudes. It seems most plausible that those features formed analogously to terrestrial eskers, as a result of infillings of ice-walled river channels by subglacial, englacial, or supraglacial drainage networks. Recent MOLA measurements of the heights and widths of the largest Martian features are consistent with the esker hypothesis (Head, 2000a; b). Kargel and Strom (1990) argued that only glaciation can account for features covering wide areas in Argyre, Hellas, and the south polar region in a simple and unifying way.

2.6. LAKES AND OCEANS

Numerous craters in the uplands of Mars show flat floors and channels entering the crater, and apparently sediment was deposited there (Figure 1). Numerous workers have proposed that these regions were the sites of standing bodies of water or lakes (e.g., Carr, 1996). Recent MOC data show evidence for layered deposits in many impact craters, suggesting that standing bodies of water occurred in these locations (e.g., Malin and Edgett, 2000). Such locations have been cataloged and are abundant and widespread (e.g., Cabrol and Grin, 2001). Lake sediments are also thought to characterize the floor of Valles Marineris (Lucchitta *et al.*, 1992).

Large outflow channels emptied into the northern lowlands and their floods (e.g. Lucchitta *et al.*, 1986) must have left extended deposits, large lakes, and possibly oceans. The water released by the outflow channels into the northern plains is estimated to amount to at least 6×10^6 km^3 but probably more (Carr *et al.*, 1987).

Baker *et al.* (1991) calculated that 6×10^7 km^3 of water was needed to fill up the northern plains. The northern plains cover an older rougher terrain that survived as hills and knobs commonly outlining old pre-plains impact craters. The plains are complex deposits probably formed by many processes, such as sedimentation from outflow channels, volcanism and mass wasting from the adjacent highlands modified by impact craters. Recent analyses using altimetry data suggest that the present surface deposits (mostly the Hesperian-aged Vastitas Borealis Formation) are underlain by early Hesperian ridged plains of volcanic origin, and the overlying Vastitas Borealis Formation is at least 100 meters thick (Head *et al.*, 2001). Many of the detailed surface features seen in MOC images seem to be recent and formed by action of ground ice and debris mantles (see Kreslavsky and Head, 2000).

Some features surrounding the northern plains form contacts interpreted as ancient shorelines (Parker *et al.*, 1989, 1993) marking the boundaries of former lakes or a northern ocean. Analysis of the elevations of the contacts and surface roughness using MOLA data show that the oldest contact interpreted to be a shoreline varies widely and is not a good candidate for an equipotential line. The younger contact also shows variations but is closer to an equipotential line (Head *et al.*, 1999). In addition, the plains inside these contacts are extremely smooth. Thus, some of these data are consistent with the presence of a large standing body of water, while others are not (Head *et al.*, 1999).

If there were standing bodies of water in the northern lowlands in the past, under current climate conditions such lakes or oceans would rapidly freeze and form an ice cover (e.g., Kargel *et al.*, 1996; Clifford and Parker, 1999). However they could stabilize if fed by meltwater or groundwater (McKay and Davis, 1991). Outflow channels, their source regions, and termination areas in the northern lowlands, provide the best evidence for surface water on Mars and a widespread groundwater system. The outflow channels could form by running water under current climatic conditions because the volume and flux of released water can change the atmosphere sufficiently to prevent freezing and sublimation at least during the relatively short periods of flooding (Carr, 1979, 1995; Gulick *et al.*, 1997). Again, for this topic we also refer to Masson *et al.* (2001).

2.7. AEOLIAN PROCESSES

Wind processes dominate the current Martian environment in the absence of known or abundant active volcanism, tectonism, and running water on the surface (Greeley *et al.*, 1993). Despite the tenuous CO_2 atmosphere (average surface pressure is 6.5 mb), winds are capable of setting large quantities of dust into motion, at times obscuring the surface from view from orbit. A century of earthbased observations as well as surface monitoring from orbit show that aeolian processes, such as dust storms, occur predominantly in the southern hemisphere summer. However, data obtained from the Viking orbiters and MGS, and measurements made by the Pathfinder and Viking landers show that aeolian activity can also occur at other seasons.

Particles are transported by the wind in three modes: suspension (dust particles, which on Mars are a few μm in diameter), saltation (which involves sand, or grains $\sim$0.6 − 2 mm in diameter), and creep (grains larger than a few mm), depending upon wind strength. According to Greeley *et al.* (2001), fine sand $\sim$80 − 100 μm in diameter is most easily moved by the lowest strength winds.

A wide variety of wind related features is seen from orbit and from landers (Figure 1; Greeley *et al.*, 2001). These include wind depositional features, such as dunes, and wind erosional features, including sculpted hills termed yardangs. The most common aeolian feature seen from orbit are various albedo patterns (such as wind streaks) which change with time, while common features seen on the surface include deposits associated with rocks, called wind tails. Both wind tails and wind streaks are considered to represent the prevailing winds at the time of their formation and can be mapped to infer wind directions.

Wind streaks occur in several forms, including bright features which appear to be stable over periods of years and dark streaks, some observed to have changed in as little as 38 days (Greeley *et al.*, 2001). Many bright streaks are thought to be dust deposited in the waning stages of dust storms, while dark streaks appear to result from erosion of windblown particles, exposing darker substrate, or leaving a lag deposit of lower albedo material on the surface. General circulation models (GCMs) of the atmosphere show that bright streaks correlate with predicted regional wind directions (Greeley *et al.*, 1993). Dark streaks appear to be influenced more by local topography than by regional wind patterns.

Images from MGS (Edgett and Malin, 2000) show that mantles of windblown materials, inferred to be dust deposits settled from suspension, can form both bright and dark surfaces. Similarly, dunes and duneforms also occur as both bright and dark features. These observations suggest a variety of source material for windblown sand and dust, including weathered and unweathered mafic rocks.

As long as Mars has had an atmosphere and loose particles on its surface, it is likely that aeolian processes have operated. Deposition of windblown sediments to form mantling blankets provides a mechanism of "resurfacing", while aeolian deflation of sediments has the potential to exhume formerly buried surfaces. The resurfacing and exhumation involve depths up to at least a few km (Malin and Edgett, 2000). Greeley *et al.* (2001) emphasize that resurfacing and exhumation must be taken into account in estimating surface ages based on impact crater statistics.

On Earth, eolian features and deposits provide clues to past climates. A similar potential exists on Mars. For example, recent analysis of duneforms, windstreaks, and eroded craters seen from orbit and ventifacts studied on images from the Mars Pathfinder lander suggest the presence of a paleowind regime (Kuzmin *et al.*, 2001). Because eolian activity is dependent on the wind regime, potential periods of more active eolian processes in the past could signal the presence of a higher density atmosphere because the threshold winds for sand and dust transport would be lower. Thus, analyses of present and past eolian activity can give new insight into the evolution of the Martian surface and climate.

3. Integrated Geological History and Relation to the Evolution of Mars

Initial differentiation, core and crustal formation occurred very early in the history of Mars prior to the time of the visible record seen in presently preserved surface geological units as confirmed from geochemical/geophysical data by Halliday *et al.* (2001). The magnetic field persisted for a long enough time period to form the magnetic fabric of the southern uplands, apparently prior to the Hellas and Argyre basins (Connerney *et al.*, 1999). The Noachian Period (Figure 2) was the time of significant impact bombardment modifying this crust, the formation of large basins, and the emplacement of highland plains in lows and plateaus. The origin of these smooth plains is uncertain because of poor preservation and mantling, but it is likely that many of them are of volcanic origin because of the higher thermal fluxes typical of the earlier history of Mars.

The northern lowlands formed during the early part of the Noachian, but the nature of its origin (external impacts or internal processes) still remains to be determined. Any theory for formation of the northern plains must reconcile the major geological differences between the northern lowlands and the southern uplands as well as distinctive differences in average crustal thickness (e.g., Zuber *et al.*, 2000). One possibility is that the lowlands result from thinning associated with magma ocean cumulate overturn (Hess and Parmentier, 2001).

Recent analyses of the northern lowlands suggests the presence of early Hesperian ridged plains lying below the late Hesperian sedimentary veneer of the Vastitas Borealis Formation (Frey *et al.*, 2001; Head *et al.*, 2001). Detection of numerous large, shallow basins in the northern plains (Frey *et al.*, 2001) suggests that the early Hesperian rigded plains are no more than 1 km thick and underlain by an early Noachian heavily cratered surface.

Tectonic activity in the Noachian is seen in the form of tectonic structures south of Hellas and in Acheron and Claritas Fossae. There is also growing evidence that the Tharsis rise underwent considerable construction and tectonic activity during the Noachian and may have been almost fully developed in shape and scale by the end of the Noachian. Emplacement of ridged plains of apparent volcanic origin had begun by the late Noachian. Extensive erosion of the heavily cratered terrain occurred during this time and extensive blankets of debris were emplaced at high southern latitudes.

A fundamental question about Mars is whether it was warm and wet in its early history. Evidence for water on the surface may be seen in the form of the extensive debris mantles and the presence of valley networks. Originally thought to represent the result of precipitation and surface runoff, valley networks are now considered by a number of researchers to have formed as a result of groundwater sapping processes. If this interpretation is correct, then it places the formation of the aquifer from which the sapping originates further back in the history of Mars. Presently, there is no compelling direct surface geological evidence to support a warm, wet early Mars.

The Hesperian Period (Figure 2) marks a critical transition from the Noachian early history characterized by high thermal fluxes and heavy cratering rates, to the later low volcanic and impact fluxes of the Amazonian. During this period some of the most important geological features and processes occurred, but ironically, we have little information about its duration, with estimates ranging from a few hundred million years to over a billion years.

Volcanic activity was extremely widespread in the Early Hesperian. Smooth plains, subsequently deformed by wrinkle ridges (mapped as Hr, Hesperian, ridged plains), form numerous regional patches such as Hesperia Planum and are testimony to global volcanism, followed by regional to global contractional deformation. Recent work suggests that the northern lowlands were also resurfaced by Hr (ridged plains) during the Early Hesperian (Head *et al.*, 2001), bringing the total resurfacing to over 40% of the planet. Individual volcanic edifices and flows are rare in these early deposits, but several centers of volcanism with unusual structure and morphology (e.g., Tyrrhena, Apollinaris, and Hadriaca Paterae) suggest that rising magma may have interacted with ground ice and ground water to produce explosive eruptions (e.g., Greeley and Spudis, 1981). Volcanic activity continued in the later Hesperian with the emplacement of Syrtis Major deposits, which differ from the Early Hesperian ridged plains in the preservation of flow fronts. A significant part of the volcanic activity during the later Hesperian shifted from broad plains to central-vent volcanism at Alba Patera, in Tharsis and in Elysium (Hecates Tholus, Albor Tholus).

Water is one of the hallmarks of the Hesperian Period. Midway through the Hesperian, the Dorsa Argentea Formation (Tanaka and Scott, 1987) was emplaced around the present south polar region. This fragmental and apparently ice-rich mantling material shows evidence of being an ice sheet that underwent meltback and drainage into surrounding low areas such as the Argyre and Hellas basins (Head and Pratt, 2001). The cause of the meltback, the fate of this water and the influence it had on the subsurface groundwater table (e.g., Clifford, 1993) are not presently known. There is evidence that volcanic eruptions interacted with this extensive ice sheet, causing melting and drainage (Ghatan and Head, 2001).

During the middle and later parts of the Hesperian, the valley networks characterizing the late Noachian period (and extending into the Hesperian, but in reduced number) transitioned to the major outflow channels. Numerous channels were emplaced along the southern uplands- northern lowlands boundary, emerging from the uplands and debouching into the lowlands. Accompanying this period of channel formation was the modification and retreat of the boundary scarp, particularly in the Deuteronilus Mensae area. Channels also entered the Hellas basin from the Hesperia Planum region. Evidence exists that the Hellas and Argyre basins were volcanically resurfaced in the Early Hesperian, and that sediments of fluvial, lacustrine and eolian origin were later emplaced on their floors.

A key question is the fate of the water involved in the formation of the outflow channels. Did the water pond in the northern lowlands to produce large bodies of water (e.g., the oceans of Parker *et al.*, 1989; 1993) or was there insufficient water to create large standing bodies? If there were large quantities of water, what was its fate? Did it persist for millennia in a liquid form, did it quickly soak back into the groundwater table, or did it rapidly freeze and sublime? Was there an ocean in the northern lowlands prior to the Hesperian? If so, when and for how long? The answers to these questions are not presently available and a key element is the duration of this period of channel formation, a number that is very poorly constrained (Figure 3). The Vastitas Borealis Formation (Tanaka and Scott, 1987), a Late Hesperian fragmental unit overlying the ridged plains (Head *et al.*, 2001), is temporally equivalent to the outflow channel deposits. A likely scenario is that the sediments of the Vastitas Borealis Formation are part of the deposits of the outflow channels, perhaps a residue remaining after the sublimation of a water-sediment mixture emplaced during one or more channel-forming events.

The Amazonian Period (Figure 2) saw the continuing emplacement of volcanic units such as the Elysium Formation, a veneer of volcanic material radial to Elysium Mons, late-stage flows from Tharsis Montes and other Tharsis volcanoes, and flow units emplaced on the upper margins of the northern lowlands (the Arcadia Formation). Volcanism apparently continued in the Elysium region until very recently, as very young flows in the Elysium-Marte Valles region have been detected (Keszthelyi *et al.*, 2000). Most volcanic activity in the Amazonian was associated with the major rises, Elysium and Tharsis, where it continued throughout this period. Outflow channel activity and even some valley networks continued into the Amazonian. Volcanic activity in the Elysium region was associated with significant quantities of groundwater. Lahar-like channels and deposits are seen debouching into the Utopia Basin, while to the east, Amazonian fluvial channels are closely related to the very recent volcanism with crater retention ages of the order 10 Myr or less (Hartmann and Berman, 2000; Hartmann and Neukum, 2001). Minor valley networks are seen on the upper flanks of Alba Patera.

In the Late Amazonian, the polar layered terrain was emplaced and modified. The crater retention ages for these deposits suggest a very young age (less than a few Myr; Herkenhoff and Plaut, 2000). There is a major hiatus between these very young Amazonian polar deposits and underlying Hesperian-aged materials at both poles (Dorsa Argentea Formation at the South Pole and the Vastitas Borealis Formation at the North Pole). Does the formation of these Late Amazonian deposits represent a major change in the atmospheric environment of Mars where it continued throughout this period? Were similar deposits formed and destroyed at several times in the Amazonian during the hiatus? Important to the understanding of these questions is a firm measurement of the duration of the Amazonian; current uncertainties in Martian cratering rates (Ivanov, 2001) mean that the duration could range from ~1.8 Gyr to ~3.5 Gyr. Hartmann and Neukum (2001), using the best estimates of crater rate, place the duration at ~2.9 to 3.2 Gyr.

Late-stage Amazonian Period activity includes abundant eolian modification of the polar deposits (Tanaka and Scott, 1987; Fishbaugh and Head, 2000) and formation of a latitudinally distributed belt of mantling material (e.g., Squyres and Carr, 1986; Kreslavsky and Head, 2000), probably related to depositional variations accompanying obliquity cycles.

4. Outstanding Chronological Questions and Measurement Requirements

Two independent systems for dating Martian units complement each other, as discussed earlier in this volume by Nyquist *et al.* (2001), Neukum *et al.* (2001), and Hartmann and Neukum (2001). Dating of rock samples is believed to be relatively exact (barring any unknown systematic errors), but in the absence of sample return missions, we have samples only from unknown locations on Mars; thus, their provenance, context, and relation to geological units and history are unknown.

The cratering method has the advantage of being able to date any chosen terrain of sufficient area, but at present the uncertainties are large. As discussed by Ivanov (2001) the main uncertainty is in the ratio of Mars/moon cratering rates, used to calibrate the system. The formal uncertainties may be of the order 20% (in terms of the estimated asteroid impact rate, crater scaling, etc), but a more realistic uncertainty may be a factor 2 (taking into account unknown effects of comets, various main belt resonances, etc). This uncertainty propagates directly into uncertainty on age. This, in turn, means that features with crater retention ages less than, say, 200 Myr are probably formed in the last 10% of Martain time and have value in terms of geological and geophysical processes. But a crater retention age of, say 2 Gyr, is less useful in geophysical terms, because the true age might lie between 1 and 4 Gyr (Figure 3). At the oldest extreme, a crater retention age around 4.0 Gyr is relatively constrained because the high crater densities are associated only with high cratering rates before about 3.5 Gyr.

Currently, Martian meteorites provide the only basis for an absolute chronology of Martian processes. Accepting the Martian origin of the meteorites, methods must be developed to generalize observations about them to observations about Mars. Attempts to do so are hampered to variable degrees by lack of knowledge of the geologic settings from whence the meteorites came. Interpretations of some types of observations, such as the composition of Martian atmospheric gases trapped in the meteorites, are independent of the location on Mars from which the meteorites came. Interpretations of other types of observations, such as the mineralogy of the meteorites, benefit from and inform the knowledge of the general environment in which rocks parental to the meteorites crystallized, but do not require exact knowledge of those environments. Interpretations of still other types of observations, such as calibrating the Martian cratering rate, would benefit greatly from exact knowledge of the places of origin of the meteorites. Some aspects of the radiometric age data for Martian meteorites fall into all three of these categories.

Because the places of origin of the Martian meteorites are unknown, use of their ages for calibrating the cratering rate is distinctly limited. Nevertheless, the observation of young igneous crystallization ages, down to ~165 Myr, among the meteorites shows that Martian volcanism continues essentially until the present day. The observation of a high proportion of young ages moreover suggests that Mars has been volcanically relatively active at recent times: 10 of 15 meteorites for which radiometric ages are summarized in Nyquist *et al.* (2001) have ages <500 Myr, 8 of those have ages in the range ~165 − 185 Myr. Only 1 in 15 has an age older than 1.3 Gyr. However, there are three outstanding questions: (a) Do the Martian meteorites give a statistically reliable sampling of Martian surface ages? (b) If not, how many Martian surface areas really have been sampled? (c) What are the potential bias factors in sampling them?

Because ~40% of the Martian surface is highlands belonging to the oldest stratigraphic unit, the Noachian, whereas only 1 of 15 (~7%) of the meteorites has a correspondingly old radiometric age, the answer to the first question appears to be negative. This conclusion invites comparison to the lunar case. Of 20 lunar meteorites, 7 (35%) are from mare areas. The mare surfaces of the Moon represent about 17% of the lunar surface, thus the highlands again appear to be underrepresented by about a factor of two. In both cases there is a need for a better statistical sampling, but there is an implication that some bias factor may discriminate against older, brecciated, and therefore more friable samples, as has been suggested by a number of authors.

However, the grouping of Martian meteorite ages around certain preferred values emphasizes the need to correct for "launch-pairing" among them. If the crystallization ages of the meteorites are used to group the meteorites, the number of apparent ejection events is reduced to 4–5, and the number of Noachian age events (1) is the statistically expected number. If instead, the cosmic ray exposure ages are used, there are seven apparent ejection events, also statistically acceptable. The two approaches are combined by Nyquist *et al.* (2001). Better understanding will be facilitated via acquisition of new data and better statistics. If the lower number of events implied by the crystallization ages is confirmed, it may imply that secondary collisions in space contribute to the distribution of cosmic ray exposure ages, increasing their apparent number relative to the actual number.

The shock metamorphic histories of the meteorites can make a more important contribution to determining launch conditions and pre-launch sample locations than perhaps has been recognized. The currently known Martian meteorites appear to populate a "launch window" of peak shock pressures in the interval ~15 − 45 GPa. Qualitatively, this appears to be systematically higher than peak shock pressures experienced by the lunar meteorites, for example. This observation needs to be quantified, and its implications assessed for models of meteorite ejection, as well as possible strength-related biases in meteorite yields. Variations in peak shock pressures among launch-paired samples can give information on the relative depths at launch for the samples.

Finally, reliable launch-pairing of the meteorites will enable better interpretation of a variety of mineralogical, geochemical and isotope geochemical data obtained for the samples. Current data provide a strong suggestion that some mantle-derived basaltic magmas have assimilated Martian crustal materials (see Nyquist *et al.*, 2001, among others). Geochemical and isotopic variations among some of the meteorites appear to reflect variable degrees of this crustal contamination. If the assimilation processes can be adequately described, it may be possible to infer the geochemical processes of a trace-element-enriched Martian "crustal" component, which has not otherwise been sampled. This objective requires the correct grouping of samples for investigation of candidate assimilation processes and scenarios.

Finally, it must be noted that the degree of reliability currently achieved for radiometric dating of lunar samples has required the experience and improvement in laboratory techniques acquired over two decades. Not all of the problems encountered in dating these samples have been analytical. Martian rocks, as evidenced by Martian meteorites, bear the record of a complex series of processes, both primary igneous processes and secondary processes (weathering, deposition of secondary minerals, cf. Bridges *et al.*, 2001). The return of actual Martian samples to terrestrial laboratories is likely to be required to answer many of the outstanding questions of Martian chronology. Doubtless they, too, will hold surprises for unwary analysts.

For the above reasons, it is critical to determine absolute ages for several broad, homogeneous Martian surface units. This could be done by sample return, and even crude (20–50% uncertainty) in situ methods would be valuable, but the surface unit must be chosen carefully to reduce the error bar in ages of currently unknown units.

To calibrate the system in this way, Noachian units are less useful; their high crater densities already give some confidence that they formed before about 3.5 or 3.7 Gyr ago (see Hartmann and Neukum, 2001). The cratering rate is believed to have been changing rapidly before that time, so units with a wide range of crater densities may have about the same age, just as on the Moon. Sampling of regional and globally widespread units such as Hesperian-aged ridged plains (Hr) would provide absolute chronological information on the key transition between the Noachian and the Hesperian (Figure 2). The very widespread nature of this unit and the well preserved craters and crater ages would have the benefit of providing a global datum somewhat analogous to the Imbrium basin ejecta on the Moon. Also extremely useful would be dating of units near the Hesperian/Amazonian boundary (e.g. to address the question does this boundary fall at 1.8 Gyr, 3.5 Gyr, or somewhere in between) and broad units, such as lava flows, that are clearly in a specific Amazonian epoch. Widespread and accessible volcanic units of the Tharsis region offer such opportunities. Interestingly, a sample return from such a unit may seem less desirable on other scientific grounds, such as the search for water or for evidence of life. For example, there is much interest in testing environmental and biological conditions represented by Noachian samples. However, examination of Figure 2 shows that exploration of Mars requires a coherent scientific strategy which must include the establishment of a firm global absolute chronology. Too narrow a focus could result in no improvement over the present understanding of the absolute age scale for the geologic history of Mars, and thus no improved context for the rates and timing of fundamental geologic processes such as volcanism, river and lake formation and tectonism. Therefore, sample return missions and in situ dating methods, need to include assessment of these critical relative time markers.

In summary, the geologic history of Mars is characterized by rock sequences (geologic units) which are derived from the topographic, geomorphologic and spectral characteristics of remotely sensed surface features (Tanaka *et al.*, 1992). The stratigraphic position of geologic units is estimated by the means of superposition and intersection and by the concentration of impact craters superposed on geologic units. Due to the lack of samples from Mars of known provenance and context, the assignment of absolute ages to the epochs based on crater densities is dependent on cratering rates (Tanaka *et al.*, 1992; Neukum and Hiller, 1981; Hartmann, 1978; Neukum and Wise, 1976) and thus is model dependent. Different models define the Amazonian-Hesperian boundary between 1.8–3.5 Gyr. and the Hesperian-Noachian boundary between 3.5–3.8 Gyr. (Hartmann *et al.*, 1981; Neukum and Wise, 1976; Neukum and Hiller, 1981).

Hartmann and Neukum (2001) have reduced the formal uncertainty and disagreement between their two systems, but they and Ivanov (2001) emphasize that the ages are still proportional to 1/(Mars cratering rate) which is still uncertain by as much as a factor of two. These large uncertainties in the Martian absolute chronology prevent a correct interpretation of the initial stage and duration of geologic processes such as volcanism, tectonism, erosion, formation of channels and valley networks, glaciation and resurfacing by wind and their implications for the climate evolution. The next stage of the exploration of Mars must strive to resolve these fundamental uncertainties.

Acknowledgements

This paper is dedicated to the memory of David H. Scott, whose careful and diligent work and unassuming sharing of his knowledge has significantly helped to provide a first-order understanding of the geological evolution of Mars. We also thank Elizabeth R. Fuller, Mikhail Ivanov, Anne Cote, William Hartmann, Baerbel Lucchitta, Ken Tanaka, Reinald Kallenbach, and Ursula Pfander, for contributing directly and indirectly to the preparation and review of this contribution.

References

Aharonson, O., Zuber, M.T., and Rothman, D.H.: 2001, "Statistics of Mars' Topography from MOLA: Slopes, Correlations, and Physical Models', *J. Geopys. Res.* **106**, in press.
Anderson, R.C., and Dohm, J.M.: 2000, 'Magmatic-Tectonic Evolution of Tharsis', *31st Lunar Planet. Sci.*, LPI, Houston, abstract #1607 (CD-ROM).

Anderson, R.C., Golombek, M.P., Franklin, B.J., Tanaka, K.L., Dohm, J.M., Lias, J.H., and Peer, B.: 1998, 'Centers of Tectonic Activity Through Time for the Western Hemisphere of Mars', *29th Lunar Planet. Sci. Conf.*, LPI, Houston, abstract #1881 (CD-ROM).

Anderson, R.C., *et al.*: 2000, 'Primary Centers and Secondary Concentrations of Tectonic Activity Through Time in the Western Hemisphere of Mars', *J. Geophys. Res*, submitted.

Baker, V.R., and Kochel, R.C.: 1979, 'Morphometry of Streamlined Forms in Terrestrial and Martian Channels', *Proc. 9th Lunar Planet. Sci. Conf.* **9**, 3181–3193.

Baker, V.R., Strom, R.G., Gulick, V.C., Kargel, J.S., Komatsu, G., and Kale, V.S.: 1991, 'Ancient Oceans, Ice Sheets and the Hydrological Cycle on Mars', *Nature* **352**, 589–594.

Baker, V.R., Carr, M.H., Gulick, V.C., Williams, C.R., and Marley, M.S.: 1992, 'Channels and Valley Networks', *in* H.H. Kieffer, B.M. Jakosky, C.W. Snyder, and M.S. Matthews (eds.), *Mars*, Univ. Arizona Press, Tucson, pp. 493–522.

Banerdt, W.B., and Golombek, M.P.: 2000, 'Tectonics of the Tharsis Region, Insights from MGS Topography and Gravity', *31st Lunar Planet. Sci. Conf.*, abstract #2038.

Banerdt, W.B., Phillips, R.J., Saunders, R.S., and Sleep, N.H.: 1982, 'Thick Shell Tectonics on One-Plate Planets: Application to Mars', *J. Geophys. Res.* **87**, 9723–9733.

Banerdt, W.B., Golombek, M.P., and Tanaka, K.L.: 1992, 'Stress and tectonics on Mars', *in* H.H. Kieffer *et al.* (eds.), *Mars*, Univ. Arizona Press, Tucson, pp. 249–297.

Barlow, N.: 1988, 'Parameters affecting formation of Martian impact crater ejecta morphology', 19th Lunar Planet. Sci., LPI, Houston, 29–30.

Bibring, J.-P., and Erard, S.: 2001, 'The Mars Surface Composition', *Space Sci. Rev.*, this volume.

Bridges, J.C., Catling, D.C., Saxton, J.M., Swindle, T.D., Lyon, I.C., and Grady, M.M.: 2001, 'Alteration Assemblages in Martian Meteorites: Implications for Near-Surface Processes', *Space Sci. Rev.*, this volume.

Cabrol, N.A. and Grin, E.A.: 2001, 'The evolution of lacustrine environemnts on Mars: Is Mars only hydrologically dormant?', *Icarus* **149**, 291–328.

Carr, M.H.: 1973, 'Volcanism on Mars', *J. Geophys. Res.* **78**, 4049–4062.

Carr, M.H.: 1974a, 'The Role of Lava Erosion in the Formation of Lunar Rilles and Martian Channels', *Icarus* **22**, 1–23.

Carr, M.H.: 1974b, 'Tectonism and Volcanism of the Tharsis Region of Mars', *J. Geophys. Res.* **79**, 3943–3949.

Carr, M.H.: 1979, 'Formation of Martian Flood Features by Release of Water from Confined Aquifers', *J. Geophys. Res.* **84**, 2995–3007.

Carr, M.H.: 1981, *The Surface of Mars*, Yale Univ. Press, New Haven.

Carr, M.H.: 1983, 'The Stability of Streams and Lakes on Mars', *Icarus* **56**, 476–495.

Carr, M.H.: 1986, 'Mars: A Water-rich Planet?', *Icarus* **56**, 187–216.

Carr, M.H.: 1995, 'The Martian Drainage System and the Origin of Valley Networks and Fretted Channels', *J. Geophys. Res.* **100**, 7479–7507.

Carr, M.H.: 1996, *Water on Mars*, Oxford Univ. Press, New York, 229 pp.

Carr, M.H.: 2001, 'Mars Global Surveyor observations of Martian fretted terrain', *J. Geophys. Res.* in press.

Carr, M.H., and Clow, G.D.: 1981, 'Martian Channels and Valleys: Their Characteristics, Distribution and Age', *Icarus* **48**, 91–117.

Carr, M.H., and Malin, M.C.: 2000, 'Meter-scale Characteristics of Martian Channels and Valleys', *Icarus* **146**, 366–386.

Carr, M.H., Greeley, R., Blasius, K.R., Guest, J.E. and Murray, J.B.: 1977b, 'Some Martian Volcanic Features as Viewed from the Viking Orbiters', *J. Geophys. Res.* **82**, 3985–4015.

Carr, M.H. *et al.*: 1980, *Viking Orbiter Views of Mars*, U.S. Government Printing Office, Washington, D.C., 136 pp.

Carr, M.H., Wu, S.S.C., Jordan, R., and Schafer, F.J.: 1987, 'Volumes of Channels, Canyons and Chaos in the Circum-Chryse Region of Mars', *Proc. 18th Lunar Planet. Sci. Conf.*, 155 (abstract).

Cattermole, P.: 1987, 'Sequence, Rheological Properties, and Effusion Rates of Volcanic Flows at Alba Patera, Mars', *J. Geophys. Res.* **92**, 553–560.

Chapman C.R.: 1974, 'Cratering on Mars. I. Cratering and Obliteration History', *Icarus* **22**, 272–291.

Chicarro, A.F., Schultz, P.H., and Masson, P.: 1985, 'Global and Regional Ridge Patterns on Mars', *Icarus* **63**, 153–174.

Christiansen, E.H.: 1989, 'Lahars in the Elysium Region of Mars', *Geology* **17**, 203–206.

Christensen, P.R., Bandfield, J.L., Smith, M.D., Hamilton, V.E. and Clark, R.N.: 2000, 'Identification of a Basaltic Component on the Martian Surface from Thermal Emission Spectrometer Data', *J. Geophys. Res.* **105**, 9609–9621.

Clifford, S.M.: 1993, 'A model for the Hydrologic and Climatic Behaviour of Water on Mars', *J. Geophys. Res.* **98**, 10,973–11,016.

Clifford, S.M., and Parker, T.J.: 1999, 'The Evolution of the Martian Hydrosphere: Implications for the Fate of a Primordial Ocean and the Current State of the Northern Plains', *Fifth Int. Conf. on Mars, July 19-24, 1999, Pasadena, CA*, abstract #6236.

Connerney, J.E.P., *et al.*: 1999, 'Magnetic Lineations in the Ancient Crust of Mars', *Science* **284**, 794–798.

Crown, D.A. and Greeley, R.: 1993, 'Volcanic geology of Hadriaca Patera and the esstern Hellas region of Mars', *J. Geophys. Res.* **98**, 3431–3451.

Crumpler, L.S., Head, J.W., and Aubele, J.C.: 1996, 'Calderas on Mars: Characteristics, Structure, and Associated Flank Deformation', *in* W.J. McGuire *et al.*(eds.), *Volcano Instability on Earth and Other Planets*, Geol. Soc. London Spec. Pub. 110, pp. 307–348.

Cutts, J.A., and Blasius, K.R.: 1981, 'Origin of Martian Outflow Channels: The Eolian Hypothesis' *J. Geophys. Res.* **86**, 5075–5102.

Dohm, J.M. and Tanaka, K.L.: 1999, 'Geology of the Thaumasia Region, Mars: Plateau Development, Valley Origins, and Magmatic Evolution', *Planet. Space Sci.* **47**, 411–431.

Edgett, K.S. and Malin, M.C.: 2000, 'New Views of Mars Eolian Activity, Materials, and Surface Properties: Three Vignettes from the Mars Global Surveyor Mars Obiter Camera', *J. Geophys. Res.* **105**, 1623–1650.

Fishbaugh, K.E., and Head, J.W.: 2000, 'North Polar Region of Mars: Topography of Circumpolar Deposits from Mars Orbiter Laser Altimeter (MOLA) Data and Evidence for Asymmetric Retreat of the Polar Cap', *J. Geophys. Res.* **105**, 22,455–22,486.

Frey, H.: 1979, 'Thaumasia: A Fossilized Early Forming Tharsis Uplift', *J. Geophys. Res.* **84**, 1009–1023.

Frey, H., and Schultz, R.A.: 1988, 'Large Impact Basins and the Mega-Impact Origin for the Crustal Dichotomy on Mars', *Geophys. Res. Lett.* **15**, 229–232.

Frey, H., Lowry, B.L., and Chase, S.A.: 1979, 'Pseudocraters on Mars', *J. Geophys. Res.* **84**, 8075–8068.

Frey, H.V., Shockey, K.M., Frey, E.L., Roark, J.H., and Sakimoto, S.E.H.: 2001, 'A Very Large Population of Likely Buried Impact Basins in the Northern Lowlands of Mars Revealed by MOLA Data', *Proc. 32nd Lunar Planet. Sci. Conf.*, abstract #1680.

Ghatan, G.J., and Head, J.W.: 2001, 'Candidate Subglacial Volcanoes in the South Polar Region of Mars', *32nd Lunar Planet. Sci. Conf.*, abstract #1039 (CD-ROM).

Golombek, M.P.: 1989, 'Geometry of Stresses around Tharsis on Mars', *20th Lunar Planet. Sci. Conf.*, LPI, Houston, 345–346.

Golombek, M.P. and Banerdt, W.B.: 1999, 'Recent Advances in Mars Tectonics', *Fifth Int. Mars Conf.* (abstract #6020), LPI Contribution No. **972**, LPI, Houston, abstract #6020 (CD-ROM).

Golombek, M.P., Plescia, J.B., and Franklin, B.J.: 1991, 'Faulting and Folding in the Formation of Planetary Wrinkle Ridges', *Proc. 21st Lunar Planet. Sci. Conf.*, 679–693.

Greeley, R.: 1973, 'Mariner 9 Photographs of Small Volcanic Structures on Mars', *Geology* **1**, 175–180.

Greeley, R., and Guest, J.E.: 1987, *Geologic Map of the Eastern Equatorial Region of Mars*, scale 1:15,000,000, U.S.G.S. Misc. Inv. Series Map I-1802-B

Greeley, R., and Schneid, B.D.: 1991, 'Magma Generation on Mars: Amounts/rates, and Comparisons with Earth, Moon, and Venus', *Science* **254**, 996–998.

Greeley, R., and Spudis, P.D.: 1978, 'Volcanism in the Cratered Terrain Hemisphere of Mars', *J. Geophys. Res.* **5**, 453–455.

Greeley, R., and Spudis, P.D.: 1981, 'Volcanism on Mars', *Rev. Geophys. Space Phys.* **19**, 13–41.

Greeley, R., Skypeck, A., and Pollack, J.B.: 1993, 'Martian Aeolian Features and Deposits: Comparisions with General Circulation Model Results', *J. Geophys. Res.* **98**, 3183–3193.

Greeley, R., Bridges, N.T.,Crown, D.A., Crumpler, L.S., Fagents, S.A., Mouginis-Mark, P.J. and Zimbleman, J.R.: 2000, 'Volcanism on the Red Planet Mars', *in* J.R. Zimbelman and T.K.P. Gregg (eds.), *Environmental Effects on Volcanic Eruptions: from Deep Oceans to Deep Space*, Kluwer Academic/Plenum Publ., New York, pp. 75–112.

Greeley, R., Kuzmin, R.O., and Haberle, R.M.: 2001, 'Aeolian Processes and Their Effects on Understanding the Chronology of Mars', *Space Sci. Rev.*, this volume.

Gulick, V.C., and Baker, V.R.: 1990, 'Origin and Evolution of Valleys on Martian Volcanoes', *J. Geophys. Res.* **95**, 14,325–14,344.

Gulick, V.C., Tyler, D., McKay, C.P., and Haberle, R.M.: 1997, 'Episodic ocean-induced CO_2 greenhouse on Mars: Implications for fluvial valley formation', *Icarus* **130**, 68–86.

Halliday, A.N., Wänke, H., Birck, J.-L., and Clayton, R.N.: 2001, 'The Accretion, Composition and Early Differentiation of Mars', *Space Sci. Rev.*, this volume.

Hamlin, S.E., Kargel, J.S., Tanaka, K.L., Lewis, K.J., and MacAyeal, D.R.: 2000, 'Preliminary Studies of Icy Debris Flows in the Martian Fretted Terrain', *31st Lunar and Planet. Sci. Conf.*, abstract #1785 (CD-ROM).

Hartmann, W.K.: 1971, 'Martian Cratering III: Theory of Crater Obliteration', *Icarus* **15**, 410–428.

Hartmann, W.K.: 1973, 'Martian Surface and Crust: Review and Synthesis', *Icarus* **19**, 550–575.

Hartmann, W.K.: 1978, 'Martian Cratering V: Toward an Empirical Martian Chronology, and its Implications', *Geophys. Res. Lett.* **5**, 450–452.

Hartmann, W.K.: 2001, 'Martian Seeps and Their Relation to Youthful Geothermal Activity', *Space Sci. Rev.*, this volume.

Hartmann, W.K., and Berman, D.C.: 2000, 'Elysium Planitia Lava Flows: Crater Count Chronology and Geological Implications', *J. Geophys. Res.* **105**, 15,011–15,025.

Hartmann, W.K., and Neukum, G.: 2001, 'Cratering Chronology and the Evolution of Mars', *Space Sci. Rev.*, this volume.

Hartmann, W.K., *et al.*: 1981, 'Chronology of Planetary Volcanism by Comparative Studies of Planetary Cratering', *Basaltic Volcanism on the Terrestrial Planets*, Pergamon, New York, pp. 1049–1127.

Head, J.W.: 2000a, 'Tests for Ancient Polar Deposits on Mars: Morphology and Topographic Relationships of Esker-like Sinuous Ridges (Dorsa Argentea) Using MOLA Data', *Proc. 31st Lunar Planet. Sci. Conf.*, abstract #1117 (CD-ROM).

Head, J.W.: 2000b, 'Tests for Ancient Polar Deposits on Mars: Assessment of Morphology and Topographic Relationships of Large Pits (Angusti and Sisyphi Cavi) using MOLA Data', *Proc. 31st Lunar Planet. Sci. Conf.*, abstract #1118 (CD-ROM).

Head, J.W., and Pratt, S.: 2001, 'Extensive South Polar Cap in Middle Mars History: Analysis using Mars Orbiter Laser Altimeter (MOLA) Data', *J. Geophys. Res.*, in review.

Head, J.W., Kreslavsky, M.A., and Pratt, S.: 2001, 'Northern Lowlands of Mars: Evidence for Widespread Volcanic Flooding and Tectonic Deformation in the Hesperian Period', *J. Geophys. Res.*, in review.

Head, J.W., Hiesinger, H., Ivanov, M.A., Kreslavsky, M.A., Pratt, S., and Thomson, B.J.: 1999, 'Possible Ancient Oceans on Mars: Evidence from Mars Orbiter Laser Altimeter Data', *Science* **286**, 2134–2137.

Herkenhoff, K.E., and Plaut, J.J.: 2000, 'Surface Ages and Resurfacing Rates of the Polar Layered Deposits on Mars', *Icarus* **144**, 243–253.

Hess, P.C. and Parmentier, E.M.: 2001, 'Implications of magma ocean cumulate overturn for Mars', 32nd Lunar and Planetary Sci. Conf., abstract #1319, (CD-ROM).

Hiesinger, H. and Head, J. W.: 2000, 'Characteristidcs and origin of polygonal terrain in southern Utopia Planitia, Mars: Results from Mars Orbiter Laser Altimeter and Mars Orbiter Camera data', *J. Geophys. Res.* **105**, 11999–12022.

Ivanov, B.: 2001, 'Mars/Moon Cratering Rate Ratio Estimates', *Space Sci. Rev.*, this volume.

Ivanov, M. and Head, J. W.: 2001, 'Chryse Planitia, Mars: Topographic configuration from MOLA data and tests for hypothesized lakes and shorelines', *J. Geophys. Res.*, in press.

Jones, K.L.: 1974, 'Evidence for an Episode of Crater Obliteration Intermediate in Martian History', *J. Geophys. Res.* **79**, 3917–3931.

Kargel, J.S. and Strom, R.G.: 1990, 'Ancient Glaciation on Mars', *Proc. 21st Lunar Planet. Sci. Conf.*, 597–598.

Kargel, J.S., Baker, V.R., Begét, J.E., Lockwood, J.F., Péwé, T.L., Shaw, J.S., and Strom, R.G.: 1996, 'Evidence of Ancient Continental Glaciation in the Martian Northern Plains', *J. Geophys. Res.* **100**, 5351–5368.

Keszthelyi, L.,McEwen, A.S., and Thordarson, T.: 2000, 'Terrestrial Analogs and Thermal Models for Martian Flood Lavas', *J. Geophys. Res.* **105**, 15,027–15,049.

Kuzmin, R.O., Greeley, R., Rafkin, S.C.R., and Haberle, R.: 2001, 'Wind Related Modification of Small Impact Craters on Mars', *Icarus*, in review.

Kreslavsky, M.A., and Head, J.W.: 1999, 'Kilometer-scale Slopes on Mars and their Correlation with Geologic Units: Initial Results from Mars Orbiter Laser Altimeter (MOLA) Data', *J. Geophys. Res.* **104**, 21,911–21,924.

Kreslavsky, M.A., and Head, J.W.: 2000, 'Kilometer-scale Roughness of Mars: Results from MOLA Data Analysis', *J. Geophys. Res.* **105**, 26,695–26,711.

Lucchitta, B.K., and Anderson, D.M.: 1980, 'Martian Outflow Channels Sculptured by Glaciers', *in* Reports of Planetary Geology Program 1980, NASA TM-81776, 271–273.

Lucchitta, B.K., Ferguson, H.M. and Summers, C.: 1986, 'Sedimentary deposits in the northern lowland plains', Mars, *J. Geophys. Res.* **91**, E166–E174.

Lucchitta, B.K., McEwen, A.S., Clow, G.D., Geissler, P.E., Singer, R.B., Schultz, R.A., and Squyres, S.W.: 1992, 'The canyon system on Mars', *in* H.H. Kieffer, B.M. Jakosky, C.W. Snyder, and M.S. Matthews (eds.), *Mars*, Univ. Arizona Press, Tucson, pp. 453–492.

Malin, M.C., and Edgett, K.S.: 1999, 'MGS MOC the First Year: Geomorphic Processes and Landforms', *Proc. 30th Lunar Planet. Sci. Conf.*, abstract #1028.

Malin, M.C., and Edgett, K.S.: 2000, 'Evidence for Recent Groundwater Seepage and Surface Runoff on Mars', *Science* **288**, 2330–2335.

Malin, M.C. *et al.*: 1998, 'Early Views of the Martian Surface from the Mars Orbiter Camera of Mars Global Surveyor', *Science* **279**, 1681–1685.

Mangold, N. and Allemand, P.: 2001, 'Topographic analysis of features related to ice on Mars', *Geophys. Res. Lett.* **28**, 407–410.

Mangold, N., Allemand, P., and Peulvast, J.-P.: 2000, 'Topography of Ice-related Features on Mars', *31st Lunar Planet. Sci. Conf.*, abstract #1131 (CD-ROM).

Mars Channel Working Group: 1983, 'Channels and Valleys on Mars', *Geol. Soc. Amer. Bull.* **94**, 1035–1054.

Masson, P., Carr, M.H., Costard, F., Greeley, R., Hauber, E., and Jaumann, R.: 2001, 'Geomorphologic Evidence for Liquid Water', *Space Sci. Rev.*, this volume.

Masursky, H., Boyce, J.V., Dial, A.L., Jr., Schaber, G.G., and Strobell, M.E.: 1977, 'Classification and Time of Formation of Martian Channels based on Viking Data', *J. Geophys. Res.* **82**, 4016–4037.

Maxwell, T.A.: 1982, 'Orientation and Origin of Ridges in the Lunae Palus — Coprates Region of Mars', *J. Geophys. Res.* **87**, 97–108.

McCauley, J.F., Carr, M.H., Cutts, J.A., Hartmann, W.K., Masursky, H.,Milton, D.J., Sharp, R.P. and Wilhelms, D.E.: 1972, 'Preliminary Mariner 9 Report on the Geology of Mars', *Icarus* **17**, 289–327, 1972.

McEwen, A.S., Malin, M.C., Carr, M.H., and Hartmann, W.K.: 1999, 'Voluminous Volcanism on Early Mars Revealed in Valles Marineris', *Nature* **397**, 584–586.

McGill, G.E.: 1989, 'Buried Topography of Utopia, Mars: Persistence of a Giant Impact Depression', *J. Geophys. Res.* **94**, 2753–2759.

McKay, C.P., and Davis, W.L.: 1991, 'Duration of Liquid Water Habitats on Early Mars', *Icarus* **90**, 214–221.

McSween, H.Y., Jr., *et al.*: 1998, 'Chemical, Multispectral, and Textural Constrains on the Composition and Origin of Rocks at the Mars Pathfinder Landing Site', *J. Geophys. Res.* **104**, 8679–8716.

Mége, D. and Masson, P.: 1996a, 'Stress Models for Tharsis Formation, Mars', *Planet. Space Sci.* **44**, 1471–1497.

Mége, D. and Masson, P.: 1996b, 'A Plume Tectonics Model for the Tharsis Province, Mars.' *Planet. Space Sci.* **44**, 1499–1546.

Mellon, M.T. and Jakosky, B.M.: 1995, 'The distribution and behavior of martian ground ice during past and present epochs', *J. Geophys. Res.* **100**, 11781–11799.

Mouginis-Mark, P.J.: 1979, 'Martian Fluidized Crater Morphology: Variations with Crater Size, Latitude, Altitude, and Target Material', *J. Geophys. Res.* **84**, 8011–8022.

Mouginis-Mark, P.J., Wilson, L., and Zimbelman, R.J.: 1988, 'Polygenic eruptions on Alba Patera, Mars: Evidence of channel erosion on pyroclastic flows', *Bull. Vol.* **50**, 361–379.

Mouginis-Mark, P.J., Wilson, L. and Zuber, M.T.: 1992, 'The Physical Volcanology of Mars', *in* H.H. Kieffer *et al.* (eds.), *Mars*, Univ. Arizona Press, Tucson, pp. 424–452.

Nelson, D.M., and Greeley, R.: 1999, 'Geology of Xanthe Terra Outflow Channels and the Mars Pathfinder Landing Site', *J. Geophys. Res.* **104**, 8653–8670.

Neukum, G., and Hiller, K.: 1981, 'Martian Ages', *J. Geophys. Res.* **86**, 3097–3121

Neukum, G., and Wise, D.U.: 1976, 'Mars: Standard Crater Curve and Possible New Time Scale', *Science* **194**, 1381–1387.

Neukum, G., Ivanov, B., Hartmann, W.K.: 2001, 'Cratering Records in the Inner Solar System in Relation to the Lunar Reference System', *Space Sci. Rev.*, this volume.

Nummedal, D.: 1978, 'The Role of Liquefaction in Channel Development on Mars', Rep. Planet. Geol. Geophys. Program-1977-1988, NASA TM–79729, 257–259.

Nummedal, D., and Prior, D.B.: 1981, 'Generation of Martian Chaos and Channels by Debris Flow', *Icarus* **45**, 77–86.

Nyquist, L.E., Bogard, D.D., Shih, C.-Y., Greshake, A., Stöffler, D., and Eugster, O.: 2001, 'Ages and Geologic Histories of Martian Meteorites', *Space Sci. Rev.*, this volume.

Parker, T.J., Saunders, R.S., and Schneeberger, D.M.: 1989, 'Transitional Morphology in the West Deuteronilus Mensae Region of Mars: Implications for Modification of the Lowland/upland Boundary', *Icarus* **82**, 111–145.

Parker, T.J., Gorsline, D.S., Saunders, R.S., Pieri, D.C., and Schneeberger, D.M.: 1993, 'Coastal Geomorphology of the Martian Northern Plains', *J. Geophys. Res.* **98**, 11,061–11,078.

Phillips, R.J., Saunders, R.S., and Conel, J.E.: 1973, 'Mars: Crustal Structure Inferred from Bouguer Gravity Anomalies', *J. Geophys. Res.* **78**, 4815–4820.

Phillips, *et al.*: 2001, 'Ancient Geodynamics and Global-scale Hydrology on Mars', *Science* **291**, 2587–2591.

Pieri, D.C.: 1980, 'Martian Valleys: Morphology, Distribution, Age and Origin', *Science* **210**, 895–897.

Plescia, J.B., and Golombek, M.P.: 1986, 'Origin of Planetary Wrinkle Ridges Based on the Study of Terrestrial Analogs', *Geol. Soc. Amer. Bull.* **97**, 1289–1299.

Plescia, J.B., and Saunders, R.S.: 1982, 'Tectonic History of the Tharsis Region, Mars', *J. Geophys. Res.* **87**, 9775–9791.

Pruis, M.J., and Tanaka, K.L.: 1995, *26th Lunar Planet. Sci.*, 1147–1148.

Schonfeld, E.: 1976, 'On the Origin of Martian Channels', *Eos: Trans. AGU* **57**, 948.

Schultz, R.A.: 1991, 'Structural Development of Coprates Chasma and Western Ophir Planum, Valles Marineris Rift, Mars', *J. Geophys. Res.* **96**, 22,777–22,792.

Schultz, R.A.: 1998, 'Multiple-Process Origin of Valles Marineris Basins and Troughs, Mars', *Planet. Space Sci.* **46**, 827–834.

Schultz, P.H. and Gault, D.E.; 1979, 'Atmospheric effects on martian ejecta emplacement', *J. Geophys. Res.* **84**, 7669–7687.

Schultz, R.A., and Tanaka, K.L.: 1994, 'Lithospheric-scale Buckling and Thrust Structures on Mars: the Coprates Rise and South Tharsis Ridge Belt', *J. Geophys. Res.* **99**, 8371–8385.

Scott, D.H., and Tanaka, K.L.: 1986, Geologic Map of the Western Equatorial Region of Mars, scale 1:15,000,000, *U.S.G.S. Misc. Inv. Ser. Map* I-1802-A.

Scott, D.H., and Dohm, J.M.: 1990, 'Chronology and Global Distribution of Fault and Ridge Systems on Mars', *Proc. 20th Lunar Planet. Sci. Conf.*, 487–501; 'Faults and Ridges: Historical Development in Tempe Terra and Ulysses Patera Regions of Mars', *ibid*, 503–513.

Scott, D.H., and Dohm, J.M.: 1992, 'Mars Highland Channels: an Age Reassessment', *Proc. 23th Lunar Planet. Sci. Conf.*, #1251.

Sharp, R.P., and Malin, M.C.: 1975, 'Channels on Mars', *Geol. Soc. Amer. Bull.* **86**, 593–609.

Sleep, N.H.: 1994, 'Martian Plate Tectonics', *J. Geophys. Res.* **99**, 5639–5655.

Sleep, N.H., and Phillips, R.J.: 1979, 'An Isostatic Model for the Tharsis Province, Mars', *Geophys. Res. Lett.* **6**, 803–806.

Sleep, N.H., and Phillips, R.J.: 1985, 'Gravity and Lithospheric Stress on the Terrestrial Planets with Reference to the Tharsis Region of Mars', *J. Geophys. Res.* **90**, 4469–4489.

Smith, D.E., and Zuber, M.T.: 1999, 'The Relationship of the MOLA Topography of Mars to the Mean Atmospheric Pressure', *31st DPS meeting, Am. Astron. Soc.*, abstract #67.02.

Smith, D.E., *et al.*: 1998, 'Topography of the Northern Hemisphere of Mars from the Mars Obiter Laser Altimeter', *Science* **279**, 1686–1692.

Smith, D.E., *et al.*: 1999a, 'The Global Topography of Mars and Implication for Surface Evolution', *Science* **284**, 1495–1503.

Smith, D.E., Sjogren, W.L., Tyler, G.L., Balmino, G., Lemoine, F.G., and Konopliv, A.S.: 1999b, 'The Gravity Field of Mars: Results from Mars Global Surveyor', *Science* **286**, 94–97.

Soderblom, L., Condit, C., West, R., Herman, B., and Kreidler, T.: 1974, 'Martian Planetwide Crater Distributions: Implications for Geologic History and Surface Processes', *Icarus* **22**, 239–263.

Solomon, S.C., and Head, J.W.: 1982, 'Evolution of the Tharsis Province of Mars: The Importance of Heterogeneous Lithospheric Thickness and Volcanic Construction', *J. Geophys. Res.* **87**, 9755–9774.

Spohn, T., Acuña, M.H., Breuer, D., Golombek, M., Greeley, R., Halliday, A.N., Hauber, E., Jaumann, R., and Sohl, F.: 2001, 'Geophysical Constraints on the Evolution of Mars', *Space Sci. Rev.*, this volume.

Squyres, S.W., and Carr, M.H.: 1986, 'Geomorphic Evidence for the Distribution of Ground Ice on Mars', *Science* **231**, 249–252.

Stöffler, D., and Ryder, G.: 2001, 'Stratigraphy and Isotope Ages of Lunar Geologic Units: Chronological Standard for the Inner Solar System', *Space Sci. Rev.*, this volume.

Strom, R.G., Croft, S.K., and Barlow, N.G.: 1992, 'The Martian impact cratering record', *in* H.H. Kieffer, B.M. Jakosky, C.W. Snyder, and M.S. Matthews (eds.), *Mars*, Univ. Arizona Press, Tucson, pp. 383–423.

Tanaka, K.L.: 1986, 'The Stratigraphy of Mars', *J. Geophys. Res.* **91, Suppl.**, 139–158.

Tanaka, K.L.: 1990, 'Tectonic History of the Alba Patera-Ceraunius Fossae Region of Mars', *Proc. 19th Lunar Planet. Sci. Conf.*, 515–523.

Tanaka, K.L.: 1995, *Mercury* **24**, 11.

Tanaka, K.L., and Davis, P.A.: 1988, 'Tectonic History of the Syria Planum Province of Mars', *J. Geophys. Res.* **93**, 14,893–14,917.

Tanaka, K.L., and Scott, D.H.: 1987, Geologic Map of the Polar Regions of Mars. *U.S. Geologic Survey Miscellaneous Investigations Map* I-1802-C, scale 1:15,000,000.

Tanaka, K.L., Isbell, N., Scott, D., Greeley, R., and Guest, J.: 1987, 'The Resurfacing History of Mars: A Synthesis of Digitized Viking-based Geology', *Proc. 18th Lunar Planet. Sci. Conf.*, 665–678.

Tanaka, K.L., Golombek, M.P., and Banerdt, W.B.: 1991, 'Reconciliation of Stress and Structural Histories of the Tharsis Region of Mars', *J. Geophys. Res.* **96**, 15,617–15,633.

Tanaka, K.L., Scott, D.H., and Greeley, R.: 1992, 'Global Stratigraphy', *in* H.H. Kieffer *et al.* (eds.), *Mars*, Univ. Arizona Press, Tucson, pp. 354–382.

Tricart, J.: 1968, 'Periglacial Landscapes', *in* R.W. Fairbridge (ed.), *Encyclopedia of Geomorphology*, Reinhold Book Co., pp. 829–833.

Tricart, J.: 1969, *Geomorphology of Cold Environments*, Macmillan, London, 320 pp.

Turcotte, D.L.: 1999, 'Tectonics and Volcanism on Mars: What Do We and What Don't We Know?', *Proc. 30th Lunar Planet. Sci. Conf.*, LPI, Houston, abstract #1187.

Watters, T.R.: 1988, 'Wrinkle Ridge Assemblages on the Terrestrial Planets', *J. Geophys. Res.* **93**, 15,599–15616.

Watters, T.R.: 1993, 'Compressional Tectonism on Mars', *J. Geophys. Res.* **98**, 17,049–17,060.

Watters, T.R., and Maxwell, T.A.: 1983, 'Crosscutting Relations and Relative Ages of Ridges and Faults in the Tharsis Region of Mars', *Icarus* **56**, 278–298.

Watters, T.R., and Maxwell, T.A.: 1986, 'Orientation, Relative Age, and Extent of the Tharsis Plateau Ridge System', *J. Geophys. Res.* **91**, 8113–8125.

Wilhelms, D.E., and Squyres, S.W.: 1984, 'The Martian Hemispheric Dichotomy May Be Due to a Giant Impact', *Nature* **309**, 138–140.

Willeman, R.J., and Turcotte, D.L.: 1982, 'The Role of Lithospheric Stress in the Support of the Tharsis Rise', *J. Geophys. Res.* **82**, 9793–9801.

Wilson, L., and Head, J.W.: 1994, 'Mars: Review and Analysis of Volcanic Eruption Theory and Relationships to Observed Landforms', *Rev. Geophysics* **32**, 221–263.

Wilson, L. and Head, J. W.: 2000, 'Tharsis-radial graben systems as the surface manifestation of plume-related dike intrusion complexes: Models and implications', 31st Lunar and Planetary Sci. Conf., abstract #1371, (CD-ROM).

Wilson, L. and Head, J. W.: 2001, 'Giant dike swarms and related graben systems in the Tharsis province of Mars', 32nd Lunar and Planetary Sci. Conf., abstract #1153, (CD-ROM).

Wise, D.U., Golombek, M.P., and McGill, G.E.: 1979a, 'Tectonic Evolution of Mars', *J. Geophys. Res.* **84**, 7934–7939.

Wise, D.U., Golombek, M.P., and McGill, G.E.: 1979b, 'Tharsis Province of Mars: Geologic Sequence, Geometry and a Deformation Mechanism', *Icarus* **38**, 456–472.

Yung, Y.L., and Pinto, J.P.: 1978, 'Primitive Atmosphere and Implications for the Formation of Channels on Mars', *Nature* **288**, 735–738.

Zuber, M.T., and Smith, D.E.: 1999, 'Interpretation of New Gravity and Topography Data for Mars', *31st DPS meeting, Am. Astron. Soc.*, abstract #43.03.

Zuber, M.T., *et al.*: 2000, 'Internal Structure and Early Thermal Evolution of Mars from Mars Global Surveyor Topography and Gravity', *Science* **287**, 1788–1793.

Address for correspondence: James W. Head, Department of Geological Sciences, Brown University, Providence, RI 02912 USA; (james_head_iii@brown.edu)

THE MARTIAN SURFACE COMPOSITION

JEAN-PIERRE BIBRING and STÉPHANE ERARD

Institut d'Astrophysique Spatiale, 91405 Orsay Campus, France

Submitted: 1 December 2000; accepted: 13 February 2001

Abstract. Mars is unique to have undergone all planetary evolutionary steps, without global resets, till its geological death: this is reflected in the variety of its surface features. The determination of Mars surface composition has thus the potential to identify the processes responsible for the entire Mars evolution, from geological timescales to seasonal variations. Due to technical challenges, only few investigations have been performed so far. They are summarized in this paper, and their interpretation is discussed in terms of surface materials (minerals, ices and frosts).

1. Introduction

1.1. MARS SURFACE ANALYSES AND COMPARATIVE PLANETOLOGY

Due to its size and orbital parameters, Mars plays a unique role in comparative planetology. Unlike the Moon, for which the activity ceased soon after the first billion years, Mars is large enough to have undergone all major planet evolutionary steps, including a period of much internal activity giving rise to impressive volcanic and tectonic surface features. On the other hand, Mars is small enough not to have been remodeled by global surface resets, as the Earth was. Its intermediate mass thus led Mars to record all major events, such as: early heavy bombardment, with crater-saturated areas predominantly located in the southern hemisphere; intense volcanic activity, with extended surface reworking and lava coverage of most northern plains, and associated atmospheric recycling; tectonic surface modeling, with huge faults and canyons; and large fluvial networks, mostly located in the oldest cratered areas. Thus, all stages of the planet interiors evolution potentially may be studied through the analysis of diverse surface features.

1.2. THE VARIOUS MARS MISSIONS

The ability to decipher the Mars entire geological evolution requires the determination of the global composition, specifically, elemental (major, minor, trace elements), isotopic, molecular and mineralogical (major, minor species) abundances.

This can only be achieved by combining global spectral imaging (remote sensing) with both high spatial and spectral resolution, in situ investigations in a variety of sites, and laboratory analyses of returned samples, collected so as to represent the diversity of Martian areas. However, planetary exploration is in its infancy. So far, only very few investigations with relevance to surface composition have been successfully carried out. It is the purpose of this paper to summarize them, while suggesting routes for future research.

The very first "geological" assessments of the Mars surface were derived from terrestrial remote sensing observations, followed initially by the Mariner and Viking space missions. From the Earth, the best resolved telescopic images show only features $\sim$100 km in size. Later, the lack of spatial resolution was in part compensated by visible and near-IR spectrometers with high spectral sampling, which allowed the identification of major geologic spectral units averaged over large areas. In contrast, the early space missions provided almost global optical imaging at sub-km resolution, with almost no diagnostic spectral characterization. However, *in situ* analyses of samples from the two Viking landing sites (1976) provided the elemental composition. The measurements consisted of Rutherford backscattering and X-ray fluorescence of atoms excited by irradiation from radioactive sources (^{55}Fe and ^{109}Cd). The results were highly reliable for elements with $Z > 10$ (a few percentages in accuracy), and very weak for H, C, N and O.

The elemental composition is not as diagnostic as mineralogical composition (i.e., a variety of minerals might have similar chemical compositions). As mineral formation depends strongly on the thermodynamic conditions, minerals are direct indicators of past environments and formation processes. However, to infer mineralogical composition from either optical imaging or sample elemental composition is a highly non univocal extrapolation, in contrast with imaging spectroscopy, in particular in the near and mid-IR spectral ranges, where surface materials have strong electronic transitions and vibrational fundamental modes and overtones.

More recent space missions have added to the surface data sets as follows: the APXS instrument on the NASA Pathfinder mission, launched in 1996, determined the soil and rock elemental compositions at a third landing site. The Soviet Phobos 2 and the NASA Mars Global Surveyor (MGS) missions contributed to orbital coverage by means of both optical imaging and IR imaging spectroscopy.

To assess the mineralogic composition, the most effective method to retrieve information on a global scale is by IR imaging spectroscopy in the $0.8 - 50$ μm range. The Mars spectrum acquired from orbit is constituted of two components: the diffused solar spectrum, which dominates the short waves up to $\sim$4 μm, and the planetary thermal emission, peaked around 10 μm. The ISM/Phobos imaging spectrometer studied the Martian surface and its atmosphere through absorption features in the spectrum between 0.8 and 3.2 μm. Although this mission observed Mars from a rather large distance ($\sim$6300 km), chosen to be the Phobos orbit, and ended prematurely in March 1989 after two months of orbital operations, ISM acquired more than 40,000 spectra of Mars, with a ground resolution of $\sim$25 km, covering all major geological units of low ($<$30°) latitudes. This unique data base was recently extended in global coverage and spectral range by the TES/MGS instrument, which essentially analyses the thermal emission of Mars in the spectral range $6 - 50$ μm. It is still in operation at the time of this writing.

Finally, it should be mentioned that no in situ mineralogical analysis has been performed so far, and therefore the most direct source of information is provided by the SNC meteorites. These meteorites are believed to be igneous Martian rocks ejected during impacts, primarily on the basis of their young crystallization ages and close match between their gaseous content and the Martian atmosphere. However, since there is no knowledge of their precise origin, it is still unclear how representative they are of Martian crustal materials.

1.3. DIAGNOSTIC SIGNATURES

The $0.8 - 50$ μm spectral domain contains a large number of signatures diagnostic of most minerals, ices and frosts. In contrast with atmospheric signatures, the mineral-associated ones are generally broad (a few nm to several tens of nm), and thus do not require high spectral sampling. Even at moderate spectral resolution, one can in principle identify not only classes of minerals such as silicates, oxides or hydrated sedimentary ones, but also identify specific minerals within these classes, provided one has access to the precise spectral value of the band peak (with a 10 to 50 nm accuracy), the band shape and possibly the correlation between two features associated to the same mineral. However, the spectral signature is affected by several parameters (e.g., Salisbury, 1993), in addition to mineralogical composition. These parameters include physical properties of the surface (most notably particle size), and viewing geometry (emergence, incidence, and phase angles).

This is illustrated here (Figure 1) with the example of the identification of silicates. In the near-IR, the "1-μm" silicate feature corresponds to the electronic excitation of the Fe ions within the silicate lattice (e.g., Gaffey *et al.*, 1993). For Fe^{3+}-rich compounds (clay silicates, and a variety of oxides), the feature is peaked at $\lambda < 0.9$ μm. For Fe^{2+}-rich silicates, the band center ranges from 0.9 to 1.05 μm., the trend towards high λ values following the Ca content, with diopside peaking at ~ 1.05 μm. In addition, olivines and feldspars have 1-μm features distinctly different from those of pyroxenes. Among these, orthopyroxenes are discriminated from clinopyroxenes by the occurrence of a "2-μm feature", peaked at $\lambda < 2.2$ μm for orthopyroxenes and $\lambda > 2.2$ μm for clinopyroxenes. The precise location of these two bands is a standard indicator of pyroxene mineral species.

The O-H vibration band at $2.7 - 3.5$ μm allows both identification of hydrated minerals and estimation of their hydration level.

The thermal IR range contains signatures which are the most intense in laboratory conditions, corresponding to the fundamental vibration of the main mineralogical bonds (such as SiO_4, CO_3, SO_4, OH), usually located in the $6 - 20$ μm spectral range (reststrahlen bands). Lattice modes are present at longer wavelengths, up to 45 μm at least, although spectra of minerals have not been very well characterized yet in this domain, at least in reflectance. However, these features are generally not quite diagnostic, being almost independent of the minor constituents: for example, all silicates exhibit large vibration bands around 10 and 18 μm (fundamentals of

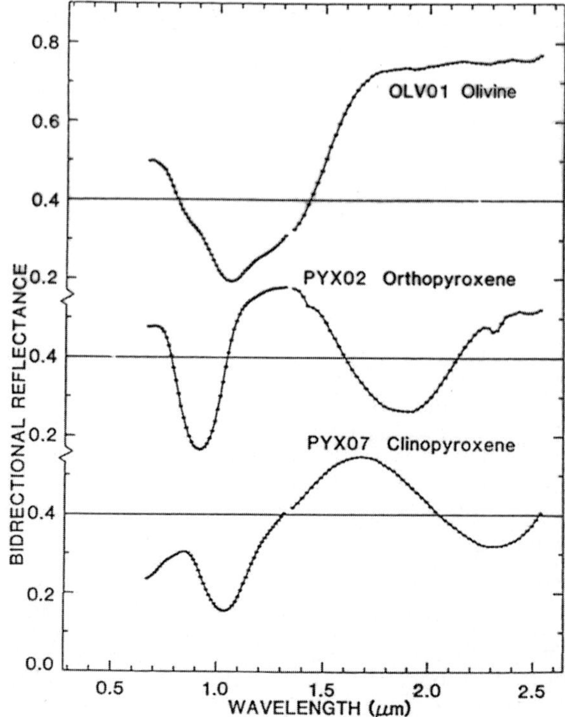

Figure 1. Laboratory NIR spectra of some silicates. Adapted from Singer (1981).

the Si-O vibration). To discriminate between the various silicate classes, one needs to identify secondary features, which usually requires higher spectral resolution.

In contrast with most minerals, pure ices have absorption bands with fine structure diagnostic of detailed composition (even at the isotopic level) and physical conditions (temperature and pressure). However, this information requires higher spectral resolution.

1.4. PIONEERING RESULTS

Thermal IR observations have been performed by IRS on board Mariner 6/7 (1.9 − 14 μm; Pimentel *et al.*, 1974), by IRIS on board Mariner 9 (5 − 50 μm; Hanel *et al.*, 1972), from the KAO in 1990 (Pollack *et al.*, 1990), and from the ground (e.g., Moersch *et al.*, 1997). Several radiometers have also observed Mars from orbital spacecraft, providing limited spectral information in the thermal range: IRTM on Viking (Kieffer *et al.*, 1977), Termoskan (Murray *et al.*, 1991) and KRMF (including a UV-visible spectrometer; Ksanfomality *et al.*, 1991) on Phobos-2.

The visible-NIR range is more easily accessed from the ground than thermal measurements. It has been used since the early 1960s (e.g., Moroz, 1964; McCord and Adams, 1969; Singer, 1982; Bell *et al.*, 1990; Pinet and Chevrel, 1990; Roush *et al.*, 1992; Martin *et al.*, 1996; Erard, 2000). Space observations have been performed by ISM (Phobos-2, $0.77 - 3.12$ μm), which is so far the only NIR imaging spectrometer that orbited Mars (Bibring *et al.*, 1989). Early results from Martian spectroscopy have been summarized by Soderblom (1992) and Roush *et al.* (1993).

1.5. THE MAJOR MARS SURFACE CONTRASTED UNITS

As viewed in the optical range, and with a very limited spatial resolution, the Mars surface is divided into large distinct areas: dark and bright surface areas and icy polar caps. The location and extent of these areas are seasonal. The precise determination of the composition of the bright and dark materials covering the non-polar areas has still not been ascertained completely.

Overall, the spectra of dark regions can be matched by basalts, whose spectral characteristics are dominated by ferromagnesian ("mafic") and andesitic volcanic materials (Bandfield *et al.*, 2000) and their alteration products. The dark areas have spectra typical of ferrous absorptions in mafic minerals such as pyroxenes, and exhibit a greater variability. Bright regions are thought to be alteration products of basalts and are spectrally very homogeneous, dominated by subtle features of hydrated ferric oxides. On the first order, this is consistent with the interpretation that bright regions are covered by a thick layer of weathered dust mixed at the planetary scale, while dark areas have large bedrock exposures and are contaminated at various levels by bright materials. Polar caps are primarily mixtures of CO_2 and H_2O ices, with ratios varying with time and location.

In the following, we discuss the measurements done so far, characterizing the dark materials, the bright soils and the polar caps.

2. Dark Materials

Although dust is widely transported seasonally, which has long been considered to prevent the observation and characterization of the underlying terrains, it has been demonstrated that a wide variety of unmantled surface materials are present, allowing a possible identification of surface geological units. These units appear much darker than the soil, with large variations in albedo observed. To date, their composition is studied by three types of space measurements: the Pathfinder in situ rock analyses, the ISM near-IR spectral images, and the TES thermal spectral images. Additional compositional information results from laboratory analyses of SNC meteorites.

2.1. PATHFINDER AND SNC ROCK ANALYSES

At the Pathfinder landing site, no sedimentary rocks were identified. Only highly Si-rich silicates were found, with an elemental composition similar to that of terrestrial andesite (Rieder *et al.*, 1997). The unusual magnetic properties measured

ISM altimetry map

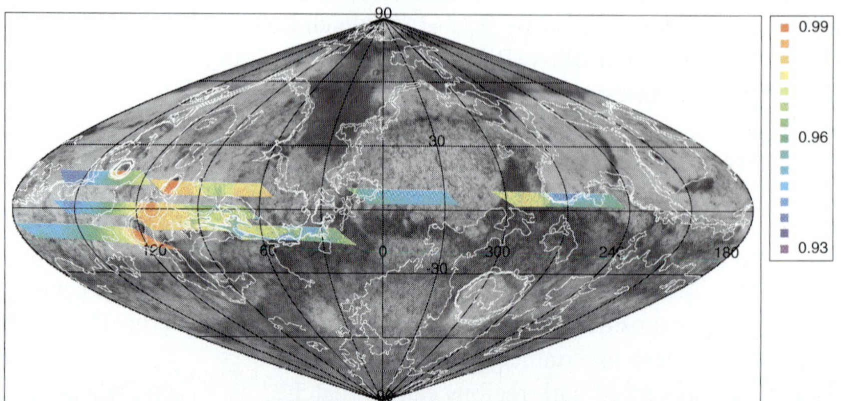

Figure 2. ISM coverage of Mars, superimposed on Viking DIM (sinusoid projection). The colour scales surface elevation from the depth of the 2.0 μm CO_2 band, with lower elevations in blue.

may indicate significant amounts of maghemite, an iron oxide (Hviid *et al.*, 1997). These rocks do not seem to represent the full diversity of Martian rocks, since they are compositionally very different from the soils (Wänke *et al.*, 2001, Table I).

SNC meteorites can be grouped into 3 broad families, according to their crystallization age (see reviews by McSween, 1994, and Treiman *et al.*, 2000). The younger family is composed of basaltic rocks dominated by two pyroxene mixtures (with Shergotty $\sim$180 Myr old, and others 300 and 420 Myr old); an older one is dominated by Ca-rich clinopyroxenes and olivine, with very low concentrations of Al (with Chassigny and Nakhla, $\sim$1.3 Gyr old). ALH84001, rich in orthopyroxenes, is much older ($\sim$4.5 Gyr). Similar to the Martian rocks analyzed in situ, all SNCs are richer in Fe than the terrestrial crust, probably reflecting a higher FeO content of the Martian mantle. However, in contrast with the rocks analyzed by Pathfinder, the SNCs are all poor in Al; hence they would derive from an Al-depleted source region. The occurrence of carbonates in SNC meteorites might suggest that these minerals were indeed crustal sinks for early CO_2. Alternatively, carbonates and other salts could have formed in restricted areas where the temperature was high enough to maintain liquid water at, or close to the surface (hydrothermal springs). Globules of carbonates contained in ALH84001 have also been presented as possible by-products of ancient life on Mars (McKay *et al.*, 1996). However, this issue is still debated. For instance, Golden *et al.* (2000) have been able to produce similar carbonate structure abiologically in the lab.

2.2. THE ISM CHARACTERIZATION

Figure 2 shows the location of all acquired ISM spectra. The very high signal to noise and the imaging capacity of ISM made it possible to identify (and map the spatial variations of) spectral features at the level of $\sim$0.5%, which is required due to the shallowness of most mineral-associated absorption features in the NIR range. In addition to the identification by ISM of major classes of silicates, subtle features have been identified in limited areas, such as possible sulfate signatures at 1.7 μm and metal-OH vibrations around $2.2 - 2.35$ μm.

Five major ISM spectral features were used to identify units of specific composition. These include the central band positions of the 1 μm and 2.2 μm silicate features (see Section I.3), the 3 μm hydration feature, the slope of the continuum, and the NIR albedo. In the following, we detail the surface heterogeneity of two large areas mapped by ISM, namely Valles Marineris and the Syrtis/Isidis region.

2.2.1. *Syrtis Isidis*

Syrtis Major (Figure 3) is the darkest region of Mars, although highly variable in albedo. It is considered as a region of seasonal deposition and removal of bright dust. High resolution images by MOC (MGS) show that the substrate itself is predominantly made up of dark volcanic material, covered with large dark sand dunes. Only the southern part of Syrtis Major was mapped by ISM, from 2° to 9° latitude. Two different units are identified in this part of Syrtis Major (Erard *et al.*, 1991; Mustard *et al.*, 1993; Murchie *et al.*, 2000). The western part (including the central area with Meroe and Nili Patera) exhibits the most typical features, which appears consistent with a mafic to ultramafic composition. This material is very dark (0.10 albedo at 1.1 μm), with the deepest pyroxenes absorptions observed in the equatorial regions (3-5% at 0.92 μm, 1-3% at 2.29 μm). Detailed analyses of these two bands showed that they are best fit by a mixture of both low and high-Ca pyroxenes, with a larger proportion of the latter (70%). The eastern part of Syrtis Major displays slightly different signatures, in particular reduced pyroxene absorptions and an elongation of the 1 μm band towards short wavelengths, indicating the presence of ferric oxides (e.g., hematite, schwertmannite or goethite) in addition to pyroxenes. The material exposed there was interpreted as dust cemented on a mafic substrate, possibly by a crystalline ferric phase.

2.2.2. *Valles Marineris*

The materials observed inside Valles Marineris (Figure 4) exhibit large variations of all spectral parameters, in contrast with the surrounding plateaus, where two distinct spectral units can be distinguished: homogeneous bright, hydrated, soils cover the western areas (on Tharsis), while darker and drier materials are exposed in the north-eastern part (Ophir Planum); they show pyroxene composition with Ca content lower than that of Syrtis Major, but higher than that of the darker materials of the floor of Valles Marineris (Erard *et al.*, 1991; Mustard *et al.*, 1997; Murchie *et al.*, 2000). Detailed analyses demonstrate a much lower content in high-Ca pyroxenes in these regions (~45% of the overall pyroxenes). All these dark materials are very dry, except for a unique area (in the ISM data set) located in the most eastern part of the canyon, in the chaotic terrains (Eos Chasma). This area is interpreted to contain high mafic, highly hydrated minerals.

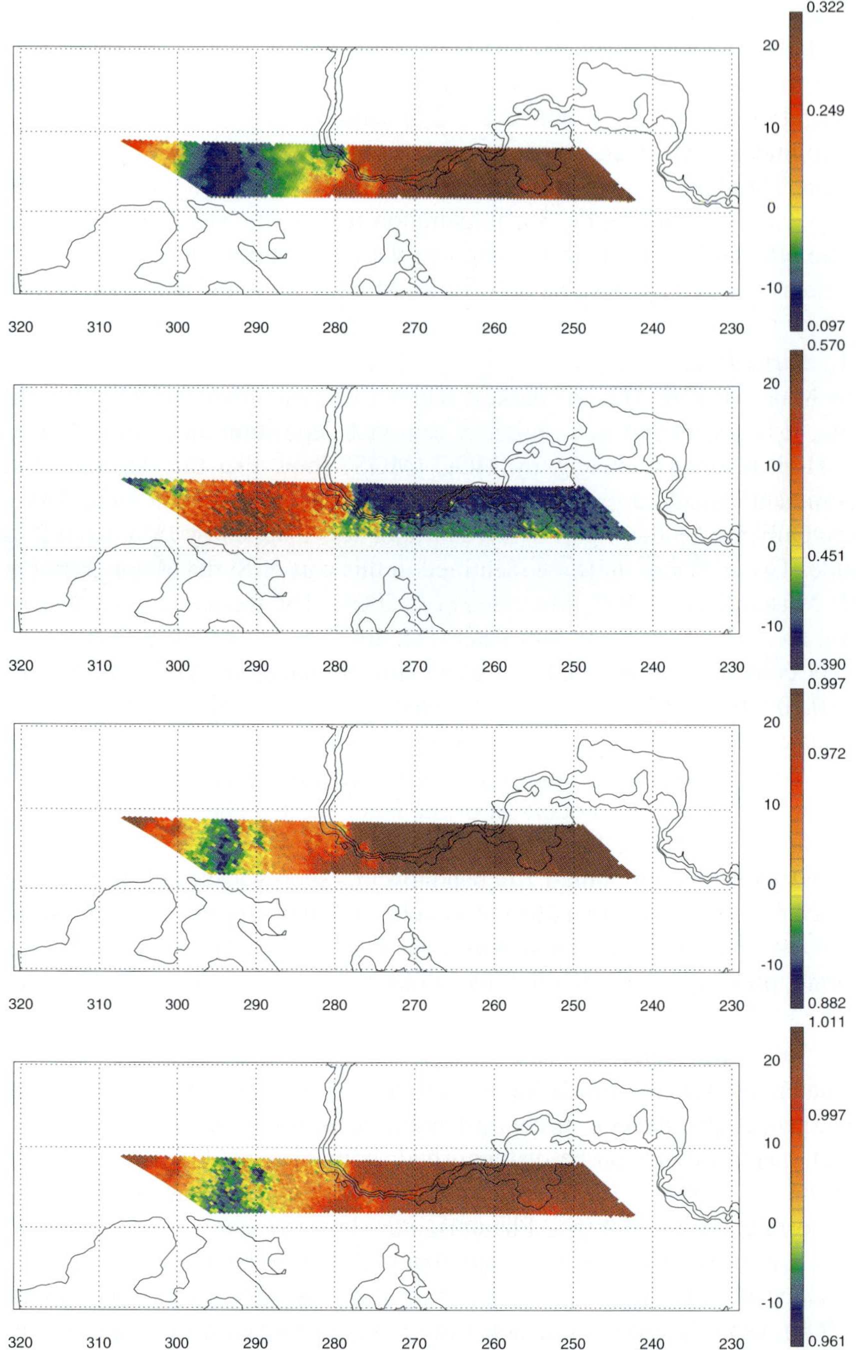

Figure 3. ISM spectral coverage of the Syrtis-Isidis area, superimposed on Viking derived altimetric contours: a) albedo map (with Minnaert correction), b) depth of the 3 μm hydration band (blue is more hydrated, red is less), c) depth of the 1.04 μm band (Fe^{2+}-rich materials) (red to blue with increasing abundance), d) depth of the 2.0 μm pyroxene band (red to blue with increasing abundance).

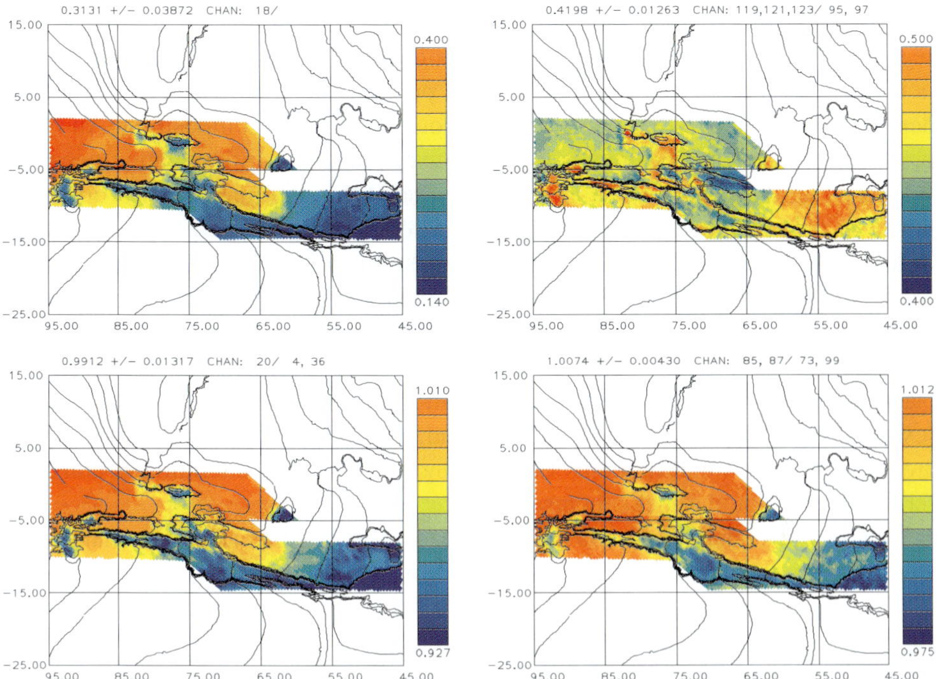

Figure 4. ISM spectral coverage of the Valles Marineris area, superimposed on Viking derived altimetric contours: a) albedo map (with Minnaert correction), b) depth of the 3 μm hydration band (blue is more hydrated, red is less), c) depth of the 1.04 μm band (Fe^{2+}-rich materials) (red to blue with increasing abundance), d) depth of the 2.0 μm pyroxene band (red to blue with increasing abundance).

The layered terrains in the central part of the canyons (Melas Chasma and Coprates Chasma) also have much stronger hydration absorptions than other materials of similar albedo. The unique spectral properties observed in the layered terrains strongly support the hypothesis of volcanic deposits restricted to the canyon, perhaps through hydrovolcanism.

2.2.3. *ISM Summary*

The high spatial resolution, spectral coverage, and high signal to noise of the ISM data permitted the determination of a mineralogical basis for the spectral properties of several distinct morphogeologic dark regions on Mars (Mustard and Sunshine, 1995; Mustard *et al.*, 1997). The freshest dark equatorial regions observed by ISM have spectral characteristics indicating a volcanic mineralogy dominated by two pyroxenes with high and low Ca content, similar to the basaltic SNC meteorites (e.g., Shergotty, Zagami), supposed to originate from Mars. However, the plateau plains (e.g. Syrtis Major) are enriched in high-Ca pyroxene relative to the floors of Valles Marineris. Within this two-pyroxene model, there exists significant diversity in the spectral properties among relatively unaltered regions on Mars, and a central question is how this spectral diversity relates to mineralogical diversity. The range of ages for these regions indicates that the SNC analyses are relevant to large regions of Mars, and to a large lapse of time (3.0–0.13 Gyr). Such materials are not common on the Earth's surface, and are probably derived from a source region depleted in Al relative to the hypothesized original mantle composition. An interpretation of these observations is that a primary planetary differentiation occurred very early in the history of Mars ($\sim$4.5 Gyr) and that the mantle reservoirs have not been remixed since. The depletion in Al could be explained by the presence of a crust enriched in Al, perhaps similar to the lunar highlands; such rocks were actually observed at the Pathfinder site, but their source area is unknown. The occurrence of significant amounts of olivine in dark areas would be inconsistent with the idea of an early differentiation. Although detection of olivine-rich materials was claimed in several occasions, these observations are not confirmed by recent data (except perhaps by TES, see below). Instead, most of the second order variations are explained by small amounts of dust, and perhaps thin alteration coatings of ferric materials on the rocks themselves.

"Anomalous" dark materials also provide important hints about the major puzzles concerning the surface. In particular, an unusual material observed in the layered deposits of Valles Marineris, with both mafic and hydration characteristics, suggests that hydrovolcanism was an important factor in the development of these areas (Murchie *et al.*, 2000).

2.3. THE TES SURFACE DETERMINATION

The presently available TES reduced data indicate three major types of surface material.

A large equatorial area (Sinus Meridiani), $350\,\mathrm{km}\times(350$–$750)\,\mathrm{km}$ wide, centered 2°S, 0–5°W, which appears gray rather than red (this may be due to a particle size effect), appears to contain crystalline hematite (α-Fe_2O_3) with an abundance $\sim$15% while it is $\ll$5% elsewhere on Mars. Through the match with laboratory spectra, it would be made of "large" grains, 5 to 10 μm at least, up to possibly 100 μm, quite different from the nano-crystalline oxides and fine-grained hematite grains postulated as constituents of the bright soil. These minerals would constitute either weathering / alteration products, or chemical precipitates, which would strongly suggest the occurrence of large amounts of stable liquid water in the past (Christensen *et al.*, 2000a).

As for the primary rocks in the dark regions, TES (Figures 5 and 6) has identified a global scale opposition on Mars, between the southern highlands, where Si-poor mafic volcanic rocks dominate (type 1 rocks, similar to basalts), and the northern plains (lowlands), made primarily of evolved Si-rich mafic rocks (type 2 rocks, similar to andesite). More precisely, the corresponding spectra have been determined to have better match with the following mixtures: feldspar (plagioclase) (50%), clinopyroxene (augite) (5%), sheet silicate (15%) for type 1 rocks; feldspar (plagioclase) (35%), clinopyroxene (augite) (10%), sheet silicate (15%), K-rich

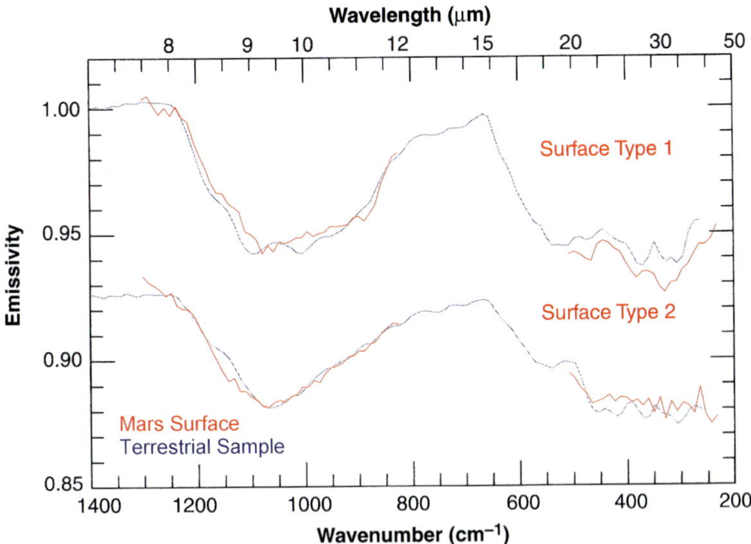

Figure 5. TES spectra of the main two surface type endmembers identified (adapted from Bandfield *et al.*, 2000), referred to Si-poor basalt (type 1) and andesite-like basalt (type 2).

glass (obsidian) with 25% for type 2 rock (Bandfield *et al.*, 2000). More detailed plagioclase and pyroxene composition was studied by Hamilton *et al.* (2000). For both types of materials, the derived plagioclase composition ranges from 40 to 55% anorthite, and high-Ca clinopyroxenes composition ranges from 30 to 45% Mg. This composition is consistent with basaltic shergottites, although more calcic than many of them. However, the TES spectra of these areas do not closely resemble those of SNC meteorites.

This contrast in volcanic materials is still challenging for our understanding of the Mars evolution: why has Mars undergone two types (one old, one later) of volcanism? Why is the "evolved type" largely spread over the planet, and coupled to the presumably thinner crust?

Recently, a very tentative detection of olivine-rich areas (as much as 30% averaged over 100 m size pixels) has been made (Hoefen *et al.*, 2000). The major concentration of olivine would be located in Nili Fossae. Large occurrence of olivine would imply that no abundant liquid water was available at the surface since the formation of these rocks, which would in turn suggest that no period of hot, wet climate has existed since then.

2.4. THE ISO SURFACE DETERMINATION

High-resolution infrared spectra of Mars have been recorded with ISO Short-Wavelength Spectrometer between 2.3 and 45 μm. Three spectra of the full Martian disk, centered respectively on Syrtis Major, Tharsis and Aeria, were recorded in July and August 1997 (Morris *et al.*, 2001; Lellouch *et al.*, 2000). At wavelengths longer

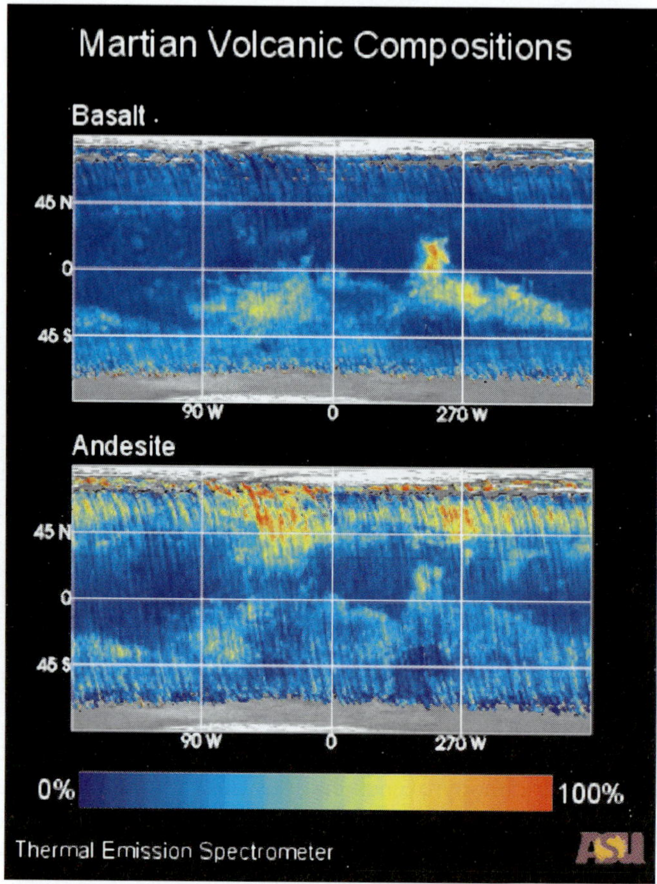

Figure 6. TES map of the two surface type endmembers identified (adapted from Bandfield *et al.*, 2000).

than about 4 μm, the main contribution to the spectrum is the thermal emission of the surface, which is dominated by dark regions and is considerably weighted towards the sub-solar point. The colder aerosols are seen in absorption, and exhibit much narrower bands. The main contributions in the ISO spectra are expected to originate from Lunae Planum, Solis Planum, Ophir Planum and Valles Marineris. Two emissivity maxima at 7.85 and 8.2 μm are interpreted as the Christiansen peaks of surface minerals. The first peak is best matched by plagioclases, particularly labradorite, which also fits most of the structures observed. Clinopyroxenes, in particular diopside, fit the second Christiansen peak and the region from 11 to 12 μm. Forsterite-poor olivines, which have a Christiansen frequency at 9.3 μm, also match this domain correctly. Crystalline oxides do not fit at all. The best fit to the $5-7$ μm domain is provided by montmorillonite in very fine grains; the weight of bright areas in the spectrum is indeed expected to increase towards the shorter wavelengths. Among rocks, the best fits are provided by dark vesicular basalts with

TABLE I

Atmospheric Mass (M_a) to Planetary Mass (M_p) ratio.

	Mars	The Earth	Venus
Present M_a/M_p	5×10^{-8}	10^{-6}	10^{-4}
Initial M_a/M_p	?	10^{-4}	10^{-4}

low Al content (7%), similar to the composition observed in SNC and at the Viking landing sites. This is consistent with analysis of pristine dark regions in the NIR range from ISM, with re-analysis of IRIS data, and with recent observations by TES (Christensen *et al.*, 2000b). The second reststrahlen band of silicates in the $16 - 25\ \mu$m region is not observed. Although olivine and feldspar bands could be undetectable with ISO calibration uncertainty at longer wavelength, pyroxenes should produce a prominent absorption in this range. This feature is not detected either in IRIS or TES spectra.

2.5. WHERE IS THE NITROGEN?

The search for surface carbonates, and more generally for sedimentary rocks, is based on the hypothesis that the atmospheric depletion of major species, and especially of CO_2, would result from its being trapped as solid compounds. The decrease of Martian internal activity would have stopped their recycling and H_2O would have been predominantly stored as subsurface ice, or permafrost. Such a scenario, in which a past, dense, CO_2-rich (and thus warm) atmosphere would progressively have faded away as Mars cooled, by efficient surface and subsurface trappings, is challenged by the depletion of N_2.

The present atmospheric N_2 is about 10^{-4} less abundant than that of Venus (Table I). This depletion is almost identical for CO_2. The present atmospheric N_2/CO_2 ratio, measured as $(3\pm0.3)\%$, is the same for Mars and Venus. Scaled to the planetary mass, the Mars atmospheric depletion factor is roughly 2000. When taking into account the present-day carbonate terrestrial reservoir of CO_2, both Venus and the Earth have an initial atmospheric to planetary mass ratio of $\sim10^{-4}$, while today's Mars value is $\sim5\times10^{-8}$. Therefore, not only the depletion of Martian N_2 and CO_2 are huge, but they have occurred with an almost identical efficiency. Besides, if carbonate formation is an easy process for removing CO_2, in particular with the concurrence of liquid water, there is no such efficient processes to convert nitrogen into rocks, apart from salt formation from acid rain and activity from living organisms. On Earth, most N_2 is still stored in the atmosphere; would terrestrial internal activity cease, and O_2 consequently disappear, the atmospheric pressure would likely decrease to 0.8 bars, with N_2 by far remaining the dominant constituent (as for Titan). One is thus left with a number of alternate mechanisms, such as atmospheric sputtering, to account for Martian nitrogen losses. It is not understood to what extent these mechanisms would also have depleted the CO_2 on Mars, to complement the potential carbonate formation. Can one derive overall identical N_2 and CO_2 depletion efficiencies, from a variety of diverse processes? Understanding the past climate and geological history of Mars requires addressing the question of where and how the Martian nitrogen has gone. The search for surface N-rich materials is probably a key part of the answer.

3. Bright Soils

3.1. OPTICAL SPECTRA

The albedo represents the brightness of a surface and can range from 0% (black) to 100% (white). The geometric albedo of the martian bright material, in the visible, can be greater than 30%, with a huge spectral variation giving the bright material its markedly reddish visible color: in violet the soils are essentially dark (geometric albedo <5% at $\lambda < 0.4$ μm) and their spectra almost featureless (Adams and McCord, 1969). At longer wavelength, the albedo increases very rapidly (Figure 7) and reaches values $\sim$0.4 in the near IR (0.7 − 2.7 μm). A convincing explanation is the occurrence of a strong Fe^{3+}-O^{2-} charge transfer absorption, centered at $\lambda = 0.34$ μm, being responsible for this spectral shape. The bright soil is widely homogeneous over the Mars surface, seasonally transported by the wind, with deposition features like dunes similar to those observed in terrestrial deserts.

3.2. ELEMENTAL COMPOSITION

As for the elemental composition of the bright soil, the Viking XRFS and Pathfinder APXS investigations provided very similar determinations, which is summarized in Table I: high Si and Fe content, MgO/Al_2O_3 of $\sim$1, very high S and Cl, while K is very low (Clark *et al.*, 1982; Banin *et al.*, 1992). This is consistent with weathering products of basalts, although richer in Fe and Mg, and poorer in K, than most terrestrial ones. The high Cl and S could be indicative of salts. All soils are very similar at the three landing sites, with high degrees of oxidation and magnetic properties.

Subsurface cemented material, identified as a duricrust by Binder *et al.* (1977), were also analyzed by the Viking landers. They have higher S content than the loose fines, suggesting possible formation by evaporation of salt-rich solutions, with sulfates as the cementing agent.

Finally, it should be recalled that the Viking biology package did not identify organic matter at levels exceeding the detection limits, in the range of <1 ppb for most compounds (see, e.g., Klein *et al.*, 1992, for a review).

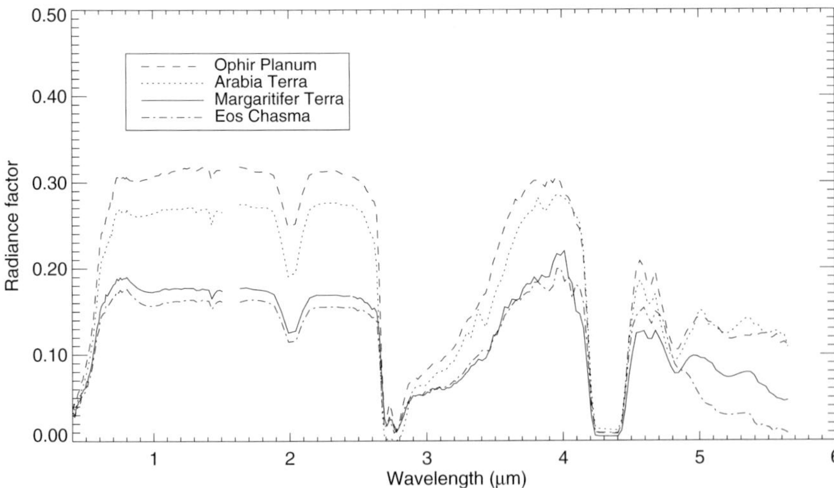

Figure 7. Composite Vis-NIR reflectance spectra of bright and dark regions. At longer wavelength, the thermal contribution is modeled to retrieve reflectance (adapted from Erard and Calvin, 1997).

3.3. NEAR-IR SPECTRAL CHARACTERISATION

The ISM spectra of typical bright regions (e.g. Tharsis) show a feature peaked at <0.9 μm. This spectral information is indicative of highly oxidized iron (ferric) rather than ferrous-rich pyroxenes. In addition, these spectra exhibit deep hydration bands at $\sim$3 μm, corresponding to an adsorbed or bound-water abundance of $\sim$1-2% (Figure 7). Actually, this "3 μm" band depth is an estimate of the hydration level. It is found to be highly correlated to reflectance (Erard *et al.*, 1991). It is more specific of the materials than albedo is, at least on transitional regions: an example is given in Lybia and Isidis, which have similar reflectance but different hydration band depth. Lybia is a dark region with highly variable albedo in the observational records, probably covered with bright dust at the time of ISM observations. There are significant variations in the strength of the 3.0 μm hydration band with layered terrains in Valles Marineris exhibiting the strongest absorptions in the ISM data set. A weak band is also observed at 2.2 μm at the threshold of the detection level ($\sim$0.5% band depth), possibly related to hydroxylated minerals (metal-OH bounds).

Spectral slope is strongly affected by atmospheric scattering. Although atmospheric scattering affects the spectra, in some specific areas variations in spectral slope may be driven by surface properties. Such variations have been studied extensively on the flanks of Olympus Mons by Fischer and Pieters (1993), who ascribe them to ferric coatings of variable thickness.

Other "anomalous" bright to intermediate soils were identified in Oxia Palus and Lunae Planum, having enhanced 3 μm band absorption and steeper spectral slope. These properties are consistent with duricrust formation process and cementation of bright dust by either water-bearing phase, sulfates, carbonates, or ferric minerals (Murchie *et al.*, 2000).

3.4. CANDIDATE MINERALS

In order to corroborate all the results acquired to date, it is thought that the bright soil predominantly could be constituted of a suite of candidate minerals of silicates (with mostly iron-rich amorphous material, clays or silica glass), iron oxides such as hematite, salts, sulfates, and (perhaps) carbonates (see review by Bell, 1996).

An interesting terrestrial analog for the iron-rich constituent is palagonite. It is formed by alteration of mafic volcanic glasses under low temperature, mainly by interaction with water ice. Large amounts of palagonites would be consistent with spectral observations, in situ analyses, thermodynamic considerations, and composition of SNC meteorites (Allen *et al.*, 1981). Most of this material could have formed during volcanic eruptions through a thick underlying permafrost (Berkley and Drake, 1981).

These materials seem to constitute alteration products of mafic minerals from the dark areas. The present rate of weathering and dust formation is unknown but is certainly small, and hence most bright materials were probably formed in a distant past.

As for the hematite, spectral features are subdued and indicate a low degree of crystallization in most areas, so that the oxide phase is thought to occur mainly as extremely fine-grained (<10 nm), or nanophase, poorly crystalline particles (Morris *et al.*, 1989). The spectra indicate a small amount ($\sim5\%$) well-crystalline (fine-grained but not nanophase) hematite to account for the spectral shape.

Although this bright material is the dominant constituent of the soil, it is not the only one. At higher spatial resolution, other units of intermediate reflectance and color can be recognized in both Viking and HST images. These units include bright crater rims and dark patches, and were thus first thought to constitute a mixture of dark and bright materials; they are currently interpreted as bright materials indurated by local processes, at least in some areas. Similarly, many "dark red" regions (e.g. Oxia Palus) exhibit ISM features inconsistent with a simple mixture between bright and dark soils. Rather, they appear to be a unique material and may contain hydrated ferric oxides and oxyhydroxides.

Independently, infrared radiometry and radar observations allow thermal inertia and rock abundance to be derived. As a rule, the finer the material, the lower the thermal inertia. On Mars, the brighter material has the lowest inertia, the darker material has the higher value. Intermediate areas are thus made of either coarse or indurated materials. Taking into account all identified physical properties, one can discriminate between four main units (e.g. Soderblom, 1992):

1. bright, fine materials interpreted as loose dust. The corresponding areas are regions of active dust accumulation several meters thick (e.g., Amazonis, Arabia);

2. bright materials with high thermal inertia, interpreted as areas of limited dust accumulation. Both Viking landing sites are located in such regions, which are rather uncommon (e.g., Chryse Planitia, Elysium Planitia);
3. intermediate materials with high thermal inertia, interpreted as indurated surfaces (cemented dust);
4. dark, coarse materials interpreted as exposed rocks with constant deposition and removal of a thin dust layer (e.g., Acidalia, Syrtis Major). In some areas however, this unit corresponds to accumulation of dark sands.

Most units appear relatively young and active, indicating that the upper surface is continually being reworked, in particular by aeolian processes and possible duricrust formation.

Subtle local variations are observed in bright to moderate regions (as in Arabia Terra, southwest Claritas Fossae, south Daedalia Planum and Sirenum Fossae, Deucalionis Regio, Sinus Meridiani, Propontis I), indicating that some bright materials are not remobilized and mixed. These variations might reflect differences in particle size or degree of crystallization, but some small areas could have high concentrations of other alteration materials such as smectite clays, oxyhydroxides, iron sulfates, chlorides, hydrous carbonates. These minerals are potentially very helpful indicators of past climates, because they are the major sinks for H_2O, CO_2 and other volatile compounds such as nitrogen. Some of them have been tentatively detected in the airborne dust, but their spatial distributions and spectral properties must be known to improve our knowledge of the surface evolution.

3.5. AIRBORNE DUST

Although the composition of airborne dust is supposed to match that of bright surface soils, the spectra themselves are different. Because airborne particles are not in contact and are much smaller than surface particles, aerosols spectra are dominated by Mie scattering, with a very steep negative spectral slope in the NIR range.

Aerosols grain size distribution is controlled by dynamical processes and seasonally varies with altitude. The ISM observations of the integrated column are consistent with a narrow particle size distribution with effective radius $r_{eff}{\sim}1.25\ \mu m$ (Drossart $et\ al.$, 1991). This particle size is only half that retrieved from Mariner 9-IRIS ($r_{eff}{\sim}2.75\ \mu m$, Toon $et\ al.$, 1977) and the distribution is twice as narrow ($v_{eff} = 0.25$ versus 0.418). However, these results agree with those of Auguste, KRFM and Termoskan on board Phobos-2 (Chassefière $et\ al.$, 1995). The difference with pre-Phobos data was ascribed to the lack of global dust storm in the late 1980s.

So far, airborne dust has been observed primarily in the mid-infrared (IRIS, TES, ISO). Analysis of IRTM photometric sequences by Clancy and Lee (1991) demonstrated that the backscattered flux can represent up to 25% of the overall signal around 1 μm, even under low opacity. This figure is consistent with detailed analyses of ISM observations by Drossart *et al.* (1991) and Erard *et al.* (1994). Erard *et al.* discussed the effect of aerosols scattering on the study of surface properties in the NIR. The main effects are to add a negative spectral slope to the spectra and to conceal the surface absorptions and shift them to longer wavelengths. Altogether, spectral slope appears equally sensitive to surface and atmospheric properties. Location of a $0.8 - 1.0$ μm band center is also quite sensitive to the amount of light scattered in the atmosphere, and can be easily shifted towards large wavelengths by 50 nm because of the apparent tilt of the continuum slope.

Absorption features were also observed in ISM continuum ratioed spectra of aerosols. Most of them were ascribed to H_2O ice (at 1.02, 1.25, 1.5, 1.9 and 2.4 μm). In addition, a small absorption was observed over Tharsis at 1.8 μm which is found in some hydrated clay minerals and in gypsum. The presence of clay minerals such as montmorillonite or kaolinite is consistent with early interpretations of IRIS observations (e.g. Hunt *et al.*, 1973), with KAO observations of suspended dust (Pollack *et al.*, 1990), with Viking biology experiments (e.g. Banin and Margulies, 1983), with SNC analyses (e.g. Gooding, 1992), and with recent thermal modeling of IRS spectra above 3.0 μm (Erard and Calvin, 1997). Crystalline clay minerals have not generally been proposed as abundant constituents on Mars essentially because the most common terrestrial species present several visible and NIR absorptions that are not observed in Martian spectra (see discussions in Roush *et al.*, 1993; Soderblom, 1992). However, recent laboratory work (e.g., Bishop and Pieters, 1995; Bishop *et al.*, 1995; Calvin and King, 1997) demonstrates that these bands are subdued and not necessarily observable under Martian conditions (low temperature and pressure), and that several less usual clays simply lack them. Consequently, it seems that the occurrence of crystalline clays in Mars bright, fine materials deserves reevaluation. Recent reanalysis of IRIS spectra suggests a composition of suspended dust dominated by feldspar, with a proposed composition of 50% albite (a feldspar), 10% sulfates, 20% pyroxenes and 20% olivines (Grassi and Formisano, 2000).

In the ISO spectra (see above, Section II.4), narrow bands at 7.2 and 11.1 μm (Figure 8) were ascribed to carbonates in the aerosols. The general band location, relative band depths and width correspond closely to fundamentals of carbonates, the best match being magnesite. However, the two bands observed are slightly more distant than those of carbonates, making the identification tentative. In the longer wavelength range, three deep absorptions at 26.5, 31 and 43.5 μm also supports the existence of carbonates. These bands are best matched by spectra of calcite and siderite, with relative intensities consistent with the $6 - 12$ μm features. The interpretation of these data remains controversial.

Spectral structures similar to those attributed to carbonates in ISO spectra are actually present in previous data sets, although little attention has been paid to them. The $7 - 7.3$ μm band is present in IRIS spectra (Christensen *et al.*, 1998), in some newly calibrated IRS spectra (Kirkland *et al.*, 1998), in KAO spectra (Pollack *et al.*, 1990) and in TES observations (Christensen *et al.*, 1998). Subtle absorptions

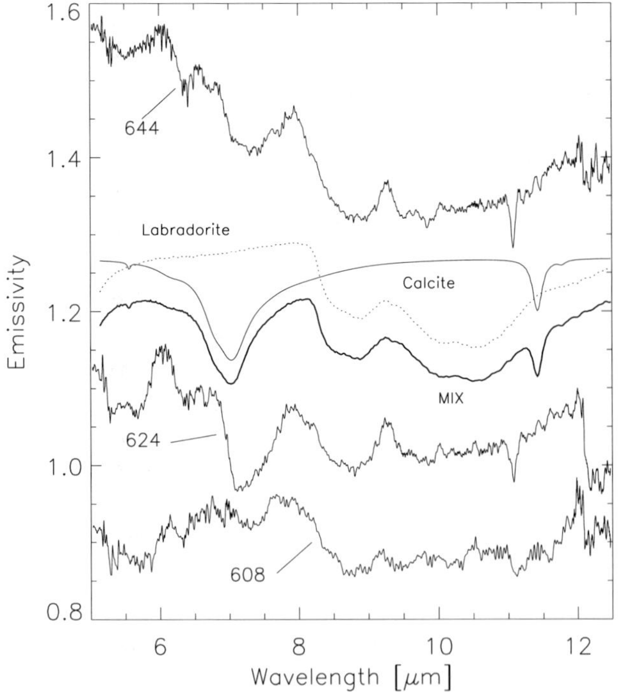

Figure 8. 5 − 12 μm portion of three disk-integrated Mars ISO-SWS spectra, compared with labradorite emission (surface-like) and carbonate transmission (aerosol-like) laboratory spectra (adapted from Morris *et al.*, 2001).

in the 11.0 − 11.5 μm domain are also present in IRIS and IRS spectra, and in telescopic spectra (Moersch *et al.*, 1997). However these bands do not necessarily imply carbonate: in particular, many minerals have transitions in the 11.0−11.5 μm domain (e.g. this feature in IRIS spectra has been ascribed to goethite by Kirkland and Herr, 2000).

4. Polar Caps, Ices and Frosts

The annual orbital motion of Mars around the Sun, on a rather eccentric orbit ($e = 0.093$), together with its obliquity (25.2°), leads to seasonal exchange of CO_2, and to a lesser extent of H_2O, between the atmosphere and polar caps, cycling up to 25% of the atmosphere. Thus, at latitudes varying with time, the surface is covered with frosts, the composition of which being entirely governed by the local thermodynamic equilibrium: composition assessments can be obtained through both remote compositional or thermal measurements.

Actually, at the local summer, none of the polar caps does completely sublimate: residual high albedo icy units remain, the composition and extent of which differing largely between the North and the South. The North residual cap is much larger, and constituted of H_2O, with no trapped CO_2 (Clark and McCord, 1982). Although its precise height is not accurately determined yet, it is a major H_2O Martian reservoir (together with the regolith), with an amount of H_2O up to 10^5 times higher than the atmospheric one (Kieffer and Zent, 1992). On the opposite, the South residual cap is predominantly constituted of CO_2, the condensation temperature of which ($= 150$ K) allows H_2O to also be trapped: the amount of H_2O, within or beneath the CO_2 ice, is still unknown.

Whenever the atmosphere starts to condense, large frosts build, predominantly made of CO_2. H_2O also condenses, with a relative abundance reflecting the atmospheric one, less than 0.1% (Calvin and Martin, 1994).

Apart from pure ice, layered deposits of variable dust content appear, at scales meters to tens of meters thick. Very few compositional information allow to discriminate between this dust and that covering a large fraction of the Martian surface (see below).

5. Future Prospects

According to the planned space missions, a number of major investigations will be conducted in the coming decade to further characterize the surface composition of Mars. At each future Mars launch window, one orbiter mission will include at least one instrument dedicated to mapping the surface. For example, the NASA Odyssey mission ('01) will have a gamma-ray spectrometer to determine the surface chemical composition, including the radioactive elements, on a ~ 100 km scale as well as THEMIS, derived from TES with a much higher spatial sampling (100 m) but much lower spectral sampling, will characterize local mineralogical species. On board the 2003 ESA Mars Express mission, OMEGA will map the entire surface, at a kilometer scale, and a few percents at 300 m, in a spectral range $0.5 - 5.2$ μm, where most potential minerals have diagnostic signatures. On the same mission, PFS will acquire spectra of the entire surface and atmosphere, at a smaller spatial resolution (5 km) but much higher spectral sampling (~ 1 cm^{-1}) and spectral domain ($1.2 - 45$ μm). Two years later, the NASA '05 mission, if confirmed, will be launched with a candidate payload including a similar visible-near infrared spectral imager, with a spatial resolution of ~ 50 m. Altogether, these orbital missions should provide global and unambiguous determination of all units, at an unprecedented resolution, so as to identify and potentially locate the presence of all minerals, including the sedimentary ones. Finally, the '03 rovers and the '07 rovers and/or landers will locally study the surface composition, down to submillimeter scales.

The next major step will undoubtedly be achieved by the analyses of Mars samples, collected in well selected areas, and returned to the Earth. The range and variety of the scientific results from these studies will surpass the mere predictions for Mars. In the future, precise dating, exhaustive elemental, isotopic, molecular and mineralogical inventory at a grain level, identification of volatile and possibly organic trapped species, detection of fossil records, if any, are among the measurements and discoveries that the present level of sophisticated analytical capabilities could make feasible. The uniqueness of Mars can contribute to understanding key issues of Solar System evolution, including possibly the formation of living organisms. Hence, the acquisition of Mars samples and their return to earth for laboratory analyses is among the most exciting scientific prospect for the early 2000s.

Acknowledgements

We are deeply indebted to the entire ISSI team for the superb meeting held, and impressive editorial support, with special thanks to Dr. Reinald Kallenbach. The paper has greatly benefited from the thorough review by an anonymous referee.

References

Adams, J.B., and McCord, T.B.: 1969, 'Mars: Interpretation of Spectral Reflectivity of Light and Dark Regions', *J. Geophys. Res.* **74**, 4851–4856.

Allen, C.C., Gooding, J.L., Jercinovic, M., and Keil, K.: 1981, 'Altered Basaltic Glass – A Terrestrial Analog to the Soil of Mars', *Icarus* **45**, 347–369.

Bandfield, J., Hamilton, V., and Christensen, P.: 2000, 'A Global View of Martian Surface Composition from MGS-TES', *Science* **287**, 1626–1630.

Banin, A., and Margulies, L.: 1983, 'Simulation of Viking Biology Experiments Suggests Smectites, not Palagonites, as Martian Soil Analogs', *Nature* **305**, 523–525.

Banin, A., Clark, B.C., and Wänke, H.: 1992, 'Surface Chemistry and Mineralogy', *in* H. Kieffer, B. Jakosky, C. Snyder, and M. Matthews (eds.), Univ. Arizona Press, Tucson, pp. 594–625.

Bell, J.: 1996, 'Iron, Sulfate, Carbonate, and Hydrated Minerals on Mars', *in* M.D. Dyar, C. McCammon, and M.W. Schaefer (eds.), *Mineral Spectroscopy: A Tribute to Roger G. Burns*, Geochemical Society, pp. 359–380.

Bell, J.F., McCord, T.B., and Owensby, P.D.: 1990, 'Observational Evidence of Crystalline Iron Oxides on Mars', *J. Geophys. Res.* **95**, 14,447–14,462.

Berkley, J.L., and Drake, M.J.: 1981, 'Weathering of Mars – Antarctic Analog Studies', *Icarus* **45**, 231–249.

Bibring, J.-P., *et al.*: 1989, 'Results from the ISM Experiment', *Nature* **341**, 591–592.

Binder, A.B., Arvidson, R.E., Guinness, E.A., Jones, K.L., Mutch, T.A., Morris, E.C., Pieri, D.C., and Sagan, C.: 1977, 'The Geology of the Viking Lander 1 Site', *J. Geophys. Res.* **82**, 4439–4451.

Bishop, J., and Pieters, C.: 1995, 'Low-temperature and Low Atmospheric Pressure Infrared Reflectance Spectra of Mars Soil Analog Materials', *J. Geophys. Res.* **100**, 5369–5379.

Bishop, J., Pieters, C., Burns, R., Edwards, J., Mancinelli, R., and Fröschl, H.: 1995, 'Reflectance Spectroscopy of Ferric Sulfate-bearing Montmorillonites as Mars Soil Analog Materials', *Icarus* **117**, 101–119.

Calvin, W.M., and King, T.: 1997, 'Spectral Characteristics of Fe-bearing Phyllosilicates: Comparison to Orgueil, Murchison and Murray', *Met. Planet. Sci.* **32**, 693–702.

Calvin, W.M., and Martin, T.Z.: 1994, 'Spectra Variability in the Seasonal South Polar Cap of Mars', *J. Geophys. Res.* **99**, 21,143–21,152.

Chassefière, E., Drossart, P., and Korablev, O.: 1995, 'Post-Phobos Model for the Altitude and Size Distribution of Dust in the Low Martian Atmosphere', *J. Geophys. Res.* **100**, 5525–5539.

Christensen, P.R., *et al.*: 1998, 'Results from the Mars Global Surveyor Thermal Emission Spectrometer', *Science* **279**, 1692–1695.

Christensen, P.R., *et al.*: 2000a, 'Detection of Crystalline Hematite Mineralization on Mars by the Thermal Emission Spectrometer: Evidence for Near-surface Water', *J. Geophys. Res.* **105**, 9623–9642.

Christensen, P.R., Bandfield, J.L., Smith, M.D., Hamilton, V.E., and Clark, R.N.: 2000b, 'Identification of a Basaltic Component on the Martian Surface from Thermal Emission Spectrometer Data', *J. Geophys. Res.* **105**, 9609–9622.

Clancy, R.T., and Lee, S.W.: 1991, 'A New Look at Dust and Clouds in the Mars Atmosphere: Analysis of Emission-phase-function Sequences from Global Viking IRTM Observations', *Icarus* **93**, 135–158.

Clark, R.N., and McCord, T.B.: 1982, 'Mars Residual Polar Cap: Earth-based Spectroscopic Confirmation of Water Ice as a Major Constituent and Evidence for Hydrated Minerals', *J. Geophys. Res.* **87**, 367–370.

Clark, B.C., Baird, A.K., Weldon, R.J., Tsusaki, D.M., Schnabel, L., and Candelaria, M.P.: 1982, 'Chemical Composition of Martian Fines', *J. Geophys. Res.* **87**, 10,059–10,067.

Drossart, P., Rosenqvist, J., Erard, S., Langevin, Y., Bibring, J.-P., and Combes, M.: 1991, 'Martian Aerosols Properties from the Phobos/ISM Experiment', *Annal. Geophys.* **9**, 754–760.

Erard, S.: 2000, 'The 1994-95 Apparition of Mars Observed from Pic-du-Midi', *Planet. Space Sci.* **48**, 1271–1287.

Erard, S., and Calvin, W.: 1997, 'New Composite Spectra of Mars, $0.4 - 5.7 \mu$m', *Icarus* **130**, 449–460.

Erard, S., *et al.*: 1991, 'Spatial Variations in Composition of the Valles Marineris and Isidis Planitia Regions of Mars Derived from the ISM Data', *Proc. 21st Lunar Planet. Sci. Conf.*, 437–455.

Erard, S., Mustard, J., Murchie, S., Bibring, J.-P., Cerroni, P., and Coradini, A.: 1994, 'Effects of Aerosols Scattering on Near-infrared Observations of the Martian Surface', *Icarus* **111**, 317–337.

Fischer, E.M., and Pieters, C.M.: 1993, 'The Continuum Slope of Mars: Bi-directional Reflectance Investigations and Applications to Olympus Mons', *Icarus* **102**, 185–202.

Gaffey, S.J., McFadden, L.A., Nash, D., and Pieters, C.: 1993, 'Ultraviolet, Visible, and Near-infrared Reflectance Spectroscopy: Laboratory Spectra of Geologic Materials', *in* C.M. Pieters and P.A. Englert (eds.), *Remote Geochimical Analysis: Elemental and Mineralogical Composition*, Cambridge University Press, pp. 43–77.

Golden, D.C., Ming, D.W., Schwandt, C.S., Morris, R.V., Yang S.V., and Lofgren, G.E.: 2000, 'An Experimental Study on Kinetically-driven Precipitation of Calcium-magnesium-iron Carbonates from Solution: Implications for the Low-temperature Formation of Carbonates in Martian Allan Hills 84001', *Met. Planet. Sci.* **35**, 457–465.

Gooding, J.L.: 1992, 'Soil Mineralogy on Mars: Possible Clues from Salts and Clay in SNC Meteorites', *Icarus* **99**, 28–41.

Grassi, D., and Formisano, V.: 2000, 'IRIS Mariner 9 Data Revisited: 2. Aerosol Dust Composition', *Planet. Space Sci.* **48**, 577–598.

Hamilton, V., Bandfield, J., and Christensen, P.: 2000, 'The Mineralogy of Martian Dark Regions from MGS TES Data: Preliminary Determination of Pyroxene and Feldspar Compositions', *Proc. 31st Lunar Planet. Sci.*, LPI, Houston, abstract #1824.

Hanel, R., *et al.*: 1972, 'Investigation of the Martian Environment by Infrared Spectroscopy on Mariner 9', *Icarus* **17**, 47–56.

Hoefen, T.M., Clark, R.N., Pearl, J.C., and Smith, M.D.: 2000, 'Unique Spectral Features in Mars Global Surveyor Thermal Emission Spectra: Implications for Surface Mineralogy in Nili Fossae', *in 32nd Annual DPS Meeting, Bull Am. Astron. Soc.*, (abstract).

Hunt, G.R., Salisbury, J.W., and Lenhoff, C.J.: 1973, 'Visible and Near-infrared Spectra of Minerals and Rocks: VI, Additional Silicates', *Mod. Geol.* **4**, 85–106.

Hviid, S.F., *et al.*: 1997, 'Magnetic Properties Experiments on the Mars Pathfinder Lander: Preliminary Results', *Science* **278**, 1768–1771.

Kieffer, H.H., and Zent: 1992, 'Quasi-periodic Climate Change on Mars', *in* H.H. Kieffer, B.M. Jakosky, C.W. Snyder and M.S. Matthews (eds.), *Mars*, Univ. Arizona Press, Tucson, pp. 1180–1218.

Kieffer, H.H., Martin, T.Z., Peterfreund, A.R., Jakosky, B.M., Miner, E.D., and Palluconi, F.D.: 1977, 'Thermal and Albedo Mapping of Mars During the Viking Primary Mission', *J. Geophys. Res.* **82**, 4249–4291.

Kirkland, L.E., and Herr, K.C.: 2000, 'Spectral Anomalies in the 11 and 12 μm Region from the Mariner Mars 7 Infrared Spectrometer', *J. Geophys. Res.* **105**, 22,507–22,516.

Kirkland, L., Forney, P., and Herr, K.: 1998, 'Mariner Mars 6/7 Infrared Spectra: New Calibration and a Search for Water Ice Clouds', *in: Lunar Planet. Sci.* **XXIX**, Lunar and Planetary Institute, abstract #1516.

Klein, H.P., Horowitz, N.H., and Biemann, K.: 1992, 'The Search for Extant Life on Mars', *in* H. Kieffer, B. Jakosky, C. Snyder, and M. Matthews (eds.), University of Arizona Press, Tucson, pp. 1221–1233.

Ksanfomality, L., *et al.*: 1991, 'Phobos: Spectrophotometry Between 0.3 and 0.6 μm and IR-radiometry', *Planet. Space Sci.* **39**, 311–325.

Lellouch, E., Encrenaz, T., de Graauw, T., Erard, S., Morris, P., Feuchtgruber, H., Crovisier, J., Girard, T., and Burgdorf, M.: 2000, 'The 2.4 − 45 μm Spectrum of Mars Observed with the Infrared Space Observatory', *Planet. Space Sci.*, in press.

Martin, P., Pinet, P., Bacon, R., Rousset, A., and Bellagh, F.: 1996, 'Martian Surface Mineralogy from 0.8 to 1.05 μm TIGER Spectro-imagery Measurements in Terra Sirenum and Tharsis Montes Formation', *Planet. Space Sci.* **44**, 859–888.

McCord, T.B., and Adams, J.B.: 1969, 'Spectral Reflectivity of Mars', *Science* **163**, 1058–1060.

McKay, D.S., Gibson, E.K., Jr., Thomas-Keprta, K.L., Vali, H., Romanek, C.S., Clemett, S.J., Chillier, X.D.F., Maechling, C.R., and Zare, R.N.: 1996, 'Search for Life on Mars: Possible Relic Biogenic Activity in Martian Meteorite ALH84001', *Science* **273**, 924–930.

McSween, H.Y., Jr.: 1994, 'What Have we Learned About Mars from SCN Meteorites', *Meteoritics* **29**, 757–779.

Moersch, J.E., Hayward, T.L., Nicholson, P.D., Suyres, S.W., Van Cleve, J., and Christensen, P.R.: 1997, 'Identification of a 10-μm Silicate Absorption Feature in the Acidalia Region of Mars', *Icarus* **126**, 183–196.

Moroz, V.I.: 1964, 'The Infrared Spectrum of Mars (1.1 − 4.1 μm)', *Soviet Astron.* **8**, 273–281.

Morris, R., Agresti, D., Lauer, H., Newcomb, J., Shelfer, T., and Murali, A.: 1989, 'Evidence for Pigmentary Hematite on Mars Based on Optical, Magnetic and Mössbauer Study of Superparamagnetic (Nanocrystalline) Hematite', *J. Geophys. Res.* **94**, 2760–2778.

Morris, P., de Graauw, T., Lellouch, E., Henderson, B.G., Erard, S., Encrenaz, T., Feuchtgruber, H., Burgdorf, M., and Davis, G.R.: 2001, 'A Detailed Assessment of Carbonates in Thermal Infrared Spectroscopy of Mars with the ISO Short Wavelength Spectrometer, *Icarus*, submitted.

Murchie, S., Kirkland, L., Erard, S., Mustard, J., and Robinson, M.: 2000, 'Near-Infrared Spectral Variations of Martian Surface Materials from ISM Imaging Spectrometer Data', *Icarus* **147**, 444–471.

Murray, B., *et al.*: 1991, 'Preliminary Assessment of Termoskan Observations of Mars', *Planet. Space Sci.* **39**, 237–265.

Mustard, J.F., and Sunshine, J.M.: 1995, 'Seeing Through the Dust: Martian Crustal Heterogeneity and Links to the SNC Meteorites', *Science* **267**, 1623–1626.

Mustard, J., Erard, S., Bibring, J.-P., Head, J.W., Hurtrez, S., Langevin, Y., Pieters, C.M., and Sotin, C.J.: 1993, 'The Surface of Syrtis Major: Composition of the Volcanic Substrate and Mixing with Altered Dust and Soil', *J. Geophys. Res.* **98**, 3387–3400.

Mustard, J., Murchie, S., Erard, S., and Sunshine, J.: 1997, 'In Situ Compositions of Martian Volcanics: Implications for the Mantle', *J. Geophys. Res.* **102**, 25,605–25,615.

Pimentel, G.C., Forney, P.B., and Herr, K.C.: 1974, 'Evidence About Hydrate Solid Water in the Martian Surface from the 1969 Mariner Infrared Spectrometer', *J. Geophys. Res.* **79**, 1623–1634.

Pinet, P., and Chevrel, S.: 1990, 'Spectral Identification of Geological Units on the Surface of Mars Related to the Presence of Silicates from Earth-based Near-infrared Telescopic Charge-coupled Device Imaging', *J. Geophys. Res.* **95**, 14,435–14,446.

Pollack, J.B., *et al.*: 1990, 'Thermal Emission Spectra of Mars (5.4 − 10.5 μm): Evidence for Sulphates, Carbonates, and Hydrates, *J. Geophys. Res.* **95**, 14,595–14,628.

Rieder, R., Economou, T., Wänke, H., Turkevich, A., Crisp, J., Brückner, J., Dreibus, G., and Mc-Sween, H.Y.J.: 1997, 'The Chemical Composition of Martian Soil and Rocks Returned by the Mobile Alpha Proton X-ray Spectrometer: Preliminary Results from the X-ray Mode', *Science* **278**, 1771–1773.

Roush, T., Roush, E., Singer, R., and Lucey, P.: 1992, 'Estimates of Absolute Flux and Radiance Factor of Localized Regions of Mars in the (2 − 4 μm) Wavelength Region', *Icarus* **99**, 42–50.

Roush, T.L., Blaney, D.L., and Singer, R.B.: 1993, 'The Surface Composition of Mars as Inferred from Spectroscopic Observations', *in* C.M. Pieters and P.A. Englert (eds.), *Remote Geochimical Analysis: Elemental and Mineralogical Composition*, Cambridge Univ. Press, New York, pp. 367–397.

Salisbury, J.W.: 1993, 'Mid-infrared Spectroscopy: Laboratory Data', *in* C.M. Pieters and P.A. Englert (eds.), 'Remote Geochimical Analysis: Elemental and Mineralogical Composition', Cambridge Univ. Press, pp. 79–98.

Singer, R.B.: 1981, Near-infrared Spectral Reflectance of Mineral Mixtures – Systematic Combinations of Pyroxenes, Olivine, and Iron Oxides', *J. Geophys. Res.* **86**, 7967–7982.

Singer, R.B.: 1982, 'Spectral Evidence for the Mineralogy of High-albedo Soils and Dust on Mars', *J. Geophys. Res.* **87**, 10,159–10,168.

Soderblom, L.: 1992, 'The Composition and Mineralogy of the Martian Surface from Spectroscopic Observations: 0.3 μm to 50 μm', *in* H. Kieffer, B. Jakosky, C. Snyder, and M. Matthews (eds.), *Mars*, Univ. Arizona Press, Tucson, pp. 557–597.

Toon, O.B., Pollack, J.B., and Sagan, C.: 1977, 'Physical Properties of the Particles Composing the Martian Dust Storm of 1971–1972', *Icarus* **30**, 663–696.

Treiman, A.H., Gleason, J.D., and Bogard, D.D.: 2000, 'The SNC Meteorites are from Mars', *Planet. Space Sci.* **48**, 1213–1230.

Wänke, H., Brückner, J., Dreibus, G., Rieder, R., and Ryabchikov, I.: 2001, 'Chemical Composition of Rocks and Soils at the Pathfinder Site', *Space Sci. Rev.*, this volume.

Address for correspondence: IAS, Institut d'Astrophysique Spatiale Batiment 121, 91405 Orsay Campus, France; (jean-pierre.bibring@ias.fr)

CHEMICAL COMPOSITION OF ROCKS AND SOILS AT THE PATHFINDER SITE

H. WÄNKE[1], J. BRÜCKNER[1], G. DREIBUS[1], R. RIEDER[1] and I. RYABCHIKOV[2]

[1]*Max-Planck-Institut für Chemie, Becher-Weg 27, D-55128, Mainz, Germany*
[2]*IGEM, Russian Academy of Sciences, 35 Staromonetny, Moscow 109017, Russia*

Received: 24 August 2000; accepted: 24 February 2001

Abstract. As Viking Landers did not measure rock compositions, Pathfinder (PF) data are the first in this respect. This review gives no proof yet whether the PF rocks are igneous or sedimentary, but for petrogenetic reasons they could be igneous. We suggest a model in which Mars is covered by about 50% basaltic and 50% andesitic igneous rocks. The soils are a mixture of the two with addition of Mg-sulfate and -chloride plus iron compounds possibly derived from the hematite deposits.

1. Pathfinder Landing and Sojourner Exploration

On July 4, 1997, the Mars Pathfinder spacecraft (see Golombek *et al.*, 1999, for mission details) landed at the mouth of Ares Vallis, $19.28°$N and $33.52°$W. The Alpha-Proton-X-ray-Spectrometer (APXS) was switched on the first day on Mars and returned an alpha-backscatter spectrum of the well-explored Martian atmosphere identical to one obtained in the laboratory from CO_2.

During the 83 days on Mars until radio contact was lost, Sojourner encircled the lander within 12 m radius driving a total distance of 104 m. During this time the sensor head of APXS was placed on 9 rocks and on soils at 7 locations. However, mainly because of electronic noise, not all of the acquired spectra were useful.

2. The APX-Spectrometer

The APXS, mounted on the rover Sojourner, returned chemical analyses of Martian rocks for the first time. Viking Landers 1 and 2 only returned in-situ X-ray fluorescence (XRF) analyses of soil samples, as neither rocks were within reach of the arms nor devices were on board to acquire and prepare solid samples.

The APXS of PF was originally designed for the Russian Mars 1996 Mission, which failed during launch. The APXS technique enables us to chemically analyze soils and rocks without any sample preparation. The spectrometer's sensor head (Figure 1) simply had to be put against the sample so that it was irradiated by alpha particles emitted by ^{244}Cu sources. The instrument had three modes of operation:

Space Science Reviews **96**: 317-330, 2001.
© 2001 *Kluwer Academic Publishers. Printed in the Netherlands.*

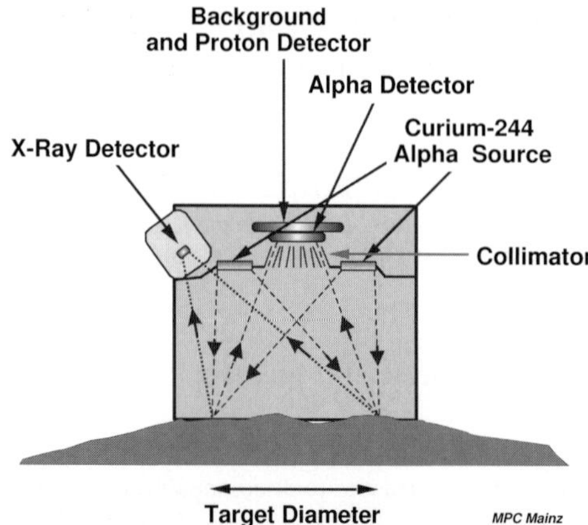

Figure 1. Functional scheme of the APXS sensor head (diameter 50 mm).

1. Alpha-back-scattering, known as Rutherford scattering. The energy of the alpha particles scattered by about 180° is a direct measure of the mass of the nucleus on which the scatter occurred; it is measured by solid-state detectors.
2. In a few cases the nuclei of the target undergo alpha-proton reactions. The protons produced have energies specific to the target nucleus. As this mode requires very long counting times to compete with the other two modes with respect to accuracy, it is only of minor importance.
3. The alpha particles also interact with the electron shell of the target nuclei generating characteristic X-rays analyzed by a solid state detector.

The alpha-back-scatter mode is especially useful to analyze the light elements C, N and O, while the X-ray mode is advantageous for all elements heavier than Na.

All measurements were performed with a sampled area of 50 mm in diameter. The depth of analysis depends on the composition of the sample but is generally on the order of a few μm. All elements, except H, can be and were analyzed. Thus, the measurement geometry does not need to be known precisely if one normalizes to 100%. For a detailed description of APXS, see Rieder *et al.* (1997a). Figure 2 illustrates the X-ray spectra obtained from the dark soil of the Mermaid Dune and the rock Half Dome. The mobility of the rover was very important to the investigation, enabling analysis of a variety of samples selected from images taken by the Imager for Mars Pathfinder (IMP) on the lander.

3. APXS Analyses at Ares Vallis

The preliminary data of the APXS analyses (Rieder *et al.*, 1997a) had to be revised because of the following reason. During tests of the APXS at JPL the electronic driver circuit for the motor activating a shutter in front of the radioactive sources failed and the project decided to discard this circuit and fly the instrument with an

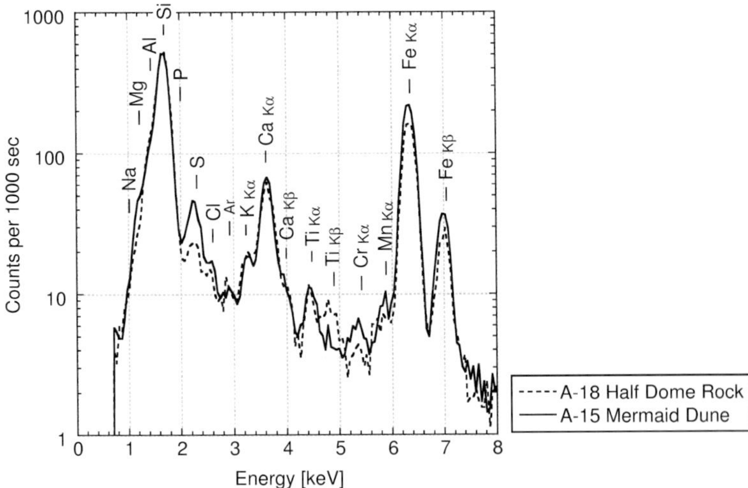

Figure 2. X-ray spectra of rock Half Dome and the dark soil of Mermaid Dune. The Ar peak is due to the 1.6% Ar in the Martian atmosphere within the APXS sensor head.

open shutter. Consequently, the thin alumina/VYNS foil in front of the sources and its support grid were also removed, because without the shutter the foil would be exposed to a high radiation dose during the flight and probably break on landing. However, all calibrations were carried out with this foil and its support grid in front of the alpha-sources. This foil had originally been installed in order to prevent contamination of calibration samples with source material, emitted by a process known as "recoil sputtering". The support grid of this foil reduces the flux of alpha particles by about 15%, whereas the attenuation of the flux of X-rays from the source is almost negligible.

This difference and other minor revisions due to more precise recalibrations lead to higher Fe concentrations for all samples by $\sim 25\%$, while the Si concentration was reduced by $\sim 10\%$ compared to the preliminary results (Rieder *et al.*, 1997b). In the case of K, appearing to the left of the Ca peak shoulder, a considerable mistake was recognized in the separation procedure of the two peaks. The data listed in Table I are the same corrected ones as in Brückner *et al.* (2001). In spite of these differences, the major observations of Rieder *et al.* (1997b) remain unchanged.

Table I contains the data for the most reliable but least disturbed measurements of the soil samples A4, A5, A10, and A15. All data were normalized to 100%, although in some cases the sums reached only about 80% because of inadequate positioning of the APXS sensor head.

The chemical composition of all the soil samples was almost identical, independent of the color and appearance of the ground. Altogether the soil composition at Ares Vallis was very similar to the landing sites of Viking 1 and 2 at Chryse and Utopia. This might well mean that the Martian soil is homogeneous on a global scale, having been distributed and mixed by impacts and storms.

WÄNKE ET AL.

TABLE I

Composition of soils and rocks at the Mars PF landing site. Corrected data (August 2000) in weight-%, normalized to 100%, based on AXPS Mainz re-calibration. The average error given in the last line includes the error due to counting statistics as well as the error from the calibration curves. For details, see Brückner *et al.* (2001).

Sample	Na_2O	MgO	Al_2O_3	SiO_2	P_2O_5	SO_3	Cl	K_2O	CaO	TiO_2	Cr_2O_3	MnO	Fe_2O_3
Soils													
A-4, soil	1.00	9.95	8.22	42.5	1.89	7.58	0.57	0.60	6.09	1.08	0.2	0.76	19.6
A-5, soil	1.05	9.20	8.71	41.0	1.55	6.38	0.55	0.51	6.63	0.75	0.4	0.34	23.0
A-10, soil	1.32	8.16	7.41	41.8	0.95	7.09	0.53	0.45	6.86	1.02	0.3	0.51	23.6
A-15, soil	0.97	7.46	7.59	44.0	1.01	6.09	0.54	0.87	6.56	1.20	0.3	0.46	23.0
Mean Soil	1.09	8.69	7.98	42.3	0.98	6.79	0.55	0.61	6.53	1.01	0.3	0.52	22.3
Mean Viking soil[a]	–	6.4	8.0	47.0	–	7.9	0.5	<0.15	6.4	0.7	–	–	19.7
Cemented Soil													
A-8, Scooby Doo	1.56	7.24	9.09	45.6	0.61	6.18	0.55	0.78	8.07	1.09	–	0.52	18.7
Rocks													FeO
A-3, Barnacle Bill	1.69	3.20	11.02	53.8	1.42	2.77	0.41	1.29	6.03	0.92	0.1	–	16.2
A-7, Yogi	1.19	6.71	9.68	49.7	0.99	4.89	0.50	0.87	7.35	0.91	–	0.47	16.7
A-16, Wedge	2.30	4.58	10.24	48.6	1.00	3.29	0.41	0.96	8.14	0.95	–	0.65	18.9
A-17, Shark	2.03	3.50	10.03	55.2	0.98	1.88	0.38	1.14	8.80	0.65	0.05	0.49	14.8
A-18, Half Dome	1.78	3.91	10.94	51.8	0.97	3.11	0.37	1.10	6.62	0.82	–	0.52	18.1
Shergotty[b]	1.29	9.28	7.07	51.4	0.80	0.33	0.011	0.16	10.0	0.87	0.20	0.53	19.4
CI[c] meteorite	0.68	15.5	1.55	22.8	0.23	13.5	0.067	0.062	1.26	0.073	0.39	0.23	23.5
Calculated soil free rock	2.46	1.51	11.0	57.0	0.95	S 0.30	0.32	1.36	8.09	0.69	–	0.55	15.7
Av. error (rel %)	40	10	7	10	20	20	15	10	10	20	50	25	5

[a] Data from Clark *et al.* (1982), normalized to 95.6% to account for Na, P, Cr, and Mn.

[b] Data from Banin *et al.* (1992).

[c] Data from Palme *et al.* (1981), CI meteorites contain in addition 3.5% C and 17% H_2O.

The agreement with Viking data is good in most cases (Table I). From the measurements of the XRF-spectrometer on board Viking 1 and 2, Clark *et al.* (1982) reported only upper limits of 0.15% K_2O, whereas from the APXS of PF 0.6±0.1% K_2O was found in soil samples. It seems likely that Clark *et al.* (1982) might have overestimated the sensitivity considerably, otherwise one has to assume that the much higher K concentrations in the PF soil are derived from local K-rich rocks.

Special emphasis was given to obtaining information on the C content of Ares Vallis soils and rocks from the alpha-mode. Unfortunately, the sensitivity for C was drastically reduced by the background due to the CO_2 of the thin Martian atmosphere ($\sim$6.5 mbar), so that only upper limits of 0.8 wt-% C (corresponding to $\sim$5% $MgCO_3$) can be given for both soils and rocks at the PF landing site.

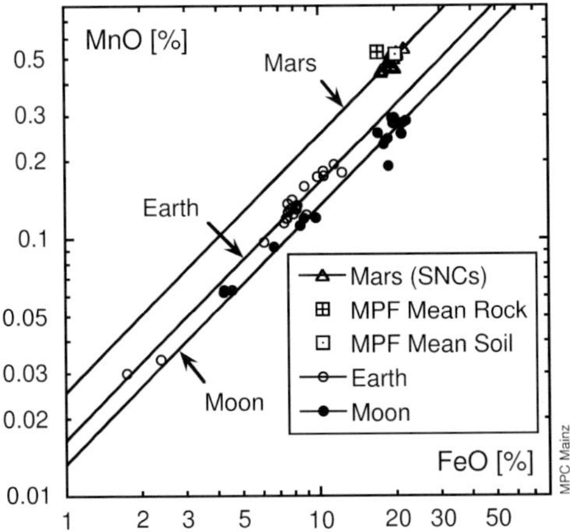

Figure 3. MnO versus FeO. While the data point for the PF soil samples fall exactly within the area of those of the Martian meteorites; the mean rock data point deviates slightly but within the error limit still agrees with the Martian meteorites.

In fact, carbonates should not be expected in the Martian soil because of the dominance of SO_3 (Wänke and Dreibus, 1994; Clark, 1999). Shergottites, the most abundant group of Martian meteorites, contain mantle derived concentrations of $\sim$200 ppm H_2O, $\sim$100 ppm CO_2, and between 1200 and 5600 ppm SO_2. Terrestrial MORB contain $\sim$2000 ppm H_2O and similar concentrations of SO_2 and CO_2. On Mars, which is much poorer in H_2O and CO_2 but similar or richer in SO_2, it is expected that SO_2 dominates the volcanic gases. At least part of SO_2 will be quickly transformed to SO_3, which together with water vapor will produce sulfuric acid which in turn will decompose carbonates and return CO_2 to the atmosphere.

One minor element, which could be analyzed with meaningful accuracy by APXS at the PF landing site on Mars is Mn. Manganese strongly correlates with Fe on the Earth and the Moon. Hence, the Fe/Mn ratio is indicative of the samples' parent body. The mean Fe/Mn ratio of PF rock and soil samples of 39 ± 11 agrees well with the Fe/Mn ratio of shergottites of 39 ± 2. This qualifies as additional proof of the parent body of the Martian meteorites (Figure 3). A similar argument is derived from the Cr concentration in Martian meteorites and the PF rocks and soil, except that the accuracy of the measured Cr concentration is lower due to the lower absolute concentrations especially in the rocks. In the case of Mn, we have to admit that rocks from any parent body not depleted in Mn in its silicate phase like eucrites will plot close to the Mars data points, too.

Another minor element of importance is P. However, the P peak sits on the shoulder of the 50-times larger Si peak and, hence, P data have errors of $\sim$30%. The observed P concentrations in the Martian samples ($\sim$0.4 wt-%) are very high when compared to terrestrial crustal concentrations of only $\sim$0.1%, but these values are in line with high P concentrations predicted for the Martian mantle by Dreibus and Wänke (1987) from data on Martian meteorites.

4. Composition of the Rocks

To derive rock compositions, the signal due to dust coatings had to be subtracted. The analyzed soils show sulfur concentrations of (2.72±0.27) wt-%. The measured S concentrations for the PF rocks, ranging from 0.75 to 1.96 wt-%, by far exceed the concentrations accommodated in magmas or igneous rocks. As evident from the PF IMP-camera images, the surfaces of the rocks are covered to varying degrees with adhering dust. When plotted against S most elements form linear arrays for the rocks analyzed (Figure 4). Extrapolation of the regression lines to zero S content yields the approximate composition of soil-free unaltered rocks (Rieder et al., 1997b). The rocks Shark and Barnacle Bill most closely approximate the composition of the soil-free rock. Minimal contamination by dust relative to other analyzed rocks is also evident from high-resolution IMP-camera images, which for these rocks exhibit higher red to blue reflectance ratios (McSween et al., 1999). As shergottites contain S in amounts between 0.13 to 0.28%, it might be more appropriate to assume for the PF rocks, which are more fractionated than shergottites, S concentrations of about 0.3 wt-%. Hence, in a new approach Brückner et al. (2001) extrapolated to 0.3% S rather to zero to find the composition of a soil-free rock. This procedure yields differences in the amounts for Si and Fe of <5% relative, but raises the MgO concentration from 0.6 to 1.5%. The good fit of all rock data to the regression lines (Figure 4) indicates an almost identical composition of the 4 rocks analyzed. Sample A8 (Scooby Doo), although of rock-like appearance, is identified by its composition as cemented soil. The lower P content of Scooby Doo compared to all other analyzed samples seems to be real.

The composition of the rocks relative to the soil yielded a surprise. The Martian soil with its considerable Mg concentration was taken as evidence for a mafic crust of Mars. The Martian meteorites have mafic to ultramafic composition, too. In contrast, the rocks at Ares Vallis turned out to represent highly fractionated crustal material, rich in SiO_2 and K, but low in Mg. This holds regardless of the not-yet-solved nature of these rocks, i.e. igneous or sedimentary. The composition of the PF rocks, together with that of the soil in which they are embedded, as well as the average composition of the Martian meteorites, is illustrated in Figure 5. The huge compositional difference between soil and rocks cannot be explained by diminution of these rocks, even considering weathering and interaction with volcanic gases SO_2 and HCl. Addition of material richer in Mg and Fe, but poorer in K and Cr as observed by the Martian meteorites, seems unavoidable.

Taking the almost identical soil composition at the three landing sites as representative for the whole surface soil of Mars, the Mainz group (Brückner et al., 1999) has shown that all elements fit into a two-component mixing diagram with the PF rocks on the high K-low Mg side and the Martian meteorites on the opposite side (Figure 6). Hence, it was concluded that large geologic units of andesitic (PF rocks) as well as of basaltic (Martian meteorites) composition must exist on Mars and cover about equal areas. Bandfield et al. (2000) recently confirmed this model

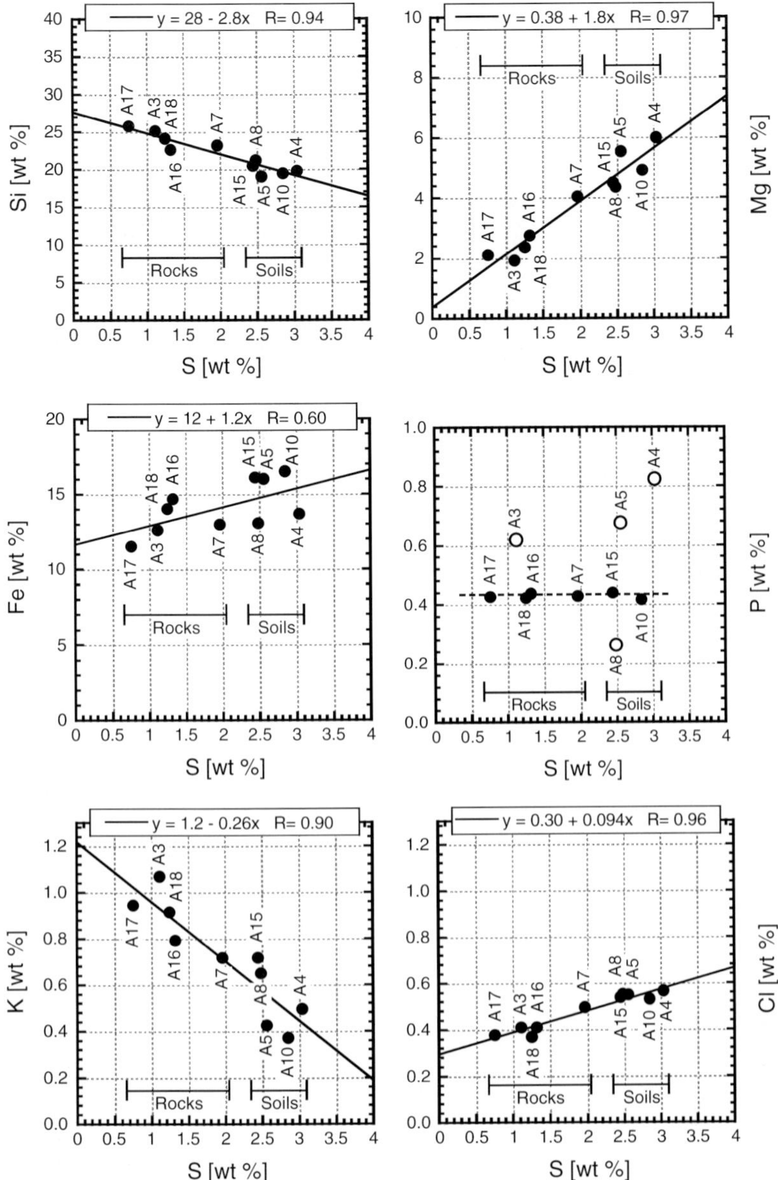

Figure 4. Top panels: Linear regression lines for Si and Mg versus S of PF rocks and soils. The good fit of all the soil samples can be taken as an indication that the S is present in form of $MgSO_4$. *Middle panels:* Linear regression lines for Fe (*left*). In the case of P (*right*) for which the uncertainties amount to about 30% for samples indicated by solid data points while those with open circles have even higher uncertainties. Six of the ten samples analyzed yield P concentrations of about 0.43%. The spectra of samples A3, A4, and A5 are difficult to analyze because of electronic noise. Hence, the true P values might be considerably lower. However P in A8, the cemented soil Scooby Doo, is definitely low. *Lower panels:* Linear regression lines for K and Cl. Although Cl in the soil is most likely to a large extent due to the presence of evaporates formed by HCl from volcanic exhalation, the PF rocks seem to contain intrinsic Cl, too.

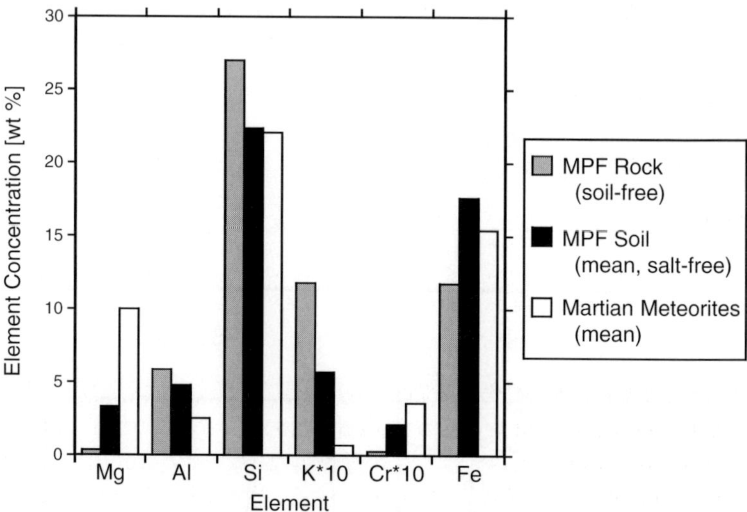

Figure 5. Histogram of some element concentrations in MPF rock (soil free), MPF soil and the mean composition of 14 Martian meteorites.

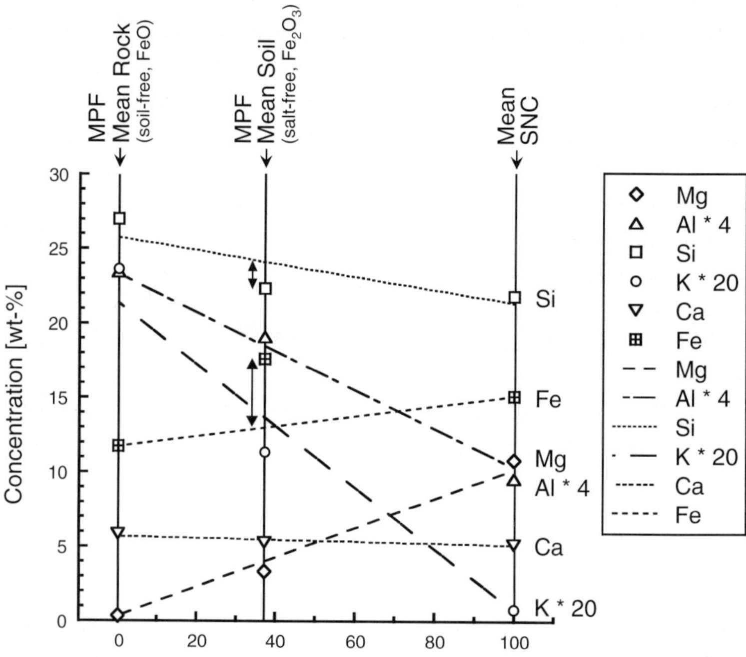

Figure 6. Two component mixing diagram of the Martian soil. Iron is assumed to be present in the soil as Fe_2O_3, but in form of FeO for the endmember rocks. MPF (Mars Pathfinder) Mean Rock is corrected for soil free.

TABLE II

Composition of the Martian soil: Comparison of measurements with the composition derived from a mixing model, which yields the following components from a least squares fit: Andesitic component (soil free PF rocks) = 54%, basaltic component (mean composition of SNC meteorites) = 28%, Mg-sulfate = 10%, hematite = 7%, and ilmenite = 0.9%.

Soil	MgO	Al_2O_3	SiO_2	SO_3	K_2O	CaO	TiO_2	Fe_2O_3
Calculated	8.6	7.4	42.6	6.9	0.7	6.2	1.0	22.3
Measured	8.7	8.0	42.3	6.8	0.6	6.5	1.0	22.3

with Thermal Emission Spectrometer (TES) data from the Mars Global Surveyor (MGS). The soil composition and its relation to specific rock types, respectively to their weathering products, has been discussed by a number of authors (e.g. Bell *et al.*, 2000; McLennan, 2000; McSween and Keil, 2000). McSween and Keil favor the view that basalts, chemically similar to basaltic shergottites, and evaporitic salts dominate the surface geology of Mars. We believe that admixture of the andesitic component is needed in addition, especially to account for K.

In the two-component mixing diagram (Figure 6), $MgSO_4$ and $MgCl_2$ have been subtracted. There is a reasonable fit for all elements except for Fe. The definite difference in the case of Fe between the concentration predicted from the two-component mixing diagram suggests that in addition to the above mentioned major components, an Fe rich component has been admixed to the soil, too. Areas rich in hematite have recently been observed at the surface of Mars (Christensen *et al.*, 2000). Hence, addition of Fe from hematite deposits seems plausible. Table II shows an attempt to account for the Martian soil by adding $MgSO_4$ and hematite (plus small amounts of ilmenite) to the basaltic and andesitic component. However, one should bear in mind that both $MgSO_4$ and hematite were originally supplied by weathering of basaltic material. Hence, the conclusion of an $\sim$1:1 abundance of andesitic and basaltic material is still justified. Pathfinder rock compositions, extrapolated to 0.3 wt-% S to account for the adhering soil, contain 57.0% SiO_2 and thus fall on the border line between basaltic andesites and andesites, according to the chemical classification of volcanic rocks after Le Bas *et al.* (1986).

5. Petrogenetic Model

Although a magmatic origin of Pathfinder rocks is far from being proven, the hypothesis of their formation as late differentiates of more primitive Martian magmas deserves to be tested. Out of different groups of SNC meteorites, widely believed to be the fragments from Martian surface (Bogard and Johnson, 1983; McSween, 1994; Walker *et al.*, 1979), shergottites belong to magmatic rocks with the largest proportion of the crystallized intercumulus melt. Compositions of Martian basaltic meteorites impose certain constraints on the mechanism of magma generation in the mantle of this planet (Ryabchikov and Wänke, 1996). It has been concluded that magmas with compositions similar to the bulk composition of Shergotty cannot be produced by direct melting of Mars primitive mantle (Wänke and Dreibus, 1988). Instead, they require a source, which may be derived from Dreibus-Wänke primitive mantle, after it had lost $\sim 5\%$ garnet (possibly at the stage of a global magma ocean), and also after subsequent extraction of $\sim 20\%$ basaltic material, which may be stored now in the crust of Mars. This source has a harzburgite composition, and the modelling of its partial melting at $1.3 - 0.8$ GPa yields magmas, which, after the fractional crystallization of $\sim 10\%$ olivine, may produce melts similar to the Shergotty bulk composition (Ryabchikov and Wänke, 1996).

On the Earth, second-stage partial melting of refractory harzburgites takes place during the subduction of young and, thus, hot suboceanic lithosphere at relatively shallow depths. It produces boninite magmas, which, being highly magnesian melts, are relatively silica-rich, and they crystallize substantial amounts of low-Ca pyroxenes (Crawford *et al.*, 1989). Like boninites, shergottites are characterized by elevated molar silica contents, and low-Ca pyroxenes in them play an important role. Terrestrial boninites are parental melts to magmatic series which in many cases include siliceous differentiates: andesites and even dacites. This evolution trend is facilitated by elevated SiO_2 contents in the initial magmas. Given this, production in Martian magma chambers of silica-rich melts similar to PF rocks during the fractional crystallization of shergottite-like primitive magmas seems quite possible. The evolution of shergottite-like melts towards silicic composition during fractional crystallization is supported by the presence of SiO_2-rich residual glass in Shergotty (Stolper and McSween, 1979). Numerical modeling of the fractional crystallization of a melt of Shergotty composition using the COMAGMAT program (Ariskin *et al.*, 1993) showed that at oxygen fugacity close to the Quartz-Magnetite-Fayalite (QMF) buffer residual melt with a composition close to the PF soil-free rock (henceforth PFSFR) may be produced (Dreibus *et al.*, 1998), whereas the MELTS program (Ghiorso and Sack, 1995) for the same purpose yielded negative results (McSween *et al.*, 1999). Neither computer program is well calibrated for compositions similar to Martian rocks, and, therefore, both results are not conclusive. In an effort to assess the crystallization trend of Martian magmas, Figure 7 plots the projections of shergottite bulk composition together with compositions of residual glass from Shergotty and the PFSFR onto the ternary joins of the system CaO-MgO-FeO-Al_2O_3-SiO_2 with $MgO + FeO$ treated as a single component. These diagrams show that more Mg-rich basaltic shergottites plot close to an olivine control line* for some common primary magma. In fact, there are likely to be a range of primary magmas with similar compositions. It follows from

---

* Olivine control line is a line, emanating from the projection of olivine composition (OL in Figure 7), which corresponds to the compositional trend of magmas or magmatic cumultes, caused by the crystallization of olivine as the only solid phase.

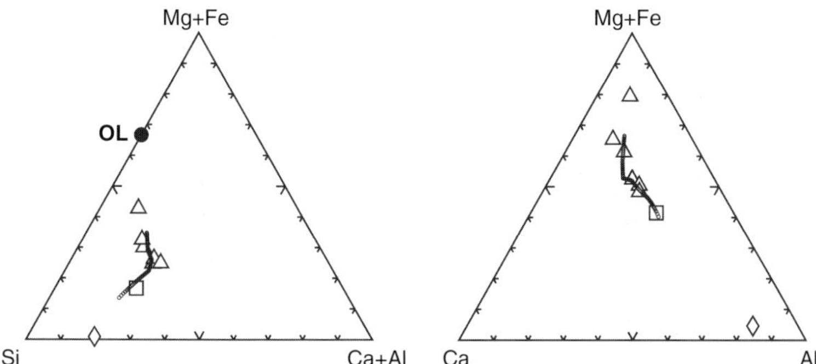

Figure 7. Projections of basaltic shergottites (*triangles*; ordered by decreasing Mg+Fe fraction on Ca-(Mg+Fe)-Al projection: Dar al Gani 476, Zagami, Shergotty, EETA B, Los Angeles, QUE 94201), late interstitial glass of Shergotty (*rhomb*; Stolper and McSween, 1979) and PF soil free rock (*square*) on Ca-(Mg+Fe)-Al and Si-(Mg+Fe)-(Ca+Al) compositional triangles (atomic proportions). The line shows the evolution of the melt composition during the fractional crystallization (at 1 atm and an oxygen fugacity 0.5 log unit above QMF buffer) of the estimated primary magma of Shergotty (Ryabchikov and Wänke, 1996) calculated using the COMAGMAT program (Ariskin *et al.*, 1993) with the revised parameters for equilibria with Fe-Ti oxides (Ariskin and Barmina, 1999).

these data that Dar al Gani 476 probably contains accumulated olivine, whereas Shergotty and Zagami are likely to have been produced by olivine fractionation from more primitive melts (Ryabchikov and Wänke, 1996). Shergotty and Zagami do not contain early olivine but probably the melts of their bulk compositions are close to olivine saturation, and they plot close to the end of an olivine control line. More evolved shergottite melts (EETA 79001B, Los Angeles, QUE 94201) deviate from an olivine control line due to the prevailing crystallization of low-Ca pyroxenes. At still later stages, the melt evolves towards the composition of residual glass in Shergotty, because the fractionation of other minerals (pyroxenes, feldspars and possibly titanomagnetite) controls the evolution of melt composition. It can be seen from Figure 7 that PFSFR plots close to the line connecting points for evolved shergottite bulk compositions and residual glass from Shergotty, which suggests that melts similar to PFSFR could have been produced during the fractional crystallization of more primitive shergottite-like melts.

During this process, a certain reduction of the Fe content should take place. This may be achieved through the crystallization of titanomagnetite. This would imply that titanomagnetite should be a liquidus mineral for PSFSR melt, which would require a certain level of oxygen fugacity. The conditions of equilibria of Fe-Ti oxides with melts of various composition has been recently formulated by Ariskin and Barmina (1999). Using their results one may estimate for 1 atm pressure, that titanomagnetite becomes a liquidus mineral for the PFSFR composition at f_{O_2} approximately 0.5 log unit above the QMF buffer. The f_{O_2} values for the late stage of Shergotty formation had been estimated also to be close to the QMF buffer (presence of late fayalite-rich olivine, cristobalite and titanomagnetite, Stolper and McSween, 1979). Therefore, the formation of PFSFR-like melts during the fractional crystallization of Shergotty-like primitive magmas at this level of oxygen fugacity, which is also typical for terrestrial magmatism (Haggerty, 1978), may be considered as a plausible mechanism of the formation of the magmatic precursors of PF rocks. During the accretion of Mars, substantial amounts of volatile-rich material must have been involved (Wänke and Dreibus, 1988) and intense degassing took place, which is likely to have resulted in relatively elevated average oxygen fugacity in the Martian mantle. In the case of more reduced conditions, initital melts of boninitic or basaltic composition may also produce Si-rich melts as late differentiates due to the loss of Fe in form of Fe-rich immiscible melt. Such a process has been observed in many terrestrial and lunar rocks (Philpotts, 1982; Roedder, 1951, 1992; Roedder and Weiblein, 1970; Ryabchikov *et al.*, 1985).

6. Summary

The first in situ analysis of Martian rocks by the Pathfinder mission provided a wealth of information not only about its specific landing site but also about the global surface chemistry of Mars and the planet's geochemical evolution:

1. Martian soil likely represents material from diminuated rocks of varied composition and their reaction products with volcanic gases, especially SO_2 and HCl, mechanically mixed and distributed on a global scale by dust storms.
2. The felsic endmember rocks have high SiO_2, high K, but low Mg contents, chemically corresponding to terrestrial andesites or icelandites. The second endmember rocks have basaltic composition similar to the mean composition of Martian meteorites.
3. The geologic units of andesitic and basaltic compositions cover similar portions of the Martian surface as suggested by Brückner *et al.* (1999), and as recently confirmed by TES data from the MGS orbiter (Bandfield *et al.*, 2000).
4. In addition, the Martian soil seems to contain $\sim$10% $MgSO_4$ and $\sim$1% chloride either in form of NaCl or $MgCl_2$, derived at least in part from evaporates, as well as about 7% hematite, plus small amounts of Ti-containing phases.
5. Carbonates are low or absent in the Martian soil as was predicted by Wänke and Dreibus (1994) because of the dominance of sulfates.
6. SNC meteorites are confirmed to originate from Mars, as they have element variation patterns identical to PF rocks and similar high abundances of P.
7. The high concentrations of P and Fe in the PF rocks confirm the prediction by Dreibus and Wänke (1987) for the Martian mantle.
8. The K concentrations in the Martian soil ($\sim$0.5%) are considerably higher than those reported by the Viking missions, but are in line with gamma-ray data as reported by Surkov *et al.* (1989).
9. At present, it is not possible to decide conclusively on the genesis – sedimentary or igneous – of the PF rocks. However, from a petrogenetic point of view, an igneous origin is theoretically possible.

7. Acknowledgements

We gratefully acknowledge the very detailed suggestions from W.K. Hartmann and H.Y. McSween, Jr., as reviewers which improved this paper considerably.

References

Ariskin, A., and Barmina, G.: 1999, 'An Empirical Model for the Calculation of Spinel-melt Equilibria in Mafic Igneous Systems at Atmospheric Pressure: 2. Fe-Ti Oxides', *Contrib. Mineral. Petrol.* **134**, 251–263.

Ariskin, A.A., Frenkel, M.Y., Barmina, G.S., and Nielsen, R.L.: 1993, 'Comagmat: A FORTRAN Program to Model Magma Differentiation Processes', *Comput. Geosci.* **19**, 1155–1170.

Bandfield, J.L., Hamilton, V.E., and Christensen, P.R.: 2000, 'A Global View of Martian Surface Compositions from MGS-TES', *Science* **287**, 1626–1630.

Banin, A., Clark, B.C., and Wänke, H.: 1992, 'Surface Chemistry and Mineralogy', *in* H.H. Kieffer *et al.* (eds.), *Mars*, Univ. Arizona Press, Tucson, pp. 594–625.

Bell, J.F., III, *et al.*: 2000, 'Mineralogic and Compositional Properties of Martian Soil and Dust: Results from Mars Pathfinder', *J. Geophys. Res.* **105**, 1721–1755.

Bogard, D.D., and Johnson, P.: 1983, 'Martian Gases in an Antarctic Meteorite', *Science* **221**, 651–654.

Brückner, J., Dreibus, G., Lugmair, G.W., Rieder, R., Wänke, H., and Economou, T.: 1999, 'Chemical Composition of the Martian Surface as Derived from Pathfinder, Viking, and Martian Meteorite Data', *Proc. 30th Lunar Planet. Sci. Conf.*, LPI, Houston, abstract #1250 (CD-ROM).

Brückner, J., Dreibus, G., Rieder, R., and Wänke, H.: 2001, 'Revised Data of APXS Analyses of Soils and Rocks at the Pathfinder Site', in preparation.

Christensen, P.R., *et al.*: 2000, 'Detection of Crystalline Hematite Mineralization on Mars by the Thermal Emission Spectrometer: Evidence for Near-surface Water', *J. Geophys. Res.* **105**, 9623–9642.

Clark, B.C., Baird, A.K., Weldon, R.J., Tsusaki, D.M., Schnabel, L., and Candelaria, M.P.: 1982, 'Chemical Composition of Martian Fines', *J. Geophys. Res.* **87**, 10,059–10,067.

Clark, B.C.: 1999, 'On the Non-observability of Carbonates on Mars', *5th Int. Conf. on Mars, LPI Contrib.* **972**, LPI, Houston, abstract #6214 (CD-ROM).

Crawford, A.J., Falloon, T.J., and Green, D.H.: 1989, 'Classification, Petrogenesis and Tectonic Setting of Boninites', *in* A.J. Crawford (ed.), *Boninites and Related Rocks*, Unwin and Hyman, pp. 1–49.

Dreibus, G., and Wänke, H.: 1987, 'Volatiles on Earth and Mars: A Comparison', *Icarus* **71**, 225–240.

Dreibus, G., Ryabchikov, I., Rieder, R., Economou, T., Brückner, J., McSween, M.Y., Jr., and Wänke, H.: 1998, 'Relationship Between Rocks and Soil at the Pathfinder Landing Site and the Martian Meteorites', *Proc. 29th Lunar Planet. Sci.*, LPI, Houston, abstract #1348 (CD-ROM).

Ghiorso, M.S., and Sack, R.S.: 1995, 'Chemical Mass Transfer in Magmatic Processes, IV. A Revised and Internally Consistent Thermodynamic Model for the Interpolation and Extrapolation of Liquid-solid Equilibria in Magmatic Systems at Elevated Temperatures and Pressures', *Contrib. Mineral. Petrol.* **119**, 197–202.

Golombek, M.P., *et al.*: 1999, 'Overview of the Mars Pathfinder Mission: Launch through Landing, Surface Operations, Data Sets, and Science Results', *J. Geophys. Res.* **104**, 8523–8553.

Haggerty, S.E.: 1978, 'The Redox State of Planetary Basalts', *Geophys. Res. Lett.* **5**, 443–446.

Le Bas, M.J., Le Maitre, R.W., Streckeisen, A., and Zanettin, B.: 1986, 'A Chemical Classification of Volcanic Rocks Based on the Total Alkali-silica Diagram', *J. Petrol.* **27**, 745–750.

McLennan, S.M.: 2000, 'Chemical Composition of Martian Soil and Rocks: Complex Mixing and Sedimentary Transport', *Geophys. Res. Lett.* **27**, 1335–1338.

McSween, H.Y., Jr.: 1994, 'What we Have Learned About Mars from SNC Meteorites', *Meteoritics* **29**, 757–779.

McSween, H.Y., Jr., *et al.*: 1999, 'Chemical, Multispectral, and Textural Constraints on the Composition and Origin of Rocks at the Mars Pathfinder Landing Site', *J. Geophys. Res.* **104**, 8679–8715.

McSween, H.Y., Jr., and Keil, K.: 2000, 'Mixing Relationships in the Martian Regolith and the Composition of Globally Homogeneous Dust', *Geochim. Cosmochim. Acta* **64**, 2155–2166.

Palme, H., Suess, H.E., and Zeh, H.D.: 1981, 'Abundances of the Elements in the Solar System', *in* K. Schaifers and H.H. Voigt (eds.), Landoldt-Börnstein **2**, Springer, Berlin, pp. 257–273.

Philpotts, A.R.: 1982, 'Compositions of Immiscible Liquids in Volcanic Rocks', *Contrib. Mineral. Petrol.* **80**, 201–218.

Rieder, R., Wänke, H., Economou, T., and Turkevich, A.: 1997a, 'Determination of the Chemical Composition of Martian Soil and Rocks: The Alpha Proton X-Ray Spectrometer', *J. Geophys. Res.* **102**, 4027–4044.

Rieder, R., Economou, T., Wänke, H., Turkevich, A., Crisp, J., Brückner, J., Dreibus, G., McSween, H.Y., Jr.: 1997b, 'The Chemical Composition of Martian Soil and Rocks Returned by the Mobile Alpha Proton X-ray Spectrometer: Preliminary Results from the X-ray Mode', *Science* **278**, 1771–1774.

Roedder, E.: 1951, 'Low Temperature Liquid Immiscibility in the System K_2O-FeO-Al_2O_3-SiO_2', *Am. Mineral.* **36**, 282–286.

Roedder, E.: 1992, 'Fluid Inclusion Evidence for Immiscibility in Magmatic Differentiation', *Geochim. Cosmochim. Acta* **56**, 5–20.

Roedder, E., and Weiblein, P.W.: 1970, 'Lunar Petrology of Silicate Melt Inclusions, Apollo 11 Rocks', *Proc. Apollo 11 Lunar Sci. Conf., Geochim. Cosmochim. Acta, Suppl.* **1**, 801–837.

Ryabchikov, I.D., Solovova, I.P., Babansky, A.D., and Borsuk, A.M.: 1985, 'Origin and Differentiation Conditions of Strongly Reduced Andesitic Magmas', *IZV. AN SSSR, ser. Geol.* **10** (in Russian).

Ryabchikov, I.D., and Wänke, H.: 1996, 'Magma Generation in Martian Mantle', *Geochemistry International* **34**, 621–627.

Stolper, E., and McSween, H.Y., Jr.: 1979, 'Petrology and Origin of the Shergottite Meteorites', *Geochim. Cosmochim. Acta* **43**, 1475–1498.

Surkov, Yu.A., Barsukov, V.L., Moskaleva, L.P., Kharyukova, V.P., Zaitseva, S.Ye., Smirnov, G.G., and Manvelyan, O.S.: 1989, 'Determination of the Elemental Composition of Martian Rocks from Phobos 2', *Nature* **341**, 595–598.

Walker, D., Stolper, E.M., and Hays, J.F.: 1979, 'Basaltic Volcanism: The Importance of Planet Size', *Geochim. Cosmochim. Acta, Suppl.* **11**, 1995–2015.

Wänke, H., and Dreibus, G.: 1988, 'Chemical Composition and Accretion History of Terrestrial Planets', *Phil. Trans. R. Soc. Lond.* **A325**, 545–557.

Wänke, H., and Dreibus, G.: 1994, 'Chemistry and Accretion History of Mars', *Phil. Trans. R. Soc. Lond.* **A349**, 295–293.

Address for correspondence: H. Wänke, Max-Planck-Institut für Chemie, Becher-Weg 27, D-55128, Mainz, Germany; (waenke@mpch-mainz.mpg.de)

III: HISTORY AND FATE OF THE MARTIAN ATMOSPHERE AND HYDROSPHERE

GEOMORPHOLOGIC EVIDENCE FOR LIQUID WATER

PHILIPPE MASSON[1], MICHAEL H. CARR[2], FRANÇOIS COSTARD[1], RONALD GREELEY[3], ERNST HAUBER[4] and RALF JAUMANN[4]

[1] *Université Paris-Sud, Equipe de Planétologie (UMR CRNS 8616), 91405 Orsay Cedex, France*
[2] *U.S. Geological Survey, MS-975, 345 Middlefield Road, Menlo Park, CA 94025, USA*
[3] *Department of Geological Sciences, Arizona State University, Box 871404, Tempe, AZ 85287-1404, USA*
[4] *DLR Institute of Space Sensor Technology and Planetary Exploration, Rutherfordstrasse 2, 12484 Berlin - Adlershof, Germany*

Received: 20 December 2000; accepted: 19 February 2001

Abstract. Besides Earth, Mars is the only planet with a record of resurfacing processes and environmental circumstances that indicate the past operation of a hydrologic cycle. However the present-day conditions on Mars are far apart of supporting liquid water on the surface. Although the large-scale morphology of the Martian channels and valleys show remarkable similarities with fluid-eroded features on Earth, there are major differences in their size, small-scale morphology, inner channel structure and source regions indicating that the erosion on Mars has its own characteristic genesis and evolution. The different landforms related to fluvial, glacial and periglacial activities, their relations with volcanism, and the chronology of water-related processes, are presented.

1. Introduction

Besides Earth, Mars is the only planet with a record of resurfacing processes and environmental circumstances that indicate the past operation of a hydrologic cycle. In 1972 for the first time the Mariner 9 photographs showed large erosion features of giant channels and branching networks of small valleys on Mars. The Viking orbiter images obtained in 1975 provided more details of these channels and valleys indicating many flow-like features which seemed to be formed by running water (Masursky *et al.*, 1977; Baker and Kochel, 1979; Pieri, 1980; Carr, 1981). However the present-day conditions on Mars with average temperatures well below freezing and atmospheric pressures at or below the 6.1 mbar triple point vapor pressure of water are far apart of supporting liquid water on the surface.

Although the large-scale morphology of the Martian channels and valleys show remarkable similarities with fluid-eroded features on Earth, there are major differences in their size, small-scale morphology, inner channel structure and source regions indicating that the erosion on Mars has its own characteristic genesis and evolution. The details of the channel forming processes are still unknown. Formerly, in analogy to similar features on Earth and with respect to their morphologic development the Martian valleys have been categorized as outflow channels, fretted channels, run off channels and quasi-dentritic networks (Sharp and Malin,

Space Science Reviews **96**: 333-364, 2001.
© 2001 *Kluwer Academic Publishers. Printed in the Netherlands.*

1975; Carr, 1981). However these names anticipate a formation process by running water that may not be justified in all cases.

The valley networks are almost entirely restricted to the old uplands and the simplest explanation is that the valleys are old themselves and the climatic requirements for valley formation were met early in the planet's history and rapidly declined during the subsequent evolution. A warmer, wet Mars with a dense atmosphere at the time after the heavy bombardment is supposed to provide the conditions for valley formation by running water. However, recent images obtained by the the Mars Observer Camera (MOC) experiment onboard Mars Global Surveyor (MGS) show relatively young small-scale alcove-like gullies combined with small channels and aprons in the walls of impact craters, thus indicating even recent groundwater seepage and probably short-term surface runoff under almost current climatic conditions (Malin and Edgett, 2000a).

2. Landform Related to Fluvial, Glacial, and Periglacial Activity

2.1. CHANNELS AND VALLEY NETWORKS

Many attempts have been made to explain the Martian erosion without including water such as formation by lava flows, winds, debris flows and liquid hydrocarbons (Carr, 1977; Schonfeld, 1976; Cutts and Blasius, 1981; Nummedal, 1978; Nummedal and Prior, 1981; Yung and Pinto, 1978), but none of these theories is able to comprehensively describe the formation of the associated flow-like features. Therefore most scientists investigating erosion landforms on Mars argue that water either in liquid form or as ice has been involved in the development of the Martian erosion (Baker *et al.*, 1992; Carr, 1996a; Lucchitta and Anderson, 1980). This led to the assumption that liquid water might have been stable on the surface in the past history of Mars (Sharp and Malin, 1975; Carr, 1981; Mars Channel Working Group, 1983; Baker *et al.*, 1992). Among several authors, glacial landforms on Mars have been reported by Lucchitta *et al.* (1981) and Lucchitta (1982), who questioned the hypothesis that outflow channels were generated by catastrophic floods of liquid water. Recent discussions about Martian erosion clearly distinguish between a fluvial channel, which is the conduit through which a river flows, and a valley, which refers to a linear depression and a fluvial valley which generally contains many channels (Mars Channel Working Group, 1983; Carr, 1996a).

Outflow Channels. These are several to tens of kilometers across, reaching length of a few hundreds to thousands of kilometers with gradients of channel floors ranging from 0 to 2.5 m/km (Baker *et al.*, 1992). Tributaries are rare, but branching downstream is common resulting in an anastomosing pattern of channels (Carr, 1996a). The channels tend to be deeper at their source than downstream and in general have a high width to depth ratio and low sinuosities (Baker *et al.*, 1992; Carr, 1996a). Isolated upland remnants separating channels, mesa-like remnants at channel borders, hanging valleys, transected divides, and expansion and constrictions of the channel beds are common. A variety of bedforms are located on the floors of outflow channels (Baker, 1979, 1982; Baker *et al.*, 1992; Carr, 1979, 1986; Lucchitta, 1982; Komar, 1983, 1984; Mars Channel Working Group, 1983) including residual or recessional headcuts, scour marks around presumed obstacles, longitudinal grooves which parallel the presumed flow direction, teardrop-shaped islands, inner channels, terraces, cataracts and plucked zones, similar to scabland topography which forms by stripping away surface materials due to high-velocity turbulent fluid flow (Baker, 1982). Small streamlined features are also present in many channels and may be deposition bars.

The distribution of outflow channels is restricted to four main areas: the vicinity of the Chryse-Acidalia basin, west of the Elysium volcano complex in Elysium Planitia, the eastern part of Hellas basin and along the western and southern border of Amazonis Planitia. By far the largest channels occur around the Chryse basin. Outflow channels in this area start in the cratered Xanthe Terra uplands or on high volcanic plains at elevations up to 4km above the datum and converge in the Chryse basin mainly from the south but also from the southeast and west. On the Chryse plains they broaden and finally disappear at about 40°N latitude. Most of the out-flow channels around Chryse emanate fully developed from circular to elliptical depressions termed chaotic terrain (Sharp, 1973a). These areas are $1 - 2$ km below the surrounding undisturbed terrain and mostly covered with large jumbled blocks which are obviously material from the former surface, indication that the chaotic terrain has formed by collapse rather than by removal of material from the above, indicating the involvement of groundwater in the channel formation process. Many of the chaotic terrain east of the Valles Marineris merge westward with the canyon system and northwards with outflow channels. In the Elysium region large outflow channels originate at fractures oriented radial to the west and northwest flanks of the Elysium volcano complex. The channels mostly start at elongated depressions that resemble oversized gullies. At the beginning the channels are rectilinear and become more sinuous and fluvial-like as the circum-Chryse channels downstream to the northwest. The rectilinear channel pattern best observed in Hephaestus Fossae at 22°N, 239°W, indicates that upstream the erosion works along lines of structural weakness voting for subsurface erosion (Carr, 1996a) and fast release of water. Outflow channels also occur at the eastern flank of the Hellas basin. Again these channels originate full sized in elongated depressions and extend downstream resembling those around Chryse into the floor of the impact basin.

Valley Networks. Open branching valleys in which tributaries merge down-stream, i.e. Valley Networks, are common in the southern highlands. Although these valleys resemble terrestrial river valleys they are far less complex than their analogues on Earth (Carr, 1995). The number of branches is low with large undis-sected areas between individual branches. The networks themselves are spaced apart leaving large areas of undissected highland between them (Pieri, 1980) indi-cating that there has been no or only little competition between adjacent drainage basins (Carr and Clow, 1981). Individual branches are characterized by their U-shaped form with flat floors and steep walls. They vary from channels by the absence of bedforms, which are indicators for fluid flow (Mars Channel Working Group, 1983). Some valley systems are several 100 km long and a few 10 km wide with short accordant tributaries such as Nirgal Vallis. However, the majority of the valley systems is typically no longer than 200 km and only a few km wide. The valley networks are mainly located in the cratered uplands, the oldest Martian terrain at elevations ranging from 2 to over 5 km. Some valley networks drain into craters, but most of them start at local heights and drain as winding streams with relatively large junction angles of branches into low areas between large impact craters, where they terminate against smooth plains (Carr, 1995). Another set of channel networks occurs on volcanoes. Although these channels usually are caused by lava or volcanic density flows ("nuées ardentes"; Reimers and Komar, 1979), there is evidence that valleys on Alba Patera, Ceraunius Tholus and Hecates Tholus as well as on Hadriaca, Thyrrena and Apollinaris Paterae are compatible with a fluvial origin (Wilson and Mouginis-Mark, 1987; Mouginis-Mark *et al.*, 1988; Gulick and Baker, 1990). These valleys are narrower and shallower than their counterparts in the uplands and show a more dense drainage pattern with small junction angles. At some places the walls of canyons are deeply incised by V-shaped valleys that, however, lack deposits at their mouth within the canyon (Carr, 1996a).

At the upland lowland boundary between about $290° - 360°$W and $30° - 50°$N fretted channels are exposed. In this area the northern plains interfere in a complex way with the highlands. Numerous islands with $1 - 2$ km high escarpments are isolated from the uplands (Sharp, 1973b). The isolated mesa-like islands are surrounded by aprons with well-developed flow fronts (Squyres, 1978). Flat floored broad (up to 20 km wide) steep-walled channels reach deep into the uplands. These fretted channels branch upstream as do valley networks. Some of these channels have longitudinal ridge-like features on their floors, indicating compressional features caused by merging aprons from the walls (Squyres, 1978) and removal of materials through the channel by mass wasting downstream (Carr, 1995). Undermining and collapse also appear to have been a process in forming the fretted channels (Lucchitta, 1984). However, the distinct differences between fretted channels and outflow channels on the one hand and valley networks on the other as well as the lack of direct evidence for fluvial erosion (Carr, 1995) indicate that fretted channels may play a major role in the processes of Martian erosion.

Valley networks in the uplands are much smaller than outflow channels indicating smaller corresponding discharges. Therefore, it must be assumed that smaller branches of the network froze very rapidly and cut off downstream motion resulting in arresting the further development of the valleys (Baker *et al.*, 1992). However, the pattern of dissection produced by valley networks, the drainage development and the general valley morphologies have been judged to be similar to those of terrestrial rivers and a similar origin has been implied. This led to the assumption that the valley networks mainly have formed by running water, and most of the discussion of the origin focused on the relative roles of surface runoff and sapping

(Sharp and Malin, 1975; Pieri, 1976, 1980; Carr, 1981; Laity and Malin, 1985; Baker, 1982, 1990). Consequently these assumption had implications on the evolution of the Martian climate (Carr, 1983, 1996b; Squyres, 1989a; Baker *et al.*, 1991, 1992; Squyres and Kasting, 1994).

Outflow channels, their source regions and termination areas in the northern lowlands provide the best evidence for surface water on Mars and a widespread groundwater system. The outflow channels can form by running water under current climatic conditions because the large amount of released water will prevent freezing and sublimation at least during the relatively short periods of flooding (Carr, 1979, 1995). The source regions of all outflow channels are directly related to tectonic and in the Elysium and Hellas region to volcanic features. This evidence together with the fluid-like erosion features of the channels indicate that there is a close correlation with fast groundwater release triggered by melting of ground ice (Costard *et al.*, 1999; Aguirre-Puente *et al.*, 1994), tectonically thinning of the cryosphere and thereby providing easy access of groundwater to the surface.

Compared to alternate hypotheses, such as erosion by lava, wind, or other fluids like liquid hydrocarbons, the hypothesis of the catastrophic release of groundwater or drainage of surface lakes accounts for most of the observed channel features (Baker *et al.*, 1992). The dimensions of the outflow channels indicate discharges in the order of $10^7 - 10^9$ m^3 s^{-1} (Carr, 1979; Komar, 1979; Baker, 1982; Robinson and Tanaka, 1990). This is about two orders of magnitude larger than the largest known flood events on Earth such as the peak discharges of the Channeled Scabland flood of eastern Washington or the Chuja Basin flood in Siberia (Baker, 1973; Baker *et al.*, 1992). Several processes are discussed that may cause large floods on Mars. Groundwater under high artesian pressure confined below a permafrost zone may break out triggered by events such as impacts or marsquakes, which disrupt the permafrost seal either by breaking the surface or sending a large pressure pulse through the aquifer (Carr, 1979, 1995, 1996a). This hypothesis requires a thick permafrost layer in order to build up the artesian pressure and thus is mainly consistent with climatic conditions similar to those of the present Mars. Another argument for this mechanism is the low elevation (<1 km) of the outflow channel source regions compared to the water level of the regional aquifer system (Carr, 1996b), which provides hydrostatic pressure. After groundwater release, the flood would have declined as the artesian pressure dropped and finally came to an end when the aquifer resealed by freezing and water could no longer reach the surface. The duration of massive flood events is short and has been estimated to last a few days to several weeks dependent on the discharge rates (Carr, 1996a). If the aquifer were recharged by groundwater flow from distant sources the flood could be repeated resulting in episodic events of catastrophic groundwater release.

Another water release process may be the melting of ground ice by volcanic heat that provides local and regional accumulation and migration of groundwater (Baker *et al.*, 1991; Hartmann, 2001). The outflow channels close to the Elysium volcanoes and Hadriaca Patera near the Hellas basin are possible examples for this mechanism. A higher thermal gradient will also be thinning the cryosphere providing easy access of the surface. Groundwater will also move preferentially along faults as indicated by the emergence of channels from fractures in Elysium and in Terra Sirrenum, where Mangala Vallis originates at about 16°S, 149°W from a graben and continues northwards for several 100 km into the Amazonis basin.

2.2. CRATER MORPHOLOGY

The particular pattern of some impact crater ejecta on Mars provides strong evidence for subsurface ice or liquid water. A combination of ballistic deposition and surface flow, caused by the presence of ground ice or ground water, was originally suggested by Carr *et al.* (1977a) to account for their unique morphology. While crater ejecta on the Moon are emplaced ballistically and their texture slowly and continuously grades radially out into secondary craters, crater ejecta on Mars exhibit flow-like morphologies, some of which include multiple flow-lobe terraces (Carr *et al.*, 1977a). These impact structures are termed ejecta flow craters or rampart craters. Most investigators attribute this morphology to the presence of liquid water or melted ice at the time of impact which mixed with the solid ejecta to create a slurry-like mass.

However, other investigators suggest that a similar morphology could result from the interaction of ejecta with the atmosphere (Schultz and Gault, 1979, 1984). Ejecta deposits from ejecta flow craters typically wrap around pre-impact terrain features, such as ridges, rather than being superposed as a mantle (Figure 1) (Carr, 1977; Carr *et al.*, 1977a; Carr, 1996b). This suggests that the ejecta was emplaced primarily as a surface or near-surface flow, in contrast to ballistically-emplaced ejecta. Moreover, estimates based on Viking Orbiter images show a greater volume of ejecta than the crater interior can accommodate, suggesting flow into the transient crater from the subsurface at the time of the impact (Greeley *et al.*, 1980). To assess the morphology of craters and ejecta deposits formed in water-rich targets, a series of experiments was conducted at the NASA-Ames Vertical Gun facility in which impacts were produced in various viscous materials (Gault and Greeley, 1978; Greeley *et al.*, 1980; Fink *et al.*, 1981). The resulting morphologies are qualitatively similar to the features observed on Mars, including multiple flow-lobes which resulted from oscillating central peaks in the experiment. Although the entrainment of atmospheric gases seems to be a possible alternative mechanism to produce fluid ejecta, most researchers view the dependence of rampart crater morphology on size and geographic latitude as a strong argument against this hypothesis and favour ejecta fluidization due to subsurface volatiles (Mouginis-Mark, 1979, 1981, 1987; Johansen, 1979; Battistini, 1984; Wohletz and Sheridan, 1983; Kargel, 1986; Costard, 1989).

According to studies on Martian impact craters by Barlow and Bradley (1990), most interior morphologies (except central peaks and peak rings) and all major ejecta patterns are controlled by the distribution of subsurface volatiles. Barlow

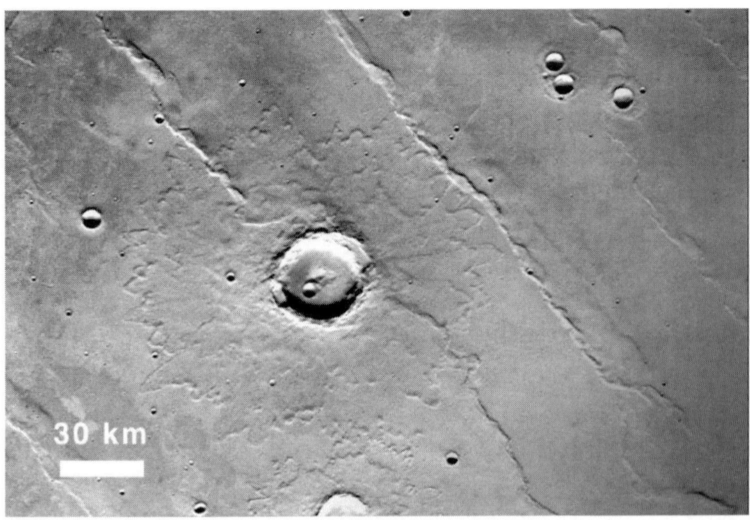

Figure 1. Rampart craters at 23°S, 79°W. The small craters, penetrating only the uppermost layers of the surface, display only one simple annular lobe (*upper right*), whereas the larger crater in the middle shows multiple lobes outlined by ramparts (Carr, 1996a). Viking orbiter image 608A45 (resolution 213 m/pixel).

et al. (2000) conclude that crater morphology is a useful tool to map the regional variations in water and ice content on Mars. Moreover, the diameters of the craters might indicate the depth to water or ice below the surface at the time of the impact. A short summary of Martian crater ejecta morphology by Boyce and Roddy (1997) confirms the earlier studies that changes in crater morphology are most simply (but not uniquely) explained by the distribution of subsurface water or ice.

Very small craters with diameters less than a few kilometers never have lobate ejecta. Assuming that ramparts also form due to groundwater or ground ice, Kuzmin *et al.* (1988) derived the global distribution of the water or ice depth from the minimum diameter of craters with rampart ejecta. The onset diameters are smaller at higher latitudes, indicating shallower levels of water-rich materials. Absolute values for the depth to this level are $\sim$300 − 400 m on the equator and 100 − 200 m at 65° latitude. For example, small craters would not penetrate to sufficient depth to reach deep subsurface water or ice. Using established values for excavation depths as a function of final crater diameter (Melosh, 1989), Kuzmin *et al.* (1988) mapped the depth to the cryolithosphere, based on the "threshold" diameter of ejecta flow craters. The results are consistent with models which predict the depth to the cryolithosphere, for an assumed thermal regime.

In another approach, the ratio of ejecta diameter to crater diameter has been used to gain information about the subsurface distribution of water. Again, a correlation to latitude has been found, the ratio being higher in higher latitudes (Blasius and Cutts, 1980; Mouginis-Mark, 1979; Kuzmin *et al.*, 1988; Costard, 1988; Squyres *et al.*, 1992). As already found for the onset diameters of rampart craters, the results seem to indicate more near-surface ice in the northern plains (Carr, 1996b). A recent global study of rampart craters confirms the earlier results and shows a correlation between high concentrations of high mobility ejecta and topography as derived from MOLA data (Costard and Kargel, 1999). The highest ratios of ejecta to crater diameter are found in large topographic depressions or basins more than 4000 m below the Martian datum near the mouth of outflow channels in Utopia and Acidalia Planitiae (a spatial relation which was already noted by Costard (1994). This is in agreement with high concentrations of fractured ground and thermokarst in the same regions and suggests the presence of a water or ice-rich subsurface, probably caused by sedimentation from the outflow channels.

2.3. PERIGLACIAL AND GLACIAL LANDFORMS

A number of Martian surface features (e.g., debris flows, polygons, thermokarst, frost mounds) seen in Mariner 9 and Viking Orbiter imagery have been attributed to periglacial or permafrost processes (Carr and Schaber, 1977; Rossbacher and Judson, 1981; Lucchitta, 1981, 1985; Squyres, 1978; Squyres and Carr, 1986; Squyres *et al.*, 1992; Carr, 1996b). The main importance of these features on Mars lies in the implications on the climatic conditions required for their formation. An additional point of interest is their potential as life-bearing environment (ESA Exobiology Study Team, 1999; Tsapin *et al.*, 1999; Brinton *et al.*, 1999).

Periglacial landforms on Earth are defined as climatically controlled surface features adjacent to the Pleistocene ice sheets (Lozinski, 1909; Washburn, 1973; Sharp, 1988). The diagnostic characteristics of periglacial domains are freezing and thawing of the ground (Tricart, 1968, 1969). Permafrost was defined by Muller (1947) as a layer of soil or bedrock in which temperatures are below the freezing point of water for a continuous period of more than two years, regardless of other parameters like texture, lithology, or water content. A unique assemblage of processes and landforms is related to this zone of permafrost that spans over 25% of Earth's solid surface (Washburn, 1973; French, 1976; Williams and Smith, 1989).

Debris Flow and Terrain Softening. The transition between the southern cratered uplands and the northern lowlands is often marked by a pronounced topographic scarp. At the base of this scarp, lobate debris aprons with well defined flow fronts and convex-upward surfaces (Figures 2 and 3) extend from the highlands and from isolated mesas into the low-lying plains (Sharp, 1973b). They have been interpreted as results from gelifluction and/or frost creep (Carr and Schaber, 1977). Squyres (1978) suggested flow of a mixture of erosional debris and ice incorporated from the atmosphere. Subsequently, Lucchitta (1984) supported most of Squyres' ideas, but preferred the idea that water might have been derived from ground ice by sapping or scarp collapse. Where the flow of material away from the scarp was confined by opposing scarp walls, as it is the case in valleys, the flow fronts collide to form linear ridges and grooves resembling terrestrial median moraines on glaciers (Figure 2). These deposits are commonly called lineated valley fill.

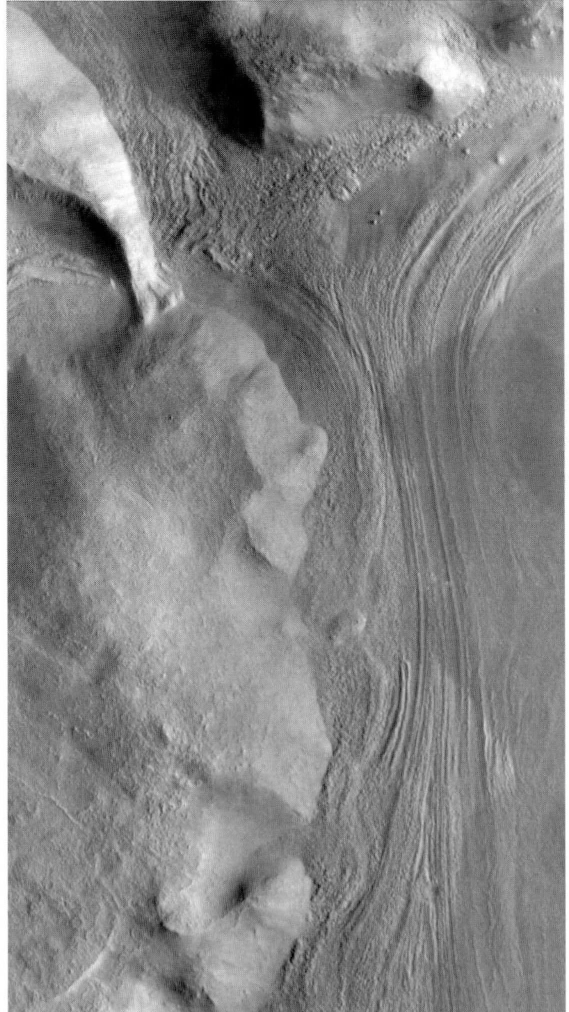

Figure 2. Lobate debris aprons and lineated valley fill in fretted terrain at 33.89°N and 290.07°W. The lobate debris aprons radiate from the left valley wall and show characteristics of plastic flow (Hamlin *et al.*, 2000). MOC image 45006, image width about 10 km (resolution 6.03 m/pixel).

In some places, the morphology of craters seems to be modified by flow of the terrain. Crater rims appear to be subdued and small craters are only sparsely visible (Jankowski and Squyres, 1992, 1993). Sometimes, material has obviously moved down the inner crater wall and forms what is called concentric crater fill (Figure 4). Calculations by Squyres (1989b) indicate that near surface flow (as opposed to flow of a >5 km deep viscous layer) was responsible for these features. All types of terrain softening are mostly seen in a latitudinal belt between 30° and 60° (Squyres, 1979; Carr, 1996a). This might be due to the fact that ice is not in equilibrium with the present atmosphere at low latitudes (Squyres and Carr,

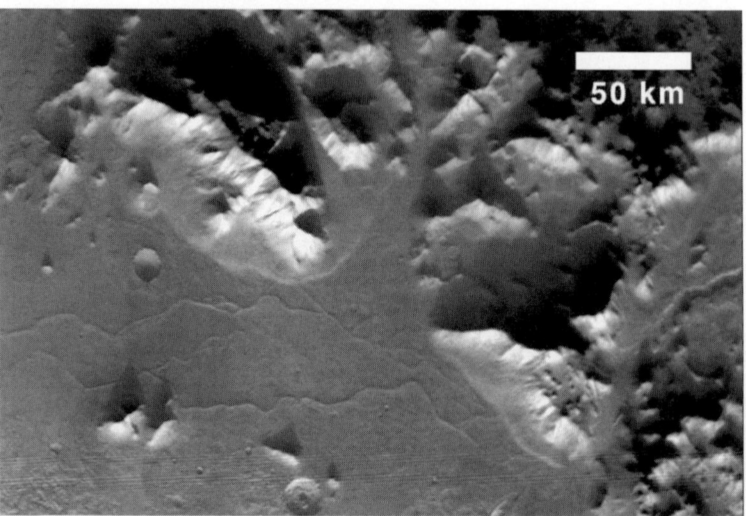

Figure 3. Glacial and periglacial features in Argyre at 52.99°S, 58.25°W. Lobate debris aprons surround the base of steep-sided mountains, esker-like sinuous ridges cover the plains of a hypothetical proglacial lake (*bottom*), and possible cirques (*e.g. upper middle part*) might have been the source regions for glaciers (Kargel and Strom, 1992). Viking orbiter image 352S34 (resolution 236 m/pixel).

1986). On the other hand, at very high latitudes $>60°$ the temperature are very low and, correspondingly, the strain rate of ice when subjected to shear stress drops significantly (Shoji and Higashi, 1978) and impedes flow.

The viscous flow features on Mars have been compared to rock glaciers (Squyres, 1989b; Colaprete and Jakosky, 1998; Rossi *et al.*, 2000), which are lobate accumulations of angular rock debris and move down due to deformation of internal ice or frozen sediments (Wahrhaftig and Cox, 1959; Barsch, 1988; Benn and Evans, 1998). Mangold *et al.* (2000a) used MOLA data to investigate the topography of lobate debris aprons. They confirmed the convex-upward shape of the lobes, and they agree with earlier assumptions that they form by creep of ground ice.

Polygons. Most polygonally fractured plains have been found in Acidalia, Elysium, and Utopia Planitiae (e.g. Pechmann, 1980), although they were identified also in the Valles Marineris (Blasius *et al.*, 1977) and in very high latitudes north of 70°N (Squyres *et al.*, 1992). Lucchitta *et al.* (1986a, 1986b) noted the close spatial relationship of the Acidalia and Utopia polygons to the mouths of the Chryse and Elysium outflow channels. While the diameters of these giant polygons were in the range of 30 km and more in Viking images (McGill, 1985, 1986; Pechmann, 1980; Lucchitta, 1983), recent observations by the MOLA and MOC experiments onboard MGS show that the average width and depth of troughs bordering the interior of polygons in Utopia Planitia is 2 km and 30 m, respectively (Hiesinger and Head, 2000). The location of polygons near the end of outflow channels lead Lucchitta and Ferguson (1983), McGill (1986), and Lucchitta *et al.* (1986a, 1986b) to the suggestion that ice-rich sediments were cooled and fractured after their deposi-

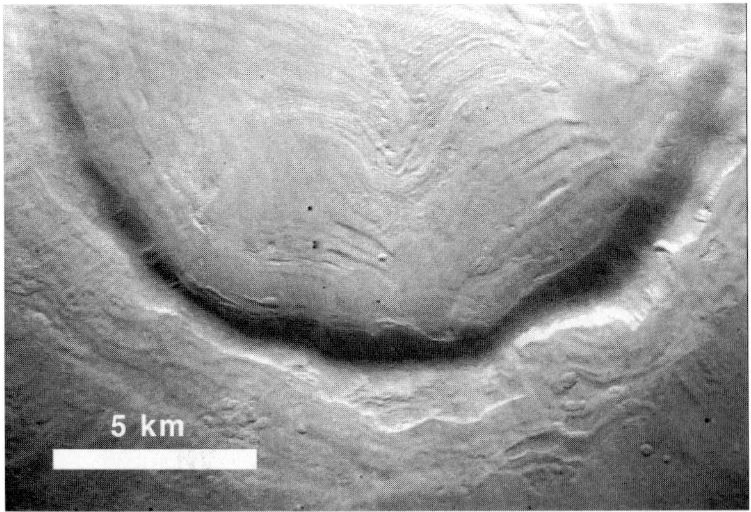

Figure 4. Concentric crater fill at 41.72°N, 272.45°W. Material has obviously flown down the inner crater wall. While an eolian origin has been suggested by Zimbelman *et al.* (1988a, 1988b) for these landforms, Squyres (1989b) and Carr (1996a) admit an eolian modification, but prefer viscous flow or creep as dominant process. Viking orbiter image 425B07 (resolution 16 m/pixel).

tion and/or self-compacted and accomodated to the underlying terrain. Dessication of wet sediments (deposited by water from the giant outflow channels) has been proposed by McGill and Hills (1992).

Pechmann (1980) applied the theory of Lachenbruch (1962) for terrestrial polygons to Martian giant polygons and found that thermal contraction is confined to the uppermost 10 m of the surface and will not produce deep cracks. The large size of giant polygons on Mars is, therefore, incompatible with a generation by thermal contraction. Instead, he preferred a tectonic origin. Hiesinger and Head (2000) also support a tectonic origin and suggest that it was caused by extension due to the uplift of the Utopia basin floor subsequent to the removal of a large body of standing water (Head *et al.*, 1998, 1999).

Another type of polygons has much smaller diameters of 10 − 100 m and was interpreted to be the result of thermal contraction in ice-rich soils based on the similar scales as compared to terrestrial ice-wedge polygons (Mutch *et al.*, 1977; Evans and Rossbacher, 1980; Brook, 1982; Lucchitta, 1983). Mellon (1997) investigated the physical processes producing thermal contraction stresses in Martian ice-cemented soils and expects the formation of contraction polygons with diameters in the 10 m range on Mars, mostly at mid and high latitudes. Many recent MOC images show polygons with relatively small diameters (Figure 5). They were investigated in detail by Yoshikawa (2000), who concludes that the formation of giant polygons is unlikely due to frost contraction cracking, while mid size (100 − 200 m diameter) and smaller polygons could be explained by ice-rich permafrost layers in the subsurface.

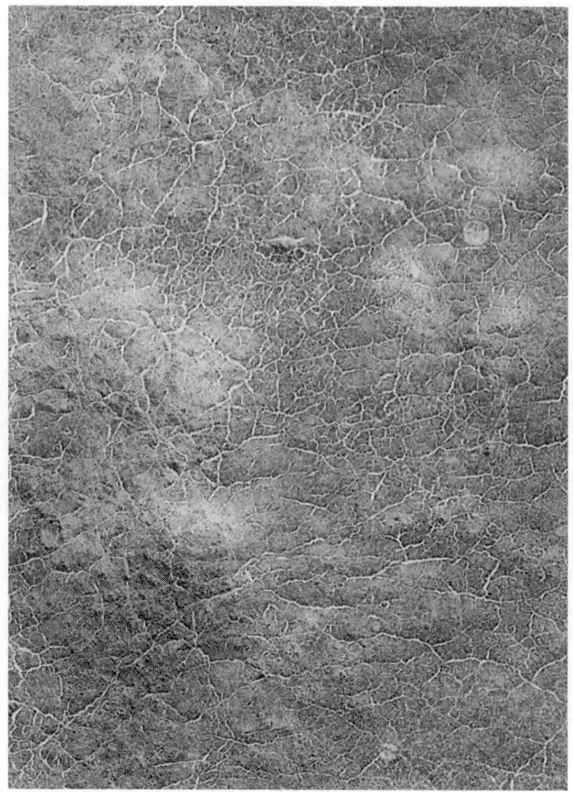

Figure 5. Polygonally fractured terrain in the plains of Malea Planum near the south pole. Remnant frost from the receding polar cap is trapped in the cracks and enhances the contrast. MOC image No. MOC2-189 (width 1.5 km, resolution 3 m/pixel) by NASA/JPL/Malin Space Science Systems.

Thermokarst. Thermokarst is defined as "karst-like features produced by the melting of ground ice and the subsequent settling or caving of the ground" (Muller, 1947; Popov, 1956). It is, therefore, the result of local or general, climatically induced permafrost degradation. Thermokarst features include linear troughs, collapsed pingos, beaded drainage patterns, thaw lakes, and alases (Kachurin, 1962; Washburn, 1973), and the term has also been applied to a variety of processes like cryoplanation, thermal abrasion, or thermal erosion (French, 1976).

Thermokarst on Mars was originally suggested by Sharp (1973a), Anderson *et al.* (1973), and Gatto and Anderson (1975), who noted the similarity of some Martian depressions to thermokarst features (thaw lakes and alases) in Alaska and Yakutia. However, in these and later works (Carr and Schaber, 1977; Theilig and Greeley, 1979; Lucchitta, 1981; Kargel and Strom, 1992) the scales of the features were mostly different, the Martian landforms being much larger than their terrestrial counterparts. In more recent studies based on high-resolution Viking Orbiter and MOC images, Costard and Kargel (1995, 1999) confirm the earlier interpretation of Martian pits and depressions as thermokarst features and find especially high concentrations in Utopia Planitia. This is the same region near the end of outflow channels where polygonally fractured ground and small rampart craters are frequently found. They attribute their generation to the melting or sublimation of ice-rich sediments with possible massive icy beds in the subsurface (Costard and Kargel, 1999). Surfaces of lobate debris aprons and lineated valley fills are also affected by thermokarst processes (Mangold *et al.*, 2000b, 2000c).

Frost Mounds. Palsas are small frost mounds containing ice lenses or a core of ice-rich frozen soil (Lundqvist, 1969; Washburn, 1973; Seppälä, 1988; Williams and Smith, 1989; Nelson *et al.*, 1992). They typically have widths of 10 − 30 m, lengths of several meters to 150 m, and heights of a few meters (Forsgren, 1968; Washburn, 1973). They are generally smaller than pingos, which can reach diameters of several hundred meters and heights of 50 − 70 m (Leffingwell, 1915; Mackay, 1973, 1977; Williams and Smith, 1989). Pingos have massive cores of clear ice. Pingos develop by freezing of taliks, i.e. unfrozen layers within or under the permafrost, or from hydraulic or artesian water pressure (Pissart, 1988; Williams and Smith, 1989). Pingo-like features on Mars have been reported by Rossbacher and Judson (1981), Lucchitta (1981), and Costard (1987), but they warn that their interpretation is not unambiguous. First, the size of Martian "pingos" is much larger than their terrestrial analogs (Coradini and Flamini, 1979). Second, the resolution of Viking images is not high enough to allow for a discrimination of pingos, pseudocraters (Frey *et al.*, 1979), craters, or secondary craters.

Eskers and Moraines. Glaciation was discussed in a limited way by several authors to account for specific landforms on Mars (Allen, 1979; Clifford, 1980; Tanaka and Scott, 1987; Scott and Underwood, 1991; Baker *et al.*, 1991). The probably most convincing morphological evidence for ancient glaciation on Mars comes from long sinuous ridges mainly found in mid and high latitudes (Figure 5). They are best developed in the Argyre region. Among the various interpretations (Kargel and Strom, 1992), a formation analogous to that of terrestrial eskers seems most plausible (Howard, 1981; Kargel and Strom, 1990, 1991, 1992; Kargel *et al.*, 1995; Metzger, 1992), while other interpretations have shortcomings (Head, 2000a). Eskers are elongated sinuous ridges of glaciofluvial sand and gravel (Banerjee and McDonald, 1975; Warren and Ashley, 1994) and form as infillings of ice-walled river channels. Their deposition may occur in subglacial, englacial, or supraglacial drainage networks (Benn and Evans, 1998), and there exist several morphological classifications of eskers (Brennand, 2000). Sizes are in the range of regular stream beds, with lengths from 5 m to 500 km, widths from 25 cm to 5 km, and heights from several centimeters to 200 m (Grout *et al.*, 1959; Lee, 1965; Wright, 1973; Shreve, 1985; Metzger, 1991). Esker-like features on Mars have dimensions near the upper limits of most terrestrial eskers (Kargel and Strom, 1992), and, therefore, their interpretation was widely regarded as problematic. On the other hand, recent MOLA measurements of the heights and widths of the largest Martian features seem to confirm the esker hypothesis (Head, 2000a, 2000b; Mangold, 2000).

2.4. VOLCANO-ICE INTERACTIONS

The potential for phreatic (steam) explosions occurs when magma or lava come into contact with surface or subsurface water (liquid or frozen). The resulting landforms include shield volcanoes composed of ash and small constructs called pseudocraters. Since there is evidence for both widespread volcanism (Greeley and Spudis, 1978; Mouginis-Mark *et al.*, 1982) and ground ice on Mars, the interaction between both processes seems inevitable.

Patera is a category of low-profile volcano with channels radiating from central calderas. Their flanks appear to be composed of easily eroded materials (Carr *et al.*, 1977b). Subsequent detailed studies (Greeley and Spudis, 1978; Crown and Greeley, 1993; Crown *et al.*, 1992) led to the proposal that these features are ash shields which formed by phreatic explosions in their early stages of eruptions. The presence of channelized lava flows superposed on the putative ash deposits in some of the features (such as Tyrrhena Patera) suggests that the water supply diminished with time, leading to effusive eruptions. If this interpretation is correct, then the locations of this class of volcano could indicate the presence of abundant surface or near-surface water at the time of the eruptions. Numerous small, often sub-kilometer sized mounds topped with craters occur in Acidalia and Utopia Planitiae. An origin as pseudo-craters formed by the explosion of steam due to the vaporization of water or ice covered by lava flows (Thorarinsson, 1953) has been suggested by Frey *et al.* (1979) and Frey and Jarosewich (1982). Images of MGS show numerous candidate features in Amazonis Planitia. Greeley and Fagents (2000) used results from field studies in Iceland and an eruptive model to investigate the formation of pseudocraters on Earth and Mars. They found that substantially less water is required to form pseudocraters on Mars than on Earth, and the Martian features are reasonably formed by phreatic processes.

Allen (1979), Frey *et al.* (1979), Hodges and Moore (1979), and Mouginis-Mark *et al.* (1982) have reported a variety of landforms related to volcano-ice or -groundwater interactions. Icelandic table mountains and Canadian tuyas resemble flat-topped mountains with steep sides mostly in northern Martian latitudes (Allen, 1979; Hodges and Moore, 1979). The table mountains (Allen, 1980) and the tuyas (Mathews, 1947; Jones, 1966, 1969) developed by subglacial volcanism as subglacial equivalents to shield volcanoes (Werner *et al.*, 1996; Bourgeois *et al.*, 1998; see Hickson, 2000, for a review of ice-contact volcanism). Ridges and mountains in the northern plains of Mars are comparable to terrestrial móberg ridges (Allen, 1979; Hodges and Moore, 1979) which consist of tephra and palagonitic tuff and represent material from subglacial fissure eruptions (Kjartansson, 1959, 1960).

The close spatial proximity of braided channels with streamlined islands to the Elysium volcanoes has lead to the idea that channel formation was due to volcano-ice interactions (Mouginis-Mark *et al.*, 1984; Mouginis-Mark, 1985; Squyres *et al.*, 1987). A detailed investigation of the Elysium region by Mouginis-Mark (1985) showed that pseudo-craters, melt water deposits or jökulhaups (formed by the

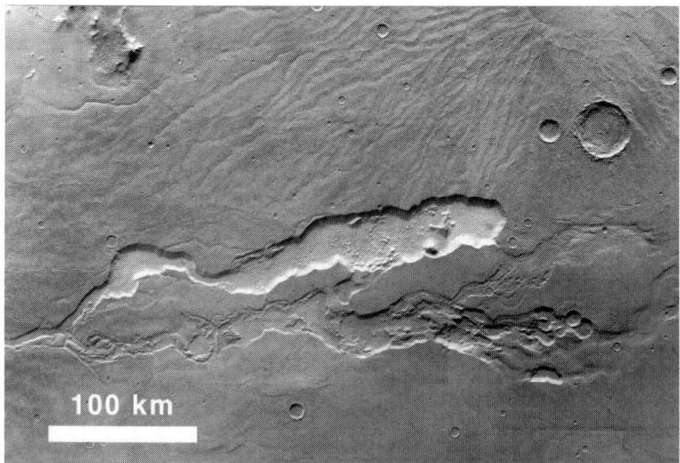

Figure 6. The source area for the outflow channel Dao Vallis near the highland volcanoe Hadriaca Patera at about 34°S, 265°W. The close proximity of the collapse features and the volcano (at the upper right of the image) suggests an origin by volcano-ice interactions (Squyres *et al.*, 1987). Viking orbiter image mosaic from orbits 408A-410A (resolutions around 160 m/pixel).

catastrophic release of water from ice-covered subglacial lakes: Björnsson, 1974, 1975, 2000; Gudmundsson, 1996; Kristmannsdóttir *et al.*, 1999; Tómasson, 2000), collapse features, and outflow channels can be attributed to water or ice-contact volcanism. Chapman (1994) found evidence for móberg ridges, table mountains, and jökulhlaup deposits, and proposed the existence of a paleoice sheet northwest of the Elysium volcanoes in Utopia Planitia. Another region heavily affected by volcano-ice interactions lies northeast of Hellas, where large outflow channels have source regions close to the ancient highland volcano Hadriaca Patera (Figure 6) (Squyres *et al.*, 1987). The same authors see further evidence for similar processes in parts of the Medusae Fossae formation (Squyres *et al.*, 1987), which had earlier been interpreted as ignimbrites by Scott and Tanaka (1982).

3. Ancient Bodies of Standing Water

Melting by volcanic heat and fracture controlled groundwater movement might also have contributed to the groundwater accumulation around Tharsis particularly in the Valles Marineris region. The canyon floors are 4 to 8 km below the surroundings. Even at a present thickness of the cryosphere of about 2.7 km at the equator (Clifford, 1993) water could have leaked into the canyons. New data from the MOC give evidence that this mechanism works also in smaller valleys (Malin and Edgett, 2000a). Given a global aquifer system it is expected that water will pool in the canyons. Layered sediments in several canyons, such as Echus Chasma and Ganges Chasma, give evidence for bodies of standing water within large depressions (McCauley, 1978; Nedell *et al.*, 1987). Catastrophic release of such lakes will also form outflow channels (McCauley, 1978). Kasei Vallis is suggested to be formed by catastrophic drainage of a former lake in Echus Chasma (Robinson and Tanaka, 1990); a large channel in Amazonis Planitia may be the drainage of a lake in southern Elysium Planitia (Scott *et al.*, 1992), and Anderson (1992) and Chapman (1994) suggest that Granicus Vallis, west of Elysium Mons, developed under thick ice deposits as a result of volcanic activity. From Ganges Chasma a broad depression trends south to the chaotic source region of Shalbatana Vallis indicating subsurface drainage of a possible former lake (Carr, 1995).

Outflow channels, their source regions and termination areas in the northern lowlands provide the best evidence for surface water on Mars. The large outflow channels emptied into the northern lowlands and their floods must have left extended deposits and probably large lakes. The water released by these channels to the northern plains is estimated to amount at least 6×10^6 km^3 but probably more (Carr *et al.*, 1987). According to Baker *et al.* (1991), it needs 6×10^7 km^3 of water to fill up the northern plains to about the 0-km contour. The northern plains cover an older rougher underground that survived as hills and knobs commonly outlining old pre-plain impact craters. The plains are complex deposits probably formed by many processes such as sedimentation from outflow channels, volcanism and mass wasting from the adjacent highlands modified by impact craters. Ridge-like features surrounding the northern plains are interpreted as ancient shorelines (Parker *et al.*, 1989, 1993), indicating the boundary of former lakes or a northern ocean. This hypothesis is supported by the fact that the shoreline-contacts altitude is close to an equipotential line and the plains inside these contacts are extremely smooth (Head *et al.*, 1999). Under current climate conditions such lakes or an ocean would rapidly freeze and form an ice cover. However they could stabilize if fed by melt- or groundwater (McKay and Davis, 1991).

Major uncertainties with the water story concern whether there were ever large bodies of water on the surface, and if so when. If Mars has a large inventory of water, as appears probable, and if climatic conditions were such the surface temperatures were above freezing then large bodies of water are inevitable. High erosion rates during the Noachian (Carr, 1992; Craddock and Maxwell, 1993) together with thick sequences of layered sediments throughout the Noachian highlands (Malin and Edgett, 2000b) strongly suggest warm conditions and an active hydrologic cycle in the Noachian. Large bodies of water may, therefore, have been present at this time, a conclusion supported by the seeming overflow of the Argyre basin (Parker, unpublished thesis). The evidence for large bodies of water (oceans) after the Noachian is, however, equivocal. Parker *et al.* (1989, 1993) identified two discontinuous contacts within the northern plains that they claim could be shorelines of former oceans. The contacts are based on a variety of observations such as breaks in slope, albedo boundaries, and textural contacts. The inner contact 2 encloses a volume of 1.4×10^7 km^3; the outer contact contains a volume of 9.6×10^7 km^3 (Head *et al.*, 1999). The MOLA data show that the outer contact 1 has a wide range of elevations ($\sim$6 km) and so is unlikely to be a shoreline. The inner contact 2 is, however at a roughly constant elevation over much of its length, and is a more plausible candidate. MOC images have, however, failed to provide any support for shorelines (Malin and Edgett, 1999). Post-Noachian oceans are questionable on other grounds. Baker *et al.* (1991) acknowledged that individual floods were likely to have been too small to produce ocean sized bodies of water so had to invoke the unlikely simultaneous occurrence of many floods. Moreover, how the oceans disappeared leaving behind almost no trace is left unexplained. It appears more likely that individual floods formed small bodies of water in low lying areas, especially in the northern plains, and that these bodies of water either sublimed or froze in place and were covered by the next flood. According to this hypothesis, therefore, much of the water that participated in the large floods, remains as ice deposits below the surface of the northern plains.

4. Water Sources and Water Budget

While the geomorphic evidence that water has played a major role in the evolution of the Martian surface is compelling, large uncertainties exist concerning how much water was involved, when liquid water was available on the surface to create the landforms, and where all the water went. The present atmosphere, although close to saturation, contains only minor amounts of water. If it all precipitated out, it would form a layer only $0.7 - 1.4$ m deep (Jakosky and Farmer, 1982). The only place on the surface where water (ice) has been unambiguously identified is at the north pole, where the layered terrains have a volume of 1.2 to 1.7×10^6 km^3 (Smith *et al.*, 1999). If they consisted entirely of water-ice, which is unlikely, the volume of their water could cover the entire planet to a depth of 10 m. For comparison, the Earth's oceans contain enough water to cover the entire Earth to a depth of $\sim$3 km. If the large outflow channels were cut by large floods, and the valley networks were cut by slow erosion of running water, as appears likely, then the planet has either lost large amounts of water since the channels and valleys formed, or large amounts of water are hidden beneath the surface as ground ice or groundwater.

Estimates of the amounts of water needed to produce the fluvial and lacustrine landforms observed on Mars are little more than educated guesses. By directly measuring canyon and channel volumes, and ignoring those volumes seemingly produced by faulting, Carr *et al.* (1987) estimated that 4×10^6 km^3 of material had been eroded away by water to form the canyons and outflow channels that drain into Chryse Planitia. If we assume, following Komar (1980), that the water carried a maximum sediment load 40 percent by volume, then 6×10^6 km^3 of water, roughly equivalent to 40 m spread over the whole planet have flowed into and across southern Chryse Planitia. If we further assume that the water in the outflow channels that entered Chryse Planitia was not recycled in some way, such as by polar basal melting (Clifford, 1987) or by precipitation (Baker *et al.*, 1991), but instead represents a one way transfer of water from the deep megaregolith to near-surface sinks (Carr, 1996b), then the 40 m represents a lower limit for the global inventory of water. The actual inventory is likely to be much larger, possibly a few hundred meters, since numerous outflow channels occur outside the Chryse region, and not all the water carried its maximum sediment load.

The floods in the Chryse region are probably the result both of the draining of lakes within the canyons (McCauley, 1978) and eruptions of groundwater under high pressure (Carr, 1979). Numerous estimates have been made of peak flood discharges from the dimensions of the channels and by extrapolating from empirical data relating discharges of terrestrial rivers to channel characteristics (Komar, 1979; Baker, 1982; Komatsu and Baker, 1997; Robinson and Tanaka, 1990). The estimates generally range from 10^7 to 10^9 m^3 s^{-1}. To calculate the volume of water involved in a flood, we need to know how the discharge changed with time. Individual floods are likely to have been short lived, a consequence of either draining of a lake or drawdown in artesian pressure. If we assume a decay time of a month, then the total volumes of individual floods for the above discharges would be 2.4×10^4 to 2.4×10^6 km^3. The upper figure appears to be unrealistically high in that it is comparable to the volume of all the rock removed around Chryse by all the floods. Alternatively, if we assume that lakes in the Ophir and Candor Chasmata, with a combined maximum volume of 7×10^5 km^3 (Carr et al., 1987), drained catastrophically to the east (McCauley, 1978) with a peak discharge of 10^9 km^3 then the decay constant of the flood would have been 8 days. The total amount of water involved in all the floods is not known but must be well in excess of the total volume of material moved. The floods appear to have formed episodically over a long period of time, mainly, but not exclusively, in the Upper Hesperian and Lower Amazonian (Tanaka et al., 1992). The above reasoning suggests that the floods around the Chryse basin likely had discharges in the 10^7 to 10^9 m^3 s^{-1} range, individual volumes in the 10^4 to 10^6 km^3 range, decay constants of the order of a week, and that they removed roughly 4×10^6 km^3 of material and deposited it in the northern plains. This discussion has just focused on the circum-Chryse channels, but similar reasoning applies to floods elsewhere, as in Hellas, Elysium and southern Amazonis Planitia.

The supposition that most valley networks did not form by floods but by slow erosion of running water (Masursky, 1973) appears supported by high resolution imaging of valleys such as Nanedi Vallis (Carr and Malin, 2000). The low drainage densities (Carr and Chuang, 1997), scarcity of fine scale dissection (Carr and Malin, 2000), and valley morphology (Baker, 1990) indicate that groundwater seepage played a prominent role in their formation. Surface runoff may have played only a minor role, because precipitation was limited or infiltration dominated over surface runoff. The climatic conditions required for valley formation remain uncertain. Theoretical models suggest that they could form under present climatic conditions if groundwater could be brought to the surface at sufficient rates (Wallace and Sagan, 1979; Carr, 1981; Goldspiel and Squyres, 2000), but the groundwater still needs to be recharged by some mechanism, and that mechanism may require much warmer climates than presently prevail. Although valley networks have a wide spread in ages (Scott and Dohm, 1992; Carr, 1995), the main period for formation was the upper Noachian, well before the main period of flood channel formation. During this time erosion rates in general much higher than in the subsequent Hesperian and Amazonian periods (Craddock and Maxwell, 1993) were suggesting very different climatic conditions.

Clifford (1987, 1993) and Clifford and Hillel (1983) have proposed a model for the history of water in which most of the planet's near-surface inventory has existed as ice within an ever-thickening, permanently frozen, cryosphere, or as groundwater below the cryosphere. He postulates that the near-surface materials are porous to considerable depths as a result of brecciation by impact, mainly during the period of heavy bombardment prior to 3.8 Gyr. He suggested, by analogy with the Moon, that the porosity decreases exponentially with depth, and estimated that the water-holding capacity of the megaregolith could range from 0.5 to 1.4 km of water spread evenly over the whole planet. We, of course, do not know what fraction is filled. Using heat flows estimated by Schubert and Spohn (1990), he suggested that the cryosphere today is roughly 2 km thick at the equator and 6 km thick at the poles. For the same surface conditions, higher heat flows at the end of heavy bombardment would have resulted in cryosphere thicknesses of 0.4 km at the equator and 1.2 km at the poles. Stability relations of water and present climatic conditions indicate that ice should be present in the near-surface cryosphere at high latitudes but not at low latitudes (Farmer and Doms, 1979). According to this model, the valley networks formed mainly by groundwater seepage, early when the cryosphere was absent or thin, and groundwater could readily access the surface; the outflow channels formed later when the cryosphere had thickened and could contain the high hydraulic pressures needed to generate large discharges.

As indicated above, if the channels and valleys are to be explained mainly by groundwater, some mechanism must exist for recharging the groundwater system. Numerous authors (Baker, 1982; Carr, 1996b) have suggested that, for the valley networks to have formed precipitation was required, if not directly to cut the valleys, then to recharge the groundwater system. For precipitation to occur, climatic conditions had to be very different from present day conditions. However, Clifford (1987) pointed to another way to recharge the groundwater system. He pointed out that any water brought to the surface, as in a seepage or flood, would, under present conditions, freeze, sublime into the atmosphere, and condense on the poles. Ice at the poles would accumulate until the deposits were thick enough for the base of the ice deposit to melt. Meltwater could then enter the groundwater system. Melting would be more readily achieved early in the planet history when heat flows were higher. The process could occur at both poles, but that at the south pole is crucial, for its high elevation would enable recharge of the groundwater system in the southern highlands where most of the valley networks are. The process is slow but hundreds of millions of years are available. Clifford and Parker (1999) suggest additionally that the outflow channels are younger than most of the valley networks because they could form only when large hydrostatic pressures could build within the aquifer system, and that this was possible only when the polar cryosphere was thick enough that north polar layered terrains were firmly anchored to bedrock and thick enough to contain the large pressures in the hydrosphere.

5. Chronology and Implications for Paleoclimatology

The stratigraphic position of geologic units on Mars is estimated by the means of superposition and intersection relations and by the concentration of impact craters superposed on geologic units (Tanaka *et al.*, 1992). Three major time systems are defined throughout which the surface has been formed: Noachian, Hesperian and Amazonian. Each system is subdivided into series that correspond to time units (epochs). Due to the lack of samples from Mars the assignment of absolute ages to the epochs based on crater densities is dependent on cratering rates (Tanaka *et al.*, 1992; Neukum and Hiller, 1981; Hartmann, 1978; Neukum and Wise, 1976) and thus is model dependent. Different early models defined the Amazonian-Hesperian boundary between 1.8 − 3.5 Gyr and the Hesperian-Noachian boundary between 3.5 − 3.8 Gyr (Hartmann *et al.*, 1981; Neukum and Wise, 1976; Neukum and Hiller, 1981). Hartmann and Neukum (2001) give new, tighter constraints as 2.9 − 3.3 Gyr and 3.5 − 3.7 Gyr, respectively, for those boundaries.

The relatively small size of Martian valleys, the modification by aeolian processes and the not sufficient coverage of high-resolution imagery makes it difficult to use crater counting for age determination. However in some places, particularly on the floors of large outflow channels it was possible to estimate crater retention ages (Masursky *et al.*, 1977; Neukum and Hiller, 1981; Marchenko *et al.*, 1998). Neukum and Hiller (1981) performed crater counts on units (surrounding and floors) of Chryse channels such as Kasei, Ares, Tiu, Maya Vallis, Elysium channels such as Hrad Vallis and Elysium Fossae, one channel NE of Hellas and Ladon Vallis north of the Agyre basin taking into account data from other authors (Malin, 1976; Masursky *et al.*, 1977; Squyres, 1978). As expected they estimated the channel surroundings to be significantly older as the measured channel ages. The number of craters with 1 km diameter $N(1)$ per square kilometer are for the surroundings either $>4 \times 10^{-2}$ km^{-2} representing the early crust or 2×10^{-3} km^{-2} to 2×10^{-2} km^{-2} representing a younger resurfacing period. This translates to crater model ages as proposed by Hartmann and Neukum (2001) of 3.3 to 3.9 Gyr and to more than 4.26 Gyr, respectively. In the following crater model ages will refer to the Neukum and Hiller (1981) Model I, however it is mentioned that these ages are model dependent and may be accurate only within a factor of two for younger units. The estimated ages of channel floors roughly fall into two groups separated by $N(1) = 8 \times 10^{-4}$ km^{-2} or about 1.1 to 1.7 Gyr the mean age of Tiu Vallis units. The older group contains most of Kasei Vallis, the mouth of Maya Vallis, Braham Vallis, and Ares Vallis in the circum-Chryse area, Ladon Vallis south of the chaotic terrain, Hrad Vallis and the Elysium Fossae. These valleys have crater retention ages ranging from $N(1) = 1 \times 10^{-2}$ km^{-2} to $N(1) = 8 \times 10^{-4}$ km^{-2} corresponding to 3.8 Gyr and 1.1 to 1.7 Gyr. The younger group contains the upper units of Tiu Vallis, Mangala Vallis, a channel NE of Hellas, the mouth of Kasei Vallis and Vedra Vallis which drains downstream into Maya Vallis being the youngest. Relative ages for this group range from $N(1) = 8 \times 10^{-4}$ km^{-2} to $N(1) = 1 \times 10^{-4}$ km^{-2} corresponding to crater model ages of 1.1 to 1.7 Gyr and $\sim$0.2 $-$ 0.3 Gyr. The data from Neukum and Hiller (1981) also show a sequence of the circum-Chryse valleys with most of the Kasei Vallis floors being oldest followed by the mouth of Maya Vallis, Ares Vallis, Tiu Vallis and the mouth of Kasei as the youngest unit in this area. Marchenko *et al.* (1998) estimated crater model ages at the mouth of Tiu and Ares Vallis ranging from 3.6 to 1.4 Gyr according to the Neukum and Hiller (1981) Model I. They suggest four major stages of fluvial activities and resurfacing in this region indicating episodic events of flooding. Some areas of Ares Vallis were formed prior to 3.5 Gyr, and have been later reworked between 2.2 and 1.6 Gyr. However, channel floors are modified by post-fluvial mass wasting and aeolian processes making it difficult to clearly address younger floor units to flood events. Neukum and Hiller (1981) concluded that the major channel formation took place between from $N(1) = 6 \times 10^{-2}$ km^{-2} to $N(1) = 7 \times 10^{-4}$ km^{-2} or 4.1 Gyr to 0.9 $-$ 1.5 Gyr ago. Channels on volcanoes are to small for crater counting on their floors but they dissect volcanic units for example at the flanks of Alba Patera measured by Neukum and Hiller (1981) to be as young as 1.2×10^{-4} km^{-2} or 0.2 $-$ 0.3 Gyr. Synthesizing the various models, the general picture is that the great majority of channel-forming activity occurred within the first third or half of Martian history, but some may have extended into the second half.

Based on the geologic maps of Scott and Tanaka (1986), Greeley and Guest (1987), Tanaka and Scott (1987) and Viking imaging data, the geologic relationships between channels and valley networks and the surrounding units they dissect were examined. For a total of 65 outflow channels and individual channel branches and 276 valley networks in the uplands the maximum relative ages have been estimated. Most of the investigated outflow channels (70%) dissect Hesperian units with about the half of the outflow channels having Upper Hesperian maximum ages. Amazonian units are eroded by about 25% of the outflow channels. The majority of valley networks in the uplands (about 63%) has an Upper Noachian maximum age. The formation of valley networks declined in the Hesperian (25%) and only about 12% of the valley networks mostly valleys on volcanoes dissect Amazonian units. Scott and Dohm (1992) measured intersection relations for several hundred networks and found about 70% to be older than Hesperian. Carr (1995) suggests that more than 90% of the valley networks have a maximum age older than Hesperian. Thus outflow channels and valley networks are not only separated spatially but are also delayed in time, indicating a major change in the erosion processes on Mars at the Noachian/Hesperian boundary.

The valley networks are almost entirely restricted to the old uplands and the simplest explanation is that the valleys are old themselves and the climatic requirements for valley formation were met early in the planet's history and rapidly declined during the subsequent evolution. A warmer, wet Mars with a dense atmosphere at the time after the heavy bombardment is supposed to provide the conditions for valley formation by running water. Carr (1995) argued that due to the drainage densities of valley networks are orders of magnitudes lower than on Earth and the short length of the valleys combined with the probably long time they needed to form, the processes which cause the valleys are extremely insufficient compared with terrestrial fluvial processes. In addition the total absence of meter-scale flow-features and dissections in the valleys support a subsurface rather than an atmospheric source for the valley formation (Malin and Carr, 1999). Based on the evaluation of high resolution MOC images Malin and Carr (1999) concluded that the valleys have been formed by fluid erosion, however, in most cases the source have been ground water. The evidences of recent groundwater seepage and probably short-term surface runoff under almost current climatic conditions (Malin and Edgett, 2000a) not necessarily contradict the hypothesis of an early warm Mars, but constrains a hydrologic cycle on Mars mostly to subsurface processes. On the other hand erosion rates declined at the end of the Noachian (Carr, 1992, 1995; Craddock and Maxwell, 1993; Hartmann, 2001) and climate change at this time is one plausible explanation. Gulick and Baker (1990) suggested that a combination of several genetic processes like water erosion, volcanic density flows and lava might have been important in the formation of those channels. Carr (1995) pointed out that mass wasting in poorly consolidated ash with only little involvement of water could have formed the channels on volcanoes. Nevertheless, the channels on volcanoes are fairly young and demonstrate that the involvement of water in Martian erosion appears to continue into the planet's recent history.

6. Future Investigations

The high resolution Mars Observer Camera (MOC) images so far demonstrate the need of recognizing small-scale features in order to fully understand the processes which caused fluid-like Martian erosion. However before high resolution stereo data with considerable areal coverage are available (as expected from future missions such as Mars Express) the problem of valley and channel formation and the implications on the climate remains debatable.

Acknowledgements

We thank William K. Hartmann for his very helpful review, and Reinald Kallenbach and the staff of the International Space Science Institute (ISSI) in Bern, Switzerland, for hospitality and assistance in preparing the manuscript.

References

Aguirre-Puente, J., Costard, F., and Posado-Cano., R.: 1994, 'Contribution to the Study of Thermal Erosion on Mars', *J. Geophys. Res.* **99**, 5657–5667.

Allen, C.C.: 1979, 'Volcano-Ice Interactions on Mars', *J. Geophys. Res.* **84**, 8048–8059.

Allen, C.C.: 1980, 'Icelandic Subglacial Volcanism: Thermal and Physical Studies', *J. Geol.* **88**, 108–117.

Anderson, D.L., Gatto, L.W., and Ugolini, F.: 1973, 'An Examination of Mariner 6 and 7 Imagery for Evidence of Permafrost Terrain on Mars', *in* F.G. Sanger (ed.), *Permafrost: 2nd Int. Conf.*, Nat. Acad. Sci., Washington, D.C., 449–508.

Anderson, W.M.: 1992, 'Glaciation in Elysium', LPI Tech. Rep. 92-08,1.

Baker, V.R.: 1973, 'Paleohydrology and Sedimentology of Lake Missoula Flooding of Eastern Washington', *Geol. Soc. Am. SP-144*, Geol. Soc. Am., Boulder.

Baker, V.R.: 1979, 'Erosional Processes in Channelized Water Flows in Mars', *J. Geophys. Res.* **84**, 7985–7993.

Baker, V.R., and Kochel, R.C.: 1979, 'Morphometry of Streamlined Forms in Terrestrial and Martian Channels', *Proc. 9th Lunar Planet. Sci. Conf.*, 3181–3193.

Baker, V.R.: 1982, *The Channels of Mars*, Univ. Texas Press, Austin.

Baker, V.R.: 1990, 'Spring Sapping and Valley Network Development', *Geol. Soc. Am.* **SP-252**, 235–265.

Baker, V.R., Strom, R.G., Gulick, V.C., Kargel, J.S., Komatsu, G.,, and Kale, V.S.: 1991, 'Ancient Oceans, Ice Sheets and the Hydrologic Cycle on Mars', *Nature* **352**, 589–594.

Baker, V.R., Carr, M.H., Gulick, V.C., Williams, C.R., and Marley, M.S.: 1992, 'Channels and Valley Networks', *in* H.H. Kieffer, B.M. Jakosky, C.W. Snyder, and M.S. Matthews (eds.), *Mars*, Univ. Arizona Press, Tucson, 493–522.

Banerjee, I., and McDonald, B.C.: 1975, 'Nature of Esker Sedimentation', *in* A.V. Jopling and B.C. McDonald (eds.), *Glaciofluvial and Glaciolacustrine Sedimentation, Spec. Publ.* **23**, Soc. Econ. Paleontol. Mineral., 132–154.

Barlow, N.G., and Bradley, T.L.: 1990, 'Martian Impact Craters: Correlations of Ejecta and Interior Morphologies with Diameter, Latitude and Terrain', *Icarus* **87**, 156–179.

Barlow, N.G., Boyce, J.M., Costard, F.M., Craddok, R.A., Garvin, J.B., Sakimoto, S.E.H., Kuzmin, R.O., Roddy, D.J., and Soderblom, L.A.: 2000, 'Standardizing the Nomenclature of Martian Impact Crater Ejecta Morphologies', *J. Geophys. Res.*, in press.

Barsch, D.: 1988, 'Rockglaciers', *in* M.J. Clark (ed.), *Advances in Periglacial Geomorphology*, Wiley, Chichester, 69–90.

Battistini, R.: 1984, 'L'Utilisation des Cratères Météoritiques à Ejectas Fluidisés Comme Moyen d'Etude Spatiale et Chronologique de l'Eau Profonde (Hydrolithosphère) de Mars', *Rev. Géom. Dyn.* **33**, 25–41.

Benn, D.I., and Evans, D.J.A.: 1998, 'Glaciers and Glaciation', Arnold, London, 734 pp.

Björnsson, H.: 1974, 'Explanation of Jökulhlaups from Grímsvötn, Vatnajökull Iceland', *Jökull* **24**, 1–26.

Björnsson, H.: 1975, 'Subglacial Water Reservoirs, Jökulhlaups and Volcanic Eruptions', *Jökull* **25**, 1–14.

Björnsson, H.: 2000, 'Jökulhlaups in Iceland: Characteristics and Impact', *2nd Mars Polar Sci. Conf.*, abstract #4064.

Blasius, K.R., Cutts, J.A., Guest, J.E., and Masursky, H.: 1977, 'Geology of the Valles Marineris: First Analysis of Imaging From the Viking 1 Orbiter Primary Mission', *J. Geophys. Res.* **82**, 4067–4091.

Blasius, K.R., and Cutts, J.A.: 1980, 'Global Patterns of Primary Crater Ejecta Morphology on Mars', *in* Reports of Planetary Geology and Geophysics Program-1980, NASA-TM 82385, Washington, D.C., 147–149.

Bourgeois, O., Dauteuil, O., and Van Vliet-Lanoë, B.: 1998, 'Pleistocene Subglacial Volcanism in Iceland: Tectonic Implications', *Earth Planet. Sci. Lett.* **164**, 165–178.

Boyce, J.M., and Roddy, D.J.: 1997, 'Martian Crater Ejecta, Emplacement and Implications for Water in the Subsurface', *Proc. 28th Lunar Planet. Sci. Conf.*, LPI, Houston, 145–146.

Brennand, T.A.: 2000, 'Deglacial Meltwater Drainage and Glaciodynamics: Inferences from Laurentide Eskers, Canada', *Geomorphology* **32**, 263–293.

Brinton, K.L.F., Tsapin, A.I., McDonald, G.D., and Gilichinsky, D.: 1999, 'Aminostratigraphy of Organisms in Antartic and Siberian Permafrost Cores', *Fifth Int. Conf. on Mars*, LPI Contribution No. 972, LPI, Houston, abstract #6137 (CD-ROM).

Brook, G.A.: 1982, 'Ice-Wedge Polygons, Baydjarakhs, and Alases in Lunae Planum and Chryse Planitia, Mars', *in* Reports of Planetary Geology Program - 1982, NASA-TM 85127, Washington, D.C., 265–267.

Carr, M.H.: 1977, 'Distribution and Emplacement of Ejecta Around Martian Impact Craters', *in* D.J. Roddy, R.O. Pepin, and R.B. Merrill (eds.), *Impact and Explosion Cratering*, Pergamon Press, New York, 593–602.

Carr, M.H.: 1979, 'Formation of Martian Flood Features by Release of Water from Confined Aquifers', *J. Geophys. Res.* **84**, 2995–3007.

Carr, M.H.: 1981, *The Surface of Mars*, Yale Univ. Press, New Haven.

Carr, M.H.: 1983, 'The Stability of Streams and Lakes on Mars', *Icarus* **56**, 476–495.

Carr, M.H.: 1986, 'Mars: A water-rich Planet?', *Icarus* **56**, 187–216.

Carr, M.H.: 1992, 'Post-Noachian Erosion Rates: Implication for Mars Climate Change', *Proc. 23rd Lunar Planet. Sci. Conf.*, 205–205.

Carr, M.H.: 1995, 'The Martian Drainage System and the Origin of Valley Networks and Fretted Channels', *J. Geophys. Res.* **100**, 7479–7507.

Carr, M.H.: 1996a, 'Channels and Valleys on Mars: Cold Climate Features Formed as a Result of a Thickening Cryosphere', *Planet. Space Sci.* **44**, 1411–1423.

Carr, M.H.: 1996b, *Water on Mars*, Oxford Univ. Press, New York, 229 pp.

Carr, M.H., and Schaber, G.G.: 1977, 'Martian Permafrost Features', *J. Geophys. Res.* **82**, 4039–4054.

Carr, M.H., and Clow, G.D.: 1981, 'Martian Channels and Valleys: Their Characteristics, Distribution and Age', *Icarus* **48**, 91–117.

Carr, M.H., and Chuang, F.C.: 1997, 'Martian Drainage Densities', *J. Geophys. Res.* **102**, 9145–9152.

Carr, M.H., and Malin, M.C.: 2000, 'Meter-scale Characteristics of Martian Channels and Valleys', *Icarus* **146**, 366–386.

Carr, M.H., Crumpler, L.S., Cutts, J.A., Greeley, R., Guest, J.E., and Masursky, H.: 1977a, 'Martian Impact Craters and Emplacement of Ejecta by Surface Flow', *J. Geophys. Res.* **82**, 4055–4065.

Carr, M.H., Greeley, R., Blasius, K.R., Guest, J.E., and Murray, J.B.: 1977b, 'Some Martian Volcanic Features as Viewed from the Viking Orbiters', *J. Geophys. Res.* **82**, 3985–4015.

Carr, M.H., Wu, S.C., Jordan, R., and Schafer, F.J.: 1987, 'Volumes of Channels, Canyons and Chaos in the Circum-Chryse Region of Mars', *Proc. 18th Lunar Planet. Sci. Conf.*, 155–156.

Chapman, M.G.: 1994, 'Evidence, Age, and Thickness of Frozen Paleolake in Utopia Planitia, Mars', *Icarus* **109**, 393–406.

Clifford, S.M.: 1980, 'A Model for the Removal and Subsurface Storage of a Primitive Martian Ice Sheet', *in* Reports of Planetary Geology and Geophysics Program-1990, NASA-TM 82385, Washington, D.C., 405–407.

Clifford, S.M.: 1987, 'Polar Basal Melting on Mars', *J. Geophys. Res.* **92**, 9135–9152.

Clifford, S.M.: 1993, 'A Model for the Hydrologic and Climatic Behaviour of Water on Mars', *J. Geophys. Res.* **98**, 10,973–11,016.

Clifford, S.M., and Hillel, D.: 1983, 'The Stability of Ground Ice in the Equatorial Regions of Mars', *J. Geophys. Res.* **88**, 2456–2474.

Clifford, S.M., and Parker, T.J.: 1999, 'The Evolution of the Martian Hydrosphere: Implications for the Fate of a Primordial Ocean and the Current State of the Northern Plains', *Fifth Int. Conf. on Mars, July 19-24, 1999, Pasadena, CA*, abstract #6236.

Colaprete, A., and Jakosky, B.M.: 1998, 'Ice Flow and Rock Glaciers on Mars', *J. Geophys. Res.* **103**, 5897–5909.

Coradini, M., and Flamini, E.: 1979, 'A Thermodynamical Study of the Martian Permafrost', *J. Geophys. Res.* **84**, 8115–8130.

Costard, F.: 1987, 'Quelques Modelés Liés à des Lentilles de Glace Fossiles sur Mars', *Z. Geomorph. N. F.* **31**, 243–251.

Costard, F.M.: 1988, 'Thickness of Sedimentary Deposits of the Mouth of Outflow Channels', *Proc. 19th Lunar Planet. Sci. Conf.*, LPI, Houston, 211–212.

Costard, F.M.: 1989, 'The Spatial Distribution of Volatiles in the Martian Hydrolithosphere', *Earth, Moon, and Planets* **45**, 265–290.

Costard, F.M.: 1994, 'Unusual Concentrations of Rampart Craters at the Mouths of Outflow Channels, Mars', *Proc. 25th Lunar Planet. Sci. Conf.*, LPI, Houston, 287–288.

Costard, F.M., and Kargel, J.S.: 1995, 'Outwash Plains and Thermokarst on Mars', *Icarus* **114**, 93–112.

Costard, F.M., and Kargel, J.S.: 1999, 'New Evidences for Ice Rich Sediments in the Northern Plains from MGS Data', *Fifth Int. Conf. on Mars*, LPI Contribution No. 972, LPI, Houston, abstract #6088 (CD-ROM).

Costard, F., Aguirre-Puente, J., Greeley, R., and Makhloufi, N.: 1999, 'Martian Fluvial-thermal Erosion: Laboratory Simulation', *J. Geophys. Res.* **104**, 14,091–14098.

Craddock, R.A., and Maxwell, T.A.: 1993, 'Geomorphic Evolution of the Martian Highlands Through Ancient Fluvial Processes', *J. Geophys. Res.* **98**, 3453–3468.

Crown, D.A., Price, K.H., and Greeley, R.: 1992, 'Geologic Evolution of the East Rim of the Hellas Basin, Mars', *Icarus* **100**, 1–25.

Crown, D.A. and R. Greeley, 1993. Volcanic geology of Hadriaca Patera and the eastern Hellas region of Mars, J. Geophys. Res. **98**, 3431-3451.

Cutts, J.A., and Blasius, K.R.: 1981, 'Origin of Martian Outflow Channels: The Eolian Hypothesis', *J. Geophys. Res.* **86**, 5075–5102.

ESA Exobiology Study Team: 1999, 'Exobiology in the Solar System and the Search for Life on Mars', *in* A. Wilson (ed.), *ESA SP* **1231**, ESA Publications Division, Noordwijk, 188 pp.

Evans, N., and Rossbacher, L.A.: 1980, 'The Last Picture Show: Small-Scale Patterned Ground in Lunae Planum', *in* Reports of Planetary Geology Program -1980, NASA-TM 82385, NASA, Washington, D.C., 376–378.

Farmer, C.B., and Doms, P.E.: 1979, 'Global and Seasonal Water Vapor on Mars and Implications for Permafrost', *J. Geophys. Res.* **84**, 2881–2888.

Fink, J.H., Greeley, R., and Gault, D.E.: 1981, 'Impact Cratering Experiments in Bingham Materials and the Morphology of Craters on Mars and Ganymede', *Proc. 12th Lunar and Planetary Sci. Conf.*, 1649–1666.

Forsgren, B.: 1968, 'Studies of Palsas in Finland, Norway and Sweden, 1964-1966', *Biuletyn Peryglacjalny* **17**, 117–123.

French, H.M.: 1976, *The Periglacial Environment*, Longman, London, 309 pp.

Frey, H., Lowry, B.L., and Chase, S.A.: 1979, 'Pseudocraters on Mars', *J. Geophys. Res.* **84**, 8075–8086.

Frey, H., and Jarosewich, M.: 1982, 'Subkilometer Martian Volcanoes: Properties and Possible Terrestrial Analogs', *J. Geophys. Res.* **87**, 9867–9879.

Gatto, L.W., and Anderson, D.M.: 1975, 'Alaskan Thermokarst Terrain and Possible Martian Analogs', *Science* **188**, 255–257.

Gault, D.E., and Greeley, R.: 1978, 'Exploratory Experiments of Impact Craters Formed in Viscous-Liquid Targets: Analogs for Martian Rampart Craters?', *Icarus* **34**, 486–495.

Goldspiel, J.M., and Squyres, S.W.: 2000, 'Groundwater Sapping and Valley Formation on Mars', *Icarus* **148**, 176–192.

Greeley, R., Fink, J., Gault, D.E., Snyder, D.B., Guest, J.E., and Schultz, P.H.: 1980, 'Impact Cratering in Viscous Targets: Laboratory Experiments', *Proc. 11th Lunar Planet Sci. Conf.*, 2075–2097.

Greeley, R., and Spudis, P.D.: 1978, 'Volcanism on Mars', *Rev. Geophys. Space Phys.* **19**, 13–41.

Greeley, R., and Guest, J.E.: 1987, Geologic Map of the Eastern Equatorial Region of Mars, scale 1:15,000,000, U.S.G.S. Misc. Inv. Series Map I-1802-B.

Greeley, R. and S.A. Fagents, 2000. Icelandic pseudocraters as analogs to some volcanic cones on Mars, J. Geophys. Res. (in process).

Grout, F.F., Sharp, R.P., and Schwartz, G.M.: 1959, 'The Geology of Cook County, Minnesota', *Minnesota Geol. Soc. Bull.* **39**, 67–69.

Gudmundsson, M.T.: 1996, 'Ice-Volcano Interaction at the Subglacial Grímsvötn Volcano, Iceland, Glaciers, Ice Sheets and Volcanoes: A Tribute to Mark F. Meier', *in* CRREL Special Report 96-27, 34–40.

Gulick, V.C., and Baker, V.R.: 1990, 'Origin and Evolution of Valleys on Martian Volcanoes', *J. Geophys. Res.* **95**, 14,325–14,344.

Hamlin, S.E., Kargel, J.S., Tanaka, K.L., Lewis, K.J., and MacAyeal, D.R.: 2000, 'Preliminary Studies of Icy Debris Flows in the Martian Fretted Terrain', *Proc. 31st Lunar Planet. Sci. Conf.*, abstract #1785 (CD-ROM).

Hartmann, W.K.: 1978, 'Martian Cratering V: Toward an Empirical Martian Chronology, and its Implications', *Geophys. Res. Lett.* **5**, 450–452.

Hartmann, W.K.: 2001, 'Martian Seeps and Their Relation to Youthful Geothermal Activity', *Space Sci. Rev.*, this volume.

Hartmann, W.K., and Neukum, G.: 2001, 'Cratering Chronology and the Evolution of Mars', *Space Sci. Rev.*, this volume.

Hartmann, W.K., *et al.*: 1981, 'Chronology of Planetary Volcanism by Comparative Studies of Planetary Cratering', *in: Basaltic Volcanism on the Terrestrial Planets*, Pergamon, New York, 1049–1127.

Head, J.W., Kreslavsky, M., Hiesinger, H., Ivanov, M., Pratt, S., Seibert, N., Smith, D.E., and Zuber, M.T.: 1998, 'Oceans in the Past History of Mars: Tests for Their Presence Using Mars Orbiter Laser Altimeter (MOLA) Data', *Geophys. Res. Lett.* **25**, 4401–4404.

Head, J.W., Hiesinger, H., Ivanov, B.A., Kreslavsky, M.A., Pratt, S., and Thomson, B.J.: 1999, 'Possible Ancient Oceans on Mars: Evidence from Mars Orbiter Laser Altimeter Data', *Science* **286**, 2134–2137.

Head, J.W.: 2000a, 'Tests for Ancient Polar Deposits on Mars: Morphology and Topographic Relationships of Esker-Like Sinuous Ridges (Dorsa Argentea) Using MOLA Data', *Proc. 31st Lunar Planet. Sci. Conf.*, LPI, Houston, abstract #1117 (CD-ROM).

Head, J.W.: 2000b, 'Tests for Ancient Polar Deposits on Mars: Assessment of Morphology and Topographic Relationships of Large Pits (Angusti and Sisyphi Cavi) Using MOLA Data', *Proc. 31st Lunar Planet. Sci. Conf.*, LPI, Houston, abstract #1118 (CD-ROM).

Hickson, C.J.: 2000, 'Physical Controls and Resulting Morphological Forms of Quaternary Ice-Contact Volcanoes in Western Canada', *Geomorphology* **32**, 239–261.

Hiesinger, H., and Head, J.W.: 2000, 'Characteristics and Origin of Polygonal Terrain in Southern Utopia Planitia, Mars: Results from Mars Orbiter Laser Altimeter and Mars Orbiter Camera Data', *J. Geophys. Res.* **105**, 11,999–12022.

Hodges, C.A., and Moore, H.J.: 1979, 'The Subglacial Birth of Olympus Mons and Its Aureoles', *J. Geophys. Res.* **84**, 8061–8074.

Howard, A.D.: 1981, 'Etched Plains and Braided Ridges of the South Polar Region of Mars: Features Produced by Basal Melting of Ground Ice?', *in* Reports of Planetary Geology Program, 1979-1980, NASA-TM 84211, 286–288.

Jakosky, B.M., and Farmer, C.B.: 1982, 'The Seasonal and Global Behavior of Water Vapor in the Mars Atmosphere: Complete Global Results of the Viking Atmospheric Water Vapor Detector Experiment', *J. Geophys. Res.* **87**, 2999–3019.

Jankowski, D.G., and Squyres, S.W.: 1992, 'The Topography of Impact Craters in 'Softened' Terrain on Mars', *Icarus* **100**, 26–39.

Jankowski, D.G. and Squyres, S.W.: 1993, 'Softened Impact Craters on Mars: Implications for Ground Ice and the Structure of the Martian Megaregolith' *Icarus* **106**, 365–379.

Johansen, L.A.: 1979, 'The Latitude Dependence of Martian Splosh Cratering and its Relationship to Water', *in* Reports of Planetary Geological Program, 1978-1979, NASA TM-80339, Washington, D.C., 123–125.

Jones, J.G.: 1966, 'Intraglacial Volcanoes of Southwest Iceland and Their Significance in the Interpretation of the Form of the Marine Basaltic Volcanoes', *Nature* **212**, 586–588.

Jones, J.G.: 1969, 'Intraglacial Volcanoes of the Laugarvatn Region, Southwest Iceland, I.', *Quart. J. Geol. Soc. London* **124**, part 3(495), 197–211.

Kachurin, S.P.: 1962, 'Thermokarst Within the Territory of the U.S.S.R.', *Biuletyn Peryglacjalny* **11**, 49–55.

Kargel, J.S.: 1986, 'Morphologic Variations of Martian Rampart Crater Ejecta and their Dependencies and Implications', *Proc. Lunar and Planetary Sci. Conf. XVII*, Lunar and Planetary Institute, Houston, 410–411.

Kargel, J.S., and Strom, R.G.: 1990, 'Ancient Glaciation on Mars', *Proc. 21st Lunar Planet. Sci. Conf.*, LPI, Houston, 597–598.

Kargel, J.S., and Strom, R.G.: 1991, 'Terrestrial Glacial Eskers: Analogs for Martian Sinuous Ridges', *Proc. 23rd Lunar Planet. Sci. Conf.*, LPI, Houston, 683–684.

Kargel, J.S., and Strom, R.G.: 1992, 'Ancient Glaciation on Mars', *Geology* **20**, 3–7.

Kargel, J., Baker, V.R., Begét, J.E., Lockwood, J.F., Péwé, T.L., Shaw, J.S., and Strom, R.G.: 1995, 'Evidence of Ancient Continental Glaciation in the Martian Northern Plains', *J. Geophys. Res.* **100**, 5351–5368.

Kjartansson, G.: 1959, 'The Móberg Formation', *in* S. Thorarinsson and G. Kjartansson (eds.), *On the Geology and Geomorphology of Iceland*, Geogr. Ann. Rep., Stockholm, 135–169.

Kjartansson, G.: 1960, 'The Móberg Formation, On the Geology and Geophysics of Iceland', *in* S. Thorarinsson (ed.), *Proc. 21st Session Int. Geol. Congr., Guide to Excursion No. A2*, Reykjavik, 21–28.

Komar, P.D.: 1979, 'Comparison of the Hydraulics of Water Flows in Martian Outflow Channels with Flows of Similar Scale on Earth', *Icarus* **37**, 156–181.

Komar, P.D.: 1980, 'Modes of Sediment Transport in Channelized Water Flows and Ramifications to the Erosion of Martian Outflow Channels', *Icarus* **42**, 317–329.

Komar, P.D.: 1983, 'Shapes of Streamlined Islands on Earth and Mars: Experiments and Analyses of Minimum-drag Form', *Geology* **11**, 651–655.

Komar, P.D.: 1984, 'The Lemniscate Loop Comparison with the Shape of Streamlined Landforms', *J. Geol.* **92**, 133–145.

Komatsu, G., and Baker, V.R.: 1997, 'Paleohydrology and Flood Geomorphology of Ares Vallis', *J. Geophys. Res.* **102**, 4151–4160.

Kristmannsdóttir, H., Björnsson, A., Pálsson, S., and Sveinbjörnsdóttir, À.E.: 1999, 'The Impact of the 1996 Subglacial Volcanic Eruption in Vatnajökull on the River Jökulsá á Fjöllum, North Iceland', *J. Volcanol. Geotherm. Res.* **92**, 359–372.

Kuzmin, R.O., Bobina, N.N., Zabalueva, E.V., and Shashkina, V.P.: 1988, 'Structural Inhomogeneities of the Martian Cryolithosphere', *Solar System Research* **22**, 121–133.

Lachenbruch, A.H.: 1962, 'Mechanics of Thermal Contraction Cracks and Ice Wedge Polygons in Permafrost', *Geol. Soc. Am., Spec. Paper* **70**, 69 pp.

Laity, J.E., and Malin, M.C.: 1985, 'Sapping Processes and the Development of Theater-headed Valley Networks on the Colorado Plateau', *Geol. Soc. Amer. Bull.* **96**, 203–217.

Lee, H.: 1965, 'Investigations of Eskers for Mineral Exploration', *Geol. Survey Can. Paper* **65-14**, 1–17.

Leffingwell, E.K.: 1915, 'Ground-Ice Wedges; the Dominant Form of Ground-Ice on the North Coast of Alaska', *J. Geol.* **23**, 635–654.

Lockwood, J.F., Kargel, J.S., and Strom, R.B.: 1992, 'Thumbprint Terrain on the Northern Plains: A Glacial Hypothesis', *Proc. 23rd Lunar Planet. Sci. Conf.*, LPI, Houston, 795–796.

Lozinski, W.: 1909, 'Über die mechanische Verwitterung der Sandsteine im gemässigten Klima', *Acad. Sci. Cracovie Bull. Internat., Cl. Sci. Math. et Naturelles* **I**, 1–25.

Lucchitta, B.K., and Anderson, D.M.: 1980, 'Martian Outflow Channels Sculptured by Glaciers', *in* Reports of Planetary Geology Program - 1980, NASA TM-81776, 271–273.

Lucchitta, B.K.: 1981, 'Mars and Earth: Comparison of Cold-Climate Features', *Icarus* **45**, 264–303.

Lucchitta, B.K.: 1982, 'Ice Sculpture in the Martian Outflow Channels', *J. Geophys. Res.* **87**, 9951–9973.

Lucchitta, B.K.: 1983, 'Permafrost on Mars: Polygonally Fractured Ground', *Proc.4th Int. Conf. on Permafrost*, Nat. Acad. Press, 744–748.

Lucchitta, B.K.: 1984, 'Ice and Debris in the Fretted Terrain, Mars', *J. Geophys. Res.* **89, suppl.**, 409–418.

Lucchitta, B.K.: 1985, 'Geomorphologic Evidence for Ground Ice on Mars', *in* J. Klinger, D. Benest, A. Dollfus, and R. Smoluchowski (eds.), *Ices in the Solar System*, D. Reidel Publ. Co., Dordrecht, 583–604.

Lucchitta, B.K., and Ferguson, H.M.: 1983, 'Chryse Basin Channels: Low Gradients and Ponded Flows', *J. Geophys. Res.* **88 Suppl.**, 553–568.

Lucchitta, B.K., Anderson, D.M., and Shoji, H.: 1981, 'Did Ice Streams Carve Martian Outflow Channels?', *Nature* **290**, 759–763.

Lucchitta, B.K., Ferguson, H.M., and Summers, C.: 1986a, 'Sedimentary Deposits in the Northern Lowland Plains, Mars', *J. Geophys. Res.* **91**, 166–174.

Lucchitta, B.K., Ferguson, H.M., and Summers, C.: 1986b, 'Northern Sinks on Mars', *Proc. 17th Lunar Planet. Sci. Conf.*, LPI, Houston, 498–499.

Lundqvist, J.: 1969, 'Earth and Ice Mounds: A Terminological Discussion', T.L. Péwé (ed.), *The Periglacial Environment*, McGill-Queen's Univ. Press, Montreal, 203–215.

Mackay, J.R.: 1973, 'The Growth of Pingos, Western Arctic Coast, Canada', *Can. J. Earth Sci.* **10**, 979–1004.

Mackay, J.R.: 1977, 'Pulsating Pingos, Tuktoyaktuk Peninsula, N.W.T.', *Can. J. Earth Sci.* **14**, 209–222.

Malin, M.C.: 1976, *J. Geophys. Res.* **81**, 4825.

Malin, M.C., and Carr, M.H.: 1999, 'Groundwater Formation of Martian Valleys', *Nature* **397**, 589–591.

Malin, M.C., and Edgett, K.S.: 1999, 'Oceans or Seas in the Martian Northern Lowlands: High Resolution Imaging Tests of Proposed Shorelines', *Geophys. Res. Lett.* **26**, 3049–3052.

Malin, M.C., and Edgett, K.S.: 2000a, 'Evidence for Recent Groundwater Seepage and Surface Runoff on Mars', *Science* **288**, 2330–2335.

Malin, M.C., and Edgett, K.S.: 2000b, 'Sedimentary Rocks of Early Mars', *Science* **290**, 927–937.

Mangold, N.: 2000, 'Giant Paleo-Eskers of Mauretania: Analogs for Martian Esker-Like Landforms', *Second Mars Polar Sci. Conf.*, abstract #4031.

Mangold, N., Allemand, P., and Peulvast, J.-P.: 2000a, 'Topography of Ice Related Features on Mars', *Proc. 31st Lunar Planet. Sci. Conf.*, LPI, Houston, abstract #1131 (CD-ROM).

Mangold, N., Costard, F., and Peulvast, J.-P.: 2000b, 'Thermokarstic Degradation of Lobate Debris Aprons and Fretted Channels', *Second Mars Polar Sci. Conf.*, abstract #4032.

Mangold, N., Costard, F., and Peulvast, J.-P.: 2000c, 'Thermokarstic Degradation of the Martian Surface', *Second Mars Polar Sci. Conf.*, abstract #4052.

Marchenko, A.G., Basilevsky, A.T., Neukum, G., Hauber, E., Hoffmann, E., and Cook, A.C.: 1998, 'The Study of the Mouth of Ares and Tiu Valles, Mars', *Proc. 29th Lunar Planet. Sci. Conf.*, abstract #1174.

Mars Channel Working Group: 1983, 'Channels and Valleys on Mars', *Geol. Soc. Amer. Bull.* **94**, 1035–1054.

Masursky, H.: 1973, 'An Overview of Geologic Results from Mariner 9', *J. Geophys. Res.* **78**, 4037–4047.

Masursky, H., Boyce, J.V., Dial, A.L., Jr., Schaber, G.G., and Strobell, M.E.: 1977, 'Classification and Time of Formation of Martian Channels Based on Viking Data', *J. Geophys. Res.* **82**, 4016–4037.

Mathews, W.H.: 1947, '"Tuyas", Flat-Topped Volcanoes in Northern British Columbia', *Am. J. Sci.* **249**, 560–570.

McCauley, J.F.: 1978, Geologic Map of the Coprates Quadrangle of Mars, scale 1:5,000,000. U.S. Geol. Surv. Misc. Inv. Series Map I-897.

McGill, G.E.: 1985, 'Age and Origin of Large Martian Polygons', *Proc. Lunar Planet. Sci. Conf.*, LPI, Houston, 534–535.

McGill, G.E.: 1986, 'The Giant Polygons of Utopia, Northern Martian Plains', *Geophys. Res. Lett.* **13**, 705–708.

McGill, G., and Hills, L.S.: 1992, 'Origin of Giant Martian Polygons', *J. Geophys. Res.* **97**, 2633–2647.

McKay, C.P., and Davis, W.L.: 1991, 'Duration of Liquid Water Habitats on Early Mars', *Icarus* **90**, 214–221.

Mellon, M.T.: 1997, 'Thermal Contraction Cracks in Martian Permafrost: Implications for Small-Scale Polygonal Features', *Proc. 28th Lunar Planet. Sci. Conf.*, LPI, Houston, 933–934.

Melosh, H.J.: 1989, *Impact Cratering, A Geologic Process*, Oxford Univ. Press, New York, 245 pp.

Metzger, S.M.: 1991, 'A Survey of Esker Morphometries, the Connection to New York State Glaciation and Criteria for Subglacial Melt-Water Channel Deposits on the Planet Mars', *Proc. 22nd Lunar Planet. Sci. Conf.*, LPI, Houston, 891–892.

Metzger, S.M.: 1992, 'The Eskers of New York State: Formation Process Implications and Esker-Like Features on the Planet Mars', *Proc. 23rd Lunar Planet. Sci. Conf.*, LPI, Houston, 901–902.

Mouginis-Mark, P.J.: 1979, 'Martian Fluidized Crater Morphology: Variations With Crater Size, Latitude, Altitude, and Target Material', *J. Geophys. Res.* **84**, 8011–8022.

Mouginis-Mark, P.J.: 1981, 'Ejecta Emplacement and Modes of Formation of Martian Fluidized Ejecta Craters', *Icarus* **45**, 60–76.

Mouginis-Mark, P.J.: 1985, 'Volcanic/Ground Ice Interactions in Elysium Planitia, Mars', *Icarus* **64**, 265–284.

Mouginis-Mark, P.J.: 1987, 'Water or Ice in the Martian Regolith?: Clues from Rampart Craters Seen at Very High Resolution', *Icarus* **71**, 268–286.

Mouginis-Mark, P.J., Wilson, L., and Head, J.W.: 1982, 'Explosive Volcanism on Hecates Tholus, Mars: Investigation of Eruption Conditions', *J. Geophys. Res.* **87**, 9890–9904.

Mouginis-Mark, P.J., Wilson, L., Head, J.W., Brown, S.H., Hall, J.L., and Sullivan, K.D.: 1984, 'Elysium Planitia, Mars: Regional Geology, Volcanology, and Evidence for Volcano - Ground Ice Interactions', *Earth, Moon, and Planets* **30**, 149–173.

Mouginis-Mark, P.J., Wilson, L., and Zimbelman, R.J.: 1988, 'Polygenic Eruptions on Alba Patera, Mars: Evidence of Channel Erosion on Pyroclastic Flows', *Bull. Vol.* **50**, 361–379.

Muller, S.W.: 1947, 'Permafrost or Permanently Frozen Ground and Related Engineering Problems', J.W. Edwards, Ann Arbor, 231 pp.

Mutch, T.A., Arvidson, R.E., Binder, Guinness, E.A., and Morris, E.C.: 1977, 'The Geology of the Viking Lander 2 Site', *J. Geophys. Res.* **82**, 4452–4467.

Nedell, S.S., Squyres, S.W., and Andersen, D.W.: 1987, 'Origin and Evolution of the Layered Deposits in the Valles Marineris, Mars', *Icarus* **70**, 409–441.

Nelson, F.E., Hinkel, K.M., and Outcalt, S.I.: 1992, 'Palsa-Scale Frost Mounds', *in* J.C. Dixon and A.D. Abrahams (eds.), *Periglacial Geomorphology*, Wiley, Chichester, 305–325.

Neukum, G., and Wise, D.U.: 1976, 'Mars: Standard Crater Curve and Possible New Time Scale', *Science* **194**, 1381–1387.

Neukum, G., and Hiller, K.: 1981, 'Martian Ages', *J. Geophys. Res.* **86**, 3097–3121.

Nummedal, D.: 1978, 'The Role of Liquefaction in Channel Development on Mars', *in* Reports of Planetary Geology and Geophysics Program-1977-1988, NSA TM-79729, 257–259.

Nummedal, D., and Prior, D.B.: 1981, 'Generation of Martian Chaos and Channels by Debris Flow', *Icarus* **45**, 77–86.

Parker, T.J., Saunders, R.S., and Sneeberger, D.M.: 1989, 'Transitional Morphology in the West Deuteronilus Mensae Region of Mars: Implications for Modification of the Plains/Upland Boundary', *Icarus* **82**, 111–145.

Parker, T.J., Gorcine, D.S., Saunders, R.S., Pieri, D.C., and Schneeberger, D.M.: 1993, 'Coastal Geomorphology of the Martian Northern Plains', *J. Geophys. Res.* **98**, 11,061–11,078.

Pechmann, J.C.: 1980, 'The Origin of Polygonal Troughs in the Northern Plains of Mars', *Icarus* **42**, 185–210.

Pieri, D.C.: 1976, 'Martian Channels: Distribution of Small Channels on the Martian Surface', *Icarus* **27**, 25–50.

Pieri, D.C.: 1980, 'Martian Valleys: Morphology, Distribution, Age and Origin', *Science* **210**, 895–897.

Pissart, A.: 1988, 'Pingos: An Overview of the Present State of Knowledge', *in* M.J. Clark (ed.), *Advances in Periglacial Geomorphology*, Wiley, Chichester, 279–297.

Popov, A.I.: 1956, 'Le Thermokarst', *Biuletyn Peryglacjalny* **4**, 319–330.

Reimers, C.E., and Komar, P.D.: 1979, 'Evidence for Explosive Volcanic Density Currents on Certain Martian Volcanoes', *Icarus* **39**, 88–110.

Robinson, M.S., and Tanaka, K.L.: 1990, 'Magnitude of a Catastrophic Flood Event at Kasei Valles', *Mars. Geology* **18**, 902–905.

Rossbacher, L.A., and Judson, S.: 1981, 'Ground Ice on Mars: Inventory, Distribution, and Resulting Landforms', *Icarus* **45**, 39–59.

Rossi, A.P., Komatsu, G., and Kargel, J.S.: 2000, 'Rock Glacier-Like Landforms in Valles Marineris, Mars', *Proc. 31st Lunar Planet. Sci. Conf.*, LPI, Houston, abstract #1587 (CD-ROM).

Schonfeld, E.: 1976, 'On the Origin of Martian Channels', *Trans. AGU, Eos* **57**, 948.

Schubert, G., and Spohn, T.: 1990, 'Thermal History of Mars and the Sulfur Content of its Core', *J. Geophys. Res.* **95**, 14,095–14,104.

Schultz, P.H., and Gault, D.E.: 1979, 'Atmospheric Effects on Martian Ejecta Emplacement', *J. Geophys. Res.* **84**, 7669–7687.

Schultz, P.H., and Gault, D.E.: 1984, 'On the Formation of Contiguous Ramparts around Martian Impact Craters', *Proc. 15th Lunar Planet. Sci. Conf.*, LPI, Houston, 732–733.

Scott, D.H., and Tanaka, K.L.: 1982, 'Ignimbrites of Amazonia Planitia Region of Mars', *J. Geophys. Res.* **87**, 1179–1190.

Scott, D.H., and Tanaka, K.L.: 1986, Geologic Map of the Western Equatorial Region of Mars, scale 1:15,000,000, U.S.G.S. Misc. Inv. Series Map I-1802-A.

Scott, D.H., and Underwood, J.R.: 1991, 'Mottled Terrain: A Continuing Martian Enigma', *in* G. Ryder and V.L. Sharpton (eds.) *Proc. 21st Lunar Planet. Sci. Conf.*, LPI, Houston, 627–634.

Scott, D.H., and Dohm, J.M.: 1992, 'Martian Highland Channels: An Age Reassessment', *Proc. 23rd Lunar Planet. Sci. Conf.*, 1251–1252.

Scott, D.H., Chapman, M.G., Rice, J.W., and Dohm, J.M.: 1992, 'New Evidence of Lakustrine Basins on Mars: Amazonis and Utopia Planitia', *Proc. 22nd Lunar. Planet. Sci. Conf.*, 53–62.

Seppälä, M.: 1988, 'Palsas and Related Forms', *in* M.J. Clerk (ed.), *Advances in Periglacial Geomorphology*, Wiley, Chichester, 247–278.

Sharp, R.P.: 1973a, 'Mars: Troughed Terrain', *J. Geophys. Res.* **78**, 4063–4072.

Sharp, R.P.: 1973b, 'Mars: Fretted and Chaotic Terrain', *J. Geophys. Res.* **78**, 4073–4083.

Sharp, R.P.: 1988, 'Living Ice', Cambridge Univ. Press, Cambridge, 225 pp.

Sharp, R.P., and Malin, M.C.: 1975, 'Channels on Mars', *Geol. Soc. Amer. Bull.* **86**, 593–609.

Shoji, H., and Higashi, A.: 1978, 'A Deformation Mechanism Map of Ice', *J. Glaciol.* **21**, 419–427.

Shreve, R.L.: 1985, 'Esker Characteristics in Terms of Glacier Physics', *Geol. Soc. Am. Bull.* **96**, 639–646.

Smith, D.E., *et al.*: 1999, 'The Global Topography of Mars and Implications for Surface Evolution', *Science* **284**, 1495–1503.

Squyres, S.W.: 1978, 'Martian Fretted Terrain: Flow of Erosional Debris', *Icarus* **34**, 600–613.

Squyres, S.W.: 1979, 'The Distribution of Lobate Debris Aprons and Similar Flows on Mars', *J. Geophys. Res.* **84**, 8087–8096.

Squyres, S.W.: 1989a, 'Early Mars: Wet and Warm or Just Wet?', *Proc. 20th Lunar Planet. Sci. Conf.*, 1044–1045.

Squyres, S.W.: 1989b, 'Water on Mars', *Icarus* **79**, 229–288.

Squyres, S.W., and Carr, M.H.: 1986, 'Geomorphic Evidence for the Distribution of Ground Ice on Mars', *Science* **231**, 249–252.

Squyres, S.W., and Kasting, J.F.: 1994, 'Early Mars: How Warm and How Wet?', *Science* **265**, 744–748.

Squyres, S.W., Wilhelms, D.E., and Moosman, A.C.: 1987, 'Large-Scale Volcano-Ground Ice Interactions on Mars', *Icarus* **70**, 385–408.

Squyres, S.W., Clifford, S.M., Kuzmin, R.O., Zimbelman, J.R., and Costard, F.M.: 1992, 'Ice In The Martian Regolith', *in* H.H. Kieffer, B.M. Jakosky, C.W. Snyder, and M.S. Matthews (eds.), *Mars*, Univ. Arizona Press, Tucson, 523–554.

Tanaka, K.L., and Scott, D.H.: 1987, Geologic Map of the Polar Region of Mars, scale 1:15,000,000, U.S.G.S. Misc. Inv. Series Map I-1802-C.

Tanaka, K.L., Scott, D.H., and Greeley, R.: 1992, 'Global Stratigraphy', *in* H.H. Kieffer *et al.* (eds.), Univ. Arizona Press, Tucson, 354–382.

Theilig, E., and Greeley, R.: 1979, 'Plains and Channels in the Lunae Planum-Chryse Planitia Region of Mars', *J. Geophys. Res.* **84**, 7994–8010.

Thorarinsson, S.: 1953, 'The Crater Groups in Iceland', *Bull. Volcanol. Ser.* **2**, 3–44.

Tómasson, H.: 2000, 'Catastrophic Floods in Iceland', *Second Mars Polar Sci. Conf.*, abstract #4117.

Tricart, J.: 1968, 'Periglacial Landscapes', *in* R.W. Fairbridge (ed.), *Encyclopedia of Geomorphology*, Reinhold Book Co., 829–833.

Tricart, J.: 1969, *Geomorphology of Cold Environments*, Macmillan, London, 320 pp.

Tsapin, A.I., McDonald, G.D., Andrews, M., Bhartia, R., Douglas, S., and Gilichinsky, D.: 1999, 'Microorganisms from Permafrost Viable and Detectable by 16SRNA Analysis: A Model for Mars', *Fifth Int. Conf. on Mars*, LPI Contribution No. 972, LPI, Houston, abstract #6104 (CD-ROM).

Wahrhaftig, C., and Cox, A.: 1959, 'Rock Glaciers in the Alaska Range', *Bull. Geol. Soc. Amer.* **70**, 383–436.

Wallace, D., and Sagan, C.: 1979, 'Evaporation of Ice in Planetary Atmospheres: Ice Covered Rivers on Mars', *Icarus* **39**, 385–400.

Warren, W.P., and Ashley, G.M.: 1994, 'Origin of the Ice-Contact Stratified Ridges (Eskers) of Ireland', *J. Sedimentary Res.* **A64**, 433–449.

Washburn, A.L.: 1973, 'Periglacial Processes and Environments', Edward Arnold, London, 320 pp.

Werner, R., Schmincke, H.-U., and Sigvaldason, G.: 1996, 'A New Model for the Evolution of Table Mountains: Volcanological and Petrological Evidence from Herdubreid and Herdubreidartogl Volcanoes (Iceland)', *Geologische Rundschau* **85**, 390–397.

Williams, P.J., and Smith, M.W.: 1989, 'The Frozen Earth', Cambridge Univ. Press, Cambridge, 306 pp.

Wilson, I., and Mouginis-Mark, P.J.: 1987, 'Volcanic Input to the Atmosphere from Alba Patera on Mars', *Nature* **330**, 354–357.

Wohletz, K.H., and Sheridan, M.F.: 1983, 'Martian Rampart Crater Ejecta: Experiments and Analysis of Melt Water Interaction', *Icarus* **56**, 15–37.

Wright, H.E.: 1973, 'Tunnel Valleys, Glacial Surges, and Subglacial Hydrology of the Superior Lobe, Minnesota', *The Wisconsinan Stage, Geol. Soc. Am. Memoir* **136**, 251–276.

Yoshikawa, K.: 2000, 'Contraction Cracking and Ice Wedge Polygons on Mars', *Second Mars Polar Sci. Conf.*, abstract #4045.

Yung, Y.L., and Pinto, J.P.: 1978, 'Primitive Atmosphere and Implications for the Formation of Channels on Mars', *Nature* **288**, 735–738.

Zimbelman, J.R., Clifford, S.M., and Williams, S.H.: 1988a, 'Terrain Softening Revisited: Photogeological Considerations', *Proc. 19th Lunar Planet. Sci. Conf.*, 1321–1322.

Zimbelman, J.R., Clifford, S.M., and Williams, S.H.: 1988b, 'Concentric Crater Fill on Mars: An Aeolian Alternative to Ice-Rich Mass Wasting', *Proc. 19th Lunar Planet. Sci. Conf.*, 397–407.

Address for correspondence: Université Paris-Sud, Equipe de Planétologie (UMR CRNS 8616), Bât. 509, 91405 Orsay Cedex, France; (masson@geol.u-psud.fr)

ALTERATION ASSEMBLAGES IN MARTIAN METEORITES: IMPLICATIONS FOR NEAR-SURFACE PROCESSES

J.C. BRIDGES[1], D.C. CATLING[2], J.M. SAXTON[3], T.D. SWINDLE[4], I.C. LYON[3]
and M.M. GRADY[1]

[1]*Department of Mineralogy, Natural History Museum, London SW7 5BD, UK*
[2]*SETI Institute/NASA Ames Research Center, Moffett Field, California, USA*
[3]*Department of Earth Sciences, Manchester University, Manchester M13 9PL, UK*
[4]*Lunar and Planetary Laboratory, University of Arizona, Tucson, Arizona 85721-0092, USA*

Received: 8 August 2000; accepted: 16 February 2001

Abstract. The SNC (Shergotty-Nakhla-Chassigny) meteorites have recorded interactions between martian crustal fluids and the parent igneous rocks. The resultant secondary minerals – which comprise up to $\sim$1 vol.% of the meteorites – provide information about the timing and nature of hydrous activity and atmospheric processes on Mars. We suggest that the most plausible models for secondary mineral formation involve the evaporation of low temperature ($25 - 150\,^{\circ}C$) brines. This is consistent with the simple mineralogy of these assemblages – Fe-Mg-Ca carbonates, anhydrite, gypsum, halite, clays – and the chemical fractionation of Ca-to Mg-rich carbonate in ALH84001 "rosettes". Longer-lived, and higher temperature, hydrothermal systems would have caused more silicate alteration than is seen and probably more complex mineral assemblages. Experimental and phase equilibria data on carbonate compositions similar to those present in the SNCs imply low temperatures of formation with cooling taking place over a short period of time (e.g. days). The ALH84001 carbonate also probably shows the effects of partial vapourisation and dehydration related to an impact event post-dating the initial precipitation. This shock event may have led to the formation of sulphide and some magnetite in the Fe-rich outer parts of the rosettes.

Radiometric dating (K-Ar, Rb-Sr) of the secondary mineral assemblages in one of the nakhlites (Lafayette) suggests that they formed between 0 and 670 Myr, and certainly long after the crystallisation of the host igneous rocks. Crystallisation of ALH84001 carbonate took place 0.5 Gyr after the parent rock. These age ranges and the other research on these assemblages suggest that environmental conditions conducive to near-surface liquid water have been present on Mars periodically over the last $\sim$1 Gyr. This fluid activity cannot have been continuous over geological time because in that case much more silicate alteration would have taken place in the meteorite parent rocks and the soluble salts would probably not have been preserved.

The secondary minerals could have been precipitated from brines with seawater-like composition, high bicarbonate contents and a weakly acidic nature. The co-existence of siderite (Fe-carbonate) and clays in the nakhlites suggests that the pCO_2 level in equilibrium with the parent brine may have been 50 mbar or more. The brines could have originated as flood waters which percolated through the top few hundred meters of the crust, releasing cations from the surrounding parent rocks. The high sulphur and chlorine concentrations of the martian soil have most likely resulted from aeolian redistribution of such aqueously-deposited salts and from reaction of the martian surface with volcanic acid volatiles.

The volume of carbonates in meteorites provides a minimum crustal abundance and is equivalent to 50–250 mbar of CO_2 being trapped in the uppermost 200–1000 m of the martian crust. Large fractionations in $\delta^{18}O$ between igneous silicate in the meteorites and the secondary minerals ($\leq 30‰$) require formation of the latter below temperatures at which silicate-carbonate equilibration could have taken place ($\sim 400^{\circ}C$) and have been taken to suggest low temperatures (e.g. $\leq 150^{\circ}C$) of precipitation from a hydrous fluid.

Space Science Reviews **96**: 365-392, 2001.
© 2001 *Kluwer Academic Publishers. Printed in the Netherlands.*

1. Introduction

Secondary mineral assemblages have been characterised in 7 of the 16 SNC (Shergotty-Nakhla-Chassigny) meteorites (Table I). To date they have been found in the 3 nakhlites Nakhla, Governador Valadares, Lafayette (olivine clinopyroxenites); ALH84001 (orthopyroxenite); Shergotty, EETA79001 (basaltic shergottites) and Chassigny (dunite). It is possible that all of the SNCs contain the secondary assemblages but they have not yet been identified or terrestrial alteration has obscured their presence. In this paper we review the mineralogy, stable isotopes, radiometric dating, associated fluid compositions and different models for formation of the SNC secondary phases. Secondary minerals have the potential to reveal compositional and isotopic information about the ancient martian atmosphere and to constrain the nature of fluid processes at the martian surface.

Radiometric dating of the secondary assemblages and their host meteorites also may constrain the times in Mars history when a thicker atmosphere was present and conditions were more favourable for the existence of liquid water. Oxygen isotopic studies provide information about chemical and isotopic fractionation processes in both the martian atmosphere and the crustal fluids. Finally, one of the underlying motivations in the study of SNCs and their secondary phases has been the search for traces of extraterrestrial life.

A description of the different SNC silicate petrographies is given in the paper by Nyquist *et al.* (2001). The main secondary mineral phases are carbonates, sulphates, halite and clay minerals (the latter particularly in the 3 nakhlites) although associated sulphides and ferric oxides are also sometimes present. There is some limited evidence for the occurrence of nitrates (Grady *et al.*, 1995). In addition, there are hydrous amphiboles associated with melt inclusions e.g. the kaersutite amphibole within Chassigny and the shergottites, that may have exchanged with or trapped martian surface water reservoirs (Treiman, 1985). Carbonate in ALH84001 occupies about 1 vol.%, heterogeneously distributed, within the meteorite (Treiman, 1995). Similarly, the proportion of secondary mineral phases can reach 1–2% in some sections of Nakhla (Bridges and Grady, 1999). Other SNCs have lower proportions of secondary minerals. Bulk meteorite water contents, associated with the secondary minerals range from 0.04 wt% in the shergottites to 0.4 wt% in the clay-rich nakhlites (Karlsson *et al.*, 1992).

The secondary minerals in the nakhlites and ALH84001 are assumed to be martian from a combination of textural and isotopic information. For instance, clay veins within the nakhlites are truncated by fusion crust and so are clearly preterrestrial (Gooding *et al.*, 1991; Treiman *et al.*, 1993). The association of some of the ALH84001 carbonate with preterrestrial fracturing is also incompatible with a terrestrial origin (e.g. Mittlefehldt, 1994). D/H ratios in some mineral phases are too high to be of terrestrial origin (Watson *et al.*, 1994; Leshin *et al.*, 1996, Saxton *et al.*, 2000a) and the highest values are comparable to values in the martian atmosphere. The D/H ratio in the martian atmosphere is $(8.1 \pm 0.3) \times 10^{-4}$ (i.e. δD

TABLE I

Summary of secondary and hydrous mineral assemblages in SNC meteorites

Meteorite (type)	Mineral assemblage	Carbonate composition (mol %)	Main textures	Other analyses	Interpretations
Lafayette (nakhlite, find)	clay, sd, po, fh, gyp (1,2,3,4)	cc22-37 rh4-35 mg0-2 sd27-67 (4)	sd + clay within fractured ol (iddingsite) (4). Carbonate grains < 50μm.	K-Ar age (5), O-isotopes (3,6,7, 8), C-isotopes (7), S-isotopes, (9) TEM smectite (2), trace elements LREE>HREE (4).	Low-T fluids (2), evaporitic, related to other nakhlites (4).
Governador Valadares (nakhlite, find)	clay, sd, gyp (4)	cc4-11 rh1-2 mg9-29 sd64-78 (4)	clay veins within ol (iddingsite), sd + gyp interstitial (4).	S-isotopes (9), C-isotopes (1), trace elements LREE>HREE (4).	Evaporitic, related to other nakhlites (4).
Nakhla (nakhlite, fall)	clay, sd, gyp, an, hal, ep, go (1,4,10,11)	cc0-6 rh1-40 mg2-41 sd23-87 (4)	Clay veins truncated by fusion crust (10), clay, gyp veins within ol (iddingsite), sd + anh + hal mainly interstitial (11).	Trace elements LREE>HREE (4,11), D/H of carbonate (12), O-isotopes (7,8,12, 13), C-isotopes (7,13,14), S-isotopes (9), TEM smectite (1).	Liquid water in parent rocks (1), evaporitic, related to other nakhlites (4), hydrothermal sulphides (9).
Chassigny (chassignite, fall)	cc, mg, gyp (15), amphiboles (kaersutite) in melt inclusions (16)		Discontinuous veins (of clay?) within ol. Amphiboles in melt inclusions suggest some water in melt (16).	D/H of amphiboles and biotite (17), S-isotopes (9).	Salts from brine (15).
Shergotty (shergottite, fall)	gyp, hal, other chlorides, sulphates, phyllosilicates (18) amphiboles in melt inclusions (19)		Salts on fracture surfaces and veins. Melt undersaturated in water (19).	D/H of amphiboles (17).	Episodic weathering (18). Extraterrestrial origin less firmly established for shergottites than nakhlites, ALH84001.
EETA79001 (shergottite, find)	cc, gyp(?) (20), amphibole in melt inclusions (21)		Present as druse vug fillings in lithology C glass.	O, C, N isotopes of druse (22,23,24).	Weathering followed by shock implantation (20).
ALH84001 (orthopyroxe nite, find)	ank, mica, pyr, Fe-sulph, mag, silica	average cc11.5 rh1.1 mg58.0, sd29.4 (25); zoned core from cc to mg end members (26)	Crack fillings and globules/rosettes (< 250μm) in interstitial areas with maskelynite.	O, C isotopes (27-32), carbonate has LREE $\leq$ HREE (33), carbonate age (Rb-Sr, U-Th-Pb) (34), D/H (35).	Evaporitic (36), hydrothermal, (25,37), impact meta-somatism (26), impact remelting/ remobilisation (38), biogenic (39).
LEW88516 (shergottite, find)	amphibole in melt inclusions(21)				
Zagami shergottite, fall)				D/H of apatite (17), O-isotopes (27).	

an anhydrite, cc calcite, ep epsomite, fh ferrihydrite, go goethite, gyp gypsum, hal halite, mag magnetite, mg magnesite, ol olivine, po unidentified 'porous oxide', rh rhodochrosite, sd siderite. 1. Reid and Bunch (1975), 2. Treiman *et al.* (1993), 3. Vicenzi and Eiler (1998), 4. Bridges and Grady (2000), 5. Swindle *et al.* (2000), 6. Romanek *et al.* (1998), 7. Wright *et al.* (1992), 8. Farquhar and Thiemens (2000), 9. Greenwood *et al.* (2000), 10. Gooding *et al.* (1991), 11. Bridges and Grady (1999), 12. Saxton *et al.* (2000b), 13. Jull *et al.* (1995), 14. Carr *et al.* (1985), 15. Wentworth and Gooding (1994), 16. Floran *et al.* (1978), 17. Watson *et al.* (1994), 18. Wentworth and Gooding (2000), 19. Treiman (1985), 20. Gooding *et al.* (1988), 21. Treiman (1998b), 22. Clayton and Mayeda (1988), 23. Wright *et al.* (1988), 24. Douglas *et al.* (1994), 25. Mittlefehldt (1994), 26. Harvey and McSween (1996), 27. Jull *et al.* (1997), 28. Saxton *et al.* (1998), 29. Leshin *et al.* (1998), 30. Farquhar *et al.* (1998), 31. Valley *et al.* (1997), 32. Romanek *et al.* (1994), 33. Wadhwa and Crozaz (1995), 34. Borg *et al.* (1999), 35. Sugiura and Hoshino (2000), 36. Warren (1998), 37. Kring *et al.* (1998), 38. Scott (1999), 39. McKay *et al.* (1996).

= 4200‰), which is about 5.2 times the ratio of 1.6×10^{-4} for terrestrial seawater (Owen, 1992). These meteorites also contain phases whose carbon (Romanek *et al.*, 1994), sulphur and oxygen isotopes (Farquhar *et al.*, 1998, 2000; Farquhar and Thiemens, 2000) are inconsistent with a terrestrial origin. Finally, as we will discuss later, the radiometric ages determined on some of the secondary mineral assemblages must predate the meteorites' residence times on Earth. However, the evidence used to establish a martian origin for the secondary mineral assemblages in the nakhlites and ALH84001 has not yet been as conclusively demonstrated for the other SNCs. Terrestrial salts can be present and are abundant in the desert finds Dar al Gani 476 and 489 (Wadhwa *et al.*, 2000). Minor efflorescence of salt from meteorite interiors onto the fusion crusts can also occur (Gooding *et al.*, 1991) and so care is necessary in distinguishing martian from terrestrial mineral assemblages. Isotopic data suggests that some of the shergottites may contain traces of terrestrial salts along with martian secondary assemblages. Therefore, we concentrate on the information obtained from the nakhlites and ALH84001.

2. Research on Secondary Minerals in SNCs and Models for Their Formation

The first description of extraterrestrial alteration assemblages in these meteorites was made by Reid and Bunch (1975) who described fibrous material ("iddingsite") associated with olivine grains within Nakhla. On the basis of a TEM study, Gooding *et al.* (1991) noted that the iddingsite veins within Nakhla olivine contained a probable smectite/illite, K-bearing mixed layer clay. Similarly, Treiman *et al.* (1993) described smectite in Lafayette. Carbonate and sulphate minerals were first identified within the EETA79001 martian meteorite (Gooding *et al.*, 1988) although their presence had earlier been inferred through pyrolysis or stepped combustion experiments (Carr *et al.*, 1985; Gooding and Muenow, 1986).

Subsequent studies (summarised in Table I) have enabled the mineralogical and stable isotope characteristics to be understood in more detail. The three nakhlites (Lafayette, Governador Valadares and Nakhla) contain smectite/illite, siderite, sulphates and halite (Chatzitheodoridis and Turner, 1990; Gooding *et al.*, 1991; Treiman *et al.*, 1993; Bridges and Grady, 1999, 2000). The clay veins in these meteorites cross-cut the olivine and, in Lafayette (Figure 1a), they are intergrown with Ca-rich siderite (Vicenzi and Eiler, 1998; Bridges and Grady, 2000). In contrast most of the siderite in Governador Valadares and Nakhla is located in interstitial positions and has lower Ca contents. These two meteorites contain gypsum, Nakhla also containing anhydrite (Chatzitheodoridis and Turner, 1990; Bridges and Grady, 1999), halite and epsomite (Gooding *et al.*, 1991). Figure 1b shows halite and anhydrite within a section of Nakhla. Chassigny contains calcite, Mg-carbonate and gypsum (Wentworth and Gooding, 1994) in veins within the silicate minerals. Shergotty has sulphate and halite with minor phyllosilicates (Gooding *et al.*,

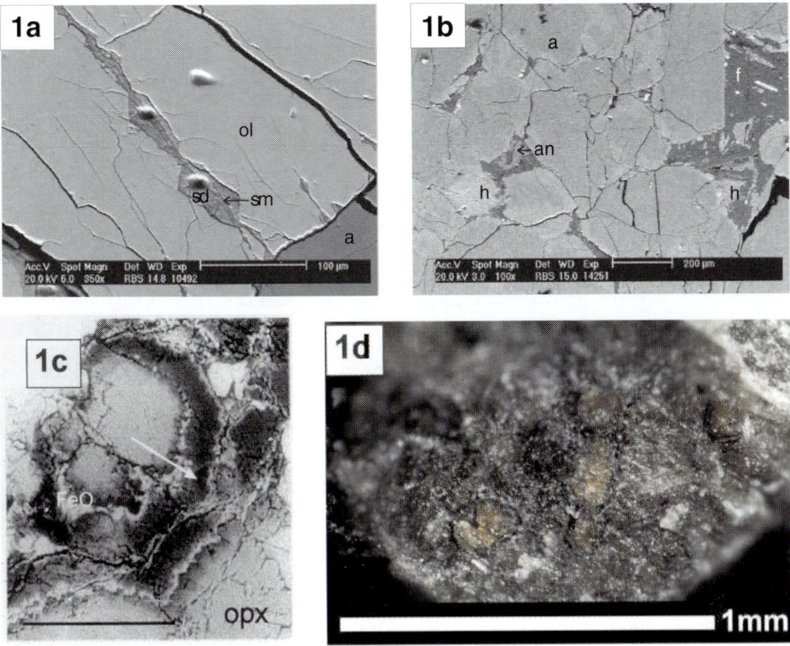

Figure 1. a) Back-scattered electron image of "iddingsite" alteration of olivine in the nakhlite Lafayette. The alteration assemblage consists of siderite and smectite along fractures. The smectite clay is found in the centres of the fractures (*arrowed*). The two spots in the fractures are ion probe analysis points (Bridges and Grady, 2000). b) Back-scattered electron image of salt minerals in Nakhla, where they mainly occupy interstitial areas. ol olivine, sd siderite, sm smectite, a cumulus augite, h halite, an anhydrite, f intergown feldspar, feldspathic glass and silica polymorphs. c) Back-scattered electron image of carbonate rosette in ALH84001. Carbonate, with increasing Mg/Ca ratio towards rim (arrowed). The white circular area (FeO) within the carbonate is a mixture of iron oxide (magnetite) and iron sulphide. opx orthopyroxene. Scale bar 100 μm. d) Carbonate 'rosettes' exposed on split surface of ALH84001 meteorite. The rounded, yellowish brown carbonate grains have dark, magnetite-rich rims. The surrounding meteorite is mainly composed of orthopyroxene grains.

1988; Treiman, 1985; Wentworth and Gooding, 2000) in fractures. The shergottite EETA79001 contains calcite and sulphate but located within druse cavities in the glassy parts of that meteorite (Gooding *et al.*, 1988).

ALH84001 has the most abundant carbonate and the largest grains or "rosettes" ($\leq$250 μm, Figure 1c, d). The average composition of the carbonate (Figure 2) is cc 11.5 rh 1.1 mg 58.0 sd 29.4 mol % and the "rosettes" are zoned from Ca-rich cores to Fe- and, on their outsides, Mg-rich (nearly pure magnesite) rims (Mittlefehldt, 1994; Harvey and McSween, 1996). Magnetite and Fe-sulphides (including pyrrhotite) are associated with the Fe-rich domains (McKay *et al.*, 1996). Minor phyllosilicate and silica veins are located in isolated pyroxene grain fractures, more extensive fracture zones in the meteorite, interstitial sites beside plagioclase shock glass and occasionally with carbonate (e.g. Mittlefehldt, 1994; Scott, 1999; Brearley, 2000).

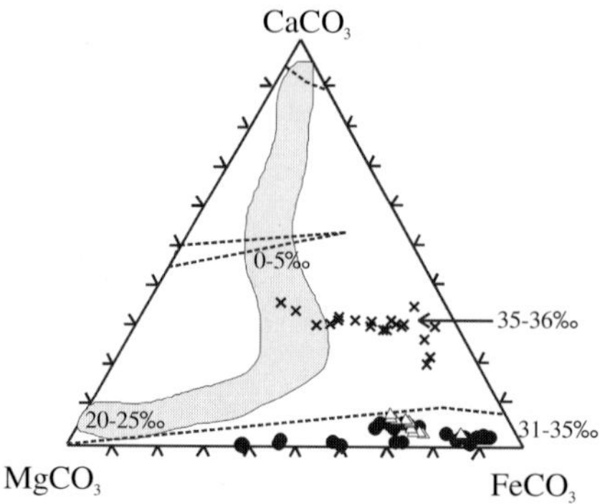

Figure 2. Fields for carbonate (mol %) compositions in ALH84001 (grey field) and the 3 nakhlites Lafayette (crosses), Governador Valadares (triangles) and Nakhla (circles). The order of decreasing Ca-carbonate contents between the nakhlites is Lafayette > Governador Valadares > Nakhla. Also shown in dashed areas are the one-phase solid solution fields for this system at 550 °C (Anovitz and Essene, 1987). With decreasing temperature the size of these fields decreases. The Lafayette and ALH84001 compositions are metastable. Further carbonate compositions are given in Bridges and Grady (2000). The correlation of δ^{18}O enrichment with Mg contents in ALH84001 carbonate is shown with isotopic values beside the chemical compositions. Carbonate data from Harvey and McSween (1996) and Bridges and Grady (1999, 2000). Oxygen isotope data from Leshin *et al.* (1998), Saxton *et al.* (1998), Vicenzi and Eiler (1998).

The first isotopic analyses of secondary minerals in SNCs were obtained through the stepped combustion experiments (Carr *et al.*, 1985). Analysis of EETA79001 suggested the possibility of two components of ^{13}C-enriched carbon, one from carbonate, the second with δ^{13}C $\sim$+36‰, from trapped martian atmospheric CO$_2$. Similarly, heavy N and O isotopic components and D/H enrichments have been described in components of martian meteorites (e.g. Wright *et al.*, 1988; Clayton and Mayeda, 1988; Watson *et al.*, 1994). These isotopic features have also been tentatively related to martian atmospheric compositions and some (heavy C, O isotopes) have been shown to be carried by secondary minerals.

Three main models to explain the formation of the mineral assemblages have been proposed: formation from (1) brines, (2) shock processes, and (3) hydrothermal processes. Each may explain some of the features seen in the meteorites. The general nature of the mineral assemblages in the SNCs – consisting largely of carbonate, sulphate, clay and chloride – suggests a link with brines and evaporite formation. However, there is also evidence for an association with shock remobilisation, particularly in ALH84001 (e.g. Scott *et al.*, 1997) and possibly EETA79001 (Gooding *et al.*, 1988). It has also been suggested that hydrothermal processes (by which we mean fluid activity at relatively high temperatures >150 °C, compared to those of groundwater), may also have left traces within the meteorites (e.g. Greenwood *et al.*, 2000). In the following subsections we attempt to summarise the supporting evidence for the main models, review evidence for martian biogenic traces in ALH84001, and suggest which of the mineralogical features may be most readily explained by each of the models.

2.1. EXPERIMENTAL CONSTRAINTS ON SNC SECONDARY MINERAL STABILITIES

Experimental studies designed to constrain the formation conditions of carbonate-bearing assemblages in SNCs have shown that they could have formed over a period of a few days from mixed CO_2-H_2O fluids interacting with basaltic rock at low temperatures (Baker *et al.*, 2000). This work demonstrated that none of the SNC meteorites need have undergone protracted interaction with aqueous fluids. In a separate set of experiments Golden *et al.* (2000) analysed the products precipitated from Ca-Mg-Fe-CO_2-H_2O-Cl solutions at various temperatures. They showed that carbonate globules, similar in terms of chemical zonation and size to those of ALH84001, could be formed at 150 °C whereas at 25 °C amorphous Fe-rich carbonate formed after 24 h and more Mg-rich carbonate after 96 h.

Experimental studies on phase equilibria in the $CaCO_3$-$MgCO_3$-$FeCO_3$ system (Anovitz and Essene, 1987) suggest that the Ca-Fe-rich carbonates of Lafayette (Bridges and Grady, 2000) and ALH84001 (Mittlefehldt, 1994; Harvey and Mc-Sween, 1996) are metastable i.e. if they had crystallised slowly then different, stable assemblages would have formed. The phase equilibria and experimental constraints point towards fairly low temperature (≤ 150 °C) precipitation of the original carbonate-bearing assemblages in SNCs over short periods of time (days).

2.2. EVAPORATION AND EVAPORITES ON MARS

The martian soil is thought to be composed of weathered rock (basalt to andesite) with a lesser Mg-rich component that may be associated with the salts, which are otherwise principally S, Cl, Br, K, Na (Bell *et al.*, 2000). The Viking and Pathfinder landing sites showed signs of salt hardpans and Christensen and Moore (1992) suggested, on the basis of thermal inertia, albedo and radar measurements, that large areas of the martian soil were partially cemented by salts. This has helped prompt speculation about whether evaporite deposits exist on Mars. Craters have preserved possible evidence for saltpan formation resulting from an enclosed drainage pattern. Some of the most detailed characterisation of an area where such deposits could have formed has been done on an unnamed, 35 km diameter, crater within the Memnonia region of the ancient highlands (Forsythe and Zimbelman, 1995) which has well marked terracing consistent with the past presence of ponded water. Similarly, high albedo patches identified in photographs of the martian surface (e.g. "White Rock" near the Schiaparelli crater and other areas in the Valles Marineris system) have been interpreted as relict evaporites (Russell *et al.*, 1999).

However, evaporite layers have not been positively identified with the Thermal Emission Spectrometer (TES) on the Mars Global Surveyor either at "White Rock" or elsewhere (Ruff *et al.*, 2000). It is possible the perceived absence reflects the constraints of pixel and spectral resolution for the current TES (Moersch *et al.*, 2000). In any case, there is little doubt that floodwaters and groundwater flow (Carr, 1996; Baker *et al.*, 1992 have been important processes, at least intermittently, in shaping the channels which dissect much of the martian surface. When such water pooled at the ends of valleys or melted into channels, plains, or fissures, salts must have precipitated in the terminal phases by solute concentration from evaporation or freezing. Forsythe and Blackwelder (1998) described 144 craters in the ancient highlands which had enclosed drainage where water would have ponded.

Results from the MOC camera have shown that surface water may have flowed intermittently on Mars in geologically recent times, creating gullies (Malin and Edgett, 2000a; Hartmann, 2001). Concentrated brines are able to exist in a liquid state at temperatures of down to $-63\,°C$ (Brass, 1980), and as mean surface temperatures on Mars are typically in the range $-33\,°C$ to $-68\,°C$ between $60\,°N$ and $60\,°S$, it seems likely that brines are associated with these features.

Brines in SNCs and floodwater models: The evaporite mineral assemblages of the 3 nakhlite meteorites have been modelled as products of brine evaporation by Bridges and Grady (2000). The similarity in cosmic ray exposure ages, crystallisation ages and silicate petrographies suggests that the 3 nakhlites were derived from the same source region on Mars, and so their secondary mineral assemblages may share a common origin. Lafayette contains minerals (Ca-rich siderite, clay) which in the model were derived from a relatively low extent (25% brine remaining) of evaporation from a brine; Governador Valadares with siderite, gypsum and clay (20%); Nakhla – which also contains anhydrite and halite – trapped the final products of evaporation through the nakhlite parent rocks ($<10\%$). The trace element abundances (Zr, Y, LREE > HREE) of the individual mineral grains reflect the dissolution of LREE-enriched mesostasis by the brine. Vicenzi and Eiler (1998) also noted the presence of distinct mineral phases and intricate banding within the Lafayette alteration assemblages.

Warren (1998) proposed that the carbonates in ALH84001 could have formed as a calcrete, where groundwater was saturated with carbonate concentrated through evaporation. Alternatively, the parent rock could have been covered by a deep layer of floodwater, with evaporation and percolation of the water leading to carbonate precipitation. The parent rocks were only briefly exposed to the envisaged floodwaters due to rapid groundwater flow or evaporation. This scenario would explain the lack of extensive alteration of the silicate phases.

Evaporation models also have the advantage of being consistent with the experimental constraints described above, especially for the siderite. The presence of gypsum e.g. in the nakhlite Governador Valadares also implies temperatures $<100°C$. However, if ALH84001 salts are evaporitic in origin, one may have to invoke physical mechanisms such as the occlusion of pore spaces (McSween and Harvey, 1998) or water stratification (Warren, 1998) to account for the absence of other salts, notably sulphates. Alternatively, the most soluble salts might have been leached on Mars if ALH84001 was not isolated from subsequent aqueous activity. Leaching of highly soluble halides or sulphate during the meteorite's residence in Antarctica is also possible. The process of evaporation related to brine evolution is considered in more detail in a later section.

2.3. HIGH TEMPERATURE: SHOCK REMOBILISATION AND HYDROTHERMAL/CO_2-CHARGED FLUIDS

All of the SNC meteorites show some of the signs of having experienced high levels of shock (e.g. 30 − 45 GPa for the shergottites, Stöffler *et al.*, 1986). These signs include melt pockets, deformation features in the silicate minerals and the presence of maskelynite. Gooding *et al.* (1988) suggested that the salts within the "lithology C" of EETA79001, which is largely composed of shock melt, might have been entrained from the martian surface into the parent rock during the impact melting event. Evaporite sediments might have been assimilated into the cooling Nakhla parent lava (Bridges and Grady, 1999), although subsequently a direct evaporation model, described in the previous section, was found to be more consistent with the mineralogical features in the 3 nakhlites (Bridges and Grady, 2000).

Harvey and McSween (1996) argued that the carbonates in ALH84001 formed through rapid reaction between a CO_2-rich, H_2O-poor fluid and the parent rock at >650 °C. Subsequently, Kring *et al.* (1998) suggested that CO_2-charged fluids at <300 °C had been active in the ALH84001 parent rock for at least a few years. This timescale was based on predicted dissolution rates of plagioclase, with replacement by carbonate, in alkaline fluids. However, as we argue in this paper, the fluids associated with ALH84001 and the nakhlites were probably acidic so this relatively long time scale for fluid activity may be inaccurate. Models involving CO_2-rich, H_2O-poor fluids do have the advantage of explaining the low abundance of hydrous minerals in ALH84001 compared to the nakhlites. However some phlogopitic mica intergrown with carbonate has since been described in ALH84001 (Brearley, 2000). Brearley suggested that the phlogopite was derived through the breakdown of pre-existing clay during shock heating at <500 °C. Treiman (1998a) documented fracturing episodes within ALH84001 and proposed that the carbonate had experienced 4 discrete deformation events. Bradley *et al.* (1998) also carefully documented the magnetite structures around the carbonate rosettes and suggested that whisker-shaped grains they found were most consistent with precipitation at high temperatures (500 − 800 °C) from a vapour. This was taken by Scott (1999) to be a result of the shock remobilisation of pre-existing carbonates and resultant break down of Fe-carbonate to a magnetite-bearing assemblage. However, the origin of the magnetite still remains controversial because some of the grains with euhedral outlines have also been suggested by other authors to be biological in origin (see next section). High temperature processes may partly explain the dehydrated nature of ALH84001 but they leave open the question of the initial origin of the salt minerals, prior to the shock event. Scott (1999) and Warren (1998) proposed that they might have been deposited during intermittent floods.

Greenwood *et al.* (2000) suggested that the nakhlites had undergone hydrothermal alteration during cooling of the igneous assemblage. This was based on the observed fractionation of $\delta^{34}S$ between the 3 meteorites (see section on stable isotopes) which was considered to be too great for low temperature processes.

However as Farquhar and Thiemens (2000) subsequently demonstrated, fractionation of the S isotopes through photolysis probably occurred in the martian atmosphere. Therefore, a high temperature origin is not required to explain the S isotopic variations in the SNCs.

Warren (1998) noted some critical problems with hydrothermal models such as that proposed in the initial description of carbonates in ALH84001 by Mittlefehldt (1994) who suggested a temperature of 700 °C. Hydrothermal systems on earth typically last for tens of thousands of years. This is implausible for the SNCs because the silicate minerals, especially in ALH84001, are relatively unaltered. The simple salt and clay mineral assemblages are not consistent with highly differentiated fluids which evolved over time. The chemical zonation preserved in the "rosettes" would have homogenised if it had been held at elevated temperatures for any protracted (> days) time (Treiman, 1998a). Another argument put against high temperature models is that the oxygen isotope compositions of silicate and carbonate are not in equilibrium (Farquhar *et al.*, 1998; Romanek *et al.*, 1994). If the carbonate is of high temperature origin then the oxygen isotopic disequilibrium could only be explained as a result of very rapid crystallisation. Temperature constraints based on oxygen isotopes are considered further in a later section.

2.4. SUGGESTED RELIC BIOLOGICAL ACTIVITY IN ALH84001

McKay *et al.* (1996) suggested that the carbonate-magnetite-sulphide rosettes in ALH84001 were of martian, biogenic origin (for a further discussion see Nyquist *et al.*, 2001). This provocative idea stimulated much interest in martian meteorites and the search for traces of life on Mars. Two of the main lines of evidence put forward by McKay *et al.* were the morphology of the magnetite and the presence of rod-like structures resembling fossil bacteria on carbonate surfaces.

The current balance of evidence does not strongly support the biogenic theory of origin for ALH84001 carbonates (Scott, 1999; Steele *et al.*, 2000; Farquhar and Thiemens, 2000; Kathie *et al.*, 2000; Bradley *et al.*, 1998; Farquhar and Thiemens, 2000). The best argument for a biological origin rests with the morphology of some of the magnetite grains. However, the association which McKay *et al.* made for the carbonates with low temperature precipitation from aqueous fluids remains valid even if the process was not directly linked to biological activity.

3. Stable Isotopes

Stable isotope ratios (O, H, C, S determined for the secondary minerals offer insights into formation conditions, fluid reservoirs and the evolution of the martian atmosphere. On the Earth, oxygen isotope ratios are frequently used to derive formation temperatures or to study mass-dependent fractionation of the isotopes between fluids and minerals. Accordingly, large fractionations in $\delta^{18}O$ compositions of the secondary minerals in martian samples may indicate low temperature precipitation. In most natural (especially geological) processes, oxygen isotopes exhibit this mass-dependent fractionation, in which $\Delta^{17}O = \delta^{17}O - 0.52\delta^{18}O =$ constant, but the SNCs contain evidence for mass-independent fractionation, with $\Delta^{17}O$ varying between different phases in the same rock. Atmospheric loss can also fractionate stable isotopes: for instance, in the current martian atmosphere, H and N are enriched in their heavy isotopes (see Bogard *et al.*, 2001, for a review).

3.1. OXYGEN

The carbonate data in ALH84001 reflect the fractionation of fluid composition, with progressively increasing Mg/Ca ratios and $\delta^{18}O$, and the temperature of carbonate precipitation. Three independent ion probe studies (Valley *et al.*, 1997; Saxton *et al.*, 1998; Leshin *et al.*, 1998) show that the large range of cation chemistry is accompanied by a correspondingly large $(20 - 25\permil)$ variation in $\delta^{18}O$. The isotopically heaviest carbonate $(\delta^{18}O = +20 - 25\permil)$ is the Mg-rich rims. The lowest carbonate $\delta^{18}O$ found by Leshin *et al.* (1998) was $\sim 5\permil$, but isotopically lighter $(\delta^{18}O \geq -10\permil)$, Ca-rich carbonate has been reported in ALH84001 by Saxton *et al.* (1998), Holland *et al.* (2000) and Eiler *et al.* (1998). The relationship of carbonate Mg/Ca composition to $\delta^{18}O$ for ALH84001 is shown on Figure 2.

Ion probe measurements of $\delta^{18}O$ have also been made on siderite in two nakhlites, Nakhla (Saxton *et al.*, 1998, 2000a) and Lafayette (Vicenzi and Eiler, 1998). The results are indistinguishable within error, both showing great enrichments in $\delta^{18}O$: Nakhla 4 grains, $+31 - 35\permil$; Lafayette, 2 grains, $+35$ and $+36\permil$.

Farquhar and Thiemens (2000) determined $\Delta^{17}O$ of secondary phases and water in ALH84001, Nakhla and Lafayette through acid extraction and fluorination. They found that $\Delta^{17}O$ of carbonate, sulphate and water released exceeded that of the silicate by up to $+1.3\permil$, demonstrating the lack of equilibrium between products of the martian hydrosphere and martian silicate. Martian silicate itself has $\Delta^{17}O = +0.3\permil$ compared to the terrestrial fractionation line (Franchi *et al.*, 1999). The data of Farquhar and Thiemens is similar to that of Karlsson *et al.* (1992) who showed that water in the SNC meteorites had $\Delta^{17}O$ greater than that of the anhydrous silicates by values up to $\Delta^{17}O = +0.6\permil$ (Nakhla) although Shergotty and EETA 79001 showed no $\Delta^{17}O$ excesses over their whole rock values.

Temperature and isotopic fractionation constraints from oxygen: Carbonate equilibrated with silicate at $500 - 700\,^{\circ}C$ would have $\delta^{18}O$ compositions $2 - 3\permil$ higher than the silicate $(\sim +4\permil)$ and $\Delta^{17}O$ equal to the silicate. Thus, the $\delta^{18}O$ values determined for most of the carbonates are too high to represent equilibrium at igneous temperatures with the silicates, and so formation from an aqueous fluid has been considered. Here we outline the models used to constrain fluid temperatures.

A common starting point for consideration of carbonate oxygen isotope data (Clayton and Mayeda, 1988; Wright *et al.*, 1992; Romanek *et al.*, 1994; Saxton *et al.*, 1998, 2000b) has been to consider water (or a water-CO_2 mixture) equilibrated with the silicate at high temperature (at which isotopic fractionation is small), and then cooled. This process can be viewed on either a global (producing an atmosphere and hydrosphere) or local scale. In the case of a water-CO_2 mixture, the final composition of the water depends on the temperature and CO_2:H_2O ratio; assuming $T = 0\,°C$, Clayton and Mayeda (1988) give $\delta^{18}O_{H_2O}(\permil) = 44.0X - 37.6$, where X is the molar fraction of oxygen present as water. In this model, the water has the same $\Delta^{17}O$ as the silicate. However, the discovery that water and secondary minerals in ALH84001, Nakhla and Lafayette have $\Delta^{17}O$ in excess of the host rock (Karlsson *et al.*, 1992; Farquhar and Thiemens, 2000) implies that not only the carbonates, but also the waters from which they formed were not equilibrated with the parent rocks. Saxton *et al.* (1998) considered several models for the formation of the ALH84001 carbonates. Models involving isotopic equilibration of fluid with the ALH84001 silicates suggest that the later Mg-rich carbonate formed at $<150\,°C$ but are difficult to reconcile with the observation of a $\Delta^{17}O$ excess in the carbonate.

The latter authors also considered a model in which no exchange with silicate takes place and the fluids are derived from a martian hydrosphere and atmosphere, approximate compositions for which were estimated from the Clayton-Mayeda model, with the requirement that it be able to generate the range of $\delta^{18}O$ values observed in martian carbonates. In this scenario, the earliest carbonates formed from a hot ($300 - 350\,°C$), water-rich fluid. The later Mg-rich carbonates would have formed at $<90°C$ if the fluid was water-rich, or at higher temperatures if it contained an appreciable mole fraction of CO_2.

The trend of increasing $\delta^{18}O$ from the Ca- to Mg-rich carbonates in ALH84001 is partially consistent with the pattern expected from brine evaporation. Enrichments in $\delta^{18}O$ are $+6\permil$ due to evaporation of terrestrial sea water and $+17\permil$ within evaporating brines in areas of low ambient humidity such as salt flats (Lloyd, 1966). However, Saxton *et al.* (1998) pointed out the isotopic zonation within this carbonate is inconsistent with a *single* evaporation episode, since $\delta^{18}O$ becomes approximately constant on reaching the magnesite rim, whilst Rayleigh fractionation would cause $\delta^{18}O$ to increase more rapidly as deposition proceeded.

3.2. CARBON, HYDROGEN AND SULPHUR

Carbon: Almost all martian meteorites have been analysed by acid dissolution to search for carbonates. Carbonate concentrations are generally low with between $30 - 60$ ppm for the nakhlites (Grady *et al.*, 1997). Carbon isotope results fall into two groups: ALH84001 and the nakhlites all contain ^{13}C-enriched carbonates, with $\delta^{13}C$ ranging from $+10\permil$ to $+55\permil$ (Carr *et al.*, 1985; Wright *et al.*, 1992; Romanek *et al.*, 1994; Jull *et al.*, 1995), whereas Chassigny and the shergottites have acid-extractable carbon, presumably from the carbonate anion, with $\delta^{13}C$ ranging from $0\permil$ to $-30\permil$. The most extensive study has been undertaken on ALH84001. Selective dissolution of the carbonate indicates that the early-forming calcium-rich cores of the rosettes are slightly less ^{13}C-enriched ($\delta^{13}C \sim +39.5\permil$)

than the later-forming, more Mg-rich mantles, with $\delta^{13}C \sim +41.8‰$ (Romanek *et al.*, 1994). Data on calcite from EETA79001 (+6.8‰, Wright *et al.*, 1988) is more ambiguous and it is possible that this phase is an effect of terrestrial contamination.

The two clear isotopic groups of carbonate are interpreted to represent formation of the minerals in two different environments: the lighter $\delta^{13}C$ values are from carbonate formed at depth within the martian crust by interaction of primary magmatic fluids with host rock, whereas carbonates in ALH84001 and the nakhlites formed either at, or close to Mars' surface, where fluid exchanged with martian atmospheric CO_2 (Carr *et al.*, 1985; Wright *et al.*, 1992; Romanek *et al.*, 1994). The carbon isotopic composition of martian atmospheric CO_2 is known to be ^{13}C-enriched, but a precise measurement of its $\delta^{13}C$ is still awaited.

Hydrogen: Several workers have found evidence of a D-enriched reservoir in the martian meteorites. Leshin *et al.* (1996) analysed D/H values in whole rock samples of several martian meteorites and found a wide variation of both δD and water content. The highest δD values were found in the shergottites, with δD up to +2140‰, the water content was ~ 0.05 wt%. The highest water contents (0.11 – 0.39 wt%) are in the nakhlites, but these have lower δD (up to +900‰ in a high temperature extraction from Lafayette).

Using an ion microprobe, Sugiura and Hoshino (2000) found an indigenous δD value of $\sim +2000‰$ for carbonate and maskelynite in ALH84001, with considerable scatter due to terrestrial contamination. Boctor *et al.* (1998) also used an ion microprobe to measure H isotopes in ALH84001 carbonate and phosphate, finding δD up to +500‰ in whitlockite. Saxton *et al.* (2000a) studied alteration phases in Lafayette – also by ion microprobe – and found δD up to $\sim +1500‰$. Watson *et al.* (1994) used an ion microprobe to measure D/H in amphibole, biotite and apatite in Chassigny, Shergotty and Zagami, finding variable, but higher than terrestrial, ratios (up to $\sim +4400‰$). They interpreted their results as the result of exchange between low D/H primary igneous minerals with a crustal fluid having high D/H. The high D/H enrichments are generally assumed to record atmospheric loss processes (see Section 5 below).

Sulphur: Greenwood *et al.* (1997, 2000) published ion probe determinations of $\delta^{34}S$ in sulphides from the nakhlites, Chassigny and the shergottites EETA79001, LEW88516 and QUE94201. Sulphides within the latter 3 shergottites had $\delta^{34}S =$ −2.6 to +3.5‰. ALH84001 pyrite, including that associated with carbonate, was +2.0 to +7.3‰. This range is inconsistent with the known sulphur isotopic fractionations associated with bacteria, which produce large enrichments in the light isotope ^{32}S. Nakhla, Governador Valadares and Lafayette had mean $\delta^{34}S$ values of +1.5‰, +0.7‰ and −3.2‰; Chassigny had isotopically light pyrite of −2.9 to −1.5‰. This was interpreted as being the result of progressive oxidation of a fluid, yielding isotopically lighter sulphides with time, which interacted with the nakhlites and Chassigny. Farquhar *et al.* (2000) also analysed $\delta^{33}S$ and $\delta^{36}S$ for SO_2 and H_2S which were mainly released through acid extraction of a range of

SNCs. The correlations between $\delta^{33}S$ and $\delta^{34}S$ that they identified could not be explained through mass fractionation.

3.3. MASS-INDEPENDENT FRACTIONATION PROCESSES

The variations in $\Delta^{17}O$ and the correlations between $\delta^{33}S$ and $\delta^{36}S$ decribed above cannot be explained by processes of mass fractionation alone (e.g. evaporation or condensation). Farquhar and Thiemens (2000) suggested generating an atmospheric reservoir of enhanced $\Delta^{17}O$ through mass-independent fractionation processes such as exchange between different gas phases (e.g. atomic oxygen and CO_2). This could then be transferred to crustal fluids and ultimately secondary mineral assemblages in the SNCs. Similarly, Farquhar *et al.* (2000) suggested that photolysis of atmospheric SO_2 and H_2S were the most likely explanation for the S isotopic anomalies. High D/H enrichments in meteorites, however, are due to atmospheric loss processes.

Previously, the presence of $\Delta^{17}O$ excesses was attributed to the influx of cometary material (Karlsson *et al.*, 1992) but as no chondritic material with suitably high $\Delta^{17}O$ has been identified this idea remains unsubstantiated.

4. Radiometric Dating: Implications for the Timing of Fluid Activity

Determining when alteration or evaporation processes occurred is difficult, particularly when many of the alteration products are as fine-grained as those in martian meteorites. There have been only five studies that have tried to address the timing of secondary mineral formation, and those focussed on only two assemblages, the carbonates in ALH84001 and the secondary mineral assemblages in Lafayette.

4.1. ALH84001 CARBONATE

Three studies of the age of these carbonates have reached three different conclusions. Wadhwa and Lugmair (1996) suggested an age of 1.41 ± 0.10 Gyr based on a two-point carbonate-plagioclase Rb-Sr "isochron", and Knott *et al.* (1997) reported an Ar-Ar age of 3.6 Gyr. However, the first was a model age with inherent uncertainties, and the second was probably actually dating plagioclase (Turner *et al.*, 1997). The most detailed chronological study of ALH84001 carbonate is that of Borg *et al.* (1999), who performed selective leaching that gave concordant Rb-Sr and U-Th-Pb ages of $3.9 - 4.0$ Gyr. Since the latter number is comparable to most ^{40}Ar-^{39}Ar ages reported (Turner *et al.*, 1997; Ilg *et al.*, 1997; Bogard and Garrison, 1999; Nyquist *et al.*, 2001) it is likely to be the age of carbonate formation.

4.2. THE NAKHLITE LAFAYETTE

In a similar approach to that used by Borg *et al.* (1999), Shih *et al.* (1998) performed leaching experiments on the nakhlite Lafayette, and obtained a two-point "isochron" from two iddingsite leachates corresponding to an age of 679 ± 66 Myr. Swindle *et al.* (2000) used a different approach. They extracted individual $1 - 20$ μg samples of alteration products from Lafayette and analyzed them using the K-Ar system. A preliminary experiment had shown that the fine-grained nature of the clay and siderite veins within olivine ('iddingsite') led to a loss of ^{39}Ar, the isotope produced by irradiation of potassium, because the recoil energy involved in the neutron capture was sufficient to expel many of the ^{39}Ar atoms from the samples. So instead, they determined K abundances on each sample via neutron activation analysis, and then measured the amount of argon. There was not a single date at which the ages of their samples clustered. Although many of their samples had demonstrably non-zero ages, the ages ranged from zero to 670 ± 91 Myr. Their oldest age was, within the uncertainty, identical to that of Shih *et al.* (1998).

5. Fluid Compositions and Environmental Conditions

In this section we discuss: (a) ancient water composition and weathering processes (b) the origin of the waters related to evaporation (c) the physical and climatic regime in which they formed.

5.1. AQUEOUS WEATHERING AND FLUID COMPOSITIONS

In a CO_2-rich environment where liquid water appeared, it is inevitable that igneous rocks would be weathered, and that the ions so released would eventually be precipitated to maintain mass balance. For instance, glacial melt or groundwater, perhaps released through local igneous activity (Gulick, 1998), would consume atmospheric CO_2 to produce carbonic acid. The aqueous weathering processes should produce bicarbonate (HCO_3^-) as the main anion, unless significant sulphurous gases are dissolved (for example in subsurface hydrothermal vents), which would cause sulphate to predominate. The bicarbonate molal concentration (i.e. mol/kg) relative to the cation concentration depends on the mole ratios in rock dissolution reactions (Garrels, 1967). The idealized case of the dissolution of iron end-members, ferrosilite (pyroxene) and fayalite (olivine), by carbonic acid exemplifies weathering under low oxygen conditions:

$$FeSiO_3 + 3H_2O + 2CO_2 = Fe^{2+} + 2HCO_3^- + Si(OH)_4$$
$$Fe_2SiO_4 + 4H_2O + 4CO_2 = 2Fe^{2+} + 4HCO_3^- + Si(OH)_4 \tag{1}$$

Given more realistic igneous rocks, other cations are released, principally Mg^{2+}, Ca^{2+}, K^+, and Na^+. The other main anions, Cl^- and SO_4^{2-} are particularly important in hydrothermal systems due to the presence of HCl and SO_2 gases. Hydrolyzed silica ($Si(OH)_4$) is a product of igneous rock weathering (see above) and silica's fate can take several paths. In a surface zone, some may combine with Al derived from plagioclase and cations to form clay minerals. Alternatively, silica can precipitate as amorphous silica, which can diagenetically change to quartz or other pure silica forms. Today on Earth, utilization of silica by microscopic organisms greatly suppresses silica concentrations in natural waters. On a non-biological Mars, waters presumably contain silica concentrations either close to temperature-dependent saturation levels or controlled by formation of clays. Smectite-type clays, like those of the nakhlites, sequester silica as well as the group I cations, particularly K^+ via cation exchange. As silica or clay minerals are early precipitates, they should be associated with other early precipitates like siderite, as seen in Lafayette. Meteorite assemblages affected by evolved waters that percolated into the pore space may be free of silica, which precipitated elsewhere. This might apply to the assemblage in Chassigny, for example. Whereas gibbsite/kaolinite clays form in dilute waters, smectite/illite clays are formed in concentrated solutions (Drever, 1997, p. 282). Consequently, the presence of smectite/illite in SNCs is consistent with solutions concentrated by evaporation in the absence of microbial life.

In general one would expect briny water on Mars to be of type Na-Mg-Fe-Ca-HCO_3-SO_4-Cl-H_2O, similar to a marine composition except subject to high levels of CO_2, making it acidic with high levels of bicarbonate. Water-soluble ions measured by leaching salt deposits directly from Nakhla (Sawyer *et al.*, 2000) provide evidence for this. The Nakhla data suggest that ion concentrations may have been in roughly similar relative proportions to terrestrial seawater, except with a greater relative abundance of Ca^{2+} and possibly Mg^{2+}. On Earth, the most abundant ions in seawater in order of decreasing concentration are Cl^-, Na^+, Mg^{2+}, SO_4^{2-}, Ca^{2+} and K^+. Because these ions are so soluble their residence times are long and their proportions are controlled by long-term geochemical processes, such as slow loss to evaporites or hydrothermal sinks at mid-ocean ridges, rather than biology. Smaller components in terrestrial seawater such as Si, C, N, P and trace elements are largely biologically regulated. The measurements of salts at two locations in Nakhla by Sawyer *et al.* included the major ions found in terrestrial seawater but not minor components. These measurements are insufficient to deduce the processes that controlled elemental abundances in martian subsurface brines. However, a lower molar ratio of Mg/Ca$\sim$0.8 $-$ 1.5 in Nakhla salts (Sawyer *et al.*, 2000) compared to a terrestrial value of $\sim$5.2 suggests hydrothermal activity controlled the brine composition at some point. Mg tends to be removed from saline water passing through hot rocks and replaced by Ca (Holland, 1978). In very high temperature hydrothermal exchange with basalt, K^+ is transferred to saline water but K^+ is actually depleted in the Sawyer *et al.* measurements relative to calcium compared to seawater, suggesting moderate temperatures $<100\,°C$.

5.2. EVAPORATION MODELS

As discussed in previous sections, it is likely that the secondary minerals are associated with brines. Numerical models of chemical evaporation can help us to understand their nature. In such models, the effect of concentrating the solution by removing water is calculated while maintaining the principles of aqueous chemistry: mass conservation, electroneutrality and equilibria. It is clear, for instance from the oxygen isotopic data discussed previously, that the silicate igneous assemblages did not equilibrate with the fluids. However, gaseous CO_2 can rapidly reach equilibrium with fluids. This presumed equilibrium is consistent with the ^{13}C-enrichments in carbonate discussed previously. The relatively shallow depth of origin of the nakhlites (estimated at $20 - 100$ m depth on the basis of comparisons with terrestrial lava flows, Friedman *et al.*, 1999) would also have facilitated brine-atmosphere exchange.

Evaporation models are, however, limited by this assumption of thermodynamic equilibrium and standard models also typically lack iron chemistry. Using a model that did incorporate iron chemistry, Catling (1999) showed that the theoretical sequence of carbonate precipitation under high CO_2 conditions follows the sequence of siderite, calcite and then magnesite. An important factor in carbonate precipitation, besides solubility, is the formation of soluble ion pairs or "complexes" in solution. For example, Fe^{2+}, tends to bind to CO_3^{2-} to form the ion pair $FeCO_3^0(aq)$ in solution, which suppresses the level of free CO_3^{2-} available and prevents the precipitation of other carbonates prior to siderite. Carbonate solid-solution equilibria (Woods and Garrels, 1992) determine the cationic composition of carbonates. The degree of cation substitution depends on the concentration ratios of cations in solution, e.g. cation displacement between high-Ca siderite and solution is mediated through the equilibrium relation

$$(a_{CaCO_3}/a_{FeCO_3})_{\text{siderite}} = K(a_{Ca^{2+}}/a_{Fe^{2+}})aq$$

where a_i is the activity of species i and K is an equilibrium constant (nonlinear with respect to species ratios). Although such a relationship neglects kinetics, it may help explain the cation composition seen in the meteorite carbonates. A Ca-rich siderite is expected early in evaporation, followed by magnesium-rich carbonate after much of the Ca and Fe have been exhausted. After carbonate precipitation, gypsum, anhydrite and halite are expected. This is consistent with the carbonate zoning sequence in ALH84001 and the variation in the mineralogies of alteration products between the 3 nakhlites (Bridges and Grady, 2000), suggesting that thermodynamic models are applicable to modelling of the meteorites' secondary assemblages.

5.3. AQUEOUS ENVIRONMENTAL AND ATMOSPHERIC CONDITIONS

Evaporite mineral precipitation is affected by temperature, pCO_2, pH, and redox conditions, which we discuss in turn.

Temperature. There is a major difference in the salts that form when water is removed from a brine by evaporation compared to removal by freezing. For example, gypsum or anhydrite followed by halite tends to form when seawater evaporates whereas mirabilite ($Na_2SO_4 \cdot 10H_2O$) followed by hydrohalite ($NaCl \cdot 2H_2O$) tends to form in freezing (Herut *et al.*, 1990). The salts identified in the nakhlites, Chassigny and the shergottites suggest that evaporation was the process for water removal rather than concentration and precipitation through freezing. As discussed in previous sections, upper temperatures are more difficult to constrain but point towards initial precipitation $\leq 150\,^{\circ}C$ and in a thermodynamic model of progressive evaporation in the nakhlites Bridges and Grady (2000) found that mineral precipitation temperatures of $10 - 25^{\circ}C$ were consistent with the evaporite assemblages identified in the 3 meteorites.

pH. Salts such as various sodium carbonates tend to appear in highly alkaline brines. These are absent from the meteorites, suggesting that solutions never reached high pH. Instead the presence of iron minerals, such as siderite, containing Fe^{2+} in the nakhlites and ALH84001, indicate that the precursor fluid had pH< 7 in order to carry sufficient Fe^{2+} cations.

pCO$_2$. The presence of siderite in the nakhlites and ALH84001 suggests high pCO$_2$ conditions. This line of argument follows the same geochemical constraint used in determining pCO$_2$ of the early Earth: for example, pCO$_2$ in the late Archaean is deduced as <50 mbar based on the absence of siderite and the presence of hydrous iron silicates in paleosols (Rye *et al.*, 1995). Essentially, in the expected presence of dissolved silica, siderite tends to form at pCO$_2 > 50 - 100$ mbar, whereas hydrous iron silicates will form at lower pCO$_2$ (Catling, 1999). Based on the presence of siderite, this would seem to suggest that the salts in nakhlites and ALH84001 were formed at pCO2 much higher than the current surface pressure on Mars of ~ 6 mbar. However, the coexistence of smectites and siderite in the nakhlites could argue for an intermediate level close to ~ 50 mbar. The equivalent amount of CO_2 locked up in the near surface can be calculated if we assume that typical carbonate amounts in the meteorites represent the whole of Mars. Taking ~ 0.5 wt% carbonate, and assuming a crustal density of 3000 kgm^{-3}, 1000 m of crust would contain 2.15×10^{18} kg of carbonate, which, using calcite as the carbonate, is equivalent to 2.15×10^{19} mol CO_2. If this were all released, a pCO$_2$ of ~ 250 mbar would result. Alternatively, 200 m of crust containing this concentration of carbonate would be equivalent to 50 mbar of CO_2 being sequestered. Supporting evidence for a martian atmosphere intermittently > 6 mbar up to geologically recent times is given by gullies identified with very low overlying crater densities and hence of young age (Malin and Edgett, 2000a; Hartmann, 2001). A periodically thicker atmosphere would almost certainly be necessary to stabilise liquid water on the martian surface.

Redox state. The presence of siderite in the nakhlites suggests a weakly reducing environment because of the need to maintain Fe^{2+} in solution prior to its precipitation in carbonate. Goethite, which is of relatively minor occurrence compared to the other secondary minerals, could have been derived from siderite or magnetite as a later oxidation product. In ALH84001, the assemblage pyrite, siderite, magnetite suggests reducing conditions although this evidence may be complicated by the possible association with shock vapourisation, after the initial carbonate precipitation. Reducing conditions are generally consistent with an early Noachian environment which experienced more volcanic outgassing than the later times in martian history in which the younger meteorites' parent rocks formed. This is also a possible explanation for the absence of sulphate in ALH84001, in favour of pyrite.

5.4. CARBONATES IN THE MARTIAN CRUST AND ATMOSPHERIC LOSS PROCESSES

The martian atmosphere consists of volatiles that were acquired from smaller bodies that either coalesced into the original bulk of the planet or impacted at a later stage. However, Mars later lost volatiles by various physical and chemical processes. Two of the main processes that would have been effective at removing an early thick CO_2 atmosphere are: (1) CO_2 reacting with surface rocks and water to form carbonates (2) impacts from asteroids or comets blowing away the atmosphere to space Melosh and Vickery (1989). In addition, preferential loss of H, compared to D occurred due to upper atmospheric loss processes, particularly that of hydrodynamic escape.

Data from Mars Global Surveyor's TES shows that carbonates are not abundant on the present surface and are loosely constrained to $< 10\%$ (Christensen $et\ al.$, 1998). As calculated in the previous section, the abundance of carbonates in meteorites, if extrapolated globally, suggests the presence of no more than ~ 0.25 bar of CO_2 in the upper kilometer. Such CO_2, if released, would not be sufficient to raise the global mean surface temperature to above freezing (Forget and Pierrehumbert, 1997) although it might contribute to seasonal tropical warming above freezing in combination with obliquity cycles (Haberle $et\ al.$, 2000). More concentrated subsurface deposits of carbonate sediments are needed to explain sequestration of a thick, early martian atmosphere. Parts of the ancient highlands are composed of meter-scale thick layers (Malin and Edgett, 2000b) and these might include carbonate sediments.

In addition to sequestration of CO_2 in carbonate the early atmosphere was thinned by impact erosion (e.g. Melosh and Vickery, 1989). Evidence for this process is given by the striking abundances of radiogenic gases ^{40}Ar and ^{129}Xe (produced after the formation of Mars by the decay of internal ^{40}K and ^{129}I, respectively) relative to the primordial gases ^{36}Ar and ^{132}Xe in the martian atmosphere (Owen, 1992). ^{40}Ar/^{36}Ar is 296 for Earth and 3000 ± 500 for Mars (Owen, 1992). The most probable scenario is that Mars' early atmosphere prior to ~ 4 Gyr was lost by a combination of subsurface precipitation of carbonates and impact erosion.

Evidence for isotopically selective atmospheric loss to space is given by D/H enrichments found within the meteorites. On present day Mars, the relative loss rates of D and H in proportion to their atmospheric density is 0.32, but fractionation on very early Mars, when there was sufficient hydrogen to cause hydrodynamic escape, may have been in the range $0.8 - 0.9$ (Zahnle *et al.*, 1990). Hydrodynamic escape is the process whereby extreme fluxes of UV radiation in the early Solar System acted to form outwardly directed fluxes of H atoms (e.g. Carr, 1996).

Jakosky (1993) modelled the relative loss rates for the different isotopes, loss flux, size of remaining reservoir, and degree of heavy isotope enrichment, which are all quantitatively related. Sugiura and Hoshino (2000) used their ALH84001 results to suggest that the fractional increase in D/H of the martian atmosphere from 4.56 to 4 Gyr was 3.0/1.0 or 2000/0‰. The fraction of original hydrogen remaining on Mars at 4 Gyr was estimated at 2.4%. Subsequent fractionation up to recent times must have taken place at a more gradual rate. Atmospheric loss effects are less clear for oxygen, as Farquhar and Thiemens (2000) noted, the lack of a correlation between $\Delta^{17}O$ and $\delta^{18}O$ in the SNC secondary minerals is not consistent with this process being the dominant control. However, a loss of $\sim7\%$ from the atmospheric oxygen reservoir corresponds to increases of $\sim20‰$ in atmospheric $\delta^{18}O$ and several tenths of a permil in $\Delta^{17}O$ (Saxton *et al.*, 2000b). Therefore, the $\Delta^{17}O$ excesses recorded by the secondary minerals probably also include some effects of atmospheric loss in addition to mass-independent fractionation.

6. Synthesis and Conclusions

6.1. RELATIONSHIP OF SNC SECONDARY ASSEMBLAGES TO CURRENT MARTIAN SURFACE DEPOSITS

Most of the martian surface between the poles is partly covered by a mixture of dust and rocks that is inferred to be salt-rich. This partly cemented "soil" is believed to have a fairly homogeneous composition across the planet as a result of redistribution by aeolian activity. Viking and Pathfinder soils have a similar basaltic nature, although the former have higher SO_3 and Cl contents (Newsom *et al.*, 1999; Wänke *et al.*, 2001). The basaltic silicate composition is similar to that of the shergottites.

Thermal emission spectrometry (e.g. at the Pathfinder site) also shows the basaltic nature of the dust and rocks (Bibring and Erard, 2001). In addition, the characteristic ferric absorption edge at $445-800$ nm of the martian surface suggests the presence of palagonite i.e. poorly crystalline ferric oxides of uncertain identity and state of hydration (Morris *et al.*, 2000). Clay and carbonate minerals have not, however, been firmly identified (Banin *et al.*, 1992). Thus, there are significant differences between the weathering products associated with the soil and those of the SNCs. In the latter, ferric oxides form a relatively minor part of the secondary mineral assemblages. The secondary minerals in the meteorites cannot be clearly related to the processes associated with the palagonite-type alteration inferred for the global martian soil. Whether the anions in the soil salts originated from brines, gas-solid reactions (Gooding, 1978), acid fog (Banin *et al.*, 1997), hydrothermal fluids (Newsom *et al.*, 1999), or some combination of these processes is not certain. The research on salt minerals in the SNCs does point towards the importance of low temperature brines at or near the martian surface. There is insufficient soil mineralogical data to make direct comparisons to the meteorite data about fluid compositions. However, the low Mg/Ca ratio and K abundances of nakhlite salts may indicate that the prior to evaporation, the brine underwent some hydrothermal exchange $<150\,°C$ with surrounding igneous rocks.

The possibility of an ancient ocean (>3.5 Gyr) on Mars has been raised (e.g. Carr, 1996; Head *et al.*, 2001). It seems probable that the ocean, if it existed, predated some ancient highland surface based on layering relationships (e.g. Malin and Edgett, 2000b). The evaporites in the nakhlites are later and are better explained by more short-lived events rather than through association with an ocean. For instance, the volume of water in the meteorite parent rocks may have been large, but its presence would have been ephemeral given that evaporite minerals comprise no more than ~ 1 vol.% of martian meteorites and that there is no evidence for prolonged weathering of the surrounding igneous rocks.

6.2. FORMATION OF THE SECONDARY ASSEMBLAGES IN SNCs

The available experimental and phase equilibrium data which are most relevant to the salt minerals in the SNCs suggest relatively low upper temperatures of formation ($<150°C$) and, in the case of much of the carbonate, as metastable phases indicating fairly rapid crystallisation, perhaps over days. Siderite may have crystallised at around $25°C$. These temperatures are consistent with the modeling based on stable isotope compositions. Although such an approach has limitations because it is now clear that the secondary assemblages have preserved mass independent fractionations of H, S and O isotopes, the ratios are clearly inconsistent with high temperature equilibration of silicate and carbonate or sulphate minerals. The large fractionation in $\delta^{18}O$ between the igneous silicate and secondary minerals ($> +30‰$) shows that the latter must have formed at $<400\,°C$. The relatively simple salt assemblages and limited amount of silicate alteration also suggest short-lived pulses of fluid in the parent rocks and a lack of extensive, long-lived hydrothermal systems.

The mineralogical features of the secondary minerals e.g. Ca-rich followed by Mg-rich carbonate, and the sequence of salt mineral crystallisation recorded in the 3 nakhlites are consistent with an origin through brine evaporation. Initial formation from an evaporating brine is the model most consistent with the data on these minerals and such brines were probably acidic and reducing. Some of the secondary minerals (e.g. the siderite located within fractures of Lafayette) are clearly filling fracture/void space. The association of carbonate with feldspathic glass, especially in ALH84001 is more ambiguous. There is also firm evidence that high temperature effects, associated with shock, have affected some of the SNCs. In particular, research on a subset of the magnetite grains within the carbonate rosettes of ALH84001 demonstrates the likelihood of shock-induced recrystallisation and partial vapourisation of pre-existing salt assemblages. This process may have partially obscured the nature of the original fluids in the parent rock.

6.3. TIMING AND ORIGIN OF AQUEOUS ACTIVITY

The carbonates in ALH84001 appear to be about 3.9 − 4.0 Gyr old. On the other hand, it is still not clear when the weathering products in the nakhlites formed. At least some of them formed several hundred Myr ago. The spread in K-Ar ages could represent partial loss of Ar from the fine-grained samples. However, some clay samples on Earth, with its warmer temperatures, have retained Ar for up to 450 Myr (Dong *et al.*, 1997). Thus, the samples shouldn't have lost their Ar in martian ambient conditions, and Swindle *et al.* (2000) argued that partial gas loss during processing and irradiation is unlikely. Another alternative is that there is not a single age for these weathering products. If the intimate banding within the Lafayette siderite and clay veins reflected deposition from a sequence of fluids (Vicenzi and Eiler, 1998), there might actually be a wide spread in ages.

The 650 − 700 Myr formation age implied by both the Rb-Sr model age and the oldest K-Ar age may well be an upper limit, or it might possibly be an average. This means that the alteration occurred and continued long after the formation of the rocks, since their crystallization ages are about 1.3 Gyr (Nyquist *et al.*, 2001). The younger crystallization ages of the shergottites (165 − 470 Myr, Nyquist *et al.*, 2001) may well indicate more recent fluid activity associated with the secondary mineral assemblages in those meteorites. However, as the distinction between martian and terrestrial salts is not as clearly established as for those in the nakhlites and ALH84001 this important issue remains to be clarified in future work.

A consideration of ALH84001 and Lafayette ages leads to a further implication. Although ALH84001 was in a location favourable to carbonate formation 3.9 Gyr ago, there is little evidence for any other alteration products. The lack of more recent weathering products in ALH84001 argues against a continuously warm and wet martian history. Similarly, as the salt assemblages, in the nakhlites particularly, are highly soluble, their preservation over hundreds of millions of years means it is likely that the fluid activity was not prolonged and recurring but instead took place during one or a few rapid and isolated events. It is possible that sporadic, localised brine migration and evaporation has continued until geologically recent times and is responsible for the gullies seen in recent MOC images on young terrains. The mechanisms associated with the formation of brines and their subsequent evaporation are not certain. An obvious source of energy to melt ice and establish a flow of water is igneous activity. However, if that is the case, the igneous activity that led to the alteration of the nakhlites was not associated with the magmatic activity in which the parent rocks formed.

Another process which could have led to ice melting is variation in the solar insolation. This occurred on timescales of $10^5 - 10^7$ years as a result of obliquity changes (Bills, 1990; Haberle *et al.*, 2000). Heating alone is not, however, sufficient to explain the formation and nature of the brines from which the SNC secondary minerals probably crystallised. For instance, the siderite and Fe-rich phyllosilicates seen in the nakhlites form together at about 50 mbar pCO_2 and not the current martian average of 6 mbar. Calculations on the amount of CO_2 trapped in the upper few hundred meters of the martian crust in the form of carbonate also suggest that an ancient atmosphere could have had $50 - 250$ mbar pCO_2. Therefore thicker atmospheres, with an associated increased chance of liquid water being stabilised were present – at least periodically – during the secondary minerals' formation $\leq 650 - 700$ Myr.

Acknowledgements

JCB and JS are supported by PPARC grants. DC is supported by NASA's Planetary Geology and Geophysics Program. This paper benefited from the reviews of W.K. Hartmann, G. Turner and an anonymous reviewer. E.J. Essene is thanked for a stimulating discussion on SNC secondary mineral compositions.

References

Anovitz, L.M., and Essene, E.J.: 1987, 'Phase Equilibria in the System $CaCO_3$-$MgCO_3$-$FeCO_3$', *J. Petrology* **28**, 389–414.

Baker, V.R., Carr, M.H., Gulick, V.C., Williams, C.R., Marley, M.S.: 1992, 'Channels and Valley Networks', *in* H.H. Kieffer, B.M. Jakosky, C.W. Snyder and M.S. Matthews (eds.), *Mars*, Univ. Arizona Press, Tucson, pp. 493–522.

Baker, L.L., Agenbroad, D.J., and Wood, S.J.: 2000, 'Experimental Hydrothermal Alteration of a Martian Analog Basalt: Implications for Martian Meteorites', *Met. Planet. Sci.* **35**, 31–38.

Banin, A., Clark, B.C. and Wänke, H.: 1992, 'Surface Chemistry and Mineralogy', *in* H.H. Kieffer, B.M. Jakosky, C.W. Snyder and M.S. Matthews (Eds.), *Mars*, Univ. Arizona Press, Tucson, pp. 594–625.

Banin, A., Han, F.X., Kan, I. and Cicelsky A.: 1997, 'Acidic Volatiles in the Mars Soil', *J. Geophys. Res.* **102**, 13,341–13,356.

Bell, J.F. III, *et al.*: 2000, 'Mineralogic and Compositional Properties of Martian Soil and Dust: Results from Pathfinder', *J. Geophys. Res.* **105**, 1721–1755.

Bibring, J.-P., and Erard, S.: 2001, 'The Martian Surface Composition', *Space Sci. Rev.*, this volume.

Bills, B.G.: 1990, 'The Rigid Obliquity History of Mars', *J. Geophys. Res.* **95**, 14,137–14,153.

Boctor, N.Z., Wang, J., Alexander, C.M.O'D., Hauri, E., Bertka, C.M. and Fei, Y.: 1998, 'Hydrogen Isotope Studies of Carbonate and Phosphate in Martian Meteorite Allan Hills 84001' *Met. Planet. Sci.* **33**, A18 (abstract).

Bogard, D.D., and Garrison, D.H.: 1999, 'Argon-39-argon-40 "Ages" and Trapped Argon in Martian Shergottites, Chassigny, and Allan Hills 84001', *Met. Planet. Sci.* **34**, 451–473.

Bogard, D.D., Clayton, R.N., Marti, K., Owen, T., and Turner, G.: 'Martian Volatiles: Isotopic Composition, Origin, and Evolution', *Space Sci. Rev.*, this volume.

Borg, L.E., Connelly, J.N., Nyquist, L.E., Shih, C.-Y., Wiesmann, H., and Reese, Y.: 1999, 'The Age of the Carbonates in Martian Meteorite ALH84001', *Science* **286**, 90–94.

Bradley, J.P., McSween, H.P., and Harvey, R.P.: 1998, 'Epitaxial Growth of Nanophase Magnetite in Martian Meteorite Allan Hills 84001: Implications for Biogenic Mineralization', *Met. Planet. Sci.* **33**, 765–773.

Brass, G.W.: 1980, 'Stability of Brines on Mars', *Icarus* **42**, 20–28.

Brearley, A.J.: 2000, 'Hydrous Phases in ALH84001: Further Evidence for Preterrestrial Alteration and a Shock-induced Thermal Overprint', *Proc. 31st Lunar Planet. Sci. Conf.*, abstract #1203 (CD-ROM).

Bridges, J.C., and Grady, M.M.: 1999, 'A Halite-siderite-anhydrite-chlorapatite Assemblage in Nakhla: Mineralogical Evidence for Evaporites on Mars', *Met. Planet. Sci.* **34**, 407–416.

Bridges, J.C., and Grady, M.M.: 2000, 'Evaporite Mineral Assemblages in the Nakhlite (Martian) Meteorites', *Earth Planet. Sci. Lett.* **176**, 267–279.

Carr, M.H.: 1996, 'Water on Mars', Oxford Univ. Press, Oxford, 229 pp.

Carr, R.H., Grady, M.M., Wright, I.P., and Pillinger, C.T.: 1985, 'Martian Atmospheric Carbon Dioxide and Weathering-products in SNC Meteorites', *Nature* **314**, 248–250.

Catling, D.C.: 1999, 'A Chemical Model for Evaporites on Early Mars: Possible Sedimentary Tracers of the Early Climate and Implications for Exploration', *J. Geophys. Res.* **104**, 16,453–16,469.

Chatzitheodoridis, E., and Turner, G.: 1990, 'Secondary Minerals in the Nakhla Meteorite' *Meteoritics* **25**, 354 (abstract).

Christensen, P.R., and Moore, H.J.: 1992, 'The Martian Surface Layer', *in* H.H. Kieffer *et al.* (eds.), *Mars*, Univ. Arizona Press, Tucson, pp. 686–729.

Christensen, P.R., *et al.*: 1998, 'Results from the Mars Global Surveyor Thermal Emission Spectrometer', *Science* **279**, 1692–1698.

Clayton, R.N., and Mayeda, T.K.: 1988, 'Isotopic Composition of Carbonate in EETA 79001 and its Relation to Parent Body Volatiles', *Geochim. Cosmochim. Acta* **52**, 925–927.

Dong, H., Hall, C.M., Halliday, A.N., and Peacor, D.R.: 1997, 'Laser ^{40}Ar-^{39}Ar Dating of Microgram-size Illite Samples and Implications for Thin Section Dating', *Geochim. Cosmochim. Acta* **61**, 3803–3808.

Douglas, C., Wright, I.P., and Pillinger, C.T.: 1994, 'A Search for Further Concentrations of Organic Materials in EETA79001', *Proc. 25th Lunar Planet. Sci. Conf.*, 339 (abstract).

Drever, J.I.: 1997, 'The Geochemistry of Natural Waters', Prentice-Hall, London. 436 pp.

Eiler, J.M., Valley, J.W., Graham, C.M., and Fournelle, J.: 1998, 'Geochemistry of Carbonates and Glass in ALH84001', *Met. Planet. Sci.* **33**, A4 (abstract).

Farquhar, J., and Thiemens, M.H.: 2000, 'Oxygen Cycle of the Martian Atmosphere-regolith System: Δ^{17}O of Secondary Phases in Nakhla and Lafayette', *J. Geophys. Res.* **105**, 11,991–11,997.

Farquhar, J., Thiemens, M.H., and Jackson, T.: 1998, 'Atmosphere-surface Interactions on Mars: Δ^{17}O Measurements of Carbonate from ALH84001', *Science* **275**, 1580–1582.

Farquhar, J., Savarino, J., Jackson, T.I., and Thiemens, M.H.: 2000, 'Evidence of Atmospheric Sulphur in the Martian Regolith from Sulphur Isotopes in Meteorites.', *Nature* **404**, 50–52.

Floran, R.J., Prinz, M., Hlava, P.F., Keil, K., Nehru, C.E., and Hinthorne, J.R.: 1978, 'The Chassigny Meteorite: a Cumulate Dunite with Hydrous Amphibole-bearing Melt Inclusions', *Geochim. Cosmochim. Acta* **42**, 1213–1230.

Forget, F., and Pierrehumbert, R.T.: 1997, 'Warming Early Mars with Carbon Dioxide Clouds that Scatter Infrared Radiation', *Science* **278**, 1273.

Forsythe, R.D., and Zimbelman, J.R.: 1995, 'A Case for Ancient Evaporite Basins on Mars', *J. Geophys. Res.* **100**, 5553–5563.

Forsythe, R.D., and Blackwelder, C.R.: 1998, 'Closed Drainage Crater Basins of the Martian Highlands: Constraints on the Early Martian Hydrologic Cycle', *J. Geophys. Res.* **103**, 31,421–31,431.

Franchi, I.A., Wright, I.P., Sexton, A.S., and Pillinger, C.T.: 1999, 'The Oxygen-isotopic Composition of Earth and Mars', *Met. Planet. Sci.* **34**, 657–661.

Friedman, Lentz, R.C., Taylor, G.J., and Treiman, A.H.: 1999, 'Formation of a Martian Pyroxenite: A Comparative Study of the Nakhlite Meteorites and Theo's Flow', *Met. Planet. Sci.* **34**, 919–932.

Garrels, R.M.: 1967, 'Genesis of Some Ground Waters from Igneous Rocks', *in* P.H. Abelson (ed.), *Researches in Geochemistry Vol. 2*, John Wiley, New York, pp. 405–420.

Golden, D.C., Ming, W., Schwandt, C.S., Morris, R.V., Yang, S.V., and Lofgren, G.E.: 2000, 'An Experimental Study on Kinetically-driven Precipitation of Calcium-magnesium-iron Carbonates from Solution: Implications for the Low-temperature Formation of Carbonates in Martian Meteorite Allan Hills 84001.' *Met. Planet. Sci.* **35**, 457–465.

Gooding, J.L.: 1978, 'Chemical Weathering on Mars: Thermodynamic Stabilities of Primary Minerals and Their Alteration Products from Mafic Igneous Rocks', *Icarus* **33**, 483–513.

Gooding, J.L., and Muenow, D.W.: 1986, 'Martian Volatiles in Shergottite EETA79001: New Evidence from Oxidised Sulfur and Sulfur-rich Alumino-silicates', *Geochim. Cosmochim. Acta* **50**, 1049–1059.

Gooding, J.L., Wentworth, S.J. and Zolensky, M.E.: 1988, 'Calcium Carbonate and Sulfate of Possible Extraterrestrial Origin in the EETA79001 Meteorite', *Geochim. Cosmochim. Acta* **52**, 909–916.

Gooding, J.L, Wentworth, S.J., and Zolensky, M.E.: 1991, 'Aqueous Alteration of the Nakhla Meteorite', *Meteoritics* **26**, 135–143.

Grady, M.M., Wright, I.P., and Pillinger, C.T.: 1995, 'A Search for Nitrates in Martian Meteorites', *J. Geophys. Res.* **100**, 5449–5455.

Grady, M.M., Wright, I.P., and Pillinger, C.T.: 1997, 'A Carbon and Nitrogen Isotope Study of Zagami', *J. Geophys. Res.* **102**, 9165–9173.

Greenwood, J.P., Riciputi, L.R., and McSween, H.Y.: 1997, 'Sulfide Isotopic Compositions in Shergottites and ALH84001, and Possible Implications for Life on Mars', *Geochim. Cosmochim. Acta* **61**, 4449–4453.

Greenwood, J.P., Riciputi, L.R., McSween, H.Y., and Taylor, L.A.: 2000, 'Modified Sulfur Isotopic Compositions of Sulfides in the Nakhlites and Chassigny', *Geochim. Cosmochim. Acta* **64**, 1121–1131.

Gulick, V.C.: 1998, 'Magmatic Intrusions and a Hydrothermal Origin for Fluvial Valleys on Mars', *J. Geophys. Res.* **103**, 19,365–19,388.

Haberle, R.M., McKay, C.P., Schaeffer, J., Joshi, M., Cabrol, N.A. and Grin, E.A.: 2000, 'Meteorological Control on the Formation of Martian Paleolakes', *Proc. 31st Lunar Planet. Sci. Conf.*, abstract #1509 (CD-ROM).

Hartmann, W.K.: 2001, 'Martian Seeps and Their Relation to Youthful Geothermal Activity', *Space Sci. Rev.*, this volume.

Harvey, R.P., and McSween, H.Y., Jr.: 1996, 'A Possible High-temperature Origin for the Carbonates in the Martian Meteorite ALH84001', *Nature* **382**, 49–51.

Head *et al.*: 2001, 'Geological Processes and Evolution', *Space Sci. Rev.*, this volume.

Herut, B., Starinsky, A., Katz, A., and Bein, A.: 1990, 'The Role of Seawater Freezing in the Formation of Subsurface Brines', *Geochim. Cosmochim. Acta* **54**, 13–21.

Holland, H.D.: 1978, 'The Chemistry of the Atmosphere and Oceans', John Wiley, New York, pp. 190–200.

Holland, G., Lyon, I.C., Saxton, J.M., and Turner, G.: 2000, 'Very Low Oxygen-isotopic Ratios in Allan Hills 84001 Carbonates: A Possible Meteoric Component?', *Met. Planet. Sci.* **35**, A76–77 (abstract).

Ilg, S., Jessberger, E.K., and El Goresy, A.: 1997, '^{40}Ar/^{39}Ar Laser Extraction Dating of Individual Maskelynites in SNC Pyroxenite Allan Hills 84001', *Met. Planet. Sci.* **33**, A65 (abstract).

Jakosky, B.M.: 1993, 'Mars Volatile Evolution: Implications of the Recent Measurement of ^{17}O in Water from SNC Meteorites', *Geophys. Res. Lett.* **20**, 1591–1594.

Jull, A.J.T., Eastoe, C.J., Xue, S., and Herzog, G.F.: 1995, 'Isotopic Composition of Carbonates in the SNC Meteorites ALH84001 and Nakhla', *Meteoritics* **30**, 311–318.

Jull, A.J.T., Eastoe, C.J., and Cloudt, S.: 1997, 'Isotopic Composition of Carbonates in the SNC Meteorites, Allan Hills 84001 and Zagami', *J. Geophys. Res.* **102**, 1663–1669.

Karlsson, H.R., Clayton, R.N., Gibson, E.K., Jr., and Mayeda, T.K.: 1992, 'Water in SNC Meteorites: Evidence for a Martian Hydrosphere', *Science* **255**, 1409–1411.

Kathie, L., Thomas-Keptra, D.A., Bazylinski, D.A., Kirschvink, J.L., Clemett, S.J., McKay, D.S., Wentworth, S.J., Hojatollah, V., Gibson, E.K., Jr., and Romanek, C.S.: 2000, 'Elongated Prismatic Magnetite Crystals in ALH84001 Carbonate Globules: Potential Martian Magnetofossils', *Geochim. Cosmochim. Acta* **64**, 3933–4096.

Knott, S.F., Ash, R.D., and Turner, G.: 1997, ^{40}Ar-^{39}Ar Dating of ALH84001: Evidence for the Early Bombardment of Mars', *Proc. 27th Lunar Planet. Sci. Conf.*, 765–766 (abstract).

Kring, D.A., Swindle, T.D., Gleason, J.D., and Grier, J.A.: 1998, 'Formation and Relative Ages of Maskelynite and Carbonate in ALH84001', *Geochim Cosmochim. Acta* **62**, 2155–2166.

Leshin, L.A., Epstein, S., and Stolper, E.M.: 1996, 'Hydrogen Isotope Geochemistry of SNC Meteorites', *Geochim. Cosmochim. Acta* **60**, 2635–2650.

Leshin, L.A., McKeegan, K.D., Carpenter, P.K. and Harvey, R.P.: 1998, 'Oxygen Isotope Constraints on the Genesis of Carbonates from Martian Meteorite ALH84001', *Geochim. Cosmochim. Acta* **62**, 3–13.

Lloyd, R.M.: 1966, 'Oxygen Isotope Enrichment of Seawater by Evaporation', *Geochim. Cosmochim. Acta* **30**, 801–814.

Malin, M.C., and Edgett, K.S.: 2000a, 'Evidence for Recent Groundwater Seepage and Surface Runoff on Mars', *Science* **288**, 2330–2335.

Malin, M.C., and Edgett, K.S.: 2000b, 'Sedimentary Rocks of Early Mars', *Science* **290**, 1927–1937.

McKay, D.S., Gibson, E.K., Jr., Thomas-Keptra, K.L., Vali, H., Romanek, C.S., Clemett, S.J., Chiller, X.D.F., Maechling, C.R., and Zare, R.N.: 1996, 'Search for Past Life on Mars: Possible Biogenic Activity in Martian Meteorite ALH84001', *Science* **273**, 924–930.

McSween, H.Y., and Harvey, R.P.: 1998, 'An Evaporation Model for the Formation of Carbonates in the ALH84001 Martian Meteorite', *Intern. Geol. Rev.* **40**, 774–783.

Melosh, H.J., and Vickery, A.M.: 1989, 'Impact Erosion of the Primordial Atmosphere of Mars', *Nature* **338**, 487–489.

Mittlefehldt, D.W.: 1994, 'ALH84001, a Cumulate Orthopyroxenite Member of the Martian Meteorite Clan', *Meteoritics* **29**, 214–221.

Moersch, J.E., Farmer, J., and Hook, S.J.: 2000, 'Detectability of Martian Evaporites – Terrestrial Analog Studies with MASTER Data', *Proc. 31st Lunar Planet. Sci. Conf.*, abstract #2054 (CD-ROM).

Moore, H.J., Bickler, D.B., Crisp, J.A., Eisen, H.J., Gensler, A., Haldemann, A.F.C., Matijevic, J.R., Reid, L.K. and Pavlics, F.: 1999, 'Soil-like Deposits Observed by Sojourner, and Pathfinder Rover', *J. Geophys. Res.* **104**, 8729–8746.

Morris, R.V., *et al.*: 2000, 'Mineralogy, Composition, and Alteration of Mars Pathfinder Rocks and Soils: Evidence from Multispectral, Elemental, and Magnetic Data on Terrestrial Analogue, SNC Meteorite, and Pathfinder Samples', *J. Geophys. Res.* **105**, 1757–1817.

Newsom, H.E., Hagerty, J.J., and Goff, F.: 1999, 'Mixed Hydrothermal Fluids and the Origin of the Martian Soil', *J. Geophys. Res.* **104**, 8717–8728.

Nyquist, L.E., Bogard, D.D., Shih, C.-Y., Greshake, A., Stöffler, D., and Eugster, O.: 2001, 'Ages and Geologic Histories of Martian Meteorites', *Space Sci. Rev.*, this volume.

Owen, T.: 1992, The Composition and Early History of the Atmosphere of Mars, *in* H.H. Kieffer *et al.* (eds.), *Mars*, Univ. Arizona Press, Tucson, pp. 818–834.

Reid, A.M., and Bunch, T.E.: 1975, The Nakhlites, Part II. Where, When and How?', *Meteoritics* **10**, 317–324.

Romanek, C.S., Grady, M.M., Wright, I.P., Mittlefehldt, D.W., Socki, R.A., Pillinger, C.T., and Gibson, E.K., Jr.: 1994, 'Record of Fluid-rock Interactions on Mars from the Meteorite ALH84001', *Nature* **372**, 655–657.

Romanek, C.S., Perry, E.C., Treiman, A.H., Sockim, R.A., Jones, J.H., and Gibson, E.K.: 1998, 'Oxygen Isotopic Record of Silicate Alteration in the Shergotty-Nakhla-Chassigny Meteorite Lafayette', *Met. Planet. Sci.* **33**, 775–784.

Ruff, S.W., *et al.*: 2000, 'Mars "White Rock" Feature Lacks Evidence of an Aqueous Origin', *Proc. 31st Lunar Planet. Sci. Conf.*, abstract #1945 (CD-ROM).

Russell, M.J., Ingham, J.K., Zadef, V., Maktav, D., Sunar, F., Hall, A.J., and Fallick, A.E.: 1999, 'Search for Signs of Ancient Life on Mars: Expectations from Hydromagnesite Microbialites, Salda Lake, Turkey', *J. Geol. Soc.* **156**, 869–888.

Rye, R., Kuo, P.H., and Holland, H.D.: 1995, 'Atmospheric Carbon Dioxide Concentration Before 2.2 billion Years Ago', *Nature* **378**, 603–605.

Sawyer, D.J., McGehee, M.D., Canepa, J., and Moore, C.B.: 2000, 'Water Soluble Ions in the Nakhla Martian Meteorite', *Met. Planet. Sci.* **35**, 743–747.

Saxton, J.M., Lyon, I.C., and Turner, G.: 1998, 'Correlated Chemical and Isotopic Zoning in Carbonates in the Martian Meteorite ALH84001', *Earth Planet. Sci. Lett.* **160**, 811–822.

Saxton, J.M., Lyon, I.C., and Turner, G.: 2000a, 'Ion Probe Studies of Deuterium/hydrogen in the Nakhlite Meteorites', *Met. Planet. Sci.* **35**, A142–A143.

Saxton, J.M., Lyon, I.C., Chatzitheodoridis, E., and Turner, G.: 2000b, 'Oxygen Isotopic Composition of Carbonate in the Nakhla Meteorite: Implications for the Hydrosphere and Atmosphere of Mars', *Geochim. Cosmochim. Acta* **64**, 1299–1309.

Scott, E.R.D.: 1999, 'Origin of Carbonate-magnetite-sulfide Assemblages in Martian Meteorite ALH84001', *J. Geophys. Res.* **104**, 3803–3813.

Scott, E.R.D., Yamaguchi, A., and Krot, A.N.: 1997, 'Petrological Evidence for Shock Melting of Carbonates in the Martian Meteorite ALH84001', *Nature* **387**, 377–379.

Shih, C.-Y., Nyquist, L.E., Reese, Y., and Wiesmann, H.: 1998, 'The Chronology of the Nakhlite, Lafayette: Rb-Sr and Sm-Nd Isotopic Ages', *Proc. 29th Lunar Planet. Sci. Conf.*, abstract #1145 (CD-ROM).

Steele, A., Goddard, D.T., Stapleton, D., Toporski, J.K.W., Peters, V., Bassinger, V., Sharples, G., Wynn-Williams, D.D., and McKay, D.S.: 2000, 'Investigations Into an Unknown Organism on the Martian Meteorite Allan Hills 84001', *Met. Planet. Sci.* **35**, 237–242.

Stöffler, D., Ostertag, R., Jammes, C., Pfannschmidt, G., Sen Gupta, P.R., Simon, S.B., Papike, J.J., and Beauchamp, R.H.: 1986, 'Shock Metamorphism and Petrography of the Shergotty Achondrite', *Geochim. Cosmochim. Acta* **50**, 889–903.

Sugiura, N., and Hoshino, H.: 2000, 'Hydrogen Isotopic Composition of Allan Hills 84001 and the Evolution of the Martian Atmosphere', *Met. Planet. Sci.* **35**, 373–380.

Swindle, T.D., Treiman, A.H., Lindstrom, D.J., Burkland, M.K., Cohen, B.A., Grier, J.A., Li, B., and Olson, E.K.: 2000, 'Noble Gases in Iddingsite from the Lafayette Meteorite: Evidence for Liquid Water on Mars in the Last Few Hundred Million Years', *Met. Planet. Sci.* **35**, 107–115.

Treiman, A.H.: 1985, 'Amphibole and Hercynite Spinel in Shergotty and Zagami: Magmatic Water, Depth of Crystallization, and Metasomatism', *Meteoritics* **20**, 229–243.

Treiman, A.H.: 1995, 'A Petrographic History of Martian Meteorite ALH84001: Two Shocks and an Ancient Age', *Meteoritics* **30**, 294–302.

Treiman, A.H.: 1998a, 'The History of Allan Hills 84001 Revised: Multiple shock events', *Met. Planet. Sci.* **33**, 753-764.

Treiman, A.H.: 1998b, 'Amphiboles in More Martian Meteorites: Elephant Moraine 79001B, Elephant Moraine 79001X, and Lewis Cliff 88516', *Met. Planet. Sci.* **33**, A156 (abstract).

Treiman, A.H., Barrett, R.A., and Gooding, J.L.: 1993, 'Preterrestrial Alteration of the Lafayette (SNC) Meteorite', *Meteoritics* **28**, 86–97.

Turner, G., Knott, S.F., Ash, R.D., and Gilmour, J.D.: 1997, 'Ar-Ar Chronology of the Martian Meteorite ALH84001: Evidence for the Timing of the Early Bombardment of Mars', *Geochim. Cosmochim. Acta* **61**, 3835–3850.

Valley, J.W., Eiler, J.M., Graham, C.M., Gibson, E.K., Romanek, C.S., and Stolper, E.M.: 1997, 'Low-temperature Carbonate Concretions in the Martian Meteorite ALH84001: Evidence from Stable Isotopes and Mineralogy', *Science* **275**, 1633–1637.

Vicenzi, E.P., and Eiler, J.: 1998, 'Oxygen-isotopic Composition and High Resolution Secondary Ion Mass Spectrometry Imaging of Martian Carbonate in Lafayette Meteorite', *Met. Planet. Sci.* **33**, A159–A160 (abstract).

Wadhwa, M., and Crozaz, G.: 1995, 'Constraints on the Rare Earth Element Characteristics of Metasomatizing Fluids in the Martian Meteorite ALH84001' *Proc. 26th Lunar Planet. Sci. Conf.*, 1451–1452 (abstract).

Wadhwa, M., and Lugmair, G.W.: 1996, 'The Formation Age of Carbonates in ALH84001', *Met. Planet. Sci.* **31**, A145 (abstract).

Wadhwa, M., Lentz, R.C.F., McSween, H.Y., and Crozaz, G.: 2000, 'Dar al Gani 476 and Dar al Gani 489, Twin Shergottites from Mars', *Proc. 31st Lunar Planet. Sci. Conf.*, abstract #1413 (CD-ROM).

Wänke, H., Brückner, J., Dreibus, G., Rieder, R., and Ryabchikov, I.: 2001, 'Chemical Composition of Rocks and Soils at the Pathfinder Site', *Space Sci. Rev.*, this volume.

Warren, P.H.: 1998, 'Petrologic Evidence for Low-temperature, Possibly Flood Evaporitic Origin of Carbonates in the ALH84001 Meteorite', *J. Geophys. Res.* **103**, 16,759–16,773.

Watson, L.L., Hutcheon, I.D., Epstein, S. and Stolper, E.M.: 1994, 'Water on Mars: Clues from D/H and Water Contents of Hydrous Phases in SNC Meteorites', *Science* **265**, 85–90.

Wentworth, S.J., and Gooding, J.L.: 1994, 'Carbonates and Sulfates in the Chassigny Meteorite: Further Evidence for Aqueous Chemistry on the SNC Parent Planet', *Meteoritics* **29**, 861–863.

Wentworth, S.J., and Gooding, J.L.: 2000, 'Weathering and Secondary Minerals in the Martian Meteorite Shergotty', *Proc. 31st Lunar Planet. Sci. Conf.*, abstract #1888 (CD-ROM).

Woods, T.L., and Garrels, R.M.: 1992, 'Calculated Aqueous Solution Solid-solution Relations in the Low Temperature System CaO-MgO-FeO-CO$_2$-H$_2$O', *Geochim. Cosmochim. Acta* **56**, 3031–3043.

Wright, I.P., Grady, M.M., and Pillinger, C.T.: 1988, 'Carbon, Oxygen and Nitrogen Isotopic Compositions of Possible Martian Weathering Products in EETA79001.' *Geochim. Cosmochim. Acta* **52**, 917–924.

Wright, I.P., Grady, M.M., and Pillinger, C.T.: 1992, 'Chassigny and the Nakhlites: Carbon Bearing Components and Their Relationship to Martian Environmental Conditions', *Geochim. Cosmochim. Acta* **56**, 817–826.

Zahnle, K., Kasting, J.F., and Pollack, J.B.: 1990, 'Mass Fraction of Noble Gases in Diffusion-limited Hydrodynamic Escape', *Icarus* **84**, 502–527.

Address for correspondence: J.C. Bridges, Department of Mineralogy, Natural History Museum, Cromwell Road, London SW7 5BD, UK; (j.bridges@nhm.ac.uk)

AEOLIAN PROCESSES AND THEIR EFFECTS ON UNDERSTANDING THE CHRONOLOGY OF MARS

RONALD GREELEY[1], RUSLAN O. KUZMIN[2] and ROBERT M. HABERLE[3]

[1] *Department of Geological Sciences, Arizona State University, Tempe, AZ, 85287-1404, USA*
[2] *Vernadsky Institute, Russian Academy of Sciences, Kosygin St 19, Moscow, 117975 GSP-1, Russia*
[3] *NASA Ames Research Center, Mail Stop 245-3, Moffett Field, CA, 94035-1000, USA*

Received: 19 September 2000; accepted: 2 November 2000

Abstract. Aeolian (wind) processes can transport particles over large distances on Mars, leading to the modification or removal of surface features, formation of new landforms, and mantling or burial of surfaces. Erosion of mantling deposits by wind deflation can exhume older surfaces. These processes and their effects on the surface must be taken into account in using impact crater statistics to derive chronologies on Mars. In addition, mapping the locations, relative ages, and orientations of aeolian features can provide insight into Martian weather, climate, and climate history.

1. Introduction

Aeolian processes involve the interaction of the atmosphere and planetary surfaces (reviewed by Greeley and Iversen, 1985). Winds can erode and transport large quantities of fine particles and deposit them elsewhere on the surface. Consequently, existing landforms, such as craters, can be modified, erased, or buried, while new landforms, such as dunes, can be created. Aeolian processes operate on all scales, from wind-abraded grooves a millimeter across on rock faces, to rocks and kilometer-size hills sculpted by windblown sand, to dune fields tens of meters thick and covering thousands of square kilometers. These and other aeolian features have been found on Mars.

In the absence of liquid water on Mars and the lack of direct evidence for active volcanism and tectonic deformation, aeolian processes are probably the dominant agent of surface modification in the current geological regime (Wells and Zimbelman, 1989; Greeley *et al.*, 1992). High resolution views of Mars from the Mars Global Surveyor (MGS) Mars Orbiter Camera (MOC) are showing the dominance of aeolian features on the surface, as reported by Edgett and Malin (2000). Aeolian processes are likely to have operated throughout much of Mars' history. Aeolian features such as wind streaks (Thomas *et al.*, 1981; see Figure 1) enable inferences to be drawn concerning the dominant wind directions at the time of their formation. Systematic mapping of these features through space and time on Mars' surface provides clues to current and past weather systems and climate. In this chapter, we outline the basics of aeolian processes, review the styles of potential resurfacing, and consider the consequences for understanding Martian chronology and history.

Space Science Reviews **96**: 393-404, 2001.
© 2001 *Kluwer Academic Publishers. Printed in the Netherlands.*

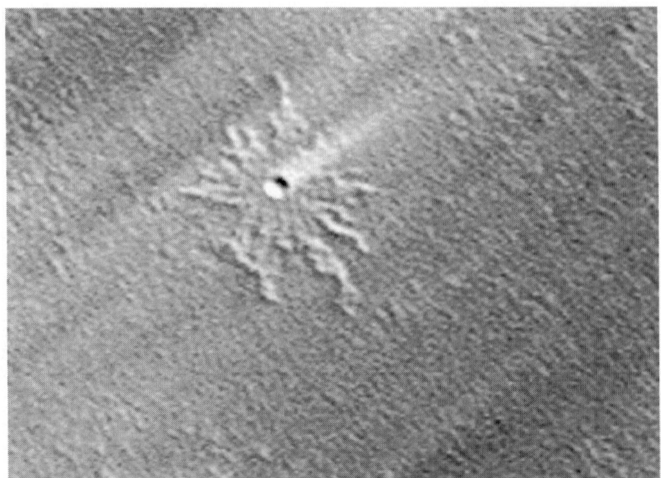

Figure 1. High and low albedo patterns forming bright and dark streaks are seen in this high resolution view of the Medusae Fossae region of Mars. The prominent bright streak associated with the impact crater is typical of those seen on Mars at all scales up to tens of kilometers long. This and other *wind streaks* are inferred to represent the prominent wind direction at the time of their formation; in the case shown here, the wind would have been blowing from the lower left to the upper right. The area shown is about 500 m wide (*NASA MGS image M00-00534; Malin Space Sciences System*).

2. Aeolian Processes

Two factors are involved in aeolian processes, a dynamic atmosphere (i.e., wind) and a supply of small, loose particles. As reviewed by Zurek (1992), the Martian atmosphere is very dynamic, with near-surface winds measured by the Viking landers in excess of 28 m s^{-1}. General circulation models (GCMs) of the atmosphere (Pollack *et al.*, 1990) show systematic wind patterns as a function of season and location on the planet. For example, some of the strongest winds are modeled to occur in the southern highlands during the southern hemisphere summer, a season and location of frequent, observed dust storms. In many areas there is a direct correlation between the orientation of bright wind streaks and directions of strongest winds predicted by the GCM, suggesting that the streaks reflect the current wind system (Greeley *et al.*, 1993).

Most windblown particles are less than a few millimeters in diameter, both on Earth and Mars (Greeley and Iversen, 1985). Particles of this size are generated by a wide variety of geological processes, including chemical weathering, impact cratering, volcanism, tectonism, and other agents of gradation, such as running water, all of which have occurred on Mars.

As shown in Figure 2, winds transport particles in three basic modes: 1) *suspension* (typically involving grains $\lesssim 60\ \mu$ in diameter, or *dust*); 2) *saltation* (for grains ~ 60 to $2000\ \mu$ in diameter, or *sand*): and 3) surface *creep* for grains $\gtrsim 2000\ \mu$ across. *Threshold curves* define the minimum *wind friction velocity* (Bagnold,

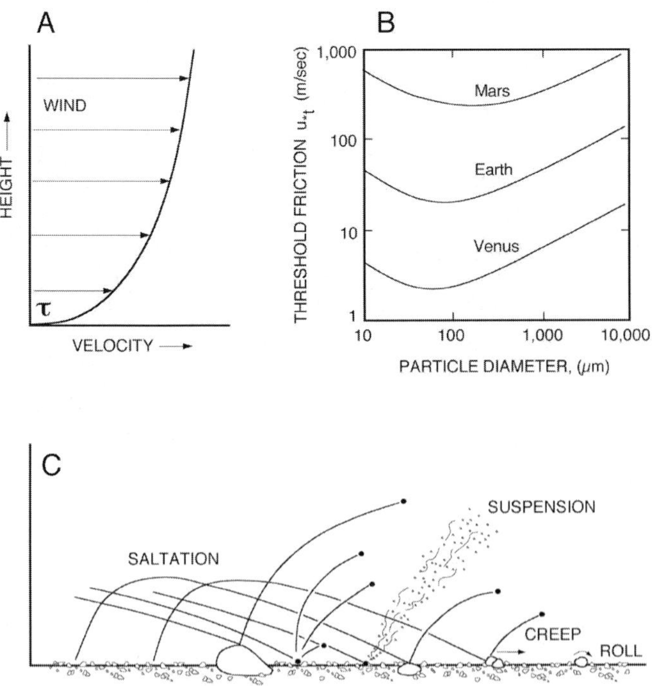

Figure 2. A) typical wind velocity profile above the surface though the boundary layer; friction along the surface generates a *shear stress* (τ) which lifts particles into the atmosphere; B) threshold curve for Mars showing the *threshold wind friction velocities* (U_{*t}, which is a function of the shear stress exerted by the wind) for particles of different sizes; differences among Mars, Earth, and Venus result from the differences in atmospheric densities (Greeley and Iversen, 1985); C) diagram showing the three modes of particle transport by the wind; although suspension and surface creep/roll can occur independently, these modes are often initiated or enhanced by saltation impact.

1954) needed to set grains into motion as a function of diameter. Extrapolation to the low-density, carbon dioxide atmosphere of Mars shows that minimum frictional velocities to set sand into saltation are about 10 times greater than on Earth, depending on the surface roughness (Greeley *et al.*, 1980). It should be noted that friction velocities are related to the shear stress exerted by the wind boundary layer above the surface, and are not equal to the wind speeds that one would experience standing on the surface. Grains $\sim$100 μ in diameter (fine sand) have the lowest threshold friction velocity, with both smaller (e.g., dust) and larger grains requiring higher velocities for movement due to interparticle forces and aerodynamic effects for the smaller grains and the higher mass of the larger grains. In addition to simple wind shear, more complicated vortical atmospheric motions, termed *dust devils*, are effective in setting grains into motion. First discovered on Mars in Viking orbiter images (Thomas and Gierasch, 1985), numerous dust devils have been observed in Mars Pathfinder data (Smith and Lemmon, 1998; Renno *et al.*, 2000) and Mars

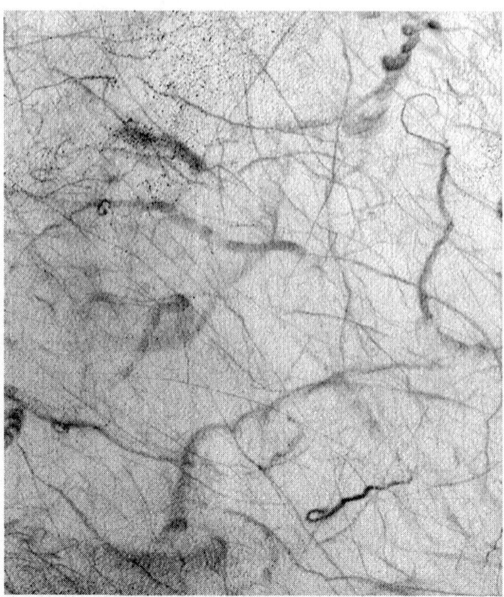

Figure 3. These dark meandering and looping patterns on the surface of Mars are inferred to be the trails left by dust devils which lifted fine particles into the atmosphere, exposing a darker substrate. The area shown is about 2 by 2.4 km (*NASA PIA02377; Malin Space Sciences System*)

Global Surveyor images (Malin *et al.*, 1998), including traces of inferred dust devil tracks (Figure 3).

Grains in saltation "hop" along the surface; as the grains impact, they can inject smaller particles into suspension and push larger grains along the surface in creep (Figure 2c). Thus, saltating sand is critical in the aeolian regime because it is 1) the grain size moved by lowest winds and, thus, will be the most common material moved (provided they are available), 2) capable of setting dust and larger grains into motion under wind conditions otherwise too low for threshold, and 3) the size commonly forming dunes, ripples, and wind-abraded rocks.

3. Aeolian Resurfacing

Resurfacing of Mars by aeolian processes includes erosion, burial or mantling, and exhumation of older surfaces which have been mantled by deposits of aeolian, fluvial, or volcanic origins.

3.1. Wind Erosion

Wind erosion occurs in two principal forms, *deflation* and *abrasion*. In deflation, relatively loose materials are picked up by the wind and removed. Such material can be unconsolidated sediments, including previously wind-transported deposits, or grains weathered free from consolidated or crystalline parent rocks. Landforms on Earth resulting from deflation include wind-stripped surfaces and depressions (commonly called *blowouts*) ranging in size from a few meters across to basins hundreds of kilometers across, as reviewed by Cooke *et al.* (1993).

Various areas on Mars have been attributed to wind deflation, as first noted by McCauley (1973) in the analyses of Mariner 9 images. These include surfaces surrounding *pedestal craters* which were assumed to have been lowered by wind erosion. The "pedestals" coincide approximately with the ejecta zone and was assumed to have been left high-standing because it was rockier than the surrounding plains, forming a type of armor-plating less subject to wind erosion. The crater shown in Figure 1 is one type of pedestal crater, although its "pedestal" has a more serrated outer boundary than typical forms. If pedestal craters do reflect deflational areas, then their presence would signal significant amounts of material removed by the wind, and lowering of the surfaces by tens of meters (the heights of the pedestals). Such deflation might be capable of removing or substantially eroding craters <100 m. On the other hand, if pedestal craters are remnants from wind erosion, one would expect the outline of the pedestal to be asymmetric in response to prevailing wind directions, as is the case with other aeolian features observed on Mars. Such asymmetry, however, is not seen around pedestal craters and their formation is open to question.

Abrasion occurs when windblown grains strike surfaces, rocks, or other grains and cause fragmentation. On the scale of millimeters to centimeters, *pits*, *flutes*, and *grooves* can be cut into rock, while on scales of centimeters to meters, rocks can be faceted and eroded into distinctive shapes called *ventifacts*. On scales of meters to kilometers, wind-sculpted hills, called *yardangs*, can be developed by a combination of abrasion and deflation. These features have been identified in images of Mars in many areas (Binder *et al.*, 1977; Mutch *et al.*, 1977; Ward, 1979; Bridges *et al.*, 1999).

It is difficult to determine the amounts and rates of wind erosion on Mars. Arvidson *et al.* (1979) estimated the age of the surface around the Viking 1 landing site from the preservation of small craters and extrapolation from lunar crater chronologies. They estimated the rates of erosion by all processes, including wind, to be relatively low. Similarly, Golombek and Bridges (2000) estimated the rate of erosion at the Pathfinder site to be 0.01 to 0.04 nm/yr. On the other hand, theoretical considerations by Sagan (1973) suggested that rates of wind abrasion should be very high on Mars because of the high wind speeds needed for particle entrainment. Laboratory experiments by Greeley *et al.* (1982) showed that the low rates of erosion could be explained if there were a paucity of effective agents of abrasion (such as holocrystalline sand grains) on Mars. They also showed that, while wind speeds are high on Mars in comparison to Earth, particles reach a smaller percentage of the wind speed on Mars because there is less effective "coupling" of the wind to the grains.

There is a fundamental consideration which must be taken into account in assessing wind erosion. Very little work has been done on the rates of landform

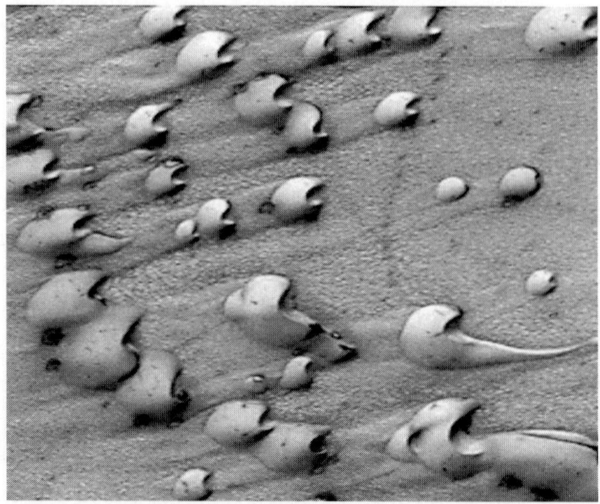

Figure 4. High resolution image of sand dunes in the Martian north polar area (*Borealis Chasma*) showing classic *slip faces* (the steeper, right-hand sides of the dunes) the orientations of which indicate that the prevailing wind direction at the time of their formation was from the left to the right. When this image was taken (September 1998) most of the area was covered with bright frost, including the dunes. Area shown is about 1.4 by 1.8 km (*NASA PIA02069; Malin Space Sciences Systems*).

erosion by the wind on the scale of hundreds of meters and larger, and extrapolation of wind abrasion from smaller features, such as rocks, probably is not appropriate. Thus, rates of wind erosion and eventual "erasure" of large landforms such as impact craters is poorly constrained on Earth and Mars.

3.2. WIND DEPOSITION

There appears to be abundant windblown material on Mars, as evidenced by the frequent dust storms and widespread occurrence of sand dunes. Viking orbiter images reveal extensive dune fields, the most prominent of which is in the north polar area (Cutts *et al.*, 1976; Tsoar *et al.*, 1979). More recently, Edgett and Malin (2000) document sand dunes in many parts of Mars imaged in high resolution, including the north polar area (Figure 4). Although few estimates of the thickness of the dune deposits have been made, one study suggests that parts of the north polar dune field have an "equivalent sediment thickness" (the thickness of all the material if it were spread out as a uniform layer) ranging from 0.5 to 6.1 m, with an average of 1.8 m (Lancaster and Greeley, 1990). Although this thickness of material could bury craters tens of meters in diameter, it is unlikely to mask larger craters. On the other hand, this estimate applies only to the dunes on the surface and does not take into account the possibility of older dune deposits which could exist below the observed dune field.

In addition to windblown materials organized into dunes, sand sheets could also occur on Mars and be unrecognized in images from orbit. For example, Viking infrared thermal mapping (IRTM) data suggest the presence of sand-size deposits in many areas (Edgett and Christensen, 1994) where obvious dunes are not seen at available orbital image resolution. In addition, most of the duneforms seen at the Pathfinder site are at the limit of detection by MOC, yet this surface suggests the presence of sand-size material in the MGS Thermal Emission Spectrometer data.

Perhaps the greatest volume of windblown material is in bright and dark dust deposits. For example, Soderblom *et al.* (1973) noted that large areas of Mars have a subdued appearance and suggested that mantling deposits were derived by wind erosion in the polar areas and were transported to lower latitudes, burying smaller craters (<1200 m in diameter) and subduing larger craters. This material was likely to be dust carried by suspension. Viking IRTM data suggest mantles of dust in many areas, including the southern highlands (Christensen, 1986), but it is not possible to determine the thickness of the deposits.

Low albedo dust is also probably present, as reviewed briefly by Edgett and Malin (2000). They show MGS MOC images supporting earlier suggestions by Dollfus *et al.* (1993) that some low albedo zones, such as Mare Erythraeum, include coarse dust which might have been emplaced in short-term suspension. Other low albedo areas probably include this type of dust which is fixed within duricrust and is immobile to transport, and sand-size particles of dark minerals.

3.3. EXHUMATION

Some areas show evidence for exhumation of older surfaces by wind erosion. For example, in Figure 5 a series of craters is seen in the Amazonis Planitia region (Greeley *et al.*, 1985); crater 1 appears to be mantled, crater 2 appears to be half mantled and half exhumed, and crater 3 is superposed on the mantle. As the material is stripped, linear streamlined hills, or yardangs, are left behind. Nearby, small dunes found along the margin of the exhumation boundary suggest that sand-size material is being stripped from the mantle. Were these relationships not seen in this area, it is doubtful that crater 2 would be recognized as being formerly buried. The pristine morphology of the exhumed part gives little evidence of the extensive deflation that must have taken place, and is very similar to the ejecta morphology of crater 3.

4. Wind Regime, Past and Present

Several lines of evidence suggest that the wind regime on Mars has been variable through time in terms of wind and windblown particles. For example, Edgett and Malin (2000) analyzed MGS MOC images and showed evidence for both active and inactive dunes, suggesting that aeolian processes might have been more vigorous (or at least variable) in the past.

Figure 5. a) Viking Orbiter image of the Medusae Fossae area showing deposits (top 3/4 of the image) which mantle cratered terrain. Crater 1 is mostly mantled, crater 2 is half- mantled and half-exhumed, and crater 3 is superposed on the mantle deposits. Without seeing the relationships portrayed here, it is doubtful that the cratered terrain in the lower part of the image would have been recognized as formerly buried, as evidenced by the similar appearance of the ejecta for crater 3 and the exhumed part of crater 2. Area shown is about 52 by 75 km. (Greeley *et al.*, 1985; *NASA Viking Orbiter image 438S01*). b) Photograph of an exhumed basaltic surface north of Askja, Iceland, showing the well-preserved surface features, such as pahoehoe "ropes" seen in the foreground; in the background (near the figure) is the remnant of the ~1.5-m-thick deposit of silt and clay which covered the entire area, but is being stripped by wind deflation. This is a smaller version of the case illustrated on Mars in a) (photograph by R. Greeley, August 1980).

The geometry of most aeolian features, including wind streaks, drift deposits, yardangs, and many duneforms, indicates the prevailing wind direction at the time of their formation. Comparing the orientations of these features with meteorological data taken from landers on the surface and predictions from GCM simulations gives insight into the aeolian regime on Mars. In addition, comparing the orientations of wind features which might reflect paleowind regimes can provide clues to changes in climate and Martian history in general. Bright wind streaks correlate well with GCM predictions for strongest winds in most parts of Mars, including the Viking 1 site where there is also a correlation with measured strongest winds. Dark wind streaks do not correlate very well with GCM runs, and it was suggested that these features result from local winds influenced by topography, which is not modeled in the GCM (Greeley *et al.*, 1993).

A wide variety of aeolian features are seen at the Pathfinder site in images taken from orbit and the surface (Greeley *et al.*, 1999, 2000), including bright wind streaks, wind tails (oriented drift deposits associate with rocks), duneforms, wind-modified craters, and wind-abraded rocks (Bridges *et al.*, 1999). Greeley *et al.* (1999) analyzed the orientations of these features for comparisons with GCM runs. Results show excellent correlations among the bright wind streaks, the wind tails, and the duneforms (seen both from orbit and the surface) with the strongest winds predicted by the GCM. The wind-abraded rocks and the wind-eroded parts of small craters, however, do not correlate with these features, suggesting that they represent a paleowind regime (Greeley *et al.*, 2000). GCM runs for all Martian seasons under current conditions and for changes in Mars' obliquity cannot account for the inferred orientation of the paleowind regime suggested by the anomalous aeolian features. This leads to the suggestion that either there are anomalous meteorological patterns not modeled by the GCM, or that Mars' spin axis was at a different geographic position than it is at present (Kuzmin *et al.*, 2000).

5. Summary Implications for Chronology and History

Aeolian processes are probably the current primary agent of surface modification on Mars today and are likely to have operated throughout much of the evolution of the surface. Landforms are both destroyed and created by these processes, as shown in Figure 6, leading to resurfacing through erosion, deposition and exhumation of older surfaces. Unfortunately, rates of resurfacing are poorly constrained because "absolute" Martian chronologies are poorly known and because the techniques for extrapolating wind abrasion rates from scales of < a few meters to larger landforms are not well developed, even on Earth.

Aeolian processes can have a significant influence on the record of impact cratering. Unlike the Moon, the Martian surface is subjected to erosion and redistribution of materials by wind and water, leading to degradation, removal, or burial of impact craters. If such modifications are not recognized, then crater statistics for the areas under study will be compromised; for the most part, age determinations based on crater counts would be underestimates. High resolution images, however, might reveal such modifications and perhaps some future technique could be developed to adjust the estimated ages.

Surfaces which have been exhumed, however, might pose a more difficult problem for obtaining impact crater statistics. Such surfaces, in effect, would have been removed from the impact crater regime for an unknown period of time corresponding to the burial. Moreover, as shown in Figure 5, it might not be possible to recognize a completely exhumed surface because small features can be perfectly preserved, even when viewed at extremely high resolution from the surface. For example, one could imagine a lava flow which has emplaced early in Mars' history, then buried by sediments before very many impacts occurred on its surface, yielding erroneously young dates. Although this scenario is conceivable, careful analysis of geological relations of all the units in the immediate area and in the regional context would likely reveal the anomaly.

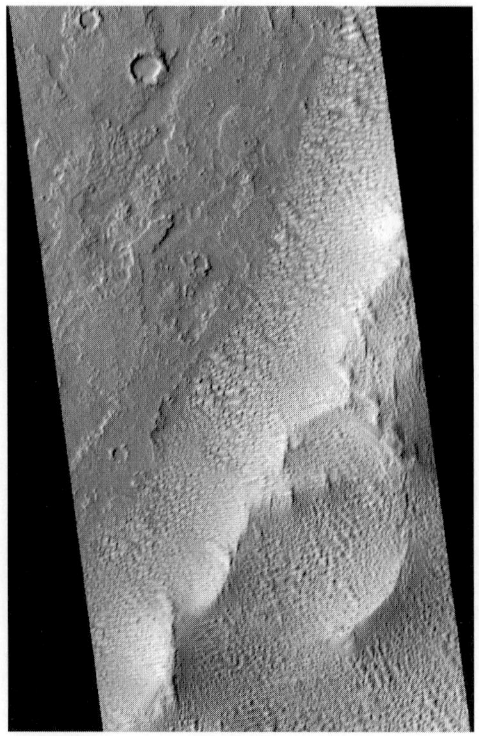

Figure 6. This high resolution MOC image illustrates the complex interplay of volcanic, impact, and aeolian processes on Mars. These lava flows northwest of the volcano, *Pavonis Mons*, show a record of impact cratering, but most of the craters are degraded. The complex sand dune field to the right includes both large dunes (the prominent ridges) and smaller duneforms which appear to encroach the lava flows. Although parts of the flow margins are clearly visible, the crater morphologies suggest partial mantling and degradation, possibly by windblown sand (*MOC image M0003198-part Malin Space Sciences System*; courtesy of W. Hartmann).

Future work should focus on quantifying rates of erosion, burial, and exhumation of large landforms by aeolian processes on Mars, drawing on terrestrial analogs to the extent reasonable to validate the approaches and to provide a means for extrapolation to Mars. Concurrently, studies should be undertaken to quantify the potential influence of aeolian processes on chronologies derived from impact crater statistics, beyond the qualitative assessment outlined here.

Acknowledgements

This work was supported by the NASA *Planetary Geology Program* and the *Mars Data Analysis Program*. We are grateful to the organizers of the *International Space Science Institute* workshop on the *Chronology and Evolution of Mars* for the opportunity to participate in the preparation of this book and acknowledge the support and encouragement of J. Geiss, R. Kallenbach, W. Hartmann, and G. Turner to submit this chapter. This manuscript benefited from thoughtful reviews provided by the editors and N. Barlow, for which we are grateful. We thank S. Selkirk for graphics, D. Nelson for image processing, and D. Burstein for LaTex processing.

References

Arvidson, R.E., Guinness, E.A., and Lee, S.: 1979, 'Differential aeolian redistribution rates on Mars', *Nature* **278**, 533–535.

Bagnold, F.R.S.: 1954, *The Physics of Blown Sand and Desert Dunes*, Methuen, New York.

Binder, A.B., *et al.*: 1977, 'The Geology of the Viking Lander 1 Site', *J. Geophys. Res.* **82**, 4439–4451.

Bridges, N.T., Greeley, R., Haldemann, A.F.C., Herkenhoff, K.E., Kraft, M., Parker, T.J., and Ward, A.W.: 1999, 'Ventifacts at the Pathfinder Landing Site', *J. Geophys. Res.* **104**, 8595–8615.

Christensen, P.R.: 1986, 'Regional Dust Deposits on Mars: Physical Properties, Age, and History', *J. Geophys. Res.* **91**, 3533–3545.

Cooke, R., Warren, A., and Goudie, A.: 1993, *Desert Geomorphology*, UCL Press, London.

Cutts, J.A., Blasius, K.R., Briggs, G.A., Carr, M.H., Greeley, R., and Masursky, H.: 1976, 'North Polar Region of Mars: Imaging Results From Viking 2', *Science* **194**, 1329–1337.

Dollfus, A., Deschamps, M., and Zimbelman, J.R.: 1993, 'Soil Texture and Granulometry at the Surface of Mars', *J. Geophys. Res.* **98**, 3413–3429.

Edgett, K.S., and Christensen, P.R.: 1994, 'Mars Aeolian Sand: Regional Variations Among Dark-hued Crater Floor Features', *J. Geophys. Res.* **99**, 1997–2018.

Edgett, K.S., and Malin, M.C.: 2000, 'New Views of Mars Aeolian Activity, Materials, and Surface Properties: Three Vignettes From the Mars Global Surveyor Mars Orbiter Camera', *J. Geophys. Res.* **105**, 1623–1650.

Golombek, M.P., and Bridges, N.T.: 2000, 'Erosion Rates on Mars and Implications for Climate Change: Constraints From the Pathfinder Landing Site', *J. Geophys. Res.* **105**, 1841–1853.

Greeley, R., and Iversen, J.: 1985, *Wind as a Geological Process*, Cambridge Univ. Press, New York.

Greeley, R., Leach, R., White, B., Iversen, J., and Pollack, J.: 1980, 'Threshold Windspeeds for Sand on Mars: Wind Tunnel Simulations', *Geophys. Res. Lett.* **7**, 121–124.

Greeley, R., Leach, R.N., Williams, S.H., White, B.R., Pollack, J.B., Krinsley, D.H., and Marshall, J.R.: 1982, 'Rate of Wind Abrasion on Mars', *J. Geophys. Res.* **87**, 10,009–10,024.

Greeley, R., Williams, S.H., White, B.R., Pollack, J.B., and Marshall, J.R.: 1985, 'Wind Abrasion on Earth and Mars', *in* M.J. Woldenberg (ed.), *Models in Geomorphology*, Allen and Unwin, Boston, 373-422.

Greeley, R., Lancaster, N., Lee, S., and Thomas, P.: 1992, 'Martian Aeolian Processes, Sediments, and Features', *in* H.H. Kieffer *et al.* (eds.), *Mars*, Univ. Arizona Press, Tucson, 730–766.

Greeley, R., Skypeck, A., and Pollack, J.B.: 1993, 'Martian Aeolian Features and Deposits: Comparisons With General Circulation Model Results', *J. Geophys. Res.* **98**, 3183–3196.

Greeley, R., *et al.*: 1999, 'Aeolian Features and Processes at the Mars Pathfinder Landing Site', *J. Geophys. Res.* **104**, 8573–8584.

Greeley, R., Kraft, M.D., Kuzmin, R.O., and Bridges, N.T.: 2000, 'Mars Pathfinder Landing Site: Evidence for a Change in Wind Regime From Lander and Orbiter Data', *J. Geophys. Res.* **105**, 1829–1840.

Kuzmin, R.O., Greeley, R., Rafkin, S.C.R., and Haberle, R.M.: 2000, 'Wind Related Modification of Small Impact Craters on Mars', *Icarus*, submitted.

Lancaster, N., and Greeley, R.: 1990, 'Sediment Volume in the North Polar Sand Seas of Mars', *J. Geophys. Res.* **95**, 10,921–10,927.

Malin, M.C., *et al.*: 1998, 'Early Views of the Martian Surface From the Mars Orbiter Camera of Mars Global Surveyor', *Science* **279**, 1681–1685.

McCauley, J.F.: 1973, 'Mariner 9 Evidence for Wind Erosion in the Equatorial and Mid-latitude Regions of Mars', *J. Geophys. Res.* **78**, 4123–4137.

Mutch, T.A., Arvidson, R.E., Binder, A.B., Guinness, E.A., and Morris, E.C.: 1977, 'The Geology of the Viking 2 Landing Site', *J. Geophys. Res.* **82**, 4452–4467.

Pollack, J.B.D., Haberle, R.M., Schaeffer, J., and Lee, H.: 1990, 'Simulations of the General Circulation of the Martian Atmosphere I: Polar Processes', *J. Geophys. Res.* **95**, 1447–1474.

Renno, N.O., Nash, A.A., Lunine, J., and Murphy, J.: 2000, 'Martian and Terrestrial Dust Devils: Test of a Scaling Theory Using Pathfinder Data', *J. Geophys. Res.* **105**, 1859–1865.

Sagan, C.: 1973, 'Sandstorms and Aeolian Erosion on Mars', *J. Geophys. Res.* **78**, 4155–4161.

Smith, P.H. and Lemmon, M.T.: 1998, 'Dust Properties at the Pathfinder Site', *Eos* **78**, F-549.

Soderblom, L.A., Kreidler, T.J., and Masursky, H.: 1973, 'Latitudinal Distribution of a Debris Mantle on the Martian Surface', *J. Geophys. Res.* **78**, 4117–4122.

Thomas, P.C., and Gierasch, P.J.: 1985, 'Dust Devils on Mars', *Science* **278**, 1758–1765.

Thomas, P.C., Veverka, J., Lee, S., and Blom, A., 1981: 'Classification of Wind Streaks on Mars', *Icarus* **45**, 124–153.

Tsoar, H., Greeley, R., and Peterfreund, A.R.: 1979, 'Mars: The North Polar Sand sea and Related Wind Patterns', *J. Geophys. Res.* **84**, 8167–8182.

Ward, A.W.: 1979, 'Yardangs on Mars: Evidence of Recent Wind Erosion', *J. Geophys. Res.* **84**, 8147–8166.

Wells, G. L. and Zimbelman, J.R.: 1989, 'Extraterrestrial Arid Surface Processes', *in* D.S.G. Thomas (ed.), *Arid Zone Geomorphology*, Belhaven Press, London, 335–358.

Zurek, R.W.: 1992, 'Comparative Aspects of the Climate of Mars: An Introduction to the Current Atmosphere', *in* H.H. Kieffer *et al.* (eds.), *Mars*, Univ. Arizona Press, Tucson, 799–817.

Address for correspondence: Department of Geological Sciences, Arizona State University Box 871404, Tempe, AZ 85287-1404, USA (greeley@asu.edu)

MARTIAN SEEPS AND THEIR RELATION
TO YOUTHFUL GEOTHERMAL ACTIVITY

WILLIAM K. HARTMANN

Planetary Science Institute, 620 N. 6th Avenue, Tucson AZ 85705-8331, USA

Received: 4 October 2000; accepted: 14 November 2000

Abstract. Gullies found on Martian hillsides by Malin and Edgett (2000) appear in many cases to be formed by water seeps produced by underground aquifers. It is proposed that these aquifers result from geologically recent melting of permafrost ice by sporadic, localized geothermal activity. This is consistent with evidence from crater counts and Martian meteorites that much higher-temperature geothermal activity has produced volcanic activity and lava flows within the last 200 Myr, and perhaps within the last 10 Myr. This hypothesis explains an aspect initially described as surprising, namely concentration of the gullies at high latitudes and on shadowed slopes. Similar features are found on Icelandic basaltic hillsides, which may be ideal analogs for further studies that may clarify the Martian phenomena.

Malin and Edgett (2000) found evidence of geologically young Martian seeps, in the form of gullies eroded on Martian hillsides. These features (Figure 1a) originate in disturbances part way down hillsides, consist of gullies running downhill, and sometimes have deltaic deposits at the foot of the gully. As proposed by Malin and Edgett, these characteristics suggest that the origin is erosion by liquid water. This process has produced virtually identical features on similar basaltic hillsides in Iceland, as shown in Figure 1b. The Martian examples, as well as many Icelandic examples, originate in resistant layers cropping out some $100 - 300$ m below the surrounding surface, as defined by the cliff top. The Martian seeps appear geologically young. Many are sharply defined, lack dune cover, and in several cases found by Malin and Edgett, have a basal apron that is deposited atop a dune field, as seen in Figure 2.

The Martian seeps are consistent with the evidence for geologically young Martian volcanism, developed elsewhere in this volume. Martian meteorites prove that molten rock has been produced on or near the surface of Mars within the last $200-300$ Myr (Nyquist *et al.*, 2001). These rocks probably represent samples from no more than three to five random impact sites on Mars, and another one of the sites has produced rocks 1.3 Gyr in age, strongly indicating that such geological units are not rare on Mars. The youngest such rocks are dated at ca. 170 Myr. Additional studies of Martian crater densities have identified geological young lava flows with inferred ages of <100 Myr, and possibly <10 Myr (Hartmann and Berman, 2000; Hartmann and Neukum, 2001).

Those ages indicate that strong geothermal heating, with temperatures of the order 1200°C has occurred on and near the surface of Mars in the last 100 Myr

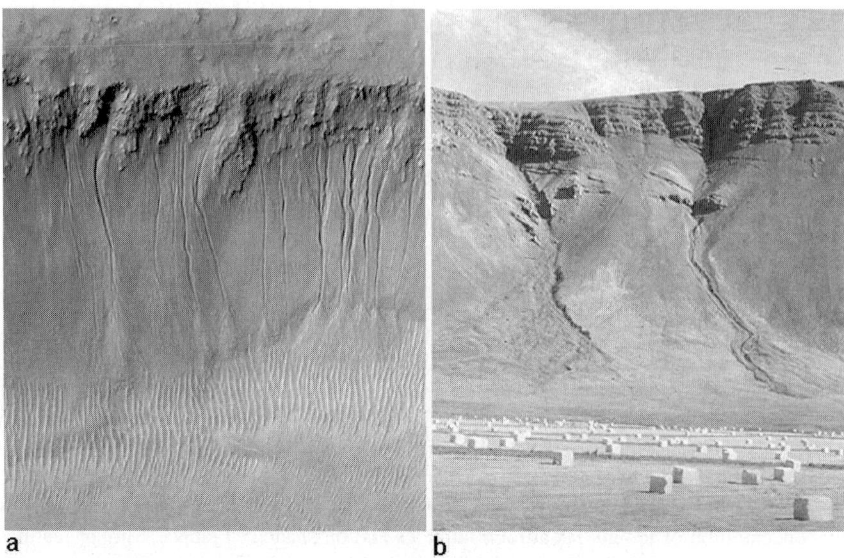

Figure 1. Comparison of Martian and Icelandic erosion gullies on hillsides. *a:* Martian example at latitude 29.7S, and longitude 38.6W. Geologic youthfulness is suggested by sharp outlines, lack of impact craters, and lack of dune or dust cover. (Mars Global Surveyor image MOC3-02990, Malin Space Science Systems and JPL). *b:* Icelandic example, north of Reykjavik. The hillside is formed in bedded basalts, with scale and morphology similar to the Martian example. (*Photo by W. Hartmann*).

or so, possibly more recently. The crater densities and inferred ages of various geologic units (Tanaka, 1986), as revealed by Mars Global Surveyor, Viking, and Mariner 9 images, imply that such igneous units have been produced sporadically throughout Martian history, from the earliest to the most recent times.

Malin and Edgett proposed that the seeps arise from "groundwater moving within and along bedrock layers," but did not discuss the source or history of the groundwater. The present paper suggests that the cause of the seeps is the melting of the underside of Martian permafrost ice layers by sporadic and scattered geothermal heating, creating underground aquifers which occasionally crop out on hillsides, producing transient water flows that erode gullies. Most gullies are in the old, cratered southern highlands, an area that was probably deeply gardened by cratering, and provided an ideal porous medium for extensive permafrost accumulation (Hartmann *et al.*, 2000). Malin and Edgett proposed outbursts of a few thousand m^3 of water, fast enough to cause erosion before sublimation or freezing could end the event. The geothermal heating events that would melt ice at around 0°C (or less in the case of brines) would be much less intense than those needed to produce magmas, and hence would be more frequent. This would explain the somewhat subjective impression that many seeps are geologically extremely young, an impression buttressed by lack of sand dune cover (Malin and Edgett, 2000). This model also explains Malin and Edgett's finding of clusters of sites with limited spatial extent, since these would be associated with regions of mild

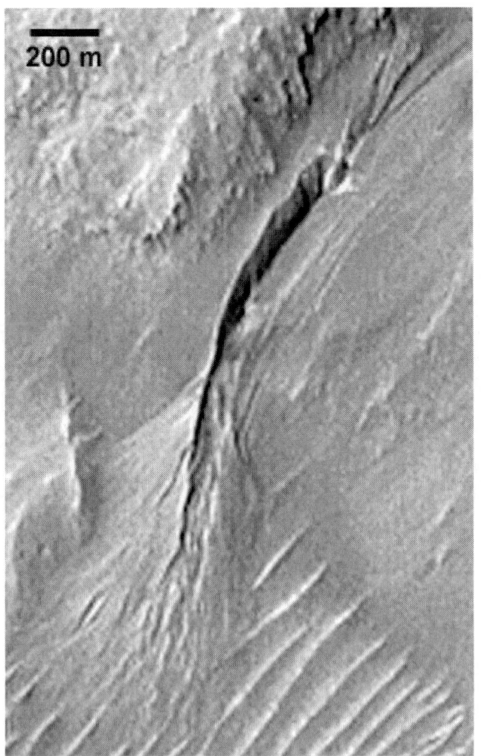

Figure 2. This gully, at 29.4S, 39.1W, produced a deltaic deposit on top of dunes in Nirgal Vallis, suggesting a very young age. This example was first pointed out by Malin and Edgett (2000). (*MGS image MO7-00752*).

subsurface heating. The model is also consistent with lack of obvious geothermal or volcanic features at the sites, because mild heating to around 0°C would not produce macro-scale topographic expressions.

Squyres *et al.* (1992) have reviewed abundant evidence that substantial ground ice deposits exist from near-surface down to kilometer scale depths at higher latitudes, affecting surface morphology. Carr *et al.* (1977), Gault and Greeley (1978), Kuzmin (1980), Boyce (1980), and later workers (notably Costard, 1989) established that the so-called rampart ejecta blankets, which resemble a muddy slurry in certain Martian impact craters, are very different from lunar ejecta blankets, and imply the presence of ice in near-surface layers at high latitudes (see review by Masson *et al.*, 2001). Figure 3 shows such a crater in a high-resolution MGS view. Kuzmin's work established that the permafrost ice exists closer to the surface at higher latitudes. Inferred depths from the surface to the ice-rich layer are <200 m in most regions poleward of 40° latitude, and <100 m in some regions poleward of 50° (Squyres *et al.*, 1992).

Tanaka (2000) and other commentators indicated surprise that the features were at high latitudes, instead of at the equator, apparently based on the supposition

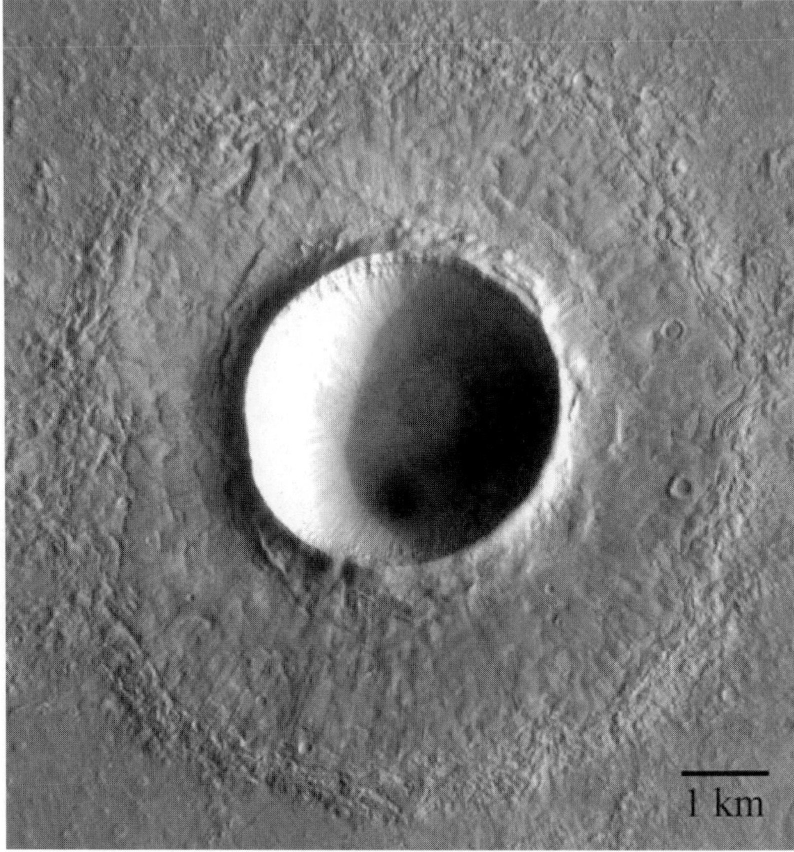

Figure 3. Martian impact crater with rampart ejecta blanket at latitude 33.2N and 238.6W longitude. Such ejecta blanket morphologies have been interpreted as evidence of near-surface ice deposits (Kuzmin, 1980). (*MGS image SPO2-43704*).

that the features should require maximum solar heating to produce the water. The present hypothesis explains the data better: the gullies naturally occur preferentially at high latitudes because that is where the ground ice deposits are known to reside in the upper 200 m and can provide a water source, whereas empirical evidence for shallow ice is lacking at low latitude.

Malin and Edgett (2000) also noted that the gullies were located preferentially on colder, pole-facing slopes. This could be a natural consequence of the present hypothesis, as illustrated schematically in Figure 4. As Malin and Edgett noticed, if an aquifer appeared on a shaded, cold, pole-facing slope, the low soil temperatures could ensure freezing of the water, causing a plug to build up at the egress of the water conduit. This could result in pressure buildup until the ice plug was breached, producing the high volume flow, as suggested by Malin and Edgett. However, if the aquifer encountered a warm, sun-facing slope, sublimation would be maximized

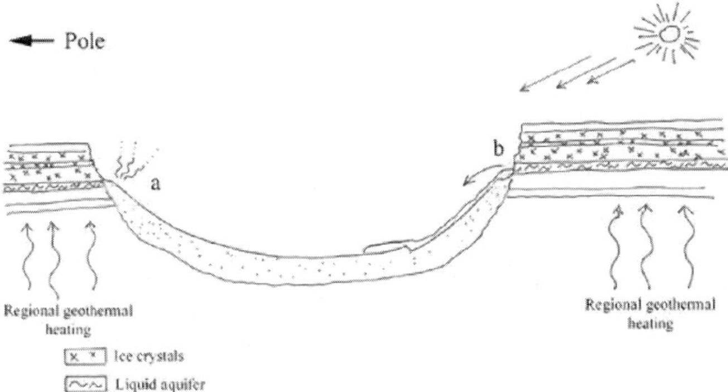

Figure 4. Sketch showing proposed features of permafrost ice hypothesis for origin of gullies. Regional geothermal heating creates an aquifer that encounters a hillside slope. High latitude is favored because near-surface ice deposits are located there. As noted by Malin and Edgett, gully formation is not favored on sun-warmed slopes (a) because sublimation is rapid there. Poleward-facing slopes (b) are favored to allow formation of an ice plug and buildup of water volumes under high pressure. Buildup of pressure leads to blowout of the ice plug and release of sufficient water volume to erode a gully. Outcropping of aquifers on sunward slopes leads to enhanced sublimation, unfavorable to buildup of sufficient water volume needed to erode a gully.

during sunlit hours, preventing build-up of a large ice plug and thus preventing accumulation of a hydrostatic head to feed a major outburst of liquid water.

In summary, the newly-discovered, geologically recent gullies are likely to be intimately associated with the recent evidence for geologically recent geothermal heating and magma production on Mars. Comparison of Figures 1a and 1b shows that the generally basaltic Icelandic hillsides, with craggy bedding of porous layers outcropping at the top, and smooth talus slopes with gullies in the lower portion, are strikingly similar to the Martian examples. The Icelandic examples may thus offer excellent opportunities for further studies that clarify Martian erosion phenomena.

Acknowledgements

The writer thanks Daniel Berman and Gilbert Esquerdo for assistance in acquiring and preparing Mars Global Surveyor images used here, and Ronald Greeley for a very helpful review. The writer also thanks Gayle Hartmann and Reinald Kallenbach for help in preparing the manuscript, and the staff of the International Space Science Institute (ISSI) in Bern, Switzerland, for hospitality and assistance. This work was supported by the Mars Global Surveyor program through the Jet Propulsion Laboratory and Malin Space Science Systems, as well as by the NASA Mars Data Analysis Program.

References

Boyce, J.P.: 1980, 'Distribution of Thermal Gradient Values in the Equatorial Region of Mars Based on Impact Crater Morphology', *in: Reports of Planetary Geology Program - 1980, NASA TM* **82385**, pp. 140–143.

Carr, M.H., Crumpler, L.S., Cutts, J.A., Greeley, R., Guest, J.E., and Masursky, H.: 1977, 'Martian Impact Craters and Emplacement of Ejecta by Surface Flow', *J. Geophys. Res.* **82**, 4055–4065.

Costard, F.: 1989, 'The Spatial Distribution of Volatiles in the Martian Hydrolithosphere', *Earth, Moon and Planets* **45**, 265–290.

Gault, D.E. and Greeley, R.: 1978, 'Exploratory Experiments of Impact Craters Formed in Viscous-liquid Targets: Analogs for Martian Rampart Craters?', *Icarus* **33**, 483–513.

Hartmann, W.K. and Berman, D.C.: 2000, 'Elysium Planitia Lava Flows: Crater Count Chronology and Geological Implications', *J. Geophys. Res.* **105**, 15011–15026.

Hartmann, W.K. and Neukum, G.: 2001, 'Cratering Chronology and the Evolution of Mars', *Space Sci. Rev.*, this volume.

Hartmann, W.K. Anguita, J., de la Casa, M.A., Berman, D.C., and Ryan, E.V.: 2000, 'Martian Cratering 7: The Role of Impact Gardening', *Icarus* **148**, in press.

Keszthelyi, K., McEwen, A.S., and Thordarson, T.: 2000, 'Terrestrial Analogs and Thermal Models for Martian Flood Lavas', *J. Geophys. Res.* **105**, 15027–15050.

Kuzmin, R.O.: 1980, 'Determination of the Bedding Depth of Ice Rocks on Mars From the Morphology of Fresh Craters', *Dokl. ANSSSP* **252**, 1445–1448.

Malin, M.C., and Edgett, K.S.: 2000, 'Evidence for Recent Groundwater Seepage and Surface Runoff on Mars', *Science* **288**, 2330–2335.

Masson, P., Carr, M.H., Costard, F., Greeley, R., Hauber, E., and Jaumann, R.: 2000, 'Geomorphic Evidence for Liquid Water', *Icarus* **148**, in press.

Nyquist, K., Bogard, D., Shih, C.-Y., Greshake, A., Stöffler, D., and Eugster, O.: 2001, 'Ages of Martian Meteorites', *Space Sci. Rev.*, this volume.

Squyres, S.W., Clifford, S., Kuzmin, R., Zimbelman, J., and Costard, F.: 1992, 'Ice in the Martian Regolith', *in* H. Kieffer *et al.* (eds.), *Mars*, Univ. Arizona Press, Tucson.

Tanaka, K.L.: 1986, 'The Stratigraphy of Mars', *Proc. 17th Lunar Planet. Sci. Conf., Part 1, J. Geophys. Res.* **91, suppl.**, 139–158.

Tanaka, K.L.: 2000, 'Fountains of Youth', *Science* **288**, 2325.

Address for correspondence: Planetary Science Institute, 620 N. 6th Avenue, Tucson AZ 85705-8331, USA (hartmann@psi.edu)

THE ATMOSPHERE OF MARS AS CONSTRAINED BY REMOTE SENSING

THERESE ENCRENAZ

DESPA, Observatoire de Paris, 92195 Meudon, France

Received: 1 July 2000; accepted: 27 September 2000

Abstract. In addition to the Viking in-situ mass spectrometry measurements, our knowledge of the Martian atmosphere comes from remote sensing spectroscopy from the ground and from space. In particular, infrared measurements from the Mariner 9, Viking, Phobos and MGS orbiters have provided information upon the thermal profile, the chemical composition, the stratospheric winds, some isotopic ratios, and the properties of suspended dust. However, further remote sensing monitoring is still needed for a better understanding of the water cycle, a more accurate knowledge of the minor species and the aerosol composition, an improved measurement of the hydrogen and oxygen isotopic ratios, and for a full mapping of the middle altitude winds. Some of these information will be provided with the Mars Express mission.

1. Introduction

The atmosphere of Mars, as compared to the one of the other telluric planets Venus and the Earth, shows two main characteristics. First, its surface pressure is very low (less than 10 mbars), to be compared with 1 bar at the surface of the Earth and almost 100 bars at Venus surface. Second, due to the inclination of the polar axis over the ecliptic plane, the Martian atmosphere shows significant seasonal variations. Because the main atmospheric constituent, CO_2, does condense at the temperatures of the Martian poles (about 180 K), CO_2 is transported from one pole to another where it forms CO_2 ice, and the total pressure of the atmosphere varies by about 30% over the seasonal cycle.

The chemical composition of the Martian atmosphere (Table I) is dominated by carbon dioxide with a small percentage of molecular nitrogen and argon, and traces of O_2, CO and H_2O. This composition shows strong similarities to the one of Venus, and also to the primitive atmosphere of the Earth, before most of the CO_2 was trapped in the oceans in the form of carbonates. The main difference, however, lies in the water abundance, very low on Venus and Mars but very high on Earth, as the total water trapped in the oceans, if it were gaseous, would correspond to more than hundred times the present nitrogen surface pressure.

However, the past atmosphere of Mars is likely to have been more abundant than today, as well as its water content (Masson *et al.*, 2001; Head *et al.*, 2001). The understanding of the past history of the Martian atmosphere raises several questions. First, if the past atmosphere was denser than today, where did CO_2 disappear? It might have been trapped in the surface, as in the case of the Earth.

Space Science Reviews **96**: 411–424, 2001.

TABLE I

The composition of the Martian lower atmosphere (from Owen, 1992)

Gaseous species	Abundance	Gaseous species	Abundance
CO_2	95.32 %	$^{36+38}Ar$	5.3 ppm
N_2	2.7 %	Ne	2.5 ppm
^{40}Ar	1.6 %	Kr	0.3 ppm
O_2	0.13 %	Xe	0.08 ppm
CO	0.07 %	O_3	0.04 - 0.2 ppm
H_2O	10 - 1000 ppm		

However, carbonates have been searched for on the Martian surface for decades but have never been firmly identified in spite of several tentative detections (Pollack *et al.*, 1990a; Clark *et al.*, 1990; Lellouch *et al.*, 2000). The search for carbonates and other mineralogic species on the Martian surface thus remains a major challenge in the future exploration program of Mars. Next, if water was so abundant in the past, how did it disappear? It has been suggested that water might be trapped today under the Martian surface in the form of permafrost. Identifying this water reservoir will be also a key question for future studies.

Another problem deals with the present water cycle on Mars. Along the Martian seasonal cycle, water, like CO_2, is transported from one polar cap to the other. But because H_2O is much less abundant and more easily condensible than CO_2, the seasonal variations of water abundance are not 30%, as for CO_2, but as much as a factor 100. In order to learn more about the history of water on Mars, we need to better understand today's cycle of H_2O, and to identify its sources and sinks.

In this paper, the present status of the Martian atmosphere is discussed, as derived from the Viking in situ measurements and from remote sensing techniques from ground and space. Most of our knowledge, since the Viking exploration, has come from the analysis of the spectrum of Mars, from the visible to the radio range. This spectrum is composed of two main components: the reflected solar spectrum, which peaks at 0.5 μm and dominates the observed spectrum of Mars up to a wavelength of $\sim$4 μm, and the thermal emission, which is the blackbody emission of the Martian surface and atmosphere. At short wavelengths, the atmospheric gases are observed through the absorption of their molecular bands in front of the solar continuum; a direct information is obtained upon their column density, i.e. the number of molecules along the line of sight. At long wavelengths, in the thermal regime, the outgoing flux is mostly a function of the temperature. The thermal vertical profile is characterized by a troposphere where the temperature decreases with increasing altitude. Above $\sim$50 km, the temperature stays more or less constant. The shape of the infrared thermal spectrum of Mars strongly depends upon the contrast between the surface temperature and the temperature of the troposphere. In

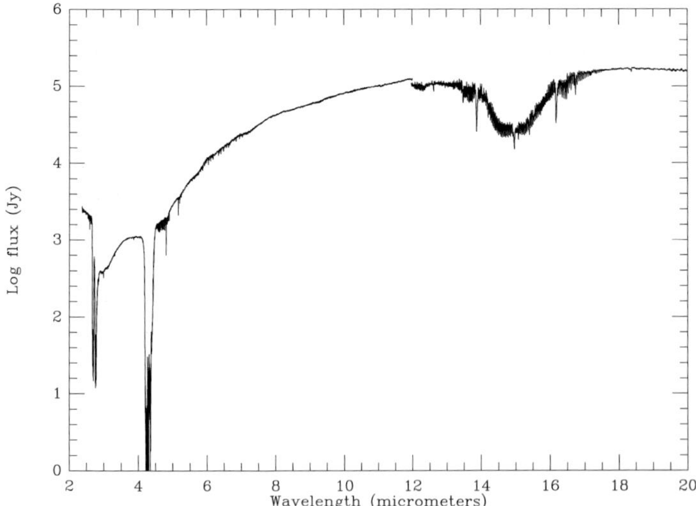

Figure 1. The spectrum of Mars (whole disk) as observed by ISO-SWS. The gap at 12 μm corresponds to a change in the grating bands. Most of the absorption features are due to atmospheric gases, in particular CO_2 at 2.7, 4.3 and 15 μm.

most of the cases, at mid-latitudes, the surface is warmer than the atmosphere and the molecular bands are observed in absorption. However, in the case of the polar caps, the surface is colder than the atmosphere and the observed spectrum is seen in emission. Thermal spectra provide information about the vertical distribution of minor constituents, when the temperature profile is known. This profile can be retrieved from the inversion of the strong bands of CO_2, which is the main atmospheric gas. This technique has been successfully used with the IRIS data of Mariner 9 (Hanel *et al.*, 1972). A typical infrared spectrum of Mars is shown in Figure 1. It corresponds to the whole disk, as observed by the Short-Wavelength Spectrograph of the Infrared Space Observatory (Lellouch *et al.*, 2000). Most of the features, attributed to gaseous CO_2, CO and H_2O, are of atmospheric origin.

2. The Data Base

Two space missions have provided in-situ data related to the Martian atmosphere. The Viking landers gave a direct measurement of the temperature profile (Seiff and Kirk, 1977), as well as the surface temperature and pressure over a seasonal cycle (Zurek, 1992); the atmospheric composition was retrieved from the mass spectrometry experiment (Owen *et al.*, 1977). The Mars Pathfinder experiment, in July 1997, provided direct information about the surface temperature and pressure, and on the water vapor abundance (Smith *et al.*, 1997).

In addition to in-situ measurements, our knowledge of th Martian atmosphere is based on a data set composed of remote sensing observations, both from space and from the ground. Space remote sensing observations of the Martian atmosphere started with the IRS data obtained during the Mariner 6 and 7 flybys (1969). Later, the IRIS infrared interferometer aboard Mariner 9 provided an important data set, still used today, which allowed to derive thermal maps, information on the water content, the dust composition and the surface mineralogy (Hanel *et al.*, 1972). Two Viking remote sensing instruments were very important for atmospheric studies: IRTM, an infrared radiometer which determined the temperature field (Kieffer *et al.*, 1977), and MAWD, which measured the water abundance from the profile of a near-IR H_2O line (Farmer *et al.*, 1977). Later, the ISM near-infrared imaging spectrometer aboard the Phobos mission provided data about the dust distribution and the CO and H_2O abundances (Rosenqvist *et al.*, 1992). Finally, the TES instrument aboard Mars Global Surveyor is able, in addition to surface mineralogy, to monitor the thermal atmospheric structure (Conrath *et al.*, 1999).

Ground-based images have been obtained in the visible range, leading to a monitoring of the polar caps evolution and the dust storms. The water abundance has been monitored through the analysis of a weak H_2O band at 0.82 μm (Barker, 1976) and oxygen was detected at 0.76 μm (Barker, 1972; Trauger and Lunine, 1983). Spectroscopy in the near-infrared has provided information about CO (Kaplan *et al.*, 1969; Billebaud *et al.*, 1998) and HDO (Owen *et al.*, 1988), in addition to the mineralogic properties of the surface. Observations in the microwave and radio range have been most successful for measuring the CO mixing ratio and monitoring the temperature profile (Clancy *et al.*, 1990), for determining the H_2O vertical distribution (Clancy *et al.*, 1996), and for measuring winds (Lellouch *et al.*, 1991a).

3. Thermal Structure and Dynamics

The main method used to retrieve the thermal atmospheric profile, up to an altitude of $\sim$50 km, is the inversion of the strong CO_2 ν_2 band at 15 μm (Figure 1). This study has been performed, in particular, by the IRIS spectrometer aboard Mariner 9 (Hanel *et al.*, 1972); in addition, the IRTM radiometers aboard the Viking orbiters, working at several infrared frequencies, provided maps of the atmospheric temperature at different altitudes over a full seasonal cycle (Martin, 1981). These maps were then used to retrieve velocity fields of thermal winds (Zurek, 1992) which were a precious tool to validate global circular models (Pollack *et al.*, 1990b; Forget *et al.*, 1999). In the upper atmosphere of Mars ($z = 100 - 200$ km), the thermal profile was retrieved from radio-occultation measurements aboard the Mariner and Viking spacecraft (Kliore *et al.*, 1972; Lindal *et al.*, 1979).

Another powerful technique for retrieving the thermal profile is the ground-based observation of CO transitions in the millimeter range by heterodyne spectroscopy. This technique is well suited to the analysis of the atmospheric Martian lines (which are very narrow due to the low pressure), as it provides a very high resolving power (10^6) over a narrow spectral interval (typically 0.5 GHz). The observation of CO transitions of different intensities (the ^{12}CO (1-0) and (2-1) lines at 115 GHz and 230 GHz, and the corresponding ^{13}CO transitions at 110 GHz and 220 GHz) allows a simultaneous retrieval of the CO abundance and the thermal profile up to an altitude of about 70 km (Clancy *et al.*, 1990, 1996; Lellouch *et al.*, 1991b). Observations show that the CO mixing ratio is constant with altitude with a value of 7×10^{-4}, and that it shows little temporal or spatial variation (Lellouch *et al.*, 1991b), as expected from its long lifetime. It can be noted that the atmospheric temperatures obtained from the millimeter technique tend to be systematically lower than the maps retrieved at 15 μm by the Mariner 9 and the Viking missions. Clancy *et al.* (1996) interpret this difference by the presence of strong dust storms at the time of Mariner 9 and Viking which had the effect of warming up the atmosphere.

Heterodyne spectroscopy was also powerful to measure stratospheric winds in the $40-70$ km altitude range, from a ground-based mapping of the CO (2-1) transition, at the time of opposition, and a measurement of its Doppler shift (Lellouch *et al.*, 1991a; Théodore *et al.*, 1993). Winds are retrieved with an uncertainty of about 20 m/s. This method, which extends the results of the infrared measurements up to higher altitudes, provide important constraints to climate models and will benefit, in the future, from the use of large interferometers.

4. Minor Constituents

4.1. CO AND THE STABILITY OF THE MARTIAN ATMOSPHERE

CO was first detected on high-resolution ground-based spectra through its (2-0) band centered at 2.35 μm (Kaplan *et al.*, 1969). The study of this band was later refined to search for possible local variations (Billebaud *et al.*, 1998), but none were detected at the observed scale ($\sim$1000 km), which thus confirmed the millimeter results mentioned above. Another experiment, however, gave more surprising results. The ISM imaging spectrometer, aboard the Phobos mission, was able to measure the CO abundance on localized areas of the Martian disk with high spatial resolution (20 km; Rosenqvist *et al.*, 1992). While the standard value of CO was measured over low altitude regions, a significant depletion was apparently found over the volcanoes. Such a result is puzzling, considering the long lifetime of CO.

A possible explanation, however, could be linked to heterogeneous chemistry Such an hypothesis was proposed by Atreya and Blamont (1990) as a tentative explanation of the stability of the atmosphere. The problem is the following: CO_2 is photodissociated into CO and O, but the inverse reaction is spin-forbidden. The CO_2 photodissociation should thus lead to an accumulation of CO and O_2 on a time scale of a few thousand years. The low contents of CO and O_2 presently observed (Table I) imply the existence of another recycling mechanism for CO_2.

Reactions involving OH have been proposed, but they seem to imply either a too high eddy diffusion coefficient (McElroy and Donahue, 1972) or a too high water content (Parkinson and Hunten, 1972). Heterogeneous chemistry, as proposed by Atreya and Blamont (1990) could favor the recombination of CO and O into CO_2 through catalysis over the surface of aerosols, and such a mechanism could lead to fluctuations of the CO abundance on a local scale. However, more recent analyses, using improved laboratory data for reaction rates and CO_2 extinction cross sections, have shown that the CO_2 production rate might exceed its loss rate even without catalytic recombination (Atreya and Gu, 1994; Rosenqvist and Chassefiere, 1995). In any case, a major objective for the infrared experiments aboard future space missions, in particular Mars Express, will be to confirm or infirm the reality of small-scale variations of the CO abundance.

4.2. H_2O: SEASONAL AND INTERANNUAL VARIATIONS

As mentioned above, water vapor exhibits huge seasonal effects due to its cyclic condensation on each of the polar caps. The water vapor abundance has been monitored from the ground, using the weak H_2O line at 0.82 μm (Barker, 1976; Rizk et al., 1991) and from the Viking orbiter using the 1.38 μm band (Farmer et al., 1977). Rotational H_2O lines in the thermal range (20-50 μm) have been also analysed by IRIS aboard Mariner 9 (Hanel et al., 1972). The Viking data, obtained over more than a seasonal cycle, showed evidence for a significant asymmetry between the northern and southern hemisphere: the maximum H_2O abundance is about 90 pr-μm around the north pole during northern summer while it is only 15 pr-μm in the southern region during southern summer. This asymmetry in the water vapor content, as well as the appearance of global dust storms during southern summers, are interpreted as a result of the ellipticity of the Martian orbit, which translates into northern summer temperatures being about 20 K colder than the southern summer temperatures (Zurek and Martin, 1993; Clancy et al., 1996).

The water vapor vertical distribution was also inferred from microwave and radio heterodyne spectroscopy measurements, using transitions of H_2O, HDO and $H_2^{18}O$ (Clancy et al., 1992, 1996; Encrenaz et al., 1991, 1995a, 1998). Water was found to be confined in the lower atmosphere, presumably by condensation, especially during northern summer, another effect of the Martian ellipticity. These results were confirmed by other H_2O measurements performed by the ISO satellite in July 1997, using the whole $2-40$ μm spectral range (Burgdorf et al., 2000; Lellouch et al., 2000) as well as in-situ measurements obtained by the Mars Pathfinder camera (Smith et al., 1997; Titov et al., 1998). Generally, the monitoring of the H_2O abundance over about 3 decades shows a seasonal cycle which is globally consistent with the Viking data and the climate models (Jakosky and Haberle, 1992); however there are also interannual variations which remain to be understood.

Observations of the water vapor content at high spatial resolution (20 km) were performed by the ISM-Phobos instrument, using the 1.8 μm and 2.6 μm bands. Data were obtained as a function of local time and latitude (Rosenqvist *et al.*, 1992). Surprisingly, an factor $3 - 5$ increase of the water vapor content was found over the Tharsis volcanoes (Titov *et al.*, 1994). This variation was interpreted as evidence for a regolith-atmosphere exchange which translates into a higher water vapor content above high-adsorbing surfaces like clay regolith, as compared to basaltic terrains (Titov *et al.*, 1995). In future space missions, the study of small-scale variations of the water vapor abundance will be crucial for better understanding the water sources and sinks on Mars and the surface-atmosphere interactions.

4.3. OTHER OXYGEN COMPOUNDS: O_2, O_3 AND NO

Molecular oxygen has been observed by high-resolution spectroscopy of two forbidden bands, the first one at 0.76 μm (Barker, 1972; Carleton and Traub, 1972) and the second one (airglow excitation coming from ozone photolysis) at 1.27 μm (Traub *et al.*, 1979). O_2 was also detected by the Viking mass spectrometers (Owen *et al.*, 1977), with a mixing ratio of 1.3×10^{-3}. Although marginally detectable in the microwave range (Encrenaz *et al.*, 1995b), O_2 has not been detected yet by ground-based heterodyne spectroscopy.

The detection of ozone was achieved by UV spectrometry, first with Mariner 6 and 7, then with Mariner 9 (Barth *et al.*, 1973), and Mars 5 (Krasnopolsky *et al.*, 1979). A map of ozone was also recorded from ground-based heterodyne spectroscopy in the 9.7 μm band (Espenak *et al.*, 1991). Ozone was found to show strong seasonal, latitudinal and local effects, and peaks at about 50 km with a mixing ratio of about 10^{-6} at the time of maximum abundance. The ozone abundance appears to be maximum in winter at high latitude, and anticorrelated with the water vapor content (Krasnopolsky, 1986). Although not yet detected in the millimeter range, ozone might be marginally detectable (Encrenaz *et al.*, 1995b).

NO, detected by the Viking mass spectrometer in the Martian upper atmosphere (Nier and McElroy, 1977; McElroy *et al.*, 1977), is a product of nitrogen dissociation by electron impact in the ionosphere. Its abundance peaks at an altitude of about 120 km, with a mixing ratio of 7×10^{-5}. A NO emission was tentatively reported from ground-based millimeter heterodyne spectroscopy (Encrenaz *et al.*, 1999) but this result failed to be confirmed in subsequent observations, and should thus be considered as very uncertain.

4.4. UNDETECTED MINOR SPECIES

Many species have been unsuccessfully searched for in the infrared or millimeter spectrum of Mars. A list of them is given in Table II. The upper limits come either from the IRIS-Mariner 9 data (CH_4, C_2H_2, C_2H_4, C_2H_6, NO_2, N_2O, NH_3, PH_3; Maguire, 1977; Owen, 1992), and from ground-based studies in the millimeter range (H_2S, SO_2, OCS; Encrenaz *et al.*, 1991, and in the infrared range (H_2O_2, HCl, H_2CO; Krasnopolsky, 1997).

TABLE II

Upper limits of minor atmospheric species

Gaseous species	Upper limit	Gaseous species	Upper limit
CH_4	$2\ 10^{-8}$	PH_3	$1\ 10^{-7}$
C_2H_2	$2\ 10^{-9}$	SO_2	$3\ 10^{-8}$
C_2H_4	$5\ 10^{-7}$	OCS	$7\ 10^{-8}$
C_2H_6	$4\ 10^{-7}$	H_2S	$2\ 10^{-8}$
N_2O	$1\ 10^{-7}$	H_2CO	$3\ 10^{-9}$
NO_2	$1\ 10^{-8}$	HCl	$2\ 10^{-9}$
NH_3	$5\ 10^{-9}$	H_2O_2	$3\ 10^{-8}$

A tentative detection of CH_4 was reported by Krasnopolsky (1997) with a mixing ratio of $(70 \pm 20) \times 10^{-9}$. However, this result is only marginally compatible with the upper limit derived from IRIS (Maguire, 1977; Table II), as well as the upper limit inferred from the ISO data (5×10^{-8}; Lellouch *et al.*, 2000).

Formaldehyde H_2CO was tentatively identified from the Phobos solar occultation data with a mixing ratio of 5×10^{-7}, in agreement with some theoretical predictions from photochemical models (Korablev *et al.*, 1993). However, subsequent ground-based observations in the infrared and millimeter range showed that such an abundance cannot be distributed over the whole disk.

The presence of H_2O_2 at the surface of Mars, has been suggested from photochemical calculations (Atreya and Gu, 1994). Indeed, H_2O_2 is suspected to be the oxidant responsible for the lack or organic material at the surface. The H_2O_2 abundance is expected to be correlated with the water vapor content. However, the lack of detection of H_2O_2 in both the infrared and millimeter range tends to imply a lower abundance than predicted by the models.

As pointed out by Krasnopolsky (1997), their stringent upper limit for HCl implies that chlorine chemistry is negligible on Mars as compared to hydrogen chemistry.

Finally, the lack of sulfur-bearing species and nitrogen-bearing species should be pointed out. This should be seen in connection with the absence of nitrates and sulfates which, in spite of a tentative detection of sulfates in the IR thermal range (Pollack *et al.*, 1990a), were never firmly identified on the Martian surface.

4.5. ISOTOPIC RATIOS

Isotopic ratios arc important tracers of the history of Martian volatiles; the present state of knowledge is summarized in Owen (1992). Isotopic ratios of the rare gases have been measured by the Viking mass spectrometer (Owen *et al.*, 1977) and their interpretation is discussed in this volume by Bogard *et al.* (2001). The $^{15}N/^{14}N$

ratio, also derived from mass spectrometry aboard Viking, showed a significant enrichment (1.6) with respect to the terrestrial value (Nier and McElroy, 1977). This enrichment was interpreted by differential nitrogen escape over the history of the planet, leading to a selective enrichment of the heavier isotope (McElroy *et al.*, 1977). The observed enrichment could imply an initial nitrogen partial pressure of between 1.3 and 30 mbar, which corresponds to an early atmosphere 10 to 300 times denser than today.

In the case of oxygen, the $^{17}O/^{16}O$ and $^{18}O/^{16}O$ ratios have been measured from CO_2 by IRIS (Maguire, 1977), then by mass spectrometry aboard Viking (Nier and McElroy, 1977) and were found in agreement with the terrestrial values. A surprising result came from Bjoraker *et al.* (1989) who inferred a depletion of ^{17}O and ^{18}O in H_2O from infrared high-resolution spectra aboard the Kuiper Airborne Observatory. These measurements should be repeated using infrared high-resolution spectroscopy aboard future space missions (and, in particular, the PFS instrument aboard Mars Express) to confirm this result, which would imply a different history of the CO_2 and H_2O reservoirs. In any case, the fact that the oxygen isotopic ratios do not show the strong enrichment in heavier species observed for nitrogen seems to imply the presence of an oxygen reservoir which could exchange with the atmosphere (McElroy *et al.*, 1977; Owen, 1992; Bogard *et al.*, 2001).

Another major result is the determination of D/H from the ground-based study of HDO infrared lines at high spectral resolution (Owen *et al.*, 1988; Bjoraker *et al.*, 1989; Krasnopolsky, 1997). Deuterium was found to be enriched by a factor between 5 and 6 with respect to the terrestrial value. Although much less than in the case of Venus, where an enrichment factor of 120 has been measured (Bézard *et al.*, 1990), this value probably implies the presence of large amounts of water in the past history of Mars. It should be noted that the D/H ratio is still poorly determined. A key factor to improve this result is the simultaneous determination of the H_2O and HDO abundances. This could be achievable with the simultaneous observation of HDO and $H_2^{18}O$ transitions in the millimeter range (Encrenaz *et al.*, 1998). The D/H determination should be also possible using the PFS instrument aboard Mars Express.

5. Aerosol Properties

Martian aerosols play a key role in the climate of Mars. Their more obvious appearance lies in the dust storms which typically appear during southern summers (Kahn *et al.*, 1992). Dust storms are regularly monitored from ground-based telescopes and, more recently, with the HST. In addition, both the Mariner 9 and Viking missions occurred during strong global dust storms, so that aerosol properties, as well as the effect of dust on atmospheric parameters, could be extensively studied.

The chemical composition of aerosols is dominated by silicates, which show a strong signature around 9 μm and a weaker one at 18 μm, as observed in the IRIS data (Hanel *et al.*, 1972). A mean particle diameter of about 2 μm was inferred from the infrared measurements (Toon *et al.*, 1977). The 9-μm flux was also used by the IRTM to map the dust at the time of the Viking mission (Martin, 1981). A measurement of the dust opacity was provided by the Viking landers (Pollack *et al.*, 1977), and later by Mars Pathfinder (Smith *et al.*, 1997). Other determinations of the dust opacity came from the slope of the near-infrared continuum as measured by Phobos ISM (Drossart *et al.*, 1991) and from the residual flux scattered in the center of the strong CO_2 band at 2.7 μm observed in the ISM spectra (Titov *et al.*, 2000) and the ISO data (Lellouch *et al.*, 2000).

The composition of the dust is consistent with a mixture of basalt and clay minerals containing at least 60% of SiO_2 (Toon *et al.*, 1977); an additional 1% of iron oxyde, such as magnetite, is needed to account for the red colour of the Martian dust (Pollack *et al.*, 1977). In addition, tentative identifications of sulfates, carbonates and hydrates in the Martian dust were reported from KAO spectra of between 5 and 11 μm (Pollack *et al.*, 1990a); another tentative detection of carbonates in suspended dust was reported from the analysis of the ISO spectrum in the $6 - 11$ μm and $25 - 45$ μm spectral ranges (Lellouch *et al.*, 2000). Some of the features observed by ISO in the $6-11$ μm range were in fact also observed by IRIS, but had remained unidentified; they cannot be attributed to atmospheric species and have to come from a solid compound. However, no laboratory analog has been found to fully account for the observed signatures, so the proposed identifications are still tentative and require further confirmation. The TES experiment aboard MGS and later PFS aboard Mars Express should be able to provide an important contribution to this major question.

6. Future Studies

A complete understanding of the processes occurring in the Martian atmosphere require a monitoring of the atmospheric parameters which has not yet been performed. We need to determine the thermal profile and the stratospheric winds up to an altitude of about 100 km, which corresponds to the maximum height of the Hadley cells. Other key measurements are the monitoring of the water vapor vertical distribution and the water ice, the search for new minor species, the determination of the aerosol composition and a refined measurement of the isotopic ratios. The PFS instrument aboard Mars Express is expected to achieve part of these objectives (temperature retrieval, water abundance, aerosol composition and isotopic ratios). However, the probed region will be limited to the troposphere (below 50 km), and no measurement will be obtained on the winds. This information requires the use of in-orbit microwave heterodyne spectroscopy. A first opportunity will be provided by the Mars flyby of the Rosetta mission, which will be equipped with a microwave spectrometer (MIRO). In the future, such an instrument embarked on an orbiter will be needed to provide a full monitoring of the Martian atmospheric parameters.

References

Atreya, S.K., and Blamont, J.: 1990, 'Stability of the Martian Atmosphere: Possible Role of Heterogeneous Chemistry', *Geophys. Res. Lett.* **17**, 287–290.

Atreya, S.K., and Gu, Z.: 1994, 'Stability of the Martian Atmosphere: is Heterogeneous Chemistry Essential?', *J. Geophys. Res.* **99**, 13,133–13145.

Barker, E.: 1972, 'Detection of Molecular Oxygen in the Martian Atmosphere', *Nature* **238**, 447–448.

Barker, E.: 1976, 'Martian Atmospheric Water Vapor Observations: 1972-74 Apparitions', *Icarus* **28**, 247–268.

Barth, C.A., Hord, C.W., Stewart, A.I., Lane, A.L., Dick, M.L., and Anderson, G.P.: 1973, 'Mariner 9 Ultraviolet Spectrometer Experiment: Seasonal Variation of Ozone on Mars', *Science* **179**, 795–796.

Bézard, B., de Bergh, C., Crisp, D., and Maillard, J.P.: 1990, 'The Deep Atmosphere of Venus Revealed by High-resolution Night-side Spectra', *Nature* **345**, 508–511.

Billebaud, F., Rosenqvist, J., Lellouch, E., Maillard, J.P., Encrenaz, Th., and Hourdin, F.: 1998, 'Observations of CO in the Atmosphere of Mars in the (2-0) Vibrational Band at 2.35 microns', *Astron. Astrophys* **333**, 1092–1099.

Bjoraker, G.L., Mumma, M.J., and Larson, H.P.: 1989, 'Isotopic Abundance Ratios for Hydrogen and Oxygen in the Martian Atmosphere, *Bull. Amer. Astron. Soc.* **21**, 990–990.

Bogard, D.D., Clayton, R.N., Marti, K., Owen, T., and Turner, G.: 2001, 'Martian Volatiles: Isotopic Composition, Origin, and Evolution', *Space Sci. Rev.*, this volume.

Burgdorf, M.J., *et al.*: 2000, 'ISO Observations of Mars: an Estimate of the Water Vapor Vertical Distribution and the Surface Emissivity', *Icarus* **145**, 79–90.

Carleton, N.P., and Traub, W.A.: 1972, 'Detection of Molecular Oxygen on Mars', *Science* **177**, 988–992.

Clancy, R.T., Muhleman, D.O., and Berge, G.L.: 1990, 'Global Changes in the $0 - 70$ km Thermal Structure of the Mars Atmosphere Derived from 1975-1989 Microwave CO Spectra', *J. Geophys. Res.* **95**, 14,543–14554.

Clancy, R.T., Grossman, A.W., and Muhleman, D.O.: 1992, 'Mapping Mars Water Vapor with the Very Large Array', *Icarus* **100**, 48–59.

Clancy, R.T., *et al.*: 1996, 'Water Vapor Saturation at low Altitudes Around Mars Aphelion: a Key to Martian Climate?', *Icarus* **122**, 36–62.

Clark, R.N., Swayze, G.A., Singer, R.B., and Pollack, J.: 1990, 'High-resolution Reflectance Spectra of Mars in the 2.3 μm Region: Evidence for Mineral Scapolite', *J. Geophys. Res.* **95**, 14,463–14480.

Conrath, B.J., Pearl, J.C., Smith, M.D., and Christensen, P.R.: 1999, 'Mars Global Surveyor TES Results: Atmospheric Thermal Structure Retrieved from Limb Measurements', *Bull. Amer. Astron. Soc.* **31**, 1150–1150.

Drossart, P., Rosenqvist, J., Erard, S., Langevin, Y., Bibring, J. P., and Combes, M.: 1991, 'Martian Aerosol Properties from the Phobos/ISM Experiment', *Annales Geophysicae* **9**, 754–760.

Encrenaz, Th., *et al.*: 1991, 'The Atmospheric Composition of Mars: ISM and Ground-based Observational Data', *Annales Geophysicae* **9**, 797–803.

Encrenaz, Th., Lellouch, E., Cernicharo, J., Paubert, G., and Gulkis, S.: 1995a, 'A Tentative Detection of the 183-GHz Water Vapor Line in the Martian Atmosphere: Constraints Upon the H_2O Abundance and Vertical Distribution', *Icarus* **113**, 110–118.

Encrenaz, Th., *et al.*: 1995b, Detectability of Molecular Species in Planetary and Satellite Atmospheres from their Rotational Transitions', *Plan. Space Sci.* **43**, 1485–1516.

Encrenaz, Th., *et al.*: 1998, 'A Study of the Water Vapor Vertical Distribution on Mars from ISO and IRAM Measurements', *Annales Geophysicae* **16**, III, C1038–C1038.

Encrenaz, Th., Lellouch, E., Paubert, G., and Gulkis, S.: 1999, IAU Circ. 7186, June 4.

Espenak, F., Mumma, M., and Kostiuk, T.: 1991, 'Ground-based Infrared Measurements of the Global Distribution of Ozone in the Atmosphere of Mars', *Icarus* **92**, 252–262.

Farmer, C.B., Davies, D.W., Holland, A.L., LaPorte, D.D., and Doms, P.E.: 1977, 'Mars: Water Vapor Observations from the Viking Orbiters', *J. Geophys. Res.* **82**, 4225–4248.

Forget, F., *et al.*: 1999, 'Improved General Circulation Models of the Martian Atmosphere from the Surface to Above 80 km', *J. Geophys. Res.* **104**, 24,155–24176.

Hanel, R.A., *et al.*: 1972, 'Investigation of the Martian Environment by Infrared Spectroscopy on Mariner 9', *Icarus* **17**, 423–442.

Head, J.W., *et al.*: 2001, 'Geological Processes and Evolution', *Space Sci. Rev.*, this volume.

Jakosky, B.M., and Haberle, R.M.: 1992, 'The Seasonal Behavior of Water on Mars', *in* H.H. Kieffer, B.M. Jakosky, C.W. Snyder, and M.S. Matthews (eds.), *Mars*, Univ. Arizona Press, pp. 969–1016.

Kahn, R.A., Martin, T.Z., Zurek, R.W., and Lee, S.W.: 1992, 'The Martian Dust Cycle', *in* H.H. Kieffer *et al.* (eds.), *Mars*, Univ. Arizona Press, pp. 1017–1053.

Kaplan, L., Connes, P., and Connes, J.: 1969, 'Carbon Monoxide in the Martian Atmosphere', *Astrophys. J.* **157**, L187–L192.

Kieffer, H.H., Martin, T.Z., Peterfreund, A.R., Jakosky, B.M., Miner, E.D., and Palluconi, F.D.: 1977, 'Thermal and Albedo Mapping of Mars during the Viking Primary Mission', *J. Geophys. Res.* **82**, 4249–4291.

Kliore, A.J., Cain, D.L., Fjeldbo, G., Sidel, B.L., Sykes, M.J., and Rasool, S.I.: 1972, 'The Atmosphere of Mars from Mariner 9 Radio-occultation Measurements', *Icarus* **17**, 484–516.

Korablev, O.I., *et al.*: 1993, 'Tentative Detetion of Formaldehyde in the Martian Atmosphere', *Plan. Space Sci.* **41**, 441–451.

Krasnopolsky, V.A., Parshev, V.A., Krysko, A.A., and Rogachev, V.N.: 1979, 'Structure of the Mars Lower and Middle Atmosphere Based on Mars 5 Ultraviolet Photometry Data', Academy of Sciences (USSR), Moscow.

Krasnopolsky, V.A.: 1986, 'Photochemistry of the Atmospheres of Venus and Mars', Springer-Verlag.

Krasnopolsky, V.A.: 1997, 'High-resolution Spectroscopy of Mars at 3.7 and 8 μm: a Sensitive Search for H_2O_2, H_2CO, HCl, and CH_4, and Detection of HDO', *J. Geophys. Res.* **102**, 6525–6534.

Lellouch, E., Goldstein, J.J., Bougher, S.W., Paubert, G., and Rosenqvist, J.: 1991a, 'First Absolute Wind Measurements in the Middle Atmosphere of Mars', *Astrophys. J.* **383**, 401–406.

Lellouch, E., Paubert, G., and Encrenaz, Th.: 1991b, 'Mapping of CO Millimeter-wave Lines in Mars' Atmosphere: the Spatial Distribution of Carbon Monoxide on Mars', *Plan. Space Sci.* **39**, 219–224.

Lellouch, E., *et al.*: 2000, 'The 2.4 − 45 μm Spectrum of Mars Observed with the Infrared Space Observatory', *Plan. Space Sci.*, in press.

Lindal, G.F., *et al.*: 1979, 'Viking Radio-occultation Measurements of the Atmosphere and Topography of Mars: Data Acquired during 1 Martian Year of Tracking', *J. Geophys. Res.* **84**, 8443–8456.

Maguire, W.C.: 1977, 'Martian Isotopic Ratios and Upper Limits for Possible Minor Constituentns as Derived from the Mariner 9 Infrared Spectrometer Data', *Icarus* **32**, 85–97.

Martin, T.Z.: 1981, 'Mean Thermal and Albedo Behavior of the Mars Surface and Atmosphere over a Martian Year', *Icarus* **45**, 427–446.

Masson, P., Carr, M.H., Costard, F., Greeley, R., Hauber, E., and Jaumann, R.: 2001, 'Geomorphologic Evidence for Liquid Water', *Space Sci. Rev.*, this volume.

McElroy, M.B., and Donahue, T.M.: 1972, 'Stability of the Martian Atmosphere', *Science* **177**, 986–988.

McElroy, M.B., Kong, T.Y., and Yung, Y.L.: 1977, 'Photochemistry and Evolution of Mars' Atmosphere: a Viking Perspective', *J. Geophys. Res.* **82**, 4379–4388.

Nier, A.O., and McElroy, M.B.: 1977, 'Composition and Structure of Mars' Upper Atmosphere: Results from the Neutral Mass Spectrometers of Viking 1 and 2', *J. Geophys. Res.* **82**, 4341–4349.

Owen, T.: 1992, 'The Composition and Early History of the Atmosphere of Mars', *in* H.H. Kieffer *et al.* (eds.), *Mars*, Univ. Arizona Press, pp. 1017–1053.

Owen, T., Biemann, K., Rushneck, D.R., Biller, J.E., Howarth, D.W., and Lafleur, A.L.: 1977, 'The Composition of the Atmosphere at the Surface of Mars', *J. Geophys. Res.* **82**, 4635–4639.

Owen, T., Maillard, J.P., de Bergh, C., and Lutz, B.L.: 1988, 'Deuterium on Mars: the Abundance of HDO and the Value of D/H', *Science* **240**, 1767–1770.

Parkinson, T.D., and Hunten, D.M.: 1972, 'Spectroscopy and Aeronomy of O_2 on Mars', *J. Atmos. Sci.* **29**, 1380–1390.

Pollack, J.B., *et al.*: 1977, 'Properties of the Aerosols in the Martian Atmosphere, as Inferred from Viking Lander Imaging Data', *J. Geophys. Res.* **82**, 4479–4496.

Pollack, J.B., *et al.*: 1990a, 'Thermal Emission Spectra of Mars (5.4 − 10.5 μm): Evidence for Sulfates, Carbonates and Hydrates', *J. Geophys. Res.* **95**, 14,595–14,628.

Pollack, J.B., Haberle, R.M., Schaeffer, J., and Lee, H.: 1990b, Simulations of the General Circulation of the Martian Atmosphere, I. Polar processes', *J. Geophys. Res.* **95**, 1447–1473.

Rizk, B., *et al.*: 1991, 'Meridional Martian Water Abundance Profiles during the 1988-1989 Season', *Icarus* **90**, 205–213.

Rosenqvist, J., *et al.*: 1992, 'Minor Constituents in the Martian Atmosphere from the ISM/Phobos Experiment', *Icarus* **98**, 254–270.

Rosenqvist, J., and Chassefière, E.: 1995, 'A Reexamination of the Relationship between Eddy Mixing and O_2 in the Martian Middle Atmosphere', *J. Geophys. Res.* **100**, 5541–5551.

Seiff, A., and Kirk, D.B.: 1977, 'Structure of the Atmosphere of Mars in Summer at Mid-latitudes', *J. Geophys. Res.* **82**, 4364–4378.

Smith, P.H., *et al.*: 1997, 'Results of the Mars Pathfinder Camera', *Science* **278**, 1758–1765.

Théodore, B., Lellouch, E., Chassefière, E., and Hauchecorne, A.: 1993, 'Solstitial Temperature Inversions in the Martian Middle Atmosphere: Observational Clues and 2-D Modeling', *Icarus* **105**, 512–528.

Titov, D.V., *et al.*: 1994, 'Observations of Water Vapor Anomaly Above Tharsis Volcanoes on Mars in the ISM (Phobos 2) Experiment', *Plan. Space Sci.* **42**, 1001–1010.

Titov, D.V., Rosenqvist, R., Moroz, V.I., Grigoriev, A.V., and Arnold, G.: 1995, 'Evidences of the Regolith-atmosphere Water Exchange on Mars from the ISM (Phobos 2) Infrared Spectrometer Observations', *Adv. Space Res.* **16**, (6)23–(6)33.

Titov, D.V., *et al.*: 1998, 'Measurements of the Atmospheric Water Vapor on Mars by the Imager for Mars Pathfinder', *J. Geophys. Res.* **104**, 9019–9026.

Titov, D.V., Fedorova, A.A., and Haus, R.: 2000, 'A new Method of Remote Sounding of the Martian Aerosols by Means of Spectroscopy in the 2.7 μm CO_2 Band', *Plan. Space Sci.* **48**, 67–74.

Toon, O.B., Pollack, J.B., and Sagan, C.: 1977, 'Physical Properties of the Particles Composing the Martian Dust Storm of 1971-1972', *Icarus* **30**, 663–696.

Traub, W.A., Carleton, N.P., Connes, P., and Noxon, J.F.: 1979, 'The Latitude Variation of O_2 Dayglow and O_3 Abundance on Mars', *Astrophys. J.* **229**, 846–850.

Trauger, J.T., and Lunine, J.I.: 1983, 'Spectroscopy of Molecular Oxygen in the Atmospheres of Venus and Mars', *Icarus* **55**, 272–281.

Zurek, R.W.: 1992, 'Comparative Aspects of the Climate of Mars: an Introduction to the Current Atmosphere', *in* H.H. Kieffer *et al.* (eds.), *Mars*, Univ. Arizona Press, pp. 799–817.

Zurek, R.W., and Martin, T.Z.: 1993, 'Interannual Variability of Planet-encircling Dust Storms on Mars', *J. Geophys. Res.* **98**, 3247–3259.

Address for offprints: DESPA, Observatoire de Paris, 92195 Meudon, France;
(therese.encrenaz@obspm.fr)

MARTIAN VOLATILES: ISOTOPIC COMPOSITION, ORIGIN, AND EVOLUTION

D.D. BOGARD[1], R. N. CLAYTON[2], K. MARTI[3], T. OWEN[4] and G. TURNER[5]

[1] *Planetary Sciences SN, NASA Johnson Space Center, Houston, TX 77058, USA*
[2] *Enrico Fermi Institute, University of Chicago, 5640 S. Ellis, Chicago, IL 60637, USA*
[3] *Chemistry Department, University of California San Diego, La Jolla, CA 92093-0317, USA*
[4] *Institute for Astronomy, 2680 Woodlawn Dr., University of Hawaii, Honolulu, HI 96822, USA*
[5] *Department of Earth Sciences, University of Manchester M13 9PL, UK*

Received: 30 November 2000; accepted: 4 February 2001

Abstract. Information about the composition of volatiles in the Martian atmosphere and interior derives from Viking spacecraft and ground-based measurements, and especially from measurements of volatiles trapped in Martian meteorites, which contain several distinct components. One volatile component, found in impact glass in some shergottites, gives the most precise measurement to date of the composition of Martian atmospheric Ar, Kr, and Xe, and also contains significant amounts of atmospheric nitrogen showing elevated $^{15}N/^{14}N$. Compared to Viking analyses, the $^{36}Ar/^{132}Xe$ and $^{84}Kr/^{132}Xe$ elemental ratios are larger in shergottites, the $^{129}Xe/^{132}Xe$ ratio is similar, and the $^{40}Ar/^{36}Ar$ and $^{36}Ar/^{38}Ar$ ratios are smaller. The isotopic composition of atmospheric Kr is very similar to solar Kr, whereas the isotopes of atmospheric Xe have been strongly mass fractionated in favor of heavier isotopes. The nakhlites and ALH84001 contain an atmospheric component elementally fractionated relative to the recent atmospheric component observed in shergottites. Several Martian meteorites also contain one or more Martian interior components that do not show the mass fractionation observed in atmospheric noble gases and nitrogen. The D/H ratio in the atmosphere is strongly mass fractionated, but meteorites contain a distinct Martian interior hydrogen component. The isotopic composition of Martian atmospheric carbon and oxygen have not been precisely measured, but these elements in meteorites appear to show much less variation in isotopic composition, presumably in part because of buffering of the atmospheric component by larger condensed reservoirs. However, differences in the oxygen isotopic composition between meteorite silicate minerals (on the one hand) and water and carbonates indicate a lack of recycling of these volatiles through the interior. Many models have been presented to explain the observed isotopic fractionation in Martian atmospheric N, H, and noble gases in terms of partial loss of the planetary atmosphere, either very early in Martian history, or over extended geological time. The number of variables in these models is large, and we cannot be certain of their detailed applicability. Evolutionary data based on the radiogenic isotopes (i.e., $^{40}Ar/^{36}Ar$, $^{129}Xe/^{132}Xe$, and $^{136}Xe/^{132}Xe$ ratios) are potentially important, but meteorite data do not yet permit their use in detailed chronologies. The sources of Mars' original volatiles are not well defined. Some Martian components require a solar-like isotopic composition, whereas volatiles other than the noble gases (C, N, and H_2O) may have been largely contributed by a carbonaceous (or cometary) veneer late in planet formation. Also, carbonaceous material may have been the source of moderate amounts of water early in Martian history.

Chronology and Evolution of Mars **96** 425–458, 2001.
© 2001 *Kluwer Academic Publishers. Printed in the Netherlands.*

1. Introduction

Very early telescopic observations of Mars revealed polar caps whose waxing and waning with the seasons were interpreted as evidence that they were made of volatiles condensed from the atmosphere. As late as 1950, by analogy to Earth, it was generally assumed that the Martian atmosphere consisted mainly of nitrogen and Ar. In 1952, strong absorption bands of CO_2 were reported, and for two decades this was the only detected component of the Martian atmosphere. As early as the 1920s it was estimated that the atmospheric pressure at the Martian surface is no more than a few percent that of the Earth's pressure, but for decades our knowledge of the actual pressure did not improve. Significant advancement of information about Mars and its atmosphere came with the Mariner spacecraft flybys in the 1960s and the Viking missions in 1976. Another major advancement occurred in the early 1980s with the realization that we had meteorites from Mars in our collections, and that these contained Martian volatiles. (See e.g., Kieffer *et al.*, 1992, for a historical discussion of Mars studies.)

2. Atmospheric Composition

2.1. VIKING MEASUREMENTS

The first detailed measurement of the composition of the Martian atmosphere was made using mass spectrometers on the two Viking landers (Owen *et al.*, 1977; Nier and McElroy, 1977; Owen, 1992). In addition to consisting of $\sim$95% CO_2 and variable amounts of H_2O, the Martian atmosphere was found to contain 2.7% N_2, 1.6% Ar, 0.13% O_2, 2.5 ppm Ne, 0.3 ppm Kr, 0.08 ppm Xe, and trace amounts of other chemically reactive species. One of the more interesting observations made by Viking was an $\sim$62% enrichment in the $^{15}N/^{14}N$ isotopic ratio compared to the terrestrial ratio. Such large ^{15}N enrichment implies that considerable amounts of N_2 have been lost from the planet over time by a mass fractionating process, which enriches the atmospheric residue in the heavier isotope. It was estimated that approximately 99% of the original atmospheric N_2 may have been lost (McElroy *et al.*, 1977).

The Viking measurements also produced interesting data for the abundances and isotopic ratios of noble gases in the Martian atmosphere. Although the atmospheric pressure on Mars ($\sim$6 millibars, depending on elevation) is less than 1% that of the Earth, the relative abundances of Ne, Ar, Kr, and Xe are similar to those in the Earth's atmosphere. This abundance pattern is distinct in detail from solar gases and noble gases trapped in primitive meteorites (e.g. Pepin, 1991). The specific reason for this similarity in relative noble gas abundances between the Earth and Mars is not understood. In addition to elemental abundances, the Viking instruments also measured $^{40}Ar/^{36}Ar \cong 3000 \pm 500$ and $^{129}Xe/^{132}Xe \cong 2.5 \, (^{+2}_{-1})$, compared to the

TABLE I

Isotopic Determinations of Some Martian Volatiles. Data listed as "ratio" are as measured. Other ratios are given as deviations (% or ‰) relative to the standard indicated (terrestrial or solar), and values are positive except where noted as negative. The four columns of data represent the atmospheric composition as measured by Viking (Owen *et al.*, 1977; Nier and McElroy, 1977); the atmospheric compositions measured in shergottite impact glass; a possible ancient composition present in ALH84001; and the Martian interior composition as measured in Chassigny and some other SNC meteorites. See text for sources of meteorite data and additional discussion.

Isotopic Ratio	Comparison Standard	Atmos. Viking	Impact-Glass	ALH-84001	Interior SNCs
^{36}Ar/^{132}Xe	ratio	350[A]	900 ± 100[B]	~50	≤ 5
^{84}Kr/^{132}Xe	ratio	11[A]	20.5 ± 1.5[B]	~6	<1.1
^{2}H/^{1}H	%-terrestrial	~450[C]	~440[D]	~78[E]	~90
^{15}N/^{14}N	%-terrestrial	62 ± 16	<50	0.7	-3.0
^{20}Ne/^{22}Ne	ratio	n.r.	~10	?	?
^{36}Ar/^{38}Ar	ratio	5.5 ± 1.5	≤3.9	≥5	≥5.26
^{40}Ar/^{36}Ar	ratio	3000 ± 500	~1800	≤128[F]	≤212
^{86}Kr/^{84}Kr	%-solar	n.r.	~0	~0	~0
^{129}Xe/^{132}Xe	ratio	~2.5	2.4–2.6	2.16	<1.07
^{136}Xe/^{130}Xe	%-solar	n.r.	~27	~0	~0
^{13}C/^{12}C	‰-terrestrial	0 ± 50	n.r.	~41	-30/+41[G]
^{18}O/^{16}O	‰-terrestrial	0 ± 50	[H]	[H]	[H]

A question mark indicates that the isotopic ratio is unknown. n.r. indicates that the value was not reported:

[A] Ar, Kr, and Xe elemental abundances measured by Viking were estimated to be uncertain to ~20%, or ~28% in these elemental ratios, assuming uncorrelated errors.

[B] Ratios were calculated assuming atmospheric ^{129}Xe/^{132}Xe=2.6 For a minimum atmospheric ^{129}Xe/^{132}Xe ratio of 2.4, these elemental ratios would be ~770 and ~18, respectively.

[C] Ground-based spectra measurement

[D] The highest δD/H measured in Martian meteorites, 4358 ± 185‰, was found in Zagami apatite. D/H ratios apparently have not been reported for impact glass.

[E] Highest δD/H value reported for ALH84001.

[F] Ratio may reflect either interior or ancient atmosphere components (see text).

[G] Approximate ^{13}C/^{12}C range measured in SNCs.

[H] The ^{17}O/^{16}O/^{18}O isotopic compositions in Martian meteorite silicates and water differ slightly from each other and from the Earth's values, after accounting for effects of mass fractionation produced from chemical interactions.

terrestrial atmospheric values of 296 and 0.98, respectively. Argon-40 and ^{129}Xe were formed over time by the radioactive decay of ^{40}K ($t_{1/2} = 1.28$ Gyr) and ^{129}I ($t_{1/2} = 16$ Myr), respectively, and their presence in the atmosphere is a measure of the extent of degassing of the Martian interior. Viking also measured some other isotopic ratios of atmospheric components (e.g., ^{13}C/^{12}C, ^{18}O/^{16}O, ^{38}Ar/^{36}Ar) and, within relatively large measurement uncertainties, found them to resemble terrestrial values (Table I).

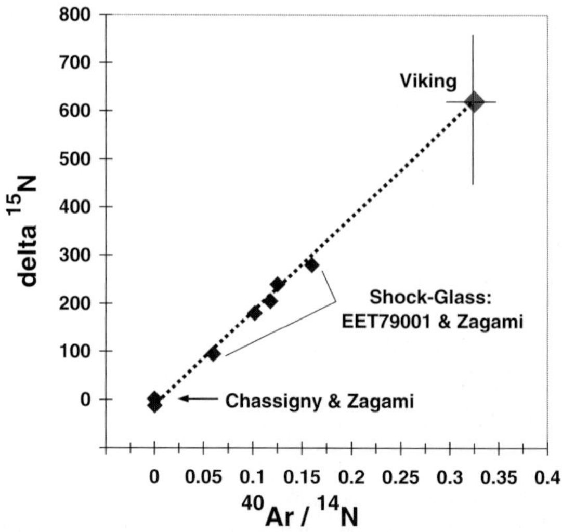

Figure 1. Values of $\delta^{15}N/^{14}N$ plotted against $^{40}Ar/^{14}N$ ratios for five samples of impact glass from the EET79001 and Zagami Martian meteorites and from bulk phases of the Chassigny and Zagami meteorites. The data indicate two-component mixing between an interior composition represented by Chassigny and the composition measured for the Martian atmosphere by Viking. Data from Owen *et al.* (1977), Becker and Pepin (1984), Marti *et al.* (1995), and Mathew and Marti (2001).

2.2. ATMOSPHERIC VOLATILES IN SNC METEORITES

A totally unexpected source of information on Martian volatiles appeared in 1983, with the recognition that impact-produced melt glass in SNC meteorite, EET79001 contained significant quantities of Martian atmospheric gases (Bogard and Johnson, 1983; Becker and Pepin, 1984). These gases were shock-incorporated into the melt when it formed by meteorite impact on the Martian surface (Bogard *et al.*, 1986; Wiens and Pepin, 1988). The presence of Martian atmospheric gases in some SNC meteorites, along with their young radiometric ages (Nyquist *et al.*, 2001) and geochemical similarity to Viking analysis of the Martian surface (McSween, 1985), became strong evidence that all of the SNC meteorites derive from Mars. Further, over the past two decades the evidence that SNC meteorites are Martian has become broader and more convincing (e.g., Treiman *et al.*, 2000), and today few workers in the field doubt their Martian origin.

Several laboratories have now measured the isotopic composition of Ne, Ar, Kr, and Xe in shock glass from several shergottite meteorites (see references in Bogard and Garrison, 1998). In addition, a significantly elevated $^{15}N/^{14}N$ ratio has been measured in shock glass from two shergottites (Wiens, 1988; Marti *et al.*, 1995; Mathew *et al.*, 1998). As shergottite shock glass can be no older than the <0.2 Gyr radiometric age of most shergottites (Nyquist *et al.*, 2001), and may well be as young as the space exposure ages of a few Myr, the gases trapped in impact glass represent the relatively recent Martian atmosphere. For the noble gases, Martian meteorite data define the atmospheric composition to a much higher precision than do the Viking data. For N_2, however, the Martian meteorites apparently do not give the pure Martian atmospheric end member measured by Viking, but rather appear to show mixing between atmospheric N_2 and a second, much less fractionated, trapped interior component. Figure 1 plots the $\delta^{15}N$ values (deviations of the $^{15}N/^{14}N$ ratio from the terrestrial ratio, in parts per thousand) against the $^{40}Ar/^{14}N$ ratios for five shock glass samples from the EET79001 and Zagami shergottites, where both N and Ar compositions were measured in the same sample (Becker and Pepin, 1984; Marti *et al.*, 1995). Corrections were applied for cosmogenic ^{15}N and radiogenic ^{40}Ar. The data define an apparent two-component mixing trend that passes through the Martian atmospheric composition measured by Viking. Although the data trend also passes near the terrestrial composition, analyses have shown that this second volatile component in Martian meteorites is different in composition from the terrestrial component and probably is characteristic of the Martian interior. This Martian interior component is better characterized through analyses of gases in two other Martian meteorites, Chassigny and ALH84001.

These results for nitrogen can be compared to the data on H_2O in the SNCs, discussed in more detail below. Water in some SNC minerals does appear to carry the strongly fractionated hydrogen found in the atmosphere by Earth-based investigations (Owen *et al.*, 1988; Krasnopolsky *et al.*, 1997), while also carrying one or more internal components that have not experienced this fractionation (Karlsson *et al.*, 1992; Watson *et al.*, 1994; Leshin *et al.*, 1996).

2.3. AR/KR/XE RELATIVE ELEMENTAL ABUNDANCES AND ^{129}XE/^{132}XE

Soon after the recognition of trapped Martian atmospheric gases in shock-produced glass of shergottite meteorite EET79001, Ott and Begemann (1985) and Ott (1988) reported very different compositions for some noble gas components trapped in Martian meteorites Chassigny, Nakhla, and non-glass samples of Shergotty. The Ar/Kr/Xe elemental ratios in Chassigny in particular indicated a very large depletion of the lighter elements, relative to both solar and Martian atmospheric compositions. The isotopic composition of Xe in some Chassigny samples, however, was found to closely resemble the solar composition, including a $^{129}Xe/^{132}Xe$ ratio of $\sim$1.03. The composition of trapped noble gases in bulk Shergotty was found to be intermediate to those of Chassigny and EET79001 shock-glass. These authors also noted that the lower ratio of radiogenic ^{129}Xe to trapped Xe in Chassigny and Nakhla, in comparison to trapped atmospheric gas in EET79001 glass, was opposite to the situation on Earth, where noble gases in the interior show larger $^{129}Xe/^{132}Xe$ compared to the Earth's atmosphere. Ott (1988) interpreted the noble gas data for Chassigny, Shergotty, and EET79001 impact glass to be variable mixtures of two Martian components, an atmospheric one represented by EET79001 glass and an interior component represented by gas in Chassigny.

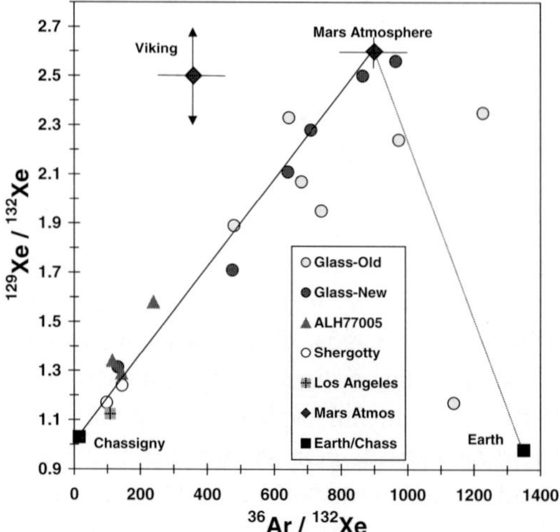

Figure 2. Plot of measured ^{129}Xe/^{132}Xe ratios against trapped ^{36}Ar/^{132}Xe ratios for various samples of shergottites. The light- and dark-tinted circles represent older and newer analyses, respectively, of impact glass. Other analyses shown are of bulk samples. Chassigny is believed to contain only a Martian interior component, whereas the impact glass contains mostly Martian atmosphere. The atmospheric composition reported by Viking and that indicated by glass data are shown. See text and Table I for relative uncertainties in these data. Two component mixing lines between Mars atmosphere and Mars interior components and between the Martian and terrestrial atmospheres are indicated. Sources of shergottite data are given in Bogard and Garrison (1998) and Garrison and Bogard (2000).

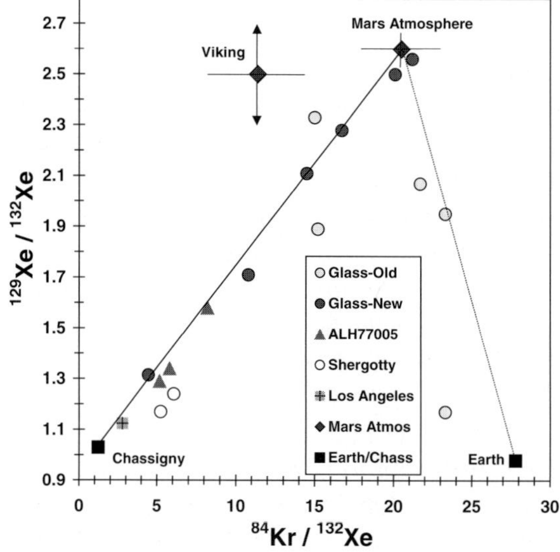

Figure 3. Plot of the measured ^{129}Xe/^{132}Xe ratio against the ^{84}Kr/^{132}Xe ratios for various samples of shergottites. See Figure 2 caption for sample explanations.

It is now apparent that both Martian atmospheric and interior components of Ar, Kr, Xe, and N, occur in Martian meteorites, with the atmospheric component being dominant in shergottite impact glass (Table I). Figures 2 and 3 illustrate these two-component mixtures for noble gases in several Martian meteorites. The elemental ratios ^{36}Ar/^{132}Xe and ^{84}Kr/^{132}Xe are plotted against the ^{129}Xe/^{132}Xe ratio because the latter is very different for the atmospheric and interior components and can generally be measured with good precision. Some data on a single impact glass inclusion, EET79001,27, and data on two Zagami glass inclusions show significant scatter on these plots. Although not always given, analytical uncertainties for most elemental abundances are of the order of $\pm10\%$ (or $\sim \pm15\%$ in elemental ratios), and thus do not explain the entire scatter in the data. Part of the data scatter is likely caused by the presence of a third component, e.g., terrestrial air. (Note that one glass inclusion (EET79001,54) plots close to the terrestrial composition in the lower right of both Figures 2 and 3.) However, recent analyses of melt glass from Shergotty, Y-793605, two different melt inclusions of EET79001, and shocked portions of ALH77005 give linear mixing relations for both the ^{36}Ar/^{132}Xe and ^{84}Kr/^{132}Xe ratios (Figures 2 and 3; Bogard and Garrison, 1998). The mixing trend for these shergottites passes through the composition measured in Chassigny (Ott, 1988), which contains the interior volatile component, but no atmospheric component. These mixing relations define Martian atmospheric ^{36}Ar/^{132}Xe $= 900 \pm 100$ and ^{84}Kr/^{132}Xe $= 20.5 \pm 2.5$, assuming an atmospheric ^{129}Xe/^{132}Xe ratio of 2.6 ± 0.05 (Bogard and Garrison, 1998).

Derivation from meteorite data of accurate Ar/Kr/Xe elemental ratios for the Martian atmosphere requires knowledge of the recent atmospheric ^{129}Xe/^{132}Xe ratio. Unfortunately, some uncertainty exists in this value. The highest ratio reported for an individual temperature extraction of EET79001 impact glass is 2.59 ± 0.03 (Bogard and Garrison, 1998), and is the basis for the atmospheric value of 2.6 assumed in deriving elemental ratios above. Other ^{129}Xe/^{132}Xe measurements in shergottite glass give somewhat lower maximum values of 2.35–2.43 (Swindle *et al.*, 1986; Marti *et al.*, 1995; Mathew *et al.*, 1998). Martian atmospheric Xe present in Nakhla has ^{129}Xe/^{132}Xe of 2.35 ± 0.03 (Gilmour *et al.*, 1999; 2001), and the maximum ratio measured in ALH84001 is 2.16 (Gilmour *et al.*, 1998; Garrison and Bogard, 1998; Mathew and Marti, 2001). It is difficult to envision how ^{129}Xe/^{132}Xe in young impact glass might be increased above the atmospheric ratio, but it might be lowered by mixing of Martian atmosphere with another component. However, if we assume that the atmospheric ^{129}Xe/^{132}Xe ratio is as low as 2.4, the mixing trend in Figures 2 and 3 imply that ^{36}Ar/^{132}Xe $\cong 770$ and ^{84}Kr/^{132}Xe $\cong 18$.

The recent Martian atmospheric composition defined by shergottite glass data in Figures 2 and 3 can be compared with the analyses made by Viking (Table I). Although the two data sources give the same ^{129}Xe/^{132}Xe ratio, the Viking measurement (2.5, $^{+2}_{-1}$ has a large uncertainty. *The ^{36}Ar/^{132}Xe and ^{84}Kr/^{132}Xe elemental ratios determined from shergottite data are higher than those measured by Viking by factors of 2.5 and 1.8, respectively, if atmospheric ^{129}Xe/^{132}Xe $= 2.6$, and higher* by factors of 2.2 and 1.6, respectively, if ^{129}Xe/^{132}Xe = 2.4. The differences in these elemental ratios from Viking and those derived for shergottites fall outside of their combined uncertainties. It seems unlikely that these ratios have changed drastically within the last $\sim$0.2 Gyr, and the shergottite data probably represent the recent Martian atmospheric composition more accurately. The shergottite impact glass values for ^{36}Ar/^{132}Xe and ^{84}Kr/^{132}Xe are within a factor of $\sim$1.5 of the terrestrial values and indicate even greater similarity in relative noble gas abundances between the two planetary atmospheres than suggested by Viking data.

Assuming the shergottite glass values of these elemental ratios are correct, this requires adjustment of the mixing ratios of noble gases in the Martian atmosphere. If we adopt the shergottite elemental ratios derived for the case that ^{129}Xe/^{132}Xe = 2.6, adopt the Viking atmospheric mixing ratio for ^{40}Ar of 1.6%, and assume atmospheric ^{40}Ar/^{36}Ar $\cong$ 1800 (see later section), we can calculate revised Martian atmospheric abundances. The Kr abundance becomes 0.36 ppm, or about 20% larger than that reported by Viking, and the Xe abundance becomes 0.05 ppm, or only 65% of that reported by Viking. The ^{36}Ar mixing ratio is $\sim$9 ppm. Similarly, the trapped ^{20}Ne/^{36}Ar ratio of $\sim$0.1 in shergottite impact glass reported by Garrison and Bogard (1998) implies that the Ne mixing ratio on Mars is only $\sim$1 ppm, a value which lies at the lower limit of the value (2.5, $^{+3.5}_{-1.5}$ ppm) reported by Viking. Given the relatively large uncertainties for much of the Viking measurements, these revised mixing ratios do not present a discrepancy.

2.4. FRACTIONATED AND ANCIENT ATMOSPHERIC GASES

The nakhlites and ALH84001 possess Martian atmospheric noble gases that appear to have been fractionated in their elemental ^{36}Ar/Xe and Kr/Xe ratios, compared to those in shergottite impact glass. In addition, at least ALH84001 contains trapped N and Xe that appear isotopically unfractionated and which may represent a trapped ancient Martian atmosphere.

Ar, Kr, and Xe data for the nakhlites and ALH84001 suggest a two component mixing trend, which passes close to the Chassigny point, but which is rotated counter-clockwise compared to the mixing trend defined by shergottite data in Figures 2 and 3 (Ott, 1988; Drake *et al.*, 1994; Swindle *et al.*, 1995; Miura *et al.*, 1995; Murty and Mohapatra, 1997; Gilmour *et al.*, 1998; Bogard and Garrison, 1998; Mathew *et al.*, 1998; Gilmour *et al.*, 2001). Although the data show some scatter, they appear to be consistent with mixing of an interior component similar to that found in Chassigny with a different type of atmosphere-like component. Swindle *et al.* (1995) concluded, based on a lack of a correlation of the ^{129}Xe/^{132}Xe and ^{136}Xe/^{132}Xe ratios, that a two-component mixture of Martian atmospheric Xe and solar-type Xe does not explain bulk ALH84001 data. The maximum ^{129}Xe/^{132}Xe ratio in Nakhla, 2.35 $\pm$ 0.03, is comparable to that in the glass of shergottites and could be largely modern, but might be as old as the 1.3 Gyr formation age of Nakhla. (ALH84001 is the only known Martian meteorite older than 1.3 Gyr.) The highest measured ^{129}Xe/^{132}Xe ratio of 2.16 in ALH84001 is outside the range of 2.35–2.59 observed for the recent Martian atmospheric component in shergottites and may represent ancient (i.e., ~4 Gyr; Turner *et al.*, 1997) trapped atmospheric gas. Other isotopic and elemental compositions in ALH84001 are ^{36}Ar/^{38}Ar $\geq$5.0, ^{36}Ar/^{132}Xe ~50, and δ^{15}N = 7‰ (Mathew and Marti, 2001; Table I).

The elementally fractionated noble gas data of ALH84001 and the nakhlites have been interpreted in two ways. The first interpretation is that the noble gas composition is an outcome of solubility controlled fractionation during direct incorporation of atmospheric gases into Martian weathering products (Drake *et al.*, 1994), as commonly occurs with terrestrial rocks (Ozima and Podosek, 1983). A second interpretation is that the elemental fractionation may have occurred during adsorption of atmospheric gases onto mineral surfaces, followed by shock implantation (Gilmour *et al.*, 2001). In contrast to Nakhla, ALH84001 silicates do not exhibit signs of alteration by liquid water, and elemental fractionation involving aqueous alteration is less likely. Using laser probe analyses of individual mineral grains, Gilmour *et al.*, (1999; 2001) demonstrated that the major carrier of the atmospheric Xe in Nakhla is actually pyroxene, with only a minor proportion being present in weathering products. Gilmour *et al.* (2001) concluded that a Chassigny-like interior component was present in Nakhla and was concentrated in feldspar. These authors argue that differences in gas release between Nakhla and ALH84001 may relate to differences in shock levels, high in ALH84001 and low in Nakhla, and that these are reflected in the corresponding (high and low) release temperatures of the atmospheric components for the two meteorites.

Most workers assume that the shock-implanted atmospheric component in shergottite glass is likely to have been much less elementally fractionated than was the component incorporated into the nakhlites and ALH84001 (e.g., see Bogard *et al.*, 1986; Drake *et al.*, 1994). If we assume that the ^{36}Ar/^{132}Xe and ^{84}Kr/^{132}Xe ratios derived above for the Martian atmosphere are correct, then the average value for this atmospheric component in the nakhlites and ALH84001 has been fractionated in favor of the heavier species by factors of approximately 15 and 2.6, respectively (Bogard and Garrison, 1998), well outside experimental uncertainties. For a ^{129}Xe/^{132}Xe ratio in the Martian atmosphere of <2.6, these fractionation factors would be proportionally less.

2.5. ARGON ISOTOPIC COMPOSITION

Analyses of shergottite impact glass indicate a lower value for the ^{40}Ar/^{36}Ar ratio of the Martian atmosphere compared to the Viking value of 3000 ± 500. Determination of this ratio in shergottites is not straightforward because of the presence of additional Ar components, including radiogenic ^{40}Ar from in situ decay of ^{40}K, cosmogenic ^{36}Ar, trapped Martian interior Ar, and sample contamination by terrestrial Ar. For example, using literature data and specific values for formation ages and space exposure ages for several Martian meteorites, Terribilini *et al.* (1998)

corrected total ^{40}Ar for *in situ* decay of ^{40}K and corrected total ^{36}Ar for cosmogenic ^{36}Ar in order to derive trapped ^{40}Ar/^{36}Ar ratios. These ^{40}Ar/^{36}Ar ratios varied widely, from $\sim$200 to $\sim$1900, indicating more than one trapped Ar component.

Two lines of evidence indicate that the ^{40}Ar/^{36}Ar ratio for trapped Martian atmosphere in shergottites actually lies in the range 1600–1900. First, a plot of ^{40}Ar/^{36}Ar against ^{129}Xe/^{132}Xe for various shergottite samples shows two component mixing between Martian atmospheric and interior components (Garrison and Bogard, 1998 and references therein). For values of ^{129}Xe/^{132}Xe approaching the maximum measured value of 2.59, the ^{40}Ar/^{36}Ar ratios approach an upper limit of $\sim$1900. The second line of evidence for a lower atmospheric ^{40}Ar/^{36}Ar than the Viking measurement comes from shergottite samples neutron-irradiated for ^{39}Ar-^{40}Ar dating (Bogard and Garrison, 1999). Higher temperature extractions of shergottites EET79001 and ALH77005 each released $\sim$61% of the total Ar and, when plotted on a modified isochron plot of ^{36}Ar/^{40}Ar versus ^{39}Ar/^{40}Ar, defined intercepts on the ^{36}Ar/^{40}Ar axis of ^{40}Ar/^{36}Ar $= 1735 \pm 85$ and 1760 ± 100, respectively. Because the abundance of trapped ^{40}Ar greatly dominates over radiogenic ^{40}Ar in these two samples, only a very short extrapolation of the mixing trend from the data to the ^{36}Ar/^{40}Ar axis is required. In addition, most extractions of a sample of Shergotty define a linear trend giving a ^{39}Ar-^{40}Ar age of $\sim$167 Myr and a ^{40}Ar/^{36}Ar trapped ratio of $\sim$1780. Lower temperature extractions of shergottites suggest release of terrestrial atmospheric Ar and possibly some Martian interior Ar. For this reason, the largest measured trapped ^{40}Ar/^{36}Ar ratios in these samples are considered to better represent the Martian atmospheric value. Although the atmospheric ^{40}Ar/^{36}Ar ratio may change with time, it is unlikely that this ratio evolved from $\sim$1800 to the Viking value of $\sim$3000 over the past 0.2 Gyr. *Thus, we conclude that an atmospheric ^{40}Ar/^{36}Ar ratio of 1800 ± 100 probably represents the present Martian atmosphere.*

Attempts have also been made to measure the ^{40}Ar/^{36}Ar ratio of the Martian interior component. The shergottite analyses by Bogard and Garrison (1999) suggest that this component is released at intermediate extraction temperatures and has a ^{40}Ar/^{36}Ar ratio of $<$500. For EET79001 glass, Wiens (1988) estimated the interior ^{40}Ar/^{36}Ar at 430–680. The ^{40}Ar/^{36}Ar ratios in the high temperature releases of Chassigny suggest interior ^{40}Ar/^{36}Ar is $<$206; the lowest ^{40}Ar/^{36}Ar ratio in ALH84001 is 128, observed at the highest release temperature (Mathew and Marti, 2001), and may indicate a much lower value for this atmospheric ratio early in Martian history. However, partitioning of atmospheric and interior Ar components in ALH84001 is not well constrained, and this low ratio may also reflect an interior Ar component. Because ^{40}Ar and ^{36}Ar in the interior have completely different sources (radiogenic and primordial, respectively), the two isotopes evolved separately over time and probably are heterogeneously distributed within Mars. Thus the interior ^{40}Ar/^{36}Ar ratio probably is not the same everywhere.

In addition to an $\sim$62% enrichment in the ^{15}N/^{14}N ratio, the isotopic compositions of Ar, Xe and H also show significant isotopic fractionation due to atmo-

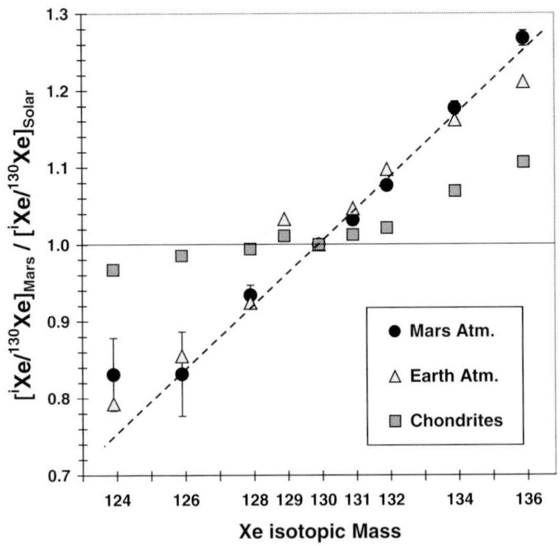

Figure 4. Isotopic composition of Xe in EET79001 impact glass (Swindle *et al.*, 1986), normalized to the solar Xe composition and mass 130. Also shown are the normalized Xe compositions for the Earth and primitive chondrites (Pepin, 1991). The dashed line indicates a possible mass fractionation trend for Martian Xe relative to solar Xe.

spheric loss. Even in the earliest analyses of impact glass in EET79001, several investigations noted *the presence of trapped Ar with a $^{36}Ar/^{38}Ar$ ratio considerably less than the terrestrial value of 5.32.* Wiens *et al.* (1986) deduced a Martian atmospheric ratio of 4.1. Swindle *et al.* (1986) derived a value of 3.60 ± 0.44. Bogard (1997) considered all shergottite data available up to that time and concluded that the ratio is less than 3.9. Deriving a precise value for Martian atmospheric $^{36}Ar/^{38}Ar$ is made difficult because of the presence of significant amounts of Ar produced by cosmic ray reactions during space exposure of Martian meteorites. The most accurate determinations derive from the EET79001 impact glass. (EET79001 has a relatively low exposure age of $\sim$0.6 Myr.) Such a low $^{36}Ar/^{38}Ar$ as that indicated for the Martian atmosphere is unique in the solar system (except for minor components produced by cosmic ray interactions). In contrast, Chassigny contains a trapped Martian interior component for which the $^{36}Ar/^{38}Ar$ ratio is $\geq$5.26 (Mathew and Marti, 2001). For comparison, this ratio is 5.32 in the Earth's atmosphere and $\sim$5.7 in the solar wind (Pepin *et al.*, 1995).

2.6. XENON ISOTOPIC COMPOSITION

The Viking Xe analyses had large uncertainties, and only the $^{129}Xe/^{132}Xe$ ratio was reported. Trapped Martian atmosphere in shergottite glass yields much more accurate isotopic data. The Xe composition of EET79001 impact glass (Swindle *et al.*, 1986) precisely agrees with shergottite glass data given by Mathew *et al.*

(1998). An informative way to consider this Xe composition is to compare it to Xe in other major volatile reservoirs and to assess the possibility that the Martian atmospheric composition was derived from one of these other reservoirs. Figure 4 plots the EET79001 Xe isotopic composition normalized to the solar wind composition (Pepin *et al.*, 1995) and to ^{130}Xe. In Figure 4 the solar composition is represented by the horizontal line and the difference between the Martian atmosphere and solar composition for a given ^{i}Xe/^{130}Xe isotopic ratio is proportional to the separation from the horizontal line. The isotopic compositions of Xe in the Earth's atmosphere and in primitive chondritic meteorites (Pepin, 1991) are also plotted relative to solar Xe. It is obvious that *the isotopic composition of Martian atmospheric Xe, as represented by EET79001 impact glass, closely resembles the terrestrial composition*, the only appreciable difference being for ^{136}Xe, which includes a fission component. (In discussing the Xe isotopic composition here we ignore ^{129}Xe, which contains a major component from decay of extinct ^{129}I.) Martian atmospheric Xe clearly differs from the solar and chondritic Xe compositions.

A characteristic of this type of isotope plot is that, for small mass intervals, derivation of one Xe component from a second Xe component by mass fractionation will cause the second composition to rotate about the first while maintaining an approximately linear relationship. Swindle *et al.* (1986) normalized their Martian Xe composition to chondritic Xe. Except for masses 126 and 128, Xe in EET79001 glass was generally consistent with derivation from chondritic Xe by mass fractionation, where the Martian ^{136}Xe/^{130}Xe ratio was enriched by $\sim$15%. However, Martian atmospheric Xe also closely resembles mass fractionated solar Xe (Swindle and Jones, 1997; Mathew *et al.*, 1998), as indicated by the dashed line in Figure 4. Using an initial solar composition, Mathew *et al.* (1998) derived a linear mass fractionation of 37‰ (3.7%) per amu, or twice as much as required for an initial chondritic composition. It is obvious that if the starting Xe composition for the Martian atmosphere resembled either solar or chondritic Xe, it has been strongly mass fractionated in favor of heavier isotopes.

In addition to the presence in the Martian atmosphere of excess ^{129}Xe from ^{129}I decay, one might also expect $^{131-136}$Xe from the fission of ^{235,238}U and extinct ^{244}Pu ($t_{1/2}$ = 82 Myr). Fission Xe apparently is present in the Earth's atmosphere (Ozima and Podosek, 1983; Porcelli and Wasserburg, 1995). Swindle and Jones (1997) argue that a chondritic initial Xe composition also implies an unreasonable fractionation of I from Pu+U in the interior of Mars in order to generate significant quantities of excess radiogenic ^{129}Xe in the Martian atmosphere, but little fission Xe. These authors suggest that the starting composition for the fractionated Martian atmosphere was solar Xe, which permits a somewhat higher concentration level of fission Xe in the current atmosphere. In addition, Mathew *et al.* (1998) present stepwise temperature data for Xe in shergottites that are consistent with mass fractionated solar Xe plus small additions of fission Xe. In further support of a solar Xe starting composition is the observation of Xe with a solar composition trapped in the Chassigny meteorite and solar Kr in the present atmosphere (see

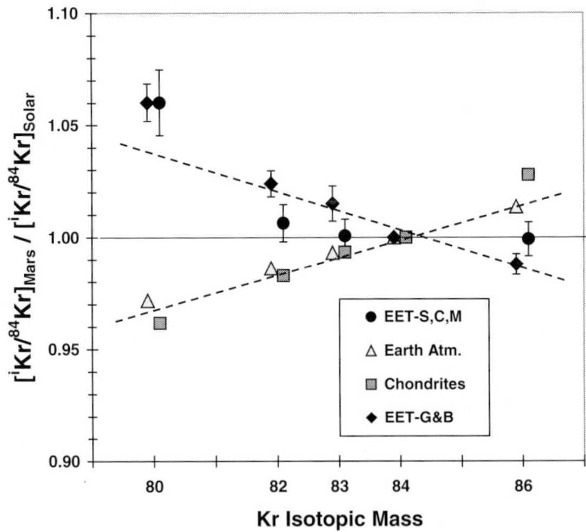

Figure 5. Isotopic composition of Kr in two samples of EET79001 impact glass (S,C,M=Swindle *et al.*, 1986; G&B=Garrison and Bogard, 1998, plotted normalized to the solar Kr composition K-1 (Pepin *et al.*, 1995) and mass 84. Also shown are the Kr compositions for the Earth and primitive chondrites (Pepin, 1991), also normalized to solar Kr and mass 84. The two dashed lines show mass fractionation trends (relative to solar K-1) for chondritic Kr and the maximum mass fractionation for Martian atmospheric Kr.

discussion below). If the dashed mass fractionation line of Martian Xe relative to solar Xe in Figure 4 had a somewhat lesser slope, it would agree better with the light Xe isotopic data of EET79001 and would permit small excesses of fission Xe at masses 132–136. In this case the ^{136}Xe/^{130}Xe mass fractionation relative to solar would be ~20%. It is not yet possible to model completely unambiguously those specific processes and starting compositions that have created the present Martian atmosphere, although a solar-like starting composition seems likely.

2.7. KRYPTON ISOTOPIC COMPOSITION

The isotopic composition of Kr in the Martian atmosphere, as measured in shergottite impact glass, presents some intriguing contrasts with the evidence for strong mass fractionation of Ar and Xe. Figure 5 plots the isotopic composition of Kr measured in two glass inclusions of EET79001, normalized to mass 84 and to the K-1 solar Kr composition derived by Pepin *et al.* (1995) from analyses of solar gases in lunar fines. The normalized compositions of Kr in the terrestrial atmosphere and in primitive chondritic meteorites are also shown (Pepin, 1991). The plotted data of Swindle *et al.* (1986) were summed over seven temperature extractions of impact glass EET79001,27. The plotted data of Garrison and Bogard (1998) are the 1550°C extraction of impact glass EET79001,8A, which released 78% of the total Kr in this sample. Garrison and Bogard (1998) suggested that the Kr released at lower extraction temperatures from both samples contained terrestrial Kr contamination, which would cause rotation of the measured shergottite composition counter-clockwise toward the terrestrial data (compare Figure 5). An average Kr composition derived from three high temperature extractions from Swindle *et al.* (1986) and three high temperature extractions from two different EET79001 glass inclusions from Garrison and Bogard (1998) is similar to the 1550°C composition of EET79001,8A plotted in Figure 5.

From Figure 5 it is obvious that both the chondritic and terrestrial Kr compositions appear mass fractionated in favor of heavier isotopes relative to solar Kr, except for a small excess in chondritic ^{86}Kr (which may be affected by fission products). Swindle *et al.* (1986) originally normalized both their Kr and Xe data for EET79001 impact glass to the chondritic composition. A chondritic normalization would imply that Martian atmospheric Kr is mass fractionated in favor of lighter isotopes by almost 1%/amu. This is opposite to the apparent enrichment of heavier isotopes of Xe when normalized to a chondritic composition and to the observed enrichment of ^{38}Ar over ^{36}Ar. Clearly such opposite mass fractionation for Kr cannot be produced by the process that produced Xe and Ar fractionation. This, along with the evidence for solar Xe inside Mars (discussed below), suggests that the Martian Kr composition is more appropriately compared to solar Kr, as is done in Figure 5.

Analyses of Kr in EET79001 impact glass indicate that a small excess exists at mass 80 (and possibly mass 82) relative to the fractionation trend defined by either a chondritic or solar normalization (Becker and Pepin, 1984; Swindle *et al.*, 1986; Garrison and Bogard, 1998). This excess ^{80}Kr is probably caused by neutron-capture by ^{79}Br at the Martian surface, where the neutrons are produced from cosmic ray interactions.

Pepin (1991) noted that the Kr composition reported by Swindle *et al.* (1986) and the K-1 solar composition are essentially identical (when a neutron-capture component has been removed) and concluded that Martian atmospheric Kr is unfractionated solar Kr. On the other hand, if the Kr composition given by Bogard and Garrison (1998) is used to represent the Martian atmosphere, then Martian atmospheric Kr appears to be mass fractionated relative to K-1 solar Kr by ~0.5% per mass unit in favor of lighter isotopes. A possible explanation of this fractionation is that the composition of solar Kr is not precisely defined. Different lunar and meteorite samples containing solar Kr show variations of as much as ~2% at individual masses (Pepin *et al.*, 1995), and Pepin (1991) considered a possible uncertainty in the solar Kr composition of ~0.5%/mass. We tend to reject the occurrence of isotopic fractionation during gas implantation into the impact melt, because laboratory experiments do not suggest isotopic fractionation during shock implantation (Bogard *et al.*, 1986). It is possible, however, that solar wind Kr implanted into lunar and meteorite samples has been slightly mass fractionated relative to solar Kr initially acquired by Mars. Thus, we conclude that *within present uncertainties, the isotopic compositions of Martian atmospheric Kr and solar Kr do not differ.*

2.8. Neon Isotopic Composition

Viking reported only 2.5 ($^{+3.5}_{-1.5}$) ppm Ne in the Martian atmosphere and was unable to measure its isotopic composition (Owen *et al.*, 1977). (We noted above that the shergottite glass data suggest only $\sim$1 ppm Ne.) Given the large mass fractionation observed in D/H, $^{15}N/^{14}N$, and $^{36}Ar/^{38}Ar$, one would expect the $^{20}Ne/^{22}Ne$ ratio also to be strongly mass fractionated. Neon compositions measured in impact glass of shergottites are ambiguous on this point, however. Garrison and Bogard (1998) plotted $^{20}Ne/^{22}Ne$ ratios against $^{21}Ne/^{22}Ne$ ratios for stepwise temperature extractions of impact glasses from three shergottites as measured in three different laboratories. In general, the data indicate a two-component mixture between cosmogenic Ne and a trapped component, but the data show significant scatter. Scatter in the cosmogenic composition exists because of the possible presence of Ne produced by energetic solar protons and because different Na/Mg concentration ratios in some samples produced different cosmogenic $^{21}Ne/^{22}Ne$ ratios. It is not clear why apparent variations exist in the $^{20}Ne/^{22}Ne$ ratio of the trapped component. Most data suggest a trapped ratio somewhat similar to the terrestrial ratio of 9.8, but some data are consistent with lower values. The trapped component tends to be released at lower extraction temperatures, and part of this release may consist of terrestrial Ne contamination in the samples. The degree of mass fractionation of $^{20}Ne/^{22}Ne$ in the Martian atmosphere, if any, is thus poorly constrained. If solar Ne with $^{20}Ne/^{22}Ne \geq 13$ were the starting composition, significant fractionation is suggested. However, if chondritic-like Ne with $^{20}Ne/^{22}Ne \cong 8.2$ were the starting composition, no significant fractionation is indicated.

3. Interior Nitrogen and Noble Gas Components

As pointed out above, the Chassigny meteorite contains primarily a Martian interior component that is very different in composition from the Martian atmospheric component. Derivation of the composition of this interior component is more difficult compared to the atmospheric component, and the interior composition reported in Table I has been deduced from several meteorite sources. In comparisons of relative abundances of trapped Ar, Kr, and Xe and the $^{129}Xe/^{132}Xe$ and $^{15}N/^{14}N$ isotopic ratios, the shergottites apparently contain variable mixtures of the Martian atmospheric component and the Martian interior component, (Figures 2 and 3). Some samples also contain varying amounts of terrestrial noble gas contamination. Noble gases from shergottites EET79001, Shergotty, Zagami, ALH77005, Yamato-793605, LEW-88516, DaG-476, Los Angeles, Dhofar-019, and Sayh al Uhaymir-005 are consistent with variable mixtures of these three gas components (Zipfel *et al.*, 2000; Shukolyukow *et al.*, 2000; Garrison and Bogard, 2000; Mohapatra and Ott, 2000; earlier data references in Bogard and Garrison, 1998). Analyses of some of these shergottites indicate the presence of the Martian interior component, but little of the Martian atmospheric component.

Recent investigations of Xe isotopes released at different temperatures show substantial variations and indicate that the Martian interior gas component actually consists of two or more components (Mathew and Marti, 2001). The Chass-S (Solar) gases in Chassigny (Mathew and Marti, 2001) represent an elementally heavily fractionated reservoir. These authors noted that a second composition, Chass-E (Evolved), can be related to Chass-S by addition of ^{244}Pu fission Xe component. However, the Chass-E xenon defines a uniform trapped signature and implies that the fission Xe component was well mixed with the solar Xe component at the time of incorporation and that in-situ decay of ^{244}Pu can be ruled out. Therefore, both Chass-S and Chass-E represent interior reservoirs of Mars and both are characterized by low ^{129}Xe/^{132}Xe ratios (<1.07). The ^{36}Ar/^{132}Xe and ^{84}Kr/^{132}Xe ratios associated with Chass-S and Chass-E are constrained to be, <5 and <1.1, and >130 and >1.8, respectively. The amount of ^{244}Pu-Xe that needs to be added to Chass-S to generate Chass-E is 8.4% of the total ^{136}Xe in Chassigny. The implied ^{244}Pu/^{238}U ratio at the time of system closure is 0.002. It is not clear why Chassigny contains Xe with a solar-like composition while showing such strong fractionation in Ar/Kr/Xe elemental abundances.

The bulk Chassigny data of Mathew and Marti (2001) show light N isotopic signatures in the ≤300°C extractions (δ^{15}N ≤ −21‰). The heaviest signature these authors observed in the low-temperature range is δ^{15}N= +15‰ at 400°C. Interestingly, the N signatures observed in the intermediate temperature steps of ALH84001 and Chassigny are rather similar. Further, the light N is enriched in Chassigny olivine inclusions, and the signature of the light nitrogen (δ^{15}N= −30‰) observed in the olivine is consistent with light N measured in ALH84001. This light nitrogen signature apparently represents the primitive interior nitrogen reservoir, which did not exchange with the evolving atmospheric nitrogen.

4. Isotopes of O, C, H, and S

On Earth, all waters and almost all rocks have oxygen isotopic compositions that lie on a single fractionation line in a three-isotope graph of δ^{17}O versus δ^{18}O. (Here the δ notation indicates deviation in the ^{17}O/^{16}O and ^{18}O/^{16}O ratios in parts per thousand relative to standard mean ocean water, SMOW.) The known exceptions are some sulfate rocks with a ^{17}O-excess ultimately derived from atmospheric ozone, which has a large isotopic anomaly (Thiemens *et al.*, 1995). The observation derived from Martian meteorites that carbonates and hydroxyl minerals have oxygen isotopic compositions displaced from the principal fractionation line defined by silicates implies the existence on Mars of volatile reservoirs (e.g., atmosphere and hydrosphere) that are not isotopically buffered by the major rock reservoir. Thus, the terrestrial model of recycling of water and carbon dioxide through the rocky interior is not applicable to Mars. The isotopic composition of oxygen (and probably hydrogen and carbon) must be dominated by sources and/or processes outside the rock reservoir. These include: (1) continued or late addition of cometary volatiles to the atmosphere (Owen and Bar-Nun, 1995a), (2) anomalous isotopic fractionation during escape of gases from the atmosphere (Jakosky, 1993), or (3) photochemical processes analogous to those occurring today in the terrestrial stratosphere (Thiemens *et al.*, 1995).

Most carbonates in SNC meteorites have $\delta^{18}O$ values of $+20$ to $+40‰$, which are much higher than the average $\delta^{18}O$ of $+4$ to $+5‰$ for the silicate minerals. Clayton and Mayeda (1984) attempted to account for this large difference by postulating high-temperature planetary outgassing of both CO_2 and H_2O, followed by low-temperature re-equilibration of the volatiles with one another, leading to ^{18}O-enrichment in the CO_2. This mechanism is inconsistent with the more recent observations that oxygen in carbonates and hydroxyl minerals does not lie on the rock fractionation line (Karlsson *et al.*, 1992; Farquhar *et al.*, 1998).

The oxygen isotopic compositions of carbonates in ALH 84001 are quite remarkable, ranging from $-5‰$ to $+30‰$ over distances of <1 mm (Leshin *et al.*, 1998; Saxton *et al.*, 1998; Holland *et al.*, 2000). Various scenarios have been proposed to account for the isotopic zoning, including temperature variations and precipitation in closed or open systems. Ion microprobe measurements, with good spatial resolution, show that the first-formed carbonates have low $\delta^{18}O$, and later-formed carbonates have progressively higher $\delta^{18}O$. This behavior suggests precipitation in a closed system from a CO_2-rich fluid, in a Rayleigh process (Romanek *et al.*, 1994; Leshin *et al.*, 1998). As noted by Leshin *et al.* (1998), the progressive increase in $\delta^{13}C$ of the carbonates in ALH 84001 requires temperatures above 200°C. On the basis of the preservation of large gradients in chemical and isotopic compositions, Valley *et al.* (1997) estimate an upper limit of about 300°C for carbonate formation (see also Kent *et al.*, 2001).

Carbon isotopes may provide some clues as to the Martian behavior of this important element. The visible inventory of carbon on Mars is only 10^{-3} of that on Earth or Venus. No evidence has been found by instrumented spacecraft for carbonate rocks at the surface. The $^{13}C/^{12}C$ ratio measured by Viking in atmospheric CO_2 resembled the typical terrestrial value, but the measurement uncertainties were very large (Owen *et al.*, 1977). The $^{13}C/^{12}C$ ratio measured in Martian meteorites gives a relatively wide range of $\sim 7.5\%$, but the reasons for these variations are not well understood.

Hydrogen isotopes also show a large difference in abundances between an internal reservoir and the atmosphere/hydrosphere. Minerals from SNC meteorites appear to contain a mixture of a mantle-derived component with $\delta D \sim +900‰$ (Watson *et al.*, 1994; Leshin, 2000 and an atmosphere-derived component with $\delta D \sim +4500$, as measured directly in the Martian atmosphere (Owen *et al.*, 1988; Krasnopolsky *et al.*, 1997). The large deuterium enrichment in the atmosphere is attributed to isotopic fractionation due to Jeans escape of hydrogen (Owen *et al.*, 1988; Yung *et al.*, 1988; Jakosky and Jones, 1997). The two-fold enrichment of D/H in the interior, relative to the terrestrial mean, may also be due to an early loss of hydrogen, or may result from accretion of deuterium-rich material, as is found in comets (Owen, 1997; Owen and Bar-Nun, 2000; Leshin, 2000) and in organic matter of carbonaceous chondrites (Robert and Epstein, 1982). In both cases, the deuterium enrichment probably resulted from low-temperature ion-molecule reactions in the interstellar cloud preceding formation of the solar system (Geiss and Reeves, 1981; Meier and Owen, 1999).

Sulfur plays several important roles in the geochemistry of Mars. As a moderately volatile element, it is probably depleted in the bulk chemical composition of the planet relative to chondritic or solar abundances. As in the case of potassium (Humayun and Clayton, 1995), planetary accretion processes seem to have brought about elemental depletions of sulfur by perhaps a factor of 5–10, without concomitant stable isotope fractionation. The planetary abundance of sulfur in Mars is unknown, and a major fraction of it may be in the core (Schubert and Spohn, 1990). On the Martian surface, chlorine and sulfur are much more abundant in soils and dust than in the underlying rocks (Rieder *et al.*, 1997), implying an origin by volcanism, some chemical transformation in the atmosphere, and eolian or aqueous deposition. Sulfur isotope measurements on returned Martian samples should have high priority.

On Earth, sulfur isotope fractionation in igneous processes produces a range of $\delta^{34}S$ of a few permil (Ohmoto, 1986), with the average of mantle-derived samples being indistinguishable from the Canyon Diablo meteorite standard. Much larger variations are found in crustal rocks, which have experienced low-temperature processing and oxidation-reduction chemistry. These low-temperature processes are commonly facilitated by biological systems; hence, large sulfur isotope variations are sometimes used to implicate biological processes in geological environments. A simple counter-example shows that this is not an infallible tool: sulfur isotopes are strongly fractionated in lunar soils due to micrometeorite bombardment and consequent vapor loss (Clayton *et al.*, 1974, and references therein). Sulfur isotope studies of SNC meteorites have had two goals: (1) a search for evidence of biological processes, and (2) a search for evidence of atmospheric chemical processes. Shearer *et al.* (1996) and Greenwood *et al.* (1997) used ion microprobe techniques to study sulfides in ALH 84001 and found a modest range of $\delta^{34}S$, from +2.0 to +7.3‰. The range found for shergottites was -1.9 to +2.7‰. These values are all compatible with origins in igneous and hydrothermal processes, and show no large effects such as are seen on Earth in bacterial processes.

On Earth, atmospheric oxidation of volcanic sulfur-bearing gases may lead to non-mass-dependent isotopic fractionation, which manifests itself as excesses or deficiencies of ^{17}O and/or ^{33}S in the sulfate products (Bao *et al.*, 2000; Farquhar *et al.*, 2000a). Farquhar *et al.* (2000b) analyzed sulfides and sulfates from three shergottites and two nakhlites, and found small ^{33}S deficits in two samples of Nakhla, whereas all other samples showed no non-mass-dependent effects. The mass-dependent variations in $\delta^{34}S$ ranged from -5 to +5‰ with both ends of this range representing sulfates. Thus the isotopic signature of Martian atmospheric chemistry in SNC meteorites is very subtle. A measurement of a returned soil or dust sample should be much more informative.

5.   Models for Atmospheric Evolution and Mass Fractionation

A common assumption made about the Martian and terrestrial atmospheres is that the volatiles initially derived from some major reservoir, such as the solar nebula or gases modified during early planetary formation and now present in primitive meteorites and comets. The early planetary atmosphere was then modified by fractionation loss processes and addition of specific components, such as those produced by radioactive decay. In detail, however, the specific sources of these volatiles and the processes that have modified them are not well known. As discussed earlier, we see evidence of appreciable mass fractionation of several volatile species in the Martian atmosphere, with accompanying enrichment of the heavier isotopes. Other elements (e.g., C, O, and S) are apparently buffered by surface reservoirs, and isotopic enrichment is unlikely to exceed several percent. For Kr and Xe, which should not be buffered by a surface reservoir, Kr appears not to be mass fractionated, whereas Xe is strongly fractionated. This observation is an important constraint on models of how isotopic mass fractionation of atmospheric species occurred. In this section we briefly discuss some models that have been presented to explain isotopic fractionation of the Martian atmosphere. Time evolution of the Martian atmosphere through degassing of radiogenic components is discussed in a later section.

5.1.   LOSS OF NITROGEN AND NOBLE GASES

Escape from a planetary gravity field often depends on the mass of an atmospheric species. Lighter species have greater velocities and higher scale heights, which gives them larger relative concentrations in the upper atmosphere compared to heavier species. Giant impacts late in the formation of Mars have been postulated to remove part of the early atmosphere (Melosh and Vickery, 1989; Benz and Cameron, 1990; Ahrens, 1990; Pepin, 1997), but such a process alone would not likely produce significant mass fractionation. Noble gases in the atmospheres of the terrestrial planets might have been mass fractionated during thermal escape from smaller planetesimals, which later accreted to form these planets (Donahue, 1986). Hunten *et al.* (1987) presented a model involving hydrodynamic escape of a massive hydrogen atmosphere very early in the history of the terrestrial planets. In this model, the mass outflow of hydrogen from the planetary atmosphere was sufficiently large that it entrained heavier species and isotopically fractionated them in the loss process. On early Mars a hydrogen atmosphere may have been produced by dissociation of abundant water through intense bombardment by energetic radiation from the early sun.

Because of its relatively large atomic mass, Xe is the species least easily lost from the Martian atmosphere. Pepin (1991) proposed that the hydrodynamic escape of an early hydrogen atmosphere on Mars caused loss of almost the entire initial inventory of atmospheric species. Part of the Xe remained and was strongly mass fractionated. Pepin (1991) proposed that most of the present atmospheric inventory for species lighter than Xe was added by planetary degassing during the latter stages of this process, and that a lesser degree of hydrodynamic fractionation of these volatiles could produce the isotopic fractionations observed in Ne and Ar. Later, Pepin (1994) described a more complex atmospheric evolution model that considered loss by hydrodynamic escape, ion sputtering, photochemical dissociation, impact erosion, and carbonate formation. Because addition of Xe with other gases could mask the isotopically fractionated Xe in the atmosphere, the model suggested that a large fraction of the Xe in the interior of Mars was retained in the core and not degassed with other species. However, no experimental evidence for Xe solubility suggests that this is likely.

To explain the similar, non-chondritic ratios of $^{36}Ar/^{84}Kr/^{132}Xe$, but very different total pressures observed in the atmospheres of Mars and Earth today requires a non-fractionating process to deplete the Martian atmosphere. Owen and Bar-Nun (1995a; 1995b; 2000) suggested that the heavy noble gases were delivered by icy planetesimals similar to the comets observed today. This model is supported by laboratory studies of the composition of noble gases trapped in amorphous ice forming at temperatures near 50 K, a reasonable value for the formation temperature of comets in the solar nebula between Uranus and Neptune. Owen and Bar-Nun pointed out that a simple mixing line drawn through the ($^{36}Ar/^{132}Xe$, $^{84}Kr/^{132}Xe$) points for the SNC data passes through the Earth's atmosphere point and the 50 K icy planetesimal point. This model gives a value of $\sim$75 mb for the post-bombardment atmospheric pressure on Mars, with a total volatile inventory of at least 7.5 bars of CO_2 and a 750 m deep layer of water over the planet. The model explains the striking similarity in isotopic composition of Martian and terrestrial xenon by postulating that the xenon trapped in comets exhibits the same fractionation from solar xenon that we find in the two atmospheres. However, there is no model yet for how this fractionation occurred (Notesco et al., 1999). Furthermore, we have no measurements of noble gases in comets, although that should change in the next decade as a result of the CONTOUR and Rosetta missions.

In order to explain the Viking discovery that the $^{15}N/^{14}N$ ratio in the Martian atmosphere is enriched by $\sim$62%, McElroy et al. (1976) and McElroy et al. (1977) proposed that solar-induced photochemical reactions produce energetic electrons that dissociate the nitrogen molecule. This results in loss of atomic nitrogen over time, with a planetary enrichment of the heavier isotope. The authors proposed that approximately 99% of the original atmospheric N_2 inventory on Mars may have been lost. Nitrogen escape from Mars has been reconsidered by a number of authors, most recently by Fox (1997) and Fox and Hac (1997). Considering all possible pathways, Fox and Hac (1997) conclude that without an early, dense CO_2

atmosphere on Mars, the enrichment of ^{15}N would be even larger than the value of 1.6 observed today. This model is specific for nitrogen and gives initial surface pressures of $\sim$114–500 mb and N_2 column abundances 4–6 times the present value (Fox and Hac, 1997). Uncertainties in the reaction rates for the production of odd nitrogen species in the Martian atmosphere remain a major factor in limiting the precision of these calculations, in addition to uncertainties in the amount of nitrogen being added to the atmosphere through volcanic outgassing and cometary bombardment. The present ^{15}N/^{14}N value on Mars may therefore be accepted as an additional indicator of a dense, early atmosphere, even though the required increase in nitrogen by itself would not produce such an atmosphere. Unfortunately, the models considered to date have treated N_2 and CO_2 as independent gases whose initial abundances can be varied at will. It seems more appropriate to use a ratio in the range of $CO_2/N_2 = 20 \pm 10$, encompassing the values in the atmosphere of Venus and the Earth's volatile reservoir (Owen and Bar-Nun, 1995a).

Jakosky *et al.* (1994) considered the role of atmospheric sputtering after the early period of heavy bombardment in removing Ar and Ne and the combined role of sputtering and photochemical mechanisms in the removal of N_2. Sputtering is defined as ionization of species in the upper atmosphere, their acceleration by local magnetic fields, and subsequent collisions with other atmospheric species, resulting in sufficient momentum transfer that atoms and molecules escape the planet. Because this loss model alone does not explain the current abundances, the authors also invoked selective planetary degassing. Whether or not the isotopic compositions of Ar and Ne changed with time over the past $\sim$3.8 Gyr depends to some extent on any assumed changes in CO_2 pressure. In a more detailed application of the basic model to Ar, Hutchins and Jakosky (1996) estimated that 85–95% of the ^{36}Ar and 70–88% of the ^{40}Ar has been lost from the atmosphere. Nitrogen loss in the basic model was $\sim$99%, but adding more complexity to the model implies even greater amounts of N_2 loss and an initial pressure of 60 mbar or higher. Jakosky *et al.* (1994) conclude that a very early period of hydrodynamic escape of hydrogen is not required to produce the fractionated N, Ar, and Ne composition in the present Martian atmosphere, although this hydrodynamic mechanism is apparently still required to explain Xe fractionation. Kr and Xe are too heavy to escape through the sputtering mechanism.

5.2. SPACECRAFT OBSERVATIONS

Spacecraft studies at Mars indicate that sputtering by itself was probably not responsible for the loss of an early dense atmosphere. Measurements by the ASPERA experiment on the PHOBOS mission suggest an O^+ outflow of about 2×10^{25} ions/s $(1.4 \times 10^7$ ions cm^{-2} s$^{-1})$, leading to a total ion escape rate of the order of 1 kg/s (Lundin *et al.*, 1989, 1990; Zakharov, 1992). This rate would lead to the loss of the present atmosphere in about 10^9 years. Lammer and Bauer (1991) and Fox (1993) have shown that the O escape flux generated by dissociative recombination is 3–

6×10^6 cm^{-2} s^{-1}. Luhmann *et al.* (1992) considered the loss of gases by sputtering caused by re-entering O$^+$ ions. Estimates of O escape from this process range from 4×10^5 cm^{-2} s^{-1} (Zhang *et al.*, 1993) to 3×10^6 cm^{-2} s^{-1} (Kass and Yung, 1995). Luhmann (1997) has argued that loss of oxygen caused by oxygen pick-up ions will dominate the loss from dissociative recombination over reasonable changes in the solar EUV flux.

Fox (1997) has modeled the escape fluxes of 11 ions on Mars, establishing the limits imposed by ionospheric production. She found that O$_2^+$ dominated O$^+$ and that "if ions are being swept away at or near their maximum rates, the loss rates inferred are of the same order as or larger than many other non- thermal mechanisms and should be accounted for in models of the history of Martian volatiles". The *upper limit* to the total escape flux from these calculations is 4.4–14.2×10^{25} s^{-1}, about 4 times higher than the value measured by ASPERA. Yet Fox's upper limit is about an order of magnitude *smaller* than the loss rate calculated by Luhmann *et al.* (1992; 1997) that would produce a serious loss of CO$_2$ (0.14 bar) and water ($\sim$50 m) from early Mars if there were enhanced solar EUV.

Although Mars Global Surveyor (MGS) does not directly measure outflowing ions, it does detect a signature of ions being formed in the Martian exosphere. These ions are picked up by the solar wind and are subsequently lost from the Martian inventory. The Electron Reflectometer onboard MGS observed an attenuation in electron flux at the magnetic pileup boundary of Mars (Acuña *et al.*, 1998), which can be reproduced by a model of solar wind electrons impact-ionizing Martian exospheric H and O (Crider *et al.*, 2000). The O loss rate due to electron impact ionization ranges from 2–20×10^{24} s^{-1}, depending on the assumptions made about the solar wind flow geometry and exospheric density. This estimate is comparable to that of Fox (1997) and slightly less than the Phobos ASPERA measurements Lundin *et al.* (1989; 1990).

Likewise, any neutral particles existing above $\sim$300 km altitude are subject to electron impact ionization and subsequent loss. H and O, being the dominant species at high altitude, suffer the greatest loss. Isotopic fractionation occurs naturally by this process because for any species, the loss rates are proportional to the exospheric density. As the atmospheric scale height is mass dependent, electron impact ionization will have an effectively higher rate for lower mass isotopes with larger scale heights.

We conclude from this brief review that the measured values of ion escape are well within theoretical upper limits which themselves are well below the levels required for removal of an early, dense atmosphere. Future results on the escape of ions and neutrals will be produced by the NOZOMI mission, scheduled to arrive at Mars in 2004.

5.3. H, C, AND O FRACTIONATION

If we assume that the large observed D/H enrichment in the Martian atmosphere (Table I) occurred by photodissociation of water during the last $\sim$3.5 Gyr, this implies that at least 90% of the original water inventory near the surface of Mars has been lost. On the other hand, the amount of H_2O lost from Mars by this process has been calculated to be only a few meters averaged over the planet's surface (Yung *et al.*, 1988; Kass and Yung, 1995). This is far less than the reservoir of water expected on Mars from the geological evidence of erosion (Carr, 1986) or volcanism (Greeley, 1987), which is tens to hundreds of meters. The resolution of this apparent paradox may lie in a decoupled cometary source for the near-surface water (Owen, 1997; Owen and Bar-Nun, 1998; 2000), which may have had a starting δD of $\sim$900‰ (Watson *et al.*, 1994; Leshin, 2000). This value is similar to the cometary D/H value for H_2O (Meier and Owen, 1999). The small size deduced for the amount of water fractionated by atmospheric escape would in turn derive from the fact that the water involved in this process was only the water left near the surface after $\sim$3.5 Gyr ago, which could be a much smaller amount than the total original inventory. Another possibility is that part of the D/H fractionation occurred during the early hydrodynamic escape of hydrogen (Carr, 1996; Leshin, 2000), if this process indeed took place. In either case, the time at which the main reservoir of water was either segregated or lost is obviously of critical importance. The key constraints on this time may come from dating the erosional features in the Martian landscape.

As noted above, mass fractionation of carbon and oxygen isotopes during loss from the upper Martian atmosphere is mitigated by buffering and exchange with surface reservoirs of CO_2. To explain an observed range in $^{13}C/^{12}C$ in Martian meteorites of $\sim$7.5%, Jakosky (1991) estimates that 40–70% of the carbon exchanging with the atmosphere has been lost to space, but this value would be higher if significant late planetary degassing occurred over time. The $^{18}O/^{16}O$ ratios in Martian meteorite carbonates imply a ratio in the atmosphere that is enriched by $\sim$3.5% in ^{18}O. With this assumption, Jakosky (1991) estimates that 20–30% of the oxygen in the surface-atmosphere exchangeable system (primarily H_2O and CO_2) has been lost to space. If we take the existing Martian polar caps and regolith to be the only exchangeable surface reservoir, this implies that the amounts of CO_2 and water lost to space cannot have exceeded 0.25 bar and 5 m, respectively (Jakosky and Jones, 1997). This assumes that any additional water in the Martian crust and oxygen in surface silicates have not been part of this exchangeable system. Without more precise knowledge of the isotopic compositions of C and O in major Martian reservoirs and the size and nature of the near-surface reservoirs in equilibrium with the atmosphere, it is difficult to model in detail the extent of mass fractionation of these isotopes due to atmospheric loss.

6. Time Evolution of Radiogenic Components

Additional constraints on the evolution of Martian volatiles in time are provided by the distribution of the noble gas products from radioactive decay of ^{40}K, ^{129}I and ^{244}Pu. The 16 Myr half-life of extinct ^{129}I provides a chronometer for evolutionary events during the very earliest history, and a fundamental constraint is provided by the radiogenic ^{129}Xe content of the atmosphere. Based on estimates of 30 ppb for the global iodine content of Mars (Wänke and Dreibus, 1988), the ratio of radiogenic ^{129}Xe in the atmosphere to the total ^{127}I in Mars is around 10^{-7}. This provides a minimum estimate for ^{129}I/^{127}I at the time when a major global fractionation of iodine and xenon must have occurred and corresponds to a maximum interval of 160 Myr after the formation of the solar system (^{129}I/^{127}I $\sim 10^{-4}$). A further constraint on the details of this early fractionation is provided by the observation that ^{129}Xe/^{132}Xe for the interior component is indistinguishable from the solar value, within an uncertainty of around ± 0.02. In order to preserve the solar value, in the presence of 30 ppb iodine with an initial ^{129}I/^{127}I ratio $\sim 10^{-4}$, the initial concentration of ^{132}Xe in Mars must have been very high ($\geq 2 \times 10^{-7}$ ccSTP/g, based on the Chassigny data), comparable to the highest concentrations observed in some C-chondrites. If the isotopic composition of Chassigny xenon is typical of the Martian mantle, it is easy to calculate that the residual concentration of ^{132}Xe after the I-Xe fractionation would also have been substantial, between 10^{-7} and 10^{-10} cc/g depending on whether the fractionation was early or late.

Details of the I-Xe fractionation are unknown but are assumed to be a combination of outgassing, global melting, and the formation of a crust and depleted mantle. Specific models have been proposed in which the fractionation of xenon arose as a result of outgassing and differential solubility in a Martian hydrosphere (Musselwhite *et al.*, 1991), or based on differences in solubility of I and Xe in silicate melts during the process of outgassing (Musselwhite and Drake, 2000). In these models the large enrichment of ^{129}Xe in the atmosphere is achieved by a temporary storage of ^{129}I in the crust while the early atmosphere is drastically eroded and isotopically fractionated during hydrodynamic escape. The very low atmospheric abundance of fission xenon from the decay of extinct ^{244}Pu ($t_{1/2} = 82$ Myr) implies that a fractionation of Pu and I occurred while ^{244}Pu was 'live', and before escape of the hypothesized early dense atmosphere. It seems likely that this was part of the early global differentiation responsible for the I-Xe fractionation but it is not yet possible to add to the time constraints imposed by the ^{129}Xe excesses.

Argon-40 in the atmosphere provides clues to the more recent outgassing history of Mars. Mars contrasts with the Earth, where much higher ^{40}Ar atmospheric abundances result from extensive and continuous outgassing of the upper mantle. Based on estimates of ~ 500 ppm for the mean K abundance in Mars, the total amount of atmospheric ^{40}Ar represents less than 2% of that produced in the last 4.0 Gyr by radioactive decay of ^{40}K. This implies either very low rates of outgassing since the presumed loss of the early proto-atmosphere (Pepin, 1994), or higher rates of outgassing combined with continuous loss by sputtering and impact erosion from the upper atmosphere (Scambos and Jakosky, 1990). Simple models involving degassing associated with partial melting, indicate a semi-quantitative consistency between the implied low outgassing rates of ^{40}Ar and the low rates of volcanism on Mars (e.g. Pepin, 1994; Greeley and Schneid, 1991).

Retention of significant primordial ^{36}Ar in the interior of Mars is implied by the recognition of an interior component with low ^{40}Ar/^{36}Ar $\sim$200 (Mathew and Marti, 2001). The concentration of radiogenic ^{40}Ar resulting from 4.0 Gyr decay and a mean K concentration of 500 ppm is 3.2×10^{-5} ccSTP/g which would require a concentration of ^{36}Ar of $\sim$$10^{-7}$ cc/g to maintain ^{40}Ar/^{36}Ar below 300. This conclusion is consistent with the elemental ratios observed in Chassigny (^{36}Ar : ^{84}Kr : ^{132}Xe = 5 : 1 : 1; Mathew and Marti, 2001). However, such a high ^{36}Ar concentration contrasts with the terrestrial situation, where ^{36}Ar concentrations in the upper mantle are of the order of 10^{-10} cc/g. A conclusion that emerges from the above discussions is that Mars is a relatively undegassed planet in comparison to the Earth.

Direct evidence for changes in the isotopic composition of the Martian atmosphere comes from analyses of the meteorite ALH84001. Its Sm-Nd age of 4.5 Gyr (Nyquist *et al.*, 2001) indicates formation early in Mars history when ^{244}Pu was live, but after the decay of ^{129}I (Gilmour *et al.*, 1998). A major shock event at $\sim$3.9 Gyr (Turner *et al.*, 1997) was probably responsible for incorporating ancient Martian atmospheric gases but the interior component could be a relict of its original crystallization. Marti and Mathew (2000) noted a relationship between nitrogen isotopic signatures and the ^{129}Xe/^{132}Xe ratio, which they proposed using as a time marker. These authors suggest that atmospheric nitrogen may have evolved from an initial light signature of δ^{15}N $= -30‰$ to a value of δ^{15}N$= 4 \pm 2‰$ at the time of gas implantation into ALH84001. This evolutionary model suggests that nitrogen was present in the atmosphere and was only slightly evolved, but that $\sim$85% of the radiogenic ^{129}Xe had been released into the atmosphere at this time, or that subsequent release of primordial ^{132}Xe has kept almost in step with radiogenic ^{129}Xe. Further, since Chassigny and ALH84001 reveal identical light nitrogen (δ^{15}N $= -30‰$) components, it appears that multiple reservoirs of volatiles with different nitrogen and radiogenic xenon signatures do exist in the Martian mantle and attest to limited exchanges between these reservoirs with the modern Martian atmosphere. A relevant aspect of the isotopic signature of the atmospheric Xe in ALH84001 is that it appears to show no fractionated characteristic of the modern atmosphere, and, apart from ^{129}Xe, still preserves a solar signature (Mathew and Marti, 2001). The time of incorporation, hence epoch, of the trapped atmospheric xenon into the nakhlites (formation ages of 1.3 Gyr) is less certain, although Gilmour *et al.* (2001) argue that it was introduced recently by mild shock associated with the ejection from Mars.

7. Overview and Implications

Information about the isotopic abundances of volatile elements on Mars derives from ground-based measurements, spacecraft instrumentation, and study of Martian meteorites in terrestrial laboratories. The major component in the Martian atmosphere is CO_2, followed by N_2 and ^{40}Ar. All other components are $\ll 1\%$. Several volatile species in the Martian atmosphere display strong isotopic fractionation, i.e., N and the noble gases, with the exception of Kr. Isotopic fractionation seen in Xe has been modeled to reflect massive and early loss of the atmosphere, which included total loss of all species save Xe. The atmosphere was later rejuvenated by internal degassing and possibly by addition of a late volatile-rich veneer. Isotopic fractionation seen in H, N, Ar and possibly Ne is explained by loss from this rejuvenated atmosphere over much longer geological times. Importantly, atmospheric Kr has very close to a solar composition and apparently has not been isotopically fractionated. The atmospheric D/H ratio is enriched over the mean terrestrial value by a factor of 5.5. Several models exist to explain various aspects of isotopic fractionation of atmospheric species, but the number of variables in these models is large, and thus we cannot be certain of their detailed applicability.

Most condensable volatiles are not expected to show appreciable isotopic fractionation effects from atmospheric loss. This is because of the existence of chemically exchangeable reservoirs on the Martian surface (e.g., the polar caps for H_2O and CO_2), which are much larger than the atmospheric reservoir for some species. Unfortunately, precise measurements of the $^{13}C/^{12}C$ and $^{18}O/^{16}O$ isotopic ratios of atmospheric CO_2 do not yet exist. However, measurements of these ratios in secondary phases (e.g., carbonates, hydrated minerals, etc) in Martian meteorites show some significant isotopic variations of a few percent. Further, real differences exist in the $\delta^{17}O$ value between these secondary phases and silicates in the same meteorite. This difference implies that O in the Martian atmosphere and hydrosphere is not in chemical equilibrium with silicates or near-surface rocks. The D/H ratio in Martian meteorites shows significant variation from a low value of $\delta D =\sim 900\%o$ up to nearly the atmospheric value of $\delta D \cong 4200\%o$. This suggests that some secondary phases in these meteorites have exchanged with atmospheric H.

The noble gases and N in Martian meteorites also reveal one or more Martian interior components that are very different isotopically from the atmospheric one. Generally, this difference is a result of the large mass fractionation effects that have occurred in the atmospheric component. The interior component appears to have a solar composition for Xe and a $^{15}N/^{14}N$ ratio similar to that of enstatite meteorites. The isotopic composition of Ar in this interior component is poorly constrained, but is consistent with a solar composition. The existence of solar-like Xe in the Martian interior and solar-like Kr in the atmosphere strongly indicates that the heavier noble gases initially acquired by Mars had a solar composition, rather than compositions similar to those seen in primitive meteorites. However, the elemental Ar/Kr/Xe ratios of this Martian interior component (as measured in Chassigny) are very strongly fractionated relative to solar compositions and are even more strongly fractionated than these elements in primitive meteorites. No good explanation exists for this contrast in elemental and isotopic fractionation, except possibly temperature-controlled adsorption.

Several isotopic ratios in the Martian atmosphere, e.g., D/H, $^{15}N/^{14}N$, $^{36}Ar/^{38}Ar$, $^{40}Ar/^{36}Ar$, and $^{129}Xe/^{132}Xe$, are expected to have evolved over part or all of Martian history due to atmospheric loss or addition of radiogenic components. In principle, if the time evolution of these isotopes could be documented, e.g. by measuring trapped gases in Martian meteorites or returned Martian rocks having different ages, much could be learned about the nature of the Martian atmosphere and crustal degassing throughout Martian history. In practice, however, an incomplete data base and the presence of interior volatile components in Martian meteorites complicate this evaluation.

Although the pressure of the Martian atmosphere is less than 1% that of the Earth's atmosphere, Martian volatiles show some similarities and some differences to those of Earth. Both planets likely had early fractionation loss of noble gases of solar-like composition. This may account for the similarity in the relative abundances of their Ar, Kr, and Xe, in spite of the large differences in elemental masses and absolute atmospheric abundances. Whether early Mars also was depleted in H_2O and CO_2 compared to the Earth is difficult to evaluate. The crust of the Earth obviously contains relatively large quantities of H_2O and of CO_2 in the form of carbonates. However, to the extent that Martian meteorites are representative of the Martian interior, Martian rocks are relatively poor in water. Further, spectroscopic measurements of Mars have yet to reveal any deposits of carbonate on the surface. Those isotopic fractionations measured in Martian meteorites suggest that the volatile reservoir involved in isotopic exchange with the Martian atmosphere is limited. Yet, the morphology of the Martian surface implies that early in its history Mars may have possessed abundant surface water (Masson *et al.*, 2001). All this suggests that Mars may contain a substantial quantity of interior volatiles, which have never outgassed into the atmosphere and which are not in isotopic exchange with the atmosphere. This would be consistent with the low ($\sim$2%) fractional degassing into the atmosphere implied for radiogenic ^{40}Ar. Part of the surface water on early Mars may now reside in the deeper crust.

However, it may also be the case that early Mars received more than one source of volatiles. The ratio of noble gases to C, N, and water is smaller by many orders of magnitude in primitive carbonaceous meteorites than in gas of solar composition. Thus, much of the water, C, and N on early Mars may have been added as a late carbonaceous veneer, possibly after loss of the early atmosphere. The $^{15}N/^{14}N$ ratio in the present atmosphere suggests that $\sim$99% of the N has been lost to space, and it seems unlikely that a significant reservoir of N exists in the Martian surface. The C/N ratio of primitive meteorites is $\sim$30 by weight, and the quantity of N_2 currently in the atmosphere is 1.4×10^{16} moles. Thus, we might deduce that late addition of meteoritic volatiles would have added $\sim 8 \times 10^{19}$ moles of CO_2 to all of Mars. This is ~ 57 moles of CO_2 per cm^2 of the surface, which is equivalent to a layer of $CaCO_3$ ~ 20 meters thick. Even considering the likelihood that cratering in the first 1 Gyr of Martian history gardened and mixed the surface to depths of up to hundreds of meters (Hartmann *et al.*, 2001), this amount of $CaCO_3$ should, in principle, be observable from orbit using spectroscopic techniques. Either most of an early inventory of Martian carbon was deeply buried by basalt flows or cratering ejecta, or removed from the planet by impacts, or the predicted quantity of carbon on Mars, based on addition by a carbonaceous veneer and estimated amounts of nitrogen, is considerably too large.

The atmospheric abundance of ^{36}Ar also may not be consistent with the late meteorite veneer model. The $N/^{36}Ar$ ratio of primitive meteorites is $\sim 3 \times 10^6$ and the solar ratio is 28, whereas the $N/^{36}Ar$ ratio in the present Martian atmosphere is $\sim 6 \times 10^3$. Although large amounts of N and Ar have been lost from Mars, the $^{15}N/^{14}N$ and $^{38}Ar/^{36}Ar$ ratios suggest somewhat comparable fractions have been lost. Thus, if a late veneer of carbonaceous chondrite material was added to Mars, either the atmospheric loss of N has actually exceeded the loss of ^{36}Ar by orders of magnitude, or a significant surface reservoir exists for N (unlikely), or most of the ^{36}Ar in the atmosphere is a solar-derived component (contributing relatively little N) that has been degassed from the interior. A solar component of atmospheric Ar is consistent with the solar composition of atmospheric Kr. This consideration implies that most of the N and the noble gases on Mars had different origins.

We also can consider the possibility of addition of water to early Mars by a carbonaceous veneer. If early Mars had water to a global depth of at least 50 m, as suggested by some, this implies a water/C molar ratio of >5. Because primitive carbonaceous meteorites have C concentrations of several percent, this water/C ratio would require primitive meteorites to contain $\sim 20\%$ or more water, a value that is about a factor of two higher than water contents of carbonaceous chondrites. An alternative explanation may be that the late veneer added to Mars was even richer in volatiles, such as is the case with comets. Comets have measured C/N values greater than solar. Owen and Bar-Nun (1995a) suggested a cometary C/N ratio of 20 ± 10 a value not much different from that in carbonaceous chondrites. If we assume the solar C/N ratio of 4.9 as a lower limit and 99% loss of N over time, only ~ 3.3 m of $CaCO_3$ would be contained within the Martian surface. This smaller amount of carbonate should be easier to dilute by impact gardening of the surface. However, the C/O ratio in comets seems to be solar, and Owen and Bar-Nun (1995a) suggested an upper limit to the CO_2/H_2O ratio of 0.6. Thus, a postulated cometary source of volatiles would seem to require that substantial amounts of C be accreted to Mars in order to produce >50 m of water planet wide. It may be that water accreted to Mars was incorporated into the Martian surface prior to a time when atmospheric gases N_2, CO_2, and Ar were being significantly lost from the planet, e.g., by impact erosion of the atmosphere. Thus, a much greater fraction of initial water may have been retained on Mars, compared to the remnants of these other gases that we observe today.

Finally, we point out that in spite of the considerable knowledge about the composition of Martian volatiles gained from the study of Martian meteorites, uncertainties remain. Is the noble gas component trapped in shergottite shock glass an accurate representation of the recent Martian atmosphere, or does it contain minor amounts of other components? How broadly representative of the Martian interior are the trapped N and noble gas components measured in Chassigny? Do the elementally fractionated Ar/Kr/Xe ratios observed in ALH84001 and the nakhlites represent fractionation of the recent Martian atmosphere during some simple absorption process? What has been the time evolution of the ^{129}Xe/^{132}Xe, ^{40}Ar/^{36}Ar, and ^{15}N/^{14}N ratios in the Martian atmosphere? What are the precise isotopic compositions of C and O in atmospheric volatiles, and with what surface reservoirs are these in isotopic equilibrium? Although we have presented above some reasonable answers to these questions, definitive answers may require study of Martian meteorites not yet found and precise, direct analyses of Martian volatiles, either in situ or in atmospheric and surface samples returned to Earth.

Acknowledgements

We thank the International Space Science Institute for hosting the workshop on the Chronology and Evolution of Mars, which was the impetus for the writing of this paper. M. Acuna contributed the discussion of Global Surveyor results. T. Encrenaz, J. Gilmour, and J. Jones provided constructive review comments.

References

Acuña, M.H. *et al.*: 1998, 'Magnetic Field and Plasma Observations at Mars: Initial Results of the Mars Global Surveyor Mission', *Science* **279**, 1676–1680.

Ahrens T.J.: 1990, 'Earth Accretion', *in* H.E. Newsom and J.H. Jones (eds.), *Origin of the Earth*, Oxford Univ. Press, New York, pp. 211–227.

Bao H., Thiemens, M.H., Farquhar, J., Campbell, D.A., Lee, C.C., Heine, K., and Loope, D.B.: 2000, 'Anomalous ^{17}O Compositions in Massive Sulphate Deposits on the Earth', *Nature* **406**, 176–178.

Becker, R.H., and Pepin, R.O.: 1984, 'The Case for a Martian Origin of the Shergottites: Nitrogen and Noble Gases in EETA79001', *Earth Planet. Sci. Lett.* **69**, 225–242.

Benz, W., and Cameron, A.G.W.: 1990, 'Terrestrial Effects of the Giant Impact', *in* H.E. Newsom and J.H. Jones (eds.), *Origin of the Earth*, Oxford Univ. Press, New York, pp. 61–67.

Bogard, D.D.: 1997, 'A Reappraisal of the Martian ^{36}Ar/^{38}Ar Ratio', *J. Geophys. Res.* **102**, 1653–1661.

Bogard, D.D. and Johnson, P.: 1983, 'Martian Gases in an Antarctic Meteorite', *Science* **221**, 651–654.

Bogard, D.D., and Garrison, D.H.: 1998, 'Relative Abundances of Argon, Krypton, and Xenon in the Martian Atmosphere as Measured in Martian Meteorites', *Geochim. Cosmochim. Acta* **62**, 1829–1835.

Bogard, D.D., and Garrison, D.H.: 1999, 'Argon-39-argon-40 "Ages" and Trapped Argon in Martian Shergottites, Chassigny, and Allan Hills 84001', *Met. Planet Sci.* **34**, 451–473.

Bogard, D.D., Hörz, F., and Johnson, P.: 1986, 'Shock-implanted Noble Gases: An Experimental Study with Implications for the Origin of Martian Gases in Shergottite Meteorites', *J. Geophys. Res.* **91 suppl.**, 99–114.

Carr, M.H.: 1986, 'Mars: A water-rich Planet?', *Icarus* **68**, 187–216.

Carr, M.H.: 1996, *Water on Mars*, Oxford Univ. Press, New York.

Carr, M.H.: 1999, 'Retention of an Atmosphere on Early Mars', *J. Geophys. Res.* **104**, 21,897–21,909.

Clayton, R.N., and Mayeda, T.K.: 1984, 'The Oxygen Isotope Record in Murchison and Other Carbonaceous Chondrites', *Earth Planet. Sci. Lett.* **67**, 151–161.

Clayton, R.N., Mayeda, T.K., and Hurd, J.M.: 1974, 'Loss of Oxygen, Silicon, Sulfur and Potassium from the Lunar Regolith', *Geochim. Cosmochim. Acta Suppl.* **5**, 1801–1809.

Crider, D.H. *et al.*: 2000, 'Evidence of Electron Impact Ionization in the Magnetic Pile-up Boundary of Mars', *Geophys. Res. Lett.* **27**, 45–48.

Donahue, T.M.: 1986, 'Fractionation of Noble Gases by Thermal Escape from Accreting Planetesimals', *Icarus* **66**, 195–210.

Drake, M.J., Swindle, T.D., Owen, T., and Musselwhite, D.S.: 1994, 'Fractionated Martian Atmosphere in the Nakhlites?', *Meteoritics* **29**, 854–859.

Farquhar, J., Thiemens, M.H., and Jackson, T.: 1998, 'Atmosphere-surface Interactions on Mars: $\delta^{17}O$ Measurements of Carbonate from ALH 84001', *Science* **280**, 1580–1582.

Farquhar, J., Bao, H., and Thiemens, M.H.: 2000a, 'Atmospheric Influence of Earth's Earliest Sulfur Cycle', *Science* **289**, 756–758.

Farquhar, J., Savarino, J., Jackson, T.L. and Thiemens, M.H.: 2000b, 'Evidence of Atmospheric Sulphur in the Martian Regolith from Sulphur Isotopes in Meteorites', *Nature* **404**, 50–52.

Fox, J.L.: 1993, 'On the Escape of Oxygen and Hydrogen from Mars', *Geophys. Res. Lett.* **20**, 1847.

Fox, J.L.: 1997, 'Upper Limits to the Outflow of Ions at Mars: Implications for Atmospheric Evolution', *Geophys. Res. Lett.* **24**, 2901.

Fox, J.L., and Hac, A.: 1997, 'The $^{15}N/^{14}N$ Isotope Fractionation in Dissociative Recombination of N_2^+', *J. Geophys. Res.* **102**, 9191–9204.

Garrison, D.H., and Bogard, D.D.: 1998, 'Isotopic Composition of Trapped and Cosmogenic Noble Gases in Several Martian Meteorites', *Met. Planet. Sci.* **33**, 721–736.

Garrison, D.H., and Bogard, D.D.: 2000, 'Cosmogenic and Trapped Noble Gases in the Los Angeles Martian Meteorite', *Met. Planet. Sci.* **35**, A58 (abstract).

Geiss, J., and Reeves, H.: 1981, 'Deuterium in the Solar System', *Astron. Astrophys.* **93**, 189–199.

Gilmour, J.D., Whitby, J.A., and Turner, G.: 1998, 'Xenon Isotopes in Irradiated ALH84001: Evidence for Shock-induced Trapping of Ancient Martian Atmosphere', *Geochim. Cosmochim. Acta* **62**, 2555–2571.

Gilmour, J.D., Whitby, J.A., and Turner, G.: 1999, 'Martian Atmospheric Xenon Contents of Nakhla Mineral Separates: Implications for the Origin of Elemental Mass Fractionation', *Earth Planet. Sci. Lett.* **166**, 139–147.

Gilmour, J.D., Whitby, J.A., and Turner, G.: 2001, 'Disentangling Xenon Components in Nakhla: Martian Atmosphere, Spallation and Martian Interior', *Geochim. Cosmochim. Acta* **65**, 343–354.

Greeley, R.: 1987, 'Release of Juvenile Water on Mars: Estimated Amounts and Timing Associated with Volcanism', *Science* **236**, 1653–1654.

Greeley, R., and Schneid, B.D.: 1991, 'Magma Generation on Mars: Amounts, Rates and Comparisons with Earth, Moon and Venus', *Science* **254**, 996–998.

Greenwood, J.P., Riciputti, L.R. and McSween, H.Y., Jr.: 1997, 'Sulfide Isotopic Compositions in Shergottites and ALH 84001, and Possible Implications for Life on Mars', *Geochim. Cosmochim. Acta* **61**, 4449–4454.

Hartmann, W.K., Anguita, J., de la Casa, M.A., Berman, D.C., and Ryan, E.V.: 2001, 'Martian Impact Cratering 7: The Role of Impact Gardening', *Icarus* **149**, 37–53.

Holland, G., Lyon, I.C., Saxton, J.M., and Turner, G.: 2000, 'Very Low Oxygen-isotopic Ratios in Allan Hills 84001 Carbonates: a Possible Meteoritic Component?', *Met. Planet. Sci.* **35**, A76–A77.

Humayun, M., and Clayton, R.N.: 1995, 'Potassium Isotope Cosmochemistry: Genetic Implications of Volatile Element Depletion', *Geochim. Cosmochim. Acta* **59**, 2131–2148.

Hunten, D.M., Pepin, R.O., and Walker, J.C.G.: 1987, 'Mass Fractionation in Hydrodynamic Escape', *Icarus* **69**, 532–549.

Hutchins, K.S., and Jakosky, B.M.: 1996, 'Evolution of Martian Atmospheric Argon: Implications for Sources of Volatiles', *J. Geophys. Res.* **101**, 14,933–14,949.

Jakosky, B.M.: 1991, 'Mars Volatile Evolution: Evidence from Stable Isotopes', *Icarus* **94**, 14.

Jakosky, B.M.: 1993, 'Mars Volatile Evolution: Implications of the Recent Measurement of ^{17}O in Water from the SNC Meteorites', *Geophys. Res. Lett.* **20**, 1591–1594.

Jakosky, B.M., and Jones, J.H. : 1997, 'The History of Martian Volatiles', *Rev. Geophys.* **35**, 1–16.

Jakosky, B.M., Pepin, R.O., Johnson, R.E., and Fox, J.L.: 1994, 'Mars Atmospheric Loss and Isotopic Fractionation by Solar-wind-induced Sputtering and Photochemical Escape', *Icarus* **111**, 271–281.

Karlsson, H.R., Clayton, R.N., Gibson, E.K., Jr., and Mayeda, T.K.: 1992, 'Water in SNC Meteorites: Evidence for a Martian Hydrosphere', *Science* **255**, 1409–1411.

Kass, D.M., and Yung, Y.L.: 1995, 'Loss of Atmosphere from Mars due to Solar Wind-Induced Sputtering', *Science* **268**, 697.

Kent, A.J.R., Hutcheon, I.D., Ryerson, F.J., and Phinney, D.L.: 2001, 'The Temperature of Formaion of Carbonate in Martian Meteorite ALH84001: Constraints from Cation Diffusion', *Geochim. Cosmochim. Acta* **65**, 311–321.

Kieffer, H.H., Jakosky, B.M., and Snyder, C.W.: 1992, 'The Planet Mars: From Antiquity to the Present', *in* H.H. Kieffer, B.M. Jakosky, C.W. Snyder, and M.S. Matthews, (eds.), *Mars*, Univ. Arizona Press, Tucson, pp. 1–33.

Krasnopolsky, V.A., Bjoraker, G.L., Mumma, M.J., and Jennings, D.E.: 1997, 'High Resolution Spectroscopy of Mars at 3.7 and 8 μm : A Sensitive Search for H_2O_2, H_2CO, HCl and CH_4, and Detection of HDO', *J. Geophys. Res.* **102**, 6524–6534.

Lammer, H., and Bauer, S.J.: 1991, 'Non-thermal Atmospheric Escape from Mars and Titan', *J. Geophys. Res.* **96**, 1819.

Leshin, L.A.: 2000, 'Insights into Martian Water Reservoirs from Analysis of Martian Meteorite QUE 94201', *Geophys. Res. Lett.* **27**, 2017–2020.

Leshin, L.A., Epstein, S., and Stolper, E.M.: 1996, 'Hydrogen Isotope Geochemistry of SNC Meteorites', *Geochim. Cosmochim. Acta* **60**, 2635–2650.

Leshin, L.A., McKeegan, K.D., Carpenter, P.K., and Harvey R.P.: 1998, 'Oxygen Isotopic Constraints on the Genesis of Carbonates from Martian Meteorite ALH 84001', *Geochim. Cosmochim. Acta* **62**, 3–13.

Luhman, J.G.: 1997, 'Correction to "The Ancient Oxygen Exosphere of Mars: Implications for Atmosphere Evolution", by Zhang *et al.*, *J. Geophys. Res.* **102**, 1637.

Luhmann, J.G., Johnson, R.E., and Zhang, M.H.G.: 1992, 'Evolutionary Impact of Sputtering of the Martian Atmosphere by O^+ Pick-up Ions', *Geophys. Res. Lett.* **19**, 2151.

Lundin, R., *et al.*: 1989, 'First Measurements of the Ionsphere Plasma Escape from Mars', *Nature* **341**, 609.

Lundin, R., *et al.*: 1990, 'Aspera/Phobos Measurements of the Ion Outflow from the Martian Atmosphere', *Geophys. Res. Lett.* **17**, 873–876.

Marti, K., and Mathew, K.J.: 2000, 'Ancient Martian Nitrogen', *Geophys. Res. Lett.* **27**, 1463–1466.

Marti, K., Kim, J.S., Thakur, A.N., McCoy, T.J., and Keil, K.: 1995, 'Signatures of the Martian Atmosphere in Glass of the Zagami Meteorite', *Science* **267**, 1981–1984.

Masson, P., Carr, M.H., Costard, F., Greeley, R., Hauber, E., and Jaumann, R.: 2001, 'Geomorphologic Evidence for Liquid Water', *Space Sci. Rev.*, this volume.

Mathew, K.J., and Marti, K.: 2001, 'Early Evolution of Martian Volatiles: Nitrogen and Noble Gas Components in ALH84001 and Chassigny', *J. Geophys. Res.*, **106**, 1401–1422.

Mathew, K.J., Kim, J.S., and Marti, K.: 1998, 'Martian Atmospheric and Indigenous Components of Xenon and Nitrogen in the Shergotty, Nakhla, and Chassigny Group Meteorites', *Met. Planet. Sci.* **33**, 655–664.

McElroy, M.B., Kong, T.Y., Yung, Y.L., and Nier, A.O.: 1976, 'Composition and Structure of the Martian Upper Atmosphere. Analysis of Results from Viking', *Science*, **194**, 1295–1298.

McElroy, M.B., Kong, T.Y., and Yung, Y.L.: 1977, 'Photochemistry and Evolution of Mars' Atmosphere: A Viking Perspective', *J. Geophys. Res.* **82**, 4379–4388.

McSween, H.Y.: 1985, 'SNC Meteorites: Clues to Martian Petrologic Evolution?' *Rev. Geophys.* **23**, 391–416.

Meier, R., and Owen, T.: 1999, 'Cometary Deuterium', *Space Sci. Rev.* **90**, 33–43.

Melosh, H.J., and Vickery, A.M.: 1989, 'Impact Erosion of the Primordial Atmosphere of Mars', *Nature* **338**, 487–489.

Mittlefehldt, D.W.: 1994, 'ALH84001, a Cumulate Orthopyroxenite Member of the Martian Meteorite Clan', *Meteoritics* **29**, 214–221.

Miura, Y.N., Nagao, K., Sugiura, N., Sagawa, H., and Matsubara, K.: 1995, 'Orthopyroxene ALH84001 and Shergottite ALH7705: Additional Evidence for a Martian Origin from Noble Gases', *Geochim. Cosmochim. Acta* **59**, 2105–2113.

Mohapatra, R.K., and Ott, U.: 2000, 'Trapped Noble Gases in Sayh al Uhaymir 005: A new Martian Meteorite from Oman', *Met. Planet. Sci.* **35**, A113 (abstract).

Murty, S.V.S., and Mohapatra, R.K.,: 1997, 'Nitrogen and Heavy Noble Gases in ALH84001: Signature of Ancient Martian Atmosphere', *Geochim. Cosmochim. Acta* **61**, 5417–5428.

Musselwhite, D.S., and Drake, M.J.: 2000, 'Early Outgassing of Mars: Implications from Experimentally Determined Solubility of Iodine in Silicate Magmas', *Icarus* **148**, 160–175.

Musselwhite, D.M., Drake, M.J., and Swindle, T.D.: 1991, 'Early Outgassing of Mars Supported by Differential Water Solubility of Iodine and Xenon', *Nature* **352**, 697–699.

Nier, A., and McElroy, M.B.: 1977, 'Composition and Structure of Mars' Upper Atmosphere: Results from the Neutral Mass Spectrometers at Viking 1 and 2', *J. Geophys. Res.* **82**, 4341–4350.

Notesco, G., Laufer, D., Bar-Nun, A., and Owen, T.: 1999, 'An Experimental Study of the Isotopic Enrichment in Ar, Kr, and Xe when Trapped in Water Ice', *Icarus* **142**, 298–300.

Nyquist, L.E., Bogard, D.D., Shih, C.-Y., Greshake, A., Stöffler, D., and Eugster, O.: 2001, 'Ages and Geologic Histories of Martian Meteorites', *Space Sci. Rev.*, this volume.

Ohmoto, H.: 1986, 'Stable Isotope Geochemistry of Ore Deposits', *in* J.W. Valley, H.P. Taylor, J.R. O'Neil (eds.), *Stable Isotopes in High Temperature Geological Processes*, Rev. Mineral. **16**, Mineral. Soc. Am., Washington, pp. 491–560.

Ott, U.: 1988, 'Noble Gases in SNC Meteorites: Shergotty, Nakhla, Chassigny', *Geochim. Cosmochim. Acta* **52**, 1937–1948.

Ott, U., and Begemann, F.: 1985, 'Are All the 'Martian' Meteorites from Mars?', *Nature* **317**, 509–512.

Owen, T.: 1992, 'The Composition and Early History of the Atmosphere of Mars', *in* H.H. Kieffer *et al.*, *Mars*, Univ. Arizona Press, Tucson, pp. 818–834.

Owen, T.: 1997, 'From Planetesimals to Planets: Contributions of Icy Planetesimals to Planetary Atmospheres', *in* Y.J. Pendleton and A.G.G.M. Tielens (eds.), *From Stardust to Planetesimals*, Astron. Soc. Pac. Conf. Ser. **122**, 435–450.

Owen, T., and Bar-Nun, A.: 1995a, 'Comets, Impacts, and Atmospheres', *Icarus* **116**, 215–226.

Owen, T., and Bar-Nun, A.: 1995b, 'Comets, Impacts, and Atmospheres II, Isotopes and Noble Gases', *in* K. Farley (ed.), *Volatiles in the Earth and Solar System* AIP Conf. Proc. **34**, pp. 123–128.

Owen, T., and Bar-Nun, A.: 1998, 'From the Interstellar Medium to Planetary Atmospheres via Comets', *Chemistry and Physics of Molecules and Grains in Space*, Faraday Discussions **109**, Roy. Soc. Chem., London, p. 453.

Owen, T., and Bar-Nun, A.: 2000, 'Volatile Contributions from Icy Planetesimals', *in* R. Canup and K. Righter (eds.), *Origin of the Earth and Moon*, Univ. Arizona Press, pp. 459–471.

Owen,T., Biemann, K., Rushneck, D.R., Biller, J.E., Howarth, D.W., and Lafleur, A.L.: 1977, 'The Composition of the Atmosphere at the Surface of Mars', *J. Geophys. Res.* **82**, 4635–4639.

Owen, T., Maillard, J.P., de Bergh, C., and Lutz, B.L.: 1988, *Science* **240**, 1767.

Ozima, N., and Podosek, F.A.: 1983, *Noble Gas Geochemistry*, Cambridge Univ. Press, 367 pp.

Pepin, R.O.: 1991, 'On the Origin and Early Evolution of Terrestrial Planet Atmospheres and Meteoritic Volatiles', *Icarus* **92**, 2–79.

Pepin, R.O.: 1994, 'Evolution of the Martian Atmosphere', *Icarus* **111**, 289–304.

Pepin, R.O.: 1997, 'Evolution of Earth's Noble Gases: Consequences of Assuming Hydrodynamic Loss Driven by Giant Impact', *Icarus* **126**, 148–156.

Pepin, R.O., Becker, R.H., and Rider, P.E.: 1995, 'Xenon and Krypton Isotopes in Extraterrestrial Regolith Soils and in the Solar Wind', *Geochim. Cosmochim. Acta* **59**, 4997–5022.

Porcelli, D., and Wasserburg, G.J.: 1995, 'Mass Transfer of Xenon through a Steady-state Upper Mantle', *Geochim. Cosmochim. Acta* **59**, 1991–2007.

Rieder, R., Economou, T., Wänke, H., Turkevich, A., Crisp, J., Brückner, J., Dreibus, G. and McSween, H.Y.: 1997, 'The Chemical Compositions of Martian Soil and Rocks returned by the mobile Alpha Proton X-ray Spectrometer: Preliminary Results from the X-ray Mode', *Science* **278**, 1771–1774.

Robert, F., and Epstein, S.: 1982, 'The Concentration and Isotopic Composition of Hydrogen, Carbon, and Nitrogen in Carbonaceous Chondrites', *Geochim. Cosmochim. Acta* **46**, 81–95.

Romanek, C.S., Grady, M.M., Wright, I.P., Mittlefehldt, D.W., Socki, R.A., Pillinger, C.T., and Gibson, E.K., Jr.: 1994, 'Record of Fluid-rock Interactions on Mars from the Meteorite ALH 84001', *Nature* **372**, 655–657.

Saxton, J.M., Lyon, I.C., and Turner, G.: 1998, 'Correlated Chemical and Isotopic Zoning in Carbonates in the Martian Meteorite ALH84001', *Earth Planet. Sci. Lett.* **160**, 811–822.

Scambos, T.A., and Jakosky, B.M.: 1990, 'An Outgassing Release Factor for Nonradiogenic Volatiles on Mars', *J. Geophys. Res.* **95**, 14,779–14,787.

Schubert, G., and Spohn, T.: 1990, 'Thermal History of Mars and the Sulfur Content of its Core', *J. Geophys. Res.* **95**, 14,095–14,104.

Shearer, C.K., Layne, G.D., Papike, J.J., and Spilde, M.N.: 1996, 'Sulfur Isotope Systematics in Alteration Assemblages in Martian Meteorite Allan Hills 84001', *Geochim. Cosmochim. Acta* **60**, 2921–2928.

Shukolyukov, Y.A., Nazarov, M.A., and Schultz, L.: 2000, 'Dhofar 019: A Shergottite with an Approximate 20-million-year Exposure Age', *Met. Planet Sci.* **35**, A147 (abstract).

Swindle, T.D., and Jones, J.H.: 1997, 'The Xenon Isotopic Composition of the Primordial Martian Atmosphere: Contributions from Solar and Fission Components', *J. Geophys. Res.* **102**, 1671–1678.

Swindle, T.D., Caffee, M.W., and Hohenberg, C.M.: 1986, 'Xenon and Other Noble Gases in Shergottites', *Geochim. Cosmochim. Acta* **50**, 1001–1015.

Swindle, T.D., Grier, J.A., and Burkland, M.K.: 1995, 'Noble Gases in Orthopyroxenite ALH84001: A Different Kind of Martian Meteorite with an Atmospheric Signature', *Geochim. Cosmochim. Acta* **59**, 793–801.

Terribilini, D., Eugster, O., Burger, M., Jakob, A., and Krähenbühl, U.: 1998, 'Noble Gases and Chemical Composition of Shergotty Mineral Fractions, Chassigny, and Yamato 793605: The Trapped Argon-40/argon-36 Ratio and Ejection Times of Martian Meteorites', *Met. Planet Sci.* **33**, 677–684.

Thiemens, M.H., Jackson, T.L., and Brenninkmeijer, C.A.M.: 1995, 'Observation of a Mass Independent Oxygen Isotopic Composition in Terrestrial Atmospheric CO_2, the Link to Ozone Chemistry, and the Possible Occurrence in the Martian Atmosphere', *Geophys. Res. Lett.* **22**, 255–257.

Treiman, A.H., Gleason, J.D., and Bogard, D.D.: 2000, 'The SNC Meteorites are from Mars', *Planet. Space Sci.* **48**, 1213–1230.

Turner, G., Knott, S.F., Ash, R.D., and Gilmour, J.D.: 1997, 'Ar-Ar Chronology of the Martian Meteorite ALH84001: Evidence for the Timing of the Early Bombardment of Mars', *Geochim. Cosmochim. Acta* **61**, 3835–3850.

Valley, J.W., Eiler, J.M., Graham, C.M., Gibson, E.K., Romanek, C.S. and Stolper, E.M.: 1997, 'Low-temperature Carbonate Concretions in the Martian Meteorite ALH 84001: Evidence from Stable Isotopes and Mineralogy', *Science* **275**, 1633–1638.

Wänke, H., and Dreibus, G.: 1988, 'Chemical Compositions and Accretion History of Terrestrial Planets', *Phil. Trans. Roy. Soc. Lond.* **A325**, 545–557.

Watson, L.L., Hutcheon, J.D., Epstein, S. and Stolper, E.M.: 1994, 'Water on Mars: Clues from Deuterium/hydrogen and Water Contents of Hydrous Phases in SNC Meteorites', *Science* **265**, 86–90.

Wiens, R.C.: 1988, 'Noble Gases Released by Vacuum Crushing of EETA79001 Glass', *Earth Planet. Sci. Lett.* **91**, 55–65.

Wiens, R.C., and Pepin, R.O.: 1988, 'Laboratory Shock Emplacement of Noble Gases, Nitrogen, and Carbon Dioxide into Basalt, and Implications for Trapped Gases in Shergottite EETA79001', *Geochim. Cosmochim. Acta* **52**, 295–307.

Wiens, R.C., Becker, R.H., and Pepin, R.O.: 1986, 'The Case for a Martian Origin of the Shergottites, II. Trapped and Indigeneous Gas Components in EETA79001 Glass', *Earth Planet Sci. Lett.* **77**, 149–158.

Yung, Y.L., Wen, J.S., Pinto, J.P., Allen, M., Pierce, K., and Paulom, S.: 1988, 'HDO in the Martian Atmosphere: Implications for the Abundance of Crustal Water', *Icarus* **76**, 146.

Zakharov, A.V.: 1992, 'The Plasma Environment of Mars: Phobos Mission Results', *AGU Geophys. Monograph* **66**, 327.

Zhang, M.H.G., Luhmann, G., Bougher, S.W., and Nagy, A.F.J.: 1993, 'The Ancient Oxygen Exosphere of Mars: Implications for Atmospheric Evolution', *Geophys. Res.* **98**, 10915.

Zipfel, J., Scherer, P., Spettel, B., Dreibus, G., and Schultz, L.: 2000, 'Petrology and Chemistry of the new Shergottite Dar al Gani 476', *Met. Planet. Sci.* **35**, 95–106.

Address for offprints: Planetary Sciences SN, NASA Johnson Space Center, Houston, TX 77058, USA, (donald.d.bogard1@jsc.nasa.gov)

EPILOGUE

SUMMARY: NEW VIEWS
AND NEW DIRECTIONS IN MARS RESEARCH

W.K. HARTMANN[1], R. KALLENBACH[2], J. GEISS[2] and G. TURNER[3]

[1]*Planetary Science Institute, Tucson AZ 85719, USA*
[2]*International Space Science Institute, Bern, Switzerland*
[3]*University of Manchester, Manchester, UK*

February 2001

The articles presented in this book reflect an interaction between geochemists, geophysicists, and photogeologists, who combined their expertise to focus on the chronology and evolution of Mars. A goal of our meeting was to offset some pressures that tend to isolate these groups from each other. As a result, we hoped to provide more integrated view, revealing Mars as a planet where we have access to primordial 4.5 Gyr-old crustal material, to diverse surfaces created during major geological processing before 3 Gyr ago, probably associated with a denser atmosphere and more fluvial environment, and also to exposures of volcanism, water release, weathering, gullying, and exhumation, all operating within the last few 100 Myr. Neither Earth nor Moon offers such a long-term, complete geologic record. This is a Mars with much longer-lived and varied geological activity than many had thought two decades ago (the time of the preparation of the major Mars conference volume, from the University of Arizona Press; Kieffer *et al.*, 1992).

A Brief History of Mars, Based (mostly) on these Articles

Part I of this book gives the general chronologic framework, based on Martian meteorite dating and cratering records. Evidence for youthful, as well as ancient, geologic processes on Mars is documented. The dating of the large lunar impact basins and craters (cf. Stöffler and Ryder, 2001; Neukum *et al.*, 2001) and the ratio of Mars/moon cratering (Ivanov, 2001) are crucial to calibrating the crater-dating scheme. Isotopic dating of Martian meteorites provides the chronology of formation and early differentiation, of igneous processes and of aqueous alteration (Bridges *et al.*, 2001; Nyquist *et al.*, 2001). Parts II and III develop the petrological and volatile evolutionary histories in the context of the new dating. Here are some highlights, in "evolutionary order."

Isotopic data indicate that Mars accreted within $\sim$10 Myr and the core formation and mantle differentiation occurred within $\sim$20 Myr, perhaps by 4520 Myr ago (Halliday *et al.*, 2001; Spohn *et al.*, 2001). Sr and Hf-W isotope data (Halliday *et al.*, 2001) suggest a lack of substantial mantle mixing since then, in sharp contrast to the Earth. This view is supported by the contrast between the isotopic composition of Xe in the Martian atmosphere and interior (Bogard *et al.*, 2001), which,

Space Science Reviews **96**: 461-470, 2001.

in addition to a major fractionation of iodine and xenon within the first 160 Myr, appears to require the hydrodynamic loss of an early dense atmosphere. Mars is relatively rich in volatiles compared to the Earth (Bogard *et al.*, 2001; Halliday *et al.*, 2001), which may reflect differences in the original accreting material, or alternatively that the Earth lost volatiles during the putative giant impact that may have formed the Moon. Mars did not have such a major impact (Halliday *et al.*, 2001), but it probably did experience smaller scale impact erosion of its early atmosphere, removing some of the early CO_2, as argued by Bridges *et al.* (2001), to explain the paucity of carbonates in Mars meteorites and Martian soils.

Early core/mantle/crust formation is consistent with the Mars Global Surveyor (MGS) discovery that some of the oldest parts of the crust are magnetized, and that the magnetizing era must have been short-lived since very early basin-forming impacts such as Hellas punched holes in the magnetized crust (Spohn *et al.*, 2001). Areas of primordial crust must be exposed near the surface, in order to explain the existence of the 4.5 Gyr-old rock ALH84001, in one out of the 4 to 8 impact "samplings" of Mars. As early proto-liths from the Moon, ALH84001 was brecciated and heavily shocked 3.9 to 4.1 Gyr ago, during the Noachian cratering period.

Igneous and volcanic petrographic evolution proceeded throughout Martian time, accompanied by moderate amounts of tectonic fracturing of the early crust (Nyquist *et al.*, 2001; Hartmann and Neukum, 2001; Head *et al.*, 2001). Much of the volcanism was early, and much of the mass and load of the Tharsis bulge were in place by the late Noachian Period, $\sim$3.6 Gyr ago, based on tectonic and geophysical mapping (Head *et al.*, 2001); though the surface flows in that region extend all the way to the last few 100 Myr or less (Hartmann and Neukum, 2001). Resurfacing data from Tanaka *et al.* (1987), combined with crater densities, suggest that volcanic resurfacing rates (km^2/yr) declined by around an order of magnitude between 3 Gyr ago and the present (Hartmann and Neukum, 2001). Low rates of volcanic outgassing are also supported by the very low radiogenic ^{40}Ar content of the Martian atmosphere (Bogard *et al.*, 2001) and a small, $\sim$10%, increase in ^{129}Xe/^{132}Xe in the last 4 Gyr.

Rock compositions range from the ultrabasic and basaltic Martian meteorites to the more andesitic chemistry observed by Pathfinder (Wänke *et al.*, 2001) and suggest a model in which both andesitic and basaltic crustal material formed, with basaltic volcanism continuing to form plains into the present or recent time. The MGS thermal emission spectrometer (TES) also mapped two basic end member rock compositions. Basalts are primarily found in the southern dark regions and andesitic rocks primarily in the northern dark regions (Bibring and Erard, 2001). Average mineralogic compositions have been estimated from the spectra, but do not quite match those of existing Mars meteorites (Bibring and Erard, 2001). Erosion and weathering, including water flow and aeolian activity, has produced the present soil, primarily a mixture of the andesitic and SNC-like basaltic rock types, plus strong components of evaporite sulfates, chlorides, and similar materials (Wänke *et al.*, 2001; Greeley *et al.*, 2001).

Isotopic composition data suggest that the early Martian atmosphere was denser than at present, though estimates show substantial variation. Bogard *et al.* (2001) review models implying early total outgassing of several bars of CO_2 and initial or early total atmospheric pressures of at least $60 - 500$ mbar or more. Chemical compositions of weathering products in Martian meteorites suggest that the ambient atmosphere at the time of weathering had a CO_2 partial pressure of at least 50 mbar, compared to $\sim$7 mbar at present (Bridges *et al.*, 2001). Even the modest carbonate content of Mars meteorites, if distributed in the upper 1 km of crust, is equivalent to 250 mbar of CO_2 atmosphere. Massive early lava outpouring associated with the early Tharsis build-up might have released a major fraction of the gas building up the putative massive early atmosphere (Head *et al.*, 2001).

A significant difference between the isotopic composition of oxygen in aqueous alteration products and the 'mass fractionation line' of igneous minerals in Martian meteorites (Bogard *et al.*, 2001) indicates that the atmosphere and hydrosphere have not undergone isotopic exchange, or only limited exchange, with the crust, e.g. by high temperature hydrothermal processes. This may reflect the low levels of volcanism on Mars and the absence of extensive hydrothermal circulation systems, as they are associated with mid-ocean ridges on Earth. Furthermore, the apparent lack of major fractionation of oxygen and carbon isotopes suggests that much of the original inventory of CO_2 may still be sequestered on Mars. In contrast, the major increase in the $^{15}N/^{14}N$ ratio over the last 4 Gyr implies that loss of nitrogen from the upper atmosphere is an ongoing process (Bogard *et al.*, 2001).

Large amounts of liquid water were involved in surface evolution during the Hesperian and perhaps Noachian Periods (Head *et al.*, 2001; Masson *et al.*, 2001; Bogard *et al.*, 2001; Encrenaz, 2001). This was probably mostly prior to 2.9 Gyr ago, but the end of the Hesperian is difficult to date because of the uncertain Mars/Moon cratering rate (Ivanov, 2001) and might, with lower probability, extend closer to 2 Gyr. The rate of fluvial and periglacial resurfacing averaged about 1 to 2 orders of magnitude greater in the first 1 Gyr than in subsequent time, based on resurfacing data (Tanaka *et al.*, 1987; Hartmann and Neukum, 2001). The presence of Noachian oceans or large exposures of water-laid sediments is still unresolved. Data from the MGS Laser Altimeter confirmed an ancient north polar basin surrounded by terraces at an equipotential level, which may be a remnant of a north polar lowland sea (Head *et al.*, 2001; Masson *et al.*, 2001). Kilometers-thick sedimentary layers, deposited on the ancient, cratered, recently exhumed floors of some ancient craters, may evidence water ponding in craters (Malin and Edgett, 2000b). These sediments may also be aeolian deposits (Greeley *et al.*, 2001), although many of them are very likely water-laid, and many of these craters have channels emptying into them (Figure 1). Exhumation and deflation processes seem to be very active in some areas of Mars today and offer opportunities to sample ancient surface units that were long hidden under masses of sediments, and only recently exposed (Greeley *et al.*, 2001). Many of the old Martian sediments appear to be weak and loosely consolidated.

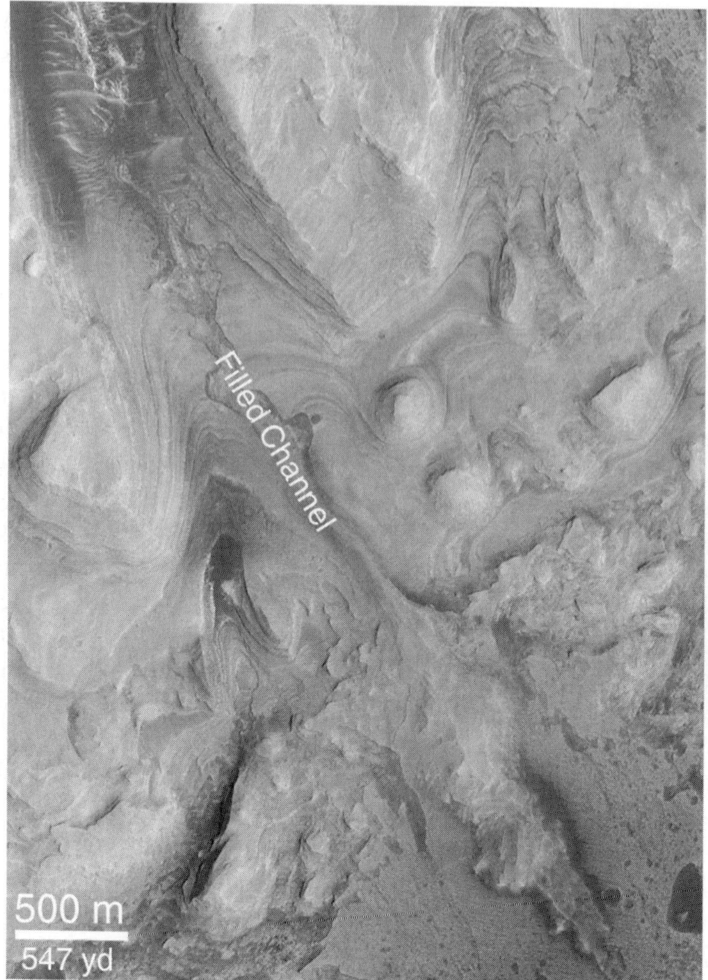

Figure 1. Stratified deposits in Gale crater symbolize the complexity of the Martian geological record. Stratified sediments (*upper left*) were deposited over an ancient cratered floor layer (*lower right*) inside Gale crater. At some point a channel cut through these sediments, but was also filled in. (This can be seen as well in other parts of the image, not shown). The sediments are currently being eroded and removed, exhuming the old surface. The youthfulness of this exhumation process is proved by the absence of young impact craters on the various surfaces. (Mars Global Surveyor MOC camera image courtesy of Malin Space Science Systems, Jet Propulsion Lab, and NASA.)

Massive volumes of carbonates were predicted to have been sequestered in the soils if long-lived early seas existed, but data from the MGS TES spectrometer did not evidence for them (Bibring and Erard, 2001). Martian sulfate chemistry may have interfered with production of the carbonates (Wänke *et al.*, 2001). Bogard *et al.* (2001) also cite the importance of sulfur chemistry in controlling carbonate formation, and discuss the need for better measurements of sulfur abundances. Alternatively, many ancient evaporite deposits may be hidden by combined effects of later sediments (Greeley *et al.*, 2001; Head *et al.*, 2001) and impact gardening (Hartmann *et al.*, 2001). Iron may be concentrated in the soils by processes that have produced patches of strong hematite concentration here and there, and most models of hematite production rely on either hydrothermally heated or cold water. The main exposed hematite concentration, in Meridiani Terra, appears to be very ancient, to have been preserved under sediments, and to have been only recently exposed by exhumation (Hartmann *et al.*, 2001); Wänke *et al.* (2001) appeal to hematite concentrations to explain Martian soil chemistry.

Various types of outflow channels and dendritic channel networks are widely accepted as establishing water flow on the surface. Although concentrated toward the beginning, channel formation and aqueous activity extended into the Amazonian period, probably including the middle or even last third of Martian history, at lower activity levels (Masson *et al.*, 2001). The dating of the outflow channels themselves remains uncertain at this writing, because (1) they have small areas, giving poor impact crater statistics, and (2) the factor 2 uncertainty in Mars impact rates (Ivanov, 2001) means that the ages are uncertain by that factor, which in turn means that middle-history model ages around 2 to 3 Gyr could, with lower probability, range from 1 to 4 Gyr – giving a poor constraint. As reviewed by Masson *et al.* (2001), most of the outflow channels and valley networks are stratigraphically associated with the late Noachian and Hesperian Periods (probably before 3 Gyr ago), but some of them extend into the Amazonian. Crater counts suggest that flow activity in certain channels extended to 1 Gyr ago, or even as recently as a few 100 Myr ago (Masson *et al.*, 2001). These dates are supported by Martian rock data showing at least brief (days) exposure to moisture as recently as 670 Myr ago (Bridges *et al.*, 2001). The Amazonian water releases, as well as the hillside gullies (Malin and Edgett, 2000b) may be associated primarily with localized volcanism, such as in Elysium Planitia (Head *et al.*, 2001), and localized mild geothermal heating that raises the bottom of the ground ice layer above $\sim -40°C$ to keep brines liquid (Hartmann, 2001). The later mild heating could produce liquid water on Mars without producing geothermal or volcanic structures. This process could also be involved in duricrust production, by cementing soils with leachates. The polar regions are also areas of continued geological processing, accumulating sediments during annual cycles of dust/ice deposition there. They show virtually uncratered, layered terrain that suggest continuing sedimentation and possible recent glacial ice flow phenomena in the polar cap regions (Head *et al.*, 2001).

Intermediate in size between the Earth and the Moon, Mars occupies a particularly interesting planetary regime. Like the Moon, its accessible surface history stretches back beyond the heavy bombardment epoch (Figure 2), but more like the Earth it has been geologically active, albeit to a much lower degree, for most or all of the intervening time. Mars will obviously continue to be a key laboratory for understanding planetary processes, including planetary and formation, crust/mantle/core evolution, basaltic volcanism, water release, and conditions that may or may not have led to biological or pre-biological chemistry.

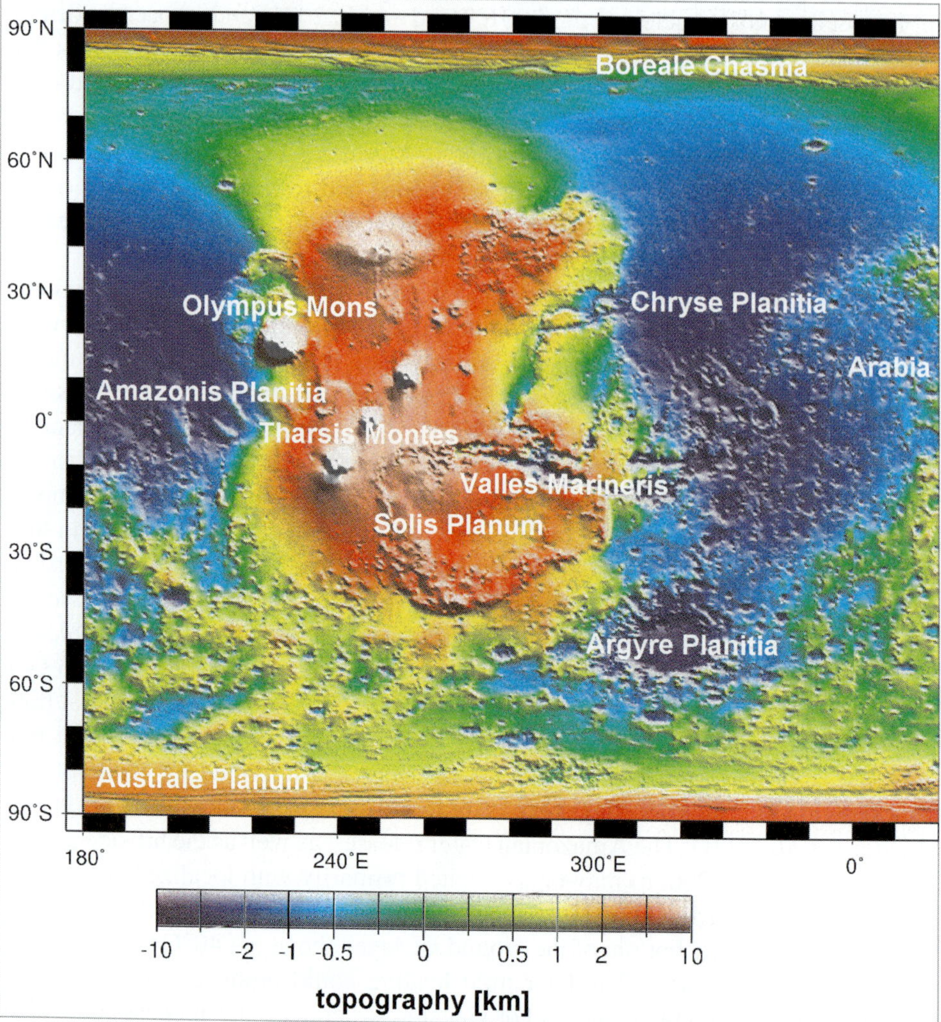

Figure 2. Topographic map of Mars, derived from MOLA (Mars Orbiter Laser Altimeter) data (Courtesy Mars Global Surveyor project, MOLA team, and NASA). The wide variation in crater density can be seen reflecting the wide range of surface ages. According to results in this volume (cf. Hartmann and Neukum, 2001; Spohn *et al.*, 2001; Head *et al.*, 2001; Bridges *et al.*, 2001; Masson *et al.*, 2001), the most heavily cratered (Noachian Period) surfaces (most of southern hemisphere plus Arabia Terra (*center-right equatorial area*) are constrained to be older than about 3.5 Gyr, while the sparsely cratered northern plains have average model ages <2 Gyr with certain areas such as Tharsis Planitia (*left center*), Elysium Planitia (*upper right*) and Amazonis Planitia (*upper right and left edges*) have individual lava flow units with model ages <100 Myr. Polar layered terrain gives even lower ages, probably due to ongoing deposition of dust trapped there seasonally. Note: MOLA topographic maps use a longitude system where longitude increases to the east. On many other maps, longitude increases to the west.

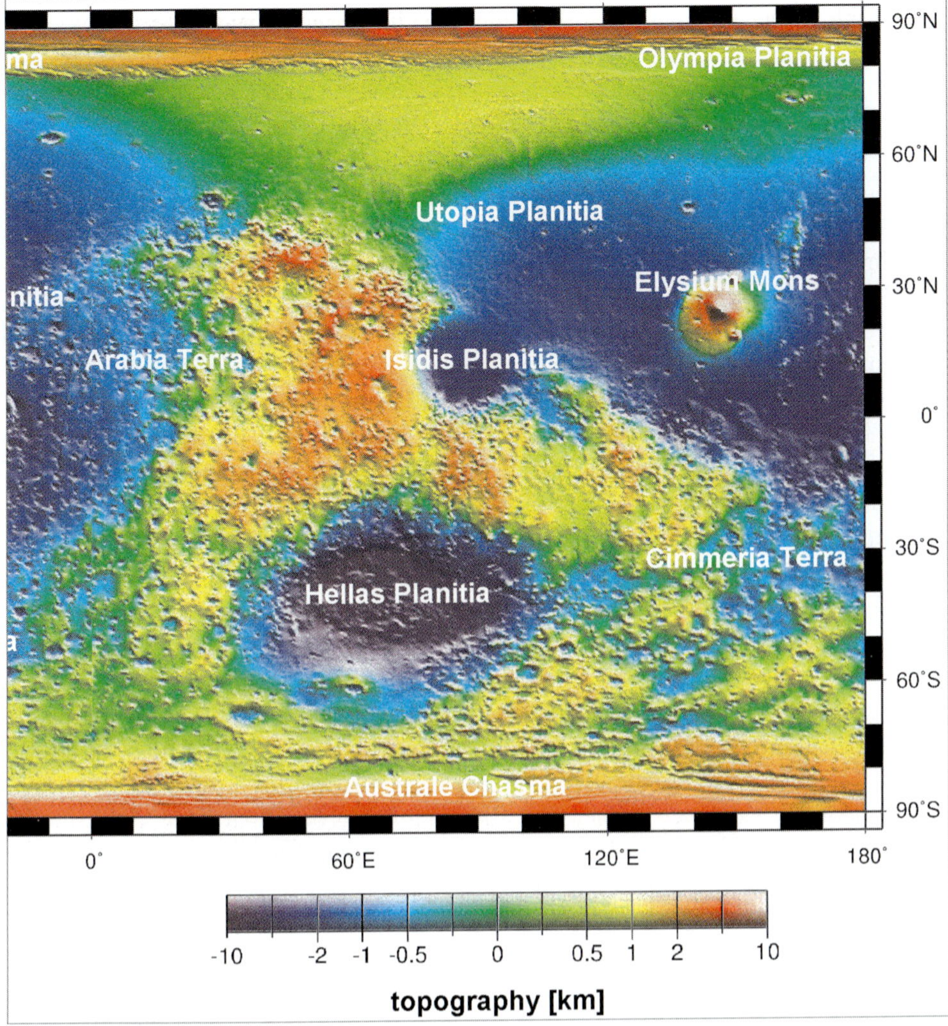

Figure 2. (continued)

Remaining Issues and Future Research

Could the emerging view of a geologically active Mars be overturned in yet another chapter of Martian exploration? Several issues have emerged that bear on this question.

1. *Radiometric Age Determinations* of Martian meteorites provide unambiguous evidence of igneous activity on Mars, within the last 1.3 Gyr to 180 Myr (Nyquist *et al.*, 2001). Since the source regions of the Martian meteorites are as yet unknown, the determination of absolute ages for cratered surface units remains a problem. Remote determination of approximate K-Ar ages of basalt flow units or cosmic ray exposure ages of outflow channels may provide a

rough answer, but it seems likely that the complexities of isotope systems will require sample return before a definitive calibration is made. If one or two young homogeneous units (lava flows?) could be sampled in such a way that the unit's age could be unambiguously determined, those units could be used to calibrate the crater count dating techniques of the other units of Mars.

Our existing knowledge on the absolute chronology of Mars has many limitations not only due to the Martian data, but also from the lunar data. The lunar crater retention age curve is well calibrated only for the time period from 3.9 to 3.2 Gyr ago. During this period we have precise radiometric ages of the mare basalts and crater counts for the corresponding mare areas. However, our estimate of the time dependence of cratering in the first 600 Myr depends strongly on our understanding of the lunar basins and upland cratering (Stöffler and Ryder, 2001; Neukum *et al.*, 2001). Similarly, the precise dating of the youngest lunar features, and hence the recent time dependence of cratering, is somewhat controversial. Great advances could be made if we had absolute dating from certain proposed "youngest" mare lava flows in Oceanus Procellarum and major craters such as Copernicus or Tycho.

2. *Cratering Chronologies.* The existence of young lava flows on Mars is also clear from crater-retention age determinations, made by measuring Martian impact crater densities and calibrating their ages with lunar radiometric ages, as discussed in the papers by Neukum *et al.* (2001), Ivanov (2001), and Hartmann and Neukum (2001). In spite of the factor two uncertainty between Martian and lunar cratering rates, the crater retention ages still offer a powerful confirmation of the existence of geologically young Martian lava flows, with model ages of the order 100 Myr and even less. To claim much older ages for the Martian features would require the Martian cratering rate to be much lower than currently estimated (cf. Ivanov, 2001).

3. *Atmospheric Evolution.* The data suggesting early high pressures are very diverse, and need to be reconciled into a firmer picture of surface pressure versus time. A better understanding of atmospheric conditions in the Noachian and Hesperian would help constrain the presently wide-ranging models of early oceans, hydrologic cycles, glaciation, outflow channels and valley networks, underground aquifers, etc. In particular, it would be valuable to have any data that could support, define, or refute short-lived episodes of enhanced atmospheric pressure. These might have occurred due to volcanic eruptions or obliquity cycles, and they could have produced transient greenhouse effects and benign periods interspersed with frozen periods. Isotopic evidence (H, C, N, O and the noble gases) from a wider range of crustal rocks and secondary minerals will be required to obtain a clearer picture of the time evolution of the atmosphere-hydrosphere-crust system on Mars. The origin of the volatile inventory of Mars will continue to be a topic for debate.

4. *Aqueous Evolution.* All evidence of aqueous activity is of high importance not only because of implications for climatic evolution but also because of implica-

tions for possible biological habitats and activity. Papers in this book generally support extensive early aqueous activity on Mars, including extensive, but perhaps short lived, bodies of water. It is clear that we need better understanding of features like the young-looking Malin/Edgett gullies, sedimentary layers, possible shorelines, and possible ocean basins (Head *et al.*, 1999, 2001; Masson *et al.*, 2001). However, the chronology of aqueous alteration products, and by inference aqueous activity on Mars, is at a very preliminary stage, and there is a need to extend this work.

5. *Crustal Evolution.* A puzzle persists that the Mars meteorite collection does not include more rocks of moderately old age, 1.5 to 4.0 Gyr, from the cratered Martian uplands. It is possible that the uplands are so fragmental that they do not easily produce meteorites, or that they produce rocks not recognized as meteorites. Wänke *et al.*, for example, regard one of the "rocks" at the Pathfinder site as a mass of "cemented soil." Intense cratering of the Mars uplands in early time not only broke through early magnetized terrain (Spohn *et al.*, 2001) but also may have pulverized hundreds of meters of early upland, creating a megaregolith that was the ideal medium for water to enter, perhaps cementing the loose materials with carbonates, salts, and other evaporates, to create weak, duricrust-like materials that can't launch meteorites. Martian meteorites, therefore, may be samples only from regions where coherent igneous near-surface rocks allow launching of meteorites, while sampling and recognition of weaker, loosely cemented sediments is inhibited. Identification of the rock types in the uplands is thus a major remaining issue in understanding Martian evolution.

6. *Martian Meteorites.* One of the strongest arguments for the derivation of the SNC meteorites from Mars was the similarity in composition of trapped gases in EET79001 and Viking analyses. Atmospheric and crustal components have now been identified in most of the SNCs but the latest data show significant differences from the Viking measurements. However the evidence that the SNCs are from Mars is now more broadly based and the assumption is that these differences arise from uncertainties in the Viking analyses. Nevertheless there is a need to obtain definitive in-situ analyses which will also provide a baseline for interpreting isotopic fractionation of carbon and oxygen.

References

Bibring, J.-P., and Erard, S.: 2001, *Space Sci. Rev.*, this volume.

Bogard, D.D., Clayton, R., Marti, K., Owen, T., and Turner, G.: 2001, *Space Sci. Rev.*, this volume.

Bridges, J.C., Catling, C.C., Saxton, J.M., Swindle, T.D., Lyon, I.C., and Grady, M.M.: 2001, *Space Sci. Rev.*, this volume.

Encrenaz, T.: 2001, *Space Sci. Rev.*, this volume.

Greeley, R., Kuzmin, R.O., and Haberle, R.M: 2001, *Space Sci. Rev.*, this volume.

Halliday, A.N., Wänke, H., Birck, J.-L., and Clayton, R.N.: 2001, *Space Sci. Rev.*, this volume.

Hartmann, W.K.:: 2001, *Space Sci. Rev.*, this volume.

Hartmann, W.K., Anguita, J., de la Casa, M., Berman, D.C., and Ryan, E.: 2001, 'Martian Cratering VI. The Role of Impact Gardening', *Icarus* **149**, 37–53.

Hartmann, W.K., and Neukum, G.: 2001, *Space Sci. Rev.*, this volume.

Head, J.W., Hiesinger, H., Ivanov, B., Kreslavsky, M., Pratt, S., and Thomson, B.: 1999, 'Possible Ancient Oceans on Mars: Evidence from Mars Orbiter Laser Altimeter Data', *Science* **286**, 2134–2137.

Head, J.W., *et al.*: 2001, *Space Sci. Rev.*, this volume.

Ivanov, B.A.: 2001, *Space Sci. Rev.*, this volume.

Kieffer, H.H., Jakosky, B.M., Snyder, C.W., and Matthews M.S. (eds.): 1992, *Mars*, Univ. Arizona Press, Tucson.

Malin, M.C., and Edgett, K.S. : 2000a, 'Evidence for Recent Groundwater Seepage and Surface Runoff on Mars', *Science* **288**, 2330–2335.

Malin, M.C., and Edgett, K.S.: 2000b, 'Sedimentary Rocks of Early Mars', *Science* **290**, 1927–1937.

Masson, P., Carr, M.H., Costard, F., Greeley, R., Hauber, E., and Jaumann, R.: 2001, *Space Sci. Rev.*, this volume.

Neukum, G., Ivanov, B.A., and Hartmann, W.K.: 2001, *Space Sci. Rev.*, this volume.

Nyquist, L.E., Bogard, D., Shih, C.-Y., Greshake, A., Stöffler, D., and Eugster, O.: 2001, *Space Sci. Rev.*, this volume.

Spohn, T., Acuña, M., Breuer, D., Golombek, M., Greeley, R., Halliday, A., Hauber, E., Jaumann, R., and Sohl, F.: 2001, *Space Sci. Rev.*, this volume.

Stöffler, D. and Ryder, G.: 2001, *Space Sci. Rev.*, this volume.

Tanaka, K.L., Isbell, N., Scott, D., Greeley, R., and Guest, J.: 1987, 'The Resurfacing History of Mars — A Synthesis of Digitized, Viking-based Geology,' *Proc. 18th Lunar Planet. Sci. Conf.*, Cambridge University Press/LPI, 1988, Cambridge and New York/Houston TX, USA, pp. 665–678.

Wänke, H., Brückner, J., Dreibus, G., Rieder, R., and Ryabchikov, I.: 2001, *Space Sci. Rev.*, this volume.

Address for correspondence: William K. Hartmann, Planetary Science Institute, 620 N. 6th Avenue, Tucson AZ 85705-8331, USA; (hartmann@psi.edu)

GLOSSARY

Geophysical, geological, petrological, and mineralogical terms used in planetary science, in particular concerning research in this book on the chronology and evolution of the Mars and the Moon, are defined.

1. Mars, Moon, and Planetary Bodies, Their Structure and Geological Units

Amazonian Period: youngest of the 3 Martian stratigraphic periods *Noachian, Hesperian, Amazonian*, defined by Tanaka (1986) in terms of crater density.

Aquifer: water-carrying rock formation below a planet's surface.

Asteroids: rocky, metallic, 1 cm – 1000 km-sized objects, orbiting the Sun, mostly in the Asteroid Main Belt between the orbits of Mars and Jupiter; consist of pristine solar system material; most likely bodies that never coalesced into a planet.

Asthenosphere: plastic zone under the *lithosphere*. In the case of Earth, this is broadly equivalent to the upper *mantle*.

Caldera: a volcanic depression usually rougher circular or oval, caused by collapse and often found on summits of volcanos.

Canal: term introduced as "canali" by Schiaparelli; Lowell ($\sim$1890) charted these hypothetical linear features on Mars, which he believed to be constructed by Martians. Not confirmed by later studies but some are associated with wind streaks and canyons; not to be confused with "channel."

Chaotic Terrain: distinctive fractured surface area on Mars usually at the head of large channels; collapsed to $\sim$1 – 2 km below the surrounding terrain presumably due to rapid release of melted subsurface ice.

Channel: apparent dry riverbeds on Mars, discovered by Mariner 9 in 1972.

Core: central portion of a differentiated planet chiefly consisting of Fe and Ni.

Crater: bowl-shaped depression with a raised rim, formed by a *meteoroid* impact (*impact or meteorite crater*); not applied to *volcanic* features in modern usage.

Crust: outer layer of a terrestrial planet.

Dentritic Network: same as *valley network*.

Dorsum: ridge (Latin).

Dunes: mound, ridge, or hill of wind-blown sand.

Duricrust: platy Martian *soil*, apparently cemented by evaporites.

Endogenic: arising from the interior of the Earth (opposed to *exogenic*).

Eskers: bouldery ridges of sediment left by melting glaciers at their margins; meandering sand lines and gravel deposited by streams beneath glaciers.

Fissure: crack extending far into a planet through which *magma* erupts.

Flow-Lobes: lobe-shaped feature in the flow field of *volcanoes*, generated by inflation of the *lava crust* as a result of magmatic overpressure in the associated lava core; larger lobate features are also found in some Martian *craters' ejecta blankets*, thought to originate by interaction with *ground ice*.

Space Science Reviews **96**: 471-482, 2001.

Fluvial Channel: the conduit through which a river flows.

Fluvial Valley: linear depression containing many channels.

Fossa: ditch (Latin); long, narrow, shallow depressions.

Fretted Channel: channel in *fretted terrain* with steep walls and a flat floor.

Fretted Terrain: fractured Martian surface area; eroded by landslides, volcanic flooding, or ice-rich debris flowing off walls of faulted valleys.

Frost Mound: area of water frost layer.

Graben: elongate, depressed crustal unit, bounded by faults on its long sides.

Hellas: one of the largest impact basins on Mars, dominating structure and flow channel orientation over a large part of the southern hemisphere.

Hesperian Period: middle of the 3 Martian stratigraphic periods defined by Tanaka (1986) in terms of crater density.

Highlands: oldest *cratered* lunar surface areas; chemically distinct from the *mare*.

Ignimbrites: deposits of *pyroclastic* flows of ash, gas and larger particles.

Lahar: Indonesian word referring to mudflows that are volcanic in origin.

Lineated Valley Fill: material in steep-walled Martian valleys, comprising their floors which have lineations (ridges and grooves); resemble alpine glaciers.

Lithosphere: rigid outer rind of a planet (*crust* and upper *mantle*), as distinct from the underlying, more fluid, *asthenosphere*.

Longitudinal Grooves: due to differential shear and lateral spreading at high velocities of landslides between mounds of interior layered deposits.

Magma Ocean: *magma* layer, covering an early planet; term derived from evidence for same on the Moon 4.5 Gyr ago.

Mantle: portion of a planet between *crust* and *core*; $\sim$2900 km thick on Earth.

Maria: dark lunar surface area; originally thought to be seas by classical observers; composed of *basaltic lava* plains; chemically distinct from the *highlands*.

Martian Meteorites: also called *SNC meteorites*; include the 3 *meteorite* classes named after the location of the fall of a prominent member of each class: Shergottites, Nakhlites, Chassignites; identified to originate from Mars because of their young ages, *basaltic* composition, and inclusion of Martian atmospheric gas.

Meteoroids: small interplanetary debris, from *asteroids* and comets.

Meteor: a *meteoroid* as it enters the atmosphere at speeds of $15 - 70$ km/s.

Meteorites: solid objects striking a planet's surface, categorized as *stony, iron,* and *stony-iron*; mainly of *asteroidal* origin, a few from the Mars and the Moon.

Möberg Mountain: see *Table Mountain*.

Noachian: oldest of the 3 Martian stratigraphic periods defined by Tanaka (1986) in terms of crater density.

Nues Ardentes: pyroclastic flow that traveled into water, separating gravitationally into a lower part containing most of the solid fragmental mass.

Obliquity: tilt of a planet's axis from perpendicular to the plane of the ecliptic.

Outflow Channel: Martian *channel* eroded by a fluid, apparently water; often starts full size in *chaotic terrain*; few, if any, tributaries; some 10 km wide and up to

2000 km long; distinct flow features such as eroded craters with teardrop-shaped tails, scour marks, "islands".

Palsas: mound of peat resulting from the formation of a number ice lenses beneath the ground surface; $2 - 30$ m wide and $1 - 10$ m high; similar to a *pingo*.

Patera: ancient Martian *volcano*; often with the semi-circular *Caldera* crater at the volcano summit ($\sim$80 km across).

Periglacial Landforms: topographic features along the margins of glaciers.

Pingo: large conical mound containing an ice core; up to $60 - 70$ m high; form in regions of permafrost; similar to *palsa*.

Planetesimals: 1 m to 100 km-sized bodies that mostly accreted to planets.

Planitia: plain (Latin); smooth low area.

Plate Tectonics: tectonic activity associated with the breakup of the *lithosphere* into moving pieces, or "plates".

Polygon: lineations, arranged in a polygonal pattern that outline flat-lying areas, scale is up to $5 - 20$ km across.

Pressure Ridges: long, narrow wavelike folds in the surface of *lava* flows.

Rampart Crater: *crater* in which the *ejecta blanket* is thick and appears to have formed from slurry due to impact into ice-rich material.

Ray: streak of material blasted out and away from an *impact crater*.

Ridged Plains: formed by crust welling up between separating plates.

Rille: long channel in lunar *maria* formed by open or underground *lava* flow.

Rock Glaciers: flowing frozen debris whose surface layer thaws in summer.

Runoff Channel: flow of water from precipitation to stream channels when the infiltration capacity of an area's soil has been exceeded.

Rupes: scarp (Latin).

Scabland Topography: created by catastrophic glacial outburst floods.

Sediment: solid rock or *mineral* fragments deposited by wind, water, gravity, or ice; precipitated by chemical reactions; accumulated as loose layers.

SNC Meteorites: see *Martian Meteorites*.

Soil: upper layers of *sediment*.

Source: location where *igneous* matter (*lava* and gases) erupts onto the surface.

Table or Möberg Mountain: special type of *volcano* (e.g. in Iceland) produced by subglacial eruptions; the discharge of meltwater is called *jökulhaup*.

Talik: island of thawed ground surrounded by permafrost.

Terrain: area of the surface with a distinctive geological character.

Tharsis: broad, domed volcanic plains of Mars dominating much of one hemisphere; major volcanoes are Olympus, Arsia, Pavonis, Ascraeus Mons.

Tidal Stress: differential gravitational force per unit area acting on a planetary body by the Sun, a moon, or another planet, resulting in periodic bulging.

Tuyas: subglacial *volcanic* deposits from eruptions beneath glacial ice.

Valles Marineris: major Martian canyon complex; similar in scale to the Red Sea.

Valley : valleys on Mars have not been filled by water to their rim and usually no river bed is observed inside them; may have formed by sapping processes.

Valley Networks: valley systems, often with complex, multiply-branched patterns of tributaries; mostly in ancient upland areas of Mars.
Vallis: sinuous valley (Latin).
Vent: opening in a planet's surface ejecting *lava*, gases, and hot particles.
Ventifact: individual pieces and pebbles etched and smoothed by *aeolian* erosion i.e. *abrasion* by wind-blown particles.
Vesicle: bubble-shaped cavity in a *volcanic* rock formed by expanding gases.
Volcano: mountain formed from the eruption of *igneous* matter.
Yardang: *aeolian* erosional rock feature formed by *deflation* and *abrasion*; elongated and aligned with the most effective wind direction.

2. Geological Processes

Abrasion: process of wearing down or rubbing away by means of mechanical friction by particles in ice, water or wind.
Aeolian Processes: wind processes; from Greek Aeolus.
Creep, Frost Creep: slow downslope movement of particles on slopes covered with loose, weathered material; directly by gravitation or by frost heaving (interstitial water freezes, surface particles are forced out and let down by melting).
Debris Flow: mixture of water-saturated rock debris flow (*lahar* or *mudflow*).
Deflation: process of wind erosion that removes and lifts individual particles, blowing away unconsolidated, dry, or noncohesive sediments.
Exhumation: Exposure of underlying terrain by removal of overburden.
Gelifluction: soil flow in *periglacial* environments; progressive, lateral movement.
General Circulation Model (GCM): time-dependent 3D model of a planet's neutral atmosphere; finite difference solution of the nonlinear equations of hydro- and thermodynamics, and coupling between dynamics and composition.
Ground Ice: subsurface water that is frozen in regions of permafrost; moisture content may vary from nearly absent to almost 100%; may be released under great pressure and flowed away so that the overlying surface collapses (*chaotic terrain*).
Hydrologic Cycle: includes evaporation, condensation, infiltration, precipitation, transpiration, runoff, groundwater on a planet.
Jökulhaup: discharge of meltwater caused by a subglacial *volcanic* eruption.
Mass Wasting: downslope transport of (*weathered*) rock, water, or soil by gravity.
Phreatic Explosion, Eruption: volcanic eruption or explosion of steam, mud, or other material that is not incandescent; caused by the heating and consequent expansion of ground water due to an adjacent *igneous* heat source.
Plinian Eruption: sustained explosive volcano eruptions which generate high-altitude eruption columns and blanket large areas with ash; after the eruption of Mount Vesuvius described by Pliny in 79 AD.
Plucking: erosive process of particle detachment by moving glacial ice when basal ice freezes in rock surface cracks.

Pyroclastic Eruption: explosive eruption of *lava* and rocks.

Saltation: transport of sand grains (usually larger than 0.2 mm) by stream or wind, bouncing the grains along the ground in asymmetrical paths.

Suspension: fine particles of given size are held in suspension in a stream of given density if the wind is faster than a threshold velocity, the wind friction velocity; finest particles are not deposited until the stream velocity nears zero.

Tectonics: fracturing and movement of large-scale rock masses in response to compressional, tensional, or shear forces causing faulting and folding (can create mountain ranges, rift zones, faults, fractured rock, and folded rock masses).

Terrain Softening: reduction in topographic relief due to the downslope movement of surficial material, probably involving ice or ground water.

Thermokarst: feature of glacial landscape; forms when the ice contained at shallow levels melts and the ground collapses.

3. Impact Craters

Cataclysm, Terminal Cataclysm: hypothetical shortlived episode of intense bombardment of the Moon, and possibly all planets, about $3.9 - 4.0$ Gyr ago; the term was introduced by Wasserburg *et al.* in the early 1970s when rock sample ages in early Apollo missions showed a strong cutoff around this time; initially, it was visualized as a very short-lived event; recent discussions imply a duration of $\sim$200 Myr (Stöffler and Ryder, 2001); also variously called early/late heavy bombardment, early intense cratering, etc. Its nature is controversial.

Central Peaks: peaks in the center of large impact craters, associated with rebound during collapse of the transient cavity.

Complex Crater: impact crater with central peak structure; generally larger than *simple craters* or larger than about $5 - 10$ km, but the transition size depends inversely on surface gravity and varies from planet to planet.

Concentric Crater Fill: multiridged intracrater deposit; possibly due to downward flow of materials from crater walls in higher latitudes on Mars.

Crater Isochron: the size distribution of craters formed on a surface in a specified period of time. In the absence of erosive losses or other effects that alter the size distribution, the observed size distribution after time T should equal the isochron for time T.

Crater Production Function: the shape of the size distribution of craters that would be formed on a surface above the atmosphere of a planet, *i.e.* in the absence of erosive or other loss effects.

Crater Retention Age: the duration of time during which impact craters of specified diameter remain visible on the surface; in the case of a homogenous surface on an atmosphereless body with no erosion and deposition, *e.g.* a *lava* flow, this may be the actual age of the unit; under erosion/deposition conditions, it may reflect the rate of obliterative processes; coined by Hartmann (1966).

Crater Saturation Equilibrium: steady state curve reached (in the absence of other geological processes) as time passes and crater density increases to the point where, in effect, new craters obliterate old craters; the observed crater density then probably fluctuates around the saturation equilibrium density, depending on statistics and locations of large impact basins and their *ejecta blankets*.

Ejecta, Ejecta Blankets or Deposits: the material thrown out from the *crater* during the *impact* that formed it.

Gardening: see *impact gardening*.

Impact Gardening: fragmentation of planetary surface layers by meteoritic impact at all sizes (or down to a cutoff size imposed by the atmosphere.)

Lobate Ejecta: *ejecta blankets* that surround a crater in the form of lobes, probably associated with impact into *ground ice*.

Megaregolith: extension of the term *regolith* (coined by Hartmann, 1973); hundreds of meters to few km deep; developed by *saturation equlibrium* among multi-kilometer scale *craters* with a size distribution such that, as *crater* time passes and density increases, craters of $D > 2$ km all begin to saturate at about the same time, producing a very rapid increase in *gardening* depth.

Peak Ring: ring of peaks on a crater floor; intermediate stage for increasing crater size (~ 100 km) in the progression from *central peaks* to multiple ring basins.

Pedestal Crater: impact crater with *ejecta blanket* standing above surrounding *terrain*.

Production Function: see *crater production function*.

Pseudo Crater: created by localized *phreatic* explosions where *lava* interacts with volatile-rich ground; note the definition of the term *crater* as *impact crater*.

Ray: streak of material blasted out and away from an *impact crater*.

Regolith: layer of loose, pulverized debris (unconsolidated rock, *minerals, glass* fragments) created on the surface of an airless or nearly airless body by *impact gardening*; measured in the lunar *mare lavas* to be $\sim 5 - 20$ m deep, grading into coarser broken materials; term probably introduced by Shoemaker after Ranger probes in the mid 1960s showed rolling terrain saturated with degraded *craters*.

Resonances: gravitational relationships, mostly with Jupiter, that force orbits of *asteroids* to change, usually toward larger eccentricity; mechanism that ejects a continuing flux of *asteroids* from the main belt onto orbits that cross those of Mars and/or the moon, creating (along with comets) the *impact crater*ing rates on those planets. Simple resonances have integer ratios (such as 2:1 and 3:2) between the *asteroid*'s and Jupiter's orbital periods.

Saturation Equilibrium: see *crater saturation equilibrium*.

Shock Wave: powerful compressional wave puls; e.g., created by a *meteorite* impact on a planetary surface at a velocity above the speed of sound in rock.

Simple Crater: bowl-shaped impact crater with no central peak; generally smaller than *complex craters* or smaller than a few km, but the transition size depends inversely on surface gravity and varies from planet to planet.

4. Meteorites

Achondrites: *meteorites* formed on *differentiated* bodies, where the complete melting removed all evidence of *chondrules*; sometimes difficult to identify because they resemble terrestrial *igneous* rocks; found in several rare types: Howardites, Eucrites, Diogenites from the *asteroid* Vesta, *SNC meteorites* from Mars, Aubrites, Ureilites, and primitive achondrites (for the latter see, *e. g.*, Graham *et al.*, 1985).

Carbonaceous (C) Chondrites: rare and primitive – though complex – *meteorites* containing organic compounds and water-bearing *minerals* that are evidence for the inclusion of water not long after formation. Chemical and *mineralogical* differences are named for the type specimens Ivuna (CI), Mighei (CM), Vigarano (CV), Ornans (CO), Karoonda (CK), and Renazzo (CR). The petrologic grade designations of C chondrites indicate increasing *metamorphism* of the *meteorite* by water (grades 1 and 2) or by heat (grades 3 to 6).

Chondrites: class of primitive *stony meteorites* containing *chondrules* inside their primary *matrix*; formed shortly after the Sun's birth (~4.56 Gyr ago); sub-classes are *ordinary*, *carbonaceous*, *enstatite*, and Rumuruti chondrites.

Chondrules: small glassy spherical inclusions formed in the solar nebula.

Enstatite (E) Chondrites: rare *chondrites* rich in *enstatite*; most of their Fe is in the form of metal or sulfide, rather than oxides in *silicates*, implying that they originate from an oxygen-poor region of the solar nebula, possibly inside the present orbit of Mars; further classified by petrologic grades 3 to 6 and according to their Fe content (EH indicates approximately 30% Fe and EL about 25% Fe).

Iron Meteorites: represent only ~5% of *meteorite* falls; some are thought to be pieces of the shattered *cores* of *differentiated asteroids*; structure variations result from the ratio of the two Ni-Fe alloys, kamacite (~27–65% Ni) and taenite ($\lesssim$7.5% Ni), that have crystallized forming the *core* of the parent *asteroids*.

Mesosiderites: a broad class of *stony-iron meteorites*, surface *regolith*, fragment mixture (*breccia*) of mantle rock and Ni-Fe alloy; much more uniform metal compositions than *iron meteorites*; stirred up and fused by repeated impacts.

Ordinary Chondrites: the most common type of *meteorite* fall on Earth; further classified by their Fe content: H denotes an Fe content of ~27% by weight, L corresponds to ~23% of Fe, and LL means "low iron" and "low metal" (~20% Fe); numbers 3 − 7 following the H, L, and LL classifications are petrologic grades indicating the degree of *chondrule* alteration by heating.

Pallasites: a class of *stony-iron meteorites*; their metallic *matrix* contains large green *olivine* crystals; probably formed at *asteroids' core-mantle* boundary.

Stony-iron Meteorites: have both stone and metal fractions; two main groups: *mesosiderites* and *pallasites*.

5. Mineralogical Terms

Agglutinates: *glass*-bonded rock, *mineral*, and *glass* fragments in lunar *sediment*.
Albite: sodium end member of the *plagioclase feldspar* group; $Na(AlSi_3O_8)$.
Amphibole: group of ferromagnesian *silicates* with a double chain (single chain: see *pyroxene*) of silicon-oxygen tetrahedra.
Andesite: medium grey rock in *lava* flows; fine-grained *plagioclase* with $52-63\%$ *silica* and smaller amounts of *biotite, hornblende,* and *augite* as small *phenocrysts*.
Anhydrite: calcium sulfate, $CaSO_4$; finely granular *mineral* found as thick beds or thinly laminated; hydrates to swell to *gypsum*; strong *weathering*.
Anorthite: calcium end member of the *plagioclase feldspar* group; $Ca(Al_2Si_2O_8)$.
Anorthosite: $>90\%$ *anorthite*, balanced by *pyroxene* and *olivine*.
Augite: *silicate mineral*, chiefly of Ca, Mg, Fe, and Al; dark-green to black variety of monoclinic *pyroxene*; characteristic of basic rocks.
Autochthonous: formed in the region where found (contrary: *allochthonous*).
Basalt: fine-grained, dark-colored *igneous* rock; contains $48-52\%$ *silica*; chiefly composed of *plagioclase, feldspar, pyroxene,* but also of other *minerals*, e.g. *olivine* and *ilmenite*; most common *volcanic* rock on the terrestrial planets.
Bedrock: solid rock underlying the *soil* and *regolith* or exposed at the surface.
Biotite: $K(Fe^{2+},Mg)_3(AlSi_3O_{10})(OH,F)_2$; *igneous rock* of the *mica* group.
Boninite: group of high MgO ($>$ 6 wt% MgO) *andesites*.
Breccia: rock consisting of broken fragments of rock (*clasts*) cemented together by a fine-grained *matrix*; formed from *regolith* during subsequent impacts.
Bronzite: a *pyroxene* with about 20% $FeSiO_3$.
Calcite: calcium carbonate, $CaCO_3$; major constituent of limestone.
Clast: rock fragment within a *breccia*.
Clay Minerals: finely crystalline, hydrous *silicates*; *weathering* or hydrothermal alteration product of, e.g., *feldspar, pyroxene,* and *amphibole*.
Dacite: light-gray *igneous* rock from *magmas* erupting at $\sim800-1000°C$; contains $63-68\%$ *silica*; composed of *plagioclase, pyroxene, amphibole* and iron oxide.
Diaplectic Glass: natural glass formed by shock pressure from any of several minerals without melting.
Diorite: speckled black and white, equigranular or porphyritic rock; mainly *plagioclase* and *hornblende*, also may contain *biotite* and *pyroxene*.
Dunite: coarse-grained *igneous* rock composed almost entirely of *olivine*.
Diopside: $CaMgSi_2O_6$; found e.g. in metamorphosed siliceous limestones.
Eclogite: extremely high-grade metamorphic *ultramafic* rock containing *garnet* and *(clino-) pyroxene*; high pressure form of mid-ocean ridge *basalts* and *picrite*.
Enstatite: belonging to the *pyroxene* group; $MgSiO_3$.
Epsomite: $MgSO_47H_2O$.
Extrusion: *igneous* rocks erupted on the surface.

Feldspar: alumino*silicate mineral*; positive ions of K, Na, and Ca fill interstices of negatively charged framework of silicon-oxygen and aluminum-oxygen tetrahedra; depending on composition monoclinic or triclinic.

Ferrosilite: belonging to the *pyroxene* group; $FeSiO_3$.

Ferrihydrite: $Fe_5O_7OH.H_2O$.

Gabbro: coarse-grained *igneous* rock rich in *olivine, pyroxene,* and *plagioclase*.

Garnet: high-pressure, deep-seated *mineral* in *eclogite*; $A_3B_2(SiO_4)_3$ where A = Ca, Mg, Fe^{+2}, and Mn^{+2}, and B = Al, Fe^{+3}, Mn^{+3}, V^{+3} and Cr.

Gibbsite: mineral formed by *weathering* of *igneous* rocks; $Al(OH)_3$.

Glass: fused from dissolved *silica* and *silicates* that also contain soda and lime.

Goethite: $FeO(OH)$.

Granite: coarse-grained *igneous* rock composed of *orthoclase, albite, quartz,* and lesser amounts of *mica, hornblende,* or *augite*; mean *silica* content is 72%.

Granulite: metamorphic rock composed of granular *minerals* of uniform size, as *quartz, feldspar,* or *pyroxene,* and showing a definite banding.

Gypsum: hydrous calcium sulfate, usually $CaSO_42H_2O$, but may occur in many different mineralogical forms depending on the degree of hydration.

Halite: NaCl.

Harzburgite: ultra*mafic* rock, mainly composed of *olivine* and *orthopyroxene*.

Hematite: Fe_2O_3; most abundant ore of iron on Earth, found as sublimation product from volcanic activity or in regionally metamorphosed rocks; only *mineral* found in a strong regional concentration on Mars by MGS/TES.

Hornblende: dark-green to black *mineral* of the *amphibole* group.

Hypersthene: a *pyroxene* with $22 - 30\%$ $FeSiO_3$.

Iddingsite: *weathered olivine*; cracks may be filled with *serpentine* or *goethite*.

Igneous: involving the solidification of hot, molten *magma* or *lava*.

Ilmenite: $FeTiO_3$; black; hexagonal crystal.

Impact Melt: rock that melts during *impact crater* formation.

Intrusion: *igneous* rocks, forced as *magma* into other rock formations underground.

Kaolinite: a *clay mineral*, $Al_2Si_2O_5(OH)_4$, formed by hydrothermal alteration or *weathering* of aluminosilicates, especially *feldspars*.

Labradorite: variety of *plagioclase feldspar*.

Lava: fluid *magma* that flows onto the surface of a planet or the Moon and the rock formed by its solidification; erupts from a *volcano* or *fissure*.

Lherzolite: ultra*mafic* rock composed of *olivine, pyroxene, spinel, plagioclase*.

Mafic: of or pertaining to rocks rich in dark, ferromagnesian *minerals*.

Magma: molten rock in a planet's interior; called *lava*, when it reaches the surface.

Magnesite: $MgCO_3$.

Magnetite: Fe_3O_4; member of the *spinel* group.

Maskelynite: diaplectic *plagioclase* glass formed by a solid-state transformation during a large *impact* event.

Matrix: solid matter in which a fossil, crystal, or rock fragment is embedded.

Metamorphic: pertaining to rocks that have recrystallized in a solid state as a result of changes in temperature, pressure, and chemical environment.

Mica: hydrous *silicates* of Al with various bases (K, Mg, Fe, and Li).

Mineral: inorganic substance of definite chemical composition and usually of definite crystal structure; some rocks consist of only one type of *mineral* such as *quartz* or *feldspar*, most rocks contain several *minerals*, however.

Montmorillonite: synonymous with *Smectite*.

Monzodiorites: heterogeneous mass of *monzonite* and *diorite*.

Monzonite: *intrusive igneous* rock composed of *plagioclase* and potash *feldspar*, and less amounts of *biotite*, *hornblende*, and sometimes *ortho-pyroxene*.

Norite: *igneous* rock of the *highlands* composed of *plagioclase* and *pyroxene*.

Obsidian: volcanic silica glass, SiO_2; glassy equivalent of *granite*.

Olivine: rock-forming *igneous silicate mineral*; $(Mg,Fe)_2SiO_4$.

Ophitic: rock *texture* in which lath-shaped *plagioclase* crystals are enclosed wholly or in part in later-formed *augite*.

Orthoclase: also called potassium *feldspar*, $KAlSi_3O_8$.

Palagonite: poorly crystalline ash, partly hydrated alterated *volcanic glass*.

Peridotite: coarse-grained *igneous* rock consisting largely of *olivine*.

Perovskite: yellow, brown, or grayish-black *ultramafic* mineral.

Phenocryst: any of the conspicuous coarse crystals in a *porphyritic* rock.

Phyllosilicates: minerals such as clay or micas which contain structural OH and whose Si atoms are arranged in a sheet structure.

Picrite: relatively rare, *olivine*-rich rock in high MgO *magma*.

Plagioclase: light-colored *feldspar* ranging from $NaAlSi_3O_8$ to $CaAl_2 Si_2O_8$.

Plutonic Rock: coarse-grained *intrusive igneous* rock that cooled slowly at depth.

Poikilitic: see *texture*.

Polymict: pertaining to multiple rock-types combined in a single rock.

Polymorph: *mineral* with the same composition as another, but with a different crystal structure; e.g., diamond and graphite are well-known carbon polymorphs.

Porosity: percentage of the total rock or *soil* volume that consists of open spaces.

Porphyry: *igneous* rock containing *phenocrysts* in a finer-grained ground mass.

Pyroxene: group of ferromagnesian *silicates* with a single chain (double chain: see *amphibole*) of silicon-oxygen tetrahedra; $(Fe,Mg,Ca)SiO_3$; crystal structure is monoclinic (Clinopyroxene) and orthorhombic (Orthopyroxene, varying from *enstatite* to *ferrosilite*); common in *basalt*;.

Pyroxenite: *igneous* rock composed largely of *pyroxene*.

Quartz: pure crystalline SiO_2.

Rhodochrosite: $MnCO_3$.

Schwertmannite: $Fe^{3+}_{16}O_{16}(OH)_{12}(SO_4)_2$.

Sedimentary: rock formed when *sediment* is compacted and lithified.

Serpentine: major rock forming *silicate* formed by *metamorphism* and hydrothermal alteration of *mafic minerals*; $(Mg,Fe)_3Si_2O_5(OH)_4$; colorless to pale green.

Siderite: $FeCO_3$; named in 1845 from the Greek sideros, for iron.

Silica: SiO_2.

Silicate: variety of *minerals* that contain Si, O, and one or more metals.

Sill: tabular, parallel-sided sheet of *igneous* layered rock, formed underground.

Smectite: *clay mineral* with $(\frac{1}{2}CaNa)_{0.7}(AlMgFe)_4Si_8O_{20}(OH)_4.nH_2O$ as approximate composition; isomorphous substitution gives the various types and causes a net permanent charge balanced by cations in such a manner that water may move between the sheets, giving very plastic properties; essentially synonymous with *Montmorillonite*.

Spinel: *mafic* rock without *plagioclase* and *peridotite*; (magnesium-iron) aluminum *silicate* material, e.g. $MgAl_2O_4$; has great hardness; forms octahedral crystals.

Textures: are characterized for *igneous* rocks on the basis of grain size and shape, and *mineral* orientation and proportions: fine grained (<1 mm), medium grained (1–5 mm), coarse grained (5–30 mm), or pegmatitic (>30 mm); *poikilitic, ophitic,* and sub*ophitic.*

Thrust Fault: low-angle fault in which rock above the fault plane moves up in relation to rock below.

Thin Section: thin rock slice, ~20 μm thick, used for studies with a microscope.

Troctolite: *igneous*, lunar *highland* rock composed of *plagioclase* and *olivine*.

Troilite: *mineral* with chemical formula FeS.

Weathering: mechanical breakdown and chemical alteration of rocks and *minerals* during exposure to air, moisture, frost, or organic matter.

6. Geochemical Processes and Related Terms

Activity: dimensionless effective concentration, a. For a solute species, i, the activity is given by $a_i = \gamma_i m_i$ where m_i is the molal concentration (mol kg^{-1}) and γ_i is an activity coefficient (kg mol^{-1}) that takes account of the interference of other ions and molecules in a real solution.

Aerosol: particles of liquid or solid dispersed as a suspension in gas.

Albedo: ratio of the outgoing solar radiation reflected by an object to the incoming solar radiation incident upon it.

Bulk Composition: chemical composition of an object averaged over its whole volume.

Chalcophile Elements: metals of the center and right-hand side of the periodic table (e.g. Cu, Zn, Sb, As, Pb, Sn, Cd, Hg, Ag); occur chiefly in sulfide *minerals*.

Core Formation: process during planetary growth that gravitationally segregates metallic iron in the molten interior to form a magnetic core; the heat sources for melting the iron can be impacts (in a first phase), radioactive nuclei, sinking of molten iron towards the center, etc.

Chemical Fractionation: see *Differentiation*.

Cumulate Layers: layers of *igneous* rocks that formed by aggregation of crystals during solidification in a *magma* chamber.

Differentiation: process of developing more than one type of *igneous* rock, in situ, from a common *magma*; also, chemical separation of materials in a planetary body during melting, allowing heavier elements to sink towards the center of the layered mass to form chemically distinct zones, e.g. *core*, *mantle*, and *crust*.

Fractional Crystallization: selective or partial crystallization of a *magma*, in which the early formed crystals are prevented from equilibrating with the parent liquid, resulting in a series of residual liquids of more extreme composition than would have resulted from continuous reaction.

Fugacity (of Oxygen): thermodynamic measure of the tendency of a substance (oxygen) to escape by some chemical process from the phase in which it exists.

Impact Erosion of Volatiles: loss of *volatiles*, especially atmospheric, during *impact* events.

Lithophile Elements: have higher chemical affinity for silicate rocks than for sulfides or for a metallic state, e.g. Al.

Mass-dependent Fractionation: fractionation of the isotopes of an element resulting from processes that depend on thermodynamic and kinetic effects such as gravity and temperature e.g. the preferential gravitational escape of H over D in the martian atmosphere, or the exchange of oxygen between a fluid and mineral.

Mass-independent Fractionation: isotope fractionation of an element due to nuclear or chemical processes (e.g. arising from photochemical reactions).

Phase Transition: transition of matter from one state with specific physical and chemical properties to another, e.g. transition of a solid from one crystal structure to another, possibly changing magnetic properties.

Radiometric Age: "absolute age" of a geologic event, feature, fossil, or rock; determined by using natural radioactive "clocks", e.g., the radioactive decay of naturally occurring ^{40}K to stable ^{40}Ar with a half-life of ~ 1.3 Gyr.

Refractory Elements: any chemical element that vaporizes at high temperatures (c.g. Ca, Al, U, and the rare earth elements, such as Ce).

Siderophile Elements: have high chemical affinity to Fe (e.g. Ir, Os, Pt, Pd); found in the metal-rich interiors of chemically segregated *asteroids* and planets .

Volatile Elements: any chemical element that vaporizes at relatively low temperatures (e.g. H_2O, CO_2, CO, CH_4, NH_3, K-, Na-, Pb-compounds).

References

Graham *et al.*: 1985, 'Catalogue of Meteorites', Univ. Arizona Press, Tucson, 460 p.

Hartmann, W. K.: 1966, 'Martian Cratering', *Icarus* **5**, 565.

Hartmann, W. K.: 1973, 'Ancient Lunar Mega-Regolith and Subsurface Structure', *Icarus* **18**, 634.

Stöffler, D., and Ryder, G.: 'Stratigraphy and Isotope Ages of Lunar Geologic Units: Chronological Standard for the Inner Solar System', this volume.

Tanaka, K.L.: 1986, 'The Stratigraphy of Mars', *J. Geophys. Res.* **91, suppl.**, 139–158.

SUBJECT INDEX

 *Space Science Reviews* **96**: 483-491, 2001.
© 2001 *Kluwer Academic Publishers. Printed in the Netherlands.*

LIST OF ACRONYMS AND BASIC MARS DATA

Acronyms

APXS	Alpha-Proton X-ray Spectrometer
ASPERA	Automatic Space Plasma Experiment with Rotating Analyzer
AU	Astronomical Unit, 1.496×10^{11} m
BSE	Bulk Silicate Earth
CAI	Ca,Al rich Inclusions
CATWG	Crater Analysis Techniques Working Group
CMB	Core-Mantle Boundary
CRA	Crater Retention Age
CRE	Cosmic Ray Exposure ages
EUV	Extreme Ultraviolet
GCM	General Circulation Model
HED	Howardite, Eucrite, and Diogenite meteorites
HREE	Heavy Rare Earth Elements
HSE	Highly Siderophile Elements
HST	Hubble Space Telescope
IMP	Imager for Mars Pathfinder
IR	Infrared
IRAS	Infrared Astronomical Satellite
IRTM	Infrared Thermal Mapping instrument
ISM	Infrared Imaging Spectrometer onboard Phobos-2
IRIS	Infrared Spectrometer on the Voyager Spacecraft
ISO	Infrared Space Observatory
JFC	Jupiter-family Comets
JPL	Jet Propulsion Laboratory
KAO	Kuiper Airborne Observatory
KREE	K for potassium, *see* REE
KRMF	Infrared Radiometer/Spectrometer onboard Phobos-2
LIL	Large Ion Lithophile
LPC	Long-period Comets
LPI	Lunar and Planetary Institute, Houston
LREE	Light Rare Earth Elements
MAG/ER	Magnetometer and Electron Reflectometer onboard MGS
MAWD	Mars Atmospheric Water Detector onboard Viking Orbiter 1
MB	Main Belt
MGS	Mars Global Surveyor
MIRO	Microwave Spectrometer Instrument for the ROSETTA Orbiter
MOC	Mars Orbital Camera

Space Science Reviews **96**: 493-494, 2001.
© 2001 *Kluwer Academic Publishers. Printed in the Netherlands.*

MoI	Moment of Inertia factor
MOLA	Mars Orbiter Laser Altimeter
MORB	Mid Ocean Ridge Basalt
NEA	Near Earth Asteroids
NIR	Near IR
OIB	Ocean Island Basalt
OMEGA	Infrared Mineralogical Mapping Spectrometer
PAH	Polycyclic Aromatic Hydrocarbons
PCA	Planet Crossing Asteroids
PSI	Planetary Science Institute
QMF	Quartz-Magnetite-Fayalite buffer
REE	Rare Earth Elements
SFD	Size Frequency Distribution
SMOW	Standard Mean Ocean Water
SNC	Shergotty, Nakhla, Chassigny (Martian meteorite classes)
TEM	Transmission Electron Microscope
TES	Thermal Emission Spectrometer onboard MGS
THEMIS	Thermal Emission Imaging System
USGS	U.S. Geological Survey
UV	Ultraviolet
XRFS	X-Ray Fluorescence Spectrometer on board Viking 1 and 2

Basic Mars Data

Diameter	6794 km
Mass	6.42×10^{23} kg
Gravitation	$3.73 \ \mathrm{m\,s^{-2}}$
Mean Distance from the Sun	1.524 AU
Orbit eccentricity	0.093
Mean speed on orbit	$24.14 \ \mathrm{km\,s^{-1}}$
Length of Martian day	24 h 39 min 35 s = 1 sol
Length of Martian year	1.88 yr = 669 sols
Magnetic field	$50 - 100$ mT
Mean air pressure	$\sim$6.5 mbar
Surface temperature range	$140 - 295$ K
Wind speed on surface	$2 - 7 \ \mathrm{m\,s^{-1}}, 5 - 10 \ \mathrm{m\,s^{-1}}$ (winter)
Wind speed in atmosphere	$60 - 80 \ \mathrm{m\,s^{-1}}$
Moons	Deimos and Phobos

AUTHOR INDEX

Space Science Reviews **96**: 495, 2001.
© 2001 *Kluwer Academic Publishers. Printed in the Netherlands.*

LIST OF PARTICIPANTS

Acuña, M. H., NASA Goddard Space Flight Center, Greenbelt, USA
    (mha@lepmom.gsfc.nasa.gov)
Albarède, F., Ecole Normale Supérieure de Lyon, France
    (albarede@ens-lyon.fr)
Arnold, G., Deutsches Zentrum für Luft- und Raumfahrt DLR, Berlin, Germany
    (gabriele.arnold@dlr.de)
Bibring, J.-P., Institut d'Astrophysique Spatiale, Orsay, France
    (bibring@ias.fr)
Birck, J. L., Institut de Physique du Globe, Paris, France
    (birck@ipgp.jussieu.fr)
Bogard, D. D., NASA Johnson Space Center, Houston, USA
    (donald.d.bogard@jsc.nasa.gov)
Breuer, Doris, Inst. für Planetologie, Universität Münster, Germany
    (breuer@uni-muenster.de)
Bridges, J. C., Natural History Museum, London, United Kingdom
    (jcb@nhm.ac.uk)
Clayton, R. N., Dept. of the Geophysical Sciences, University of Chicago, USA
    (r-clayton@uchicago.edu)
Drake, M. J., Lunar and Planetary Laboratory, University of Arizona, Tucson, USA
    (drake@lpl.arizona.edu)
Encrenaz, Th., Observatoire de Paris, Meudon, France
    (therese.encrenaz@obspm.fr)
Eugster, O., Physikalisches Institut, Universität Bern, Switzerland
    (otto.eugster@phim.unibe.ch)
Golombek, M., Jet Propulsion Laboratory, Caltech, Pasadena, USA
    (matthew.p.golombek@jpl.nasa.gov)
Greeley, R., Department of Geology, Arizona State University, Tempe, USA
    (greeley@dione.la.asu.edu)
Greshake, A., Museum für Naturkunde, Humboldt-Universität Berlin, Germany
    (ansgar.greshake@rz.hu-berlin.de)
Halliday, A., Department of Earth Sciences, ETH Zentrum, Zürich, Switzerland
    (halliday@erdw.ethz.ch)
Hartmann, W. K., Planetary Science Institute, Tucson, USA
    (hartmann@psi.edu)
Hauber, E., Deutsches Zentrum für Luft- und Raumfahrt DLR, Berlin, Germany
    (ernst.hauber@dlr.de)
Head, J.W., Dept. of Geological Sciences, Brown University, Providence, USA
    (head@pggipl.geo.brown.edu)
Jaumann, R., Deutsches Zentrum für Luft- und Raumfahrt DLR, Berlin, Germany
    (ralf.jaumann@dlr.de)
Jessberger, E.K., Inst. f. Planetologie, Universität Münster, Germany
    (ekj@uni-muenster.de)
Marti, K., Dept. of Chemistry, B-017, University of California, La Jolla, USA
    (kmarti@ucsd.edu)
Masson, Ph., LGDTP, Université Paris-Sud, Orsay, France
    (masson@geol.u-psud.fr)

Space Science Reviews **96**: 497-498, 2001.
© 2001 *Kluwer Academic Publishers. Printed in the Netherlands.*

Neukum, G., Deutsches Zentrum für Luft- und Raumfahrt DLR, Berlin, Germany
(gerhard.neukum@dlr.de)

Nyquist, L. E., NASA Johnson Space Center, Houston, USA
(laurence.e.nyquist1@jsc.nasa.gov)

Ott, U., Max-Planck-Institut für Chemie, Mainz, Germany
(ott@mpch-mainz.mpg.de)

Owen, T., Institute for Astronomy, University of Hawaii at Manoa, Honolulu, USA
(owen@uhifa.ifa.hawaii.edu)

Ryder, G., Lunar and Planetary Institute, Space Studies Center, Houston, USA
(zryder@lpi.usra.edu)

Saxton, J.M., Dept. of Geology, University of Manchester, United Kingdom
(jsaxton@fs1.ge.man.ac.uk)

Spohn, T., Inst. für Planetologie, Universität Münster, Germany
(spohn@uni-muenster.de)

Stöffler, D., Museum für Naturkunde, Humboldt-Universität Berlin, Germany
(dieter.stoeffler@museum.hu-berlin.de)

Swindle, T.D., Lunar and Planetary Laboratory, University of Arizona, Tucson, USA
(tswindle@u.arizona.edu)

Turner, G., Dept. of Geology, University of Manchester, United Kingdom
(gturner@fs1.ge.man.ac.uk)

Wänke, H., Abteilung f. Kosmochemie, Max-Planck-Institut für Chemie, Mainz, Germany
(waenke@mpch-mainz.mpg.de)

Warren, P.H., Institute of Geophysics, University of California, Los Angeles, USA
(pwarren@ucla.edu)

Space Science Series of ISSI

1. R. von Steiger, R. Lallement and M.A. Lee (eds.): *The Heliosphere in the Local Interstellar Medium.* 1996
ISBN 0-7923-4320-4

2. B. Hultqvist and M. Øieroset (eds.): *Transport Across the Boundaries of the Magnetosphere.* 1997
ISBN 0-7923-4788-9

3. L.A. Fisk, J.R. Jokipii, G.M. Simnett, R. von Steiger and K.-P. Wenzel (eds.): *Cosmic Rays in the Heliosphere.* 1998
ISBN 0-7923-5069-3

4. N. Prantzos, M. Tosi and R. von Steiger (eds.): *Primordial Nuclei and Their Galactic Evolution.* 1998
ISBN 0-7923-5114-2

5. C. Fröhlich, M.C.E. Huber, S.K. Solanki and R. von Steiger (eds.): *Solar Composition and its Evolution – From Core to Corona.* 1998
ISBN 0-7923-5496-6

6. B. Hultqvist, M. Øieroset, Goetz Paschmann and R. Treumann (eds.): *Magnetospheric Plasma Sources and Losses.* 1999
ISBN 0-7923-5846-5

7. A. Balogh, J.T. Gosling, J.R. Jokipii, R. Kallenbach and H. Kunow (eds.): *Co-rotating Interaction Regions.* 1999
ISBN 0-7923-6080-X

8. K. Altwegg, P. Ehrenfreund, J. Geiss and W. Huebner (eds.): *Composition and Origin of Cometary Materials.* 1999
ISBN 0-7923-6154-7

9. W. Benz, R. Kallenbach and G.W. Lugmair (eds.): *From Dust to Terrestrial Planets.* 2000
ISBN 0-7923-6467-8

10. J.W. Bieber, E. Eroshenko, P. Evenson, E.O. Flückiger and R. Kallenbach (eds.): *Cosmic Rays and Earth.* 2000
ISBN 0-7923-6712-X

11. E. Friis-Christensen, C. Fröhlich, J.D. Haigh, M. Schüssler and R. von Steiger (eds.): *Solar Variability and Climate.* 2000
ISBN 0-7923-6741-3

12. R. Kallenbach, J. Geiss and W.K. Hartmann (eds.): *Chronology and Evolution of Mars.* 2001
ISBN -7923-7051-1

Kluwer Academic Publishers – Dordrecht / Boston / London